SAFETY SYMBOLS

SAFETY SYMBOLS	HAZARD	EXAMPLE	[illegible]	MEDY
DISPOSAL	Special disposal procedures need to be followed.	certain chemicals, living organisms	[illegible] can.	[illegible] f wastes as [illegible] by your teacher.
BIOLOGICAL	Organisms or other biological materials that might be harmful to humans	bacteria, fungi, blood, unpreserved tissues, plant materials	Avoid skin contact with these materials. Wear mask or gloves.	Notify your teacher if you suspect contact with material. Wash hands thoroughly.
EXTREME TEMPERATURE	Objects that can burn skin by being too cold or too hot	boiling liquids, hot plates, dry ice, liquid nitrogen	Use proper protection when handling.	Go to your teacher for first aid.
SHARP OBJECT	Use of tools or glassware that can easily puncture or slice skin	razor blades, pins, scalpels, pointed tools, dissecting probes, broken glass	Practice common-sense behavior and follow guidelines for use of the tool.	Go to your teacher for first aid.
FUME	Possible danger to respiratory tract from fumes	ammonia, acetone, nail polish remover, heated sulfur, moth balls	Make sure there is good ventilation. Never smell fumes directly. Wear a mask.	Leave foul area and notify your teacher immediately.
ELECTRICAL	Possible danger from electrical shock or burn	improper grounding, liquid spills, short circuits, exposed wires	Double-check setup with teacher. Check condition of wires and apparatus.	Do not attempt to fix electrical problems. Notify your teacher immediately.
IRRITANT	Substances that can irritate the skin or mucous membranes of the respiratory tract	pollen, moth balls, steel wool, fiberglass, potassium permanganate	Wear dust mask and gloves. Practice extra care when handling these materials.	Go to your teacher for first aid.
CHEMICAL	Chemicals that can react with and destroy tissue and other materials	bleaches such as hydrogen peroxide; acids such as sulfuric acid, hydrochloric acid; bases such as ammonia, sodium hydroxide	Wear goggles, gloves, and an apron.	Immediately flush the affected area with water and notify your teacher.
TOXIC	Substance may be poisonous if touched, inhaled, or swallowed	mercury, many metal compounds, iodine, poinsettia plant parts	Follow your teacher's instructions.	Always wash hands thoroughly after use. Go to your teacher for first aid.
OPEN FLAME	Open flame may ignite flammable chemicals, loose clothing, or hair	alcohol, kerosene, potassium permanganate, hair, clothing	Tie back hair. Avoid wearing loose clothing. Avoid open flames when using flammable chemicals. Be aware of locations of fire safety equipment.	Notify your teacher immediately. Use fire safety equipment if applicable.

Eye Safety
Proper eye protection should be worn at all times by anyone performing or observing science activities.

Clothing Protection
This symbol appears when substances could stain or burn clothing.

Animal Safety
This symbol appears when safety of animals and students must be ensured.

Radioactivity
This symbol appears when radioactive materials are used.

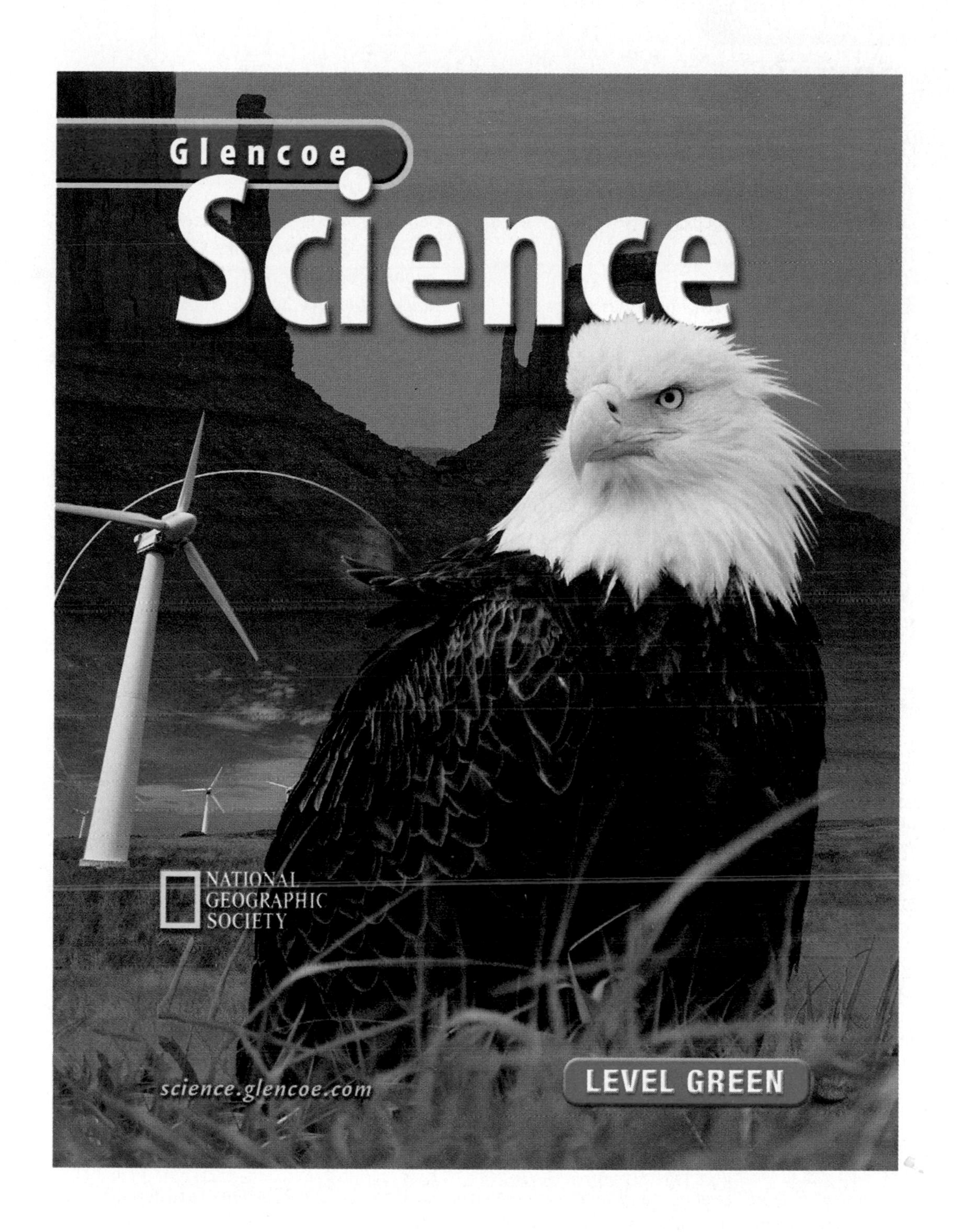

New York, New York Columbus, Ohio Woodland Hills, California Peoria, Illinois

LEVEL GREEN

Student Edition
Teacher Wraparound Edition
Interactive Teacher Edition CD-ROM
Interactive Lesson Planner CD-ROM
Lesson Plans
Content Outline for Teaching
Dinah Zike's Teaching Science with Foldables
Directed Reading for Content Mastery
Foldables: Reading and Study Skills
Assessment
- *Chapter Review*
- *Chapter Tests*
- *ExamView Pro Test Bank Software*
- *Assessment Transparencies*
- *Performance Assessment in the Science Classroom*
- *The Princeton Review Standardized Test Practice Booklet*

Directed Reading for Content Mastery in Spanish
Spanish Resources
English/Spanish Guided Reading Audio Program
Reinforcement
Enrichment
Activity Worksheets
Section Focus Transparencies
Teaching Transparencies
Laboratory Activities
Science Inquiry Labs
Critical Thinking/Problem Solving
Reading and Writing Skill Activities
Mathematics Skill Activities
Cultural Diversity
Laboratory Management and Safety in the Science Classroom
MindJogger Videoquizzes and Teacher Guide
Interactive CD-ROM with Presentation Builder
Vocabulary PuzzleMaker Software
Cooperative Learning in the Science Classroom
Environmental Issues in the Science Classroom
Home and Community Involvement
Using the Internet in the Science Classroom

"Study Tip," "Test-Taking Tip," and the "Test Practice" features in this book were written by The Princeton Review, the nation's leader in test preparation. Through its association with McGraw-Hill, The Princeton Review offers the best way to help students excel on standardized assessments.

The Princeton Review is not affiliated with Princeton University or Educational Testing Service.

Glencoe/McGraw-Hill
A Division of The McGraw-Hill Companies

Cover Images: full moon in the sky over Monument Valley, Arizona; bald eagle; wind turbines

Send all inquiries to:
Glencoe/McGraw-Hill
8787 Orion Place
Columbus, OH 43240

ISBN 0-07-828240-3
Printed in the United States of America.
2 3 4 5 6 7 8 9 10 071/055 06 05 04 03 02

Authors

National Geographic Society
Education Division
Washington, D.C.

Alton Biggs
Biology Teacher
Allen High School
Allen, Texas

Lucy Daniel, EdD
Teacher/Consultant
Rutherford County Schools
Rutherfordton, North Carolina

Ralph M. Feather Jr., PhD
Science Department Chair
Derry Area School District
Derry, Pennsylvania

Edward Ortleb
Science Consultant
St. Louis Public Schools
St. Louis, Missouri

Peter Rillero, PhD
Professor of Science Education
Arizona State University West
Phoenix, Arizona

Susan Leach Snyder
Earth Science Teacher, Consultant
Jones Middle School
Upper Arlington, Ohio

Dinah Zike
Educational Consultant
Dinah-Might Activities, Inc.
San Antonio, Texas

Contributing Authors

Cathy Ezrailson
Oak Ridge High School
Conroe ISD
Conroe, Texas

Patricia Horton
Mathematics and Science Teacher
Summit Intermediate School
Etiwanda, California

Deborah Lillie
Math and Science Writer
Sudbury, Massachusetts

Series Reading Consultants

Elizabeth Babich
Special Education Teacher
Mashpee Public Schools
Mashpee, Massachusetts

Barry Barto
Special Education Teacher
John F. Kennedy Elementary
Manistee, Michigan

Carol A. Senf, PhD
Associate Professor of English
Georgia Institute of Technology
Atlanta, Georgia

Rachel Swaters
Science Teacher
Rolla Middle Schools
Rolla, Missouri

Nancy Woodson, PhD
Professor of English
Otterbein College
Westerville, Ohio

Series Math Consultants

Michael Hopper, D.Eng
Manager of Aircraft Certification
Raytheon Company
Greenville, Texas

Teri Willard, EdD
Department of Mathematics
Montana State University
Belgrade, Montana

Content Consultants

Michelle Anderson
Community Faculty
Marion Technical College
Marion, Ohio

Jack Cooper
Adjunct Faculty Math and Science
Navarro College
Corsicana, Texas

Sandra K. Enger, PhD
Coordinator
UAH Huntsville Institute for Science Education
Huntsville, Alabama

Leanne Field, PhD
Lecturer Molecular Genetics and Microbiology
University of Texas
Austin, Texas

Michael A. Hoggarth, PhD
Department of Life and Earth Sciences
Otterbein College
Westerville, Ohio

William C. Keel, PhD
Department of Physics and Astronomy
University of Alabama
Tuscaloosa, Alabama

Linda Knight, EdD
Associate Director
Rice Model Science Lab
Houston, Texas

Lisa McGaw
Science Teacher
Hereford High School
Hereford, Texas

Lee Meadows, PhD
UAB Birmingham Education Department
Birmingham, Alabama

Robert Nierste
Science Department Head
Hendrick Middle School
Plano, Texas

Connie Rizzo, MD
Professor of Biology
Pace University
New York, New York

Dominic Salinas, PhD
Middle School Science Supervisor
Caddo Parish Schools
Shreveport, Louisiana

Carl Zorn, PhD
Staff Scientist
Jefferson Laboratory
Newport News, Virginia

Betsy Wrobel-Boerner
Department of Microbiology
Ohio State University
Columbus, Ohio

Series Activity Testers

José Luis Alvarez, PhD
Math and Science Mentor Teacher
El Paso, Texas

Nerma Coats Henderson
Teacher
Pickerington Jr. High School
Pickerington, Ohio

Mary Helen Mariscal-Cholka
Science Teacher
William D. Slider Middle School
El Paso, Texas

José Alberto Marquez
TEKS for Leaders Trainer
El Paso, Texas

Science Kit and Boreal Laboratories
Tonawanda, New York

Series Safety Consultants

Malcolm Cheney, PhD
OSHA Chemical Safety Officer
Hall High School
West Hartford, Connecticut

Aileen Duc, PhD
Science II Teacher
Hendrick Middle School
Plano, Texas

Sandra West, PhD
Associate Professor of Biology
Southwest Texas State University
San Marcos, Texas

Reviewers

Sharla Adams
McKinney High School North
McKinney, Texas

Michelle Bailey
Northwood Middle School
Houston, Texas

Maureen Barrett
Thomas E. Harrington Middle School
Mt. Laurel, New Jersey

Desiree Bishop
Baker High School
Mobile, Alabama

Janice Bowman
Coke R. Stevenson Middle School
San Antonio, Texas

Lois Burdette
Green Bank Elementary-Middle School
Green Bank, West Virginia

Marcia Chackan
Pine Crest School
Boca Raton, Florida

Anthony DiSipio
Octorana Middle School
Atglen, Pennsylvania

Sandra Everhart
Honeysuckle Middle School
Dothan, Alabama

Cory Fish
Burkholder Middle School
Henderson, Nevada

Linda V. Forsyth
Merrill Middle School
Denver, Colorado

George Gabb
Great Bridge Middle School
Chesapeake, Virginia

Annette Garcia
Kearney Middle School
Commerce City, Colorado

Nerma Coats Henderson
Pickerington Jr. High School
Pickerington, Ohio

Tammy Ingraham
Westover Park Intermediate School
Canyon, Texas

Michael Mansour
John Page Middle School
Madison Heights, Michigan

Linda Melcher
Woodmont Middle School
Piedmont, South Carolina

Amy Morgan
Berry Middle School
Hoover, Alabama

Annette Parrott
Lakeside High School
Atlanta, Georgia

Michelle Punch
Northwood Middle School
Houston, Texas

Billye Robbins
Lomax Junior High School
LaPorte, Texas

Pam Starnes
North Richland Middle school
Fort Worth, Texas

Joanne Stickney
Monticello Middle School
Monticello, New York

Delores Stout
Sterling City High School
Sterling City, Texas

Darcy Vetro-Ravndal
Middleton Middle School of Technology
Tampa, Florida

Clabe Webb
Sterling City High School
Sterling City, Texas

Contents in Brief

UNIT 1 The Nature of Science—2

CHAPTER 1 The Nature of Science—4

CHAPTER 2 Measurement—36

UNIT 2 Life's Building Blocks and Processes—66

Respiration and Excretion—152

Animal Behavior—178

Contents

CONTENTS

UNIT 3 Reproduction and Heredity—206

Chapter 8 Cell Reproduction—208

Chapter 9 Plant Reproduction—238

Chapter 10 Regulation and Reproduction—268

Heredity—298

UNIT 4 Ecology—328

Interactions of Life—330

CHAPTER 13 The Nonliving Environment—358

UNIT 5 Earth and the Solar System—388

CHAPTER 14 Plate Tectonics—390

CHAPTER 15 Earthquakes and Volcanoes—418

CHAPTER 19

Properties and Changes of Matter—544

UNIT 7 Waves, Sound, and Light—572

Waves—574

Sound—602

CHAPTER 22

Electromagnetic Waves—632

CHAPTER 23

Light, Mirrors, and Lenses—662

Field Guides

Interdisciplinary Connections

NATIONAL GEOGRAPHIC Unit Openers

VISUALIZING

TIME SCIENCE AND Society

TIME SCIENCE AND HISTORY

Oops! Accidents in SCIENCE

Science and Language Arts

Science Stats

Activities

Full Period Labs

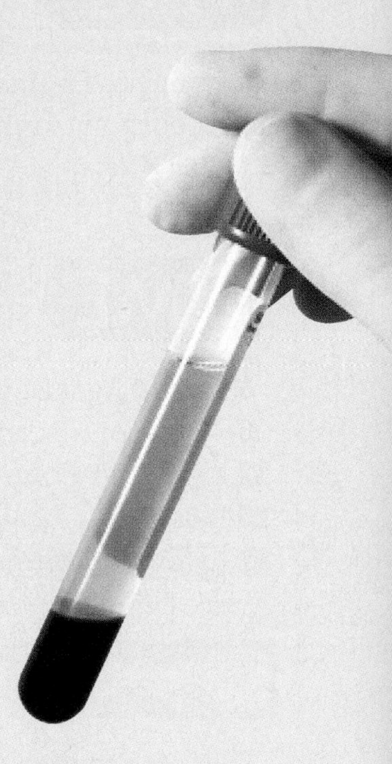

Mini LAB

Explore Activity

Problem-Solving Activities

Math Skills Activities

Activities

Skill Builder Activities

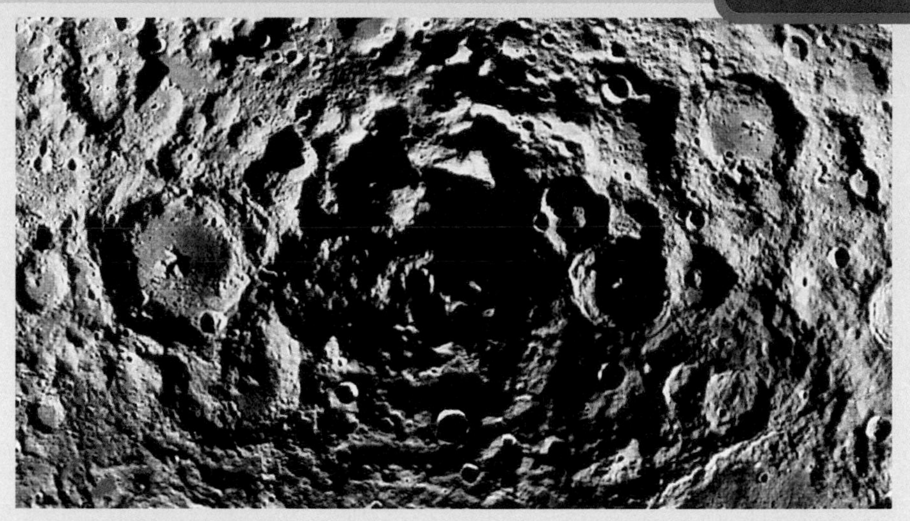

Science INTEGRATION

SCIENCE *Online*

UNIT 1

The Nature of Science

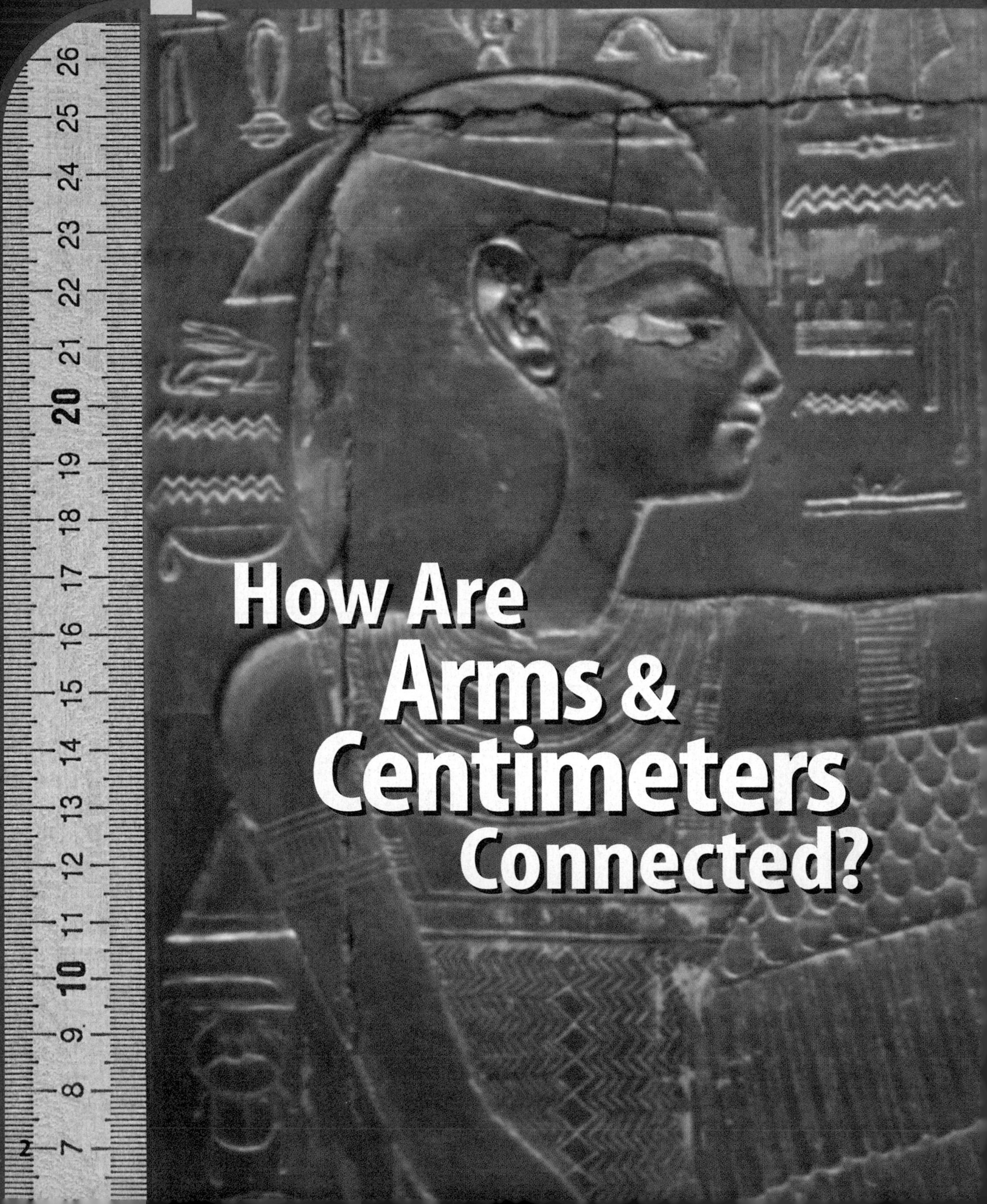

How Are Arms & Centimeters Connected?

About 5,000 years ago, the Egyptians developed one of the earliest recorded units of measurement—the cubit, which was based on the length of the arm from elbow to fingertip. The Egyptian measurement system probably influenced later systems, many of which also were based on body parts such as arms and feet. Such systems, however, could be problematic, since arms and feet vary in length from one person to another. Moreover, each country had its own system, which made it hard for people from different countries to share information. The need for a precise, universal measurement system eventually led to the adoption of the meter as the basic international unit of length. A meter is defined as the distance that light travels in a vacuum in a certain fraction of a second—a distance that never varies. Meters are divided into smaller units called centimeters, which are seen on the rulers here.

SCIENCE CONNECTION

MEASUREMENT SYSTEMS Ancient systems of measurement had their flaws, but they paved the way for the more exact and uniform systems used today. Devise your own measurement system based on parts of your body (for example, the length of your hand or the width of your shoulders) or common objects in your classroom or home. Give names to your units of measurement. Then calculate the width and height of a doorway using one or more of your units.

CHAPTER 1

The Nature of Science

Why explore outer space or dive to the depths of the ocean? How can you examine microscopic cells or study animal behavior? In this chapter you will learn about skills and tools that are used to answer scientific questions. You'll also learn about the different methods scientists use to investigate the world. You will discover that you already use many of these skills every day.

What do you think?

Science Journal Look at the picture below with a classmate. Discuss what you think this might be. Here's a hint: *Somebody found this by drawing a map.* Write your answer or best guess in your Science Journal.

Magnification: 6,000×

Ouch! That soup is hot. Your senses tell you a great deal of information about the world around you, but they can't answer every question. Scientists use tools, such as thermometers, to measure accurately. Learn more about the importance of tools in the following activity.

Measure using tools 

1. Use three bowls. Fill one with cold water, one with lukewarm water, and the third with hot water. **WARNING:** *Make sure the hot water will not burn you.*
2. Use a thermometer to measure the temperature of the lukewarm water. Record the temperature.
3. Submerse one hand in the cold water and the other in the hot water for 2 min.
4. Put both hands into the bowl of lukewarm water. What do you sense with each hand? Record your response in your Science Journal.

Observe

In your Science Journal, write a paragraph that explains why it is important to use tools to measure information.

Before You Read

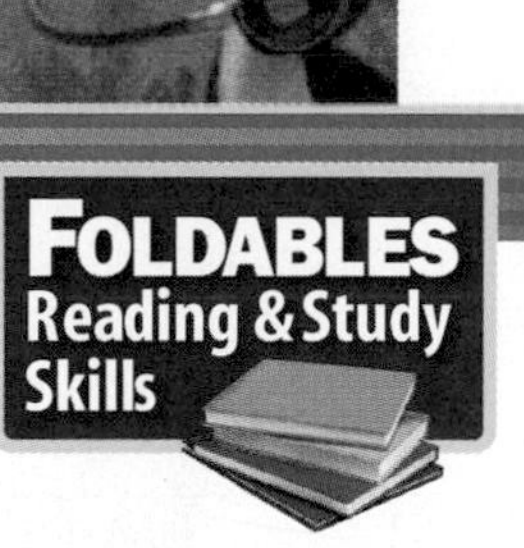

Making a Question Study Fold **Asking yourself questions helps you stay focused and better understand scientists when you are reading the chapter.**

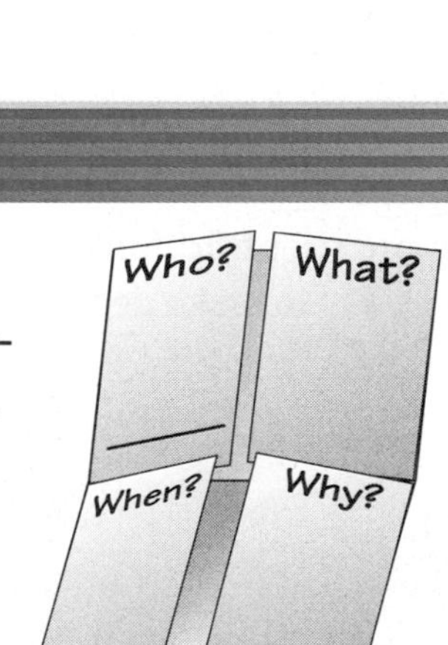

1. Place a sheet of paper in front of you so the long side is at the top. Fold the paper in half from the left side to the right side and then unfold.
2. Fold each side in to the center line, dividing the paper into fourths.
3. Fold the paper in half from top to bottom and unfold.
4. Through the top thickness of paper, cut along both middle fold lines to form four tabs, as shown. Label each tab *Who, When, What,* and *Why,* as shown.
5. Before you read, select a scientist and write their name on the front of the *Who* tab.
6. As you read the chapter, write answers to *What, When,* and *Why* under the tabs.

SECTION

What is science?

As You Read

What You'll Learn

- **Identify** how science is a part of your everyday life.
- **Describe** what skills and tools are used in science.

Vocabulary

science
technology

Why It's Important

What and how you learn in science class can be applied to other areas of your life.

Science in Society

When you hear the word *science,* do you think only of your science class, your teacher, and certain terms and facts? Is there any connection between what happens in science class and the rest of your life? You might have problems to solve or questions that need answers, as illustrated in **Figure 1. Science** is a way or a process used to investigate what is happening around you. It can provide possible answers.

Science Is Not New Throughout history, people have tried to find answers to questions about what was happening around them. Early scientists tried to explain things based on their observations. They used their senses of sight, touch, smell, taste, and hearing to make these observations. From the Explore Activity, you know that using only your senses can be misleading. What is cold or hot? How heavy is heavy? How much is a little? How close is nearby? Numbers can be used to describe observations. Tools, such as thermometers and metersticks, are used to give numbers to descriptions. Scientists observe, investigate, and experiment to find answers, and so can you.

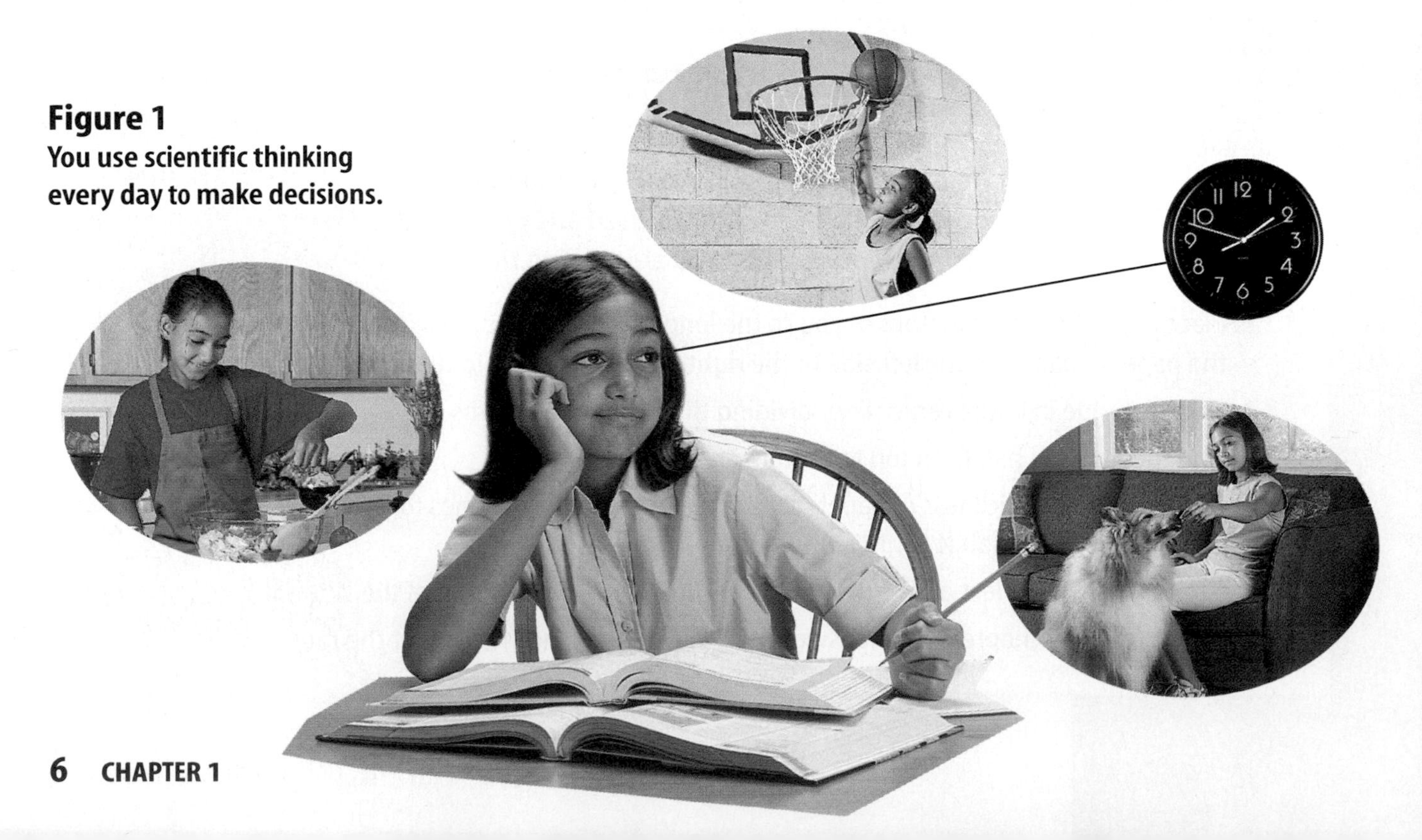

Figure 1
You use scientific thinking every day to make decisions.

Science as a Tool

As Luis and Midori walked into science class, they still were talking about their new history assignment. Mr. Johnson overheard them and asked what they were excited about.

"We have a special assignment—celebrating the founding of our town 200 years ago," answered Luis. "We need to do a project that demonstrates the similarities of and differences between a past event and something that is happening in our community now."

Mr. Johnson responded. "That sounds like a big undertaking. Have you chosen the two events yet?"

"We read some old newspaper articles and found several stories about a cholera epidemic here that killed ten people and made more than 50 others ill. It happened in 1871—soon after the Civil War. Midori and I think that it's like the *E. coli* outbreak going on now in our town," replied Luis.

"What do you know about an outbreak of cholera and problems caused by *E. coli,* Luis?"

"Well, Mr. Johnson, cholera is a disease caused by a bacterium that is found in contaminated water," Luis replied. "People who eat food from this water or drink this water have bad cases of diarrhea and can become dehydrated quickly. They might even die. *E. coli* is another type of bacterium. Some types of *E. coli* are harmless, but others cause intestinal problems when contaminated food and water are consumed."

"In fact," added Midori, "one of the workers at my dad's store is just getting over being sick from *E. coli.* Anyway, Mr. Johnson, we want to know if you can help us with the project. We want to compare how people tracked down the source of the cholera in 1871 with how they are tracking down the source of the *E. coli* now."

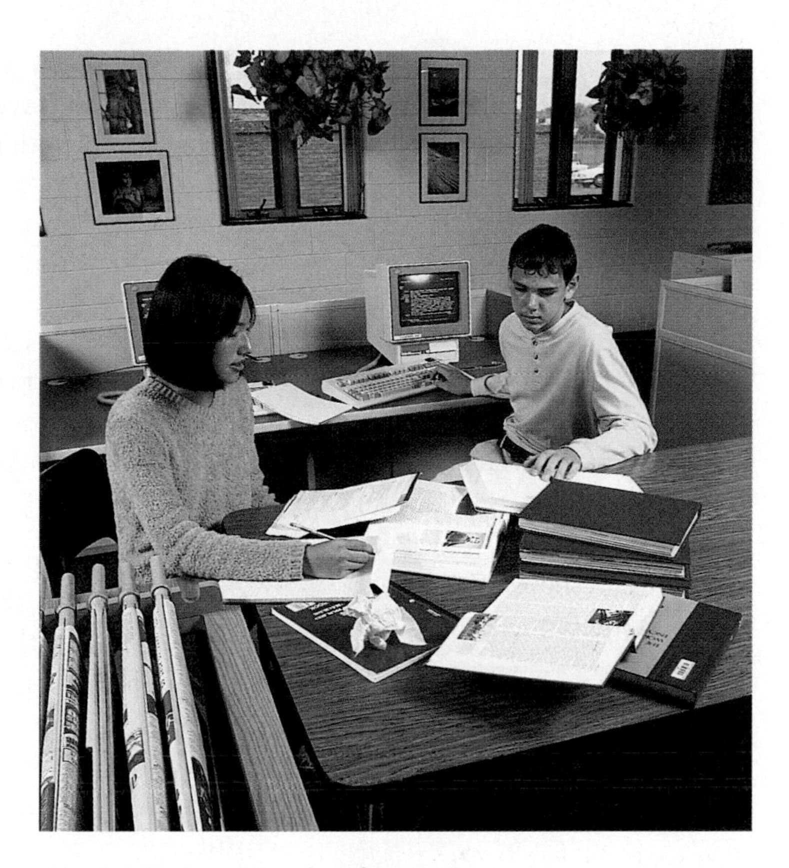

Figure 2
Newspapers, magazines, books, and the Internet are all good sources of information.

Using Science Every Day

"I'll be glad to help. This sounds like a great way to show how science is a part of everyone's life. In fact, you are acting like scientists right now," Mr. Johnson said proudly.

Luis had a puzzled look on his face, then he asked, "What do you mean? How can we be doing science? This is supposed to be a history project."

Health INTEGRATION

You can't prevent all illnesses. You can, however, take steps to reduce your chances of coming in contact with disease-causing organisms. Antibacterial soaps and cleansers claim to kill such organisms, but how do you know if they work? Read ads for or labels on such products. Do they include data to support their claims? Communicate what you learn to your class.

SCIENCE Online

Research Visit the Glencoe Science Web site at **science.glencoe.com** for more information about disease control. Report two different diseases that the Centers for Disease Control and Prevention (CDC) have tracked down and identified in the past five years. Make a poster that shows what you have learned.

Scientists Use Clues "Well, you're acting like a detective right now. You have a problem to solve. You and Midori are looking for clues that show how the two events are similar and different. As you complete the project, you will use several skills and tools to find the clues." Mr. Johnson continued, "In many ways, scientists do the same thing. People in 1871 followed clues to track the source of the cholera epidemic and solve their problem. Today, scientists are doing the same thing by finding and following clues to track the source of the *E. coli*."

Using Prior Knowledge

Mr. Johnson asked, "Luis, how do you know what is needed to complete your project?"

Luis thought, then responded, "Our history teacher, Ms. Hernandez, said the report must be at least three pages long and have maps, pictures, or charts and graphs. We have to use information from different sources such as written articles, letters, videotapes, or the Internet. I also know that it must be handed in on time and that correct spelling and grammar count."

"Did Ms. Hernandez actually talk about correct spelling and grammar?" asked Mr. Johnson.

Midori quickly responded, "No, she didn't have to. Everyone knows that Ms. Hernandez takes points away for incorrect spelling or grammar. I forgot to check my spelling in my last report and she took off two points."

"Ah-ha! That's where your project is like science," exclaimed Mr. Johnson. "You know from experience what will happen. When you don't follow her rule, you lose points. You can predict, or make an educated guess, that Ms. Hernandez will react the same way with this report as she has with others."

Mr. Johnson continued, "Scientists also use prior experience to predict what will occur in investigations. Scientists form theories when their predictions have been well tested. A theory is an explanation that is supported by facts. Scientists also form laws, which are rules that describe a pattern in nature, like gravity."

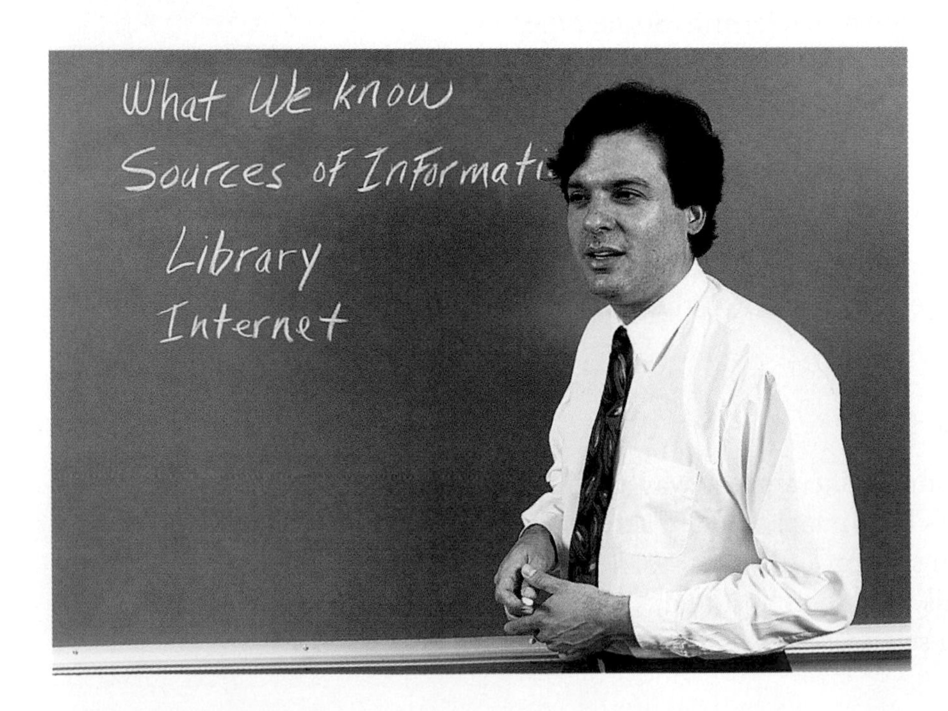

Figure 3
When solving a problem, it is important to discover all background information. Different sources can provide such information. *How would you find information on a specific topic? What sources of information would you use?*

Using Science and Technology

"Midori, you said that you want to compare how the two diseases were tracked. Like scientists, you will use skills and tools to find the similarities and differences." Mr. Johnson then pointed to Luis. "You need a variety of resource materials to find information. How will you know which materials will be useful?"

"We can use a computer to find books, magazines, newspapers, videos, and web pages that have information we need," said Luis.

"Exactly," said Mr. Johnson. "That's another way that you are thinking like scientists. The computer is one tool that modern scientists use to find and analyze data. The computer is an example of technology. **Technology** is the application of science to make products or tools that people can use. One of the big differences you will find between the way diseases were tracked in 1871 and how they are tracked now is the result of new technology."

Figure 4
Computers are one example of technology. Schools and libraries often provide computers for students to do research and word processing.

Science Skills Perhaps some of the skills used to track the two diseases will be one of the similarities between the two time periods," continued Mr. Johnson. "Today's doctors and scientists, like those in the late 1800s, use skills such as observing, classifying, and interpreting data. In fact, you might want to review the science skills we've talked about in class. That way, you'll be able to identify how they were used during the cholera outbreak and how they still are used today."

Luis and Midori began reviewing the science skills that Mr. Johnson had mentioned. Some of these skills used by scientists are described in the **Science Skill Handbook** at the back of this book. The more you practice these skills, the better you will become at using them.

Inferring from Pictures

Procedure

1. Study the two pictures to the left. Write your observations in your **Science Journal.**
2. Make and record inferences based on your observations.
3. Share your inferences with others in your class.

Analysis

1. Analyze your inferences. Are there other explanations for what you observed?
2. Why must you be careful when making inferences?

Observation and Measurement Think about the Explore Activity at the beginning of this chapter. Observing, measuring, and comparing and contrasting are three skills you used to complete the activity. Scientists probably use these skills more than other people do. You will learn that sometimes observation alone does not provide a complete picture of what is happening. To ensure that your data are useful, accurate measurements must be taken, in addition to making careful observations.

Reading Check *What are three skills commonly used in science?*

Luis and Midori want to find the similarities and differences between the disease-tracking techniques used in the late 1800s and today. They will use the comparing and contrasting skill. When they look for similarities among available techniques, they compare them. Contrasting the available techniques is looking for differences.

Communication in Science

What do scientists do with their findings? The results of their observations, experiments, and investigations will not be of use to the rest of the world unless they are shared. Scientists use several methods to communicate their observations.

Results and conclusions of experiments often are reported in one of the thousands of scientific journals or magazines that are published each year. Some of these publications are shown in **Figure 5.** Scientists spend a large part of their time reading journal articles. Sometimes, scientists discover information in articles that might lead to new experiments.

Figure 5
Scientific publications allow scientists around the world to learn about the latest research. Papers are submitted to journals. Other scientists review them before they are published.

Science Journal Another method to communicate scientific data and results is to keep a Science Journal. Observations and plans for investigations can be recorded, along with the step-by-step procedures that were followed. Listings of materials and drawings of how equipment was set up should be in a journal, along with the specific results of an investigation. You should record mathematical measurements or formulas that were used to analyze the data. Problems that occurred and questions that came up during the investigation should be noted, as well as any possible solutions. Your data might be summarized in the form of tables, charts, or graphs, or they might be recorded in a paragraph. Remember that it's always important to use correct spelling and grammar in your Science Journal.

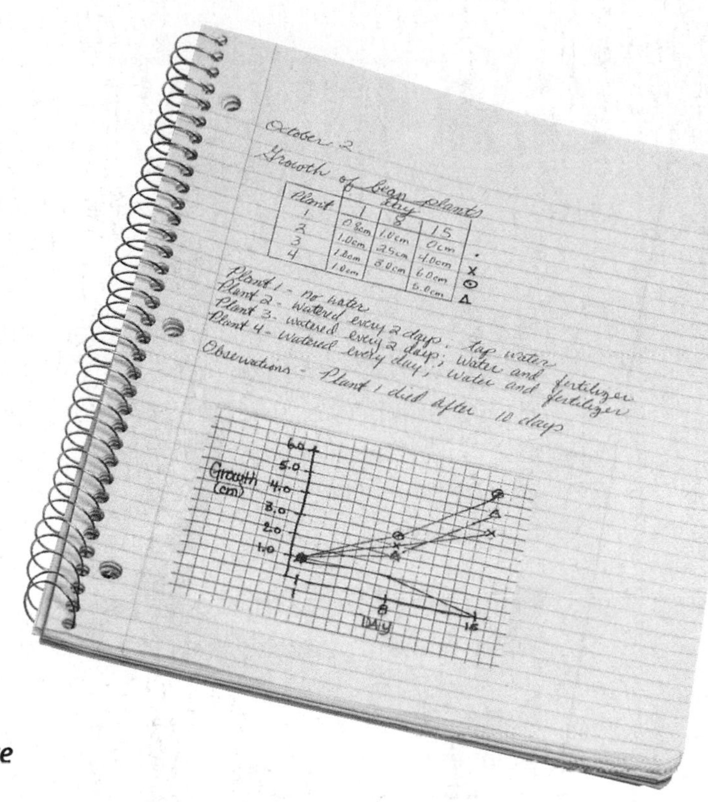

Reading Check *What are some ways to summarize data from an investigation?*

You will be able to use your Science Journal, as illustrated in **Figure 6,** to communicate your observations, questions, thoughts, and ideas as you work in science class. You will practice many of the science skills and become better at identifying problems. You will learn to plan investigations and experiments that might solve these problems.

Figure 6
Your Science Journal is used to record and communicate your findings. It might include graphs, tables, and illustrations.

Section  Assessment

1. Why do scientists use tools, such as thermometers and metersticks, when they make observations?
2. What are some sources for information about problems that you need to solve?
3. What are some skills used in science? Name one science skill that you have used today.
4. Give one example of technology. How is technology different from science?
5. **Think Critically** Why is a Science Journal used to record data? What are three different ways you could record or summarize data in your Science Journal?

Skill Builder Activities

6. **Comparing and Contrasting** Sometimes you use your senses and observations to find the answer to a question. Other times you use tools and measurements to provide answers. Compare and contrast these two methods of answering scientific questions. **For more help, refer to the Science Skill Handbook.**
7. **Communicating** In your Science Journal record five things you observe in or about your classroom. Make sure to include observations based on more than just one of your senses. **For more help, refer to the Science Skill Handbook.**

Activity

Battle of the Drink Mixes

You can use science skills to answer everyday questions or to solve problems. For example, you might know that the cheapest brand of a product is not always the best value. In this activity, you will test one aspect, or quality, of a product.

What You'll Investigate

Which brand of powdered drink mix dissolves best?

Materials

weighing paper
50-mL graduated cylinder
powdered drink mixes (3 or 4)
triple-beam balance
250-mL beaker
water
spoon

Goals

- **Determine** which brand of powdered drink mix dissolves best using science skills.

Safety Precautions

WARNING: *Never eat or drink anything during science experiments.*

Procedure

1. Copy the following data table in your Science Journal.

Drink Mix Data	
Drink Mix	Mass of Dissolved Powder (g)

2. Using the graduated cylinder, measure 50 mL of water and pour the water into the beaker.
3. **Measure** 20 g of powder from one of the drink-mix brands.
4. Gradually add the powder to the water. Stir the mixture each time you add more powder. Stop adding powder when undissolved powder begins to accumulate at the bottom of the beaker.
5. **Measure** the mass of the remaining powder. Subtract this number from 20 g to find the amount of powder that was dissolved. Record your answer in your data table.
6. Empty the drink mix into the sink, rinse out your beaker, and repeat steps 2 through 5 for the other drink-mix brands.

Conclude and Apply

1. **Identify** the drink-mix powder that dissolved best in the water.
2. Based on the data you collected, infer which drink-mix brand would taste the best. Remember, do not taste any of your samples.
3. Which drink-mix brand would you buy? Identify the science skills you used during this experiment that helped you determine the best drink mix.
4. Review promotional pamphlets for services such as landscaping or pool services. Make a list of inferences about the claims presented.

Communicating Your Data

Write the script for a 15 s advertisement that tells why people should buy your best-dissolving drink-mix brand. Perform your commercial for the class. **For more help, refer to the Science Skill Handbook.**

SECTION

Doing Science

Solving Problems

When Luis and Midori did their project, they were answering a question. However, there is more than one way to answer a question or solve a scientific problem. Every day, scientists work to solve scientific problems. Although the investigation of each problem is different, scientists use some steps in all investigations.

Identify the Problem Scientists first make sure that everyone working to solve the problem has a clear understanding of the problem. Sometimes, scientists find that the problem is easy to identify or that several problems need to be solved. For example, before a scientist can find the source of a disease, the disease must be identified correctly.

How can the problem be solved? Scientists know that scientific problems can be solved in different ways. Two of the methods used to answer questions are descriptive research and experimental research design. **Descriptive research** answers scientific questions through observation. When Louis and Midori gathered information to learn about cholera and *E. coli*, they performed descriptive research. **Experimental research design** is used to answer scientific questions by testing a hypothesis through the use of a series of carefully controlled steps. **Scientific methods,** like the one shown in **Figure 7,** are ways, or steps to follow, to try to solve problems. Different problems will require different scientific methods to solve them.

As You Read

What You'll Learn

- **Examine** the steps used to solve a problem in a scientific way.
- **Explain** how a well-designed investigation is developed.

Vocabulary

descriptive research
experimental research design
scientific methods
model
hypothesis
independent variable
dependent variable
constant
control

Why It's Important

Using scientific methods and carefully thought-out experiments can help you solve problems.

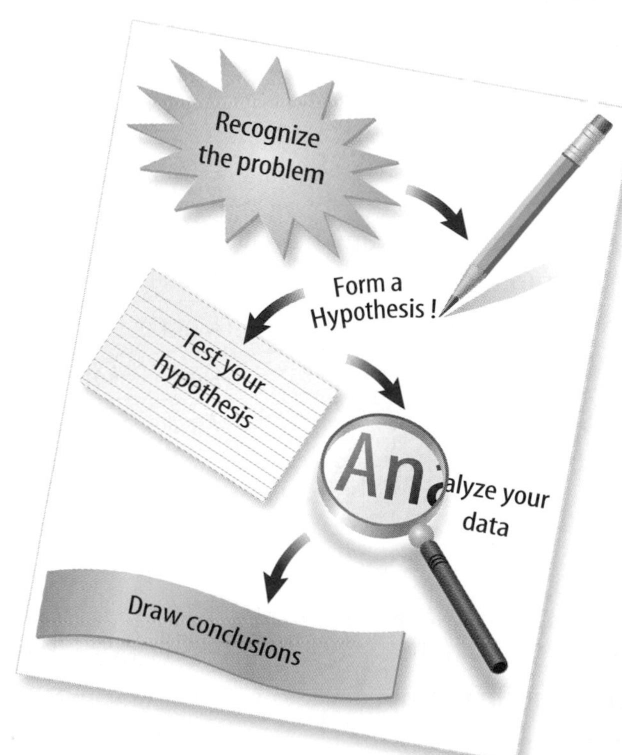

Figure 7
This poster shows one way to solve problems using scientific methods.

Figure 8
Items can be described by using words and numbers. *How could you describe these objects using both of these methods?*

Descriptive Research

Some scientific problems can be solved, or questions answered, by using descriptive research. Descriptive research is based mostly on observations. What observations can you make about the objects in **Figure 8?** Descriptive research can be used in investigations when experiments would be impossible to perform. For example, a London doctor, Dr. John Snow, tracked the source of a cholera epidemic in the 1800s by using descriptive research. Descriptive research usually involves the following steps.

State the Research Objective This is the first step in solving a problem using descriptive research. A research objective is what you want to find out, or what question you would like to answer. Luis and Midori might have said that their research objective was "to find out how the sources of the cholera epidemic and *E. coli* epidemic were tracked." Dr. John Snow might have stated his research objective as "finding the source of the cholera epidemic in London."

Problem-Solving Activity

Problem-Solving Skills

Drawing Conclusions from a Data Table

During an investigation, data tables often are used to record information. The data can be evaluated to decide whether or not the prediction was supported and then conclusions can be drawn.

A group of students conducted an investigation of the human populations of some states in the United States. They predicted that the states with the highest human population also would have the largest area of land. Do you have a different prediction? Record your prediction in your Science Journal before continuing.

Identifying the Problem

The results of the students' research are shown in this chart. Listed are several states in the United States, their human population, and land area.

State Population and Size

State	Human Population	Area (km^2)
New York	18,976,457	122,284
New Jersey	8,414,350	19,210
Massachusetts	6,349,097	20,306
Maine	1,274,923	79,932
Montana	902,195	376,978
North Dakota	642,200	178,647
Alaska	626,902	1,481,350

Source: United States Census Bureau, United States Census 2000

1. What can you conclude about your prediction? If your prediction is not supported by the data, can you come up with a new prediction? Explain.
2. What other research could be conducted to support your prediction?

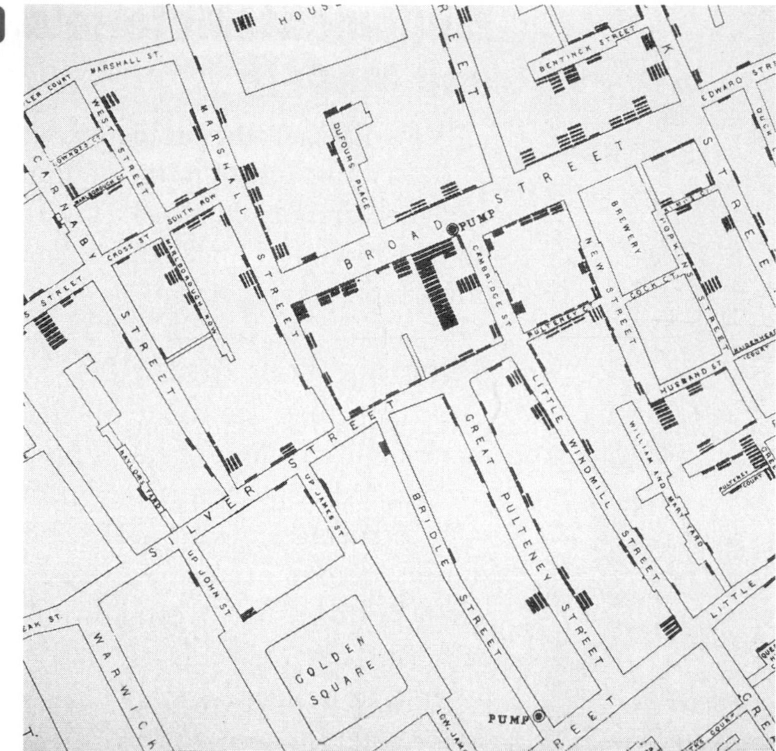

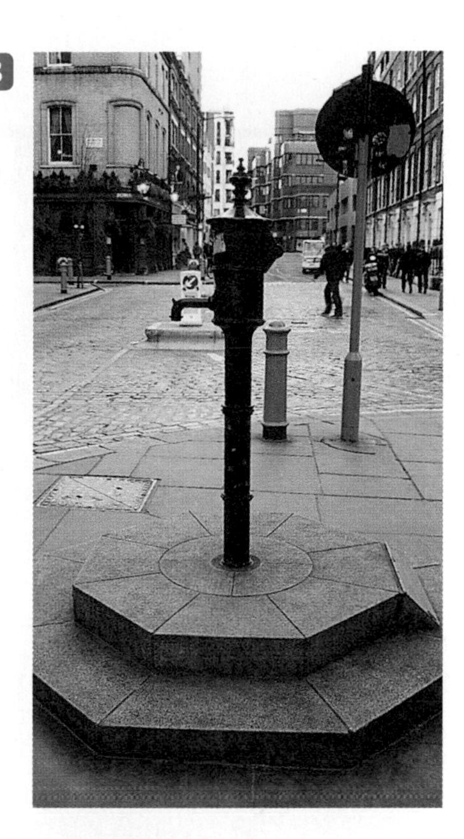

Figure 9
A Each mark on Dr. Snow's map shows where a cholera victim lived. B Dr. Snow had the water-pump handle removed, and the cholera epidemic ended.

Describe the Research Design How will you carry out your investigation? What steps will you use? How will the data be recorded and analyzed? How will your research design answer your question? These are a few of the things scientists think about when they design an investigation using descriptive research. An important part of any research design is safety. Check with your teacher several times before beginning any investigation.

✔ Reading Check ***What are some questions to think about when planning an investigation?***

Dr. John Snow's research design included the map in **Figure 9A.** The map shows where people with cholera had lived, and where they obtained their water. He used these data to predict that the water from the Broad Street pump, shown in **Figure 9B,** was the source of the contamination.

Eliminate Bias It's a Saturday afternoon. You want to see a certain movie, but your friends do not. To persuade them, you tell them about a part of the show that they will find interesting. You give only partial information so they will make the choice you want. Similarly, scientists might expect certain results. This is known as bias. Good investigations avoid bias. One way to avoid bias is to use careful numerical measurements for all data. Another type of bias can occur in surveys or groups that are chosen for investigations. To get an accurate result, you need to use a random sample.

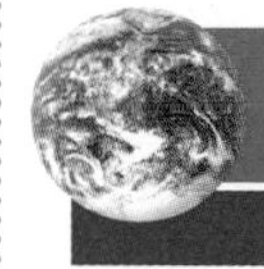

Environmental Science INTEGRATION

The U.S. Congress has passed several laws to reduce water pollution. The 1986 Safe Drinking Water Act is a law to ensure that drinking water in the United States is safe. The 1987 Clean Water Act gives money to the states for building sewage- and wastewater-treatment facilities. Find information about a state or local water quality law and share your findings with the class.

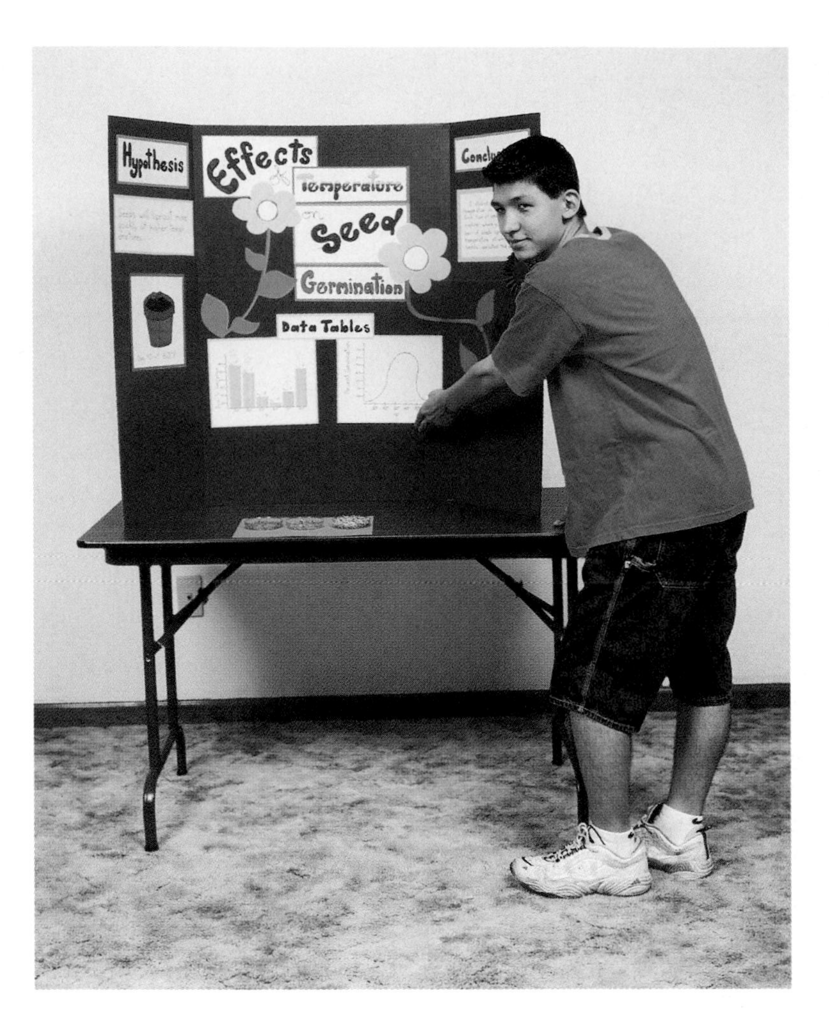

Figure 10
This presentation neatly and clearly shows experimental data.

Equipment, Materials, and Models

When a scientific problem is solved by descriptive research, the equipment and materials used to carry out the investigation and analyze the data are important.

Selecting Your Materials Scientists try to use the most up-to-date materials available to them. If possible, you should use scientific equipment such as balances, spring scales, microscopes, and metric measurements when performing investigations and gathering data. Calculators and computers can be helpful in evaluating or displaying data. However, you don't have to have the latest or most expensive materials and tools to conduct good scientific investigations. Your investigations can be completed successfully and the data displayed with materials found in your home or classroom, like paper, colored pencils, or markers. An organized presentation of data, like the one shown in **Figure 10,** is as effective as a computer graphic or an extravagant display.

Using Models One part of carrying out the investigative plan might include making or using scientific models. In science, a **model** represents things that happen too slowly, too quickly, or are too big or too small to observe directly. Models also are useful in situations in which direct observation would be too dangerous or expensive.

Dr. John Snow's map of the cholera epidemic was a model that allowed him to predict possible sources of the epidemic. Today, people in many professions use models. Many kinds of models are made on computers. Graphs, tables, and spreadsheets are models that display information. Computers can produce three-dimensional models of a microscopic bacterium, a huge asteroid, or an erupting volcano. They are used to design safer airplanes and office buildings. Models save time and money by testing ideas that otherwise are too small, too large, or take too long to build.

Table 1 Common SI Measurements

Measurement	Unit	Symbol	Equal to
Length	1 millimeter	mm	0.001 (1/1,000) m
	1 centimeter	cm	0.01 (1/100) m
	1 meter	m	100 cm
	1 kilometer	km	1,000 m
Liquid Volume	1 milliliter	mL	0.001 L
	1 liter	L	1,000 mL
Mass	1 milligram	mg	0.001 g
	1 gram	g	1,000 mg
	1 kilogram	kg	1,000 g
	1 tonne	t	1,000 kg = 1 metric ton

Scientific Measurement Scientists around the world use a system of measurements called the International System of Units, or SI, to make observations. This allows them to understand each other's research and compare results. Most of the units you will use in science are shown in **Table 1.** Because SI uses certain metric units that are based on units of ten, multiplication and division are easy to do. Prefixes are used with units to change their names to larger or smaller units. See the Reference Handbook to help you convert English units to SI. **Figure 11** shows equipment you can use to measure in SI.

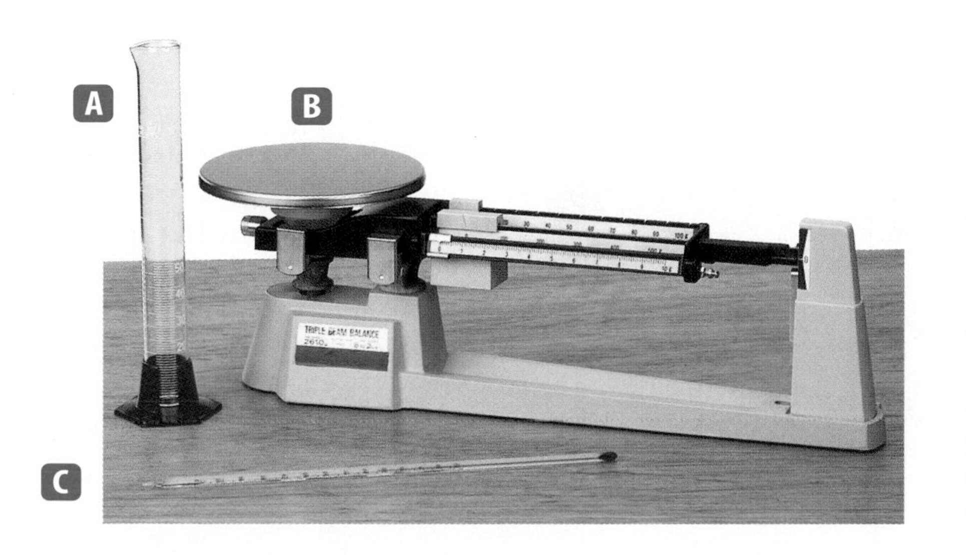

Figure 11
Some of the equipment used by scientists is shown here.
A The amount of space occupied by an object is its volume. A graduated cylinder is used to measure liquid volume. **B** Mass is the amount of matter in an object. Mass is measured with a balance. **C** A scientist would use a thermometer with the Celsius scale to measure temperature. On the Celsius scale, water freezes at 0°C and boils at 100°C.

TRY AT HOME
Mini LAB

Comparing Paper Towels

Procedure

1. Make a data table similar to the one in **Figure 12.**
2. Cut a 5-cm by 5-cm square from each of **three brands of paper towel.** Lay each piece on a level, smooth, waterproof surface.
3. Add one drop of **water** to each square.
4. Continue to add drops until the piece of paper towel no longer can absorb the water.
5. Tally your observations in a frequency table and graph your results.
6. Repeat steps 2 through 5 three more times.

Analysis

1. Did all the squares of paper towels absorb equal amounts of water?
2. If one brand of paper towel absorbs more water than the others, can you conclude that it is the towel you should buy? Explain.
3. Which scientific methods did you use to compare paper towel absorbency?

Figure 12
Data tables help you organize your observations and results.

Paper Towel Absorbency (Drops of Water Per Sheet)			
Trial	**Brand A**	**Brand B**	**Brand C**
1			
2			
3			
4			

Data

In every type of scientific research, data must be collected and organized carefully. When data are well organized, they are easier to interpret and analyze.

Designing Your Data Tables A well-planned investigation includes ways to record results and observations accurately. Data tables, like the one shown in **Figure 12,** are one way to do this. Most tables have a title that tells you at a glance what the table is about. The table is divided into columns and rows. These are usually trials or characteristics to be compared. The first row contains the titles of the columns. The first column identifies what each row represents.

As you complete a data table, you will know that you have the information you need to analyze the results of the investigation accurately. It is wise to make all of your data tables before beginning the experiment. That way, you will have a place for all of your data as soon as they are available.

Analyze Your Data

Your investigation is over. You breathe a sigh of relief. Now you have to figure out what your results mean. To do this, you must review all of the recorded observations and measurements. Your data must be organized to analyze them. Charts and graphs are excellent ways to organize data. You can draw the charts and graphs, like the ones in **Figure 13,** or use a computer to make them.

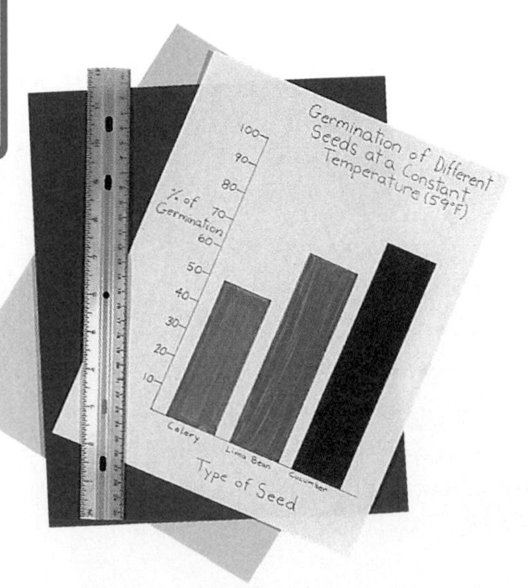

Figure 13
Charts and graphs can help you organize and analyze your data.

Draw Conclusions

After you have organized your data, you are ready to draw a conclusion. Do the data answer your question? Was your prediction supported? You might be concerned if your data are not what you expected, but remember, scientists understand that it is important to know when something doesn't work. When looking for an antibiotic to kill a specific bacteria, scientists spend years finding out which antibiotics will work and which won't. Each time scientists find that a particular antibiotic doesn't work, they learn some new information. They use this information to help make other antibiotics that have a better chance of working. A successful investigation is not always the one that comes out the way you originally predicted.

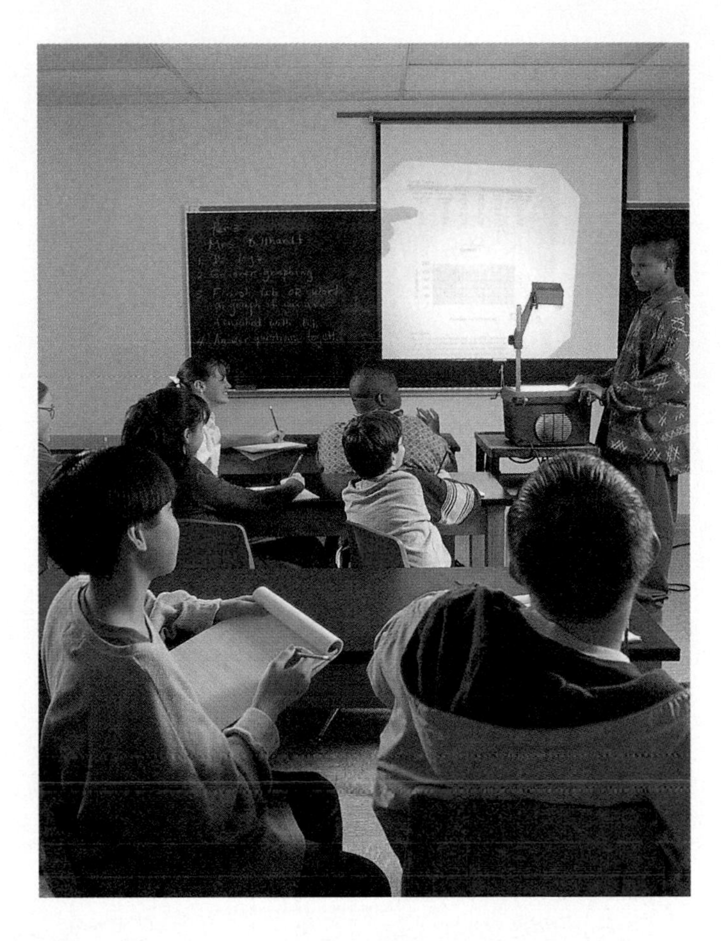

Figure 14
Communicating experimental results is an important part of the laboratory experience.

Communicating Your Results Every investigation begins because a problem needs to be solved. Analyzing data and drawing conclusions are the end of the investigation. However, they are not the end of the work a scientist does. Usually, scientists communicate their results to other scientists, government agencies, private industries, or the public. They write reports and presentations that provide details on how experiments were carried out, summaries of the data, and final conclusions. They can include recommendations for further research. Scientists usually publish their most important findings.

✔ Reading Check ***Why is it important for scientists to communicate their data?***

Just as scientists communicate their findings, you will have the chance to communicate your data and conclusions to other members of your science class, as shown in **Figure 14.** You can give an oral presentation, create a poster, display your results on a bulletin board, prepare computer graphics, or talk with other students or your teacher. You will share with other groups the charts, tables, and graphs that show your data. Your teacher, or other students, might have questions about your investigation or your conclusions. Organized data and careful analysis will allow you to answer most questions and to discuss your work confidently. Analyzing and sharing data are important parts of descriptive and experimental research, as shown in **Figure 15.**

Figure 15

Scientists use a series of steps to solve scientific problems. Depending on the type of problem, they may use descriptive research or experimental research with controlled conditions. Several of the research steps involved in determining water quality at a wastewater treatment plant are shown here.

A Gathering background information is an important first step in descriptive and experimental research.

B Some questions can be answered by descriptive research. Here, the scientists make and record observations about the appearance of a water sample.

C Some questions can be answered by experimentation. These scientists collect a wastewater sample for testing under controlled conditions in the laboratory.

D Careful analysis of data is essential after completing experiments and observations. The technician at right uses computers and other instruments to analyze data.

Experimental Research Design

Another way to solve scientific problems is through experimentation. Experimental research design answers scientific questions by observation of a controlled situation. Experimental research design includes several steps.

Form a Hypothesis A **hypothesis** (hi PAH thuh sus) is a prediction, or statement, that can be tested. You use your prior knowledge, new information, and any previous observations to form a hypothesis.

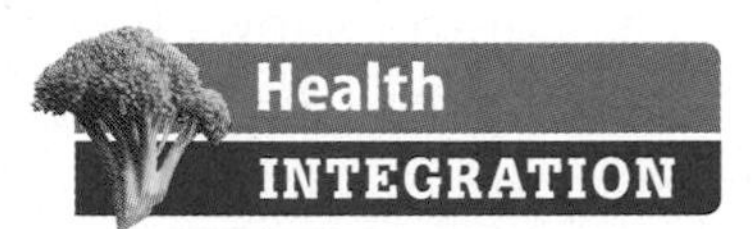

Variables In well-planned experiments, one factor, or variable, is changed at a time. This means that the variable is controlled. The variable that is changed is called the **independent variable.** In the experiment shown below, the independent variable is the amount or type of antibiotic applied to the bacteria. A **dependent variable** is the factor being measured. The dependent variable in this experiment is the growth of the bacteria, as shown in **Figure 16.**

To test which of two antibiotics will kill a type of bacterium, you must make sure that every variable remains the same but the type of antibiotic. The variables that stay the same are called **constants.** For example, you cannot run the experiments at two different room temperatures, for different lengths of time, or with different amounts of antibiotics.

Figure 16
In this experiment, the effect of two different antibiotics on bacterial growth was tested. The type of antibiotic is the independent variable.

A At the beginning of the experiment, dishes A and B of bacteria were treated with different antibiotics. The control dish did not receive any antibiotic.

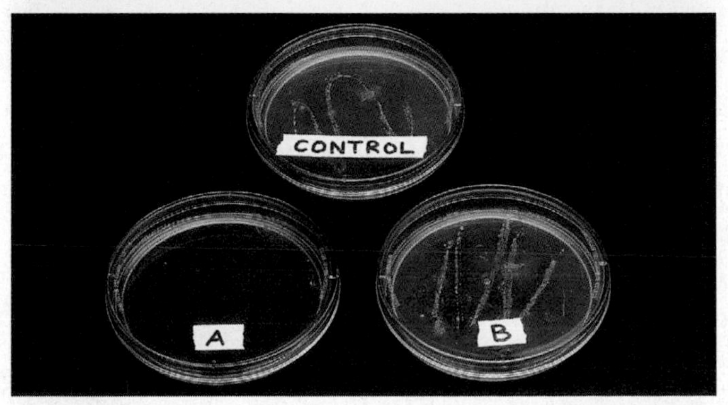

B The results of the experiment are shown. All factors were constant except the type of antibiotic applied. *Based on these photographs, what would you conclude about the effects of these antibiotics on bacteria?*

Figure 17
Check with your teacher several times as you plan your experiment.

Identify Controls Your experiment will not be valid unless a control is used. A **control** is a sample that is treated like the other experimental groups except that the independent variable is not applied to it. In the experiment with antibiotics, your control is a sample of bacteria that is not treated with either antibiotic. The control shows how the bacteria grow when left untreated by either antibiotic.

Reading Check *What is an experimental control?*

You have formed your hypothesis and planned your experiment. Before you begin, you must give a copy of it to your teacher, who must approve your materials and plans before you begin, as shown in **Figure 17.** This is also a good way to find out whether any problems exist in how you proposed to set up the experiment. Potential problems might include health and safety issues, length of time required to complete the experiment, and the cost and availability of materials.

Once you begin the experiment, make sure to carry it out as planned. Don't skip or change steps in the middle of the process. If you do, you will have to begin the experiment again. Also, you should record your observations and complete your data tables in a timely manner. Incomplete observations and reports result in data that are difficult to analyze and threaten the accuracy of your conclusions.

Number of Trials Experiments done the same way do not always have the same results. To make sure that your results are valid, you need to conduct several trials of your experiment. Multiple trials mean that an unusual outcome of the experiment won't be considered the true result. For example, if another substance is spilled accidentally on one of the containers with an antibiotic, that substance might kill the bacteria. Without results from other trials to use as comparisons, you might think that the antibiotic killed the bacteria. The more trials you do using the same methods, the more likely it is that your results will be reliable and repeatable. The number of trials you choose to do will be based on how much time, space, and material you have to complete the experiment.

Analyze Your Results After completing your experiment and obtaining all of your data, it is time to analyze your results. Now you can see if your data support your hypothesis. If the data do not support your original hypothesis, you can still learn from the experiment. Experiments that don't work out as you had planned can still provide valuable information. Perhaps your original hypothesis needs to be revised, or your experiment needs to be carried out in a different way. Maybe more background information is available that would help. In any case, remember that professional scientists, like those shown in **Figure 18,** rarely have results that support their hypothesis without completing numerous trials first.

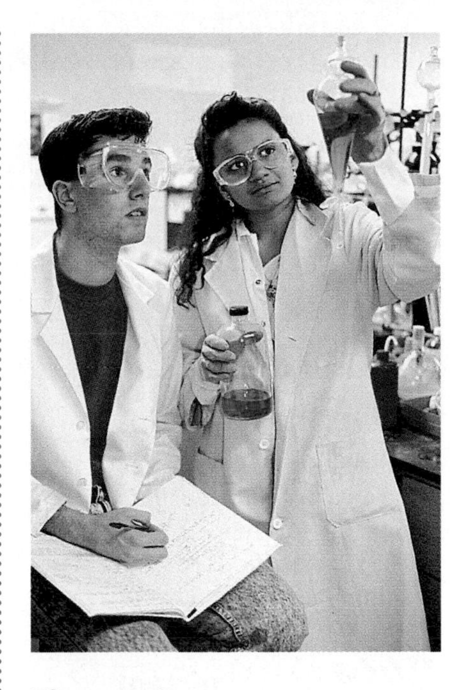

Figure 18
These scientists might work for months or years to find the best experimental design to test a hypothesis.

Reading Check ***What are some possible reasons that data are different than expected?***

After your results are analyzed, you can communicate them to your teacher and your class. Sharing the results of experiments allows you to hear new ideas from other students that might improve your research. Your results might contain information that will be helpful to other students.

In this section you learned the importance of scientific methods—steps used to solve a problem. Remember that some problems are solved using descriptive research, and others are solved through experimental research. In the next section you will learn more ways that science and technology are parts of your life.

Section  Assessment

1. Why do scientists use models? Give three examples of models.
2. What is a hypothesis?
3. Name the three steps scientists might use when designing an investigation to solve a problem.
4. Why is it important to identify carefully the problem to be solved?
5. **Think Critically** The data that you gathered and recorded during an experiment do not support your original hypothesis. Explain why your experiment is not a failure.

Skill Builder Activities

6. **Measuring in SI** Measurements can communicate experimental results. Use a meterstick to measure the length of your desktop in meters, centimeters, and millimeters. **For more help, refer to the Science Skill Handbook.**
7. **Using Percentages** A town of 1,000 people is divided into five areas, each with the same number of people. Use the data below to make a bar graph showing the number of people ill with cholera in each area. *Area: A–50%; B–5%; C–10%; D–16%; E–35%.* **For more help, refer to the Math Skill Handbook.**

SECTION

Science and Technology

As You Read

What **You'll Learn**

- **Determine** how science and technology influence your life.
- **Analyze** how modern technology allows scientific discoveries to be communicated worldwide.

Why **It's Important**

Modern communication systems allow scientific discoveries and information to be shared with people all over the world.

Science in Your Daily Life

You have learned how science is useful in your daily life. Doing science means more than just completing a science activity, reading a science chapter, memorizing vocabulary words, or following a scientific method to find answers.

Scientific Discoveries

Science is meaningful in other ways in your everyday life. New discoveries constantly lead to new products that influence your lifestyle or standard of living, such as those shown in **Figure 19.** For example, in the last 100 years, technological advances have allowed entertainment to move from live stage shows to large movie screens. Now DVDs allow users to choose a variety of options while viewing a movie. Do you want to hear English dialogue with French subtitles or Spanish dialogue with English subtitles? Do you want to change the ending? You can do it all from your chair by using the remote control.

Figure 19
New technology has changed the way people work and relax.

Technological Advances Technology also makes your life more convenient. Hand-held computers can be carried in a pocket. Foods can be prepared quickly in microwave ovens, and hydraulic tools make construction work easier and faster. A satellite tracking system in your car can give you verbal and visual directions to a destination in an unfamiliar city.

New discoveries influence other areas of your life as well, including your health. Technological advances, like the ones shown in **Figure 20,** help many people lead healthier lives. A disease might be controlled by a skin patch that releases a constant dose of medicine into your body. Miniature instruments allow doctors to operate on unborn children and save their lives. Bacteria also have been engineered to make important drugs such as insulin for people with diabetes.

Figure 20
Modern medical technology helps people have better health. The physician is studying a series of X rays. New, more complete ways of seeing internal problems helps to solve them.

Reading Check *What new scientific discoveries have you used?*

Science—The Product of Many

New scientific knowledge can mean that old ways of thinking or doing things are challenged. Aristotle, an ancient Greek philosopher, classified living organisms into plants and animals. This system worked until new tools, such as the microscope, allowed scientists to study organisms in greater detail. The new information changed how scientists viewed the living world. The current classification system will be used only as long as it continues to answer questions scientists have or until a new discovery allows them to look at information in a different way.

Research Visit the Glencoe Science Web site at **science.glencoe.com** for news about students who have made a scientific discovery or invented new technology. Communicate to your class what you learn.

Figure 21
Science and technology are the results of many people's efforts.

Who practices science? Scientific discoveries have never been limited to people of one race, sex, culture, or time period, or to professional scientists, as shown in **Figure 21.** In fact, students your age have made some important discoveries.

A Sarita M. James was a teenager when she developed a system that allows computers to recognize human speech easily.

B Stephen Hawking, a physicist, studies the universe and black holes.

C Grace Murray Hopper, a mathematician and software developer, helped pioneer the computer field.

D Fred Begay is a physicist who studies ways to produce heat energy without harming the environment.

E Ellen Ochoa is an inventor and an astronaut in NASA's space shuttle program.

F Daniel Hale Williams performed the first open-heart surgery and founded a hospital.

Use of Scientific Information The Internet quickly spreads word of new discoveries. New knowledge and technology brought about by these discoveries are shared by people in all countries. Any information gathered from the Internet must be checked carefully for accuracy.

Science provides new information every day that people use to make decisions. A new drug can be found or a new way to produce electricity can be developed. However, science cannot decide whether the new information is good or bad, moral or immoral. People decide whether the new information is used to help or harm the world and its inhabitants.

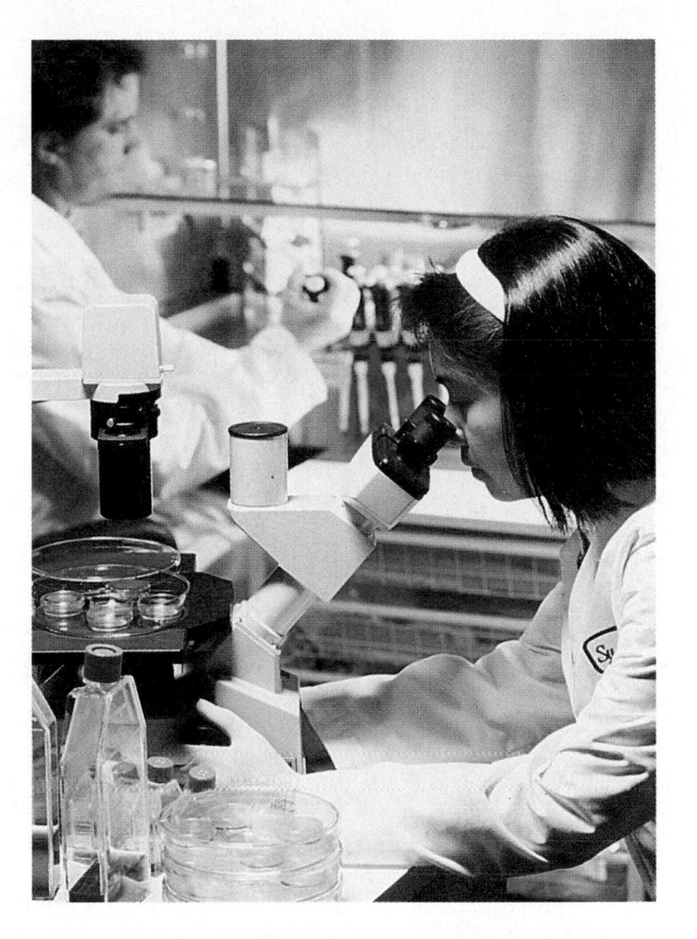

Figure 22
Modern laboratories allow scientists to track the source of a disease or solve many other scientific problems.

Looking to the Future

Midori and Luis discovered that technology has changed how modern scientists track the source of a disease. New information about bacteria and modern tools, such as those shown in **Figure 22,** help identify specific types of these organisms. Computers are used to model how the bacteria kill healthy cells or which part of a population the bacteria will infect. Today's scientists use cellular phones and computers to communicate with each other. This information technology has led to the globalization, or worldwide distribution, of information.

Section  Assessment

1. What is one way that science or technology has improved your health?
2. What might cause scientists to change a 100-year-old theory?
3. List five ways that scientists are able to communicate their discoveries.
4. Name an advance in technology that makes your life more enjoyable. What discoveries contributed to this technology?
5. **Think Critically** Explain why modern communications systems are important to scientists worldwide.

Skill Builder Activities

6. **Comparing and Contrasting** Make a drawing showing how a modern scientist and one from the 1800s would communicate their data with other scientists of their time. **For more help, refer to the Science Skill Handbook.**
7. **Using a Word Processor** Research the life of a famous scientist. Find at least two sources for your information. Take notes on ten facts about the scientist and use a word processing program to write a short biography. **For more help, refer to the Technology Skill Handbook.**

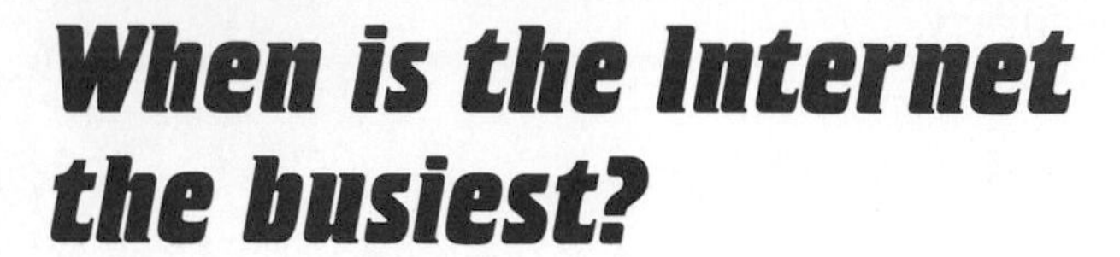

When is the Internet the busiest?

Using the Internet, you can get information any time from practically anywhere in the world. It has been called the "information superhighway." But does the Internet ever get traffic jams like real highways? Is the Internet busier at certain times?

Recognize the Problem

How long does it take data to travel across the Internet at different times of the day?

Form a Hypothesis

Think about using the Internet. When are you online most often? When do most other people use the Internet? Can you measure how busy the Internet is? Form a prediction about what time of day the Internet is the busiest.

Goals

- **Observe** when you, your friends, or your family use the Internet.
- **Research** how to measure the speed of the Internet.
- **Identify** the times of day when the Internet is the busiest in different areas of the country.
- **Graph** your findings and communicate them to other students.

Data Source

SCIENCE*Online* Go to the Glencoe Science Web site at **science.glencoe.com** for more information on how to measure the speed of the Internet, when the Internet is busiest, and data from other students.

Test Your Prediction

Plan

1. **Observe** when you, your family, and your friends use the Internet. Do you think that everyone in the world uses the Internet during the same times?
2. How are you going to measure the speed of the Internet? Research different factors that might affect the speed of the Internet. What are your variables?
3. How many times are you going to measure the speed of the Internet? What times of day are you going to gather your data?

Do

1. Make sure your teacher approves your plan before you start.
2. Visit the Glencoe Science Web site. Click on the Web Links button to view links that will help you do this activity.
3. Complete your investigation as planned.
4. **Record** all of your data in your Science Journal.
5. **Share** your data by posting it on the Glencoe Science Web site.

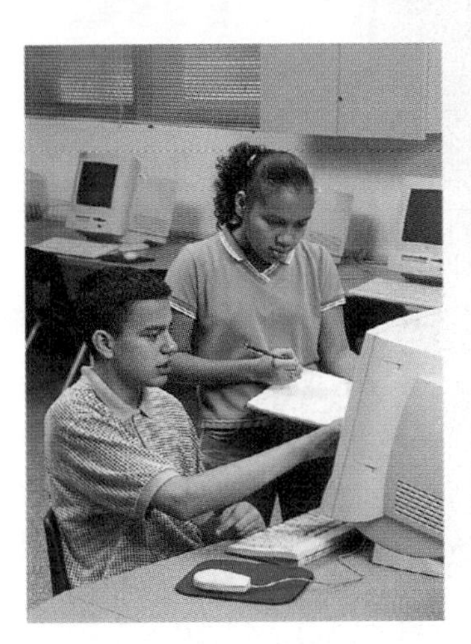

Analyze Your Data

1. **Record** in your Science Journal what time of day you found it took the most time to send data over the Internet.
2. **Compare** your results with those of other students around the country. In which areas did data travel the most quickly?

Draw Conclusions

1. **Compare** your findings to those of your classmates and other data that were posted on the Glencoe Science Web site. When is the Internet the busiest in your area? How does that compare to different areas of the country?
2. What factors could cause different results in your class?
3. How do you think your data would be affected if you had performed this experiment during a different time of the year, like the winter holidays?

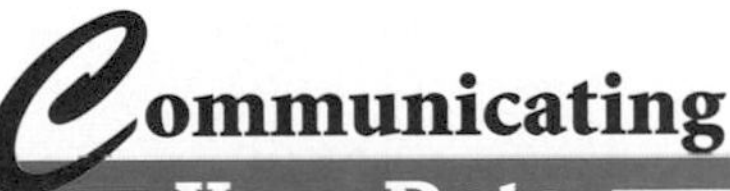

SCIENCE Online Find this *Use the Internet* activity on the Glencoe Science Web site at **science.glencoe.com.** **Post** your data in the table provided. Combine your data with those of other students and plot the combined data on a map to recognize patterns in internet traffic.

Science and Language Arts

The Everglades: River of Grass

by Marjory Stoneman Douglas

Respond to the Reading

1. How would you verify facts contained in this passage such as the construction of the dike and its location?
2. What hints does the author give you about her opinion of the dike-building project?

In this passage, Douglas writes about Lake Okeechobee, the large freshwater lake that lies in the southern part of Florida, north of the Everglades. A dike is an earthen wall usually built to protect against floods.

Something had to be done about the control of Okeechobee waters in storms. . . . A vast dike was constructed from east to south to west of the lake, within its average rim.[1] Canal gates were opened in it. It rises now between the lake itself and all those busy towns. . . .

To see the vast pale water you climb the levee[2] and look out upon its emptiness, hear the limpkins[3] crying among the islands of reeds in the foreground, and watch the wheeling creaking sea gulls flying about a man cutting bait in a boat. . . .

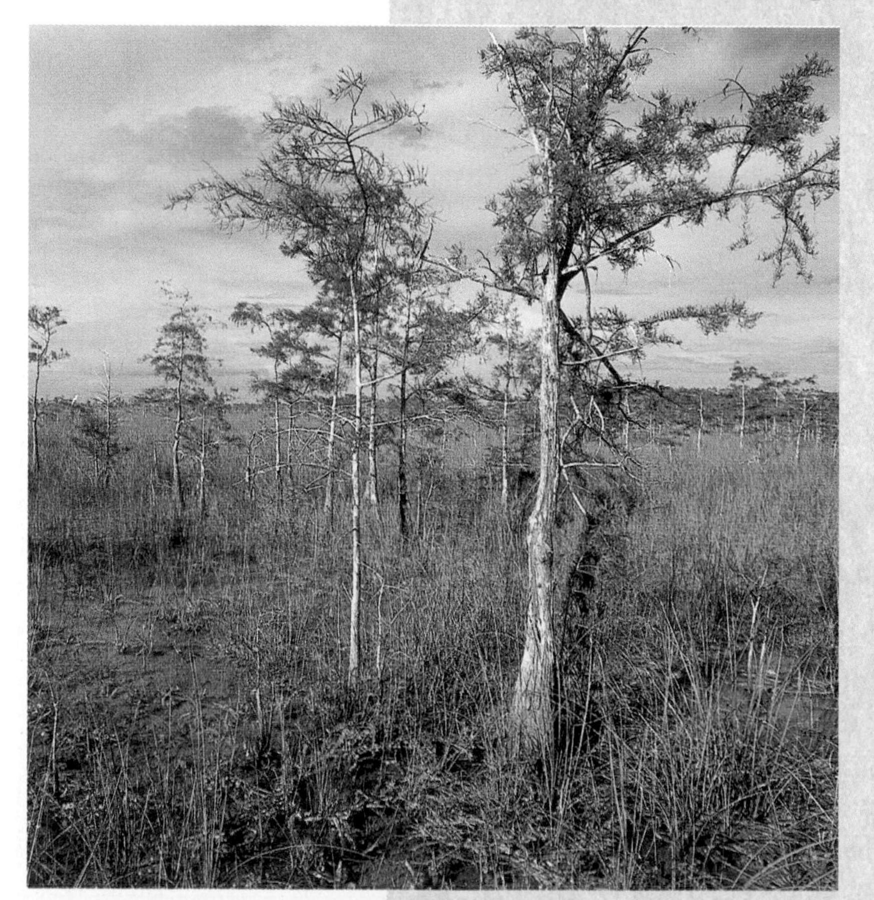

From the lake the control project extended west, cutting a long ugly canal straight through the green curving jungle and the grove-covered banks of Caloosahatchee [River].

[1] "Average rim" refers to the average location of the southern bank of the lake. Before the dike was built, heavy rains routinely caused Lake Okeechobee to overflow, emptying water over its southern banks into the Everglades. The overflowing water would carry silt and soil toward the southern banks of the lake, causing the southern banks to vary in size and location.
[2] dike
[3] waterbirds

Understanding Literature

Nonfiction *The Everglades: River of Grass* is a nonfiction book about the history of the Florida Everglades. Nonfiction stories are about real people, places, and events. Nonfiction includes autobiographies, biographies, and essays, as well as encyclopedias, history and science books, and newspaper and magazine articles. When reading nonfiction, you should ask yourself these questions: What do I know about this subject that will help me judge the accuracy of this information? Do I understand why the writer has included this specific information? Does the writer express an opinion about the subject matter?

Science Connection Because nonfiction is based upon real life, nonfiction writers must research their subjects thoroughly. As you learned in this chapter, scientific investigation involves being a detective. Author Marjory Stoneman Douglas relied upon her own observations as a long-time resident of Florida. She also thoroughly researched the history of the Florida Everglades. Douglas's book is an excellent example of the importance of communication in science. *The Everglades: River of Grass* brought the world's attention to the need to preserve the Everglades because of its unique ecosystems.

Linking Science and Writing

Nonfiction Write a one-page nonfiction account of your favorite outdoor place. You might write a description of the place, or write about an experience that you had there. Afterwards, read through your account and underline or highlight the facts that are in it. What information is factual? What information is based on your opinion?

Career Connection

Anthropologist

Jane Goodall entered the African jungle at the age of 26 to study chimps in the wild. Goodall spent five years observing the chimps, taking extensive notes, and earning their trust. Since then, she has been the world's foremost expert on chimpanzees. She was the first person to discover that chimpanzees use tools to gather food. Today she spends her time traveling and teaching young people about the environment.

SCIENCE*Online* To learn more about careers in anthropology, visit the Glencoe Science Web site at **science.glencoe.com.**

Chapter 1 Study Guide

Reviewing Main Ideas

Section 1 What is science?

1. Science is a process that can be used to solve problems or answer questions. Everyone uses science every day.
2. Scientists use tools to measure. *Why should this student use a thermometer to measure the temperature of the water?*

3. Technology is the application of science to make tools and products you use each day. Computers are a valuable technological tool.
4. Communication is an important part of all aspects of science.

Section 2 Doing Science

1. No one scientific method is used to solve all problems. Organization and careful planning are important when trying to solve any problem.
2. Scientific questions can be answered by descriptive research or experimental research.
3. Models save time and money by testing ideas that are too difficult to build or carry out. Computer models cannot completely replace experimentation. *Why might this scientist use a computer model to design aircraft?*

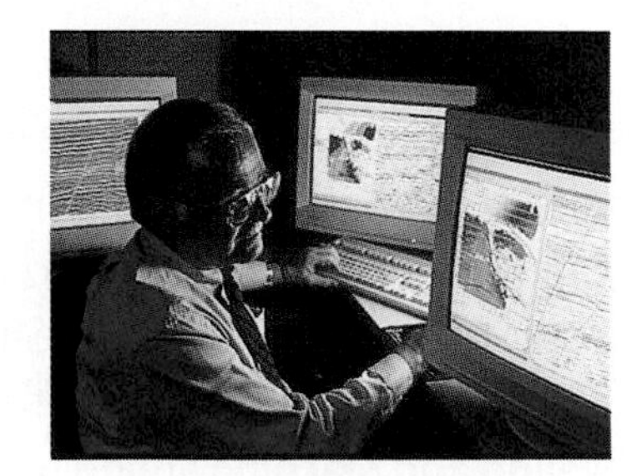

4. A hypothesis is an idea that can be tested. Sometimes experiments don't support the original hypothesis, and a new hypothesis must be formed.
5. In a well-planned experiment, there is a control and only one variable is changed at a time. All other factors are kept constant.

Section 3 Science and Technology

1. Science is part of everyone's life. New discoveries lead to new technology and products.
2. Science continues to challenge old knowledge and ways of doing things. Old ideas are kept until new discoveries prove them wrong.
3. People of all races, ages, sexes, cultures, and professions practice science.
4. Modern communication assures that scientific information is spread around the world. *When a new discovery is made in a South American rain forest, how can a scientist in Chicago find out about the data?*

After You Read

FOLDABLES Reading & Study Skills

Exchange your Question Study Fold with another classmate to learn more about other scientists. Write down the information your classmates collected on their scientists.

Chapter 1 Study Guide

Visualizing Main Ideas

Complete the following concept map with steps to solving a problem.

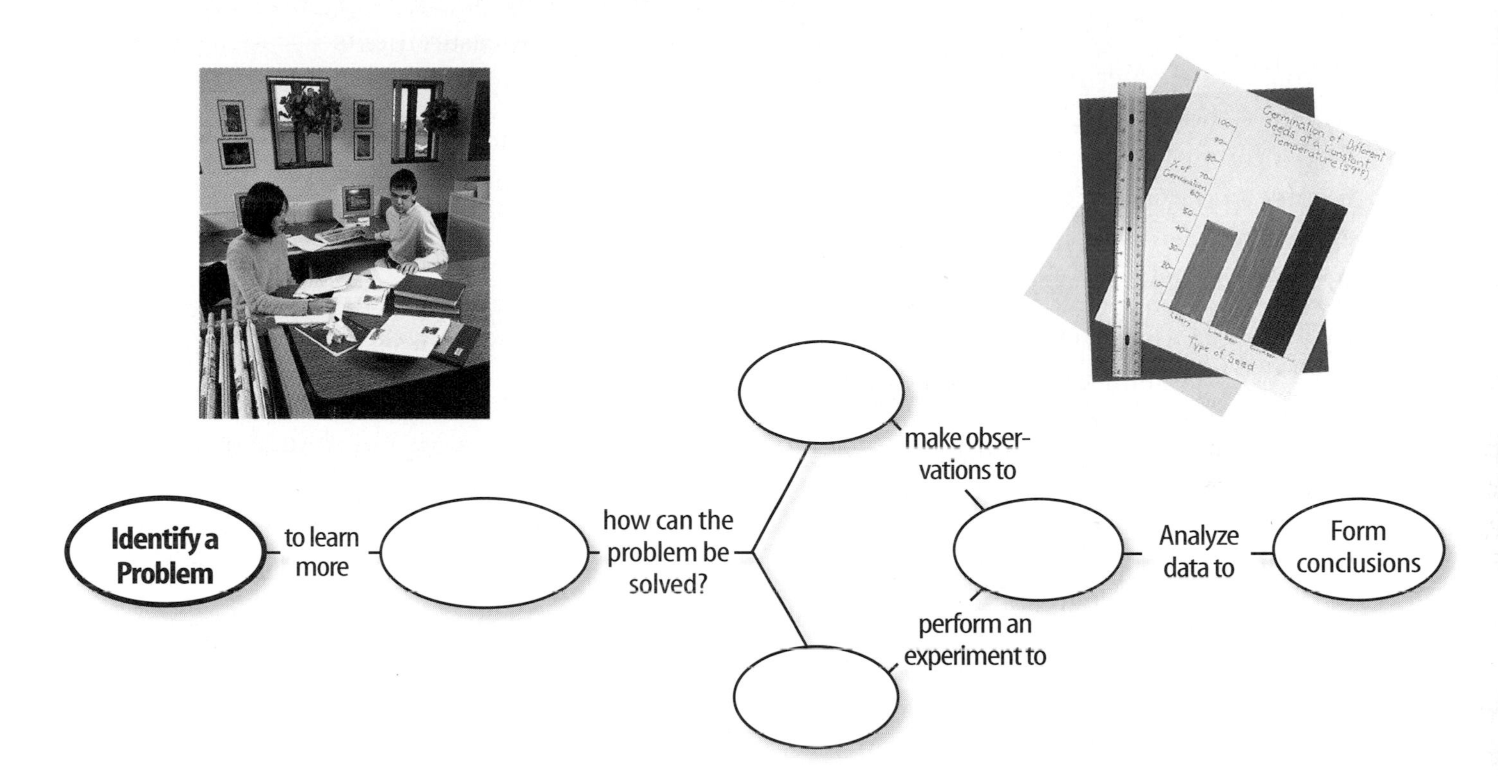

Vocabulary Review

Vocabulary Words

a. constant
b. control
c. dependent variable
d. descriptive research
e. experimental research design
f. hypothesis
g. independent variable
h. model
i. science
j. scientific methods
k. technology

THE PRINCETON REVIEW Study Tip

Practice reading tables. Devise a graph that shows the same information as a table does.

Using Vocabulary

Match each phrase with the correct vocabulary word from the list.

1. the factor being measured in an experiment
2. a statement that can be tested
3. use of knowledge to make products
4. sample treated like other experimental groups except variable is not applied
5. steps to follow to solve a problem
6. a variable that stays the same during every trial of an experiment
7. the variable that is changed in an experiment

Chapter 1 Assessment

Checking Concepts

Choose the word or phrase that best answers the question.

1. To make sure experimental results are valid, which of these procedures must be followed?
   A) conduct multiple trials
   B) pick two hypotheses
   C) add bias
   D) communicate uncertain results

2. In an experiment on bacteria, using different amounts of antibiotics is an example of which of the following?
   A) control **C)** bias
   B) hypothesis **D)** variable

3. Computers are used in science to do which of the following processes?
   A) analyze data
   B) make models
   C) communicate with other scientists
   D) all of the above

4. If you use a computer to make a three-dimensional picture of a building, it is an example of which of the following?
   A) model **C)** control
   B) hypothesis **D)** variable

5. Predictions about what will happen can be based on which of the following?
   A) controls **C)** technology
   B) prior knowledge **D)** number of trials

6. Which of the following is the greatest concern for scientists using the Internet?
   A) speed **C)** language
   B) availability **D)** accuracy

7. Which of the following can be used to record information?
   A) conclusion **C)** observation
   B) data table **D)** hypothesis

8. When scientists make a prediction that can be tested, what skill is being used?
   A) hypothesizing
   B) inferring
   C) taking measurements
   D) making models

9. Which of the following is the first step toward finding a solution?
   A) analyze data
   B) draw a conclusion
   C) identify the problem
   D) test the hypothesis

10. Which of the following terms describes a variable that does not change in an experiment?
    A) hypothesis **C)** constant
    B) dependent **D)** independent

Thinking Critically

11. How is a Science Journal a valuable tool for scientists?

12. Why is it important to record data as they are collected?

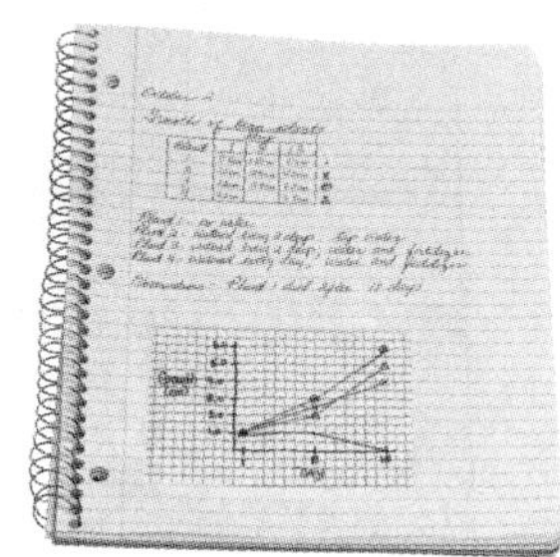

13. What is the difference between analyzing data and drawing conclusions?

14. What is the advantage of eliminating bias in experiments?

15. When trying to solve a problem, why do scientists collect information about what is already known?

Developing Skills

16. **Recognizing Cause and Effect** If three variables were changed at one time, what would happen to the accuracy of the conclusions made for an experiment?

17. Communicating Prepare a report for a kindergarten class about the result of a science experiment. How would this be different than a report prepared for a group of adults?

18. Interpreting Data You applied three different antibiotics to three bacteria samples. The control bacteria sample did not receive any antibiotics. All four of the bacteria samples grew at the same rate. How could you interpret your data?

19. Making and Using Graphs Prepare a bar graph of the data in this table. Which age group seems most likely to get the disease?

Disease Victims

Age Group (years)	Number of People
0–5	37
6–10	20
11–15	2
16–20	1
over 20	0

20. Form Hypotheses More books were checked out of the school library during April than at any other time of the year. Form a hypothesis about why this occurred. How can the hypothesis be tested?

Performance Assessment

21. Poster Create a poster showing steps in a scientific method. Use creative images to show the steps to solving a scientific problem.

Technology

Go to the Glencoe Science Web site at **science.glencoe.com** or use the **Glencoe Science CD-ROM** for additional chapter assessment.

Test Practice

Anita heated four beakers of water to boiling. Each beaker contained 100 mL of water with a certain amount of salt dissolved in it. The results of her experiment are listed in the table below.

Boiling Temperatures of Salt Solutions

Beaker	Mass of Salt Dissolved (g)	Boiling Temperature (°C)
A	0	99
B	3	102
C	6	105
D	9	?

Study the table and answer the following questions.

1. If everything remains the same, what will be the boiling temperature of the solution in beaker D?

A) 102°C
B) 106°C
C) 108°C
D) 110°C

2. Which hypothesis probably was being tested by this experiment?

F) The boiling temperature of water increases as the mass of water increases.
G) When salt is dissolved in water, the water takes longer to boil.
H) The boiling temperature of salt depends on the mass of the salt.
J) The boiling temperature of a salt-water solution increases as the mass of dissolved salt increases.

CHAPTER 2 Measurement

Does the expression "winning by a nose" mean anything to you? If you have ever "won by a nose," that means the race was close. Sometimes horse races, such as this one, are so close the winner has to be determined by a photograph. But there is more to measure than just how close the race was. How fast did the horse run? Did he break a record? In this chapter, you will learn how scientists measure things like distance, time, volume, and temperature. You also will learn how to use illustrations, pictures, and graphs to communicate measurements.

What do you think?

Science Journal Look at the picture below with a classmate. Discuss what you think this might be. Here's a hint: *How fast did you come up with an answer?* Write your answer or best guess in your Science Journal.

You make measurements every day. If you want to communicate those measurements to others, how can you be sure that they will understand exactly what you mean? Using vague words without units won't work. Do the Explore Activity below to see the confusion that can result from using measurements that aren't standard.

Measure length

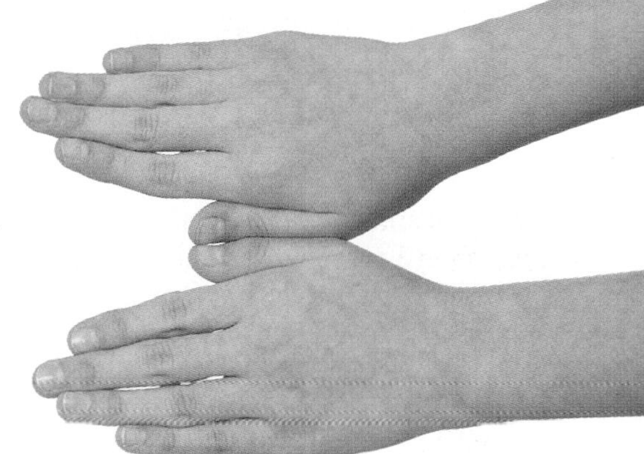

1. As a class, choose six objects to measure in your classroom.
2. Measure each object using the width of your hand and write your measurements in your Science Journal.
3. Compare your measurements to those of your classmates.

Observe

Is your hand the same width as your classmates' hands? Discuss in your Science Journal why it is better to switch from using hands to using units of measurement that are the same all the time.

Before You Read

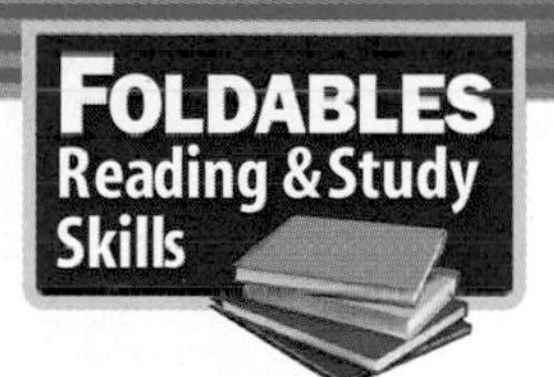

Making an Organizational Study Fold **When information is grouped into clear categories, it is easier to understand what you are learning. Before you begin reading, make the following Foldable to help you organize your thoughts about measurements.**

1. Place a sheet of paper in front of you so the short side is at the top. Fold the paper in half from the left side to the right side two times. Unfold all the folds.
2. Fold the paper from top to bottom in equal thirds and then in half. Unfold all the folds.
3. Trace over all the fold lines and label the table you created. Label the columns: *Estimate It, Measure It,* and *Round It,* as shown. Label the rows: *Length of ______, Volume of ______, Mass of ______, Temperature of ______,* and *Rate of ______,* as shown.
4. Before you read the chapter, select objects to measure and estimate their measurements. As you read the chapter, complete the *Measure It* column.

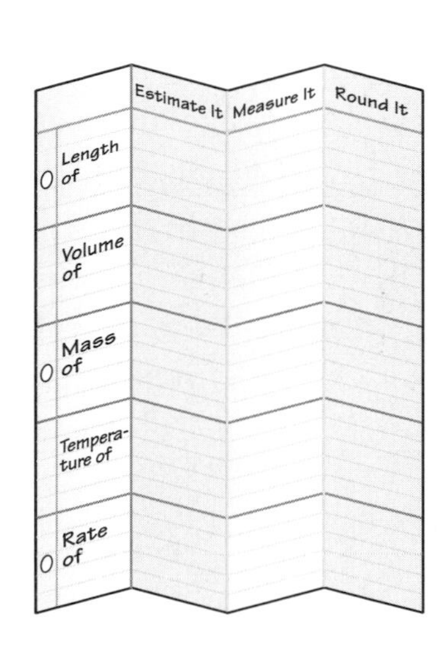

SECTION

Description and Measurement

As You Read

What You'll Learn

- **Determine** how reasonable a measurement is by estimating.
- **Identify** and use the rules for rounding a number.
- **Distinguish** between precision and accuracy in measurements.

Vocabulary

measurement
estimation
precision
accuracy

Why It's Important

Measurement helps you communicate information and ideas.

Measurement

How would you describe what you are wearing today? You might start with the colors of your outfit, and perhaps you would even describe the style. Then you might mention sizes—size 7 shoes, size 14 shirt. Every day you are surrounded by numbers. **Measurement** is a way to describe the world with numbers. It answers questions such as how much, how long, or how far. Measurement can describe the amount of milk in a carton, the cost of a new compact disc, or the distance between your home and your school. It also can describe the volume of water in a swimming pool, the mass of an atom, or how fast a penguin's heart pumps blood.

The circular device in **Figure 1** is designed to measure the performance of an automobile in a crash test. Engineers use this information to design safer vehicles. In scientific endeavors, it is important that scientists rely on measurements instead of the opinions of individuals. You would not know how safe the automobile is if this researcher turned in a report that said, "Vehicle did fairly well in head-on collision when traveling at a moderate speed." What does "fairly well" mean? What is a "moderate speed?"

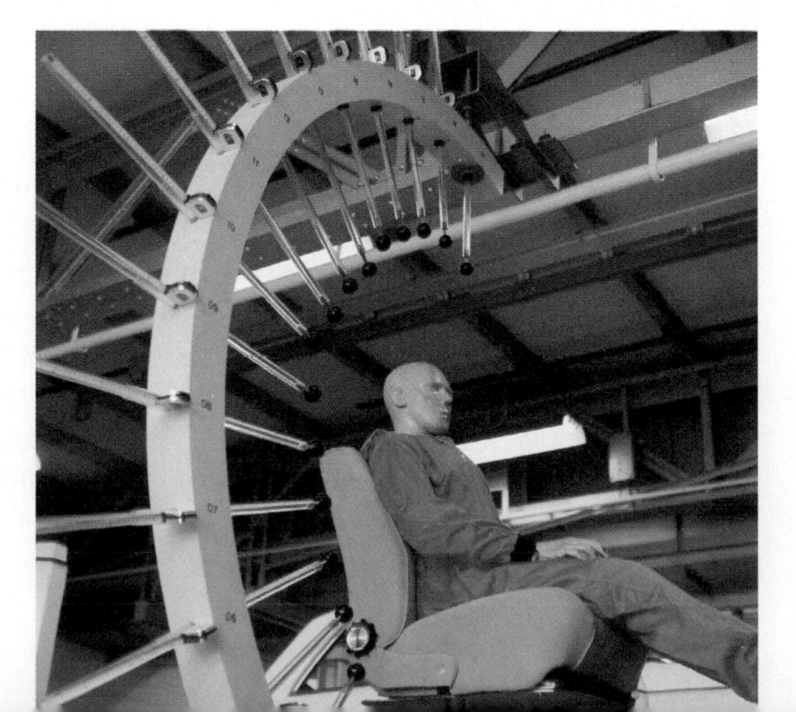

Figure 1
This device measures the range of motion of a seat-belted mannequin in a simulated accident.

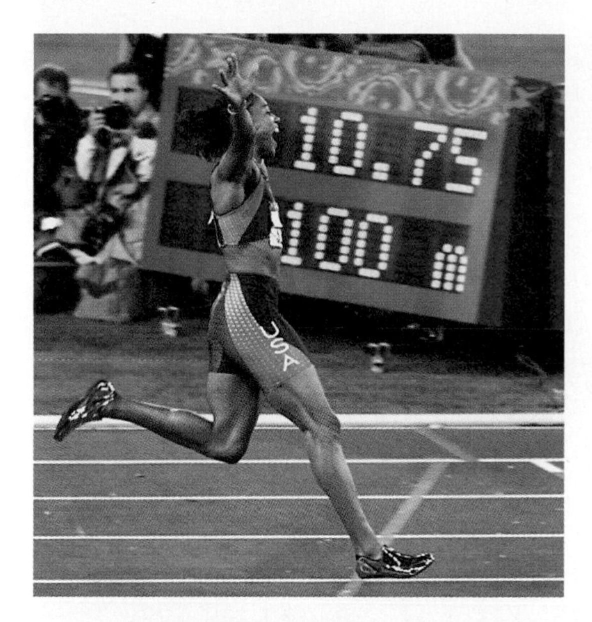

Figure 2
Accurate measurement of distance and time is important for competitive sports like track and field. *Why wouldn't a clock that measured in minutes be precise enough for this race?*

Describing Events Measurement also can describe events such as the one shown in **Figure 2.** In the 1956 summer Olympics, sprinter Betty Cuthbert of Australia came in first in the women's 200 m dash. She ran the race in 23.4 s. In the 2000 summer Olympics, Marion Jones of the United States won the 100-m dash in a time of 10.75 s. In this example, measurements convey information about the year of the race, its length, the finishing order, and the time. Information about who competed and in what event are not measurements but help describe the event completely.

Estimation

What happens when you want to know the size of an object but you can't measure it? Perhaps it is too large to measure or you don't have a ruler handy. **Estimation** can help you make a rough measurement of an object. When you estimate, you can use your knowledge of the size of something familiar to estimate the size of a new object. Estimation is a skill based on previous experience and is useful when you are in a hurry and exact numbers are not required. Estimation is a valuable skill that improves with experience, practice, and understanding.

Reading Check ***When should you not estimate a value?***

How practical is the skill of estimation? In many instances, estimation is used on a daily basis. A caterer prepares for each night's crowd based on an estimation of how many will order each entree. A chef makes her prize-winning chili. She doesn't measure the cumin; she adds "just that much." Firefighters estimate how much hose to pull off the truck when they arrive at a burning building.

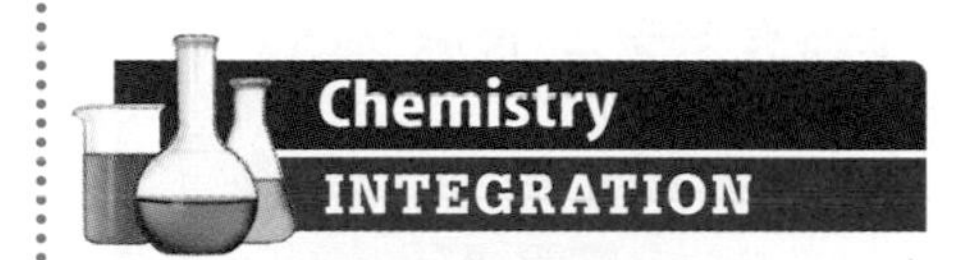

A description of matter that does not involve measurement is *qualitative*. For example, water is composed of hydrogen and oxygen. A *quantitative* description uses numbers to describe. For example, one water molecule is composed of one oxygen atom and two hydrogen atoms. Research another compound containing hydrogen and oxygen—hydrogen peroxide. Infer a qualitative and quantitative description of hydrogen peroxide in your Science Journal.

Figure 3
This student is about 1.5 m tall. *Estimate the height of the tree in the photo.*

Measuring Accurately

Procedure

1. Fill a **400-mL beaker** with **crushed ice.** Add enough **cold water** to fill the beaker.
2. Make three measurements of the temperature of the ice water using a **computer temperature probe.** Remove the computer probe and allow it to warm to room temperature between each measurement. Record the measurements in your **Science Journal.**
3. Repeat step two using an **alcohol thermometer.**

Analysis

1. Average each set of measurements.
2. Which measuring device is more precise? Explain. Can you determine which is more accurate? How?

Using Estimation You can use comparisons to estimate measurements. For example, the tree in **Figure 3** is too tall to measure easily, but because you know the height of the student next to the tree, you can estimate the height of the tree. When you estimate, you often use the word *about.* For example, doorknobs are about 1 m above the floor, a sack of flour has a mass of about 2 kg, and you can walk about 5 km in an hour.

Estimation also is used to check that an answer is reasonable. Suppose you calculate your friend's running speed as 47 m/s. You are familiar with how long a second is and how long a meter is. Think about it. Can your friend really run a 50-m dash in 1 s? Estimation tells you that 47 m/s is unrealistically fast and you need to check your work.

Precision and Accuracy

One way to evaluate measurements is to determine whether they are precise. **Precision** is a description of how close measurements are to each other. Suppose you measure the distance between your home and your school five times with an odometer. Each time, you determine the distance to be 2.7 km. Suppose a friend repeated the measurements and measured 2.7 km on two days, 2.8 km on two days, and 2.6 km on the fifth day. Because your measurements were closer to each other than your friend's measurements, yours were more precise. The term *precision* also is used when discussing the number of decimal places a measuring device can measure. A clock with a second hand is considered more precise than one with only an hour hand.

Degrees of Precision The timing for Olympic events has become more precise over the years. Events that were measured in tenths of a second 100 years ago are measured to the hundredth of a second today. Today's measuring devices are more precise. **Figure 4** shows an example of measurements of time with varying degrees of precision.

Accuracy When you compare a measurement to the real, actual, or accepted value, you are describing **accuracy.** A watch with a second hand is more precise than one with only an hour hand, but if it is not properly set, the readings could be off by an hour or more. Therefore, the watch is not accurate. On the other hand, measurements of 1.03m, 1.04m and 1.06m compared to an actual value of 1.05 m is accurate, but not precise. **Figure 5** illustrates the difference between precision and accuracy.

Reading Check *What is the difference between precision and accuracy?*

Figure 4
Each of these clocks provides a different level of precision. *Which of the three could you use to be sure to make the 3:35 bus?*

A **Before the invention of clocks, as they are known today, a sundial was used. As the Sun passes through the sky, a shadow moves around the dial.**

B **For centuries, analog clocks—the kind with a face—were the standard.**

C **Digital clocks are now as common as analog ones.**

Figure 5

From golf to gymnastics, many sports require precision and accuracy. Archery—a sport that involves shooting arrows into a target—clearly shows the relationship between these two factors. An archer must be accurate enough to hit the bull's-eye and precise enough to do it repeatedly.

A The archer who shot these arrows is neither accurate nor precise—the arrows are scattered all around the target.

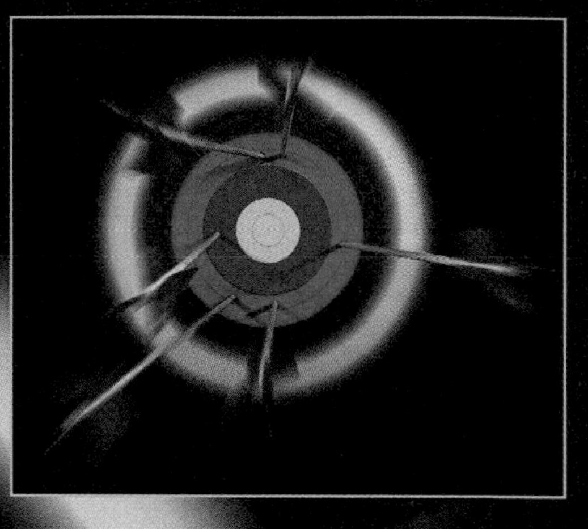

C Here we have a winner! All of the arrows have hit the bull's-eye, a result that is both precise and accurate.

B This archer's attempt demonstrates precision but not accuracy—the arrows were shot consistently to the left of the target's center.

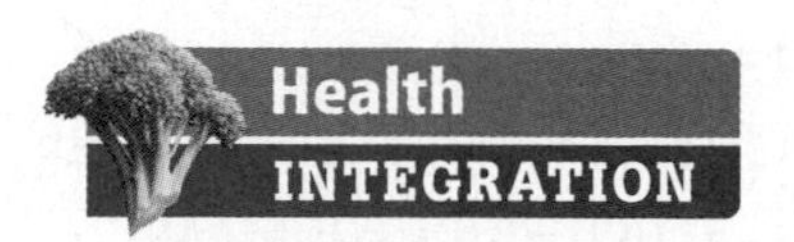

Precision and accuracy are important in many medical procedures. One of these procedures is the delivery of radiation in the treatment of cancerous tumors. Because radiation damages cells, it is important to limit the radiation to only the cancerous cells that are to be destroyed. A technique called Stereotactic Radiotherapy (SRT) allows doctors to be accurate and precise in delivering radiation to areas of the brain. The patient makes an impression of his or her teeth on a bite plate that is then attached to the radiation machine. This same bite plate is used for every treatment to position the patient precisely the same way each time. A CAT scan locates the tumor in relation to the bite plate, and the doctors can pinpoint with accuracy and precision where the radiation should go.

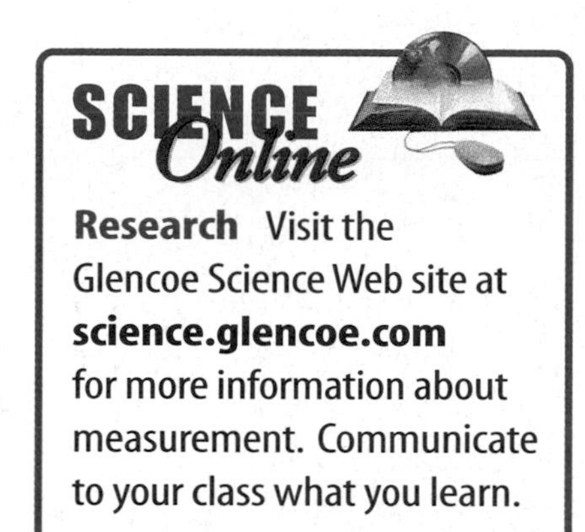

Research Visit the Glencoe Science Web site at **science.glencoe.com** for more information about measurement. Communicate to your class what you learn.

Rounding a Measurement Not all measurements have to be made with instruments that measure with great precision like the scale in **Figure 6.** Suppose you need to measure the length of the sidewalk outside your school. You could measure it to the nearest millimeter. However, you probably would need to know the length only to the nearest meter or tenth of a meter. So, if you found that the length was 135.841 m, you could round off that number to the nearest tenth of a meter and still be considered accurate. How would you round this number? To round a given value, follow these steps:

1. Look at the digit to the right of the place being rounded to.
   - If the digit to the right is 0, 1, 2, 3, or 4, the digit being rounded to remains the same.
   - If the digit to the right is 5, 6, 7, 8, or 9, the digit being rounded to increases by one.
2. The digits to the right of the digit being rounded to are deleted if they are also to the right of a decimal. If they are to the left of a decimal, they are changed to zeros.

Look back at the sidewalk example. If you want to round the sidewalk length of 135.841 to the tenths place, you look at the digit to the right of the 8. Because that digit is a 4, you keep the 8 and round it off to 135.8 m. If you want to round to the ones place, you look at the digit to the right of the 5. In this case you have an 8, so you round up, changing the 5 to a 6, and your answer is 136 m.

Figure 6
This laboratory scale measures to the nearest hundredth of a gram.

Precision and Number of Digits When might you need to round a number? Suppose you want to divide a 2-L bottle of soft drink equally among seven people. When you divide 2 by 7, your calculator display reads as shown in **Figure 7.** Will you measure exactly 0.285 714 285 L for each person? No. All you need to know is that each person gets about 0.3 L of soft drink.

Using Precision and Significant Digits The number of digits that truly reflect the precision of a number are called the significant digits or significant figures. They are figured as follows:

- Digits other than zero are always significant.
- Final zeros after a decimal point (6.54600 g) are significant.
- Zeros between any other digits (507.0301 g) are significant.
- Zeros before any other digits (0.0002030 g) are NOT significant.
- Zeros in a whole number (1650) may or may not be significant.
- An exact number, such as the number of people in a room or the number of meters in a kilometer, has infinite significant digits.

Math Skills Activity

Rounding

Example Problem

The mass of one object is 6.941 g. The mass of a second object is 20.180 g. You need to know these values only to the nearest whole number to solve a problem. What are the rounded values?

Solution

1 *This is what you know:* mass of first object = 6.941 g
mass of second object = 20.180 g

2 *This is what you need to know:* the number to the right of the one's place

first object: 9, second object: 1

3 *This is what you need to use:* digits 0, 1, 2, 3, 4 remain the same
for digits 5, 6, 7, 8, 9, round up

4 *Solution:* first object: 9 makes the 6 round up = 7
second object: 1 makes the 0 remain the same = 20

Practice Problem

What are the rounded masses of the objects to the nearest tenth of a unit?

For more help, refer to the Math Skill Handbook.

Following the Rules In the soda example you have an exact number, seven, for the number of people. This number has infinite significant digits. You also have the number two, for how many liters of soda you have. This has only one significant digit.

There are also rules to follow when deciding the number of significant digits in the answer to a calculation. They depend on what kind of calculation you are doing.

- For multiplication and division, you determine the number of significant digits in each number in your problem. The significant digits of your answer are determined by the number with fewer digits.

$$6.14 \times 5.6 = \boxed{34}.384$$

3 digits 2 digits 2 digits

- For addition and subtraction, you determine the place value of each number in your problem. The significant digits of the answer is determined by the number that is least precise.

6.14	to the hundredths
+ 5.6	to the tenths
$\boxed{11.7}$4	to the tenths

Therefore, in the soda example you are dividing and the limiting number of digits is determined by the amount of soda, 2 L. There is one significant digit there; therefore, your answer has one.

Reading Check ***What determines the number of significant digits in the answer to an addition problem?***

Figure 7
Sometimes considering the size of each digit will help you realize they are unneeded. In this calculation, the seven ten-thousandths of a liter represents just a few drops of soda.

Section  Assessment

1. Estimate the distance between your desk and your teacher's desk. Explain the method you used.
2. Measure the height of your desk to the nearest half centimeter.
3. Sarah's garden is 11.72 m long. Round to the nearest tenth of a meter.
4. John's puppy has chewed on his ruler. Will John's measurements be accurate or precise?
5. **Think Critically** Would the sum of 5.7 and 6.2 need to be rounded? Why or why not? Would the sum of 3.28 and 4.1 need to be rounded? Why or why not?

Skill Builder Activities

6. **Using Precision and Significant Digits** Perform the following calculations and express the answer using the correct number of significant digits: 42.35 + 214; 225/12. **For more help, refer to the Math Skill Handbook.**
7. **Communicating** Describe your backpack in your Science Journal. Include in your description one set of qualities that have no measurements, such as color and texture, and one set of measured quantities, such as width and mass. **For more help, refer to the Science Skill Handbook.**

SECTION 2 SI Units

As You Read

What You'll Learn

- **Identify** the purpose of SI.
- **Identify** the SI units of length, volume, mass, temperature, time, and rate.

Vocabulary

SI | kilogram
meter | kelvin
mass | rate

Why It's Important

The SI system is used throughout the world, allowing you to measure quantities in the exact same way as other students around the world.

The International System

Can you imagine how confusing it would be if people in every country used different measuring systems? Sharing data and ideas would be complicated. To avoid confusion, scientists established the International System of Units, or **SI,** in 1960 as the accepted system for measurement. It was designed to provide a worldwide standard of physical measurement for science, industry, and commerce. SI units are shown in **Table 1.**

Reading Check *Why was SI established?*

The SI units are related by multiples of ten. Any SI unit can be converted to a smaller or larger SI unit by multiplying by a power of 10. For example, to rewrite a kilogram measurement in grams, you multiply by 1,000. The new unit is renamed by changing the prefix, as shown in **Table 2.** For example, one millionth of a meter is one *micro*-meter. One thousand grams is one *kilo*gram. **Table 3** shows some common objects and their measurements in SI units.

Table 1 SI Base Units

Quantity	Unit	Symbol
length	meter	m
mass	kilogram	kg
temperature	kelvin	K
time	second	s
electric current	ampere	A
amount of substance	mole	mol
intensity of light	candela	cd

Table 2 SI Prefixes

Prefix	Multiplier
giga-	1,000,000,000
mega-	1,000,000
kilo-	1,000
hecto-	100
deka-	10
[unit]	1
deci-	0.1
centi-	0.01
milli-	0.001
micro-	0.000 001
nano-	0.000 000 001

Length

Length is defined as the distance between two points. Lengths measured with different tools can describe a range of things from the distance from Earth to Mars to the thickness of a human hair. In your laboratory activities, you usually will measure length with a metric ruler or meterstick.

The **meter** (m) is the SI unit of length. One meter is about the length of a baseball bat. The size of a room or the dimensions of a building would be measured in meters. For example, the height of the Washington Monument in Washington, D.C. is 169 m.

Smaller objects can be measured in centimeters (cm) or millimeters (mm). The length of your textbook or pencil would be measured in centimeters. A twenty-dollar bill is 15.5 cm long. You would use millimeters to measure the width of the words on this page. To measure the length of small things such as blood cells, bacteria, or viruses, scientists use micrometers (millionths of a meter) and nanometers (billionths of a meter).

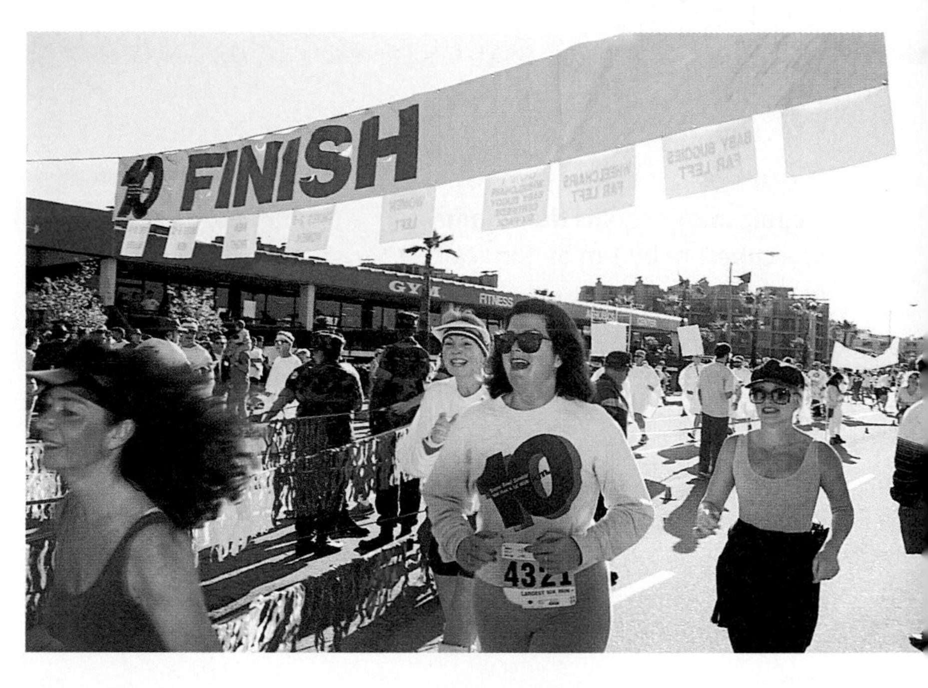

Figure 8
These runners have just completed a 10-kilometer race—known as a 10K. *About how many kilometers is the distance between your home and your school?*

A Long Way Sometimes people need to measure long distances, such as the distance a migrating bird travels or the distance from Earth to the Moon. To measure such lengths, you use kilometers. Kilometers might be most familiar to you as the distance traveled in a car or the measure of a long-distance race, as shown in **Figure 8.** The course of a marathon is measured carefully so that the competitors run 42.2 km. When you drive from New York to Los Angeles, you cover 4,501 km.

How important are accurate measurements? In 1999, the *Mars Climate Orbiter* disappeared as it was to begin orbiting Mars. NASA later discovered that a unit system error caused the flight path to be incorrect and the orbiter to be lost. Research the error and determine what systems of units were involved. How can using two different systems of units cause errors?

Table 3 Common Objects in SI Measurements

Object	Type of Measurement	Measurement
can of soda	volume	355 cm^3
bag of potatoes	mass	4.5 kg
fluorescent tube	length	1.2 m
refrigerator	temperature	276 K

Figure 9
A cubic meter equals the volume of a cube 1 m by 1 m by 1 m. *How many cubic centimeters are in a cubic meter?*

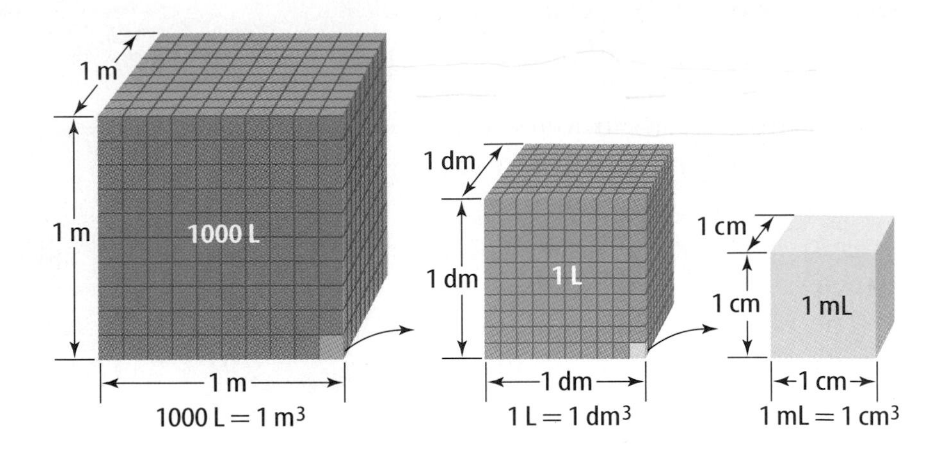

Volume

The amount of space an object occupies is its volume. The cubic meter (m^3), shown in **Figure 9,** is the SI unit of volume. You can measure smaller volumes with the cubic centimeter (cm^3 or cc). To find the volume of a square or rectangular object, such as a brick or your textbook, measure its length, width, and height and multiply them together. What is the volume of a compact disc case?

You are probably familiar with a 2-L bottle. A liter is a measurement of liquid volume. A cube 10 cm by 10 cm by 10 cm holds 1 L (1,000 cm^3) of water. A cube 1 cm on a side holds 1 mL (1 cm^3) of water.

Volume by Immersion Not all objects have an even, regular shape. How can you find the volume of something irregular like a rock or a piece of metal?

Have you ever added ice cubes to a nearly full glass of water only to have the water overflow? Why did the water overflow? Did you suddenly have more water? The volume of water did not increase at all, but the water was displaced when the ice cubes were added. Each ice cube takes up space or has volume. The difference in the volume of water before and after the addition of the ice cubes equals the volume of the ice cubes that are under the surface of the water.

The ice cubes took up space and caused the total volume in the glass to increase. When you measure the volume of an irregular object, you do the same thing. You start with a known volume of water and drop in, or immerse, the object. The increase in the volume of water is equal to the volume of the object.

TRY AT HOME
Mini LAB

Measuring Volume

Procedure

1. Fill a plastic or glass **liquid measuring cup** until half full with **water.** Measure the volume.
2. Find an **object,** such as a rock, that will fit in your measuring cup.
3. Carefully lower the object into the water. If it floats, push it just under the surface with a **pencil.**
4. Record in your **Science Journal** the new volume of the water.

Analysis

1. How much space does the object occupy?
2. If 1 mL of water occupies exactly 1 cm^3 of space, what is the volume of the object in cm^3?

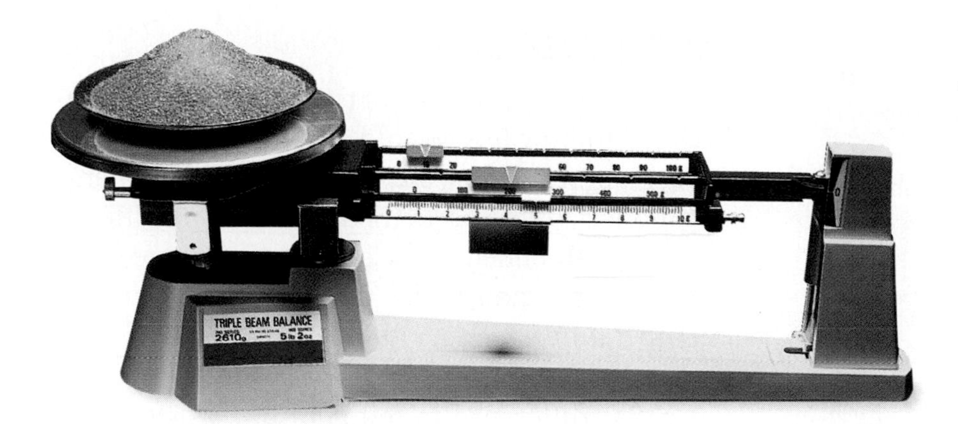

Figure 10
A triple beam balance compares an unknown mass to known masses.

Mass

The **mass** of an object measures the amount of matter in the object. The **kilogram** (kg) is the SI unit for mass. One liter of water has a mass of about 1 kg. Smaller masses are measured in grams (g). One gram is about the mass of a large paper clip.

You can determine mass with a triple beam balance, shown in **Figure 10.** The balance compares an object to a known mass. It is balanced when the known mass of the slides on the balance is equal to the mass of the object on the pan.

Why use the word *mass* instead of *weight?* Weight and mass are not the same. Mass depends only on the amount of matter in an object. If you ride in an elevator in the morning and then ride in the space shuttle later that afternoon, your mass is the same. Mass does not change when only your location changes.

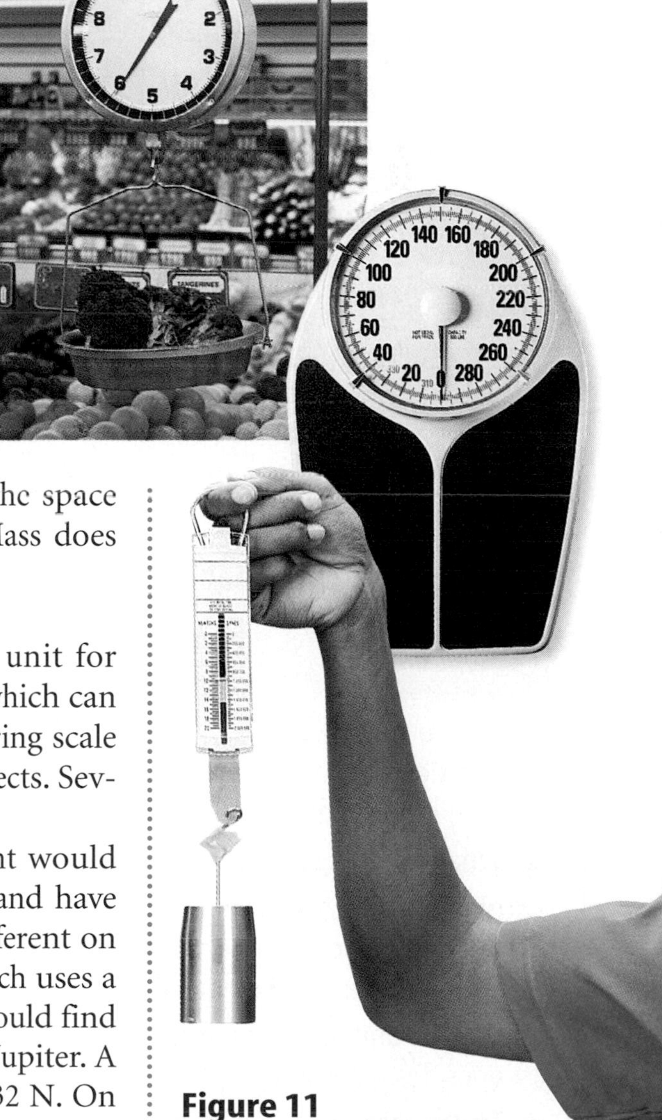

Figure 11
A spring scale measures an object's weight by how much it stretches a spring.

Weight Weight is a measurement of force. The SI unit for weight is the newton (N). Weight depends on gravity, which can change depending on where the object is located. A spring scale measures how a planet's gravitational force pulls on objects. Several spring scales are shown in **Figure 11.**

If you were to travel to other planets, your weight would change, even though you would still be the same size and have the same mass. This is because gravitational force is different on each planet. If you could take your bathroom scale, which uses a spring, to each of the planets in this solar system, you would find that you weigh much less on Mars and much more on Jupiter. A mass of 75 pounds, or 34 kg, on Earth is a weight of 332 N. On Mars, the same mass is 126 N, and on Jupiter it is 782 N.

✔ Reading Check *What does weight measure?*

Figure 12
The kelvin scale starts at 0 K. In theory, 0 K is the coldest temperature possible in nature.

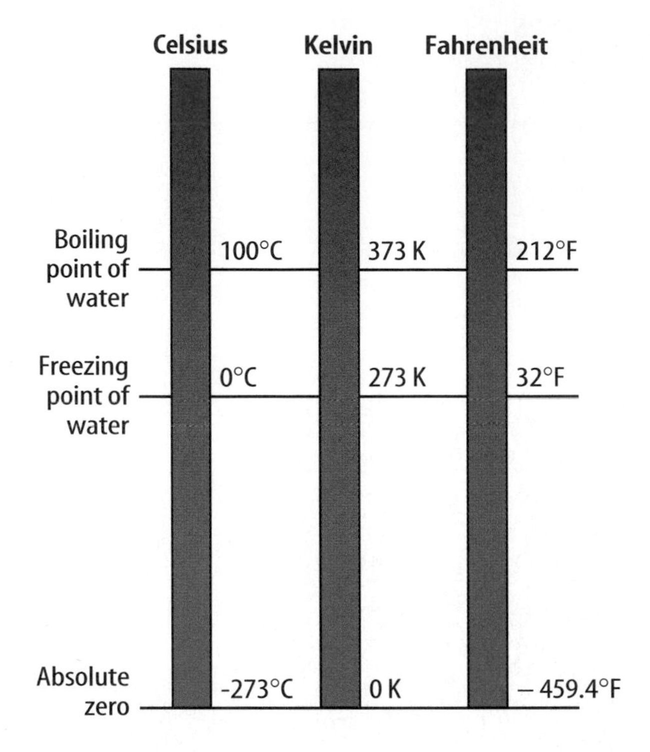

Temperature

The physical property of temperature is related to how hot or cold an object is. Temperature is a measure of the kinetic energy, or energy of motion, of the particles that make up matter.

The Fahrenheit and Celsius temperature scales are two common scales used on thermometers. Temperature is measured in SI with the **kelvin (K)** scale. A 1-K difference in temperature is the same as a 1°C difference in temperature, as shown in **Figure 12.** However, the two scales do not start at zero.

Time and Rates

Time is the interval between two events. The SI unit of time is the second (s). Time also is measured in hours (h). Although the hour is not an SI unit, it is easier to use for long periods of time. Can you imagine hearing that a marathon was run in 7,620 s instead of 2 h and 7 min?

A **rate** is the amount of change of one measurement in a given amount of time. One rate you are familiar with is speed, which is the distance traveled in a given time. Speeds often are measured in kilometers per hour (km/h).

The unit that is changing does not necessarily have to be an SI unit. For example, you can measure the number of cars that pass through an intersection per hour in cars/h. The annual rate of inflation can be measured in percent/year.

Section  Assessment

1. Describe a situation in which different units of measure could cause confusion.
2. What type of quantity does the cubic meter measure?
3. How would you change a measurement in centimeters to kilometers?
4. What SI unit replaces the pound? What does this measure?
5. **Think Critically** You are told to find the mass of a metal cube. How will you do it?

Skill Builder Activities

6. **Measuring in SI** Measure the length, volume, and mass of your textbook in SI units. Describe any tools or calculations you use. **For more help, refer to the Science Skill Handbook.**
7. **Converting Units** A block of wood is 0.2 m by 0.1 m by 0.5 m. Find its dimensions in centimeters. Use these to find its volume in cubic centimeters. Show your work. **For more help, refer to the Math Skill Handbook.**

Activity

Scale Drawing

A scale drawing is used to represent something that is too large or too small to be drawn at its actual size. Blueprints for a house are a good example of a scale drawing.

What You'll Investigate

How can you represent your classroom accurately in a scale drawing?

Materials

1-cm graph paper
pencil
metric ruler
meterstick

Goals

- **Measure** using SI.
- **Make** a data table.
- **Calculate** new measurements.
- **Make** an accurate scale drawing.

Procedure

1. Use your meterstick to measure the length and width of your classroom. Note the locations and sizes of doors and windows.
2. **Record** the lengths of each item in a data table similar to the one below.
3. Use a scale of 2 cm = 1 m to calculate the lengths to be used in the drawing. Record them in your data table.
4. **Draw** the floor plan. Include the scale.

Room Dimensions

Part of Room	Distance in Room (m)	Distance on Drawing (cm)

Conclude and Apply

1. How did you calculate the lengths to be used on your drawing? Did you put a scale on your drawing?
2. What would your scale drawing look like if you chose a different scale?
3. **Sketch** your room at home, estimating the distances. Compare this sketch to your scale drawing of the classroom. When would you use each type of illustration?
4. What measuring tool simplifies this task?

Communicating Your Data

Measure your room at home and compare it to the estimates on your sketch. Explain to someone at home what you did and how well you estimated the measurements. **For more help, refer to the Science Skill Handbook.**

SECTION

Drawings, Tables, and Graphs

As You Read

***What* You'll Learn**

- **Describe** how to use pictures and tables to give information.
- **Identify** and use three types of graphs.
- **Distinguish** the correct use of each type of graph.

Vocabulary

table
graph
line graph
bar graph
circle graph

***Why* It's Important**

Illustrations, tables, and graphs help you communicate data about the world around you in an organized and efficient way.

Scientific Illustrations

Most science books include pictures. Photographs and drawings model and illustrate ideas and sometimes make new information more clear than written text can. For example, a drawing of an airplane engine shows how all the parts fit together much better than several pages of text could describe it.

Drawings A drawing is sometimes the best choice to show details. For example, a canyon cut through red rock reveals many rock layers. If the layers are all shades of red, a drawing can show exactly where the lines between the layers are. The drawing can emphasize only the things that are necessary to show.

A drawing also can show things you can't see. You can't see the entire solar system, but drawings show you what it looks like. Also, you can make quick sketches to help model problems. For example, you could draw the outline of two continents to show how they might have fit together at one time.

A drawing can show hidden things, as well. A drawing can show the details of the water cycle, as in **Figure 13.** Architects use drawings to show what the inside of a building will look like. Biologists use drawings to show where the nerves in your arm are found.

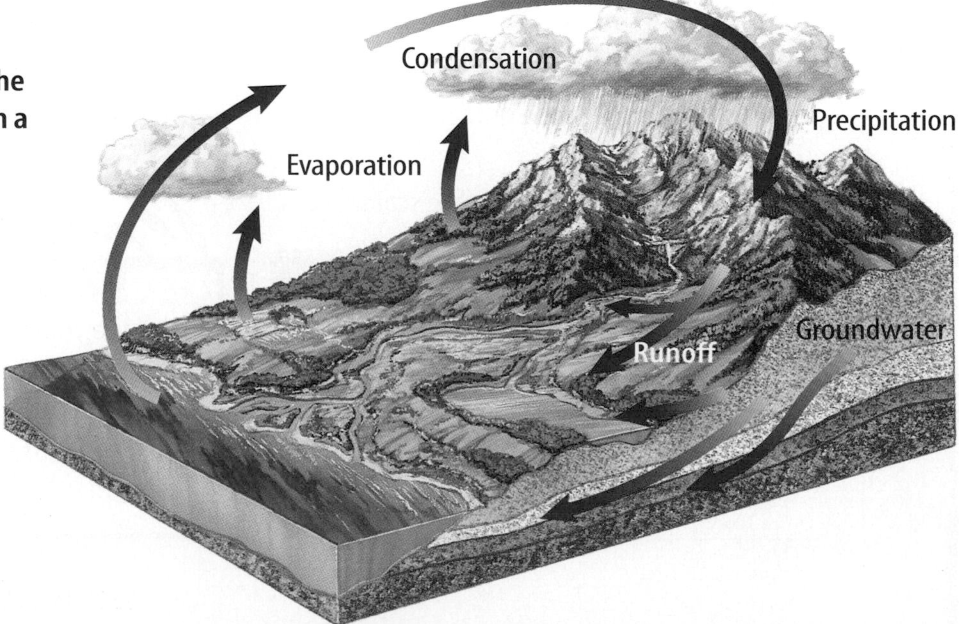

Figure 13
This drawing shows details of the water cycle that can't be seen in a photograph.

Photographs A still photograph shows an object exactly as it is at a single moment in time. Movies show how an object moves and can be slowed down or sped up to show interesting features. In your schoolwork, you might use photographs in a report. For example, you could show the different types of trees in your neighborhood for a report on ecology.

Tables and Graphs

Everyone who deals with numbers and compares measurements needs an organized way to collect and display data. A **table** displays information in rows and columns so that it is easier to read and understand, as seen in **Table 4.** The data in the table could be presented in a paragraph, but it would be harder to pick out the facts or make comparisons.

A **graph** is used to collect, organize, and summarize data in a visual way. The relationships between the data often are seen more clearly when shown in a graph. Three common types of graphs are line, bar, and circle graphs.

Line Graph A **line graph** shows the relationship between two variables. A variable is something that can change, or vary, such as the temperature of a liquid or the number of people in a race. Both variables in a line graph must be numbers. An example of a line graph is shown in **Figure 14.** One variable is shown on the horizontal axis, or *x*-axis, of the graph. The other variable is placed along the vertical axis, or *y*-axis. A line on the graph shows the relationship between the two variables.

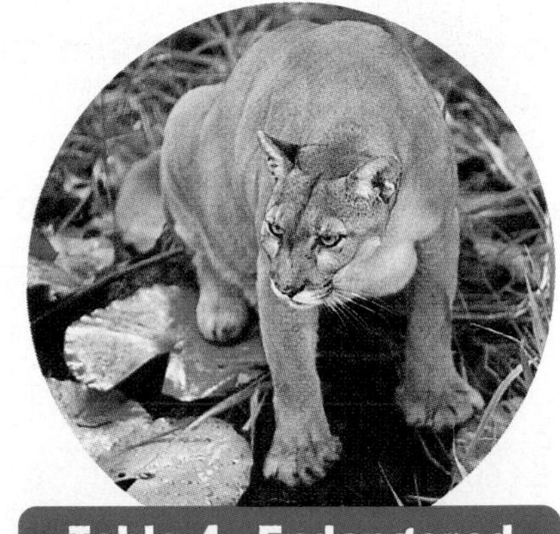

Table 4 Endangered Animal Species in the United States

Year	Number of Endangered Animal Species
1980	174
1982	179
1984	192
1986	213
1988	245
1990	263
1992	284
1994	321
1996	324
1998	357

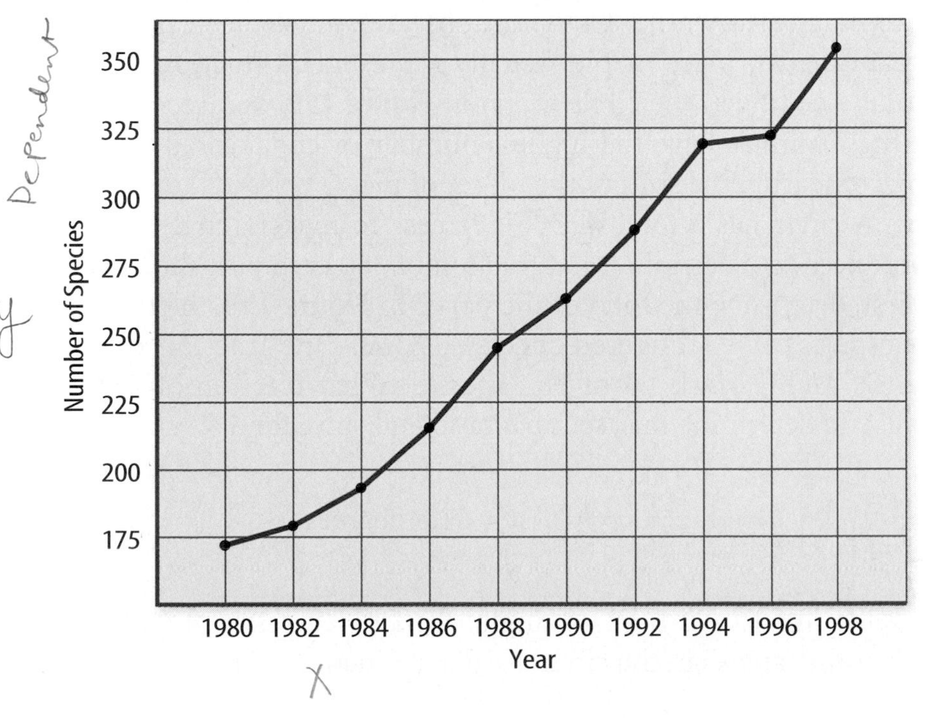

Figure 14
To find the number of endangered animal species in 1988, find that year on the *x*-axis and see what number corresponds to it on the *y*-axis.

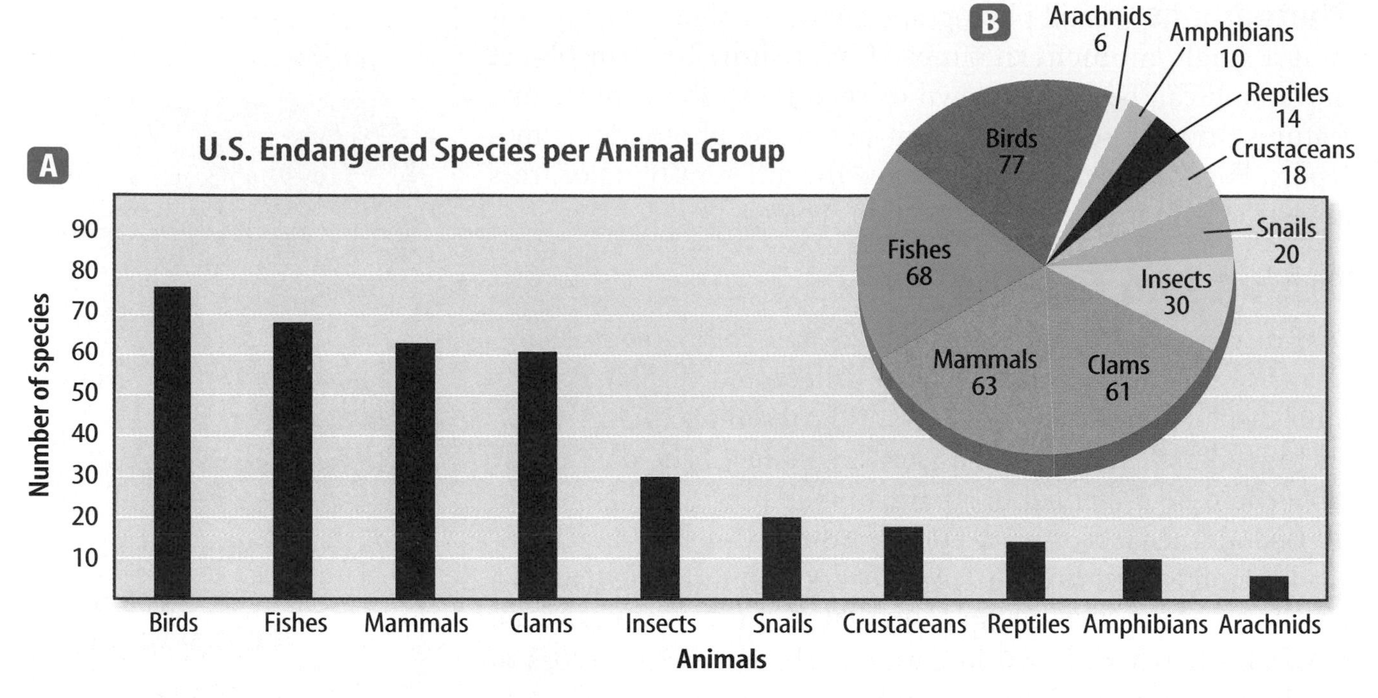

Figure 15
A Bar graphs allow you to picture the results easily. *Which category of animals has the most endangered species?* **B On this circle graph, you can see what part of the whole each animal represents.**

Bar Graph A **bar graph** uses rectangular blocks, or bars, of varying sizes to show the relationships among variables. One variable is divided into parts. It can be numbers, such as the time of day, or a category, such as an animal. The second variable must be a number. The bars show the size of the second variable. For example, if you made a bar graph of the endangered species data from **Figure 14,** the bar for 1990 would represent 263 species. An example of a bar graph is shown in **Figure 15A.**

Circle Graph Suppose you want to show the relationship among the types of endangered species. A **circle graph** shows the parts of a whole. Circle graphs are sometimes called pie graphs. Each piece of pie visually represents a fraction of the total. Looking at the circle graph in **Figure 15B,** you see quickly which animals have the highest number of endangered species by comparing the sizes of the pieces of pie.

A circle has a total of 360 degrees. To make a circle graph, you need to determine what fraction of 360 each part should be. First, determine the total of the parts. In **Figure 15B,** the total of the parts, or endangered species, is 367. One fraction of the total, *Mammals,* is 63 of 367 species. What fraction of 360 is this? To determine this, set up a ratio and solve for x:

$$\frac{63}{367} = \frac{x}{360} \qquad x = 61.8 \text{ degrees}$$

Mammals will have an angle of 61.8 degrees in the graph. The other angles in the circle are determined the same way.

Research Visit the Glencoe Science Web site at **science.glencoe.com** for more information about scientific illustrations. Communicate to your class what you learn.

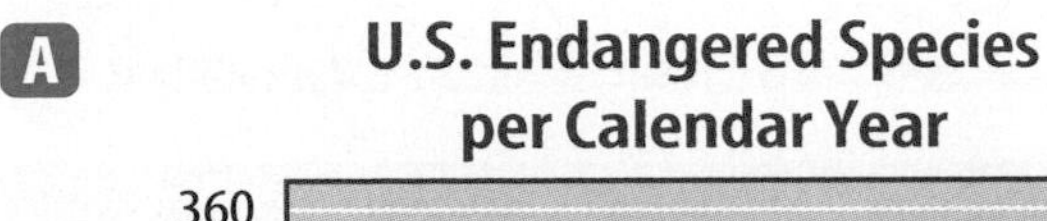

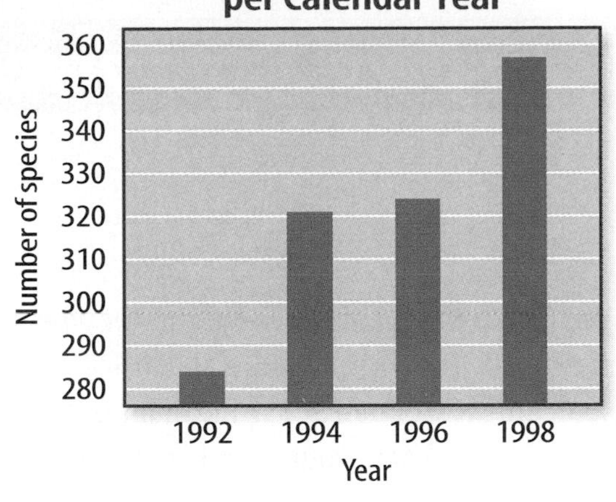

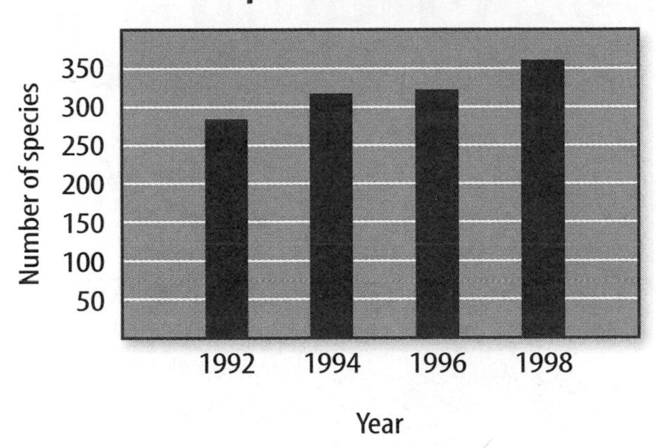

Reading Graphs When you are using or making graphs to display data, be careful—the scale of a graph can be misleading. The way the scale on a graph is marked can create the wrong impression, as seen in **Figure 16A.** Until you see that the *y*-axis doesn't start at zero, it appears that the number of endangered species has quadrupled in just six years.

This is called a broken scale and is used to highlight small but significant changes, just as an inset on a map draws attention to a small area of a larger map. **Figure 16B** shows the same data on a graph that does not have a broken scale. The number of species has only increased 22 percent from 1980 to 1986. Both graphs have correct data, but must be read carefully. Always analyze the measurements and graphs that you come across. If there is a surprising result, look closer at the scale.

Figure 16
Careful reading of graphs is important. A This graph does not start at zero, which makes it appear that the number of species has more than quadrupled from 1980 to 1986. B The actual increase is about 22 percent as you can see from this full graph. The broken scale must be noted in order to interpret the results correctly.

Section  Assessment

1. Describe a time when an illustration would be helpful in everyday activities.
2. Explain how to use **Figure 16** to find the number of endangered species in 1998.
3. Explain the difference between tables and graphs.
4. Suppose your class surveys students about after-school activities. What type of graph would you use to display your data? Explain.
5. **Think Critically** How are line, bar, and circle graphs the same? How are they different?

Skill Builder Activities

6. **Making and Using Graphs** Record the amount of time you spend reading each day for the next week. Then make a graph to display the data. What type of graph will you use? Could more than one kind of graph be used? **For more help, refer to the Science Skill Handbook.**
7. **Using an Electronic Spreadsheet** Use a spreadsheet to display how the total mass of a 500-kg elevator changes as 50-kg passengers are added one at a time. **For more help, refer to the Technology Skill Handbook.**

Activity Design Your Own Experiment

Pace Yourself

Track meets and other competitions require participants to walk, run, or wheel a distance that has been precisely measured. Officials make sure all participants begin at the same time, and each person's time is stopped at the finish line. If you are practicing for a marathon or 10K race, you need to know your speed or pace in order to compare it with those of other participants. How can your performance be measured accurately?

Recognize the Problem

How will you measure the speed of each person in your group? How will you display these data?

Form a Hypothesis

Think about the information you have learned about precision, measurement, and graphing. In your group, make a hypothesis about a technique that will provide you with the most precise measurement of each person's pace.

Goals

- **Design** an experiment that allows you to measure speed for each member of your group accurately.
- **Display** data in a table and a graph.

Possible Materials

meterstick
stopwatch
**watch with a second hand*
**Alternate materials*

Safety Precautions

Work in an area where it is safe to run. Participate only if you are physically able to exercise safely. As you design your plan, make a list of all the specific safety and health precautions you will take as you perform the investigation. Get your teacher's approval of the list before you begin.

Test Your Hypothesis

Plan

1. As a group, decide what materials you will need.
2. How far will you travel? How will you measure that distance? How precise can you be?
3. How will you measure time? How precise can you be?
4. List the steps and materials you will use to test your hypothesis. Be specific. Will you try any part of your test more than once?
5. Before you begin, create a data table. Your group must decide on its design. Be sure to leave enough room to record the results for each person's time. If more than one trial is to be run for each measurement, include room for the additional data.

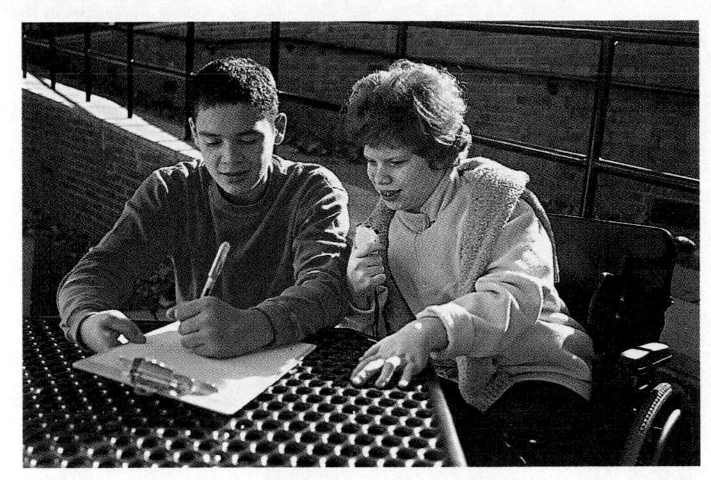

Do

1. Make sure that your teacher approves your plan before you start.
2. Carry out the experiment as planned and approved.
3. Be sure to record your data in the data table as you proceed with the measurements.

Analyze Your Data

1. **Graph** your data. What type of graph would be best?
2. Are your data table and graph easy to understand? Explain.
3. How do you know that your measurements are precise?
4. Do any of your data appear to be out of line with the rest?

Draw Conclusions

1. How is it possible for different members of a group to find different times while measuring the same event?
2. What tools would help you collect more precise data?
3. What other data displays could you use? What are the advantages and disadvantages of each?

Make a larger version of your graph to display in your classroom with the graphs of other groups. **For more help, refer to the Science Skill Handbook.**

Science Stats

Biggest, Tallest, Loudest

Did you know...

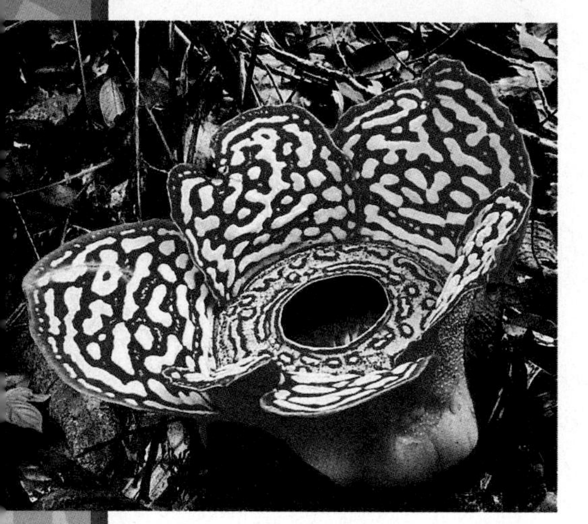

... The world's most massive flower belongs to a species called *Rafflesia* (ruh FLEE zhee uh) and has a mass of up to 11 kg. The diameter, or the distance across the flower's petals, can measure up to 1 m.

... The world's tallest building is the Petronus Towers in Kuala Lumpur, Malaysia. It is 452 m tall. The tallest building in the United States is Chicago's Sears Tower, shown here, which measures 442 m.

... The Grand Canyon is so deep— as much as 1,800 m—that it can hold more than four Empire State Buildings stacked on top of one another.

... The world's tallest tree is a coast redwood in the Montgomery Woods State Park in California. The tree stands 112.1 m high.

...The largest animal on Earth

is the blue whale. It can grow to be 33.5 m long. If 20 people who are each 1.65 m tall were lying head to toe, it would almost equal this length.

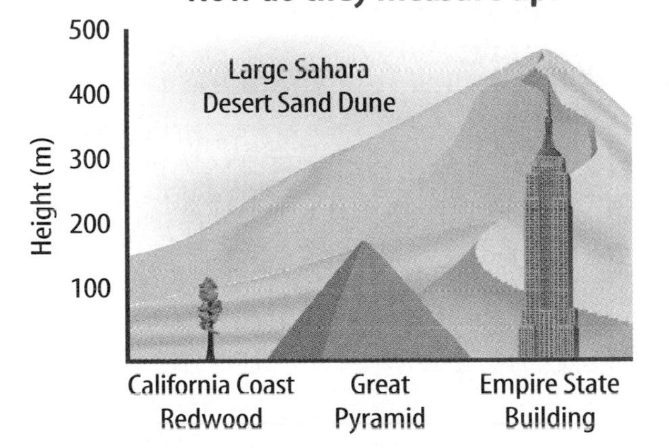

...One of the loudest explosions on Earth

was the 1883 eruption of Krakatau (krah kuh TAHEW), an Indonesian volcano. It was heard from more than 3,500 km away.

Do the Math

1. How many of the largest rafflesia petals would you have to place side by side to equal the length of a blue whale?
2. When Krakatau erupted, it ejected 18,000 km^3 of ash and rock. Other large eruptions released the following: Mount Pinatubo—7,000 km^3, Mount Katmai—13,000 km^3, Tambora—30,000 km^3, Vesuvius—5,000 km^3. Make a bar graph to compare the sizes of these eruptions.
3. Use the information provided about the Grand Canyon to calculate how many Sears Towers would have to stand end on end to equal the depth of the canyon.

Go Further

Do research on the Internet at **science.glencoe.com** to find facts that describe some of the shortest, smallest, or fastest things on Earth. Create a class bulletin board with the facts you and your classmates find.

Chapter 2 Study Guide

Reviewing Main Ideas

Section 1 Description and Measurement

1. Measurements such as length, volume, mass, temperature, and rates are used to describe objects and events.
2. Estimation is used to make an educated guess at a measurement.
3. Accuracy describes how close a measurement is to the true value.
4. Precision describes how close measurements are to each other. *Are the shots accurate or precise on the basketball hoop shown?*

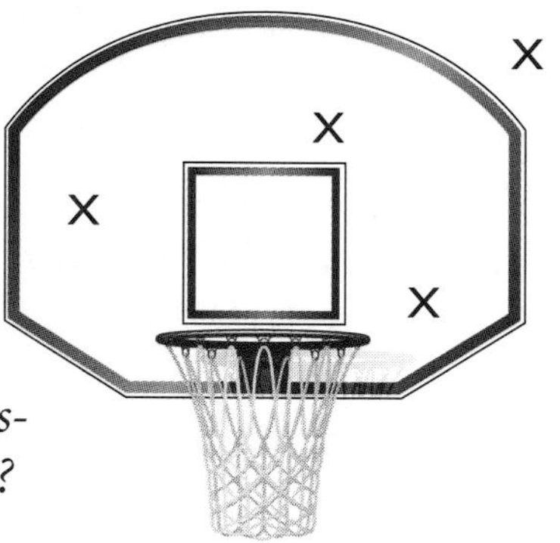

Section 2 SI Units

1. The international system of measurement is called SI. It is used throughout the world for communicating data.
2. The SI unit of length is the meter. Volume—the amount of space an object occupies—can be measured in cubic meters. The mass of an object is measured in kilograms. The SI unit of temperature is the kelvin. *What type of measurement is being made according to the sign shown?*

Section 3 Drawings, Tables, and Graphs

1. Tables, photographs, drawings, and graphs can sometimes present data more clearly than explaining everything in words. Scientists use these tools to collect, organize, summarize, and display data in a way that is easy to use and understand.
2. The three common types of graphs are line graphs, bar graphs, and circle graphs. *Which city on the line graph shown is the coldest in the fifth month?*

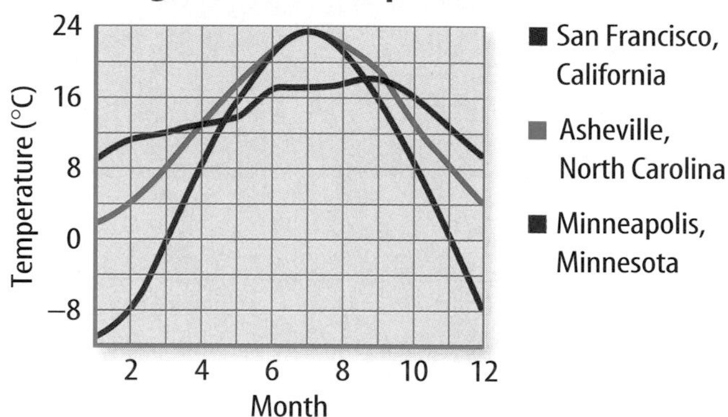

3. Line graphs show the relationship between two variables that are numbers on an x-axis and a y-axis. Bar graphs divide a variable into parts to show a relationship. Circle graphs show the parts of a whole like pieces of a pie.

After You Read

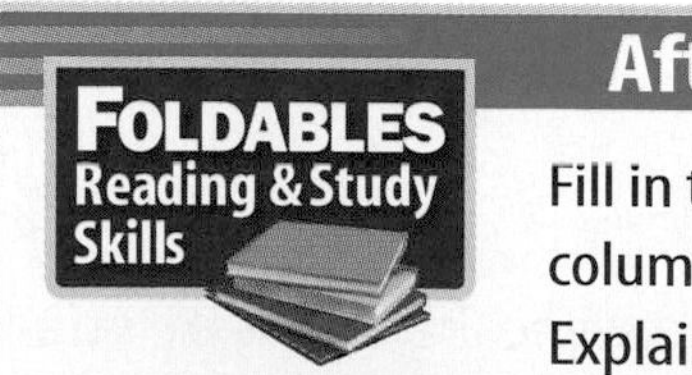

Fill in the *Round It* column on your Foldable. Explain when it is acceptable and appropriate for scientists to round measurements.

Visualizing Main Ideas

Complete the following concept map.

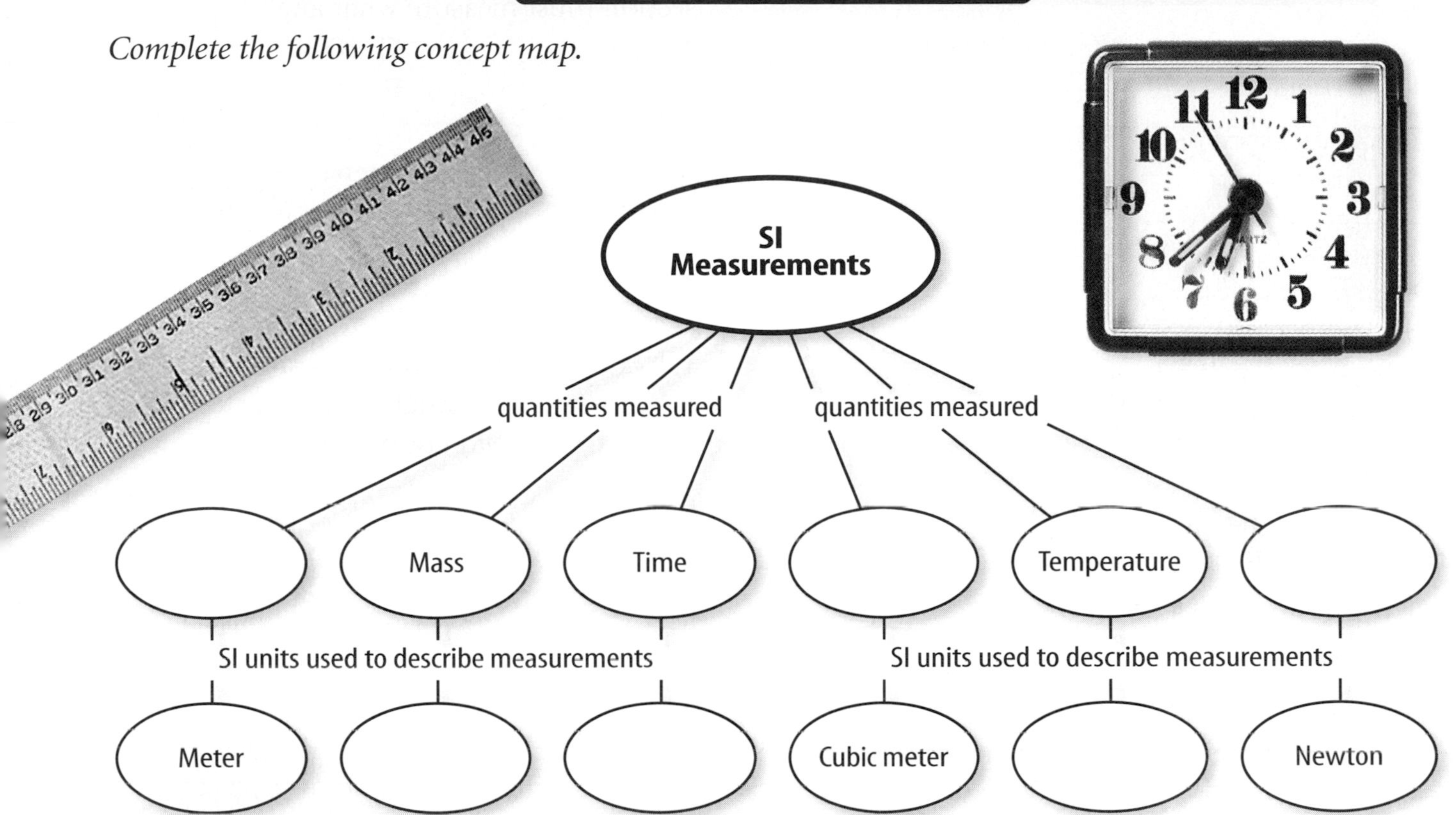

Vocabulary Review

Vocabulary Words

- **a.** accuracy
- **b.** bar graph
- **c.** circle graph
- **d.** estimation
- **e.** graph
- **f.** kelvin
- **g.** kilogram
- **h.** line graph
- **i.** mass
- **j.** measurement
- **k.** meter
- **l.** precision
- **m.** rate
- **n.** SI
- **o.** table

THE PRINCETON REVIEW Study Tip

When you encounter new vocabulary, write it down in your Science Journal. This will help you understand and remember them.

Using Vocabulary

Each phrase below describes a vocabulary word. Write the word that matches the phrase describing it.

1. the SI unit for length
2. a description with numbers
3. a method of making a rough measurement
4. the amount of matter in an object
5. a graph that shows parts of a whole
6. a description of how close measurements are to each other
7. the SI unit for temperature
8. an international system of units

Chapter 2 Assessment

Checking Concepts

Choose the word or phrase that best answers the question.

1. The measurement 25.81 g is precise to the nearest what?
A) gram
B) kilogram
C) tenth of a gram
D) hundredth of a gram

2. What is the SI unit of mass?
A) kilometer **C)** liter
B) meter **D)** kilogram

3. What would you use to measure length?
A) graduated cylinder
B) balance
C) meterstick
D) spring scale

4. The cubic meter is the SI unit of what?
A) volume **C)** mass
B) weight **D)** distance

5. Which term describes how close measurements are to each other?
A) significant digits **C)** accuracy
B) estimation **D)** precision

6. Which is a temperature scale?
A) volume **C)** Celsius
B) mass **D)** mercury

7. Which is used to organize data?
A) table **C)** precision
B) rate **D)** meterstick

8. To show the number of wins for each football team in your district, which of the following would you use?
A) photograph **C)** bar graph
B) line graph **D)** SI

9. What organizes data in rows and columns?
A) bar graph **C)** line graph
B) circle graph **D)** table

10. To show 25 percent on a circle graph, the section must measure what angle?
A) 25° **C)** 180°
B) 90° **D)** 360°

Thinking Critically

11. How would you estimate the volume your backpack could hold?

12. Why do scientists in the United States use SI rather than the English system (feet, pounds, pints, etc.) of measurement?

13. List the following in order from smallest to largest: 1 m, 1 mm, 10 km, 100 mm.

14. Describe an instance when you would use a line graph. Can you use a bar graph for the same purpose?

15. Computer graphics artists can specify the color of a point on a monitor by using characters for the intensities of three colors of light. Why was this method of describing color invented?

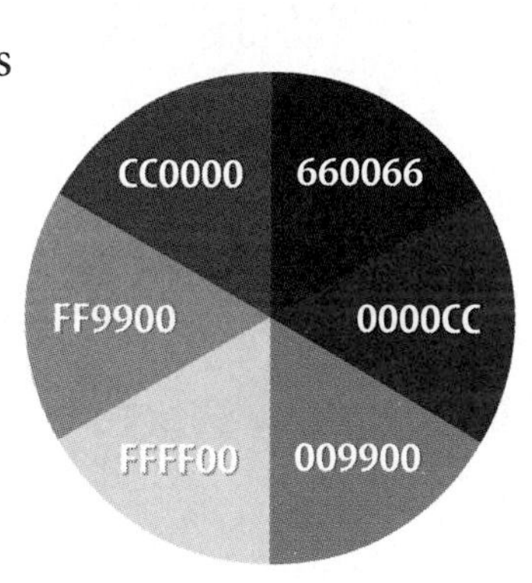

Developing Skills

16. Measuring in SI Make a fist. Use a centimeter ruler to measure the height, width, and depth of your fist.

17. Comparing and Contrasting How are volume, length, and mass similar? How are they different? Give several examples of units that are used to measure each quantity. Which units are SI?

18. Making and Using Graphs The table shows the area of several bodies of water. Make a bar graph of the data.

Areas of Bodies of Water

Body of Water	Area (km^2)
Currituck Sound (North Carolina)	301
Pocomoke Sound (Maryland/Virginia)	286
Chincoteague Bay (Maryland/Virginia)	272
Core Sound (North Carolina)	229

19. Interpreting Scientific Illustrations What does the figure show? How has this drawing been simplified?

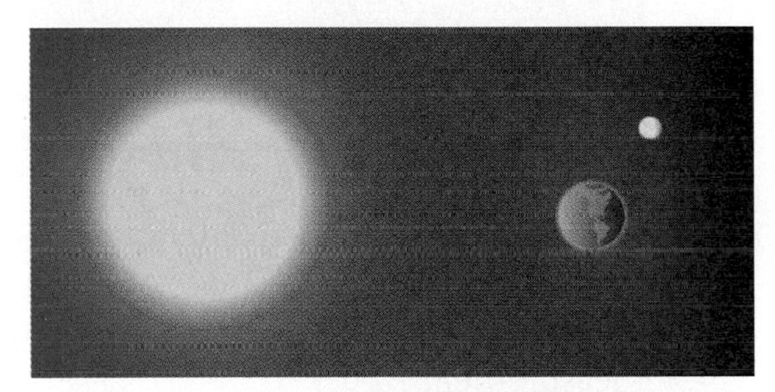

Performance Assessment

20. Poster Make a poster to alert the public about the benefits of using SI units.

21. Newspaper Search Look through a week's worth of newspapers and evaluate any graphs or tables that you find.

TECHNOLOGY

Go to the Glencoe Science Web site at **science.glencoe.com** or use the **Glencoe Science CD-ROM** for additional chapter assessment.

Test Practice

Some students in Mrs. Olsen's science class measured their masses during three consecutive months. They placed their results in the following table. Study the table and answer the following questions.

Student Masses: Sept. – Nov. 1999

Student	September	October	November
Domingo	41.13 kg	40.92 kg	42.27 kg
Latoya	35.21 kg	35.56 kg	36.07 kg
Benjamin	45,330 g	45,680 g	45,530 g
Poloma	31.78 kg	31.55 kg	31.51 kg
Frederick	50,870 g	51,880 g	51,030 g
Fiona	37.62 kg	37.71 kg	37.85 kg

1. According to the table, which shows Frederick's weight in kilograms for the three months?

A) 5.087, 5.118, 5.103
B) 50.87, 51.88, 51.03
C) 508.7, 511.8, 510.3
D) 5,087, 5,118, 5,103

2. According to this information, which lists the students from lightest to heaviest during November?

F) Poloma, Benjamin, Domingo, Frederick
G) Domingo, Latoya, Frederick, Benjamin
H) Fiona, Domingo, Benjamin, Frederick
J) Frederick, Benjamin, Domingo, Poloma

Standardized Test Practice

Reading Comprehension

Read the passage carefully. Then read each question that follows the passage. Decide which is the best answer to each question.

Test-Taking Tip After you read and think about the passage, write one or two sentences that summarize the most important points. Read your sentences out loud.

History of Measurement Units

In modern society, units of measurement that have been defined and agreed upon by international scientists are used. In ancient times, people were just beginning to invent and use units of measurement. For example, thousands of years ago, a cabinetmaker would build one cabinet at a time and measure the pieces of wood needed relative to the size of the other pieces of that cabinet. Today, factories manufacture many of the same products. Ancient cabinetmakers rarely made two cabinets that were exactly the same. Eventually, it became obvious that units of measurement had to mean the same thing to everybody.

Measurements, such as the inch, foot, and yard, began many years ago as fairly crude units. For example, the modern-day inch began as "the width of one's thumb." The foot originally was defined as "the length of one's foot." The yard was defined as "the distance from the tip of one's nose to the end of one's arm."

Although using these units of measurement was easier than not using any units of measurement, these ancient units were confusing. Human beings come in many different sizes and shapes, and one person's foot can be much larger than another person's foot. So, whose foot defines a foot? Who's thumb width defines an inch? Ancient civilizations used these kinds of measurements for thousands of years. Over time, these units were redefined and standardized, eventually becoming the exact units of measurement that you know today.

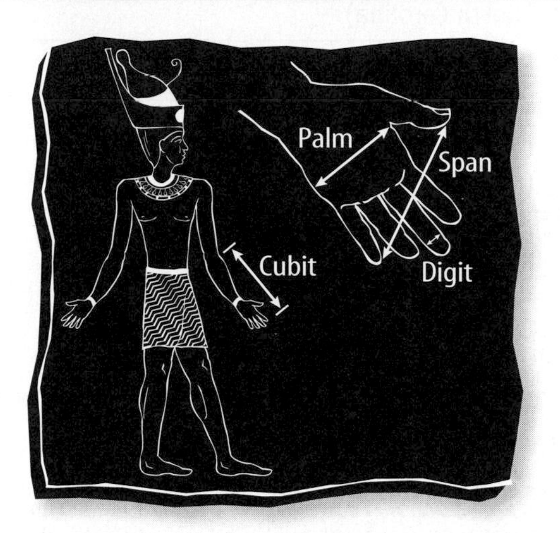

Ancient people created units of measurement.

1. Based on the passage, the reader can conclude that _____.

A) ancient cultures had no concept of measurement
B) standards of measurement developed over a long period of time
C) ancient people were probably good at communicating exacts units of measurement
D) units of measurement are needed only in modern and technologically advanced societies

2. According to the passage, which of these best describes ancient units of measurement?

F) precise
G) incorrect
H) approximate
J) irresponsible

Standardized Test Practice

Reasoning and Skills

Read each question and choose the best answer.

1. All of these are things that can be measured accurately EXCEPT _____.

A) the temperature of a human body
B) the space that a couch takes up in a living room
C) the beauty in a piece of artwork
D) the mass of a rock from the Moon

Test-Taking Tip Think about the reasons why people use measurement and the kinds of things that can and cannot be measured.

2. Jodie and William are using the graduated cylinder pictured above to measure the volume of liquid in milliliters. What multiple of a liter is a milliliter?

F) 1,000
G) 0.01
H) 0.001
J) 0.000 001

Test-Taking Tip Consider the prefixes used in SI units, as well as the amount of liquid shown.

Group S
• How does this medication work to reduce a fever?
• Why do tides occur?
• What is the melting point of iron?
• What was Earth's climate like in the past?

Group T
• Who would make the best class president?
• Is it right to compliment a friend's new shirt if you don't like the shirt?
• Shouldn't everyone like cauliflower, broccoli, turnips, and spinach?
• Why is a sunset beautiful?

3. The questions in Group S are different from the questions in Group T because only the questions in Group S _____.

A) cannot be answered by science
B) will have answers that are solely opinions
C) can be answered with absolute certainty
D) can be answered by science

Test-Taking Tip Think about the kinds of questions that scientists try to and are able to answer.

Consider this question carefully before writing your answer on a separate sheet of paper.

4. Suppose you decide to investigate this problem: Which brand of fertilizer produces the most tomatoes per plant? Identify the independent and dependent variables. List the constants in your investigation.

Test-Taking Tip Think about the procedures that scientists must go through in order to discover and explain.

UNIT 2

Life's Building Blocks and Processes

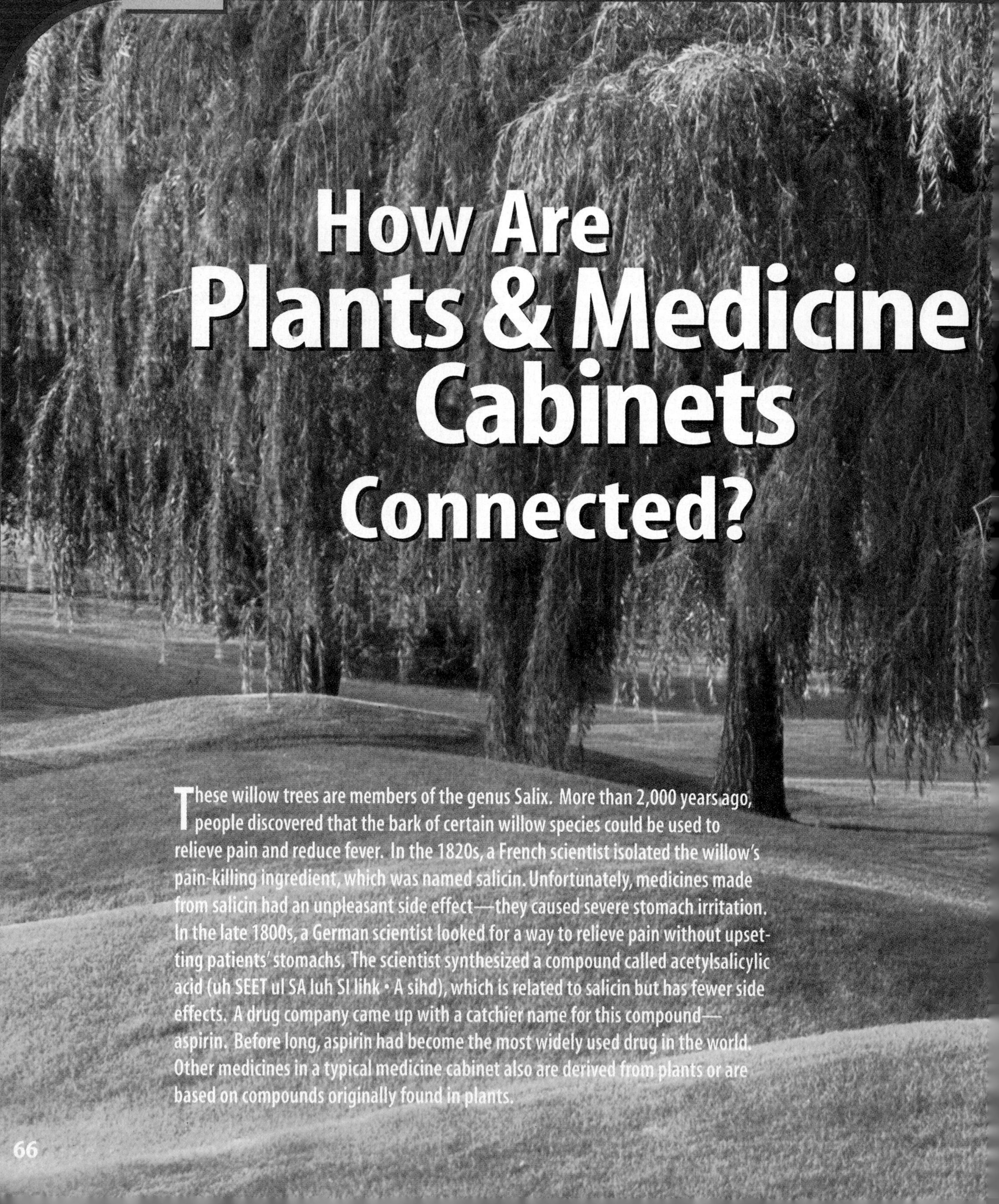

How Are Plants & Medicine Cabinets Connected?

These willow trees are members of the genus Salix. More than 2,000 years ago, people discovered that the bark of certain willow species could be used to relieve pain and reduce fever. In the 1820s, a French scientist isolated the willow's pain-killing ingredient, which was named salicin. Unfortunately, medicines made from salicin had an unpleasant side effect—they caused severe stomach irritation. In the late 1800s, a German scientist looked for a way to relieve pain without upsetting patients' stomachs. The scientist synthesized a compound called acetylsalicylic acid (uh SEET ul SA luh SI lihk • A sihd), which is related to salicin but has fewer side effects. A drug company came up with a catchier name for this compound—aspirin. Before long, aspirin had become the most widely used drug in the world. Other medicines in a typical medicine cabinet also are derived from plants or are based on compounds originally found in plants.

SCIENCE CONNECTION

PLANT COMPOUNDS Some modern medicines contain compounds extracted directly from plants. Others contain synthetic versions of plant compounds. Among the drugs with plant origins are digitalis, vincristine, and quinine. Investigate one of these three drugs to discover what plant it comes from, how it helps people, and how its medicinal properties were first discovered. Then write a newspaper article in which you relate your findings.

CHAPTER 3

Cells

The world around you is filled with organisms that you could overlook, or even be unable to see. Some of these organisms are one-celled and some are many-celled. The monster in this photograph is a louse crawling across human skin. It can be seen in great detail with a microscope that is found in many classrooms. You can study the cells of smaller organisms with other kinds of microscopes.

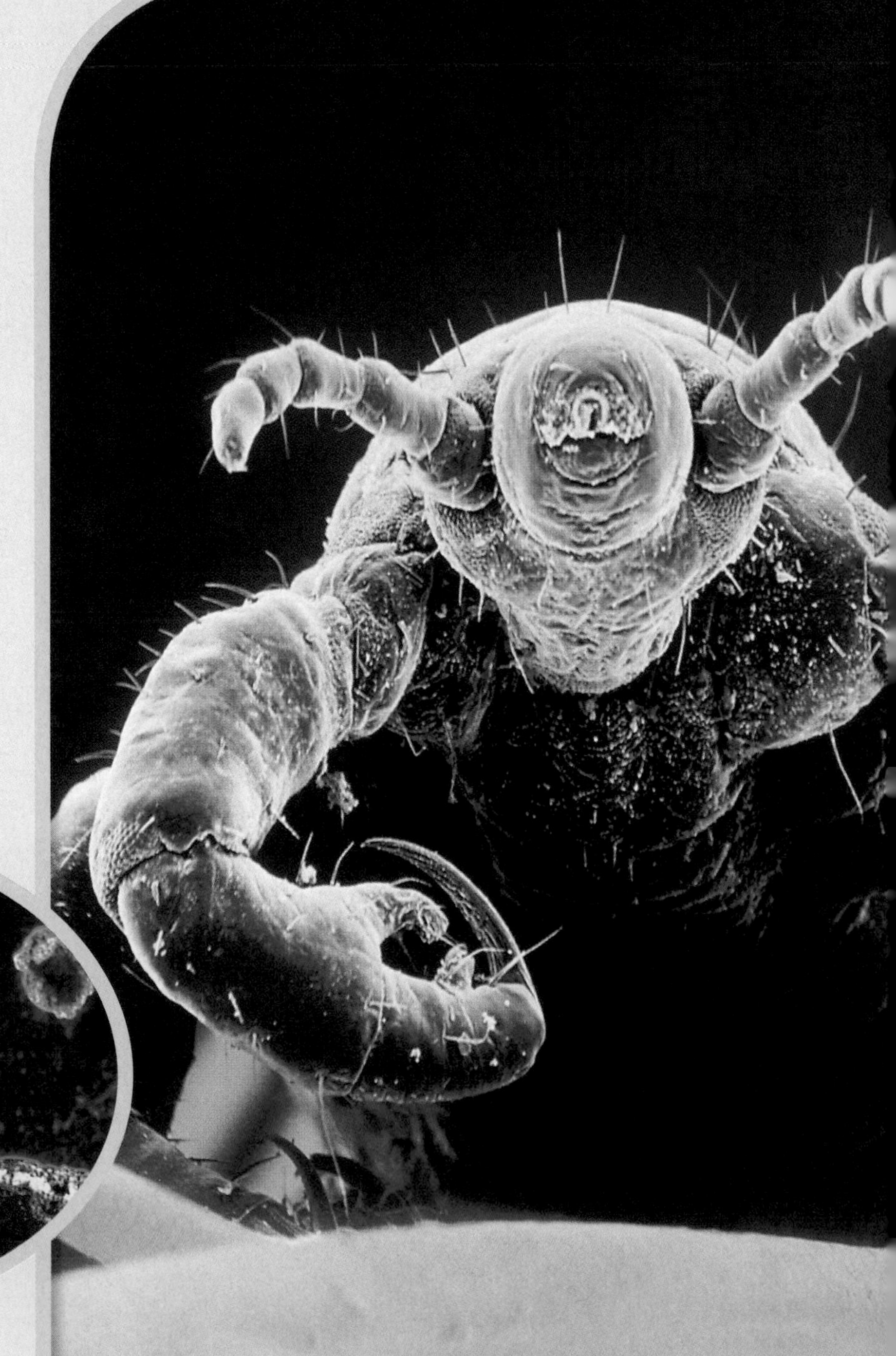

What do you think?

Science Journal Look at the picture below with a classmate. Discuss what you think is happening. Here's a hint: *Not every battlefield is found on land or at sea.* Write your answer or best guess in your Science Journal.

If you look around your classroom, you can see many things of all sizes. With the aid of a hand lens, you can see more details. You might examine a speck of dust and discover that it is a living or dead insect. In the following activity, use a hand lens to search for the smallest thing you can find in the classroom.

Measure a small object

1. Obtain a hand lens from your teacher. Note its power (the number followed by ×, shown somewhere on the lens frame or handle).
2. Using the hand lens, look around the room for the smallest object you can find.
3. Measure the size of the image as you see it with the hand lens. To estimate the real size of the object, divide that number by the power. For example, if it looks 2 cm long and the power is 10×, the real length is about 0.2 cm.

Observe

In your Science Journal, describe what you observe. Did the details become clearer? Explain.

Before You Read

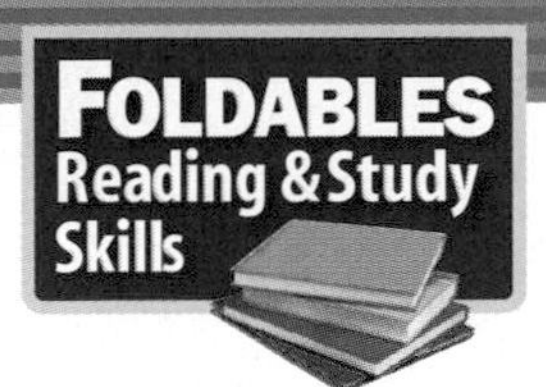

Making a Main Ideas Study Fold **Make the following Foldable to help you identify the main ideas or major topics on cells.**

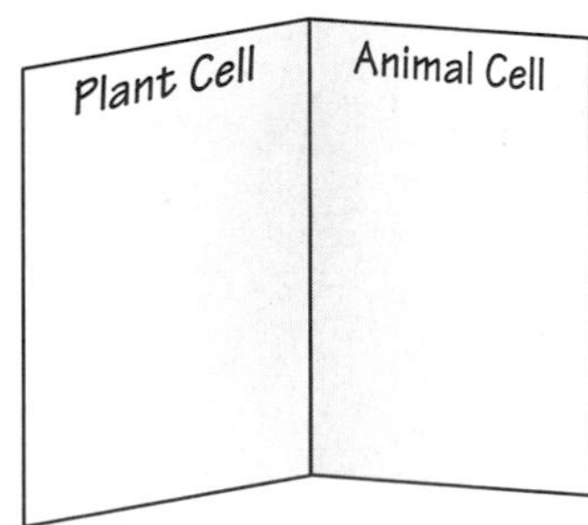

1. Place a sheet of paper in front of you so the long side is at the top. Fold the paper in half from the left side to the right side. Then unfold.
2. Label the left side of the paper *Plant Cell.* Label the right side of the paper *Animal Cell,* as shown.
3. Before you read the chapter, draw a plant cell on the left side of the paper and an animal cell on the right side of the paper.
4. As you read the chapter, change and add to your drawings.

Cell Structure

As You Read

What You'll Learn

- **Identify** names and functions of each part of a cell.
- **Explain** how important a nucleus is in a cell.
- **Compare** tissues, organs, and organ systems.

Vocabulary

cell membrane, cytoplasm, cell wall, organelle, nucleus, chloroplast, mitochondrion, ribosome, endoplasmic reticulum, Golgi body, tissue, organ

Why It's Important

If you know how organelles function, it's easier to understand how cells survive.

Common Cell Traits

Living cells are dynamic and have several things in common. A cell is the smallest unit that is capable of performing life functions. All cells have an outer covering called a **cell membrane.** Inside every cell is a gelatinlike material called **cytoplasm** (SI toh plaz uhm). In the cytoplasm of every cell is hereditary material that controls the life of the cell.

Comparing Cells Cells come in many sizes. A nerve cell in your leg could be a meter long. A human egg cell is no bigger than the dot on this **i.** A human red blood cell is about one-tenth the size of a human egg cell. A bacterium is even smaller—8,000 of the smallest bacteria can fit inside one of your red blood cells.

A cell's shape might tell you something about its function. The nerve cell in **Figure 1** has many fine extensions that send and receive impulses to and from other cells. Though a nerve cell cannot change shape, muscle cells and some blood cells can. In plant stems, some cells are long and hollow and have openings at their ends. These cells carry food and water throughout the plant.

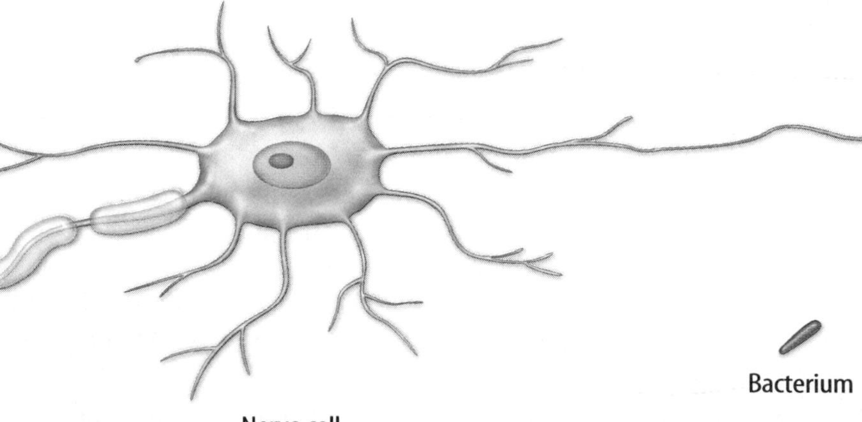

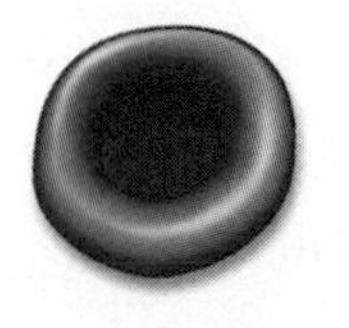

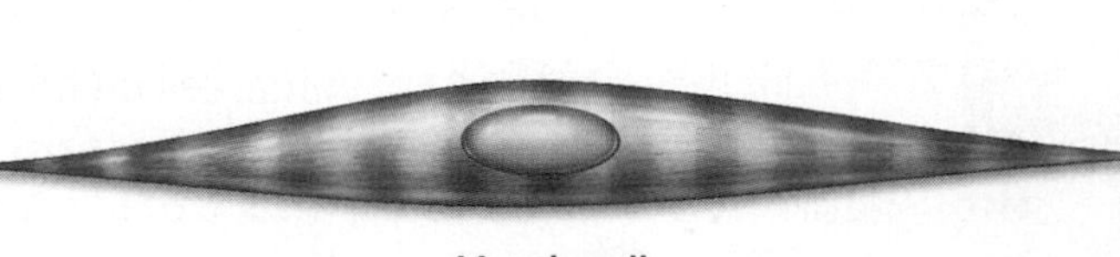

Figure 1
The shape of the cell can tell you something about its function. These cells are drawn 700 times their actual size.

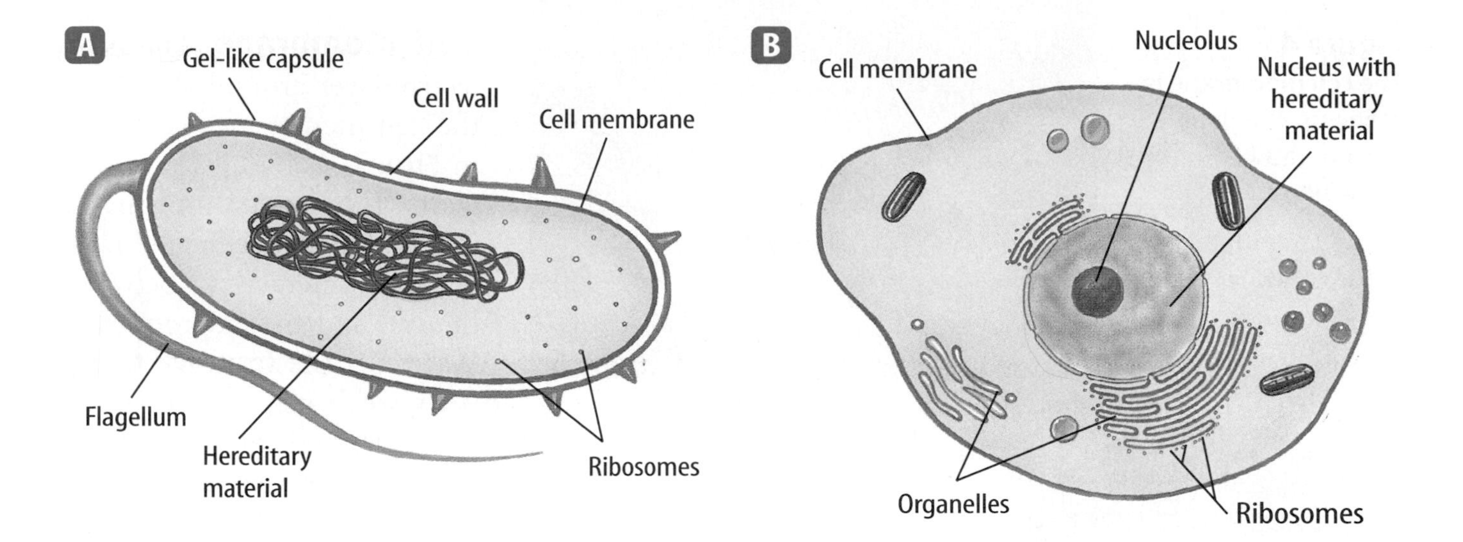

Figure 2
Examine these drawings of cells. A Prokaryotic cells are only found in one-celled organisms, such as bacteria. B Protists, fungi, plants and animals are made of eukaryotic cells. *What differences do you see between them?*

Cell Types Scientists have found that cells can be separated into two groups. One group has no membrane-bound structures inside the cell and the other group does, as shown in **Figure 2.** Cells without membrane-bound structures are called prokaryotic (proh KAYR ee yah tihk) cells. Cells with membrane-bound structures are called eukaryotic (yew KAYR ee yah tihk) cells.

Reading Check *Into what two groups can cells be separated?*

Cell Organization

Each cell in your body has a specific function. You might compare a cell to a busy delicatessen that is open 24 hours every day. Raw materials for the sandwiches are brought in often. Some food is eaten in the store, and some customers take their food with them. Sometimes food is prepared ahead of time for quick sale. Wastes are put into trash bags for removal or recycling. Similarly, your cells are taking in nutrients, secreting and storing chemicals, and breaking down substances 24 hours every day.

Cell Wall Just like a deli that is located inside the walls of a building, some cells are enclosed in a cell wall. The cells of plants, algae, fungi, and most bacteria are enclosed in a cell wall. **Cell walls** are tough, rigid outer coverings that protect the cell and give it shape.

A plant cell wall, as shown in **Figure 3,** mostly is made up of a carbohydrate called cellulose. The long, threadlike fibers of cellulose form a thick mesh that allows water and dissolved materials to pass through it. Cell walls also can contain pectin, which is used in jam and jelly, and lignin, which is a compound that makes cell walls rigid. Plant cells responsible for support have a lot of lignin in their walls.

Figure 3
The protective cell wall of a plant cell is outside the cell membrane.

Magnification: 9,000×

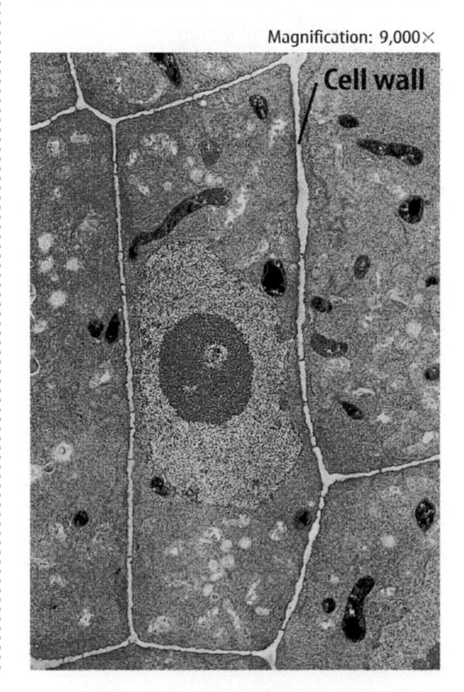

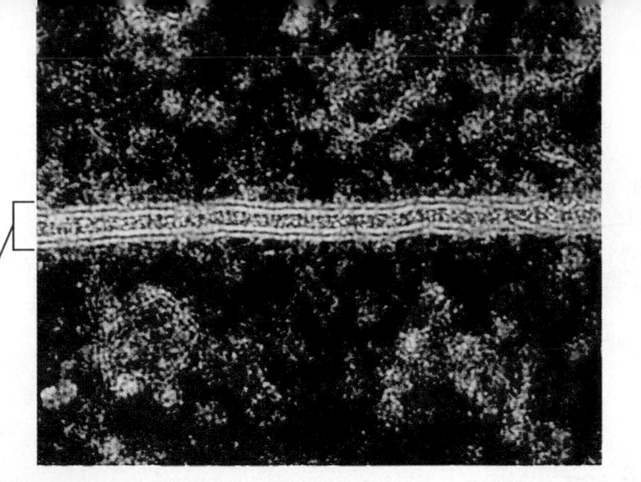

Figure 4
The cell membrane is made up of a double layer of fatlike molecules.

Cell Membrane

The protective layer around all cells is the cell membrane, as shown in **Figure 4.** If cells have cell walls, the cell membrane is inside of it. The cell membrane regulates interactions between the cell and the environment. Water is able to move freely into and out of the cell through the cell membrane. Food particles and some molecules enter and waste products leave through the cell membrane.

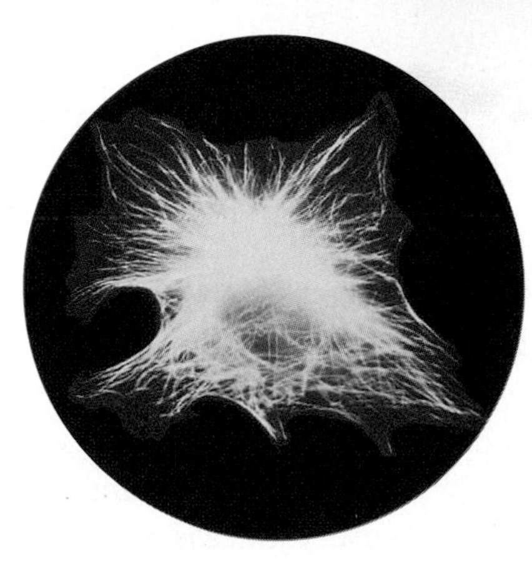

Figure 5
Cytoskeleton, a network of fibers in the cytoplasm, gives cells structure and helps them maintain shape.

Cytoplasm

Cells are filled with a gelatinlike substance called cytoplasm that constantly flows inside the cell membrane. Many important chemical reactions occur within the cytoplasm.

Throughout the cytoplasm is a framework called the cytoskeleton, which helps the cell maintain or change its shape. Cytoskeletons enable some cells to move. An amoeba, for example, moves by stretching and contracting its cytoskeleton. The cytoskeleton is made up of thin, hollow tubes of protein and thin, solid protein fibers, as shown in **Figure 5.** Proteins are organic molecules made up of amino acids.

✔ Reading Check *What is the function of the cytoskeleton?*

Most of a cell's life processes occur in the cytoplasm. Within the cytoplasm of eukaryotic cells are structures called **organelles.** Some organelles process energy and others manufacture substances needed by the cell or other cells. Certain organelles move materials, while others act as storage sites. Most organelles are surrounded by membranes. The nucleus is usually the largest organelle in a cell.

Modeling Cytoplasm

Procedure

1. Add 100 mL of **water** to a **clear container.**
2. Add **unflavored gelatin** and stir.
3. Shine a **flashlight** through the solution.

Analysis

1. Describe what you see.
2. How does a model help you understand what cytoplasm might be like?

Nucleus

The nucleus is like the deli manager who directs the store's daily operations and passes on information to employees. The **nucleus,** shown in **Figure 6,** directs all cell activities and is separated from the cytoplasm by a membrane. Materials enter and leave the nucleus through openings in the membrane. The nucleus contains the instructions for everything the cell does. These instructions are found on long, threadlike, hereditary material made of DNA. DNA is the chemical that contains the code for the cell's structure and activities. During cell division, the hereditary material coils tightly around proteins to form structures called chromosomes. A structure called a nucleolus also is found in the nucleus.

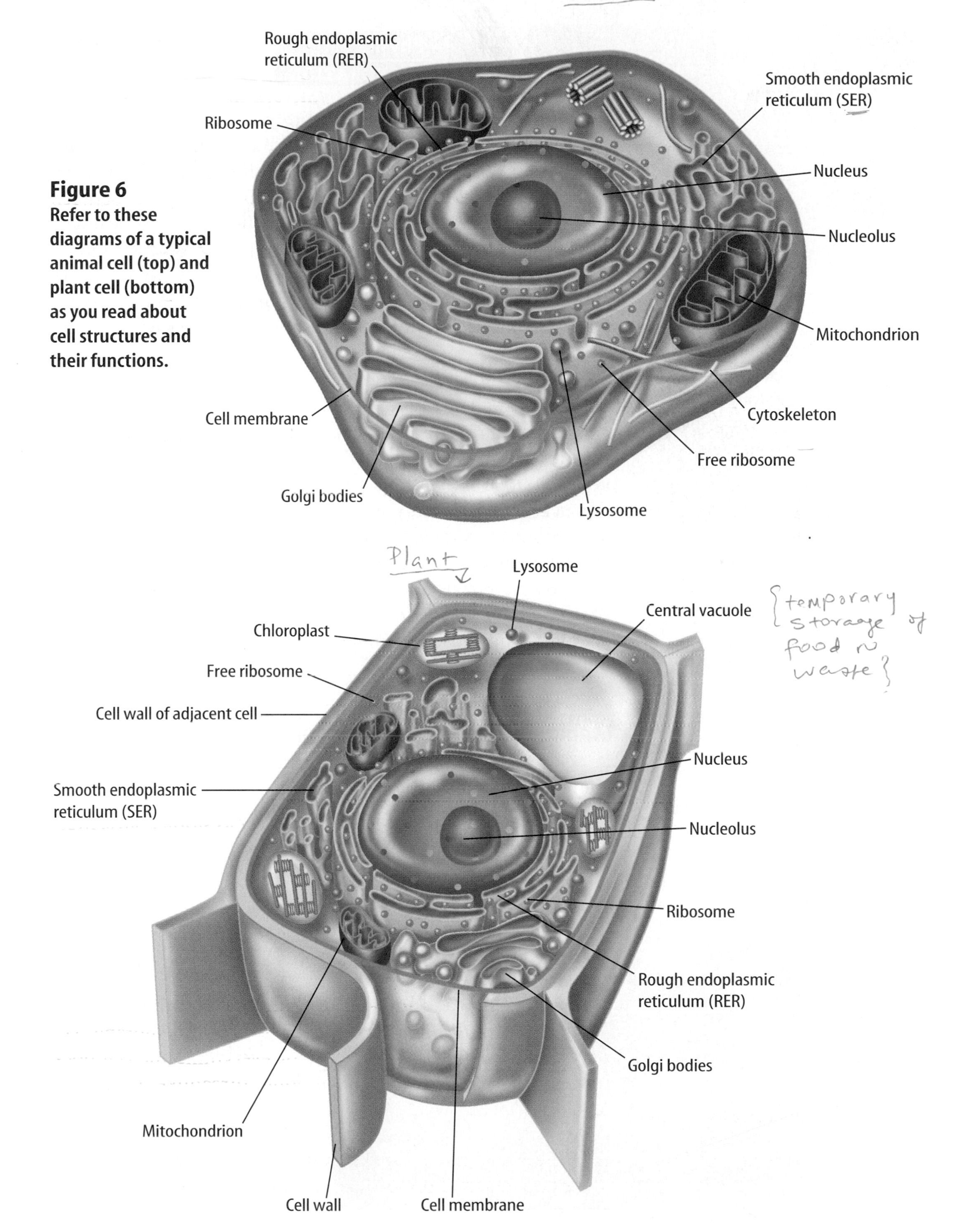

Figure 6
Refer to these diagrams of a typical animal cell (top) and plant cell (bottom) as you read about cell structures and their functions.

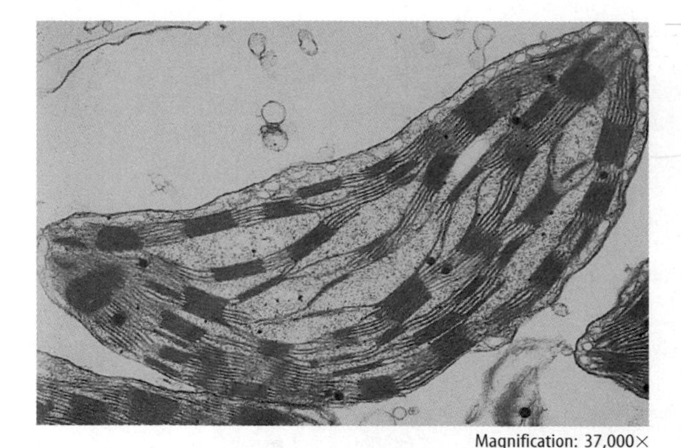

Figure 7
Chloroplasts are organelles that use sunlight to make sugar from carbon dioxide and water. They contain chlorophyll, which gives most leaves and stems their green color.

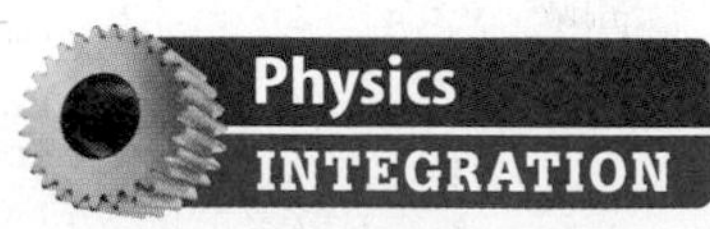

Energy-Processing Organelles Cells require a continuous supply of energy to process food, make new substances, eliminate wastes, and communicate with each other. In plant cells, food is made in green organelles in the cytoplasm called **chloroplasts** (KLOR uh plasts), as shown in **Figure 7.** Chloroplasts contain the green pigment chlorophyll, which gives leaves and stems their green color. Chlorophyll captures light energy that is used to make a sugar called glucose. Glucose molecules store the captured light energy as chemical energy. Many cells, including animal cells, do not have chloroplasts for making food. They must get food from their environment.

The energy in food is stored until it is released by the mitochondria. **Mitochondria** (mi tuh KAHN dree uh) (singular, *mitochondrion*), such as the one shown in **Figure 8,** are organelles where energy is released from breaking down food into carbon dioxide and water. Just as the gas or electric company supplies fuel for the deli, a mitochondrion releases energy for use by the cell. Some types of cells, such as muscle cells, are more active than other cells. These cells have large numbers of mitochondria. Why would active cells have more or larger mitochondria?

Figure 8
Mitochondria are known as the powerhouses of the cell because they release energy that is needed by the cell from food.
What types of cells might contain many mitochondria?

Manufacturing Organelles One substance that takes part in nearly every cell activity is protein. Proteins are part of cell membranes. Other proteins are needed for chemical reactions that take place in the cytoplasm. Cells make their own proteins on small structures called **ribosomes.** Even though ribosomes are considered organelles, they are not membrane bound. Some ribosomes float freely in the cytoplasm; and others are attached to the endoplasmic reticulum. Ribosomes are made in the nucleolus and move out into the cytoplasm. Ribosomes receive directions from the hereditary material in the nucleus on how, when, and in what order to make specific proteins.

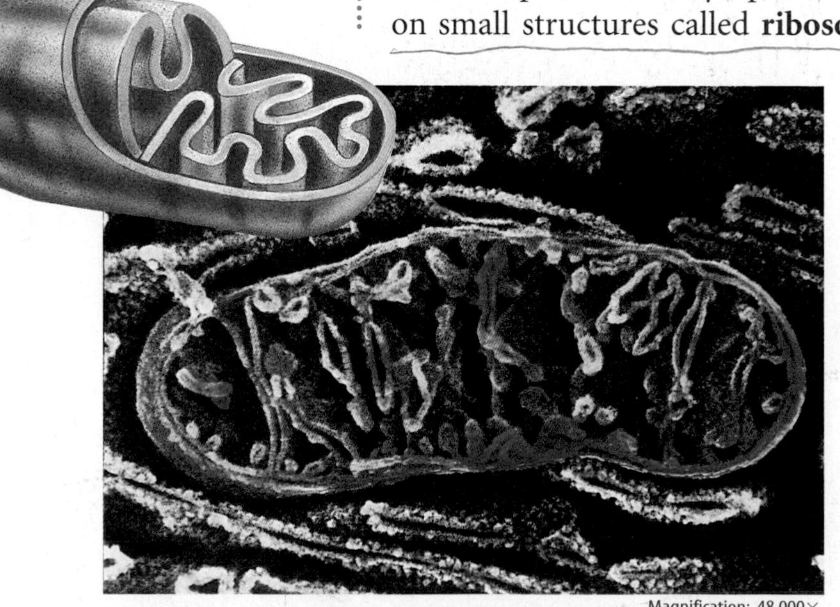

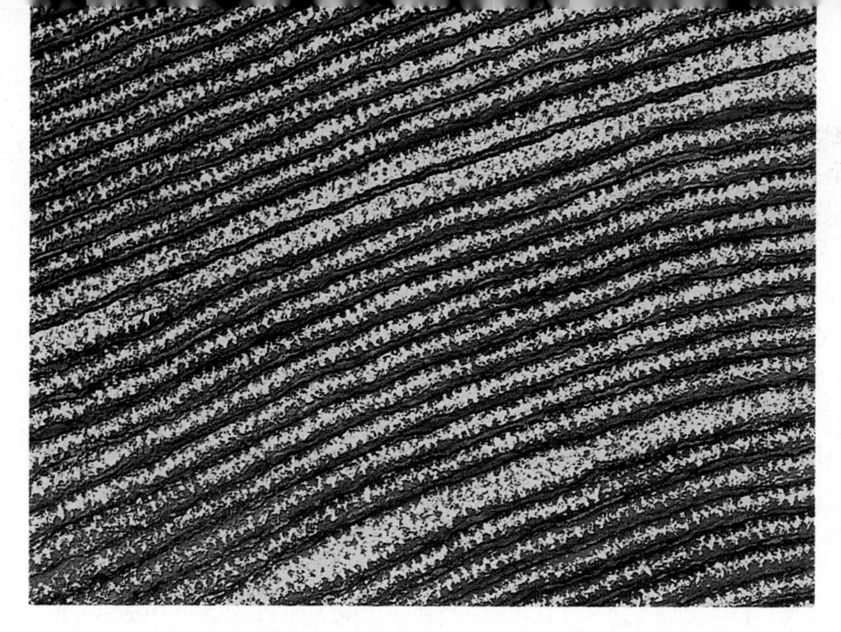

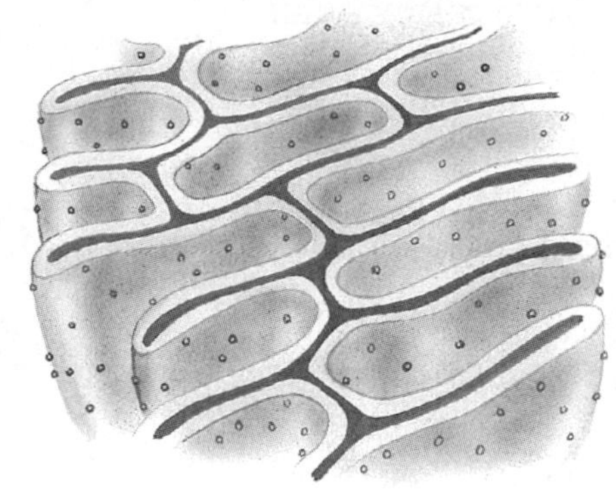

Figure 9
Endoplasmic reticulum (ER) is a complex series of membranes in the cytoplasm of the cell. *What would smooth ER look like?*

Processing, Transporting, and Storing Organelles

The **endoplasmic reticulum** (en duh PLAZ mihk • rih TIHK yuh lum) or ER, as shown in **Figure 9,** extends from the nucleus to the cell membrane. It is a series of folded membranes in which materials can be processed and moved around inside of the cell. The ER takes up a lot of space in some cells.

- Ribosomes make Protein
- ER are tubes that move material

The endoplasmic reticulum may be "rough" or "smooth." ER that has no attached ribosomes is called smooth endoplasmic reticulum. This type of ER processes other cellular substances such as lipids that store energy. Ribsomes are attached to areas on the rough ER. There they carry out their job of making proteins that are moved out of the cell or used within the cell.

Reading Check *What is the difference between rough ER and smooth ER?*

After proteins are made in a cell, they are transferred to another type of cell organelle called the Golgi (GAWL jee) bodies. The **Golgi bodies**, as shown in **Figure 10,** are stacked, flattened membranes. The Golgi bodies sort proteins and other cellular substances and package them into membrane-bound structures called vesicles. The vesicles deliver cellular substances to areas inside the cell. They also carry cellular substances to the cell membrane where they are released to the outside of the cell.

Just as a deli has refrigerators for temporary storage of some its foods and ingredients, cells have membrane-bound spaces called vacuoles for the temporary storage of materials. A vacuole can store water, waste products, food, and other cellular materials. In plant cells, the vacuole may make up most of the cell's volume.

Figure 10
The Golgi body packages materials and moves them to the outside of the cell. *Why are materials removed from the cell?*

Magnification: 28,000×

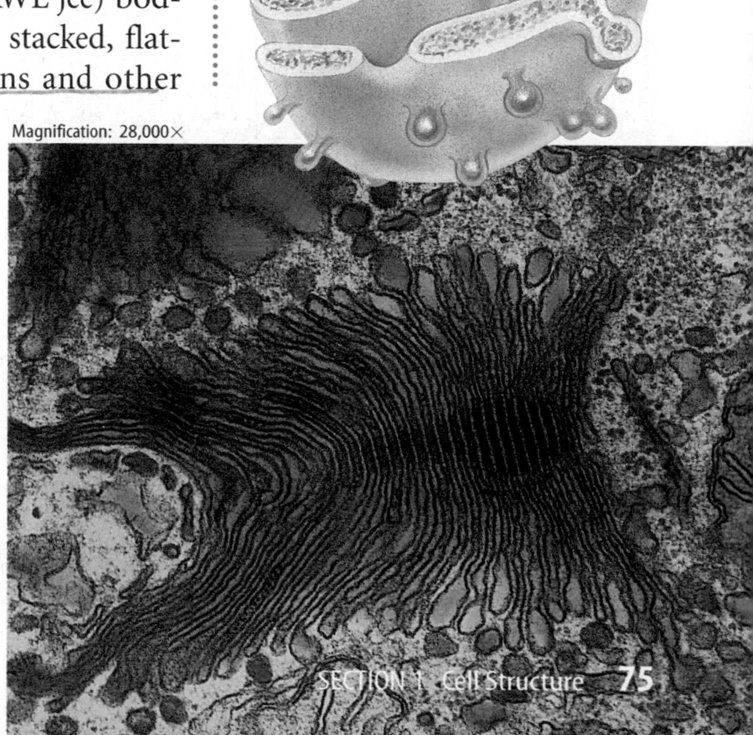

Environmental Science
INTEGRATION

Just like a cell, you can recycle materials. Paper, plastics, aluminum, and glass are materials that can be recycled into usable items. Make a promotional poster to encourage others to recycle.

Recycling Organelles Active cells break down and recycle substances. Organelles called lysosomes (LI suh sohmz) contain digestive chemicals that help break down food molecules, cell wastes, and worn-out cell parts. In a healthy cell, chemicals are released into vacuoles only when needed. The lysosome's membrane prevents the digestive chemicals inside from leaking into the cytoplasm and destroying the cell. When a cell dies, a lysosome's membrane disintegrates. This releases digestive chemicals that allow the quick breakdown of the cell's contents.

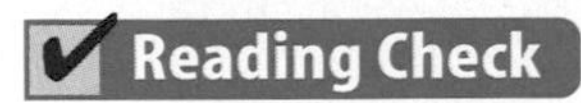

What is the function of the lysosome's membrane?

Math Skills Activity

Calculate the Ratio of Surface Area to Volume of Cells

Example Problem

Assume that a cell is like a cube with six equal sides. Find the ratio of surface area to volume for a cube that is 4 cm high.

Solution

1. *This is what you know:* A cube has 6 equal sides of 4 cm × 4 cm.
2. *This is what you want to find:* the ratio (R) of surface area to volume for each cube
3. *These are the equations you use:*
   surface area (A) = width × length × 6
   volume (V) = length × width × height
   $R = A/V$
4. *Solve for surface area and volume, then solve for the ratio:*
   $A = 4\text{ cm} \times 4\text{ cm} \times 6 = 96\text{ cm}^2$
   $V = 4\text{ cm} \times 4\text{ cm} \times 4\text{ cm} = 64\text{ cm}^3$
   $R = 96\text{ cm}^2/64\text{ cm}^3 = 1.5\text{ cm}^2/\text{cm}^3$

Check your answer by multiplying the ratio by the volume. Do you calculate the surface area?

Practice Problems

1. Calculate the ratio of surface area to volume for a cube that is 2 cm high. What happens to this ratio as the size of the cube decreases?
2. If a 4-cm cube doubled just one of its dimensions—length, width, or height—what would happen to the ratio of surface area to volume?

For more help, refer to the Math Skills Handbook.

From Cell to Organism

Many one-celled organisms perform all their life functions by themselves. Cells in a many-celled organism, however, do not work alone. Each cell carries on its own life functions while depending in some way on other cells in the organism.

In **Figure 11,** you can see cardiac muscle cells grouped together to form a tissue. A **tissue** is a group of similar cells that work together to do one job. Each cell in a tissue does its part to keep the tissue alive.

Tissues are organized into organs. An **organ** is a structure made up of two or more different types of tissues that work together. Your heart is an organ made up of cardiac muscle tissue, nerve tissue, and blood tissues. The cardiac muscle tissue contracts, making the heart pump. The nerve tissue brings messages that tell the heart how fast to beat. The blood tissue is carried from the heart to other organs of the body.

Reading Check *What type of tissues make up your heart?*

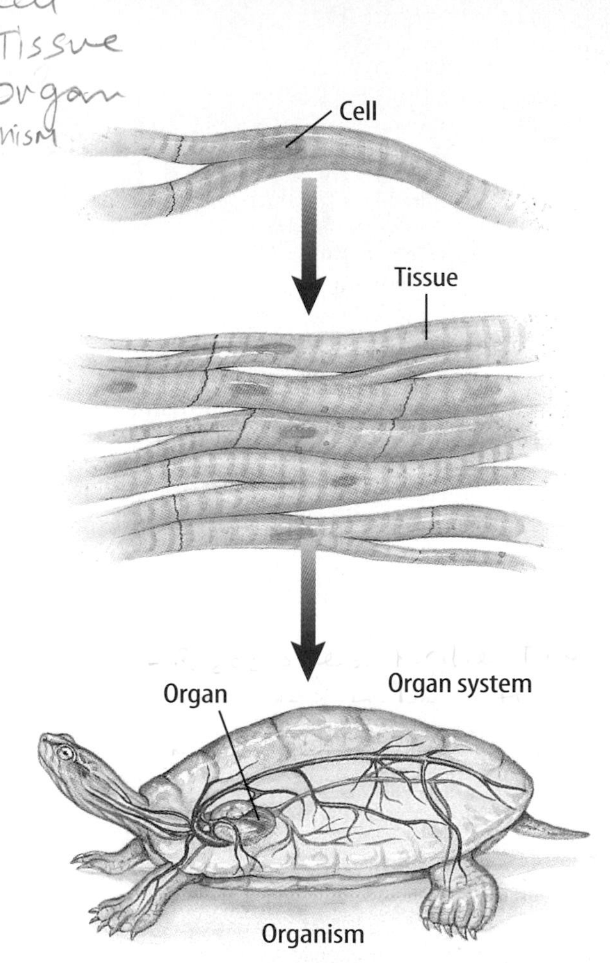

Figure 11
In a many-celled organism, cells are organized into tissues, tissues into organs, organs into systems, and systems into an organism.

A group of organs working together to perform a certain function is an organ system. Your heart, arteries, veins, and capillaries make up your cardiovascular system. In a many-celled organism, several systems work together in order to perform life functions efficiently. Your nervous, circulatory, respiratory, muscular, and other systems work together to keep you alive.

Section  Assessment

1. Explain the important role of the nucleus in the life of a cell.
2. Compare and contrast the energy processing organelles.
3. Why are digestive enzymes in a cell enclosed in a membrane-bound organelle?
4. How are cells, tissues, organs, and organ systems related?
5. **Think Critically** How is the cell of a one-celled organism different from the cells in many-celled organisms?

Skill Builder Activities

6. **Interpreting Scientific Illustrations** Examine the illustrations of the animal cell and the plant cell in **Figure 6** and make a list of differences and similarities between them. **For more help, refer to the Science Skill Handbook.**
7. **Communicating** Your textbook compared some cell functions to that of a deli. In your Science Journal, write an essay that explains how a cell is like your school or town. **For more help, refer to the Science Skill Handbook.**

Activity

Comparing Cells

If you compared a goldfish to a rose, you would find them unlike each other. Are their individual cells different also? Try this activity to compare plant and animal cells.

What You'll Investigate

How do human cheek cells and plant cells compare?

Materials

microscope
microscope slide
coverslip
forceps
tap water
dropper
Elodea plant
prepared slide of human cheek cells

Goal

- **Compare and contrast** an animal cell and a plant cell.

Safety Precautions

Procedure

1. Copy the data table in your Science Journal. Check off the cell parts as you observe them.

Cell Observations		
Cell Part	**Cheek**	***Elodea***
Cytoplasm		
Nucleus		
Chloroplasts		
Cell Wall		
Cell Membrane		

2. Using forceps, make a wet-mount slide of a young leaf from the tip of an *Elodea* plant.
3. **Observe** the leaf on low power. Focus on the top layer of cells.
4. Switch to high power and focus on one cell. In the center of the cell is a membrane-bound organelle called the central vacuole. Observe the chloroplasts—the green, disk-shaped objects moving around the central vacuole. Try to find the cell nucleus. It looks like a clear ball.
5. **Draw** the *Elodea* cell. Label the cell wall, cytoplasm, chloroplasts, central vacuole, and nucleus. Return to low power and remove the slide. Properly dispose of the slide.
6. **Observe** the prepared slide of cheek cells under low power.
7. Switch to high power and observe the cell nucleus. Draw and label the cell membrane, cytoplasm, and nucleus. Return to low power and remove the slide.

Conclude and Apply

1. **Compare and contrast** the shapes of the cheek cell and the *Elodea* cell.
2. What can you conclude about the differences between plant and animal cells?

Communicating Your Data

Draw the two kinds of cells on one sheet of paper. Use a green pencil to label the organelles found only in plants, a red pencil to label the organelles found only in animals, and a blue pencil to label the organelles found in both. **For more help, refer to the Science Skill Handbook.**

Viewing Cells

Magnifying Cells

The number of living things in your environment that you can't see is much greater than the number that you can see. Many of the things that you cannot see are only one cell in size. To see most cells, you need to use a microscope.

Trying to see separate cells in a leaf, like the ones in **Figure 12,** is like trying to see individual photos in a photo mosaic picture that is on the wall across the room. As you walk toward the wall, it becomes easier to see the individual photos. When you get right up to the wall, you can see details of each small photo. A microscope has one or more lenses that enlarge the image of an object as though you are walking closer to it. Seen through these lenses, the leaf appears much closer to you, and you can see the individual cells that carry on life processes.

Early Microscopes In the late 1500s, the first microscope was made by a Dutch maker of reading glasses. He put two magnifying glasses together in a tube and got an image that was larger than the image that was made by either lens alone.

In the mid 1600s, Antonie van Leeuwenhoek, a Dutch fabric merchant, made a simple microscope with a tiny glass bead for a lens, as shown in **Figure 13.** With it, he reported seeing things in pond water that no one had ever imagined. His microscope could magnify up to 270 times. Another way to say this is that his microscope could make the image of an object 270 times larger than its actual size. Today you would say his lens had a power of 270×. Early compound microscopes were crude by today's standards. The lenses would make an image larger, but it wasn't always sharp or clear.

As You Read

What You'll Learn

- **Compare** the differences between the compound light microscope and the electron microscope.
- **Summarize** the discoveries that led to the development of the cell theory.
- **Relate** the cell theory to modern biology.

Vocabulary
cell theory

Why It's Important

Humans are like other living things because they are made of cells.

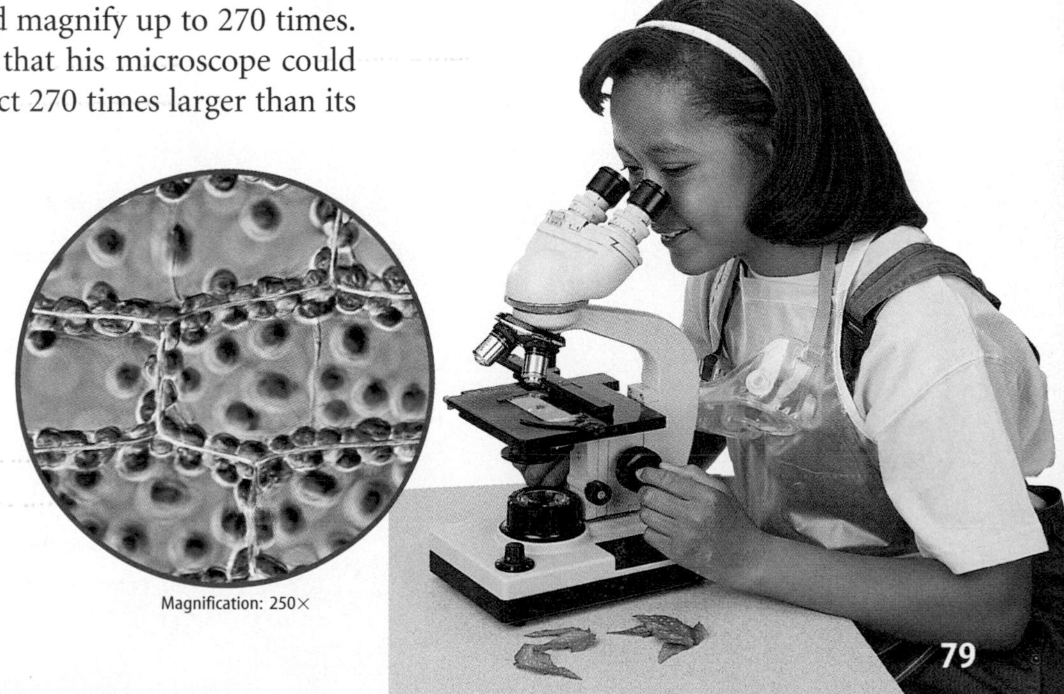

Figure 12
Individual cells become visible when a plant leaf is viewed using a microscope with enough magnifying power.

NATIONAL GEOGRAPHIC VISUALIZING MICROSCOPES

Figure 13

Microscopes give us a glimpse into a previously invisible world. Improvements have vastly increased their range of visibility, allowing researchers to study life at the molecular level. A selection of these powerful tools—and their magnification power—is shown here.

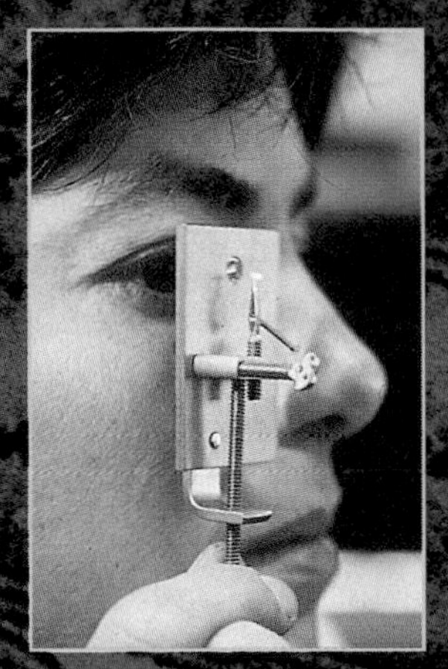

Up to 250x LEEUWENHOEK MICROSCOPE Held by a modern researcher, this historic microscope allowed Leeuwenhoek to see clear images of tiny freshwater organisms that he called "beasties."

Up to 2,000x BRIGHTFIELD / DARKFIELD MICROSCOPE The light microscope is often called the brightfield microscope because the image is viewed against a bright background. A brightfield microscope is the tool most often used in laboratories to study cells. Placing a thin metal disc beneath the stage, between the light source and the objective lenses, converts a brightfield microscope to a darkfield microscope. The image seen using a darkfield microscope is bright against a dark background. This makes details more visible than with a brightfield microscope. Below are images of a *Paramecium* as seen using both processes.

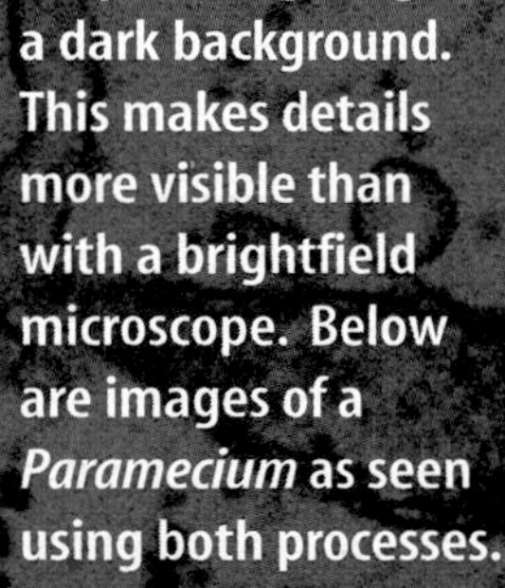

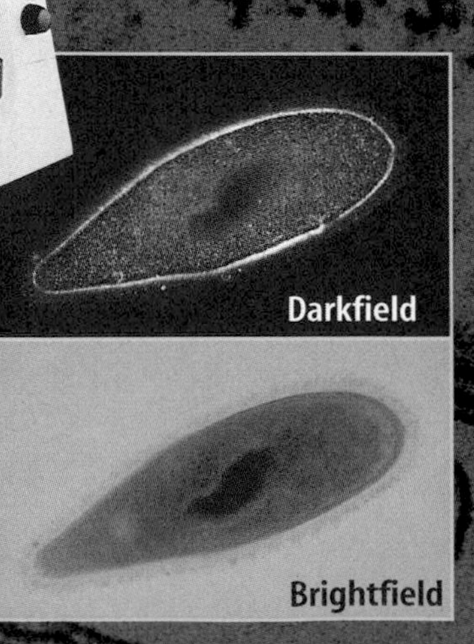

Darkfield

Brightfield

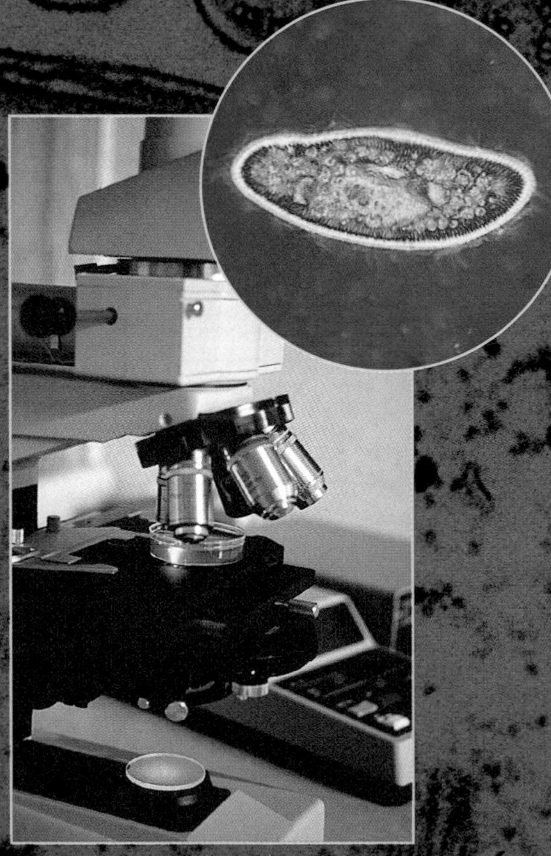

Up to 1,500x FLUORESCENCE MICROSCOPE This type of microscope requires that the specimen be treated with special fluorescent stains. When viewed through this microscope, certain cell structures or types of substances glow, as seen in the image of a *Paramecium* above.

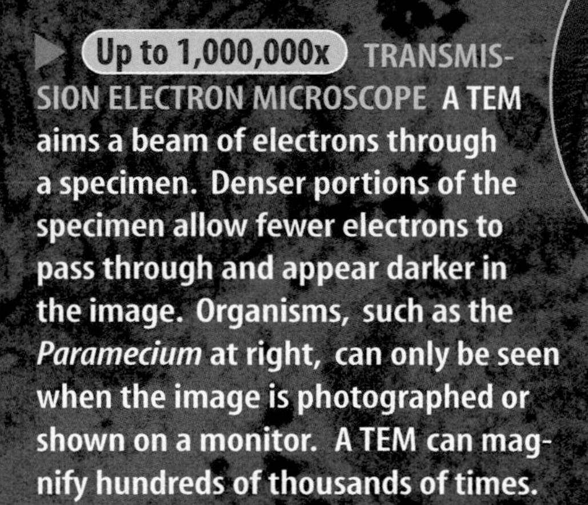

Up to 1,000,000x TRANSMISSION ELECTRON MICROSCOPE **A TEM aims a beam of electrons through a specimen. Denser portions of the specimen allow fewer electrons to pass through and appear darker in the image. Organisms, such as the *Paramecium* at right, can only be seen when the image is photographed or shown on a monitor. A TEM can magnify hundreds of thousands of times.**

Up to 1,500x PHASE-CONTRAST MICROSCOPE **A phase-contrast microscope emphasizes slight differences in a specimen's capacity to bend light waves, thereby enhancing light and dark regions without the use of stains. This type of microscope is especially good for viewing living cells, like the *Paramecium* above left. The images from a phase-contrast microscope can only be seen when the specimen is photographed or shown on a monitor.**

Up to 200,000x SCANNING ELECTRON MICROSCOPE **An SEM sweeps a beam of electrons over a specimen's surface, causing other electrons to be emitted from the specimen. SEMs produce realistic, three-dimensional images, which can only be viewed as photographs or on a monitor, as in the image of the *Paramecium* at right. Here a researcher compares an SEM picture to a computer monitor showing an enhanced image.**

TRY AT HOME
Mini LAB

Observing Magnified Objects

Procedure

1. Look at a **newspaper** through the curved side and through the flat bottom of an **empty, clear glass.**
2. Look at the newspaper through a **clear glass bowl** filled with **water** and then with a **magnifying glass.**

Analysis

In your Science Journal, compare how well you can see the newspaper through each of the objects.

Modern Microscopes Scientists use a variety of microscopes to study organisms, cells, and cell parts that are too small to be seen with the human eye. Depending on how many lenses a microscope contains, it is called simple or compound. A simple microscope is similar to a magnifying glass. It has only one lens. A microscope's lens makes an enlarged image of an object and directs light toward your eye. The change in apparent size produced by a microscope is called magnification. Microscopes vary in powers of magnification. Some microscopes can make images of individual atoms.

The microscope you probably will use to study life science is a compound light microscope, similar to the one in the Reference Handbook at the back of this book. The compound light microscope has two sets of lenses—eyepiece lenses and objective lenses. The eyepiece lenses are mounted in one or two tubelike structures. Images of objects viewed through two eyepieces, or stereomicroscopes, are three-dimensional. Images of objects viewed through one eyepiece are not. Compound light microscopes usually have two to four movable objective lenses.

Magnification The powers of the eyepiece and objective lenses determine the total magnifications of a microscope. If the eyepiece lens has a power of 10× and the objective lens has a power of 43×, then the total magnification is 430× (10× times 43×). Some compound microscopes, like those in **Figure 13,** have more powerful lenses that can magnify an object up to 2,000 times its original size.

Electron Microscopes Things that are too small to be seen with other microscopes can be viewed with an electron microscope. Instead of using lenses to direct beams of light, an electron microscope uses a magnetic field in a vacuum to direct beams of electrons. Some electron microscopes can magnify images up to one million times. Electron microscope images must be photographed or electronically produced.

Several kinds of electron microscopes have been invented, as shown in **Figure 13.** Scanning electron microscopes (SEM) produce a realistic, three-dimensional image. Only the surface of the specimen can be observed using an SEM. Transmission electron microscopes (TEM) produce a two-dimensional image of a thinly-sliced specimen. Details of cell parts can be examined using a TEM. Scanning tunneling microscopes (STM) are able to show the arrangement of atoms on the surface of a molecule. A metal probe is placed near the surface of the specimen and electrons flow from the tip. The hills and valleys of the specimen's surface are mapped.

Physics INTEGRATION

A magnifying glass is a convex lens. All microscopes use one or more convex lenses. In your Science Journal, diagram a convex lens and describe its shape.

Development of the Cell Theory

During the seventeenth century, scientists used their new invention, the microscope, to explore the newly discovered microscopic world. They examined drops of blood, scrapings from their own teeth, and other small things. Cells weren't discovered until the microscope was improved. In 1665, Robert Hooke cut a thin slice of cork and looked at it under his microscope. To Hooke, the cork seemed to be made up of empty little boxes, which he named cells.

Table 1 The Cell Theory

All organisms are made up of one or more cells.	An organism can be one cell or many cells like most plants and animals.
The cell is the basic unit of organization in organisms.	Even in complex organisms, the cell is the basic unit of structure and function.
All cells come from cells.	Most cells can divide to form two new, identical cells.

In the 1830s, Matthias Schleiden used a microscope to study plant parts. He concluded that all plants are made of cells. Theodor Schwann, after observing many different animal cells, concluded that all animals also are made up of cells. Eventually, they combined their ideas and became convinced that all living things are made of cells.

Several years later, Rudolf Virchow hypothesized that cells divide to form new cells. Virchow proposed that every cell came from a cell that already existed. His observations and conclusions and those of others are summarized in the **cell theory,** as described in **Table 1.**

✔ Reading Check *Who made the conclusion that all animals are made of cells?*

Section  Assessment

1. Explain why the invention of the microscope was important in the study of cells.
2. What is stated in the cell theory?
3. What is the difference between a simple and a compound light microscope?
4. What was Virchow's contribution to the cell theory?
5. **Think Critically** Why would it be better to look at living cells than at dead cells?

Skill Builder Activities

6. **Concept Mapping** Using a network tree concept map, compare a compound light microscope to an electron microscope. **For more help, refer to the Science Skill Handbook.**
7. **Solving One-Step Equations** Calculate the magnifications of a microscope that has an 8× eyepiece, and 10× and 40× objectives. **For more help, refer to the Math Skill Handbook.**

SECTION 3

Viruses

As You Read

What You'll Learn

- **Explain** how a virus makes copies of itself.
- **Identify** the benefits of vaccines.
- **Investigate** some uses of viruses.

Vocabulary
virus
host cell

Why It's Important
Viruses infect nearly all organisms, usually affecting them negatively yet sometimes affecting them positively.

What are viruses?

Cold sores, measles, chicken pox, colds, the flu, and AIDS are diseases caused by nonliving particles called viruses. A **virus** is a strand of hereditary material surrounded by a protein coating. Viruses don't have a nucleus or other organelles. They also lack a cell membrane. Viruses, as shown in **Figure 14,** have a variety of shapes. Because they are too small to be seen with a light microscope, they were discovered only after the electron microscope was invented. Before that time, scientists only hypothesized about viruses.

How do viruses multiply?

All viruses can do is make copies of themselves. However, they can't do that without the help of a living cell called a **host cell.** Crystalized forms of some viruses can be stored for years. Then, if they enter an organism, they can multiply quickly.

Once a virus is inside of a host cell, the virus can act in two ways. It can either be active or it can become latent, which is an inactive stage.

Figure 14
Viruses come in a variety of shapes.

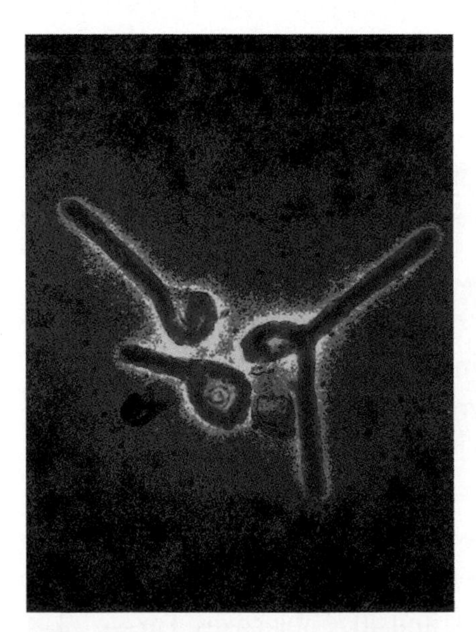

A Filoviruses do not have uniform shapes. Some of these *Ebola* viruses have a loop at one end.

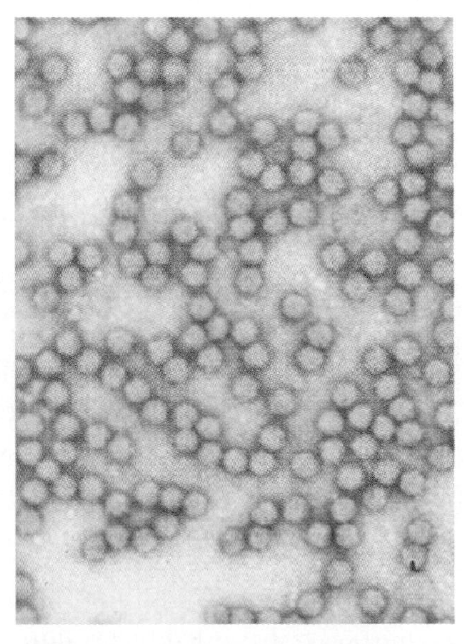

B The potato leafroll virus, *Polervirus,* damages potato crops worldwide.

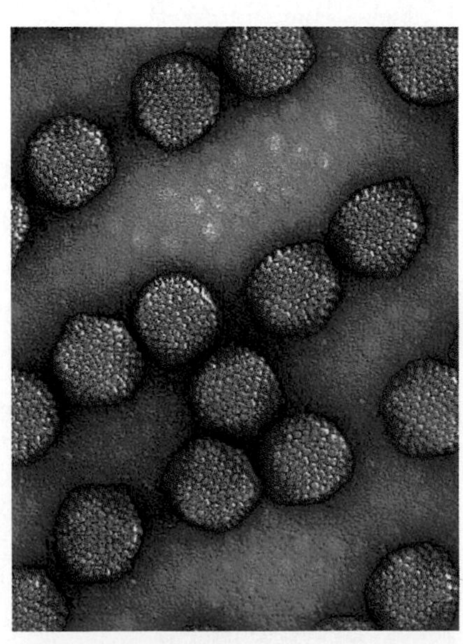

C This is just one of the many adenoviruses that can cause the common cold.

Figure 15
An active virus multiplies and destroys the host cell. A The virus attaches to a specific host cell. B The virus's hereditary material enters the host cell. C The hereditary material of the virus causes the cell to make viral hereditary material and proteins. D New viruses form inside of the host cell. E New viruses are released as the host cell bursts open and is destroyed.

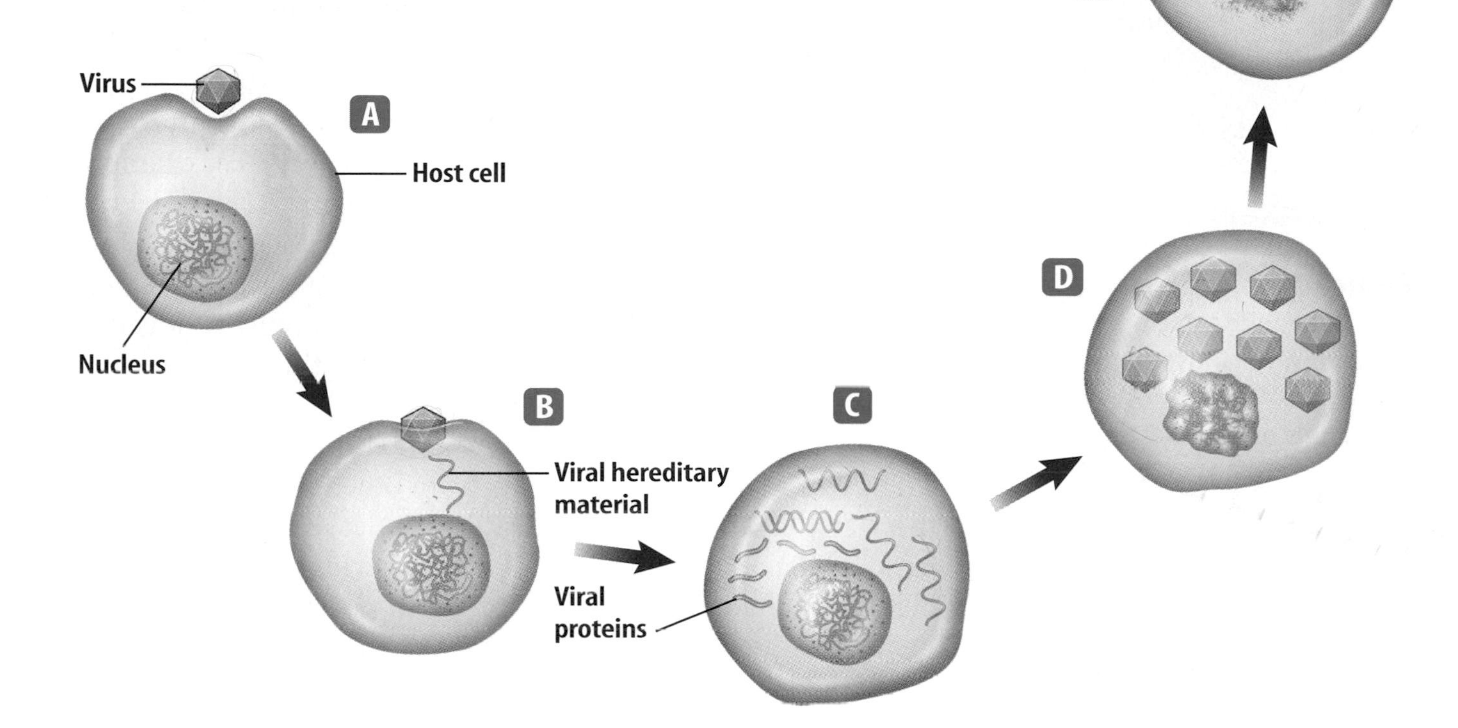

Active Viruses When a virus enters a cell and is active, it causes the host cell to make new viruses. This process destroys the host cell. Follow the steps in **Figure 15** to see one way that an active virus functions inside a cell.

Latent Viruses Some viruses can be latent. That means that after the virus enters a cell, its hereditary material can become part of the cell's hereditary material. It does not immediately make new viruses or destroy the cell. As the host cell reproduces, the viral DNA is copied. A virus can be latent for many years. Then, at any time, certain conditions, either inside or outside your body, can activate the virus.

If you have had a cold sore on your lip, a latent virus in your body has become active. The cold sore is a sign that the virus is active and destroying cells in your lip. When the cold sore disappears, the virus has become latent again. The virus is still in your body's cells, but it is hiding and doing no apparent harm.

Research Visit the Glencoe Science Web site at **science.glencoe.com** for information on viruses. What environmental stimuli might activate a latent virus? Record your answer in your Science Journal.

Virus

Cell membrane

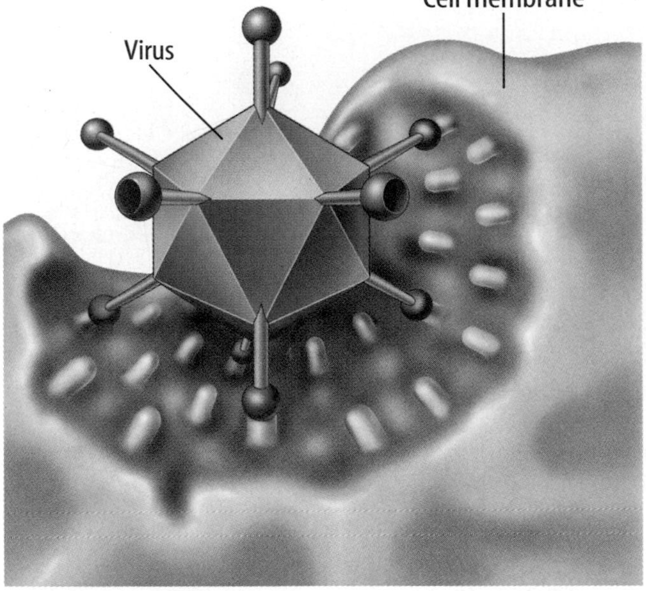

Figure 16
Viruses and the attachment sites of the host cell must match exactly. That's why most viruses infect only one kind of host cell.

How do viruses affect organisms?

Viruses attack animals, plants, fungi, protists, and all prokaryotes. Some viruses can infect only specific kinds of cells. For instance, many viruses, such as the potato leafroll virus, are limited to one host species or to one type of tissue within that species. A few viruses affect a broad range of hosts. An example of this is the rabies virus. Rabies can infect humans and many other animal hosts.

A virus cannot move by itself, but it can reach a host's body in several ways. For example, it can be carried onto a plant's surface by the wind or it can be inhaled by an animal. In a viral infection, the virus first attaches to the surface of the host cell. The virus and the place where it attaches must fit together exactly, as shown in **Figure 16.** Because of this, most viruses attack only one kind of host cell.

Viruses that infect bacteria are called bacteriophages (bak TIHR ee uh fay juhz). They differ from other kinds of viruses in the way that they enter bacteria and release their hereditary material. Bacteriophages attach to a bacterium and inject their hereditary material. The entire cycle takes about 20 min, and each virus-infected cell releases an average of 100 viruses.

SCIENCE *Online*

Data Update Scientists have determined that *Marburg* virus, *Ebola zaire*, and *Ebola reston* belong to the virus family Filoviridae. Visit the Glencoe Science Web site at **science.glencoe.com** for the latest information about these viruses. Share your results with your class.

Fighting Viruses

Vaccines are used to prevent disease. A vaccine is made from weakened virus particles that can't cause disease anymore. Vaccines have been made to prevent many diseases, including measles, mumps, smallpox, chicken pox, polio, and rabies.

What is a vaccine?

The First Vaccine Edward Jenner is credited with developing the first vaccine in 1796. He developed a vaccine for smallpox, a disease that was still feared in the early twentieth century. Jenner noticed that people who got a disease called cowpox didn't get smallpox. He prepared a vaccine from the sores of people who had cowpox. When injected into healthy people, the cowpox vaccine protected them from smallpox. Jenner didn't know he was fighting a virus. At that time, no one understood what caused disease or how the body fought disease.

Treating and Preventing Viral Diseases Antibiotics are used to treat bacterial infections. They are ineffective against any viral disease. One way your body can stop viral infections is by making interferons. Interferons are proteins that protect cells from viruses. These proteins are produced rapidly by infected cells and move to noninfected cells in the host. They cause the noninfected cells to produce protective substances.

Antiviral drugs can be given to infected patients to help fight a virus. A few drugs show some effectiveness against viruses but some have limited use because of their adverse side effects.

Public health measures for preventing viral diseases include vaccinating people, improving sanitary conditions, quarantining patients, and controlling animals that spread the disease. Yellow fever was wiped out completely in the United States through mosquito-control programs. Annual rabies vaccinations protect humans by keeping pets and farm animals free from infection. To control the spread of rabies in wild animals such as coyotes and wolves, wildlife workers place bait containing an oral rabies vaccine, as shown in **Figure 17**, where wild animals will find it.

Figure 17
This oral rabies bait is being prepared for an aerial drop by the Texas Department of Health as part of their Oral Rabies Vaccination Program. This five-year program has prevented the expansion of rabies into Texas.

Research with Viruses

You might think viruses are always harmful. However, through research, scientists are discovering helpful uses for some viruses. One use, called gene therapy, is being tried on cells with defective genes. Normal hereditary material is substituted for a cell's defective hereditary material. The normal material is enclosed in viruses. The viruses then "infect" targeted cells, taking the new hereditary material into the cells to replace the defective hereditary material. Using gene therapy, scientists hope to help people with genetic disorders and find a cure for cancer.

Section  Assessment

1. Describe the structure of viruses and explain how viruses multiply.
2. How are vaccines beneficial?
3. How might some viruses be helpful?
4. How might viral diseases be prevented?
5. **Think Critically** Explain why a doctor might not give you any medication if you have a viral disease.

Skill Builder Activities

6. **Concept Mapping** Make an events chain concept map to show what happens when a latent virus becomes active. **For more help, refer to the Science Skill Handbook.**
7. **Using a Word Processor** Make an outline of the cycle of an active virus. **For more help, refer to the Technology Skill Handbook.**

Activity Design Your Own Experiment

Comparing Light Microscopes

You're a technician in a police forensic laboratory. You use a stereomicroscope and a compound light microscope in the laboratory. A detective just returned from a crime scene with bags of evidence. You must examine each piece of evidence under a microscope. How do you decide which microscope is the best tool to use?

Recognize the Problem

Will all of the evidence that you've collected be viewable through both microscopes?

Form a Hypothesis

Compare the items to be examined under the microscopes. Which microscope will be used for each item?

Possible Materials

compound light microscope
stereomicroscope
items from the classroom—include some living or once-living items (8)
microscope slides and coverslips
plastic petri dishes
distilled water
dropper

Goals

- **Learn** how to correctly use a stereomicroscope and a compound light microscope.
- **Compare** the uses of the stereomicroscope and compound light microscope.

Safety Precautions

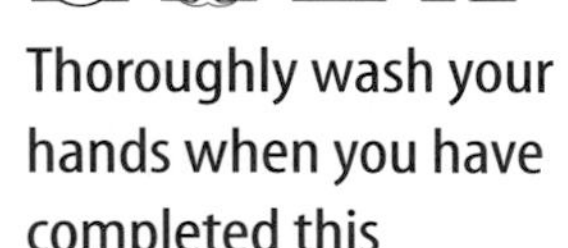

Thoroughly wash your hands when you have completed this experiment.

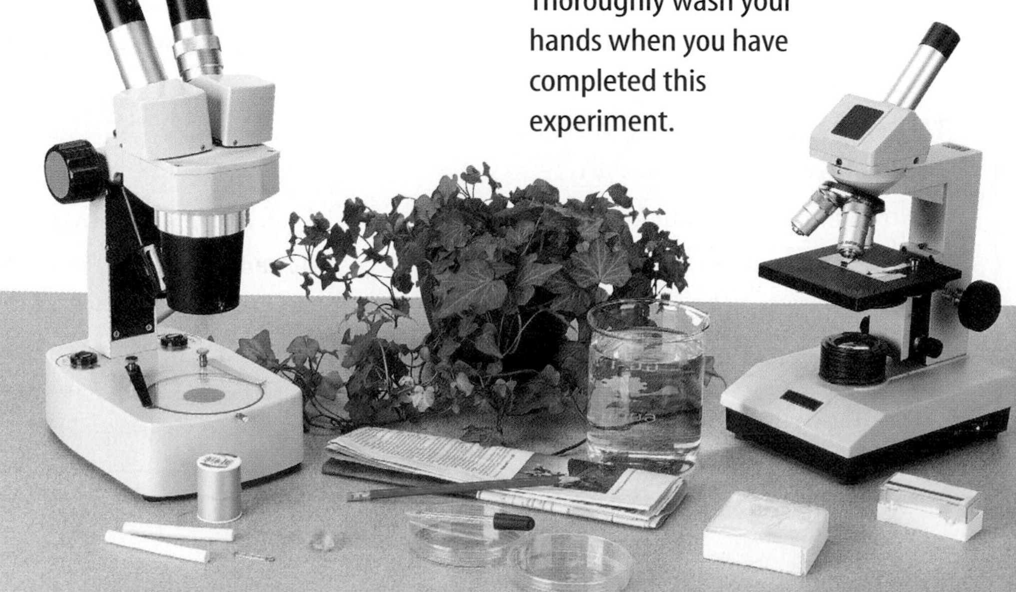

Test Your Hypothesis

Plan

1. As a group, decide how you will test your hypothesis.
2. **Describe** how you will carry out this experiment using a series of specific steps. Make sure the steps are in a logical order. Remember that you must place an item in the bottom of a plastic petri dish to examine it under the stereomicroscope and you must make a wet mount of any item to be examined under the compound light microscope. For more help, see the Reference Handbook.
3. If you need a data table or an observation table, design one in your Science Journal.

Do

1. Make sure your teacher approves the objects you'll examine, your plan, and your data table before you start.
2. Carry out the experiment.
3. While doing the experiment, record your observations and complete the data table.

Analyze Your Data

1. **Compare** the items you examined with those of your classmates.
2. Based on this experiment, classify the eight items you observed.

Draw Conclusions

1. **Infer** which microscope a scientist might use to examine a blood sample, fibers, and live snails.
2. **List** five careers that require people to use a stereomicroscope. List five careers that require people to use a compound light microscope. Enter the lists in your Science Journal.
3. If you examined an item under a compound light microscope and a stereomicroscope, how would the images differ?
4. Which microscope was better for looking at large, or possibly live items?

Communicating Your Data

In your Science Journal, **write** a short description of an imaginary crime scene and the evidence found there. **Sort** the evidence into two lists—items to be examined under a stereomicroscope and items to be examined under a compound light microscope. **For more help, refer to the Science Skill Handbook.**

Magnification: 2,000×

This colored scanning electron micrograph (SEM) shows two breast cancer cells in the final stage of cell dvision.

Cobb Against Cancer

New York City, 1950. Jewel Plummer put yet another slide onto the stage of her microscope and clipped it into place. She switched to the high power objective, looked through the eyepiece, and turned the fine adjustment a tiny bit to bring her subject—cells from a cancerous tumor—into focus. She switched back to low power and removed the slide. She had found no change in the tumor cells. The drug that doctors had used wasn't killing or slowing the growth rate of those cancer cells. Sighing, she reached for the next slide. Maybe the slightly different drug they had used on that batch of cells would be the answer....

Jewel Plummer Cobb is a cell biologist who did important background research on the use of drugs against cancer. She removed cells from cancerous tumors and cultured them in the lab. Then, in a controlled study, she tried a series of different drugs against batches of the same cells. Her goal was to find the right drug to cure each patient's particular cancer. Cobb never met that goal, but her research laid the groundwork for modern chemotherapy—the use of chemicals to treat people with cancer.

Role Model

Jewel Cobb also influenced the course of science in a different way. She served as dean or president of several universities, retiring as president of the University of California at Fullerton. In her role as a college official, she was able to promote equal opportunity for students of all backgrounds, especially in the sciences.

Light Up a Cure

Vancouver, British Columbia, 2000. While Cobb herself was only able to infer what was going on inside a cell from its reactions to various drugs, her work has helped others go further. Building on Cobb's work, Professor Julia Levy and her research team at the University of British Columbia actually go inside cells and even inside organelles to work against cancer. One technique they are pioneering is the use of light to guide cancer drugs to the right cells. First, the patient is given a chemotherapy drug that reacts to light. Next, a fiber optic tube is inserted into the tumor. Finally, laser light is passed through the tube. The light activates the light-sensitive drug—but only in the tumor itself. This technique keeps healthy cells healthy but kills sick cells on the spot.

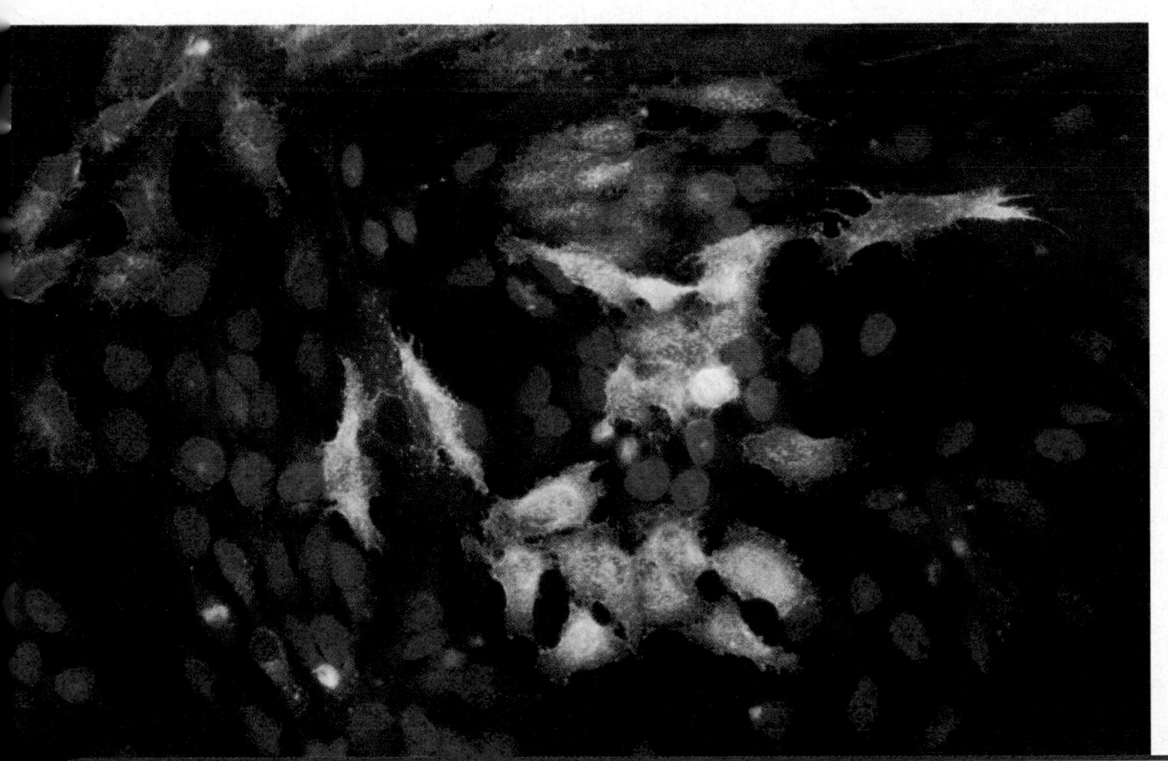

The image to the left shows human cervical cells magnified 125 times that have been attacked by cancer. The light blue areas at the center are keratin, a kind of protein. The cell nuclei are stained blue, and the red areas are fibroblasts, a kind of connective-tissue cell. These are the first human cells used to research cancer. This type of cell grows well in a lab, and is used in research worldwide.

CONNECTIONS Write **Report on Cobb's experiments on cancer cells. What were her dependent and independent variables? What would she have used as a control? What sources of error did she have to guard against? Answer the same questions about Levy's work.**

SCIENCE *Online*
For more information, visit science.glencoe.com

Chapter 3 Study Guide

Reviewing Main Ideas

Section 1 Cell Structure

1. There are two basic cell types. Cells without membrane-bound structures are called prokaryotic cells. Cells with membrane-bound structures are called eukaryotic cells.
2. Most of the life processes of a cell occur within the cytoplasm.
3. Cell functions are performed by organelles under the control of DNA in the nucleus.
4. Organelles such as mitochondria and chloroplasts process energy.
5. Proteins take part in nearly every cell activity.

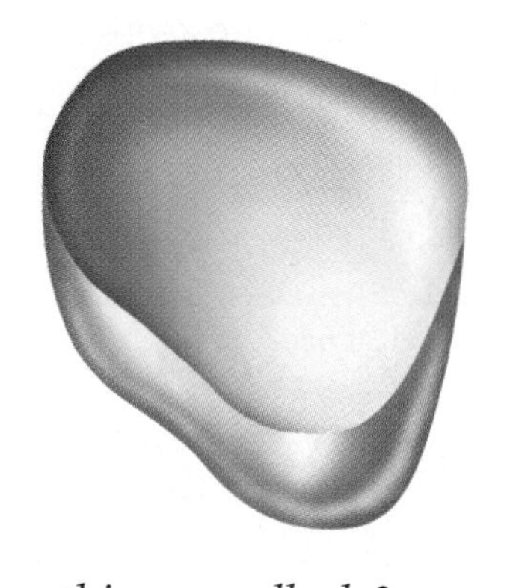

6. Golgi bodies and vacuoles transport substances, rid the cell of wastes, and store cellular materials. *What does this organelle do?*
7. Most many-celled organisms are organized into tissues, organs, and organ systems that perform specific functions to keep an organism alive.

Section 2 Viewing Cells

1. A simple microscope has just one lens. A compound light microscope has eyepiece lenses and objective lenses.
2. To calculate the magnification of a microscope, multiply the power of the eyepiece by the power of the objective lens.
3. An electron microscope uses a beam of electrons instead of light to produce an image of an object.
4. Things that are too small to be viewed with a light microscope can be viewed with an electron microscope. This is an SEM of an ant. *How do you know if the ant is alive or dead?*

5. According to the cell theory, the cell is the basic unit of life. Organisms are made of one or more cells, and all cells come from other cells.

Section 3 Viruses

1. A virus is a structure containing hereditary material surrounded by a protein coating.
2. A virus can make copies of itself only when it is inside a living host cell.
3. Viruses cause diseases in animals, plants, fungi, and bacteria. *Why don't scientists consider viruses like these in the photo to be living organisms?*

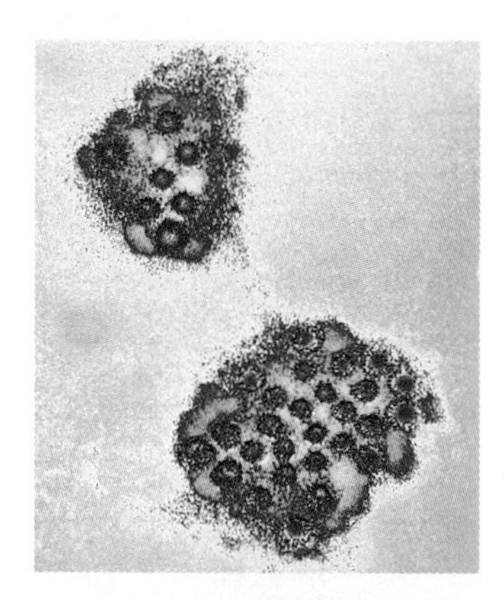

After You Read

FOLDABLES Reading & Study Skills

On the inside of the Main Ideas Study Fold you made at the beginning of the chapter describe the characteristics of each type of cell.

Chapter 3 Study Guide

Visualizing Main Ideas

Complete the following concept map of the basic units of life.

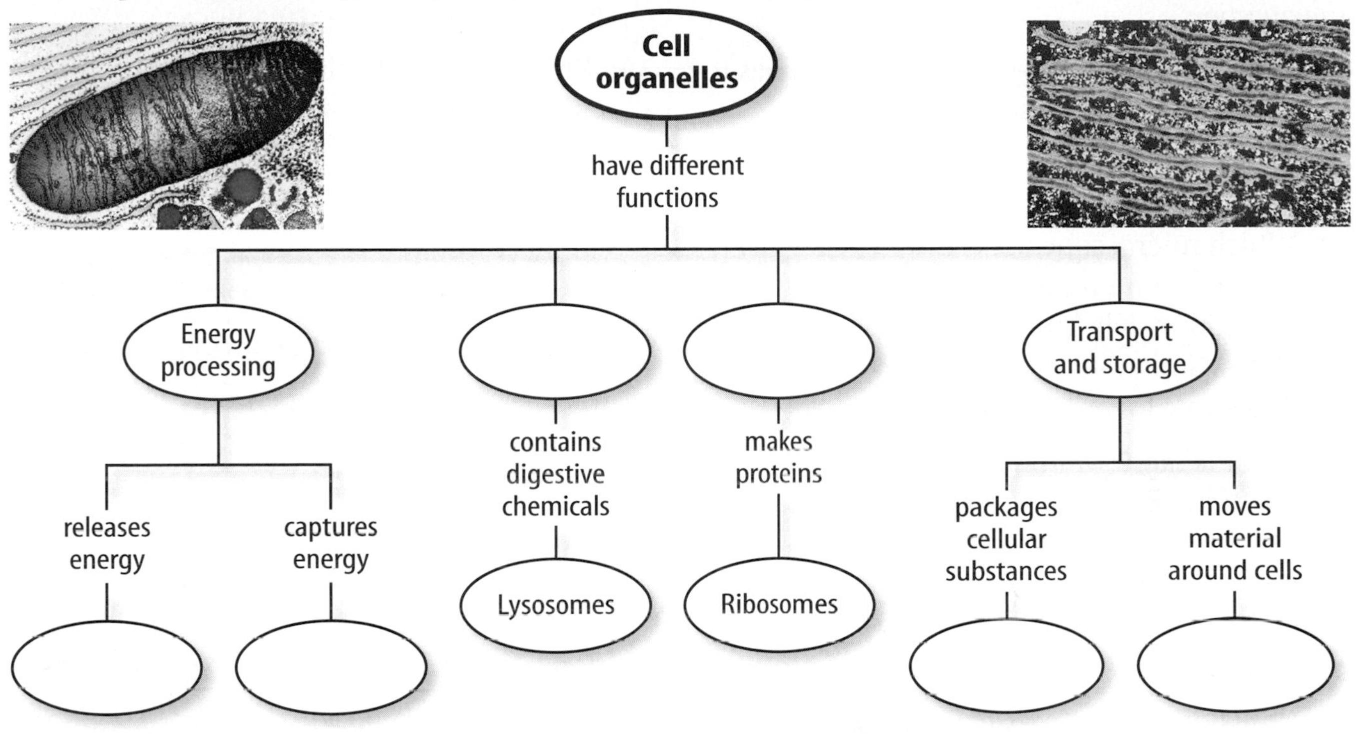

Vocabulary Review

Vocabulary Words

a. cell membrane
b. cell theory
c. cell wall
d. chloroplast
e. cytoplasm
f. endoplasmic reticulum
g. Golgi body
h. host cell
i. mitochondrion
j. nucleus
k. organ
l. organelle
m. ribosome
n. tissue
o. virus

THE PRINCETON REVIEW Study Tip

In order to understand the information that a graph is trying to communicate, write out a sentence that talks about the relationship between the *x*-axis and *y*-axis in the graph.

Using Vocabulary

Using the vocabulary words, give an example of each of the following.

1. found in every organ
2. smaller than one cell
3. a plant-cell organelle
4. part of every cell
5. powerhouse of a cell
6. used by biologists
7. contains hereditary material
8. a structure that surrounds the cell
9. can be damaged by a virus
10. made up of cells

Chapter 3 Assessment

Checking Concepts

Choose the word or phrase that best answers the question.

1. What structure allows only certain things to pass in and out of the cell?
A) cytoplasm **C)** ribosomes
B) cell membrane **D)** Golgi body

2. Which microscope uses lenses to magnify?
A) compound light microscope
B) scanning electron microscope
C) transmission electron microscope
D) atomic force microscope

3. What is made of folded membranes that move materials around inside the cell?
A) nucleus
B) cytoplasm
C) Golgi body
D) endoplasmic reticulum

4. Which scientist gave the name *cells* to structures he viewed?
A) Hooke **C)** Schleiden
B) Schwann **D)** Virchow

5. What organelle helps recycle old cell parts?
A) chloroplast **C)** lysosome
B) centriole **D)** cell wall

6. Which of the following is a viral disease?
A) tuberculosis **C)** smallpox
B) anthrax **D)** tetanus

7. What are structures in the cytoplasm of a eukaryotic cell called?
A) organs **C)** organ systems
B) organelles **D)** tissues

8. Which microscope can magnify up to a million times?
A) compound light microscope
B) stereomicroscope
C) transmission electron microscope
D) atomic force microscope

9. Which of the following is part of a bacterial cell?
A) a cell wall **C)** mitochondria
B) lysosomes **D)** a nucleus

10. Which of the following do groups of different tissues form?
A) organ **C)** organ system
B) organelle **D)** organism

Thinking Critically

11. Why is it difficult to treat a viral disease?

12. What type of microscope would be best to view a piece of moldy bread? Explain.

13. What would happen to a plant cell that suddenly lost its chloroplasts?

14. What would happen to this animal cell if it didn't have ribosomes?

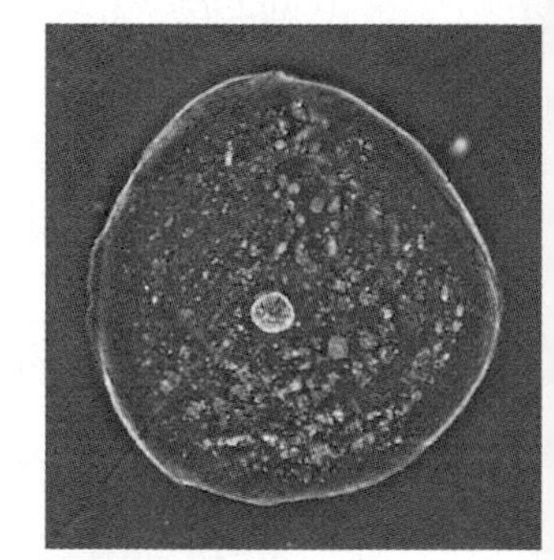

15. How would you decide whether an unknown cell was an animal cell, a plant cell, or a bacterial cell?

Developing Skills

16. Concept Mapping Make an events-chain concept map of the following from simple to complex: *small intestine, circular muscle cell, human,* and *digestive system.*

17. Interpreting Scientific Illustrations Use the illustrations in **Figure 1** to describe how the shape of a cell is related to its function.

18. Making and Using Graphs Use a computer to make a line graph of the following data. At 37°C there are 1.0 million viruses; at, 37.5°C, 0.5 million; at 37.8°C, 0.25 million; at 38.3°C, 0.1 million; and at 38.9°C, 0.05 million.

19. Comparing and Contrasting Complete the following table to compare and contrast the structures of a prokaryotic cell to those of a eukaryotic cell.

Cell Structures

Structure	Prokaryotic Cell	Eukaryotic Cell
Cell Membrane		Yes
Cytoplasm	Yes	
Nucleus		Yes
Endoplasmic Reticulum		
Golgi Bodies		

20. Making a Model Make and illustrate a time line to show the development of the cell theory. Begin with the development of the microscope and end with Virchow. Include the contributions of Leeuwenhoek, Hooke, Schleiden, and Schwann.

Performance Assessment

21. Model Use materials that resemble cell parts or that represent their functions to make a model of a plant cell or an animal cell. Make a key to the cell parts to explain your model.

22. Poster Research the history of vaccinations. Contact your local Health Department for current information. Display your results on a poster.

TECHNOLOGY

Go to the Glencoe Science Web site at **science.glencoe.com** or use the **Glencoe Science CD-ROM** for additional chapter assessment.

Test Practice

A scientist is studying living cells. Below is an image of one of the cells that is being studied. This image represents what a scientist sees when he or she uses a tool in the laboratory.

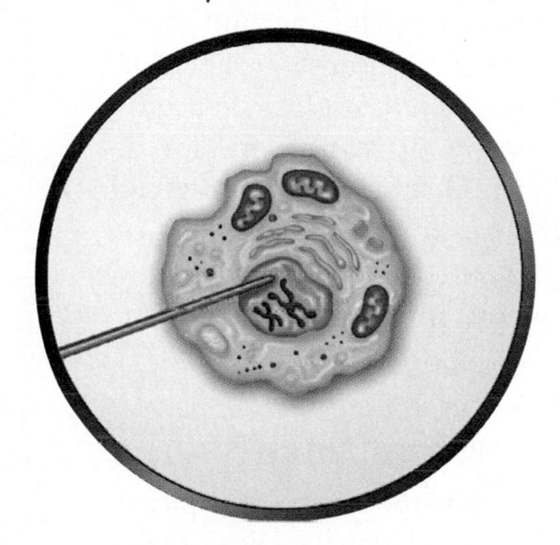

Closely examine the image above then answer the following questions.

1. If the pointer shown above with the cell is 10 micrometers in length, then about how wide is this cell?

A) 20 micrometers
B) 10 micrometers
C) 5 micrometers
D) 0.1 micrometers

2. Which of the following tools is the scientist probably using to view the living cell?

F) telescope
G) endoplasmic reticulum
H) compound light microscope
J) kaleidoscope

CHAPTER

Cell Processes

The Sun is hot. Your back aches and your hands are sore. Weeding a garden is hard work. You are sweaty, tired, thirsty, and hungry. Are the weeds having the same reactions? You may know that plants don't sweat or get tired, but they do need water and food, just like you. How do plants take in and use water and food? In this chapter you'll find the answer to this question. You'll also find out how living things get the energy that they need to survive.

What do you think?

Science Journal Look at the picture below with a classmate. Discuss what this might be or what is happening. Here's a hint: *It's sometimes called the powerhouse of the cell.* Write your answer or best guess in your Science Journal.

If you forget to water a plant, it will wilt. After you water the plant, it probably will straighten up and look healthier. Why does the plant straighten? In the following activity, find out about water entering and leaving plant cells.

Demonstrate why water leaves plant cells

1. Label a small bowl "salt water." Pour 250 mL of water into the bowl. Then add 15 g of salt to the water and stir.
2. Pour 250 mL of water into another small bowl.
3. Place two carrot sticks into each bowl. Also, place two carrot sticks on the lab table.
4. After 30 min, remove the carrot sticks from the bowls and keep them next to the bowl they came from. Examine all six carrot sticks, then describe them in your Science Journal.

Observe

Predict what would happen if you moved the carrot sticks from the plain water to the lab table, the ones from the salt water into the plain water, and the ones from the lab table into the salt water. Now try it. Write your predictions and your results in your Science Journal.

Before You Read

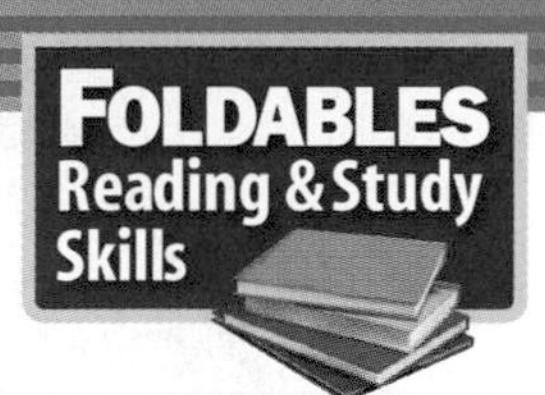

Making a Vocabulary Study Fold To help you study cell processes, make the following vocabulary Foldable. Knowing the definition of vocabulary words in a chapter is a good way to ensure that you have understood the content.

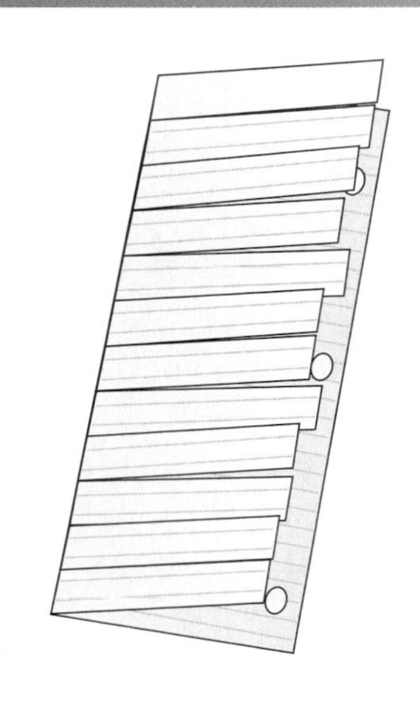

1. Place a sheet of notebook paper in front of you so the short side is at the top. Fold the paper in half from the left to the right side.
2. Through the top thickness of paper, cut along every third line from the outside edge to the center fold, forming ten tabs as shown.
3. On the front of each tab, write a vocabulary word listed on the first page of each section in this chapter. On the back of each tab, define the word.

Chemistry of Life

As You Read

What You'll Learn

- **List** the differences among atoms, elements, molecules, and compounds.
- **Explain** the relationship between chemistry and life science.
- **Discuss** how organic compounds are different from inorganic compounds.

Vocabulary

mixture
organic compound
enzyme
inorganic compound

Why It's Important

You grow because of chemical reactions in your body.

The Nature of Matter

Think about everything that surrounds you—chairs, books, clothing, other students, and air. What are all these things made up of? You're right if you answer "matter and energy." Matter is anything that has mass and takes up space. Energy is anything that brings about change. Everything in your environment, including you, is made of matter. Energy can hold matter together or break it apart. For example, the food you eat is matter that is held together by chemical energy. When food is cooked, energy in the form of heat can break some of the bonds holding the matter in food together. **Table 1** compares matter and energy and gives some examples of each.

Atoms Whether it is solid, liquid, or gas, matter is made of atoms. **Figure 1** shows a model of an oxygen atom. At the center of an atom is a nucleus that contains protons and neutrons. Although they have nearly equal masses, a proton has a positive charge and a neutron has no charge. Outside the nucleus are electrons, each of which has a negative charge. It takes about 1,837 electrons to equal the mass of one proton. Electrons are important because they are the part of the atom that is involved in chemical reactions. Look at **Figure 1** again and you will see that an atom is mostly empty space. Energy holds the parts of an atom together.

Figure 1
An oxygen atom model shows the placement of electrons, protons, and neutrons.

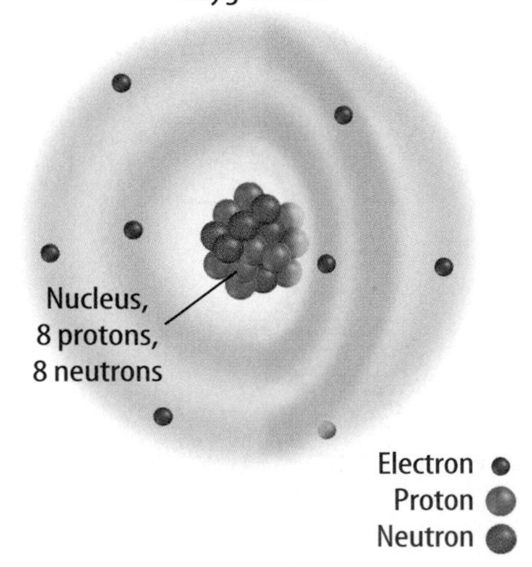

Table 1 Matter and Energy

	Definition	Examples
Matter	anything that has mass and takes up space	atoms, electrons, protons, and neutrons, living things, rocks, soil, and air
Energy	ability to cause change	sunlight, electricity, heat, chemical energy

Table 2 Elements That Make Up the Human Body

Symbol	Element	Percent
O	Oxygen	65.0
C	Carbon	18.5
H	Hydrogen	9.5
N	Nitrogen	3.2
Ca	Calcium	1.5
P	Phosphorus	1.0
K	Potassium	0.4
S	Sulfur	0.3
Na	Sodium	0.2
Cl	Chlorine	0.2
Mg	Magnesium	0.1
	Other elements	0.1

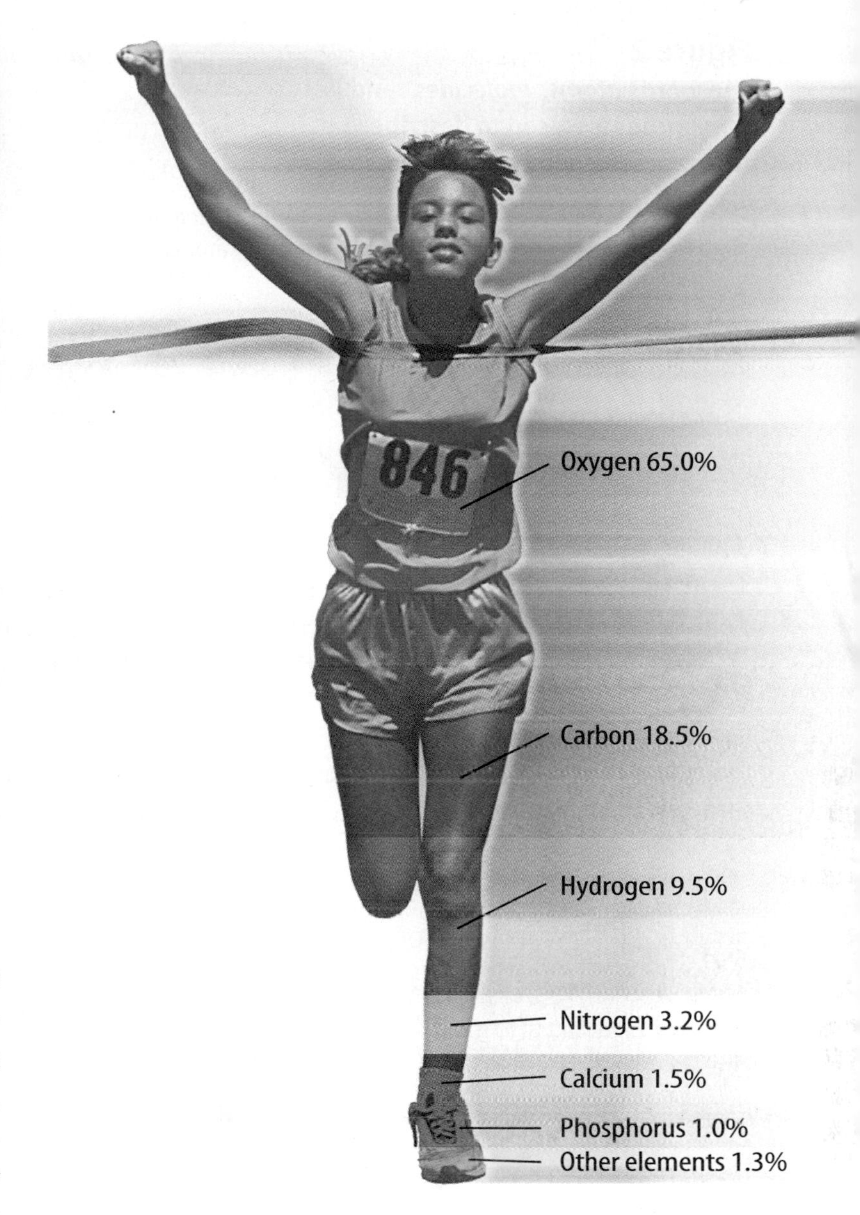

Elements When something is made up of only one kind of atom, it is called an element. An element can't be broken down into a simpler form by chemical reactions. The element oxygen is made up of only oxygen atoms, and hydrogen is made up of only hydrogen atoms. Scientists have given each element its own one- or two-letter symbol.

All elements are arranged in a chart known as the periodic table of elements. You can find this table at the back of this book. The table provides information about each element including its mass, how many protons it has, and its symbol.

Everything is made up of elements. Most things, including all living things, are made up of a combination of elements. Few things exist as pure elements. **Table 2** lists elements that are in the human body. What two elements make up most of your body?

Reading Check *What types of things are made up of elements?*

Six of the elements listed in the table are important because they make up about 99 percent of living matter. The symbols for these elements are S, P, O, N, C, and H. Use **Table 2** to find the names of these elements.

Figure 2
The words *atoms, molecules,* and *compounds* are used to describe substances. *How are they related to each other?*

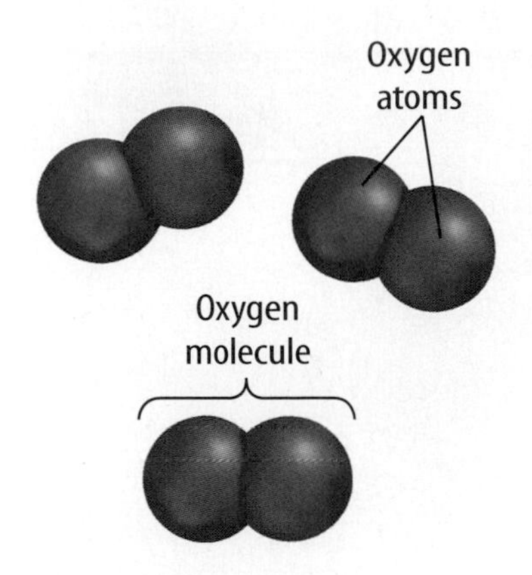

A Some elements, like oxygen, occur as molecules. These molecules contain atoms of the same element bonded together.

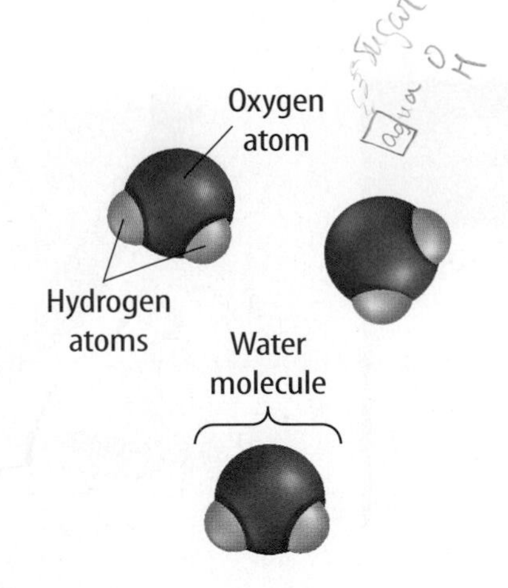

B Compounds also are composed of molecules. Molecules of compounds contain atoms of two or more different elements bonded together, as shown by these water molecules.

Compounds and Molecules

Suppose you make a pitcher of lemonade using a powdered mix and water. The water and the lemonade mix, which is mostly sugar, contain the elements oxygen and hydrogen. Yet, in one, they are part of a nearly tasteless liquid—water. In the other they are part of a sweet solid—sugar. How can the same elements be part of two materials that are so different? Water and sugar are compounds. Compounds are made up of two or more elements in exact proportions. For example, pure water, whether one milliliter of it or one million liters, is always made up of hydrogen atoms bonded to oxygen atoms in a ratio of two hydrogen atoms to one oxygen atom. Compounds have properties different from the elements they are made of. There are two types of compounds—molecular compounds and ionic compounds.

Molecular Compounds The smallest part of a molecular compound is a molecule. A molecule is a group of atoms held together by the energy of chemical bonds, as shown in **Figure 2.** When chemical reactions occur, chemical bonds break, atoms are rearranged, and new bonds form. The molecules produced are different from those that began the chemical reaction.

Molecular compounds form when different atoms share their outermost electrons. For example, two atoms of hydrogen each can share one electron on one atom of oxygen to form one molecule of water, as shown in **Figure 2B.** Water does not have the same properties as oxygen and hydrogen. Under normal conditions on Earth, oxygen and hydrogen are gases. Yet, water can be a liquid, a solid, or a gas. When hydrogen and oxygen combine, changes occur and a new substance forms.

Ions Atoms also combine because they've become positively or negatively charged. Atoms are usually neutral—they have no overall electric charge. When an atom loses an electron, it has more protons than electrons, so it becomes positively charged. When an atom gains an electron, it has more electrons than protons, so it becomes negatively charged. Electrically charged atoms—positive or negative—are called ions.

Ionic Compounds Ions of opposite charges attract one another to form electrically neutral compounds called ionic compounds. Table salt is made of sodium (Na^+) and chlorine (Cl^-) ions, as shown in **Figure 3B.** When they combine, a chlorine atom gains an electron from a sodium atom. The chlorine atom becomes a negatively charged ion, and the sodium atom becomes a positively charged ion. These oppositely charged ions then are attracted to each other and form the ionic compound sodium chloride, NaCl.

Ions are important in many life processes that take place in your body and in other organisms. For example, messages are sent along your nerves as potassium and sodium ions move in and out of nerve cells. Calcium ions are important in causing your muscles to contract. Ions also are involved in the transport of oxygen by your blood. The movement of some substances into and out of a cell would not be possible without ions.

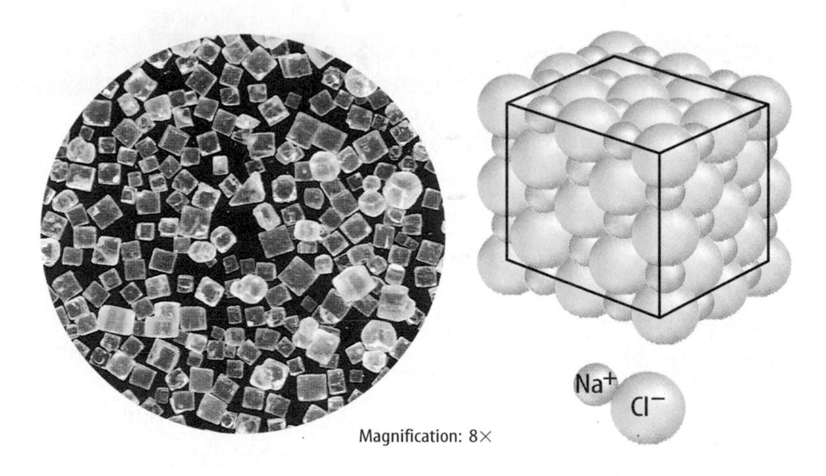

A Magnified crystals of salt look like this.

B The salt crystal is held together by the attraction between sodium ions and chlorine ions.

Figure 3
Table salt, or sodium chloride (NaCl), is a crystal composed of sodium ions and chlorine ions held together by ionic bonds.

Mixtures

Some substances, such as a combination of sugar and salt, can't change each other or combine chemically. A **mixture** is a combination of substances in which individual substances retain their own properties. Mixtures can be solids, liquids, gases, or any combination of them.

Reading Check ***Why is a combination of sugar and salt said to be a mixture?***

Most chemical reactions in living organisms take place in mixtures called solutions. You've probably noticed the taste of salt when you perspire. Sweat is a solution of salt and water. In a solution, two or more substances are mixed evenly. A cell's cytoplasm is a solution of dissolved molecules and ions.

Living things also contain mixtures called suspensions. A suspension is formed when a liquid or a gas has another substance evenly spread throughout it. Unlike solutions, the substances in a suspension eventually sink to the bottom. If blood, shown in **Figure 4,** is left undisturbed, the red blood cells and white blood cells will sink gradually to the bottom. However, the pumping action of your heart constantly moves your blood and the blood cells remain suspended.

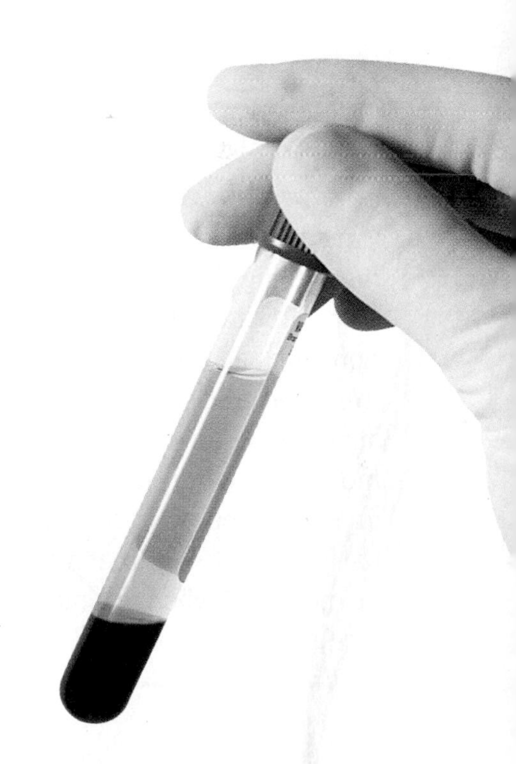

Figure 4
When a test tube of whole blood is left standing, the blood cells sink in the watery plasma.

Table 3 Organic Compounds Found in Living Things

	Carbohydrates	Lipids	Proteins	Nucleic Acids
Elements	carbon, hydrogen, and oxygen	carbon, oxygen, hydrogen, and phosphorus	carbon, oxygen, hydrogen, nitrogen, and sulfur	carbon, oxygen, hydrogen, nitrogen, and phosphorus
Examples	sugars, starch, and cellulose	fats, oils, waxes, phospholipids, and cholesterol	enzymes, skin, and hair	DNA and RNA
Function	supply energy for cell processes; form plant structures; short-term energy storage	store large amounts of energy long term; form boundaries around cells	regulate cell processes and build cell structures	carry hereditary information; used to make proteins

Organic Compounds

You and all living things are made up of compounds that are classified as organic or inorganic. Rocks and other nonliving things contain inorganic compounds, but most do not contain large amounts of organic compounds. **Organic compounds** always contain carbon and hydrogen and usually are associated with living things. One exception would be nonliving things that are products of living things. For example, coal contains organic compounds because it was formed from dead and decaying plants. Organic molecules can contain hundreds or even thousands of atoms that can be arranged in many ways. **Table 3** compares the four groups of organic compounds that make up all living things—carbohydrates, lipids, proteins, and nucleic acids.

Research Air is a mixture of many things. Weather forecasts often include information about air quality. Visit the Glencoe Science Web site at **science.glencoe.com** for more information about air quality. In your Science Journal list some things that may be measured when testing air quality.

Carbohydrates Carbohydrates are organic molecules that supply energy for cell processes. Sugars and starches are carbohydrates that cells use for energy. Some carbohydrates also are important parts of cell structures. For example, a carbohydrate called cellulose is an important part of plant cells.

Lipids Another type of organic compound found in living things is a lipid. Lipids do not mix with water. Lipids such as fats and oils store and release even larger amounts of energy than carbohydrates do. One type of lipid, the phospholipid, is a major part of cell membranes.

Proteins Organic compounds called proteins have many important functions in living organisms. They are made up of smaller molecules called amino acids. Proteins are the building blocks of many structures in organisms. Your muscles contain large amounts of protein. Proteins are scattered throughout cell membranes. Certain proteins called **enzymes** regulate nearly all chemical reactions in cells.

Nucleic Acids Large organic molecules that store important coded information in cells are called nucleic acids. One nucleic acid, deoxyribonucleic acid, or DNA—genetic material—is found in all cells. It carries information that directs each cell's activities. Another nucleic acid, ribonucleic acid, or RNA, is needed to make enzymes and other proteins.

Inorganic Compounds

Most **inorganic compounds** are made from elements other than carbon. Generally, inorganic molecules contain fewer atoms than organic molecules. Inorganic compounds are the source for many elements needed by living things. For example, plants take up inorganic compounds from the soil. These inorganic compounds can contain the elements nitrogen, phosphorus, and sulfur. Many foods that you eat contain inorganic compounds. **Table 4** shows some of the inorganic compounds that are important to you. One of the most important inorganic compounds for living things is water.

Table 4 Some Inorganic Compounds Important in Humans

Compound	Use in Body
Water	makes up most of the blood; most chemical reactions occur in water
Calcium phosphate	gives strength to bones
Hydrochloric acid	breaks down foods in the stomach
Sodium bicarbonate	helps the digestion of food to occur
Salts containing sodium, chlorine, and potassium	important in sending messages along nerves

Mini LAB

Determining How Enzymes Work

Procedure

1. Get two small cups of **prepared gelatin** from your teacher. Do not eat or drink anything in lab.
2. On the gelatin in one of the cups, place a piece of **fresh pineapple.**
3. Let both cups stand undisturbed during your class period. Wash your hands when you are done.
4. Observe what happens to the gelatin.

Analysis

1. What effect did the piece of fresh pineapple have on the gelatin?
2. What does fresh pineapple contain that caused it to have the effect on the gelatin you observed?
3. Why do the preparation directions on a box of gelatin dessert tell you not to mix it with fresh pineapple?

Importance of Water Some scientists hypothesize that life began in the water of Earth's ancient oceans. Chemical reactions might have occurred that produced organic molecules. Similar chemical reactions can take place in cells in your body.

Living things are composed of more than 50 percent water and depend on water to survive. You can live for weeks without food but only for a few days without water. **Figure 5** shows where water is found in your body. Although seeds and spores of plants, fungi, and bacteria can exist without water, they must have water if they are to grow and reproduce. All the chemical reactions in living things take place in water solutions, and most organisms use water to transport materials through their bodies. For example, many animals have blood that is mostly water and moves materials. Plants use water to move minerals and sugars between the roots and leaves.

Math Skills Activity

Calculating the Importance of Water

All life on Earth depends on water for survival. Water is the most vital part of humans and other animals. It is required for all of the chemical processes that keep us alive.

Example Problem

At least 60% of an adult human body consists of water. If an adult man weighs 90 kg, how many kilograms of water does his body contain?

Solution

1. *This is what you know:* adult human body = 60% water; man = 90 kg
2. *This is what you want to find:* 60% of 90 kg
3. *This is the equation you need to use:* $60/100 = x/90$
4. *Solve the equation for* x: $x = (60 \times 90)/100$; $x = 54$ kg

Check your answer by dividing your answer by 90, then multiplying by 100. Do you get 60%?

Practice Problem

A human body at birth consists of 78% water. This percent gradually decreases to 60% in an adult. Assume a baby weighed 3.2 kg at birth, and grew into an adult weighing 95 kg. Calculate the approximate number of kilograms of water the human gained.

For more help, refer to the Math Skill Handbook.

Characteristics of Water

The atoms of a water molecule are arranged in such a way that the molecule has areas with different charges. Water molecules are like magnets. The negative part of a water molecule is attracted to the positive part of another water molecule just like the north pole of a magnet is attracted to the south pole of another magnet. This attraction, or force, between water molecules is why a film forms on the surface of water. The film is strong enough to support small insects because the forces between water molecules are stronger than the force of gravity on the insect.

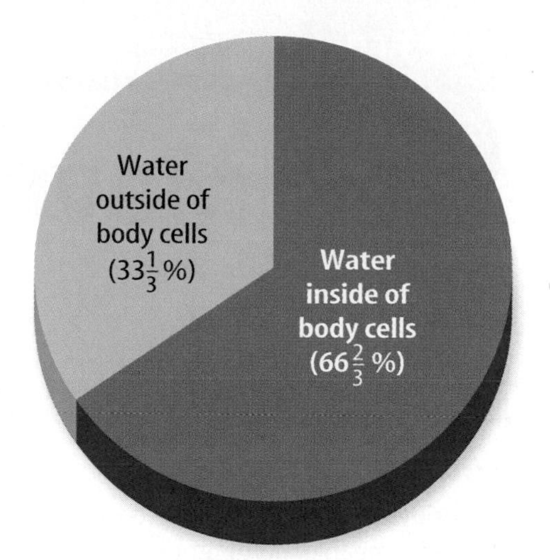

Figure 5
About two thirds of your body's water is located within your body's cells. Water helps maintain the cells' shapes and sizes. One third of your body's water is outside of your body's cells.

When heat is added to any substance, its molecules begin to move faster. Because water molecules are so strongly attracted to each other, the temperature of water changes slowly. The large percentage of water in living things acts like an insulator. The water in a cell helps keep its temperature constant, which allows life-sustaining chemical reactions to take place.

You've seen ice floating on water. When water freezes, ice crystals form. In the crystals, each water molecule is spaced at a certain distance from all the others. Because this distance is greater in frozen water than in liquid water, ice floats on water. Bodies of water freeze from the top down. The floating ice provides insulation from extremely cold temperatures and allows living things to survive in the cold water under the ice.

Section  Assessment

1. What are the similarities and differences between atoms and molecules?
2. What is the difference between organic and inorganic compounds? Give an example of each type of compound.
3. What are the four types of organic compounds found in all living things?
4. Why does life as we know it depend on water?
5. **Think Critically** If you mix salt, sand, and sugar with water in a small jar, will the resulting mixture be a suspension, a solution, or both?

Skill Builder Activities

6. **Interpreting Scientific Illustrations** Carefully observe **Figure 1** and determine how many protons, neutrons, and electrons an atom of oxygen has. **For more help, refer to the Science Skill Handbook.**
7. **Using an Electronic Spreadsheet** Research to find the percentage of elements that make up Earth's crust. Make a spreadsheet that includes this information and the information in **Table 2.** Create a circle graph for each set of percentages. **For more help, refer to the Technology Skill Handbook.**

SECTION

Moving Cellular Materials

As You Read

What You'll Learn

- **Describe** the function of a selectively permeable membrane.
- **Explain** how the processes of diffusion and osmosis move molecules in living cells.
- **Explain** how passive transport and active transport differ.

Vocabulary

passive transport
diffusion
equilibrium
osmosis
active transport
endocytosis
exocytosis

Why It's Important

Cell membranes control the substances that enter and leave the cells in your body.

Passive Transport

"Close that window. Do you want to let in all the bugs and leaves?" How do you prevent unwanted things from coming through the window? As seen in **Figure 6,** a window screen provides the protection needed to keep unwanted things outside. It also allows some things to pass into or out of the room like air, unpleasant odors, or smoke.

Cells take in food, oxygen, and other substances from their environments. They also release waste materials into their environments. A cell has a membrane around it that works for a cell like a window screen does for a room. A cell's membrane is selectively permeable (PUR mee uh bul). It allows some things to enter or leave the cell while keeping other things outside or inside the cell. The window screen also is selectively permeable based on the size of its openings.

Things can move through a cell membrane in several ways. Which way things move depends on the size of the molecules or particles, the path taken through the membrane, and whether or not energy is used. The movement of substances through the cell membrane without the input of energy is called **passive transport.** Three types of passive transport can occur. The type depends on what is moving through the cell membrane.

Figure 6
A cell membrane, like a screen, will let some things through more easily than others. Air gets through a screen, but insects are kept out.

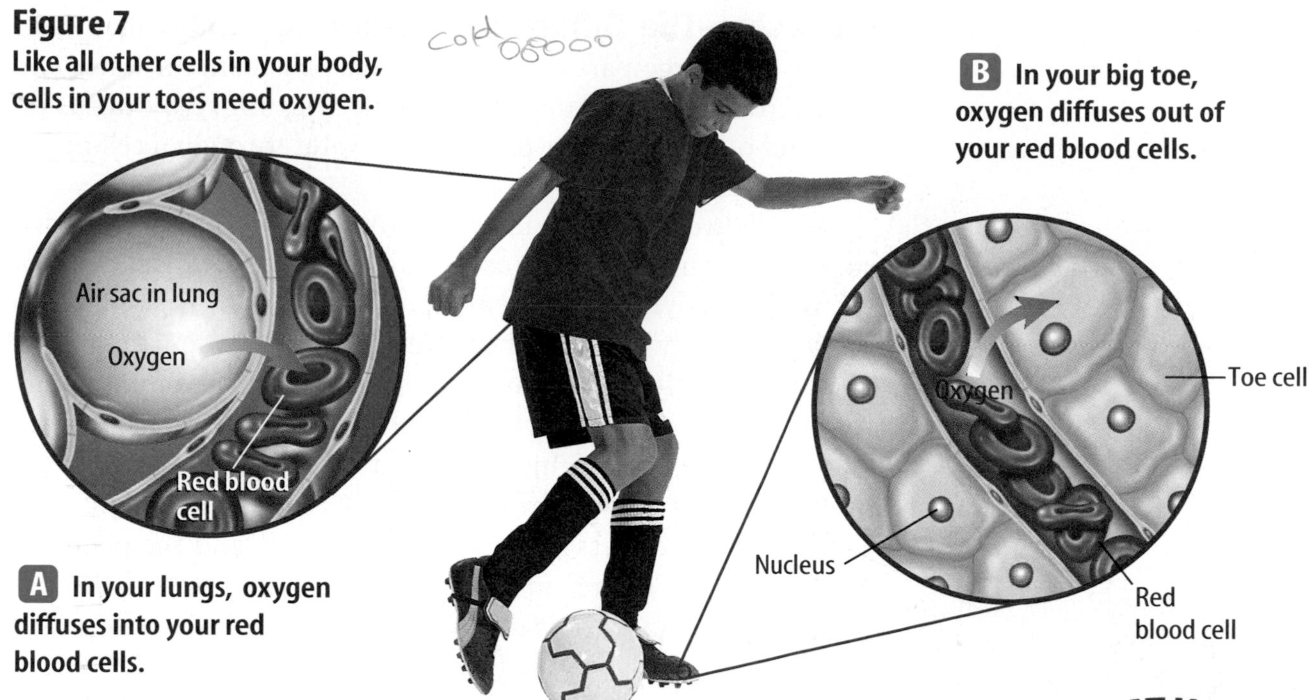

Figure 7
Like all other cells in your body, cells in your toes need oxygen.

A In your lungs, oxygen diffuses into your red blood cells.

B In your big toe, oxygen diffuses out of your red blood cells.

Diffusion Molecules in solids, liquids, and gases move constantly and randomly. You might smell perfume when you sit near or as you walk past someone who is wearing it. This is because perfume molecules randomly move throughout the air. This random movement of molecules from an area where there is relatively more of them into an area where there is relatively fewer of them is called **diffusion.** Diffusion is one type of cellular passive transport. Molecules of a substance will continue to move from one area into another until the relative number of these molecules is equal in the two areas. When this occurs, **equilibrium** is reached and diffusion stops. After equilibrium occurs, it is maintained because molecules continue to move.

✔ Reading Check *What is equilibrium?*

Every cell in your body uses oxygen. When you breathe, how does oxygen get from your lungs to cells in your big toe? Oxygen is carried throughout your body in your blood by the red blood cells. When your blood is pumped from your heart to your lungs, your red blood cells do not contain much oxygen. However, your lungs have more oxygen molecules than your red blood cells do, so the oxygen molecules diffuse into your red blood cells from your lungs, as shown in **Figure 7A.** When the blood reaches your big toe, there are more oxygen molecules in your red blood cells than in your big toe cells. The oxygen diffuses from your red blood cells and into your big toe cells, as shown in **Figure 7B.**

TRY AT HOME Mini LAB

Observing Diffusion

Procedure

1. Use **two clean glasses** of equal size. Label one "hot," then fill it until half full with **very warm water.** Label the other "cold," then fill it until half full with **cold water. WARNING:** *Do not use boiling hot water.*
2. Add one drop of **food coloring** to each glass. Carefully release the drop just at the water's surface to avoid splashing the water.
3. Observe the glasses. Record your observations immediately and again after 15 min.

Analysis

1. Describe what happens when food coloring is added to each glass.
2. How does temperature affect the rate of diffusion?

Osmosis—The Diffusion of Water Remember that water makes up a large part of living matter. Cells contain water and are surrounded by water. Water molecules move by diffusion into and out of cells. The diffusion of water through a cell membrane is called **osmosis.**

If cells weren't surrounded by water that contains few dissolved substances, water inside the cells would diffuse out of them. This is why water left the carrot cells in this chapter's Explore Activity. Because there were relatively fewer water molecules in the salt solution around the carrot cells than in the carrot cells, water moved out of the cells and into the salt solution.

Losing water from inside a plant cell causes the cell membrane to come away from the cell wall, as shown in **Figure 8A.** This reduces the pressure against the cell wall, and the plant cell becomes limp. If the carrot sticks were taken out of the salt water and put in pure water, the water around the cells would move into the cells. The cells would fill with water and their cell membranes would press against their cell walls, as shown in **Figure 8B.** Pressure would increase and the plant cells would become firm. That is why the carrot sticks would be crisp again.

✔ **Reading Check** *Why do carrots in salt water become limp?*

Osmosis also takes place in animal cells. If animal cells were placed in pure water, they too would swell up. However, animal cells are different from plant cells. Just like an overfilled water balloon, animal cells will burst if too much water enters the cell.

Figure 8
Cells respond to differences between the amount of water inside and outside the cell.

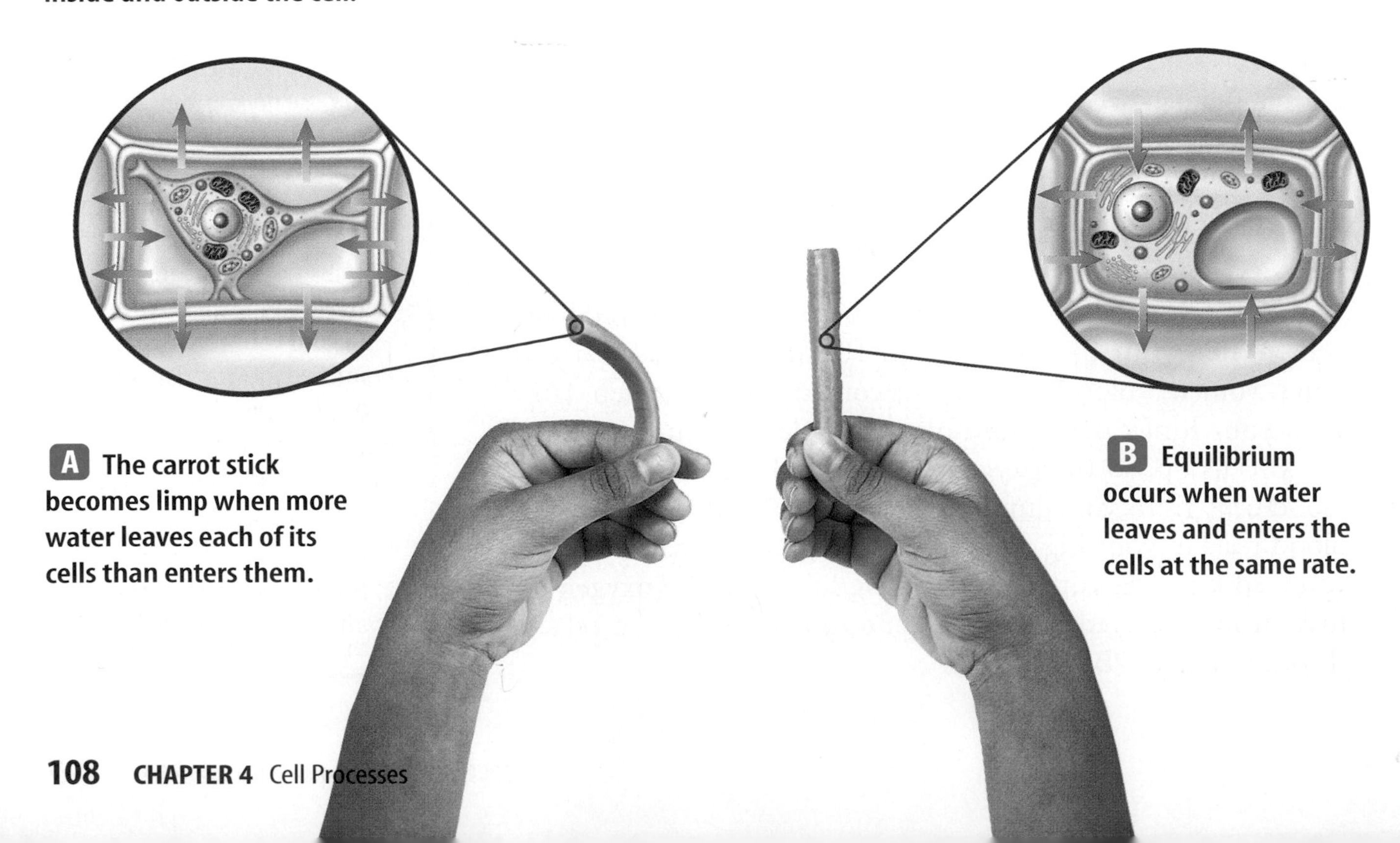

A **The carrot stick becomes limp when more water leaves each of its cells than enters them.**

B **Equilibrium occurs when water leaves and enters the cells at the same rate.**

Facilitated Diffusion Cells take in many substances. Some substances pass easily through the cell membrane by diffusion. Other substances, such as glucose molecules, are so large that they can enter the cell only with the help of molecules in the cell membrane called transport proteins. This process, a type of passive transport, is known as facilitated diffusion. Have you ever used the drive through at a fast-food restaurant to get your meal? The transport proteins in the cell membrane are like the drive-through window at the restaurant. The window lets you get food out of the restaurant and put money into the restaurant. Similarly, transport proteins are used to move substances into and out of the cell.

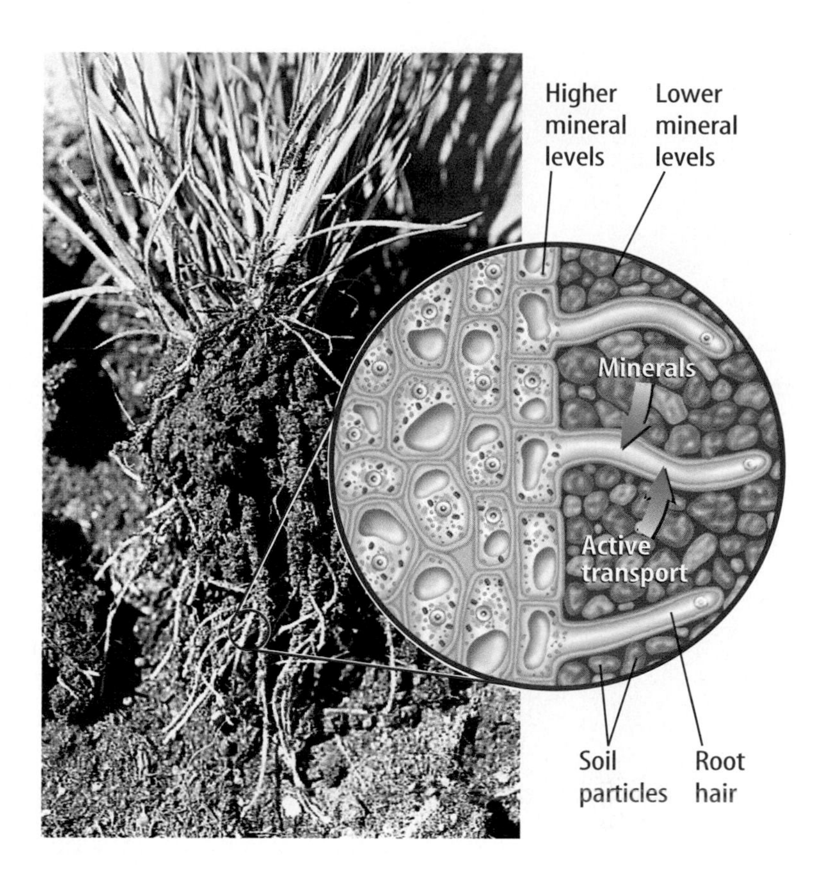

Figure 9
Some root cells have extensions called root hairs that may be 5 mm to 8 mm long. Minerals are taken in by active transport through the cell membranes of root hairs.

Active Transport

Imagine that a football game is over and you leave the stadium. As soon as you get outside of the stadium, you remember that you left your jacket on your seat. Now you have to move against the crowd coming out of the stadium to get back in to get your jacket. Which required more energy—leaving the stadium with the crowd or going back to get your jacket? Something similar to this happens in cells.

Sometimes, a substance is needed inside a cell even though the amount of that substance inside the cell is already greater than the amount outside the cell. For example, root cells require minerals from the soil. The roots of the plant in **Figure 9** already might contain more of those mineral molecules than the surrounding soil does. The tendency is for mineral molecules to move out of the root by diffusion or facilitated diffusion. But they need to move back across the cell membrane and into the cell just like you had to move back into the stadium. When an input of energy is required to move materials through a cell membrane, **active transport** takes place.

Active transport involves transport proteins, just as facilitated diffusion does. In active transport, a transport protein binds with the needed particle and cellular energy is used to move it through the cell membrane. When the particle is released, the transport protein can move another needed particle through the membrane.

Health INTEGRATION

Transport proteins are important to your health. Sometimes transport proteins are missing or do not function correctly. What would happen if proteins that transport cholesterol across membranes were missing? Cholesterol is an important lipid used by your cells. Write your ideas in your Science Journal.

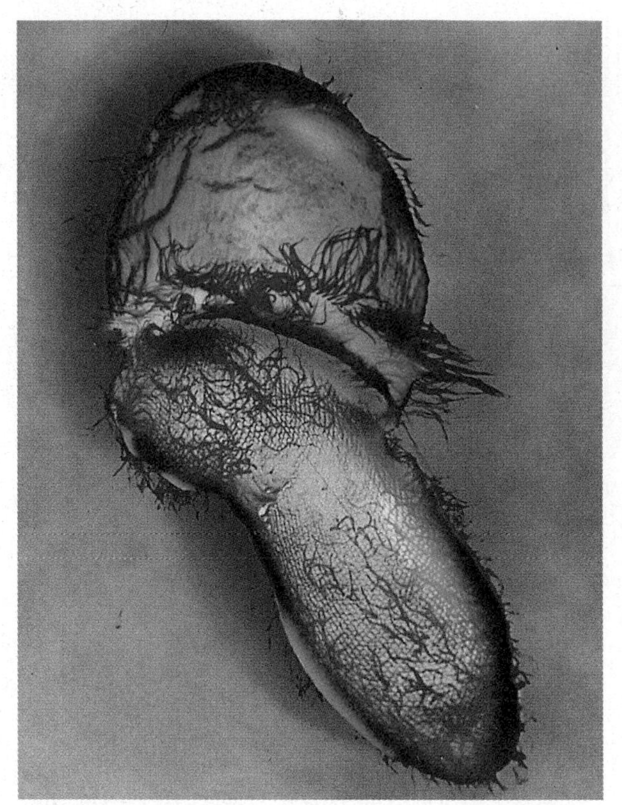

Magnification: 1400×

Figure 10
One-celled organisms like this egg-shaped one can take in other one-celled organisms using endocytosis.

Endocytosis and Exocytosis

Some molecules and particles are too large to move by diffusion or to use the cell membrane's transport proteins. Large protein molecules and bacteria, for example, can enter a cell when they are surrounded by the cell membrane. The cell membrane folds in on itself, enclosing the item in a sphere called a vesicle. Vesicles are transport and storage structures in a cell's cytoplasm. The sphere pinches off, and the resulting vesicle enters the cytoplasm. A similar thing happens when you poke your finger into a partially inflated balloon. Your finger is surrounded by the balloon in much the same way that the protein molecule is surrounded by the cell membrane. This process of taking substances into a cell by surrounding it with the cell membrane is called **endocytosis** (en duh si TOH sus). Some one-celled organisms, as shown in **Figure 10,** take in food this way.

The contents of a vesicle may be released by the cell using a process called **exocytosis** (ek soh si TOH sus). Exocytosis occurs in the opposite way that endocytosis does. The membrane of the vesicle fuses with the cell's membrane, and the vesicle's contents are released. Cells in your stomach use this process to release chemicals that help digest food. The different ways that materials may enter or leave a cell are summarized in **Figure 11.**

Section  Assessment

1. Explain how cell membranes are selectively permeable.
2. Compare and contrast the processes of osmosis and diffusion.
3. Identify the molecules that help substances move through the cell membrane during active transport and facilitated diffusion.
4. Why are endocytosis and exocytosis important processes to cells?
5. **Think Critically** Why are fresh fruits and vegetables sprinkled with water at produce markets?

Skill Builder Activities

6. **Concept Mapping** Make a network tree concept map to use as a study guide to help you tell the difference between passive transport and active transport. Begin with the phrase "Transport through membranes." **For more help, refer to the Science Skill Handbook.**
7. **Communicating** Seawater is saltier than tap water. In your Science Journal, explain why drinking large amounts of seawater would be dangerous to humans. **For more help, refer to the Science Skill Handbook.**

VISUALIZING CELL MEMBRANE TRANSPORT

Figure 11

A flexible yet strong layer, the cell membrane is built of two layers of lipids (gold) pierced by protein "passageways" (purple). Molecules can enter or exit the cell by slipping between the lipids or through the protein passageways. Substances that cannot enter or exit the cell in these ways may be surrounded by the membrane and drawn into or expelled from the cell.

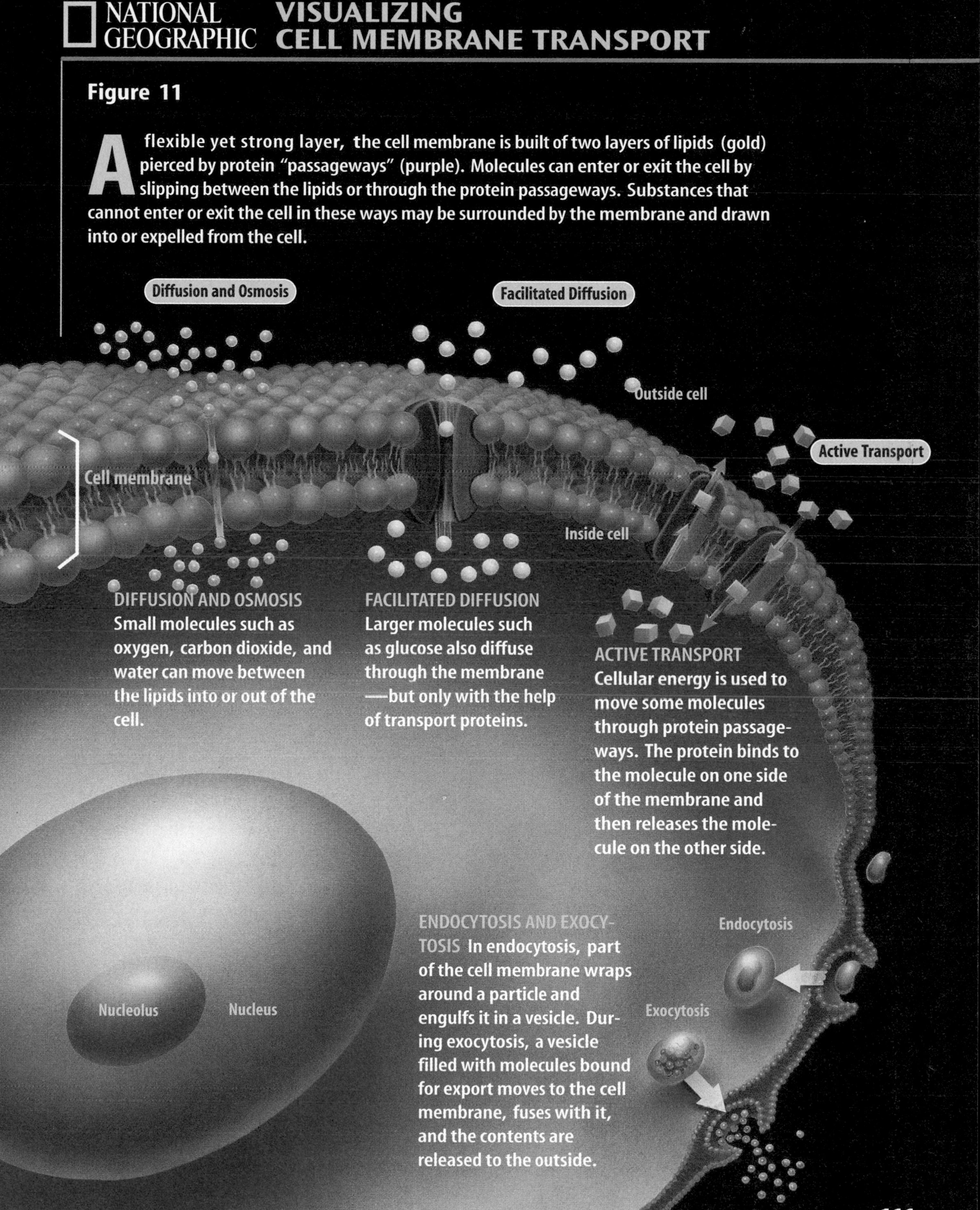

DIFFUSION AND OSMOSIS Small molecules such as oxygen, carbon dioxide, and water can move between the lipids into or out of the cell.

FACILITATED DIFFUSION Larger molecules such as glucose also diffuse through the membrane—but only with the help of transport proteins.

ACTIVE TRANSPORT Cellular energy is used to move some molecules through protein passageways. The protein binds to the molecule on one side of the membrane and then releases the molecule on the other side.

ENDOCYTOSIS AND EXOCYTOSIS In endocytosis, part of the cell membrane wraps around a particle and engulfs it in a vesicle. During exocytosis, a vesicle filled with molecules bound for export moves to the cell membrane, fuses with it, and the contents are released to the outside.

Activity

Observing Osmosis

It is difficult to see osmosis occurring in cells because most cells are so small. However, a few cells can be seen without the aid of a microscope. Try this activity to see how osmosis occurs in a large cell.

What You'll Investigate

How does osmosis occur in an egg cell?

Materials

unshelled egg*
balance
spoon
distilled water (250 mL)
light corn syrup (250 mL)
500-mL container

Goals

- **Observe** osmosis in an egg cell.
- **Determine** what affects osmosis.

Safety Precautions

Eggs may contain bacteria. Avoid touching your face. Wash your hands thoroughly when you are done.

*an egg whose shell has been dissolved by vinegar

Procedure

1. Copy the table below into your Science Journal and use it to record your data.

Egg Mass Data

	Beginning Egg Mass	Egg Mass After Two Days
Distilled water		
Corn syrup		

2. Obtain an unshelled egg from your teacher. Handle the egg gently. Use a balance to find the egg's mass and record it in the table.

3. Place the egg in the container and add enough distilled water to cover it.
4. **Observe** the egg after 30 min, one day, and two days. After each observation, record the egg's appearance in your Science Journal.
5. After day two, remove the egg with a spoon and allow it to drain. Find the egg's mass and record it in the table.
6. Empty the container, then put the egg back in. Now add enough corn syrup to cover it. Repeat steps 4 and 5.

Conclude and Apply

1. **Explain** the difference between what happened to the egg in water and in corn syrup.
2. **Calculate** the mass of water that moved into and out of the egg.
3. **Hypothesize** why you used an unshelled egg for this investigation.
4. **Infer** what part of the egg controlled water's movement into and out of the egg.

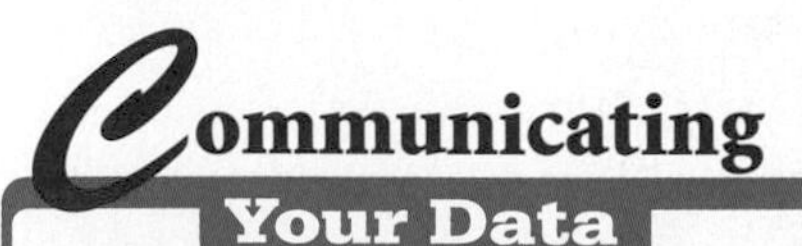

Communicating Your Data

Compare your conclusions with those of other students in your class. **For more help, refer to the Science Skill Handbook.**

Energy for Life

Trapping and Using Energy

Think of all the energy that players use in a basketball game. Where does the energy come from? The simplest answer is "from the food they eat." The chemical energy stored in food is changed in cells into forms needed to perform all the activities necessary for life. In every cell, these changes involve chemical reactions. All of the activities of an organism involve chemical reactions in some way. The total of all chemical reactions in an organism is called **metabolism.**

The chemical reactions of metabolism need enzymes. What do enzymes do? Suppose you are hungry and decide to open a can of spaghetti. You use a can opener to open the can. Without a can opener, the spaghetti is unusable. The can of spaghetti changed because of the can opener, but the can opener did not change. The can opener can be used again later to open more cans of spaghetti. Enzymes in cells work something like can openers, as shown in **Figure 12.** The enzyme, like the can opener, causes a change, but the enzyme is not changed and is reusable. Unlike the can opener, which can only break things apart, enzymes also can cause molecules to join. Without the right enzymes, chemical reactions in cells cannot take place.

As You Read

***What* You'll Learn**

- **List** the differences between producers and consumers.
- **Explain** how the processes of photosynthesis and respiration store and release energy.
- **Describe** how cells get energy from glucose through fermentation.

Vocabulary

metabolism
photosynthesis
respiration
fermentation

***Why* It's Important**

Because of photosynthesis and respiration, you use the Sun's energy.

A
B
Enzyme
Enzyme
Large molecule
Small molecules

Figure 12
Enzymes are needed for most chemical reactions that take place in cells. A The enzyme attaches to the large molecule it will help change. B The enzyme causes the larger molecule to break down into two smaller molecules. Like the can opener, the enzyme is not changed and can be used again.

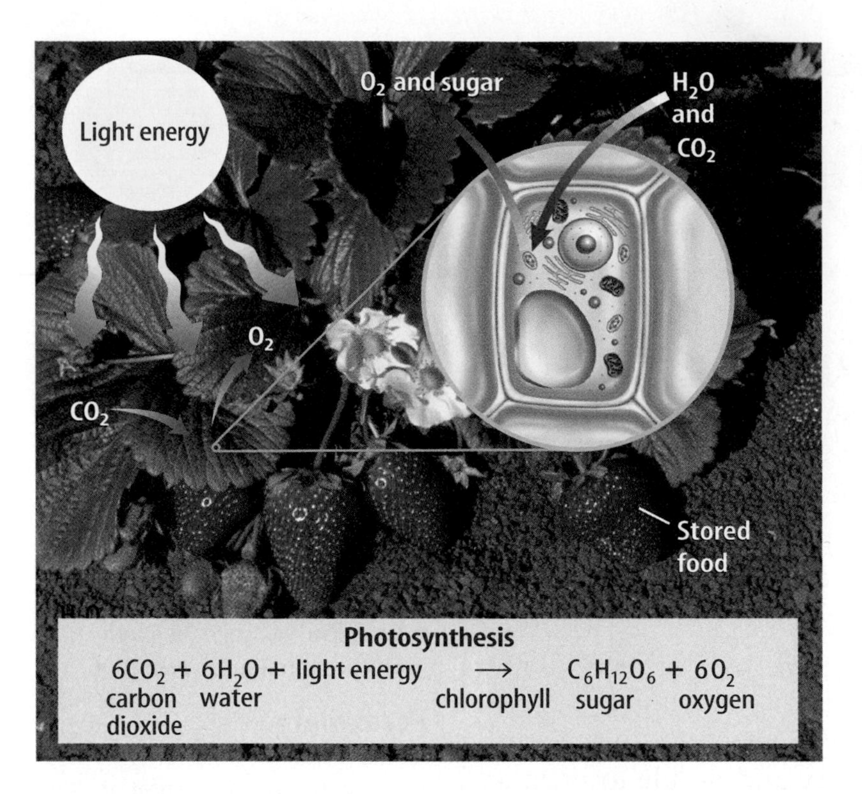

Figure 13
Plants use photosynthesis to make food. *According to the chemical equation, what raw materials would the plant pictured need for photosynthesis?*

Photosynthesis Living things are divided into two groups—producers and consumers—based on how they obtain their food. Organisms that make their own food, such as plants, are called producers. Organisms that cannot make their own food are called consumers.

If you have ever walked barefoot across a sidewalk on a sunny summer day, you probably moved quickly because the sidewalk was hot. Sunlight energy was converted into thermal energy and heated the sidewalk. Plants and many other producers can convert sunlight energy into another kind of energy—chemical energy. The process they use is called photosynthesis. During **photosynthesis,** producers use light energy to make sugars, which can be used as food.

Producing Carbohydrates Producers that use photosynthesis are usually green because they contain a green pigment called chlorophyll (KLOR uh fihl). Chlorophyll and other pigments are used in photosynthesis to capture sunlight energy. In plant cells, these pigments are found in chloroplasts.

The captured sunlight energy is used to drive chemical reactions during which the raw materials, carbon dioxide and water, are used to produce sugar and oxygen. For plants, the raw materials come from air and soil. Some of the captured sunlight energy is stored in the chemical bonds that hold the sugar molecules together. **Figure 13** shows what happens during photosynthesis in a plant. Enzymes also are needed before these reactions can occur.

Storing Carbohydrates Plants make more sugar during photosynthesis than they need for survival. Excess sugar is changed and stored as starches or used to make other carbohydrates. Plants use these carbohydrates as food for growth, maintenance, and reproduction.

Why is photosynthesis important to consumers? Do you eat apples? Apple trees use photosynthesis to produce apples. Do you like cheese? Some cheese comes from milk, which is produced by cows that eat plants. Consumers take in food by eating producers or other consumers. No matter what you eat, photosynthesis was involved directly or indirectly in its production.

Respiration Imagine that you get up late for school. You dress quickly, then run three blocks to school. When you get to school, you feel hot and are breathing fast. Why? Your muscle cells use a lot of energy when you run. To get this energy, muscle cells break down food. Some of the energy from the food is used when you move and some of it becomes thermal energy, which is why you feel warm or hot. Most cells also need oxygen to break down food. You were breathing fast because your body was working to get oxygen to your muscles. Your muscle cells were using the oxygen for the process of respiration. During **respiration,** chemical reactions occur that break down food molecules into simpler substances and release their stored energy. Just as in photosynthesis, enzymes are needed for the chemical reactions of respiration.

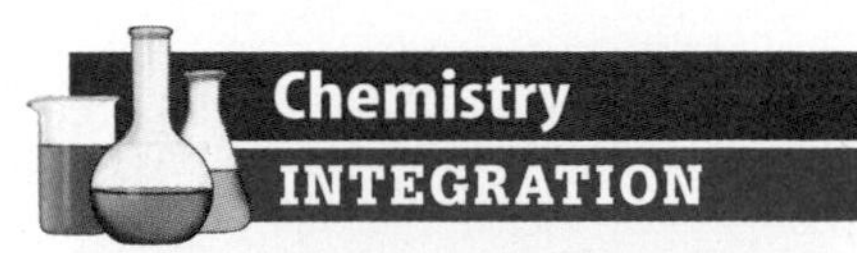

Compounds often are represented by a chemical formula. The chemical formula shows how many and what type of atoms are found in one molecule of the compound. For example, the sugar glucose has the chemical formula $C_6H_{12}O_6$. What is the total number of atoms in one glucose molecule?

What must happen to food molecules for respiration to take place?

Breaking Down Carbohydrates The type of food that is most easily broken down by cells is carbohydrates. Respiration of carbohydrates begins in the cytoplasm of the cell. The carbohydrates are broken down into glucose molecules. Each glucose molecule is broken down further into two simpler molecules. As the glucose molecules are broken down, energy is released.

The two simpler molecules are broken down again. This breakdown occurs in the mitochondria of the cells of plants, animals, fungi, and many other organisms. This process uses oxygen, releases much more energy, and produces carbon dioxide and water as wastes. When you exhale, you breathe out carbon dioxide and some of the water.

Respiration occurs in the cells of all living things. **Figure 14** shows how respiration occurs in one consumer. As you are reading this section of the chapter, millions of cells in your body are breaking down glucose, releasing energy, and producing carbon dioxide and water.

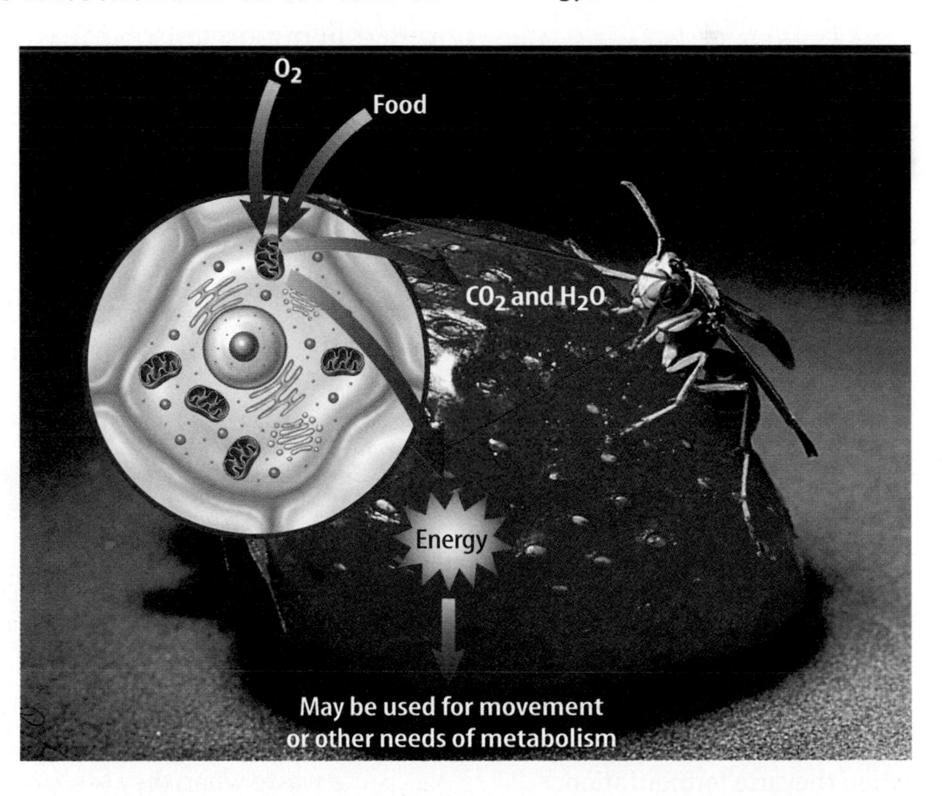

Figure 14
Producers and consumers carry on respiration that releases energy from foods.

Fermentation Remember imagining you were late and had to run to school? During your run, your muscle cells might not have received enough oxygen, even though you were breathing rapidly. When cells do not have enough oxygen for respiration, they use a process called **fermentation** to release some of the energy stored in glucose molecules.

Like respiration, fermentation begins in the cytoplasm. Again, as the glucose molecules are broken down, energy is released. But the simple molecules from the breakdown of glucose do not move into the mitochondria. Instead, more chemical reactions occur in the cytoplasm. These reactions release some energy and produce wastes. Depending on the type of cell, the wastes may be lactic acid, alcohol, and carbon dioxide, as shown in **Figure 15.** Your muscle cells can use fermentation to change the simple molecules into lactic acid while releasing energy. The presence of lactic acid is why your muscle cells might feel stiff and sore after you run to school.

SCIENCE Online

Collect Data Visit the Glencoe Science Web site at **science.glencoe.com** for more information about how microorganisms are used to produce many useful products. In your Science Journal list three products produced by microorganisms.

✔ Reading Check *Where in a cell does fermentation take place?*

Some microscopic organisms, such as bacteria, carry out fermentation and make lactic acid. Some of these organisms are used to produce yogurt and some cheeses. These organisms break down a sugar in milk to release energy. The lactic acid produced causes the milk to become more solid and gives these foods some of their flavor.

Have you ever used yeast to make bread? Yeasts are one-celled living organisms. Yeast cells use fermentation to break down sugar in bread dough. They produce alcohol and carbon dioxide as wastes. The carbon dioxide waste is a gas that makes bread dough rise before it is baked. The alcohol is lost as the bread bakes.

Figure 15
Organisms that use fermentation produce several different wastes.

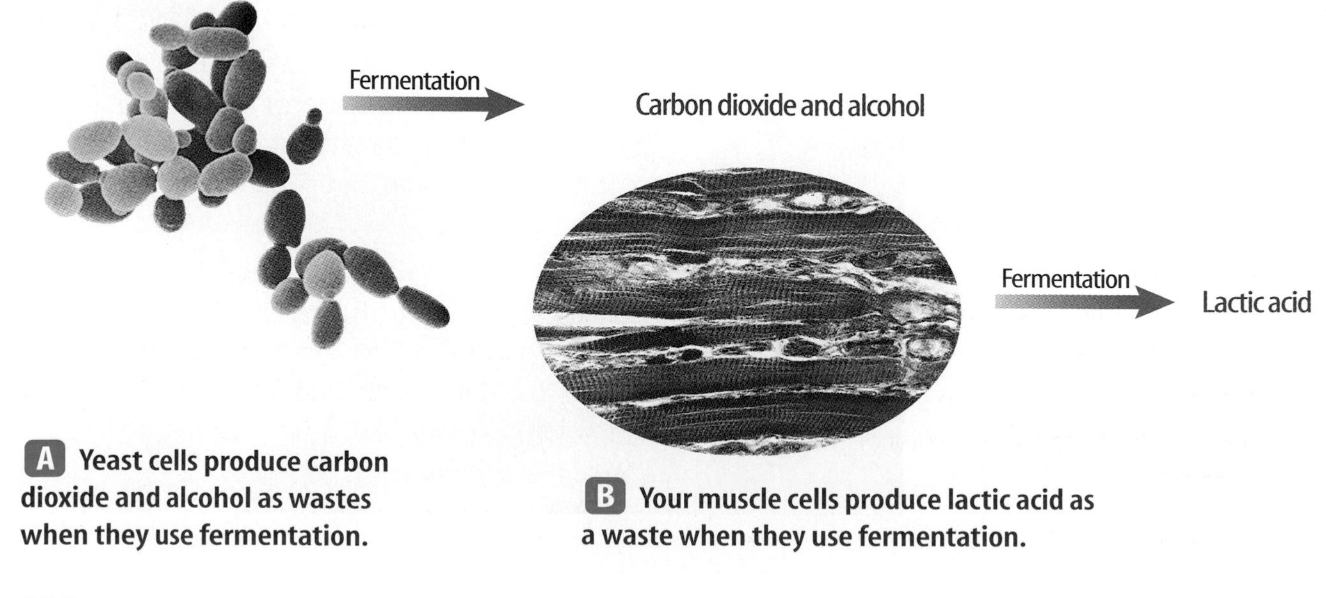

A **Yeast cells produce carbon dioxide and alcohol as wastes when they use fermentation.**

B **Your muscle cells produce lactic acid as a waste when they use fermentation.**

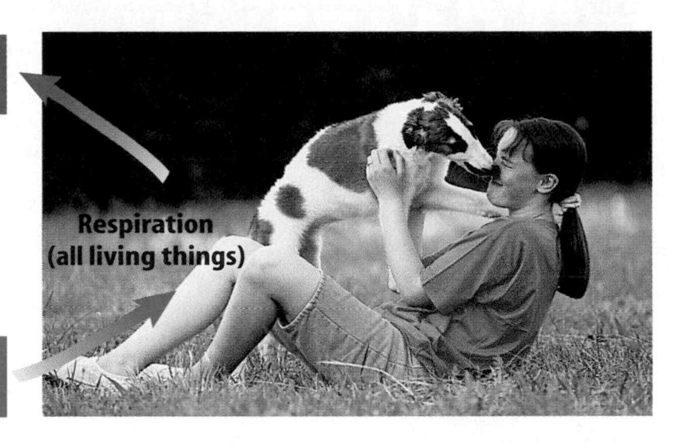

Figure 16
The chemical reactions of photosynthesis and respiration could not take place without each other.

Related Processes How are photosynthesis, respiration, and fermentation related? Some producers use photosynthesis to make food. All living things use respiration or fermentation to release energy stored in food. If you think carefully about what happens during photosynthesis and respiration, you will see that what is produced in one is used in the other, as shown in **Figure 16.** These two processes are almost the opposite of each other. Photosynthesis produces sugars and oxygen, and respiration uses these products. The carbon dioxide and water produced during respiration are used during photosynthesis. Most life would not be possible without these important chemical reactions.

Section  Assessment

1. Explain the difference between producers and consumers and give three examples of each.
2. Explain how the energy used by many living things on Earth can be traced back to sunlight.
3. Compare and contrast respiration and fermentation.
4. What condition must exist in cells for fermentation to occur?
5. **Think Critically** How can some indoor plants help improve the quality of air in a room?

Skill Builder Activities

6. **Identifying and Manipulating Variables and Controls** Design an experiment to show what happens to a plant when you limit sunlight or one of the raw materials for photosynthesis. Identify the control. **For more help, refer to the Science Skill Handbook.**
7. **Solving One-Step Equations** Refer to the chemical equation for photosynthesis. Calculate then compare the number of carbon, hydrogen, and oxygen atoms before and after photosynthesis. **For more help, refer to the Math Skill Handbook.**

Activity

Photosynthesis and Respiration

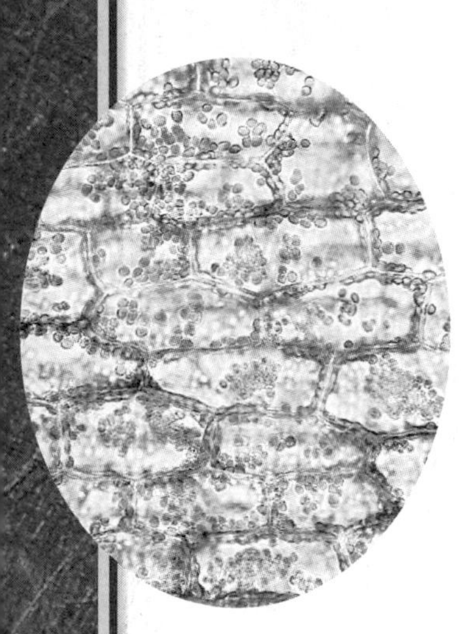

Every living cell carries on many chemical processes. Two important chemical processes are respiration and photosynthesis. All cells, including the ones in your body, carry on respiration. However, some plant cells can carry on both processes. In this experiment you will investigate when these processes occur in plant cells. How could you find out when plants were using these processes? Are the products of photosynthesis and respiration the same?

What You'll Investigate

When do plants carry on photosynthesis and respiration?

Materials

16-mm test tubes (3)
150-mm test tubes with stoppers (4)
**small, clear-glass baby food jars with lids (4)*
test-tube rack
stirring rod
scissors
carbonated water (5 mL)
bromothymol blue solution in dropper bottle
aged tap water (20 mL)
**distilled water (20 mL)*
sprig of *Elodea* (2)
**other water plants*

**Alternate materials*

Goals

- **Observe** green water plants in the light and dark.
- **Determine** whether plants carry on photosynthesis and respiration.

Safety Precautions

Wear splash-proof safety goggles to protect eyes from hazardous chemicals. Wash hands thoroughly after the activity.

Procedure

1. Label each test tube using the numbers 1, 2, 3, and 4. Pour 5 mL of aged tap water into each test tube.
2. Add 10 drops of carbonated water to test tubes 1 and 2.
3. Add 10 drops of bromothymol blue to all of the test tubes. Bromothymol blue turns green to yellow in the presence of an acid.
4. Cut two 10-cm sprigs of *Elodea*. Place one sprig in test tube 1 and one sprig in test tube 3. Stopper all test tubes.
5. In your Science Journal, copy and complete the test-tube data table.
6. Place test tubes 1 and 2 in bright light. Place tubes 3 and 4 in the dark. Observe the test tubes for 30 min or until the color changes. Record the color of each of the four test tubes.

Test Tube Data

Test Tube	Color at Start	Color After 30 Minutes
1		
2		
3		
4		

Conclude and Apply

1. What is indicated by the color of the water in all four test tubes at the start of the activity?
2. **Infer** what process occurred in the test tube or tubes that changed color after 30 min.
3. **Describe** the purpose of test tubes 2 and 4 in this experiment.
4. Do the results of this experiment show that photosynthesis and respiration occur in plants? Explain.

Communicating Your Data

Choose one of the following activities to **communicate** your data. Prepare an oral presentation that explains how the experiment showed the differences between products of photosynthesis and respiration. Draw a cartoon strip to **explain** what you did in this experiment. Use each panel to show a different step. **For more help, refer to the Science Skill Handbook.**

Science and Language Arts

from "Tulip"

by Penny Harter

Respond to the Reading

1. Why do you suppose the tulip survived the builders' abuse?
2. The poet chooses to write about a tulip rather than another kind of flower. Why do you think that is?
3. What is the yellow throat that the narrator is staring into?

I watched its first green push
through bare dirt, where the builders
had dropped boards, shingles,
plaster—
killing everything.
I could not recall what grew
there,
what returned each spring,
but the leaves looked tulip,
and one morning it arrived,
a scarlet slash against the
aluminum siding.

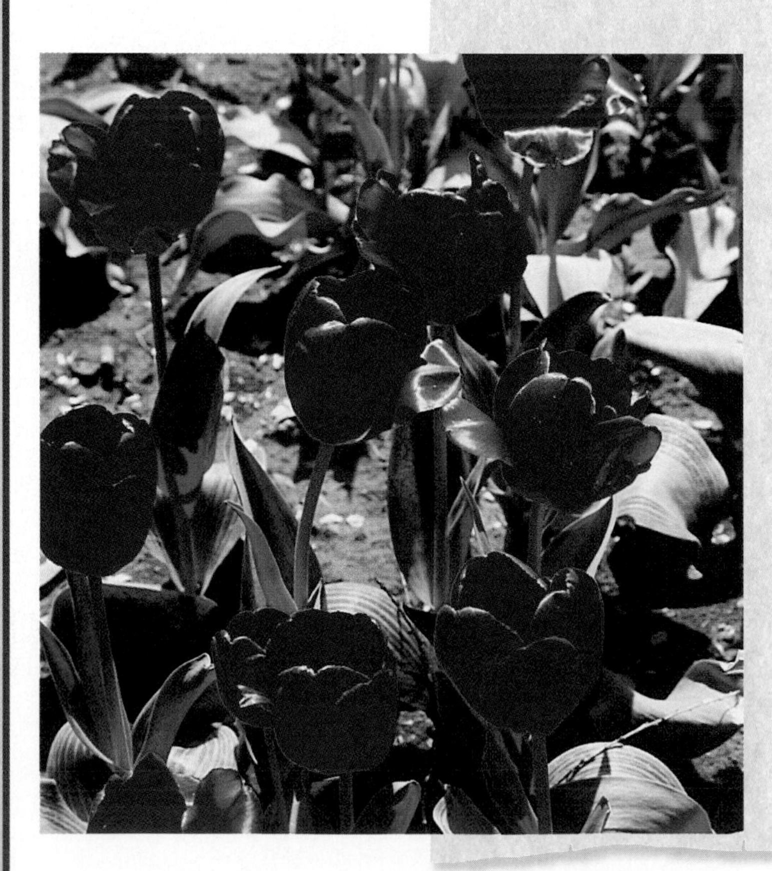

Mornings, on the way to
my car,
I bow to the still bell
of its closed petals; evenings,
it greets me, light ringing
at the end of my driveway.

Sometimes I kneel
to stare into the yellow
throat . . . .
It opens and closes my days.
It has made me weak with
love . . . .

Understanding Literature

Personification Using human traits or emotions to describe an idea, animal, or inanimate object is called personification. When the poet writes that the tulip has a "yellow throat," she uses personification. This can make the reader think of the tulip as more than just a plant. The poet also uses personification when she states that the tulip inspires love.

Science Connection Living things are made of more than 50 percent water and depend on it for their survival. Because most chemical reactions in plants take place in water, plants must have water in order to grow. In the poem, the tulip only pushes up through the ground in the spring when the tulip's underground bulb and roots absorb enough water. The water carries nutrients and minerals from the soil into the plant.

The process of active transport allows needed nutrients to enter the roots. The cell membranes of root cells contain proteins that bind with the needed nutrients. Cellular energy is used to move these nutrients through the cell membrane.

You also learned about photosynthesis in this chapter. From reading the poem, how would you know that photosynthesis had taken place?

Linking Science and Writing

Gardener's Journal Select a plant to observe. It could be a plant that you, your family, or one of your classmates grows. Depending on the season, it could be a plant growing on the grounds of your school or in a public park. Keep a gardener's observation journal of the plant for a month. Write weekly entries in your journal, describing the plant's condition, size, health, color, and other physical qualities.

Career Connection

Microbiologist

Dr. Harold Amos is a microbiologist who has studied cell processes in bacteria and mammals over the course of his career. He studied the way that sugar is transported in normal cells and cancer cells. Dr. Amos has a medical degree and a doctorate in bacteriology and immunology, which deals with the immune system and its interaction with diseases. He also has received many awards for his scientific work and his contributions to the careers of other scientists.

SCIENCE*Online* Visit the Glencoe Science Web site at **science.glencoe.com** to learn more about careers in microbiology.

Chapter 4 Study Guide

Reviewing Main Ideas

Section 1 Chemistry of Life

1. Matter is anything that has mass and takes up space.
2. Energy in matter is in the chemical bonds that hold matter together.
3. All organic compounds contain the elements hydrogen and carbon. The organic compounds in living things are carbohydrates, lipids, proteins, and nucleic acids.
4. Organic and inorganic compounds are important to living things. *What organic compounds could be found in an elephant and a pumpkin?*

Section 2 Moving Cellular Materials

1. The selectively permeable cell membrane controls which molecules can pass into and out of the cell.
2. In diffusion, molecules move from areas where there are relatively more of them into areas where there are relatively fewer of them. Osmosis is the diffusion of water through a cell membrane. *Why might these plants use osmosis?*

3. Cells use energy to move molecules by active transport but do not use energy for passive transport.
4. Cells move large particles through cell membranes by endocytosis and exocytosis.

Section 3 Energy for Life

1. Photosynthesis is the process by which some producers change light energy into chemical energy. *Why can't cells in these humans use sunlight to make food?*

2. Respiration that uses oxygen releases the energy in food molecules and produces waste carbon dioxide and water.
3. Some one-celled organisms and cells that lack oxygen use fermentation to release small amounts of energy from glucose. Wastes such as alcohol, carbon dioxide, and lactic acid are produced.

After You Read

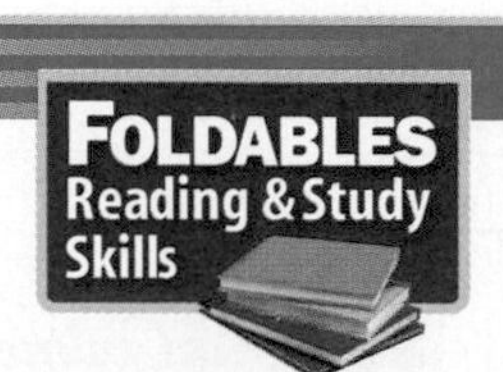

Under each tab of your Vocabulary Study Fold, write a sentence about one of the cell processes using the vocabulary word on the tab.

Chapter 4 Study Guide

Visualizing Main Ideas

Complete the following table on energy processes.

Energy Processes	Photosynthesis	Respiration	Fermentation
Energy Source		food (glucose)	food (glucose)
In plant and animal cells, occurs in			
Reactants are			
Products are			

Vocabulary Review

Vocabulary Words

a. active transport
b. diffusion
c. endocytosis
d. enzyme
e. equilibrium
f. exocytosis
g. fermentation
h. inorganic compound
i. metabolism
j. mixture
k. organic compound
l. osmosis
m. passive transport
n. photosynthesis
o. respiration

Study Tip

Make a note of anything you don't understand so that you'll remember to ask your teacher about it.

Using Vocabulary

Use what you know about the vocabulary words to answer the following questions.

1. What is the diffusion of water called?

2. What type of protein regulates nearly all chemical reactions in cells?

3. How do large food particles enter an amoeba?

4. What type of compound is water?

5. What process is used by some producers to convert light energy into chemical energy?

6. What type of compounds always contain carbon and hydrogen?

7. What process uses oxygen to break down glucose?

8. What is the total of all chemical reactions in an organism called?

Chapter 4 Assessment

Checking Concepts

Choose the word or phrase that best answers the question.

1. What is it called when cells use energy to move molecules?
A) diffusion **C)** active transport
B) osmosis **D)** passive transport

2. How might a cell take in a bacterium?
A) osmosis **C)** exocytosis
B) endocytosis **D)** diffusion

3. What occurs when the number of molecules of a substance is equal in two areas?
A) equilibrium **C)** fermentation
B) metabolism **D)** cellular respiration

4. Which of the following substances is an example of a carbohydrate?
A) enzymes **C)** waxes
B) sugars **D)** proteins

5. What is RNA an example of?
A) carbon dioxide **C)** lipid
B) water **D)** nucleic acid

6. What organic molecule stores the greatest amount of energy?
A) carbohydrate **C)** lipid
B) water **D)** nucleic acid

7. Which of these formulas is an example of an organic compound?
A) $C_6H_{12}O_6$ **C)** H_2O
B) NO_2 **D)** O_2

8. What are organisms that cannot make their own food called?
A) biodegradables **C)** consumers
B) producers **D)** enzymes

9. Which one of these cellular processes requires the presence of chlorophyll?
A) fermentation **C)** respiration
B) endocytosis **D)** photosynthesis

10. What kind of molecule is water?
A) organic **C)** carbohydrate
B) lipid **D)** inorganic

Thinking Critically

11. If you could place one red blood cell in distilled water, what would you see happen to the cell? Explain.

12. In snowy places, salt is used to melt ice on the roads. Explain what could happen to many roadside plants as a result.

13. Why does sugar dissolve faster in hot tea than in iced tea?

14. What would happen to the consumers in a lake if all the producers died?

15. Meat tenderizers contain protein enzymes. How do these enzymes affect meat?

Developing Skills

16. Concept Mapping Complete the events-chain concept map to sequence the following parts of matter from smallest to largest: *atom, electron,* and *compound.*

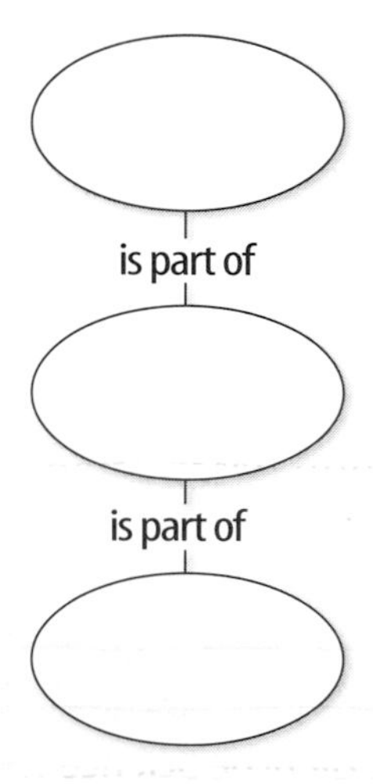

17. Forming Hypotheses Make a hypothesis about what will happen to wilted celery when placed in a glass of plain water.

18. Interpreting Data Water plants were placed at different distances from a light source. Bubbles coming from the plants were counted to measure the rate of photosynthesis. What can you say about how the distance from the light affected the rate?

Photosynthesis in Water Plants

Beaker Number	Distance from Light (cm)	Bubbles per Minute
1	10	45
2	30	30
3	50	19
4	70	6
5	100	1

19. Making and Using Graphs Using the data from question 18, make a line graph that shows the relationship between the rate of photosynthesis and the distance from light.

Performance Assessment

20. Puzzle Make a crossword puzzle with words describing ways substances are transported across cell membranes. Use the following words in your puzzle: *diffusion, osmosis, facilitated diffusion, active transport, endocytosis,* and *exocytosis.* Make sure your clues give good descriptions of each transport method.

TECHNOLOGY

Go to the Glencoe Science Web site at **science.glencoe.com** or use the **Glencoe Science CD-ROM** for additional chapter assessment.

Test Practice

Organic compounds called carbohydrates and proteins form many parts of a cell and also help connect cells to each other. Several organic compounds, along with their characteristics and where they are found, are listed below.

Cell Substances

Organic Compound	Flexibility	Found In
Keratin	Not very flexible	Hair and skin of mammals
Collagen	Not very flexible	Skin, bones, and tendons of mammals
Chitin	Very rigid	Tough outer shell of insects, crabs
Cellulose	Very flexible	Trees and flowers

Study the chart and answer the following questions.

1. According to this information, which organic compound is the least flexible?

A) keratin
B) collagen
C) chitin
D) cellulose

2. According to the chart, cellulose might be found in _____.

F) mammals
G) bones
H) insects
J) trees

CHAPTER 5

Plant Processes

From crabgrass to giant sequoias, many plants start as small seeds. Some trees may grow to be more than 20 m tall. One tree can be cut up to produce many pieces of lumber. Where does all that wood come from? You may have seen a plant on a windowsill with all its leaves growing toward the window. Why do they grow that way? In this chapter, find the answers to these questions. In addition, learn how plants are essential to the survival of all animals on Earth—including you!

What do you think?

Science Journal Look at the picture below with a classmate. Discuss what you think this might be or what is happening. Here's a hint: *This would never happen without light.* Write your answer or best guess in your Science Journal.

Plants are similar to other living things because they are made of cells, reproduce, make and use substances, and need water. If you forgot to water a houseplant, what do you think would happen? From your own experiences, you probably know that the houseplant would wilt. Do the following activity to discover one way plants lose water.

Infer how plants lose water 

1. Obtain a self-sealing plastic bag, some aluminum foil, and a small potted plant from your teacher.
2. Using the foil, carefully cover the soil around the plant in the pot. Place the potted plant in the plastic bag.
3. Seal the bag and place it in a sunny window. Wash your hands.
4. Look at the plant at the same time every day for a few days.

Observe

In your Science Journal, describe what happens in the bag. If enough water is lost by a plant and not replaced, predict what will happen to the plant.

Before You Read

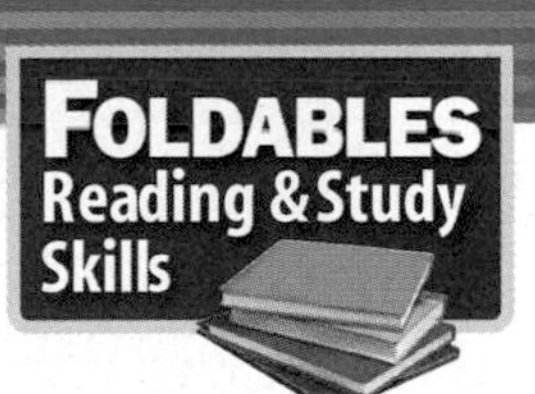

Making a Compare and Contrast Study Fold **As you study plant processes, use the following Foldable to help you compare and contrast plant respiration and animal respiration.**

1. Place a sheet of paper in front of you so the long side is at the top. Fold the paper in half from top to bottom.
2. Write *Respiration* across the front, as shown.
3. Unfold the paper. Draw a picture of an animal on the top half and a plant on the bottom half. Leave room to write below the drawings.
4. Before you read the chapter write what you know about animal respiration and plant respiration on the appropriate flaps.
5. As you read the chapter, add to or change your information.

Respiration

Photosynthesis and Respiration

As You Read

What You'll Learn

- **Explain** how plants take in and give off gases.
- **Compare and contrast** relationships between photosynthesis and respiration.
- **Discuss** why photosynthesis and respiration are important.

Vocabulary

stomata
chlorophyll
photosynthesis
respiration

Why It's Important

Understanding photosynthesis and respiration in plants will help you understand how life is maintained on Earth.

Taking in Raw Materials

Sitting in the cool shade under a tree, you finish eating your lunch. The food you eat is one of the raw materials that you need to grow. Oxygen is another. It enters your lungs and eventually reaches every cell in your body. Your cells use oxygen to help release the energy from the food that you eat. The process that uses oxygen to release the energy from food produces carbon dioxide and water as wastes. These wastes move in your blood to your lungs where they are removed as gases when you exhale. You look up at the tree and wonder, "Does a tree need to eat? Does it use oxygen? How does a tree get rid of wastes?

Movement of Materials in Plants No one packs a sack lunch for the tree. Trees and other plants don't take in foods the way you do. Plants make their own foods using the raw materials water, carbon dioxide, and inorganic chemicals in the soil. Just like you, plants also produce waste products.

Most of the water used by plants is taken in through roots, as shown in **Figure 1.** Water moves into root cells and then up through the plant to where it is used. When you pull up a plant, some of its roots are damaged. If you replant it, the plant will need extra water until new roots grow to replace those that were damaged.

Leaves, instead of lungs, are where most gas exchange occurs in plants. Most of the water taken in through the roots exits through the leaves of a plant. Carbon dioxide, oxygen, and water vapor exit and enter the plant through the leaf. The leaf's structure helps explain how it functions in gas exchange.

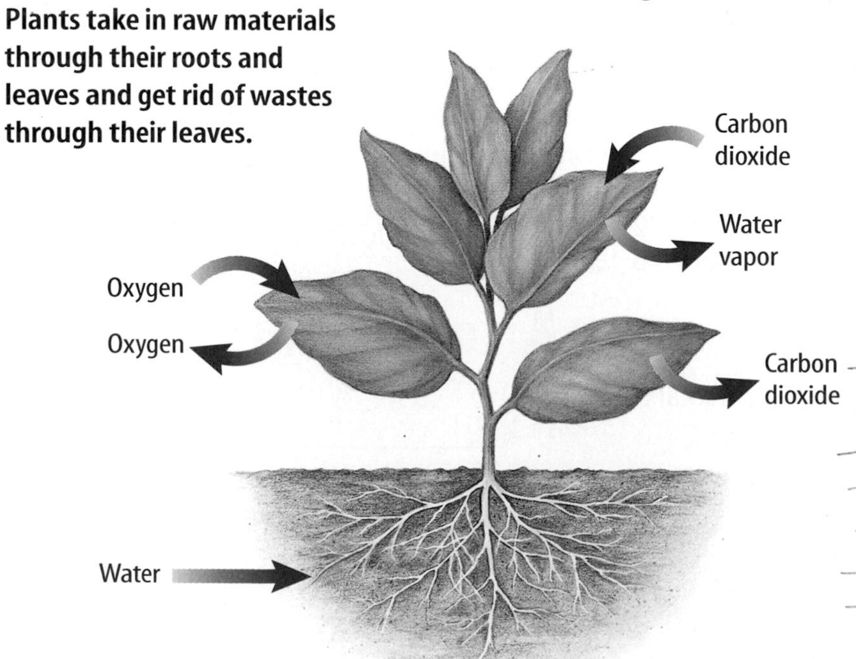

Figure 1
Plants take in raw materials through their roots and leaves and get rid of wastes through their leaves.

Figure 2
A leaf's structure determines its function. Food is made in the inner layers. Most stomata are found on the lower epidermis.

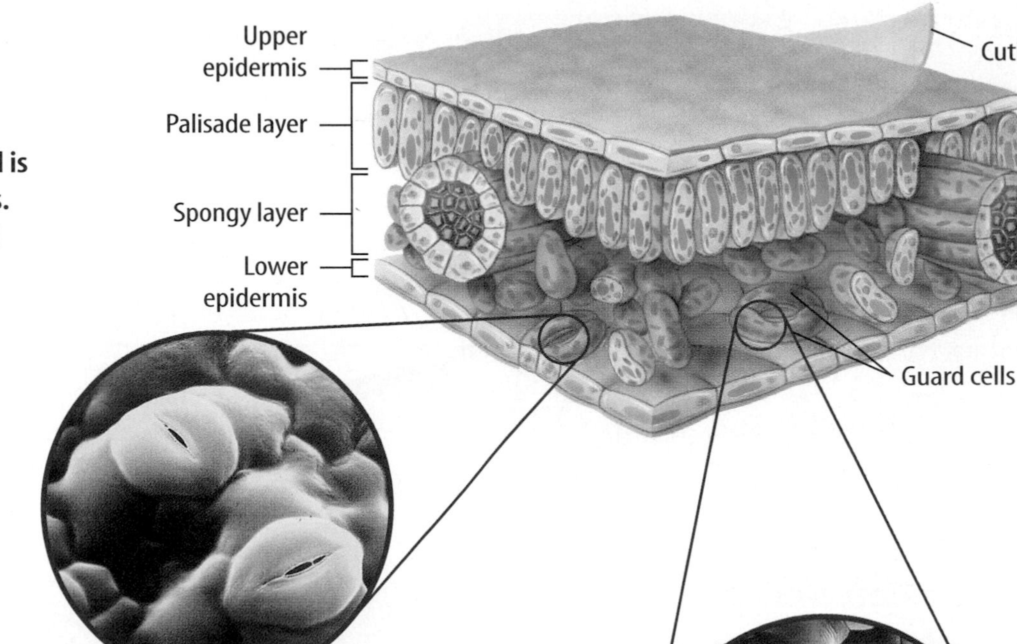

A Closed stomata

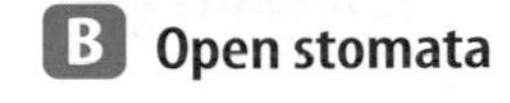

B Open stomata

Leaf Structure and Function A leaf is made up of many different layers, as shown in **Figure 2.** The outer cell layer of the leaf is the epidermis. A waxy cuticle that helps keep the leaf from drying out covers the epidermis. Because the epidermis is nearly transparent, sunlight—which is used to make food—reaches the cells inside the leaf. If you examine the epidermis under a microscope, you will see that it contains many small openings. These openings, called **stomata** (stoh MAH tuh) (singular, *stoma*), act as doorways for raw materials such as carbon dioxide, water vapor, and waste gases to enter and exit the leaf. Stomata also are found on the stems of many plants. More than 90 percent of the water plants take in through their roots is lost through the stomata. In one day, a growing tomato plant can lose up to 1 L of water.

Two cells called guard cells surround each stoma and control its size. As water moves into the guard cells, they swell and bend apart, opening a stoma. When guard cells lose water, they deflate, closing the stoma. **Figures 2A** and **2B** show closed and open stomata.

Stomata usually are open during the day when most plants need to take in raw materials to make food. They usually are closed at night when food making slows down. Stomata also close when a plant is losing too much water. This adaptation conserves water, because less water vapor escapes from the leaf.

Inside the leaf are two layers of cells, the spongy layer and the palisade layer. Carbon dioxide and water vapor, which are needed in the food-making process, fill the spaces of the spongy layer. Most of the food is made in the palisade layer.

Health INTEGRATION

Vitamins are substances needed for good health. You get most of the vitamins you need from the plants you eat. Research to learn about four vitamins and the plant foods you would need to eat to get them. Display your results on a poster.

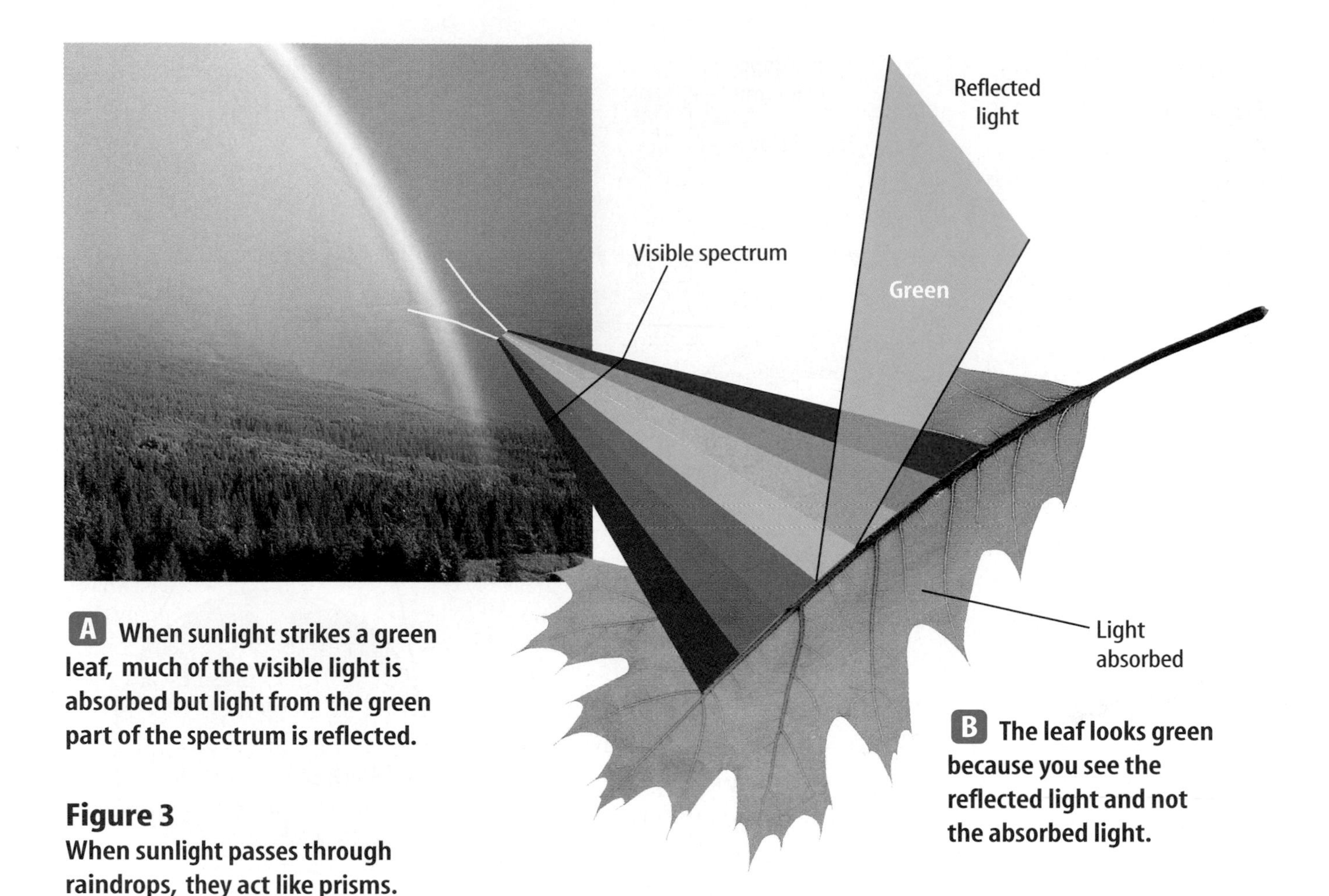

A When sunlight strikes a green leaf, much of the visible light is absorbed but light from the green part of the spectrum is reflected.

B The leaf looks green because you see the reflected light and not the absorbed light.

Figure 3
When sunlight passes through raindrops, they act like prisms. Light separates into the colors of the visible spectrum. You see a rainbow when this happens.

Chloroplasts and Plant Pigments If you look closely at the leaf in **Figure 2,** you'll see that some of the cells contain small, green structures called chloroplasts. Most leaves look green because their cells contain so many chloroplasts. Chloroplasts are green because they contain a green pigment called **chlorophyll** (KLOR uh fihl).

Reading Check *Why are chloroplasts green?*

As shown in **Figure 3,** light from the Sun contains all colors of the visible spectrum. A pigment is a substance that reflects a particular part of the visible spectrum and absorbs the rest. When you see a green leaf, you are seeing green light energy reflected from chlorophyll. Most of the other colors of the spectrum, especially red and blue, are absorbed by chlorophyll. In the spring and summer, most leaves have so much chlorophyll that it hides all other pigments. In fall, the chlorophyll in some leaves breaks down and the leaves change color as other pigments become visible. Pigments, especially chlorophyll, are important to plants because the light energy that they absorb is used to make food. For plants, this food-making process—photosynthesis—happens in the chloroplasts.

The Food-Making Process

Photosynthesis (foh toh SIHN thuh suhs) is the process during which a plant's chlorophyll traps light energy and sugars are produced. In plants, photosynthesis occurs only in cells with chloroplasts. For example, photosynthesis occurs only in a carrot plant's lacy green leaves, shown in **Figure 4.** Because a carrot's root cells lack chlorophyll and normally do not receive light, they can't perform photosynthesis. But excess sugar produced in the leaves is stored in the familiar orange root that you and many animals eat.

Besides light, plants also need the raw materials carbon dioxide and water for photosynthesis. The overall chemical equation for photosynthesis is shown below. What happens to each of the raw materials in the process?

$$6CO_2 + 6H_2O + \text{light energy} \xrightarrow{\text{chlorophyll}} C_6H_{12}O_6 + 6O_2$$

carbon dioxide, water, glucose, oxygen

Light-Dependent Reactions Some of the chemical reactions that take place during photosynthesis need light but others do not. Those that need light can be called the light-dependent reactions of photosynthesis. During light-dependent reactions, chlorophyll and other pigments trap light energy that eventually will be stored in sugar molecules. Light energy causes water molecules, which were taken up by the roots, to split into oxygen and hydrogen. The oxygen leaves the plant through the stomata. This is the oxygen that you breathe. Leftover hydrogen is used in photosynthesis reactions that occur when there is no light.

Mini LAB

Inferring What Plants Need to Produce Chlorophyll

Procedure

1. Cut two pieces of **black construction paper** large enough so that each one completely covers one leaf on a **plant.**
2. Cut a square out of the center of each piece of paper.
3. Sandwich the leaf between the two paper pieces and **tape** the pieces together along their edges.
4. Place the plant in a sunny area. Wash your hands.
5. After seven days, carefully remove the paper and observe the leaf.

Analysis

In your **Science Journal,** describe how the color of the areas covered by paper compare to the areas not covered. Infer why this happened.

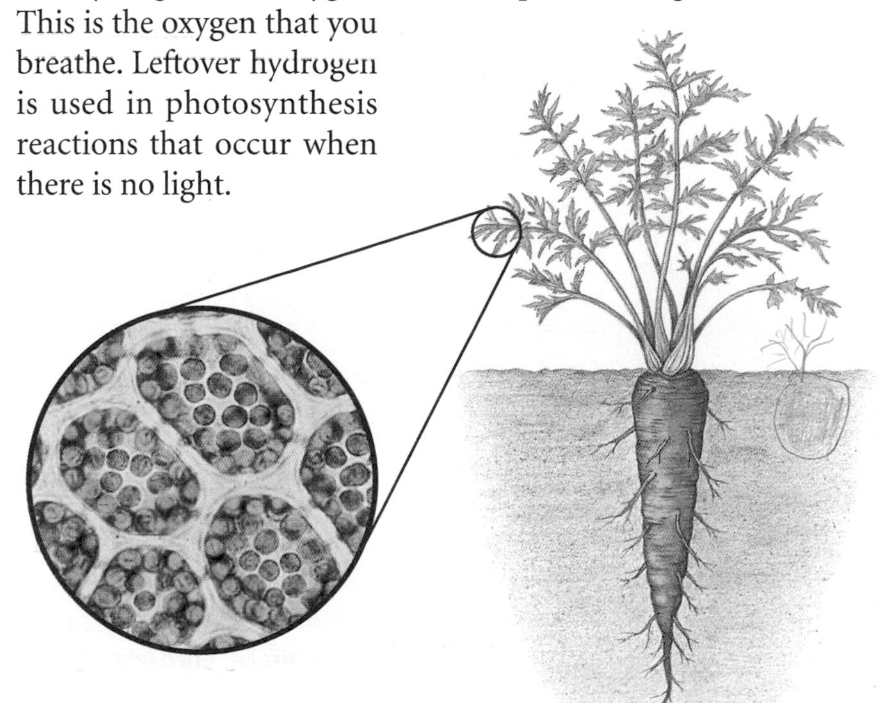

Figure 4
Because they contain chloroplasts, cells in the leaf of the carrot plant are the sites for photosynthesis.

Light-Independent Reactions Reactions that don't need light are called the light-independent reactions of photosynthesis. Carbon dioxide, the raw material from the air, is used in these reactions. The light energy trapped during the light-dependent reactions is used to combine carbon dioxide and hydrogen to make sugars. One important sugar that is made is glucose. The chemical bonds that hold glucose and other sugars together are stored energy. **Figure 5** compares what happens during each stage of photosynthesis.

What happens to the oxygen and glucose that were made during photosynthesis? Most of the oxygen from photosynthesis is a waste product and is released through stomata. Glucose is the main form of food for plant cells. A plant usually produces more glucose than it can use. Excess glucose is stored in plants as other sugars and starches. When you eat carrots, as well as beets, potatoes, or onions, you are eating the stored product of photosynthesis.

Glucose also is the basis of a plant's structure. You don't grow larger by breathing in and using carbon dioxide. However, that's exactly what plants do as they take in carbon dioxide gas and convert it into glucose. Cellulose, an important part of plant cell walls, is made from glucose. Leaves, stems, and roots are made of cellulose and other substances produced using glucose. The products of photosynthesis are used by plants to grow.

SCIENCE Online

Research Besides glucose, what other sugars do plants produce? Visit the Glencoe Science Web site at **science.glencoe.com** for more information about plant sugars. In your Science Journal list three sugars produced by plants.

Figure 5
Photosynthesis includes two sets of reactions, the light-dependent reactions and the light-independent reactions.

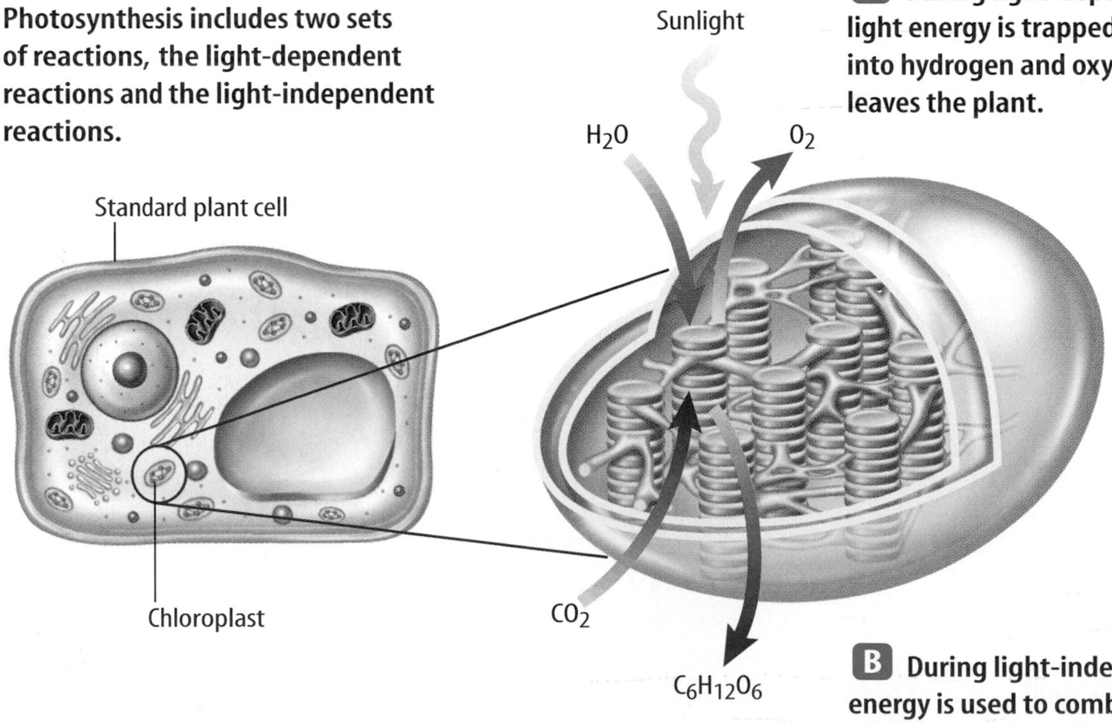

A During light-dependent reactions, light energy is trapped and water is split into hydrogen and oxygen. Oxygen leaves the plant.

B During light-independent reactions, energy is used to combine carbon dioxide and hydrogen to make glucose and other sugars.

Figure 6
Tropical rain forests contain large numbers of photosynthetic plants.

Importance of Photosynthesis Why is photosynthesis important to living things? First, photosynthesis produces food. Organisms that carry on photosynthesis provide food directly or indirectly for nearly all the other organisms on Earth. Second, photosynthetic organisms, like the plants in **Figure 6,** use carbon dioxide and release oxygen. This removes carbon dioxide from the atmosphere and adds oxygen to it. Most organisms, including humans, need oxygen to stay alive. As much as 90 percent of the oxygen entering the atmosphere today is a result of photosynthesis.

The Breakdown of Food

Look at the photograph in **Figure 7.** Do the fox and the plants in the photograph have anything in common? They don't look alike, but the fox and the plants are made of cells that break down food, and release energy in a process called respiration. How does this happen?

Respiration is a series of chemical reactions that breaks down food molecules and releases energy. Respiration occurs in cells of most organisms. The breakdown of food might or might not require oxygen. Respiration that uses oxygen to break down food chemically is called aerobic respiration. In plants and many organisms that have one or more cells, a nucleus, and other organelles, aerobic respiration occurs in the mitochondria (singular, *mitochondrion*). The overall chemical equation for aerobic respiration is shown below.

$$\underset{\text{glucose}}{C_6H_{12}O_6} + \underset{\text{oxygen}}{6O_2} \longrightarrow \underset{\text{carbon dioxide}}{6CO_2} + \underset{\text{water}}{6H_2O} + \text{energy}$$

Figure 7
You know that animals such as this red fox carry on respiration, but so do all the plants that surround the fox.

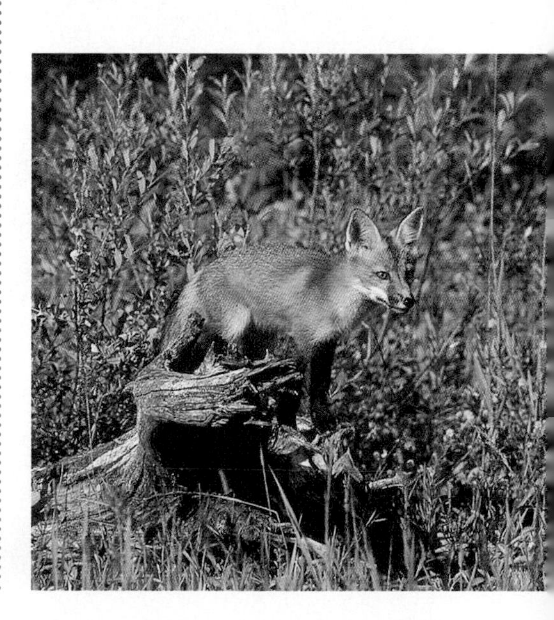

Figure 8
Aerobic respiration takes place in the mitochondria of plant cells.

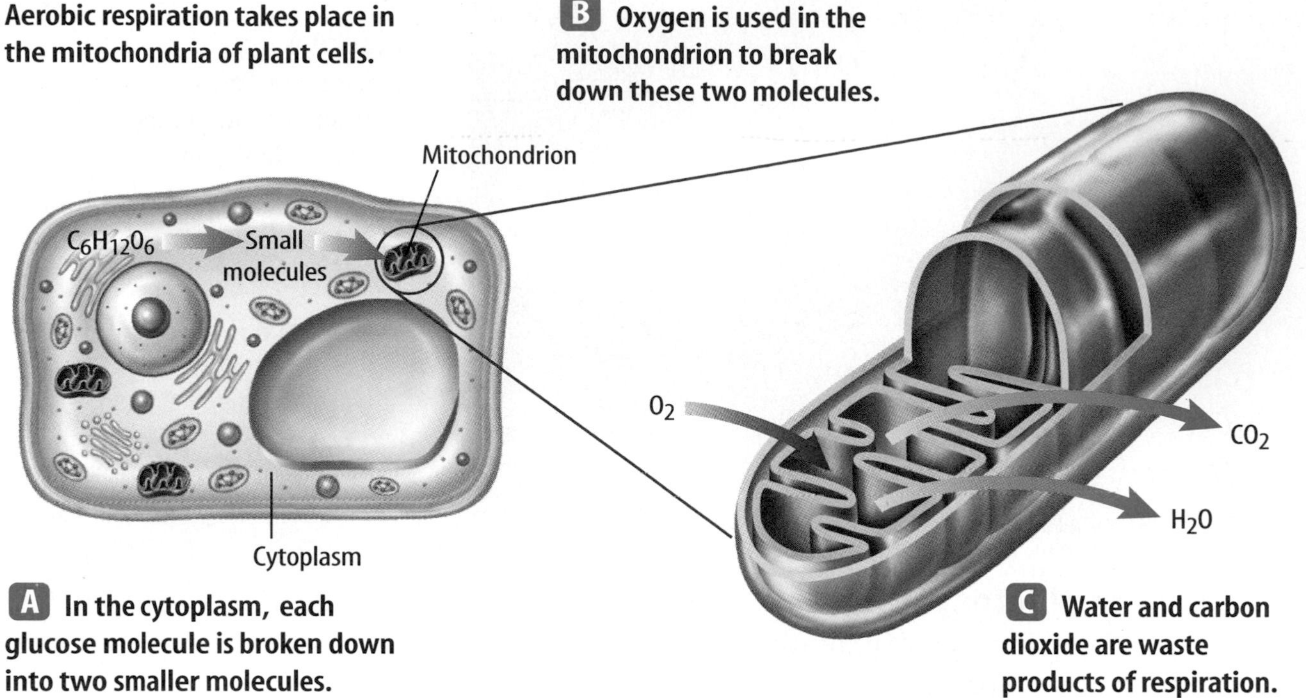

Aerobic Respiration Before aerobic respiration begins, glucose molecules are broken down into two smaller molecules. This happens in the cytoplasm. The smaller molecules then enter a mitochondrion, where aerobic respiration takes place. Oxygen is used in the reactions that break the small molecules into the waste products water and carbon dioxide. The reactions also release energy. Every cell in the organism needs this energy. **Figure 8** shows aerobic respiration in a plant cell.

Figure 9
Plants use the energy released from the respiration of food to carry out many functions.

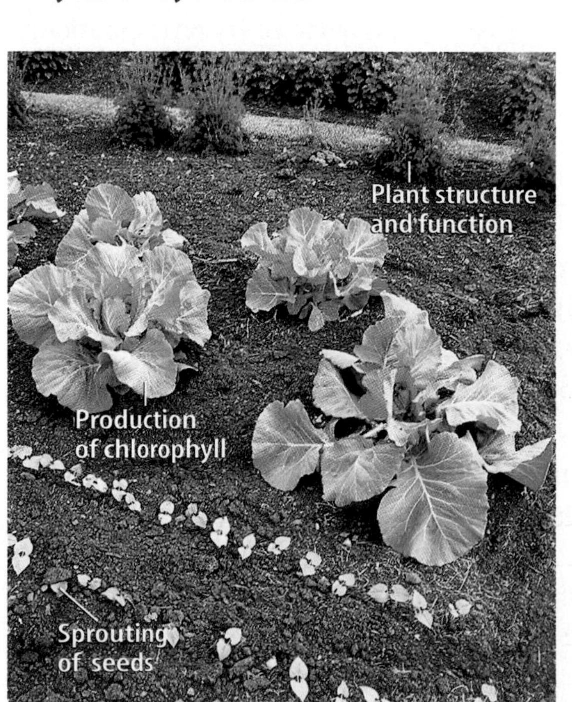

Importance of Respiration Although food contains energy, it is not in a form that can be used by cells. Respiration changes food energy into a form all cells can use. This energy drives the life processes of almost all organisms on Earth.

Reading Check *What organisms use respiration?*

Plants use energy produced by respiration to transport sugars and open and close stomata. Some of the energy is used to produce substances needed for photosynthesis, such as chlorophyll. When seeds sprout, they use energy from the respiration of stored food in the seed. **Figure 9** shows uses of energy in plants.

The waste product carbon dioxide is also important. Aerobic respiration returns carbon dioxide to the atmosphere, where it can be used again by plants and some other organisms for photosynthesis.

Table 1 Comparing Photosynthesis and Aerobic Respiration

	Energy	Raw Materials	End Products	Where
Photosynthesis	stored	water and carbon dioxide	glucose and oxygen	cells with chlorophyll
Aerobic Respiration	released	glucose and oxygen	water and carbon dioxide	cells with mitochondria

Comparison of Photosynthesis and Respiration

Look back in the chapter to find the equations for photosynthesis and aerobic respiration. Do they resemble each other? If you look closely, you can see that overall, aerobic respiration is almost the reverse of photosynthesis. Photosynthesis combines carbon dioxide and water by using light energy. The end products are glucose (food) and oxygen. During photosynthesis, energy is stored in food. Photosynthesis occurs only in cells that contain chlorophyll, such as those in the leaves of plants. Aerobic respiration combines oxygen and food to release the energy in the chemical bonds of the food. The end products of aerobic respiration are energy, carbon dioxide, and water. Because all plant cells contain mitochondria, all plant cells and any cell with mitochondria can use the process of aerobic respiration. **Table 1** compares photosynthesis and aerobic respiration.

Section  Assessment

1. Explain how a leaf exchanges carbon dioxide and water vapor.
2. Why are photosynthesis and respiration important?
3. What must happen to glucose molecules before respiration begins?
4. Compare the number of organisms that respire to those that photosynthesize.
5. **Think Critically** Humidity is water vapor in the air. How do plants contribute to the amount of humidity in the air?

Skill Builder Activities

6. **Forming Hypotheses** White potatoes sometimes have green areas on their skins. Hypothesize what process can take place in the green part but not in the white part of the potato. **For more help, refer to the Science Skill Handbook.**
7. **Solving One-Step Equations** How many CO_2 molecules result from the aerobic respiration of a glucose molecule ($C_6H_{12}O_6$)? Refer to the equation in this section. **For more help, refer to the Math Skill Handbook.**

Activity

Stomata in Leaves

Stomata open and close, which allows gases into and out of a leaf. These openings are usually invisible without the use of a microscope. Do this activity to see some stomata.

What You'll Investigate

Where are stomata in lettuce leaves?

Materials

lettuce in dish of water
coverslip
microscope
microscope slide
salt solution
forceps

Goals

- **Describe** guard cells and stomata.
- **Infer** the conditions that make stomata open and close.

Safety Precautions

WARNING: *Do not eat or taste any of the materials in the activity.*

Procedure

1. Copy the Stomata Data table into your Science Journal.
2. From a head of lettuce, tear off a piece of an outer, crisp, green leaf.
3. Bend the piece of leaf in half and carefully use a pair of forceps to peel off some of the epidermis, the transparent tissue that covers a leaf. Prepare a wet mount of this tissue.
4. **Examine** your prepared slide under low and high power on the microscope.
5. **Count** the total number of stomata in your field of view and then count the number of open stomata. Enter these numbers in the data table.

Stomata Data

	Wet Mount	Salt-Solution Mount
Total Number of Stomata		
Number of Open Stomata		
Percent Open		

6. Make a second slide of the lettuce leaf epidermis. This time place a few drops of salt solution on the leaf instead of water.
7. Repeat steps 4 and 5.
8. **Calculate** the percent of open stomata using the following equation:

$$\frac{\text{number of open stomata}}{\text{total number of stomata}} \times 100 = \text{percent open}$$

Conclude and Apply

1. Determine which slide preparation had a greater percentage of open stomata.
2. **Infer** why fewer stomata were open in the salt-solution mount.
3. What can you infer about the function of stomata in a leaf?

Communicating Your Data

Collect data from other students in your class. Compare your data to the class data. Discuss any differences you find and why these differences occurred. **For more help, refer to the Science Skill Handbook.**

SECTION 2

Plant Responses

What are plant responses?

It's dark. You're alone in a room watching a horror film on television. Suddenly, the telephone near you rings. You jump, and your heart begins to beat faster. You've just responded to a stimulus. A stimulus is anything in the environment that causes a response in an organism. The response often involves movement either toward the stimulus or away from the stimulus. A stimulus may come from outside (external) or inside (internal) the organism. The ringing telephone is an example of an external stimulus. It caused you to jump, which is a response. Your beating heart is a response to an internal stimulus. Internal stimuli are usually chemicals produced by organisms. Many of these chemicals are hormones. Hormones are substances made in one part of an organism for use somewhere else in the organism.

All living organisms, including plants, respond to stimuli. Many different chemicals are known to act as hormones in plants. These internal stimuli have a variety of effects on plant growth and function. Plants respond to external stimuli such as touch, light, and gravity. Some responses, such as the response of the Venus's-flytrap plant in **Figure 10,** are rapid. Other plant responses are slower because they involve changes in growth.

As You Read

What **You'll Learn**

- **Identify** the relationship between a stimulus and a tropism in plants.
- **Compare and contrast** long-day and short-day plants.
- **Explain** how plant hormones and responses are related.

Vocabulary

tropism
auxin
photoperiodism
long-day plant
short-day plant
day-neutral plant

Why **It's Important**

You will be able to grow healthier plants if you understand how they respond to certain stimuli.

Figure 10
A Venus's-flytrap has three small trigger hairs on the surface of its toothed leaves. When two hairs are touched at the same time, the plant responds by closing its trap in less than 1 second.

Figure 11
Tropisms are responses to external stimuli.

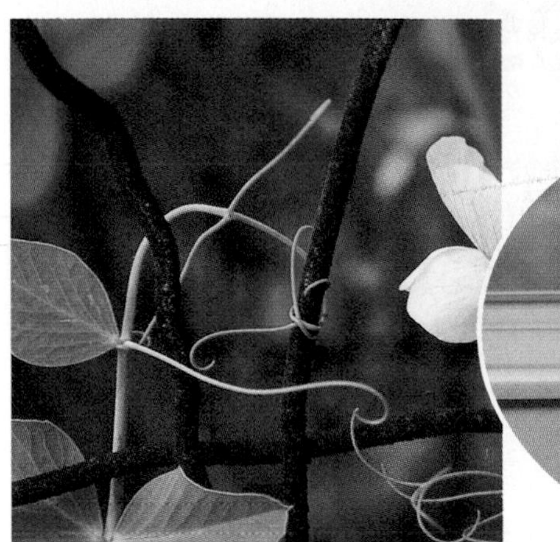

A **The pea plant's tendrils respond to touch by coiling around things.**

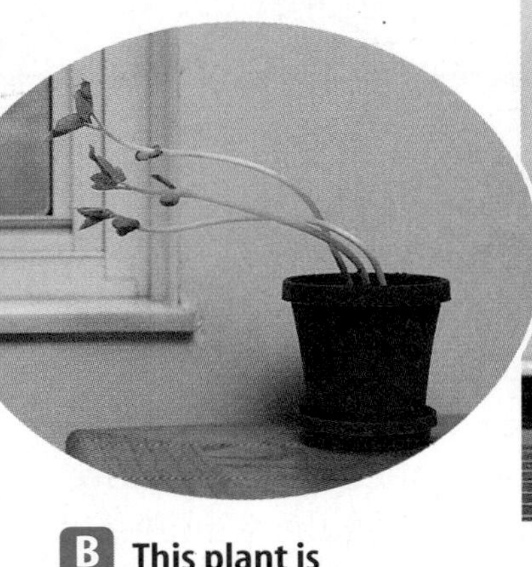

B **This plant is growing toward the light, an example of positive phototropism.**

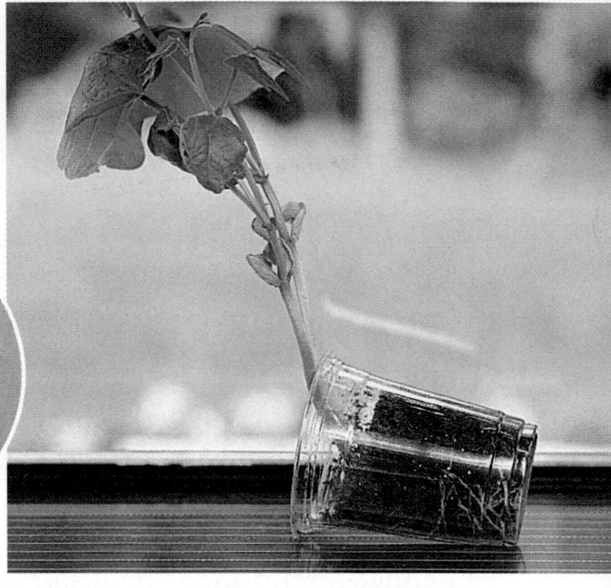

C **This plant was turned on its side. With the roots visible, you can see that they are showing positive gravitropism.**

Tropisms

Some responses of a plant to an external stimuli are called tropisms. A **tropism** (TROH pih zum) can be seen as movement caused by a change in growth and can be positive or negative. For example, plants might grow toward a stimulus—a positive tropism—or away from a stimulus—a negative tropism.

Touch One stimulus that can result in a change in a plant's growth is touch. When the pea plant, shown in **Figure 11A,** touches a solid object, it responds by growing faster on one side of its stem than on the other side. As a result the stem bends and twists around any object it touches.

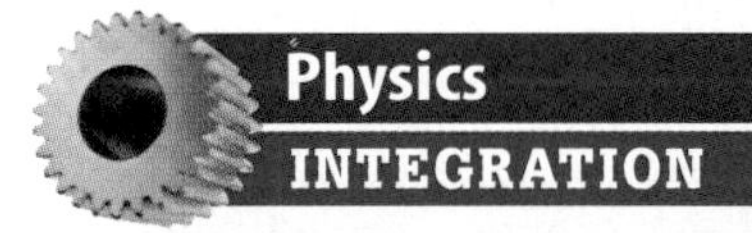

Gravity is a stimulus that affects how plants grow. Can plants grow without gravity? In space the force of gravity is low. Write a paragraph in your Science Journal that describes your idea for an experiment aboard a space shuttle to test how low gravity affects plant growth.

Light Did you ever see a plant leaning toward a window? Light is an important stimulus to plants. When a plant responds to light, the cells on the side of the plant opposite the light get longer than the cells facing the light. Because of this uneven growth, the plant bends toward the light. This response causes the leaves to turn in such a way that they can absorb more light. When a plant grows toward light it is called a positive response to light, as shown in **Figure 11B.**

Gravity Plants respond to gravity. The downward growth of plant roots is a positive response to gravity, as shown in **Figure 11C.** A stem growing upward is a negative response to gravity. Plants also may respond to electricity, temperature, and darkness.

Plant Hormones

Hormones control the changes in growth that result from tropisms and affect other plant growth. Plants often need only millionths of a gram of a hormone to stimulate a response.

Ethylene Many plants produce the hormone ethylene (EH thuh leen) gas and release it into the air around them. This means that ethylene produced by one plant can cause a response in a nearby plant. One plant response to ethylene causes a layer of cells to form between a leaf and the stem. That's why most leaves fall from plants.

Ethylene is produced in cells of ripening fruit, which stimulates the ripening process. Commercially, fruits such as oranges and bananas are picked when they are still green. During shipping the green fruits are exposed to ethylene and they ripen.

Math Skills Activity

Calculating Averages

Example Problem

What is the average height of control bean seedlings after 14 days?

Solution

1. *This is what you know:*
   height of control seedlings after 14 days
   number of control seedlings
2. *This is what you need to find:*
   average height of control seedlings after14 days
3. *This is what you must do:*
   total the heights of all control seedlings
   15 + 12 + 14 + 13 + 10 + 11 = 75 cm
4. *Divide the total height by the total number of control seedlings:*
   75 cm/6
   average height of control seedlings = 12.5 cm

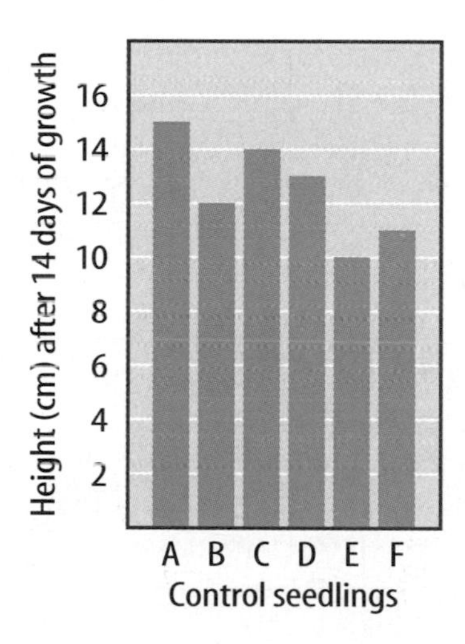

Practice Problem

Calculate the average height of seedlings treated with gibberellin.

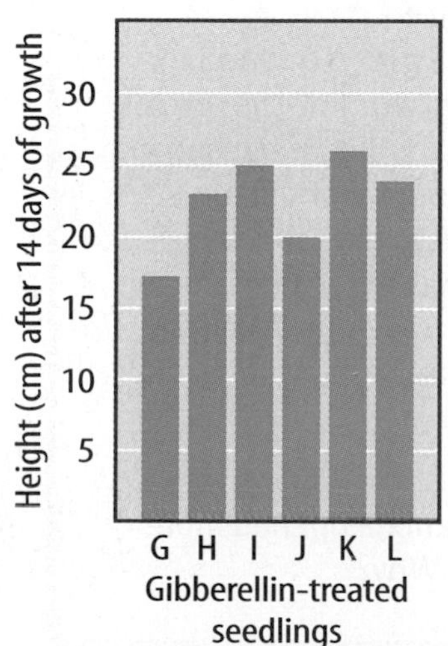

For more help, refer to the Math Skill Handbook.

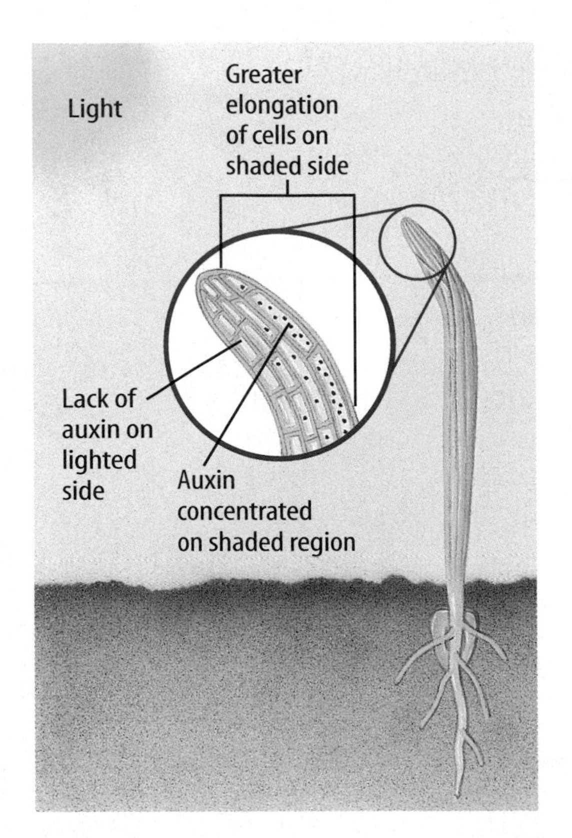

Figure 12
The concentration of auxin on the shaded side of a plant causes cells to lengthen on that side.

Auxin Scientists identified the plant hormone, **auxin** (AWK sun) more than 100 years ago. Auxin is a type of plant hormone that causes plant stems and leaves to exhibit positive response to light. When light shines on a plant from one side, the auxin moves to the shaded side of the stem where it causes a change in growth, as shown in **Figure 12.** Auxins also control the production of other plant hormones, including ethylene.

Reading Check *How are auxins and positive response to light related?*

Development of many parts of the plant, including flowers, roots, and fruit, is stimulated by auxins. Because auxins are so important in plant development, synthetic auxins have been developed for use in agriculture. Some of these synthetic auxins are used in orchards so that all plants produce flowers and fruit at the same time. Other synthetic auxins damage plants when they are applied in high doses and are used as weed killers.

Gibberellins and Cytokinins Two other groups of plant hormones that also cause changes in plant growth are gibberellins and cytokinins. Gibberellins (jih buh REH lunz) are chemical substances that were isolated first from a fungus. The fungus caused a disease in rice plants called "foolish seedling" disease. The fungus infects the stems of plants and causes them to grow too tall. Gibberellins can be mixed with water and sprayed on plants and seeds to stimulate plant stems to grow and seeds to germinate.

Like gibberellins, cytokinins (si tuh KI nunz) also cause rapid growth. Cytokinins promote growth by causing faster cell divisions. Like ethylene, the effect of cytokinins on the plant also is controlled by auxin. Interestingly, cytokinins can be sprayed on stored vegetables to keep them fresh longer.

Abscisic Acid Because hormones that cause growth in plants were known to exist, biologists suspected that substances that have the reverse effect also must exist. Abscisic (ab SIH zihk) acid is one such substance. Many plants grow in areas that have cold winters. Normally, if seeds germinate, or buds develop on plants during the winter, they will die. Abscisic acid is the substance that keeps seeds from sprouting and buds from developing during the winter. This plant hormone also causes stomata to close and helps plants respond to water loss on hot summer days. **Figure 13** summarizes how plant hormones affect plants and how hormones are used.

Observing Ripening

Procedure
1. Place a **green banana** in a **paper bag**. Roll the top shut.
2. Place another green banana on a counter or table.
3. After two days check the bananas to see how they have ripened. **WARNING:** *Do not eat the materials used in the lab.*

Analysis
Which banana ripened more quickly? Why?

VISUALIZING PLANT HORMONES

Figure 13

Chemical compounds called plant hormones help determine how a plant grows. There are five main types of hormones. They coordinate a plant's growth and development, as well as its responses to environmental stimuli, such as light, gravity, and changing seasons. Most changes in plant growth are a result of plant hormones working together, but exactly how hormones cause these changes is not completely understood.

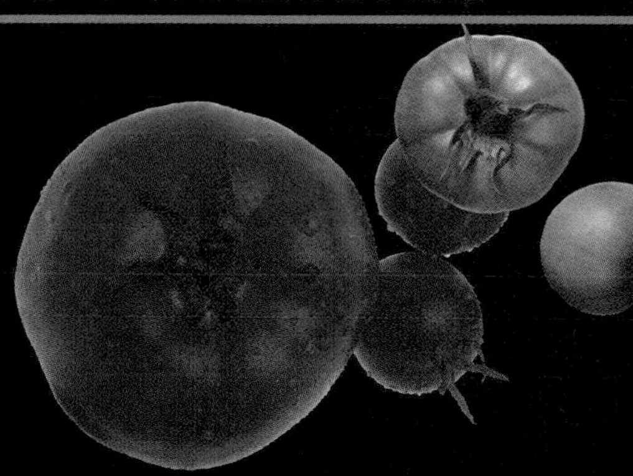

▲ ETHYLENE By controlling the exposure of these tomatoes to ethylene, a hormone that stimulates fruit ripening, farmers are able to harvest unripe fruit and make it ripen just before it arrives at the supermarket.

◀ GIBBERELLINS The larger mustard plant in the photo at left was sprayed with gibberellins, plant hormones that stimulate stem elongation and fruit development.

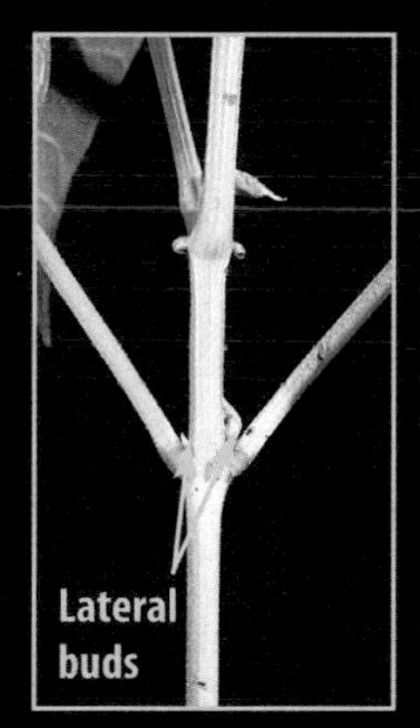

◀ CYTOKININS Lateral buds do not usually develop into branches. However, if a plant's main stem is cut, as in this bean plant, naturally occurring cytokinins will stimulate the growth of lateral branches, causing the plant to grow "bushy."

▼ AUXINS Powerful growth hormones called auxins regulate responses to light and gravity, stem elongation, and root growth. The root growth on the plant cuttings, center and right, is the result of auxin treatment.

▶ ABA (ABSCISIC ACID) In plants such as the American basswood, right, abscisic acid causes buds to remain dormant for the winter. When spring arrives, ABA stops working and the buds sprout.

Photoperiods

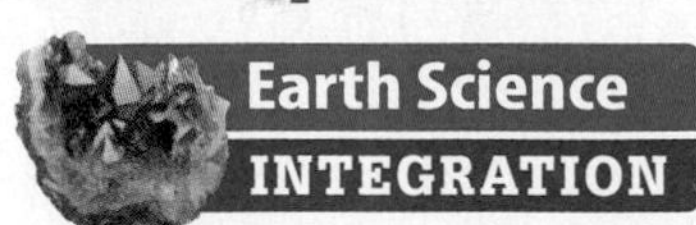

Sunflowers bloom in the summer, and cherry trees flower in the spring. Some plant species produce flowers at specific times during the year. A plant's response to the number of hours of daylight and darkness it receives daily is **photoperiodism** (foh toh PIHR ee uh dih zum).

Earth revolves around the Sun once each year. As Earth moves in its orbit, it also rotates. One rotation takes about 24 h. Because Earth is tilted about 23.5° from a line perpendicular to its orbit, the hours of daylight and darkness vary with the seasons. As you probably have noticed, the Sun sets later in summer than in winter. These changes in lengths of daylight and darkness affect plant growth.

SCIENCE Online

Research Visit the Glencoe Science Web site at **science.glencoe.com** to find out how plant pigments are involved in photoperiodism. Communicate what you learn to your class.

Darkness and Flowers Many plants require a specific length of darkness to begin the flowering process. Generally, plants that require less than 10 h to 12 h of darkness to flower are called **long-day plants.** You may be familiar with some long-day plants such as spinach, lettuce, and beets. Plants that need 12 or more hours of darkness to flower are called **short-day plants.** Some short-day plants are poinsettias, strawberries, and ragweed. **Figure 14** shows what happens when a short-day plant receives less darkness than it needs to flower.

Reading Check *What is needed to begin the flowering process?*

Day-Neutral Plants Plants like dandelions and roses are **day-neutral plants.** They have no specific photoperiod, and the flowering process can begin within a range of hours of darkness.

In nature, photoperiodism affects where flowering plants can grow and produce flowers and fruit. Even if a particular environment has the proper temperature and other growing conditions for a plant, it will not flower and produce fruit without the correct photoperiod. **Table 2** shows how day length affects flowering in all three types of plants.

Sometimes the photoperiod of a plant has a narrow range. For example, some soybeans will flower with 9.5 h of darkness but will not flower with 10 h of darkness. Farmers must choose the variety of soybeans with a photoperiod that matches the hours of darkness in the section of the country where they plant their crop.

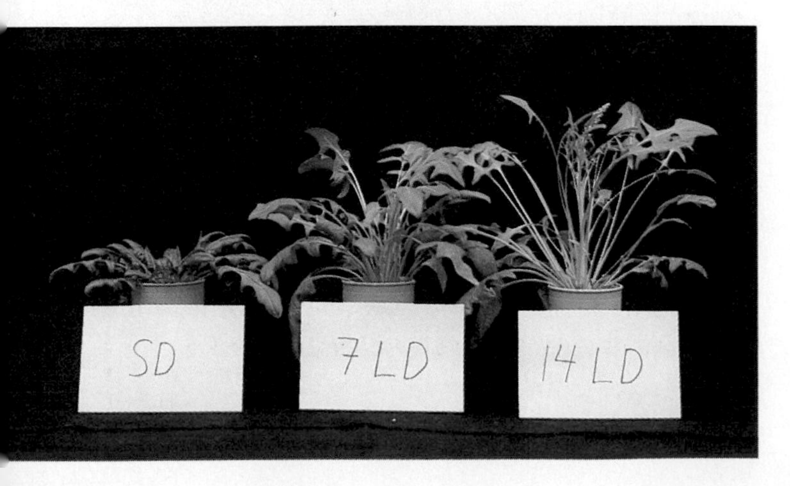

Figure 14
When short-day plants receive less darkness than required to produce flowers, they produce larger leaves instead.

Table 2 Photoperiodism

	Long-Day Plants	Short-Day Plants	Day-Neutral Plants
Early Summer Noon 6 AM 6 PM Midnight			
Late Fall Noon 6 AM 6 PM Midnight			
	An iris is a long-day plant that is stimulated by short nights to flower in the early summer.	Goldenrod is a short-day plant that is stimulated by long nights to flower in the fall.	Roses are day-neutral plants and have no specific photoperiod.

Today, greenhouse growers are able to provide any length of artificial daylight or darkness. This means that you can buy short-day flowering plants during the summer and long-day flowering plants during the winter.

Section  Assessment

1. Give an example of an internal stimulus and an external stimulus in plants.
2. Compare and contrast photoperiodism and phototropism.
3. Some red raspberries produce fruit in late spring, then again in the fall. What term describes their photoperiod?
4. How do the effects of abscisic acid differ from those of gibberellins?
5. **Think Critically** What is the relationship between plant hormones and tropisms?

Skill Builder Activities

6. **Comparing and Contrasting** Different plant parts exhibit positive and negative tropisms. Compare and contrast the responses of roots and stems to gravity. **For more help, refer to the Science Skill Handbook.**
7. **Communicating** For three years a farmer in Costa Rica grew healthy strawberry plants. But the plants never produced fruit. In your Science Journal, explain why this happened. **For more help, refer to the Science Skill Handbook.**

Activity

Tropism in Plants

Grapevines can climb on trees, fences, or other nearby structures. This growth is a response to the stimulus of touch. Tropisms are specific plant responses to stimuli outside of the plant. One part of a plant can respond positively while another part of the same plant can respond negatively to the same stimulus. Gravitropism is a response to gravity. Why might it be important for some plant parts to have a positive response to gravity while other plant parts have a negative response? You can design an experiment to test how some plant parts respond to the stimulus of gravity.

What You'll Investigate

Do stems and roots respond to gravity in the same way?

Materials

paper towel
30 cm × 30 cm sheet of aluminum foil
water
mustard seeds
marking pen
1-L clear glass or plastic jar

Safety Precautions

WARNING: *Some kinds of seeds are poisonous. Do not put any seed in your mouth. Wash your hands after handling the seeds.*

Goals

- **Describe** how roots and stems respond to gravity.
- **Observe** how changing the stimulus changes the growth of plants.

Procedure

1. Copy the following data table in your Science Journal.

Response to Gravity		
Position of Arrow on Foil Package	**Observations of Seedling Roots**	**Observations of Seedling Stems**
Arrow Up		
Arrow Down		

2. Moisten the paper towel with water so that it's damp but not dripping. Fold it in half twice.
3. In the center of the foil, place the folded paper towel and sprinkle mustard seeds in a line across the center of the towel.
4. Fold the foil around the towel and seal each end by folding the foil over. Make sure the paper towel is completely covered by the foil.
5. Use a marking pen to draw an arrow on the foil, and place the foil package in the jar with the arrow pointing upward.
6. After five days carefully open the package and record your observations in the data table. (Note: *If no stems or roots are growing yet, reseal the package and place it back in the jar, making sure that the arrow points upward. Reopen the package in two days.*)

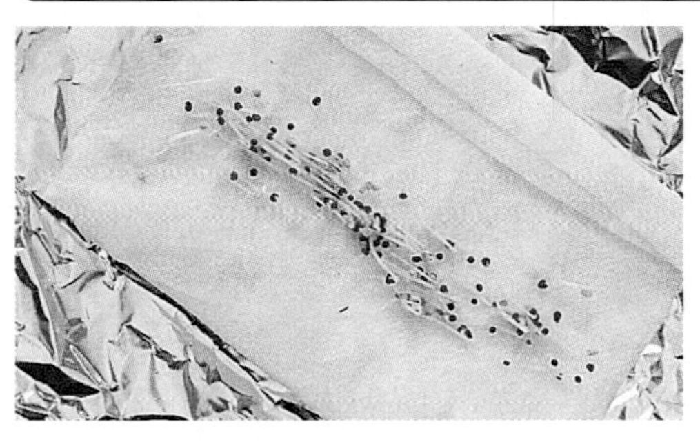

7. Reseal the foil package, being careful not to disturb the seedlings. Place it in the jar so that the arrow points downward instead of upward.
8. After five more days reopen the package and observe any new growth of the seedlings' roots and stems. Record your observations in your data table.

Conclude and Apply

1. **Classify** the responses you observed as positive or negative tropisms.
2. **Explain** why the plants' growth changed when you placed them upside down.
3. Why was it important that no light reach the seedlings during your experiment?
4. What are some other ways you could have changed the position of the foil package to test the seedlings' response?

Use drawings to **compare** the growth of the seedlings before and after you turned the package. **Compare** your drawing with those of other students in your class. **For more help, refer to the Science Skill Handbook.**

ucleus organelle marsupial mantle element ribosome primate sea otter hypertension

Science and Language Arts

Sunkissed: An Indian Legend

as told by Alberto and Patricia De La Fuente

A long time ago, deep down in the very heart of the old Mexican forests, so far away from the sea that not even the largest birds ever had time to fly that far, there was a small, beautiful valley. A long chain of snow-covered mountains stood between the valley and the sea. . . . Each day the mountains were the first ones to tell everybody that Tonatiuh, the King of Light, was coming to the valley. The meadows would see the shining white tops of the mountains and spread out their flowery skirts for the Sun.

"Good morning, Tonatiuh!" cried a little meadow.

"Hurry up and bring us warmth and light!" sang all the wild roses along the river bank together as an opening line. . . .

The wild flowers always started their fresh new day with a kiss of golden sunlight from Tonatiuh, but it was necessary to first wash their sleepy baby faces with the dew that Metztli, the Moon, sprinkled for them out of her bucket onto the nearby leaves during the night. . . .

. . . All night long, then, Metztli Moon would walk her night-field making sure that by sun-up all flowers had the magic dew that made them feel beautiful all day long.

However, much as flowers love to be beautiful as long as possible, they want to be happy too. So every morning Tonatiuh himself would give each one a single golden kiss of such power that it was possible to be happy all day long after it. As you can see, then, a flower needs to feel beautiful in the first place, but if she does not feel beautiful, she will not be ready for her morning sun-kiss. If she cannot wash her little face with the magic dew, the whole day is lost.

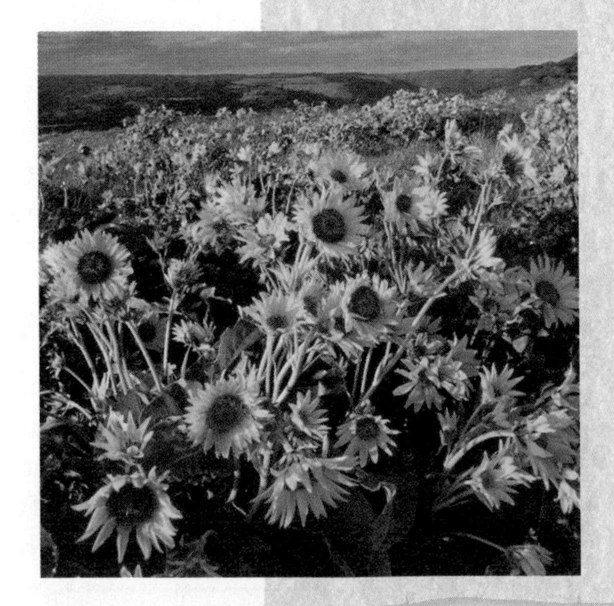

Respond to the Reading

1. What does this passage tell you about the relationship between the Sun and plants?
2. What does this passage tell you about the relationship between water and the growth of flowers?
3. Do you think this passage is effective in making the reader understand the importance of light and darkness to a growing plant? Why or why not?

Understanding Literature

Legends and Oral Traditions A legend is a traditional story often told orally and believed to be based on actual people and events. Legends are believed to be true even if they cannot be proved. Sunkissed: An Indian Legend is a legend about a little flower that is changed forever by the Sun. What in this story indicates that it is a legend? This legend also is an example of an oral tradition. Oral traditions are stories or skills that are handed down by word of mouth. They can be stories about real people and events, fictional stories, recipes, or crafts.

Science Connection In this chapter, you learned about the processes of photosynthesis and respiration. The passage from Sunkissed: An Indian Legend does not teach us the details about photosynthesis or respiration. However, it does show how sunshine and water are important to plant life. The difference between the legend and the information contained in your textbook is this— photosynthesis and respiration can be proved scientifically, and the legend, although fun to read, cannot.

Linking Science and Writing

Creating Oral Traditions
Create an idea for a fictional story that explains why the sky becomes so colorful during a sunset. Write a few short notes about your story on a piece of paper. Then retell your story to your classmates using only your short notes and your imagination. When you retell your story, remember that good storytellers are enthusiastic and entertaining. An oral tradition is started because listeners want to pass the story along.

Career Connection

Horticulturist/Landscape Designer

Jill Nokes is a horticulturist who studies plants and how to grow them. Many horticulturists work in large nurseries as managers or plant breeders. Nokes works as a landscape designer, a person who creates gardens for homes and businesses. There are two important parts to a landscape designer's job. Designers must first create attractive landscapes for their clients. They also have to be plant experts so they can choose plants that will thrive in the local climate and with other plants in the design.

SCIENCE*Online* To learn more about careers in horticulture, visit the Glencoe Science Web site at **science.glencoe.com.**

Chapter 5 Study Guide

Reviewing Main Ideas

Section 1 Photosynthesis and Respiration

1. Carbon dioxide and water vapor gases enter and leave a plant through openings in the epidermis called stomata. Guard cells cause a stoma to open and close.
2. Photosynthesis takes place in the chloroplasts of plant cells. Light energy is used to produce glucose and oxygen from carbon dioxide and water.
3. Photosynthesis provides the food for most organisms on Earth. *Why are plants called producers?*

4. All organisms use respiration to release the energy stored in food molecules. Oxygen is used in the mitochondria to complete respiration in plant cells and many other types of cells. Energy, carbon dioxide, and water are produced.
5. The energy produced from respiration is needed by most living organisms including plants. *Why is respiration important to this sprouting seed?*

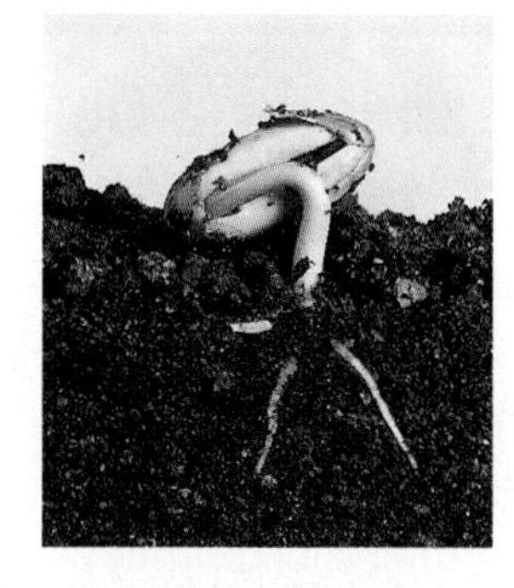

6. Photosynthesis and respiration are almost the reverse of each other. The end products of photosynthesis are the raw materials needed for aerobic respiration. The end products of aerobic respiration are the raw materials needed for photosynthesis.

Section 2 Plant Responses

1. Plants respond positively and negatively to stimuli. The response may be a movement, a change in growth, or the beginning of some process such as flowering.
2. A stimulus from outside the plant is called a tropism. Outside stimuli include such things as light, gravity, and touch. *What outside stimulus is affecting the growth of this plant?*

3. The length of darkness each day can affect flowering times of plants. *Why can day-neutral plants, such as this one, flower almost any time?*

4. Hormones control changes from inside plants. These chemicals affect plants in many ways. Some hormones cause plants to exhibit tropisms. Other hormones cause changes in plant growth.

After You Read

FOLDABLES Reading & Study Skills

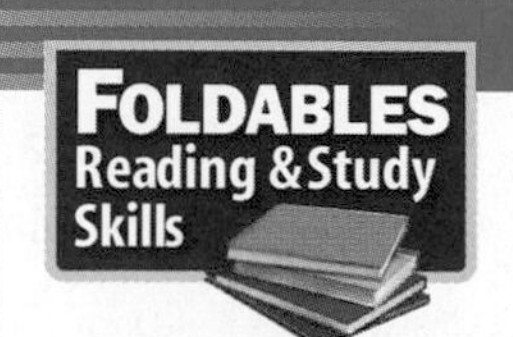

Use the information in your Compare and Contrast Study Fold to compare and contrast aerobic respiration that occurs in plants and animals.

Chapter 5 Study Guide

Visualizing Main Ideas

Complete the following cycle concept map that shows how photosynthesis and respiration are related.

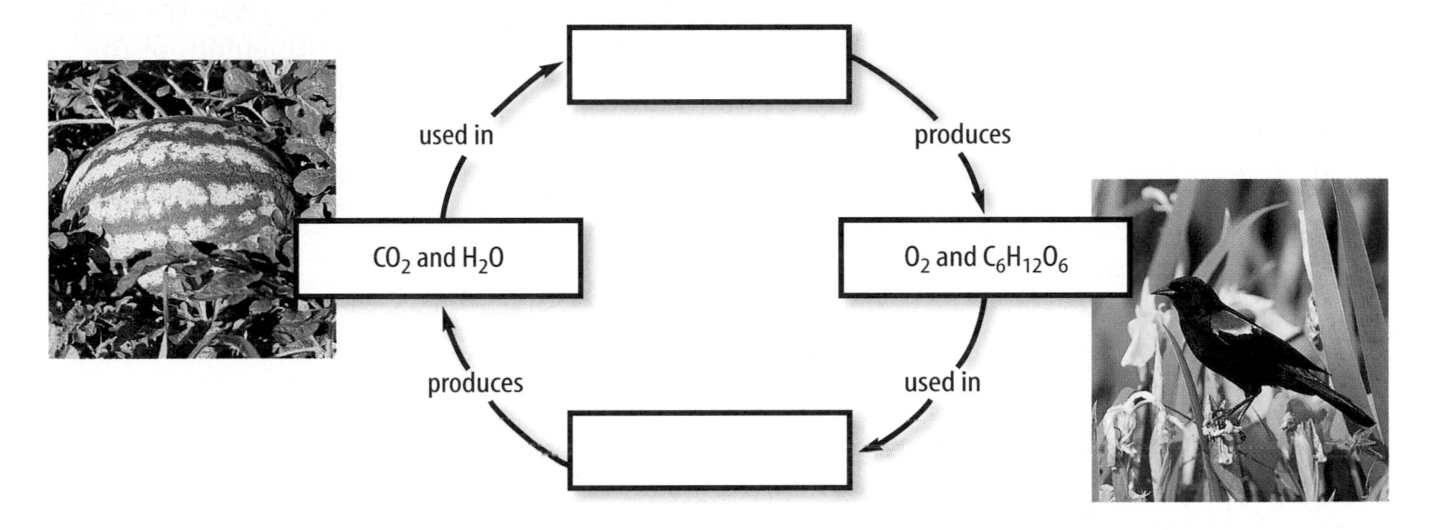

Vocabulary Review

Vocabulary Words

- **a.** auxin
- **b.** chlorophyll
- **c.** day-neutral plant
- **d.** long-day plant
- **e.** photoperiodism
- **f.** photosynthesis
- **g.** respiration
- **h.** short-day plant
- **i.** stomata
- **j.** tropism

Using Vocabulary

Replace the underlined definition with the correct vocabulary word from the list above.

1. A plant hormone causes plant stems and leaves to exhibit positive phototropism.
2. An important process of green plants is using light to make glucose and oxygen.
3. A green pigment is important in the process of photosynthesis.
4. A poinsettia, often seen flowering during December holidays, is a plant that requires long nights to flower.
5. The process of energy release from food occurs in most living things.
6. Spinach is a plant that requires only ten hours of darkness at night to flower.
7. A response of a plant to an outside stimulus can cause the plant to bend toward light.
8. Plants usually take in carbon dioxide through tiny pores in their leaves.
9. A plant's response to the number of hours of darkness it receives daily determines many plant processes.
10. Marigolds are plants that flower without regard to the length of darkness.

Study Tip

Outline the chapters to make sure that you understand the key ideas that are presented. Writing down the main points of the chapter will help you remember important details and understand larger themes.

Chapter 5 Assessment

Checking Concepts

Choose the word or phrase that best answers the question.

1. What raw material needed by plants enters through open stomata?
A) sugar **C)** carbon dioxide
B) chlorophyll **D)** cellulose

2. What is a function of stomata?
A) photosynthesis
B) to guard the interior cells
C) to allow sugar to escape
D) to permit the release of oxygen

3. What plant process produces water, carbon dioxide, and energy?
A) cell division **C)** growth
B) photosynthesis **D)** respiration

4. What type of plant needs short nights in order to flower?
A) day-neutral **C)** long-day
B) short-day **D)** nonvascular

5. What do you call things such as light, touch, and gravity that cause plant growth responses?
A) tropisms **C)** responses
B) growth **D)** stimuli

6. What are the products of photosynthesis?
A) glucose and oxygen
B) carbon dioxide and water
C) chlorophyll and glucose
D) carbon dioxide and oxygen

7. What are plant substances that affect plant growth called?
A) tropisms **C)** germination
B) glucose **D)** hormones

8. Leaves change colors because what substance breaks down?
A) hormone **C)** chlorophyll
B) carotenoid **D)** cytoplasm

9. Which of these is a product of respiration?
A) CO_2 **C)** C_2H_4
B) O_2 **D)** H_2

10. What is a plant's response to gravity called?
A) phototropism **C)** thigmotropism
B) gravitropism **D)** hydrotropism

Thinking Critically

11. You buy pears at the store that are not completely ripe. What could you do to help them ripen more rapidly?

12. Name each tropism and state whether it is positive or negative.
a. Stem grows up.
b. Roots grow down.
c. Plant grows toward light.
d. A vine grows around a pole.

13. Scientists who study sedimentary rocks and fossils suggest that oxygen was not in Earth's atmosphere until plantlike, one-celled organisms appeared. Why?

14. Explain why apple trees bloom in the spring but not in the summer.

15. Why do day-neutral and long-day plants grow best in countries near the equator?

Developing Skills

16. Forming Hypotheses Make a hypothesis about when guard cells open and close in desert plants.

17. Identifying and Manipulating Variables and Controls Plan an experiment to test your hypothesis in question 16.

18. Predicting Make a prediction about how the number and location of stomata differ in land plants and water plants whose leaves float on the water's surface.

19. Concept Mapping Complete the following concept map about photoperiodism using the following information: flower year-round—*corn, dandelion, tomato;* flower in the spring, fall, or winter—*chrysanthemum, rice, poinsettia;* flower in summer—*spinach, lettuce, petunia.*

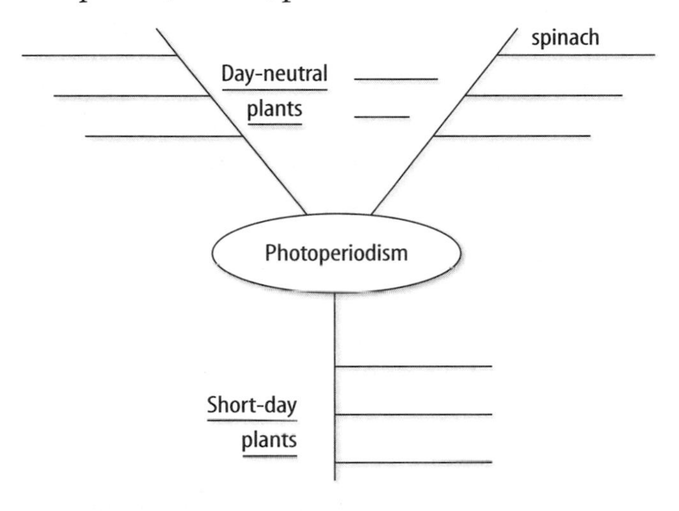

20. Comparing and Contrasting Compare and contrast the action of auxin and the action of ethylene on a plant.

Performance Assessment

21. Coloring Book Create a coloring book of day-neutral plants, long-day plants, and short-day plants. Use pictures from magazines and seed catalogs to get your ideas. Label the drawings with the plant's name and how it responds to darkness. Let a younger student color the flowers in your book.

TECHNOLOGY

Go to the Glencoe Science Web site at **science.glencoe.com** or use the **Glencoe Science CD-ROM** for additional chapter assessment.

Test Practice

Eileen and Logan wanted to learn more about the materials that plants use during photosynthesis. They designed the following data table to record the results of their investigation.

Resources Used During Photosynthesis

Plant	Water Used	Carbon Dioxide Used	Light Absorbed	Oxygen Produced
Plant X				
Plant Y				
Plant Z				

Study the table and answer the following questions.

1. Using your knowledge of photosynthesis, which of the data columns would not be needed for recording results from the investigation?

A) water used
B) carbon dioxide used
C) light absorbed
D) oxygen produced

2. The most likely source of energy for the plants during this investigation is _____.

F) water
G) carbon dioxide
H) light
J) oxygen

CHAPTER 6

Respiration and Excretion

How do you feel when you've just finished running a mile, or sliding into home base, or slamming a soccer ball into the goal past your opponent? If you're like most people, you probably breathe hard and perspire. Maybe you have even felt your lungs would burst. You need a constant supply of oxygen to keep your body cells functioning. Your body is adapted to meet that need.

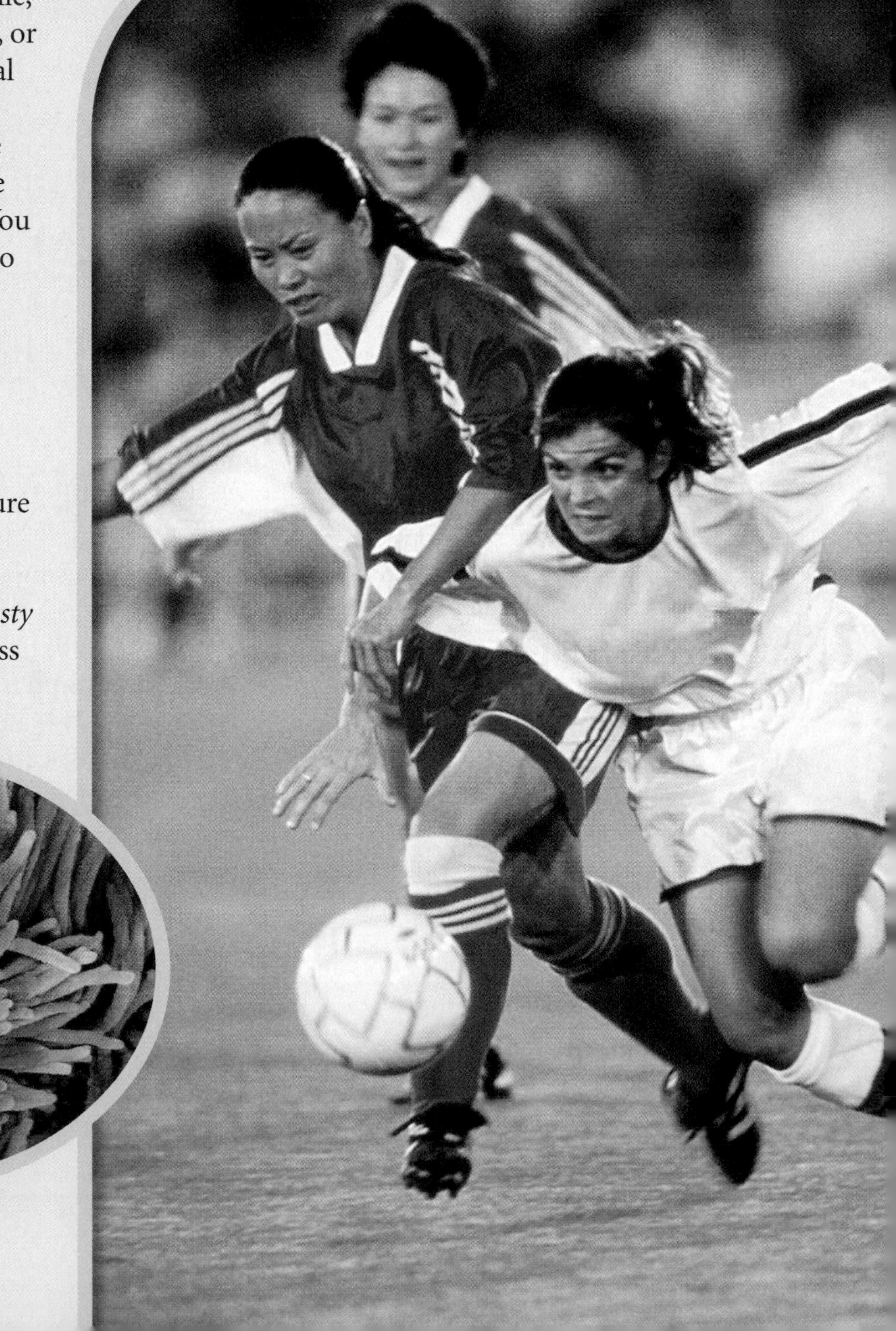

What do you think?

Science Journal Look at the picture below. What do you think these wormlike things are? Here's a hint: *You are glad you have them on a dusty day.* Write your answer or best guess in your Science Journal.

Your body can store food and water, but it cannot store much oxygen. Breathing brings oxygen into your body. In the following activity, find out about one factor that can change your breathing rate.

Measure breathing rate

1. Put your hand on the side of your rib cage. Take a deep breath. Notice how your rib cage moves out and upward when you inhale.
2. Count the number of breaths you take for 15 s. Multiply this number by four to calculate your normal breathing rate for 1 min.
3. Repeat step 2 two more times, then calculate your average breathing rate.
4. Do a physical activity described by your teacher for 1 min and repeat step 2 to determine your breathing rate now.
5. Time how long it takes for your breathing rate to return to normal.

Observe

How does breathing rate appear to be related to physical activity? Write your answer in your Science Journal.

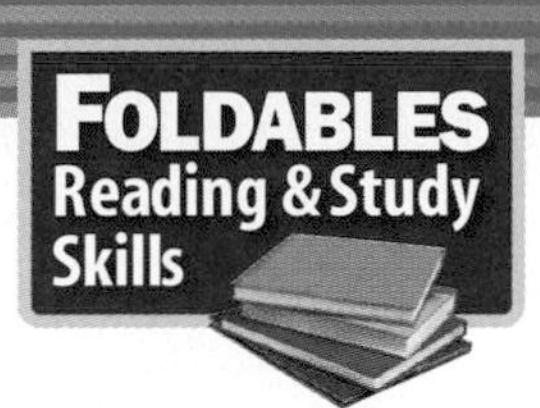

Before You Read

Making a Know-Want-Learn Study Fold Make the following Foldable to help identify what you already know and what you want to know about respiration.

1. Place a sheet of paper in front of you so the long side is at the top. Fold the paper in half from top to bottom.
2. Fold in both sides to divide the paper into thirds. Unfold the paper.
3. Cut through the top thickness of paper along each of the fold lines to the top fold to form three tabs. Label each tab as shown.
4. Before you read the chapter, write *I breathe* under the left tab. Write *Why do I breathe?* under the middle tab.
5. As you read the chapter, write the answer you learn under the right tab.

SECTION

The Respiratory System

As You Read

What You'll Learn

- **Describe** the functions of the respiratory system.
- **Explain** how oxygen and carbon dioxide are exchanged in the lungs and in tissues.
- **Identify** the pathway of air in and out of the lungs.
- **Explain** the effects of smoking on the respiratory system.

Vocabulary

pharynx
larynx
trachea
bronchi
alveoli
diaphragm
emphysema
asthma

Why It's Important

Your body's cells depend on your respiratory system to supply oxygen and remove carbon dioxide.

Functions of the Respiratory System

Can you imagine an astronaut walking on the Moon without a space suit or a diver exploring the ocean without scuba gear? Of course not. You couldn't survive in either location under those conditions because you need to breathe air. Earth is surrounded by a layer of gases called the atmosphere (AT muh sfihr). You breathe atmospheric gases that are closest to Earth. As shown in **Figure 1,** oxygen is one of those gases.

For thousands of years people have known that air, food, and water are needed for life. However, the gas in the air that is necessary for life was not identified as oxygen until the late 1700s. At that time, a French scientist experimented and discovered that an animal breathed in oxygen and breathed out carbon dioxide. He measured the amount of oxygen that the animal used and the amount of carbon dioxide produced by its bodily processes. After his work with animals, the French scientist used this knowledge to study the way that humans use oxygen. He measured the amount of oxygen that a person uses when resting and when exercising. These measurements were compared, and he discovered that more oxygen is used by the body during exercise.

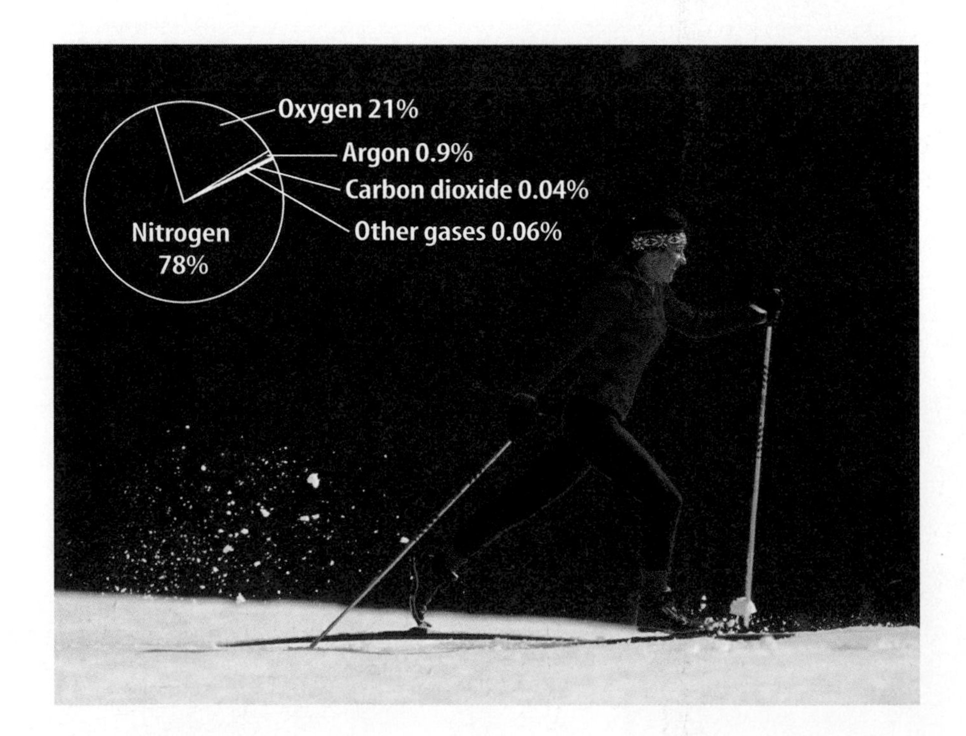

Figure 1
Air, which is needed by most organisms, is only 21 percent oxygen.

Figure 2
Several processes are involved in how the body obtains, transports, and uses oxygen.

$$C_6H_{12}O_6 + 6O_2 \longrightarrow 6CO_2 + 6H_2O + \text{Energy}$$

Glucose + Oxygen ⟶ Carbon dioxide + Water + Energy

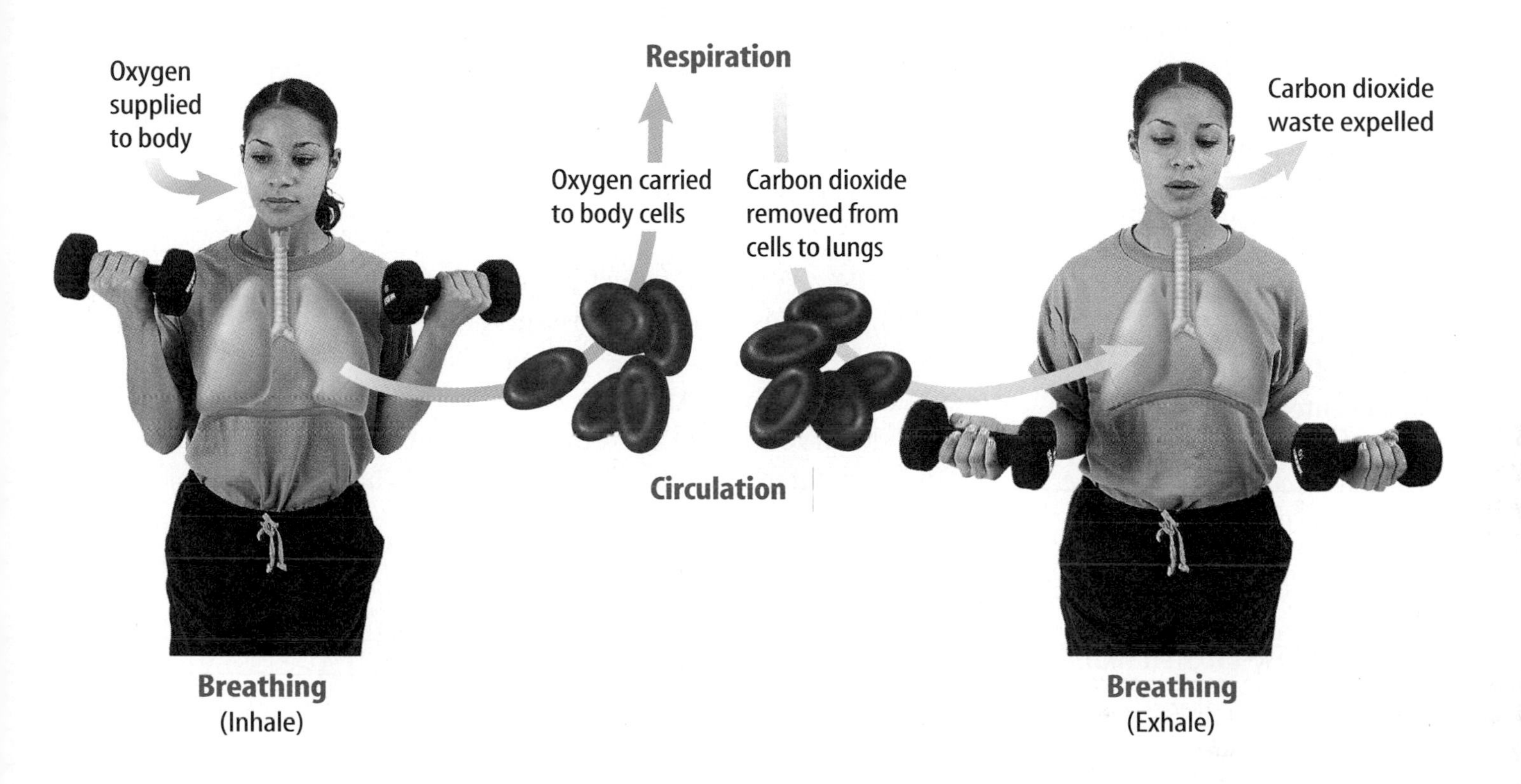

Breathing and Respiration People often confuse the terms *breathing* and *respiration.* Breathing is the movement of the chest that brings air into the lungs and removes waste gases. The air entering the lungs contains oxygen. It passes from the lungs into the circulatory system because there is less oxygen in the blood than in cells of the lungs. Blood carries oxygen to individual cells. At the same time, the digestive system supplies glucose from digested food to the same cells. The oxygen delivered to the cells is used to release energy from glucose. This chemical reaction, shown in the equation in **Figure 2,** is called cellular respiration. Without oxygen, this reaction would not take place. Carbon dioxide and water molecules are waste products of cellular respiration. They are carried back to the lungs in the blood. Exhaling, or breathing out, eliminates waste carbon dioxide and some water molecules.

What is respiration?

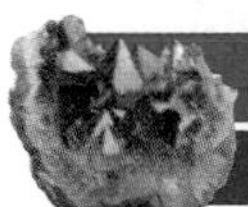
Earth Science INTEGRATION

The amount of water vapor in the atmosphere varies from almost none over deserts to nearly four percent in tropical rain forest areas. This means that every 100 molecules that make up air include only four molecules of water. In your Science Journal, infer how breathing dry air can stress your respiratory system.

Organs of the Respiratory System

The respiratory system, shown in **Figure 3,** is made up of structures and organs that help move oxygen into the body and waste gases out of the body. Air enters your body through two openings in your nose called nostrils or through the mouth. Fine hairs inside the nostrils trap dust from the air. Air then passes through the nasal cavity, where it gets moistened and warmed by the body's heat. Glands that produce sticky mucus line the nasal cavity. The mucus traps dust, pollen, and other materials that were not trapped by nasal hairs. This process helps filter and clean the air you breathe. Tiny, hairlike structures, called cilia (SIHL ee uh), sweep mucus and trapped material to the back of the throat where it can be swallowed.

Pharynx Warmed, moist air then enters the **pharynx** (FER ingks), which is a tubelike passageway used by food, liquid, and air. At the lower end of the pharynx is a flap of tissue called the epiglottis (ep uh GLAHT us). When you swallow, your epiglottis folds down to prevent food or liquid from entering your airway. The food enters your esophagus instead. What do you think has happened if you begin to choke?

Figure 3
Air can enter the body through the nostrils and the mouth. *What is an advantage of having air enter through the nostrils?*

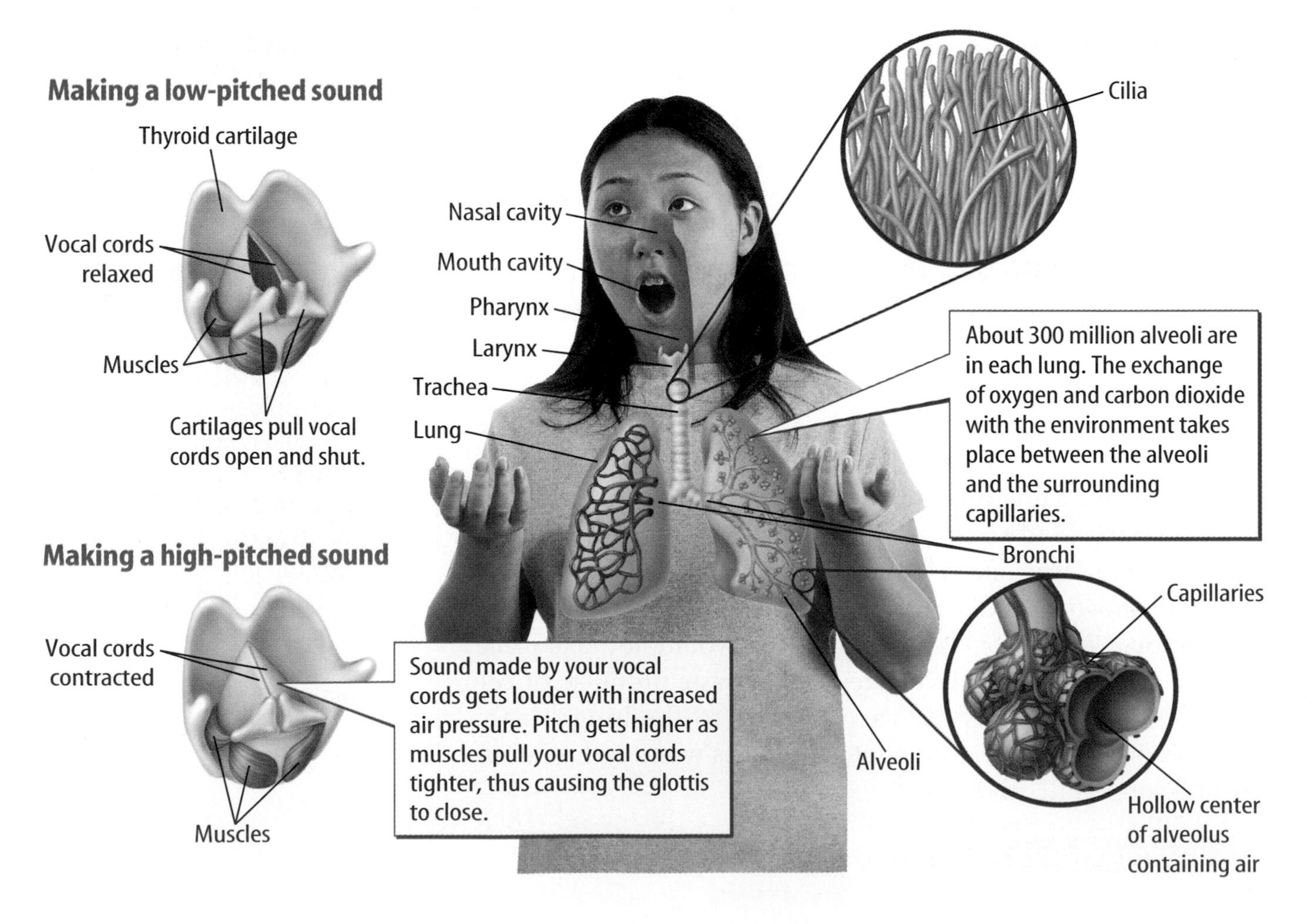

Larynx and Trachea Next, the air moves into your larynx (LER ingks). The **larynx** is the airway to which two pairs of horizontal folds of tissue, called vocal cords, are attached as shown in **Figure 3.** Forcing air between the cords causes them to vibrate and produce sounds. When you speak, muscles tighten or loosen your vocal cords, resulting in different sounds. Your brain coordinates the movement of the muscles in your throat, tongue, cheeks, and lips when you talk, sing, or just make noise. Your teeth also are involved in forming letter sounds and words.

From the larynx, air moves into the **trachea** (TRAY kee uh), which is a tube about 12 cm in length. Strong, C-shaped rings of cartilage prevent the trachea from collapsing. The trachea is lined with mucous membranes and cilia, as shown in **Figure 3,** that trap dust, bacteria, and pollen. Why must the trachea stay open all the time?

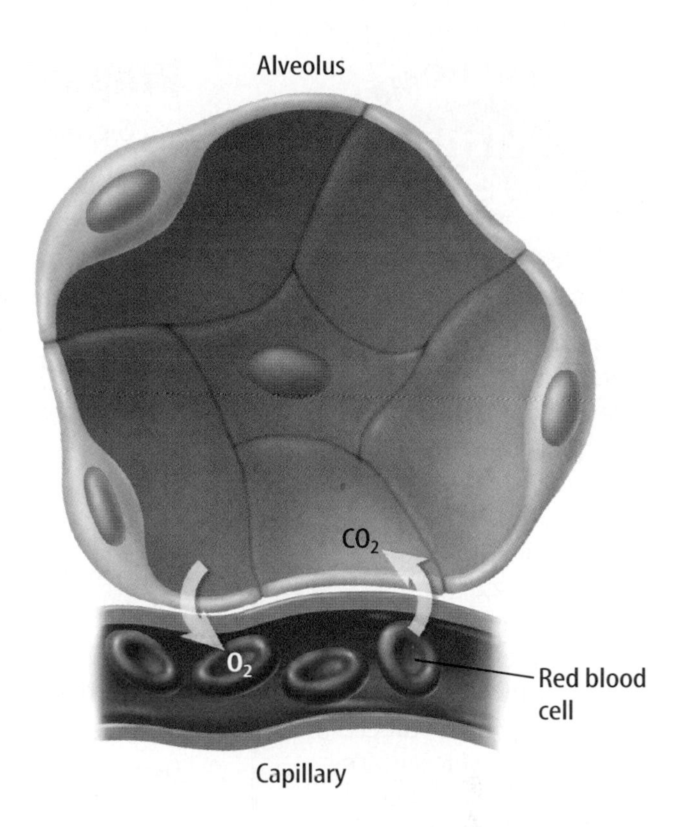

Figure 4
The thin capillary walls allow gases to be exchanged easily between the alveoli and the capillaries.

Bronchi and the Lungs Air is carried into your lungs by two short tubes called **bronchi** (BRAHN ki) (singular, *bronchus)* at the lower end of the trachea. Within the lungs, the bronchi branch into smaller and smaller tubes. The smallest tubes are called bronchioles (BRAHN kee ohlz). At the end of each bronchiole are clusters of tiny, thin-walled sacs called **alveoli** (al VEE uh li). Air passes into the bronchi, then into the bronchioles, and finally into the alveoli. As shown in **Figure 3,** lungs are masses of alveoli arranged in grapelike clusters. The capillaries surround the alveoli like a net.

The exchange of oxygen and carbon dioxide takes place between the alveoli and capillaries. This easily happens because the walls of the alveoli (singular, *alveolus)* and the walls of the capillaries are each only one cell thick, as shown in **Figure 4.** Oxygen moves through the cell membranes of the alveoli and then through the cell membranes of the capillaries into the blood. There the oxygen is picked up by hemoglobin (HEE muh gloh bun), a molecule in red blood cells, and carried to all body cells. At the same time, carbon dioxide and other cellular wastes leave the body cells. The wastes move through the cell membranes of the capillaries. Then they are carried by the blood. In the lungs, waste gases move through the cell membranes of the capillaries and through the cell membranes of the alveoli. Then waste gases leave the body during exhalation.

Research Visit the Glencoe Science Web site at **science.glencoe.com** for more information about how speech sounds are made. Report to your class what you learn.

Comparing Surface Area

Procedure

1. Stand a **bathroom-tissue cardboard tube** in an **empty bowl.**
2. Drop **marbles** into the tube, filling it to the top.
3. Count the number of marbles used.
4. Repeat steps 2 and 3 two more times. Calculate the average number of marbles needed to fill the tube.
5. The tube's inside surface area is approximately 161.29 cm^2. Each marble has a surface area of approximately 8.06 cm^2. Calculate the surface area of the average number of marbles.

Analysis

1. Compare the inside surface area of the tube with the surface area of the average number of marbles needed to fill the tube.
2. If the tube represents a bronchus, what do the marbles represent?
3. Using this model, explain what makes gas exchange in the lungs efficient.

Why do you breathe?

Signals from your brain tell the muscles in your chest and abdomen to contract and relax. You don't have to think about breathing to breathe, just like your heart beats without you telling it to beat. Your brain can change your breathing rate depending on the amount of carbon dioxide present in your blood. If a lot of carbon dioxide is present, your breathing rate increases. It decreases if less carbon dioxide is in your blood. You do have some control over your breathing—you can hold your breath if you want to. Eventually, though, your brain will respond to the buildup of carbon dioxide in your blood. The brain's response will tell your chest and abdomen muscles to work automatically, and you will breathe whether you want to or not.

Inhaling and Exhaling Breathing is partly the result of changes in air pressure. Under normal conditions, a gas moves from an area of high pressure to an area of low pressure. When you squeeze an empty, soft-plastic bottle, air is pushed out. This happens because air pressure outside the top of the bottle is less than the pressure you create inside the bottle when you squeeze it. As you release your grip on the bottle, the air pressure inside the bottle becomes less than it is outside the bottle. Air rushes back in, and the bottle returns to its original shape.

Your lungs work in a similar way to the squeezed bottle. Your **diaphragm** (DI uh fram) is a muscle beneath your lungs that contracts and relaxes to help move gases into and out of your lungs. **Figure 5** illustrates breathing.

✔ **Reading Check** *How does your diaphragm help you breathe?*

When a person is choking, a rescuer can use abdominal thrusts, as shown in **Figure 6,** to save the life of the choking victim.

Figure 5
Your lungs inhale and exhale about 500 mL of air with an average breath. This increases to 2,000 mL of air per breath when you do strenuous activity.

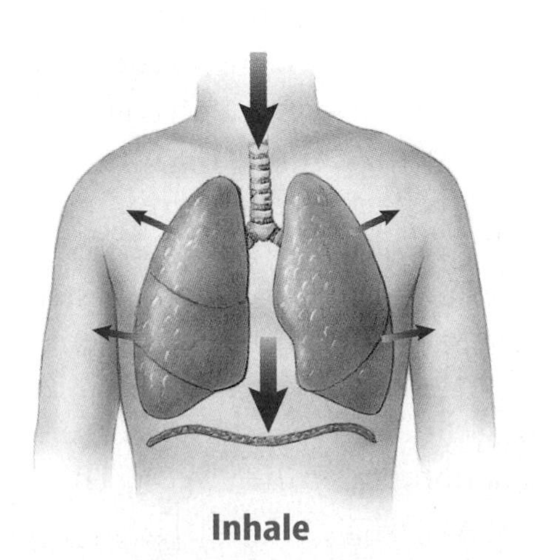

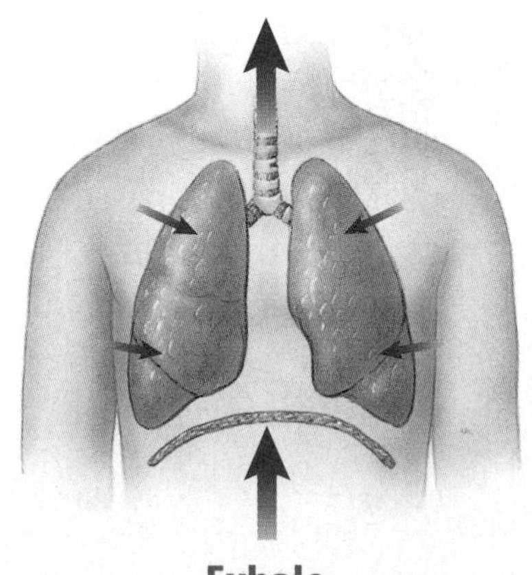

NATIONAL GEOGRAPHIC

VISUALIZING ABDOMINAL THRUSTS

Figure 6

When food or other objects become lodged in the trachea, airflow between the lungs and the mouth and nasal cavity is blocked. Death can occur in minutes. However, prompt action by someone can save the life of a choking victim. The rescuer uses abdominal thrusts to force the victim's diaphragm up. This decreases the volume of the chest cavity and forces air up in the trachea. The result is a rush of air that dislodges and expels the food or other object. The victim can breathe again. This technique is shown at right and should only be performed in emergency situations.

Food is lodged in the victim's trachea.

The rescuer places her fist against the victim's stomach.

The rescuer's second hand adds force to the fist.

A The rescuer stands behind the choking victim and wraps her arms around the victim's upper abdomen. She places a fist (thumb side in) against the victim's stomach. The fist should be below the ribs and above the navel.

An upward thrust dislodges the food from the victim's trachea.

B With a violent, sharp movement, the rescuer thrusts her fist up into the area below the ribs. This action should be repeated as many times as necessary.

Table 1 Smokers' Risk of Death from Disease

Disease	Smokers' Risk Compared to Nonsmokers' Risk
Lung Cancer	23 times higher for males, 11 times higher for females
Chronic Bronchitis and Emphysema	5 times higher
Heart Disease	2 times higher

Diseases and Disorders of the Respiratory System

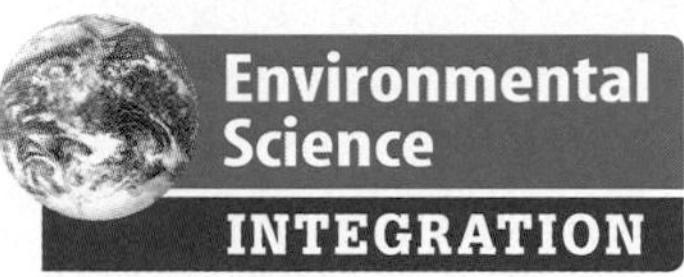

If you were asked to list some of the things that can harm your respiratory system, you probably would put smoking at the top. As you can see in **Table 1,** many serious diseases are related to smoking. The chemical substances in tobacco—nicotine and tars—are poisons and can destroy cells. The high temperatures, smoke, and carbon monoxide produced when tobacco burns also can injure a smoker's cells. Even if you are a nonsmoker, inhaling smoke from tobacco products—called secondhand smoke—is unhealthy and has the potential to harm your respiratory system. Smoking, polluted air, coal dust, and asbestos (as BES tus) have been related to respiratory problems such as bronchitis (brahn KITE us), emphysema (em fuh SEE muh), asthma (AZ muh), and cancer.

Research Visit the Glencoe Science Web site at **science.glencoe.com** for more information about the health aspects of secondhand smoke. Make a poster explaining what you learn.

Respiratory Infections Bacteria, viruses, and other microorganisms can cause infections that affect any of the organs of the respiratory system. The common cold usually affects the upper part of the respiratory system—from the nose to the pharynx. The cold virus also can cause irritation and swelling in the larynx, trachea, and bronchi. The cilia that line the trachea and bronchi can be damaged. However, cilia usually heal rapidly. A virus that causes influenza, or flu, can affect many of the body's systems. The virus multiplies in the cells lining the alveoli and damages them. Pneumonia is an infection in the alveoli that can be caused by bacteria, viruses, or other microorganisms. Before antibiotics were available to treat these infections, many people died from pneumonia.

What parts of the respiratory system are affected by the cold virus?

Chronic Bronchitis When bronchial tubes are irritated and swell and too much mucus is produced, a disease called bronchitis develops. Sometimes, bacterial infections occur in the bronchial tubes because the mucus there provides nearly ideal conditions for bacteria to grow. Antibiotics are effective treatments for this type of bronchitis.

Many cases of bronchitis clear up within a few weeks, but the disease sometimes lasts for a long time. When this happens, it is called chronic (KRAHN ihk) bronchitis. A person who has chronic bronchitis must cough often to try to clear the excess mucus from the airway. However, the more a person coughs, the more the cilia and bronchial tubes can be harmed. When cilia are damaged, they cannot move mucus, bacteria, and dirt particles out of the lungs effectively. Then harmful substances, such as sticky tar from burning tobacco, build up in the airways. Sometimes, scar tissue forms and the respiratory system cannot function properly.

Emphysema A disease in which the alveoli in the lungs enlarge is called **emphysema** (em fuh SEE muh). When cells in the alveoli are reddened and swollen, an enzyme is released that causes the walls of the alveoli to break down. As a result, alveoli can't push air out of the lungs, so less oxygen moves into the bloodstream from the alveoli. When blood becomes low in oxygen and high in carbon dioxide, shortness of breath occurs. Some people with emphysema require extra oxygen as shown in **Figure 7C.** Because the heart works harder to supply oxygen to body cells, people who have emphysema often develop heart problems, as well.

Figure 7
Lung diseases can have major effects on breathing.

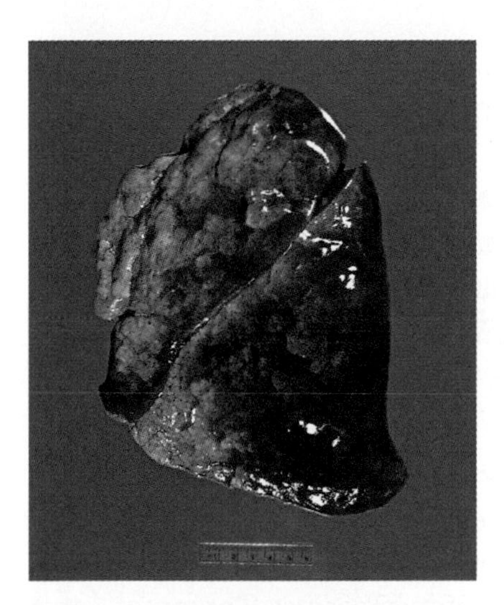

A **A normal, healthy lung can exchange oxygen and carbon dioxide effectively.**

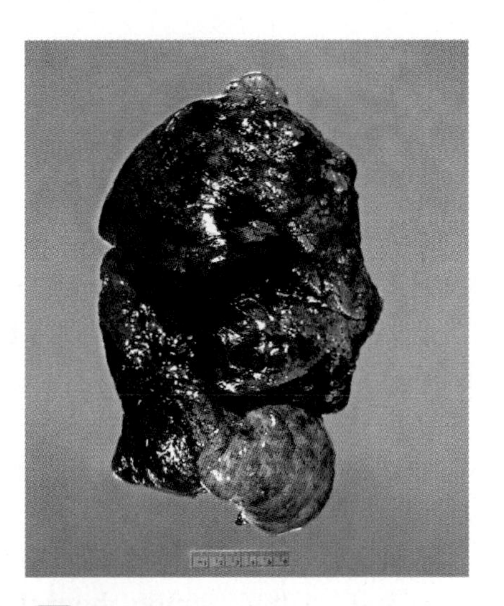

B **A diseased lung carries less oxygen to body cells.**

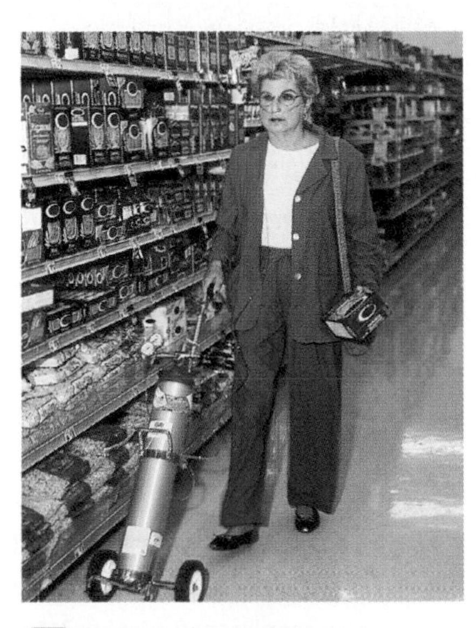

C **Emphysema may take 20 to 30 years to develop.**

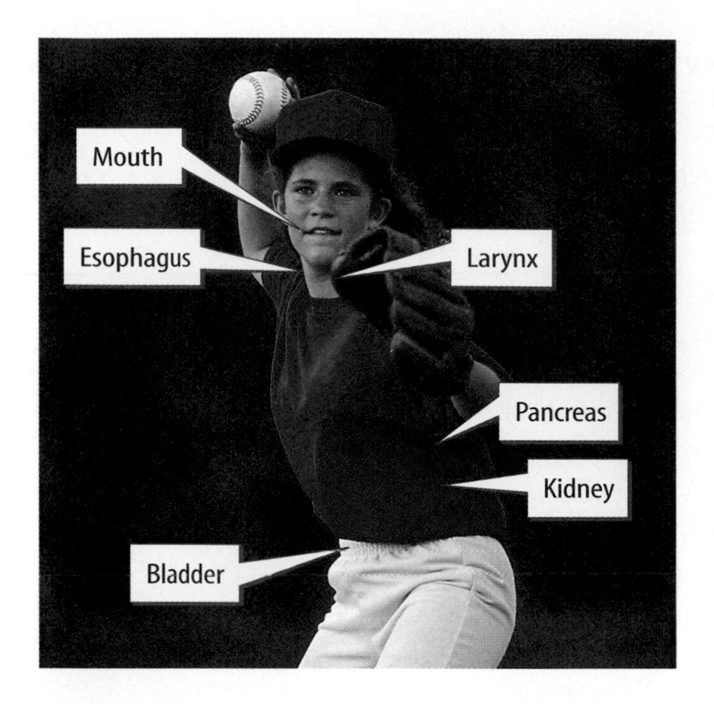

Figure 8
More than 85 percent of all lung cancer is related to smoking. Smoking also can play a part in the development of cancer in other body organs indicated above.

Lung Cancer The third leading cause of death in men and women in the United States is lung cancer. Inhaling the tar in cigarette smoke is the greatest contributing factor to lung cancer. Tar and other ingredients found in smoke act as carcinogens (kar SIHN uh junz) in the body. Carcinogens are substances that can cause an uncontrolled growth of cells. In the lungs, this is called lung cancer. Lung cancer is not easy to detect in its early stages. Smoking also has been linked to the development of cancers of the mouth, esophagus, larynx, pancreas, kidney, and bladder. See **Figure 8.**

Reading Check ***What do you think will happen to the lungs of a young person who begins smoking?***

Asthma Shortness of breath, wheezing, or coughing can occur in a lung disorder called **asthma.** When a person has an asthma attack, the bronchial tubes contract quickly. Inhaling medicine that relaxes the bronchial tubes is the usual treatment for an asthma attack. Asthma is often an allergic reaction. An allergic reaction occurs when the body overreacts to a foreign substance. An asthma attack can result from breathing certain substances such as cigarette smoke or certain plant pollen, eating certain foods, or stress in a person's life.

Section  Assessment

1. What is the main function of the respiratory system?
2. How are oxygen, carbon dioxide, and other waste gases exchanged in the lungs and body tissues?
3. What causes air to move into and out of a person's lungs?
4. How does smoking affect the respiratory and circulatory systems?
5. **Think Critically** How is the work of the digestive and circulatory systems related to the respiratory system?

Skill Builder Activities

6. **Researching Information** Nicotine in tobacco is a poison. Using library references, find out how nicotine affects the body. **For more help, refer to the Science Skill Handbook.**
7. **Communicating** Use references to find out about lung disease common among coal miners, stonecutters, and sandblasters. Find out what safety measures are required now for these trades. In your Science Journal, write a paragraph about these safety measures. **For more help, refer to the Science Skill Handbook.**

SECTION

The Excretory System

Functions of the Excretory System

It's your turn to take out the trash. You carry the bag outside and put it in the trash can. The next day, you bring out another bag of trash, but the trash can is full. When trash isn't collected, it piles up. Just as trash needs to be removed from your home to keep it livable, your body must eliminate wastes to remain healthy. Undigested material is eliminated by your large intestine. Waste gases are eliminated through the combined efforts of your circulatory and respiratory systems. Some salts are eliminated when you sweat. These systems function together as parts of your excretory system. If wastes aren't eliminated, toxic substances build up and damage organs. If not corrected, serious illness or death occurs.

The Urinary System

The **urinary system** rids the blood of wastes produced by the cells. **Figure 9** shows how the urinary system functions as a part of the excretory system. The urinary system also controls blood volume by removing excess water produced by body cells during respiration.

As You Read

What **You'll Learn**

- **Distinguish** between the excretory and urinary systems.
- **Describe** how the kidneys work.
- **Explain** what happens when urinary organs don't work.

Vocabulary

urinary system
urine
kidney
nephron
ureter
bladder
urethra

Why **It's Important**

The urinary system helps clean your blood of cellular wastes.

Figure 9
The urinary system, along with the digestive and respiratory systems, and the skin make up the excretory system.

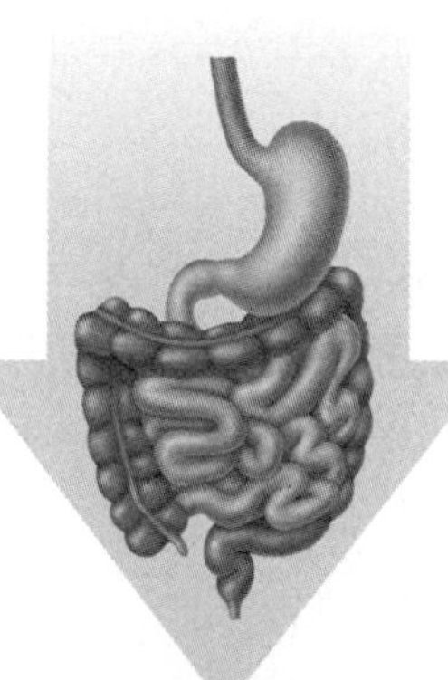

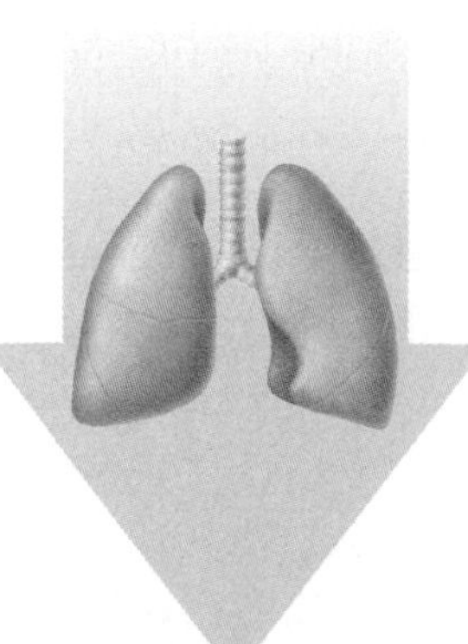

Figure 10
The amount of urine that you eliminate each day is determined by the level of a hormone that is produced by your hypothalamus.

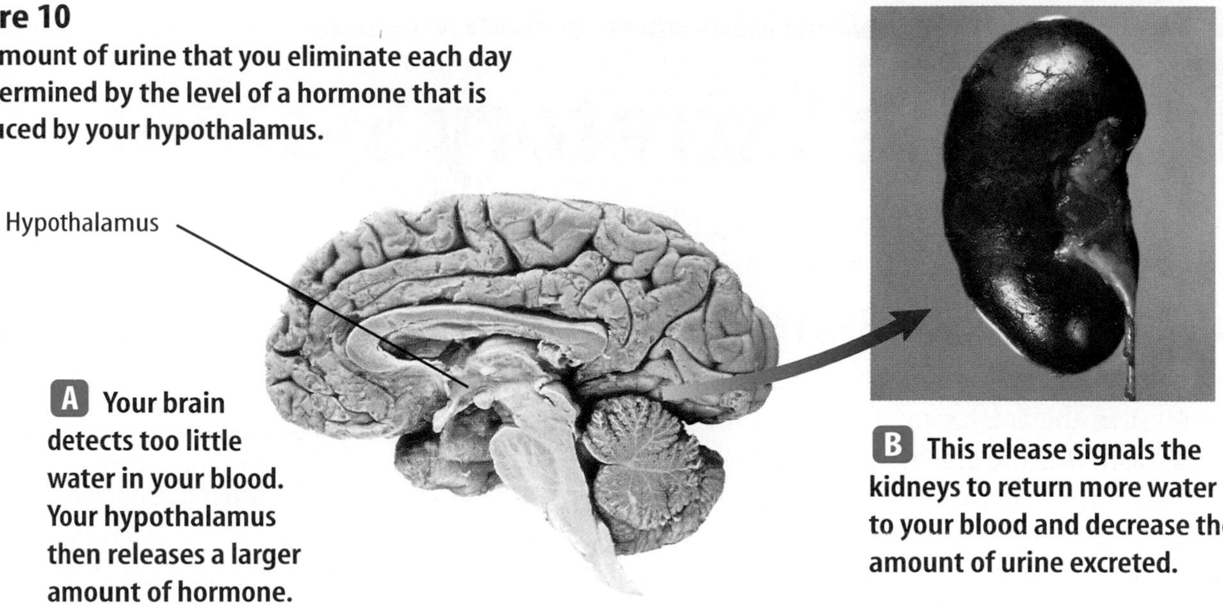

Hypothalamus

A Your brain detects too little water in your blood. Your hypothalamus then releases a larger amount of hormone.

B This release signals the kidneys to return more water to your blood and decrease the amount of urine excreted.

Regulating Fluid Levels To stay in good health, the fluid levels within the body must be balanced and normal blood pressure must be maintained. An area in the brain, the hypothalamus (hi poh THAL uh mus), constantly monitors the amount of water in the blood. When the brain detects too much water in the blood, the hypothalamus releases a lesser amount of a specific hormone. This signals the kidneys to return less water to the blood and increase the amount of wastewater, called **urine,** that is excreted. **Figure 10** shows what happens when too little water is in the blood.

Reading Check *How does the urinary system control the volume of water in the blood?*

A specific amount of water in the blood is also important for the movement of gases and excretion of solid wastes from the body. The urinary system also balances the amounts of certain salts and water that must be present for all cell activities to take place.

Organs of the Urinary System Excretory organs is another name for the organs of the urinary system. The main organs of the urinary system are two bean-shaped **kidneys.** Kidneys are located on the back wall of the abdomen at about waist level. The kidneys filter blood that contains wastes collected from cells. In approximately 5 min, all of the blood in your body passes through the kidneys. The red-brown color of the kidneys is due to their enormous blood supply. In **Figure 11A,** you can see that blood enters the kidneys through a large artery and leaves through a large vein.

Filtration in the Kidney The kidney, shown in **Figure 11B,** is a two-stage filtration system. It is made up of about 1 million tiny filtering units called **nephrons** (NEF rahnz), shown in **Figure 11C.** Each nephron has a cuplike structure and a tubelike structure called a duct. Blood moves from a renal artery to capillaries in the cuplike structure. The first filtration occurs when water, sugar, salt, and wastes from the blood pass into the cuplike structure. Left behind in the blood are the red blood cells and proteins. Next, liquid in the cuplike structure is squeezed into a narrow tubule. Capillaries that surround the tubule perform the second filtration. Most of the water, sugar, and salt are reabsorbed and returned to the blood. These collection capillaries merge to form small veins, which merge to form a renal vein in each kidney. Purified blood is returned to the main circulatory system. The liquid left behind flows into collecting tubules in each kidney. This wastewater, or urine, contains excess water, salts, and other wastes that are not reabsorbed by the body. An average-sized person produces about 1 L of urine per day.

Mini LAB

Modeling Kidney Function

Procedure

1. Mix a small amount of **soil** and **fine gravel** with **water** in a **clean cup.**
2. Place the **funnel** into a **second cup.**
3. Place a small piece of **wire screen** in the funnel.
4. Carefully pour the mud-water-gravel mixture into the funnel. Let it drain.
5. Remove the screen and replace it with a piece of **filter paper.**
6. Place the funnel in **another clean cup.**
7. Repeat step 4.

Analysis

1. What part of the blood does the gravel represent?
2. How does this experiment model the function of a person's kidneys?

Figure 11
The urinary system removes wastes from the blood.

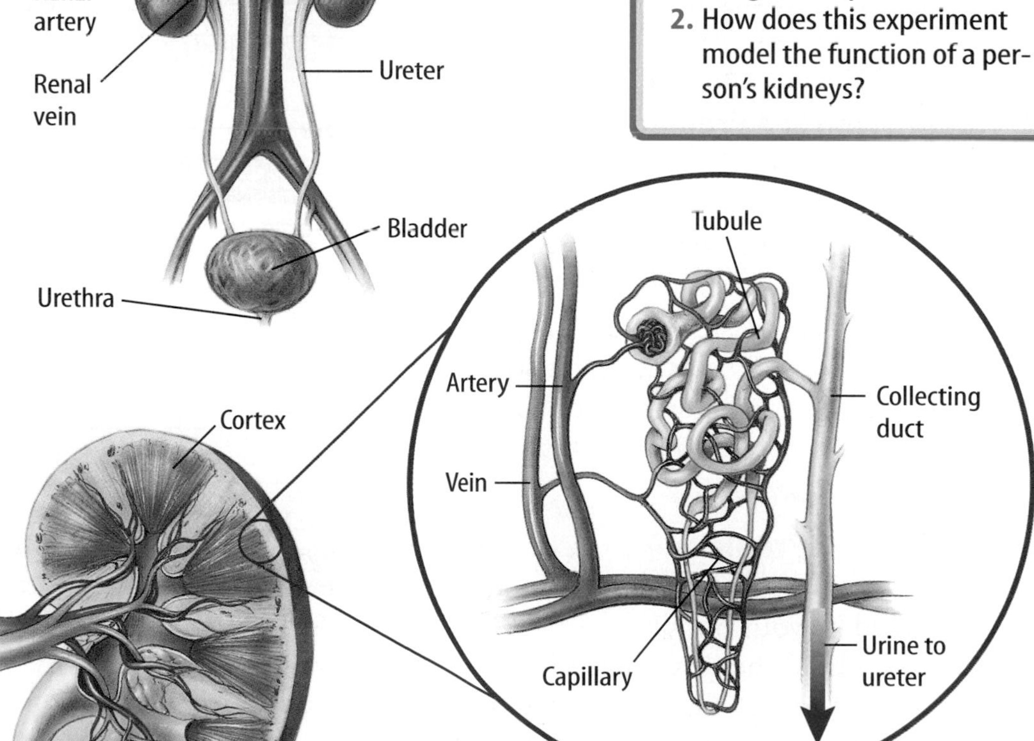

A **The urinary system includes the kidneys, the bladder, and the connecting tubes.**

B **Kidneys are made up of many nephrons.**

C **A single nephron is shown in detail.**
What is the main function of the nephron?

Urine Collection and Release The urine in each collecting tubule drains into a funnel-shaped area of each kidney that leads to the ureter (YER ut ur). **Ureters** are tubes that lead from each kidney to the bladder. The **bladder** is an elastic, muscular organ that holds urine until it leaves the body. The elastic walls of the bladder can stretch to hold up to 0.5 L of urine. When empty, the bladder looks wrinkled and the cells lining the bladder are thick. When full, the bladder looks like an inflated balloon and the cells lining the bladder are stretched and thin. A tube called the **urethra** (yoo REE thruh) carries urine from the bladder to the outside of the body.

Problem-Solving Activity

How does your body gain and lose water?

Your body depends on water. Without water, your cells could not carry out their activities and body systems could not function. Water is so important to your body that your brain and other body systems are involved in balancing water gain and water loss.

Identifying the Problem

Table A shows the major sources by which your body gains water. Oxidation of nutrients occurs when energy is released from nutrients by your body's cells. Water is a waste product of these reactions. **Table B** lists the major sources by which your body loses water. The data show you how daily gain and loss of water are related.

Table A

Major Sources by Which Body Water Is Gained		
Source	**Amount (mL)**	**Percent**
Oxidation of Nutrients	250	10
Foods	750	30
Liquids	1,500	60
Total	**2,500**	**100**

Table B

Major Sources by Which Body Water Is Lost		
Source	**Amount (mL)**	**Percent**
Urine	1,500	60
Skin	500	20
Lungs	350	14
Feces	150	6
Total	**2,500**	**100**

Solving the Problem

1. What is the greatest source of water gained by your body?
2. How would the percentages of water gained and lost change in a person who was working in extremely warm temperatures? In this case, what organ of the body would be the greatest contributor to water loss?

Other Organs of Excretion

Large amounts of liquid wastes are lost every day by your body in other ways, as shown in **Figure 12.** The liver also filters the blood to remove wastes. Certain wastes are converted to other substances. For example, excess amino acids are changed to urea (yoo REE uh), which is a chemical that ends up in urine. Hemoglobin from broken-down red blood cells becomes part of bile, which is the digestive fluid from the liver.

Figure 12
On average, the volume of water lost daily by exhaling is a little more than the volume of a soft-drink can. The volume of water lost by your skin each day is about the volume of a 20-ounce soft-drink bottle.

Urinary Diseases and Disorders

What happens when someone's kidneys don't work properly or stop working? Waste products that are not removed build up and act as poisons in body cells. Water that normally is removed from body tissues accumulates and causes swelling of the ankles and feet. Sometimes these fluids also build up around the heart, and it has to work harder to move blood to the lungs.

Without excretion, an imbalance of salts occurs. The body responds by trying to restore this balance. If the balance isn't restored, the kidneys and other organs can be damaged. Kidney failure occurs when the kidneys don't work as they should. This is always a serious problem because the kidneys' job is so important to the rest of the body.

Infections caused by microorganisms can affect the urinary system. Usually, the infection begins in the bladder. However, it can spread and involve the kidneys. Most of the time, these infections can be cured with antibiotics.

Because the ureters and urethra are narrow tubes, they can be blocked easily in some disorders. A blockage of one of these tubes can cause serious problems because urine cannot flow out of the body properly. If the blockage is not corrected, the kidneys can be damaged.

✔ Reading Check ***Why is a blocked ureter or urethra a serious problem?***

Detecting Urinary Diseases Urine can be tested for any signs of a urinary tract disease. A change in the urine's color can suggest kidney or liver problems. High levels of glucose can be a sign of diabetes. Increased amounts of a protein called albumin (al BYEW mun) indicate kidney disease or heart failure. When the kidneys are damaged, albumin can get into the urine, just as a leaky water pipe allows water to drip.

Nearly 80 percent of Earth's surface is covered by water. Ninety-seven percent of this water is salt water. Humans cannot drink salt water, so they depend on the less than one percent of freshwater that is available for use. In your Science Journal, infer how your kidneys would need to be different for you to be able to drink salt water.

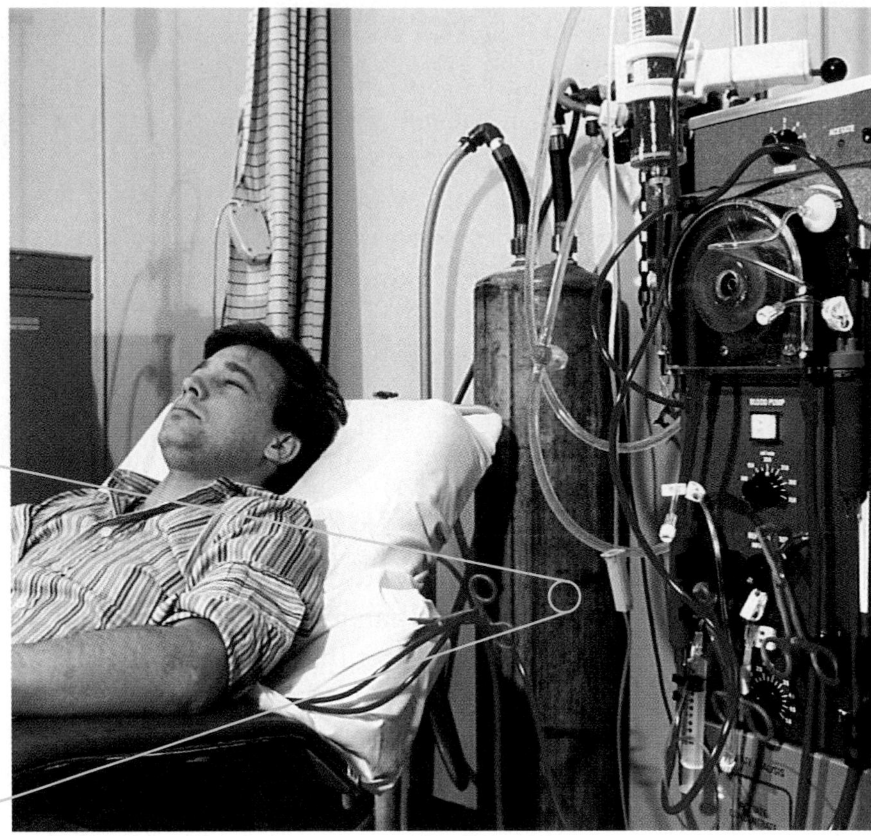

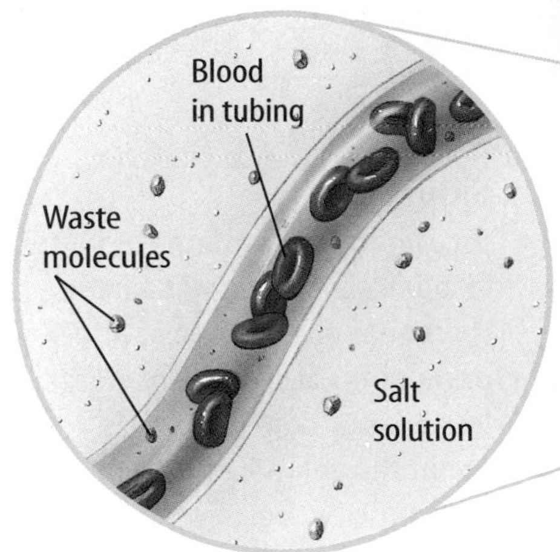

Figure 13
A dialysis machine can replace or help with some of the activities of the kidneys in a person with kidney failure. Like the kidney, the dialysis machine removes wastes from the blood.

Dialysis A person who has only one kidney still can live normally. The remaining kidney increases in size and works harder to make up for the loss of the other kidney. However, if both kidneys fail, the person will need to have his or her blood filtered by an artificial kidney machine in a process called dialysis (di AL uh sus), as shown in **Figure 13.**

Section  Assessment

1. Describe the functions of a person's urinary system.
2. Explain how the kidneys remove wastes and keep fluids and salts in balance.
3. Describe what happens when the urinary system does not function properly.
4. Compare the excretory system and urinary system.
5. **Think Critically** Explain why reabsorption of certain materials in the kidneys is important to your health.

Skill Builder Activities

6. **Concept Mapping** Using a network tree concept map, compare the excretory functions of the kidneys and the lungs. **For more help, refer to the Science Skill Handbook.**
7. **Solving One-Step Equations** In approximately 5 min, all 5 L of blood in the body pass through the kidneys. Calculate the average rate of flow through the kidneys in liters per minute. **For more help, refer to the Math Skill Handbook.**

Activity

Kidney Structure

As your body uses nutrients, wastes are created. One role of the kidneys is to filter waste products out of the bloodstream and excrete this waste outside the body. How can these small structures filter all the blood in the body in 5 min?

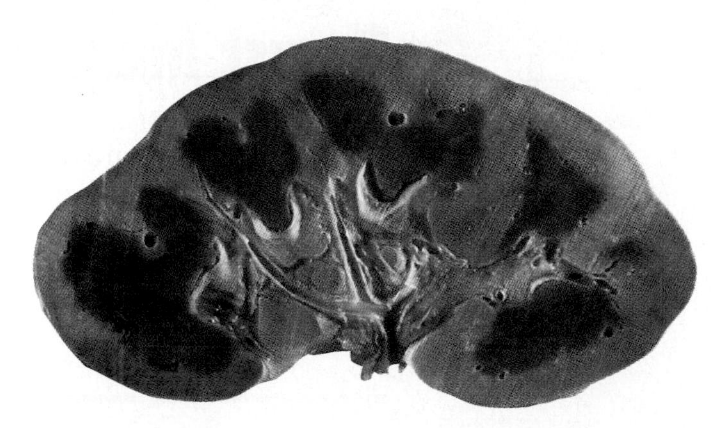

What You'll Investigate

How does the structure of the kidney relate to the function of a kidney?

Materials

large animal kidney
**model of a kidney*
scalpel
hand lens
disposable gloves
**Alternate material*

Goal

- **Observe** the external and internal structures of a kidney.

Safety Precautions

WARNING: *Use extreme care when using sharp instruments. Wear disposable gloves. Wash your hands with soap after completing this activity.*

Procedure

1. **Examine** the outside of the kidney supplied by your teacher.
2. If the kidney still is encased in fat, peel off the fat carefully.
3. Using a scalpel, carefully cut the tissue in half lengthwise around the outline of the kidney. This cut should result in a section similar to the illustration on this page.
4. **Observe** the internal features of the kidney using a hand lens, or view these features in a model.
5. **Compare** the specimen or model with the kidney in the illustration.
6. **Draw** the kidney in your Science Journal and label its structures.

Conclude and Apply

1. What part makes up the cortex of the kidney? Why is this part red?
2. What is the main function of nephrons?
3. The medulla of the kidney is made up of a network of tubules that come together to form the ureter. What is the function of this network of tubules?
4. How can the kidney be compared to a portable water-purifying system?

Communicating Your Data

Compare your conclusions with those of other students in your class. **For more help, refer to the Science Skill Handbook.**

Activity Model and Invent

Simulating the Abdominal Thrust Maneuver

Have you ever taken a class in CPR or learned about how to help a choking victim? Using the abdominal thrust maneuver, or Heimlich maneuver, is one way to remove food or another object that is blocking someone's airway. What happens internally when the maneuver is used? How can you simulate the internal effects of the abdominal thrust maneuver?

Recognize the Problem

How can you simulate the removal of an object from the trachea when the abdominal thrust maneuver is used?

Thinking Critically

What can you use to make a model of the trachea? How can you simulate what happens during an abdominal thrust maneuver using your model?

Goals

- **Construct** a model of the trachea with a piece of food stuck in it.
- **Demonstrate** what happens when the abdominal thrust maneuver is performed on someone.

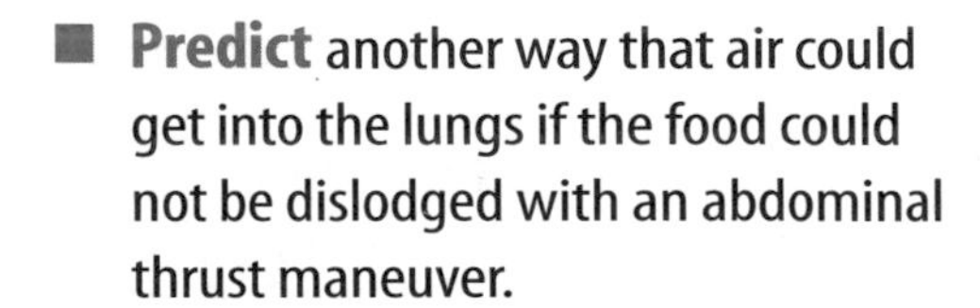

- **Predict** another way that air could get into the lungs if the food could not be dislodged with an abdominal thrust maneuver.

Possible Materials

paper towel roll or other tube
paper (wadded into a ball)
clay
bicycle pump
sports bottle
scissors

Safety Precautions

Always be careful when you use scissors.

Planning the Model

1. **List** the materials that you will need to construct your model. What will represent the trachea and a piece of food or other object blocking the airway?
2. How can you use your model to simulate the effects of an abdominal thrust maneuver?
3. Suggest a way to get air into the lungs if the food could not be dislodged. How would you simulate this method in your model?

Check the Model Plans

1. **Compare** your plans for the model and the abdominal thrust maneuver simulation with those of other students in your class. Discuss why each of you chose the plans and materials that you did.
2. Make sure your teacher approves your plan and materials for your model before you start.

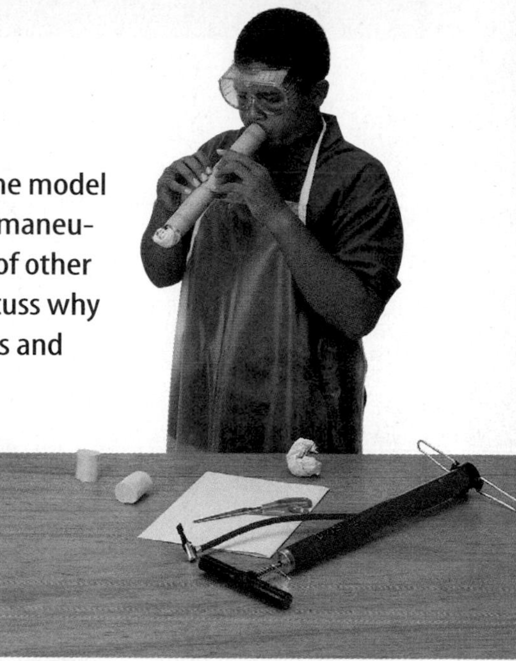

Making the Model

1. **Construct** your model of a trachea with an object stuck in it. Make sure that air cannot get through the trachea if you try blowing softly through it.
2. Simulate what happens when an abdominal thrust maneuver is used. Record your observations. Was the object dislodged? How hard was it to dislodge the object?
3. Replace the object in the trachea. Use your model to simulate how you could get air into the lungs if an abdominal thrust maneuver did not remove the object. Is it easy to blow air through your model now?
4. Model a crushed trachea. Is it easy to blow air through the trachea in this case?

Analyzing and Applying Results

1. **Describe** how easy it was to get air through the trachea in each step in the Making the Model section above. Include any other observations that you made as you worked with your model.
2. Think about what you did to get air into the trachea when the object could not be dislodged with an abdominal thrust maneuver. How could this be done to a person? Do you know what this procedure is called?
3. **Explain** why the trachea has cartilage around it to protect it. What might happen if it did not?

Explain to your family or friends what you have learned about how the abdominal thrust maneuver can help choking victims.

Overcoming the Odds

Guts and determination helped one pioneering doctor to save the lives of thousands

Overcoming the odds—especially when the odds seem stacked against you—is a challenge that many people face. Dr. Samuel Lee Kountz, Jr. (photo, right) had the odds stacked against him. Thanks to his determination he beat them.

Samuel Kountz decided at age eight to become a doctor. He faced his first challenge when he failed the entrance exam to his local Arkansas college. That didn't stop him, though. He asked the college president to give him another chance, and the president did. Kountz got into school and earned As and Bs. Kountz went on to get a graduate degree in biochemistry and was admitted to the University of Arkansas's medical school. For many, these achievements would be more than enough. But for Dr. Kountz, it was just the beginning of his quest to improve medicine—and to change history.

Dr. Kountz was especially interested in a process that was still brand new in the 1950s—the kidney transplant. For many patients, a kidney transplant added months or a year to one's life. But then a patient's body would reject the kidney, and the patient would die. Dr. Kountz was determined to see that kidney transplants saved lives and kept patients healthy for years.

Fixing the Problem

Kountz discovered the root of the problem—why and how a patient's body rejected the transplanted kidney. He discovered that the patient's cells attacked and destroyed the small blood vessels of the transplanted kidney. So the new kidney would die from lack of blood-supplied oxygen. He and others at Stanford University developed a way for doctors to watch the flow of the kidney's blood supply following surgery. Then doctors can give patients the right kinds of drugs at the right time, so that their bodies can overcome the rejection process.

In 1959, Kountz performed the first successful kidney transplant. He went on to develop a procedure to keep body organs healthy for up to 60 hours after being taken from a donor. He also set up a system of organ donor cards through the National Kidney Foundation. And in his career, Dr. Kountz transplanted more than 1,000 kidneys himself—and paved the way for thousands more.

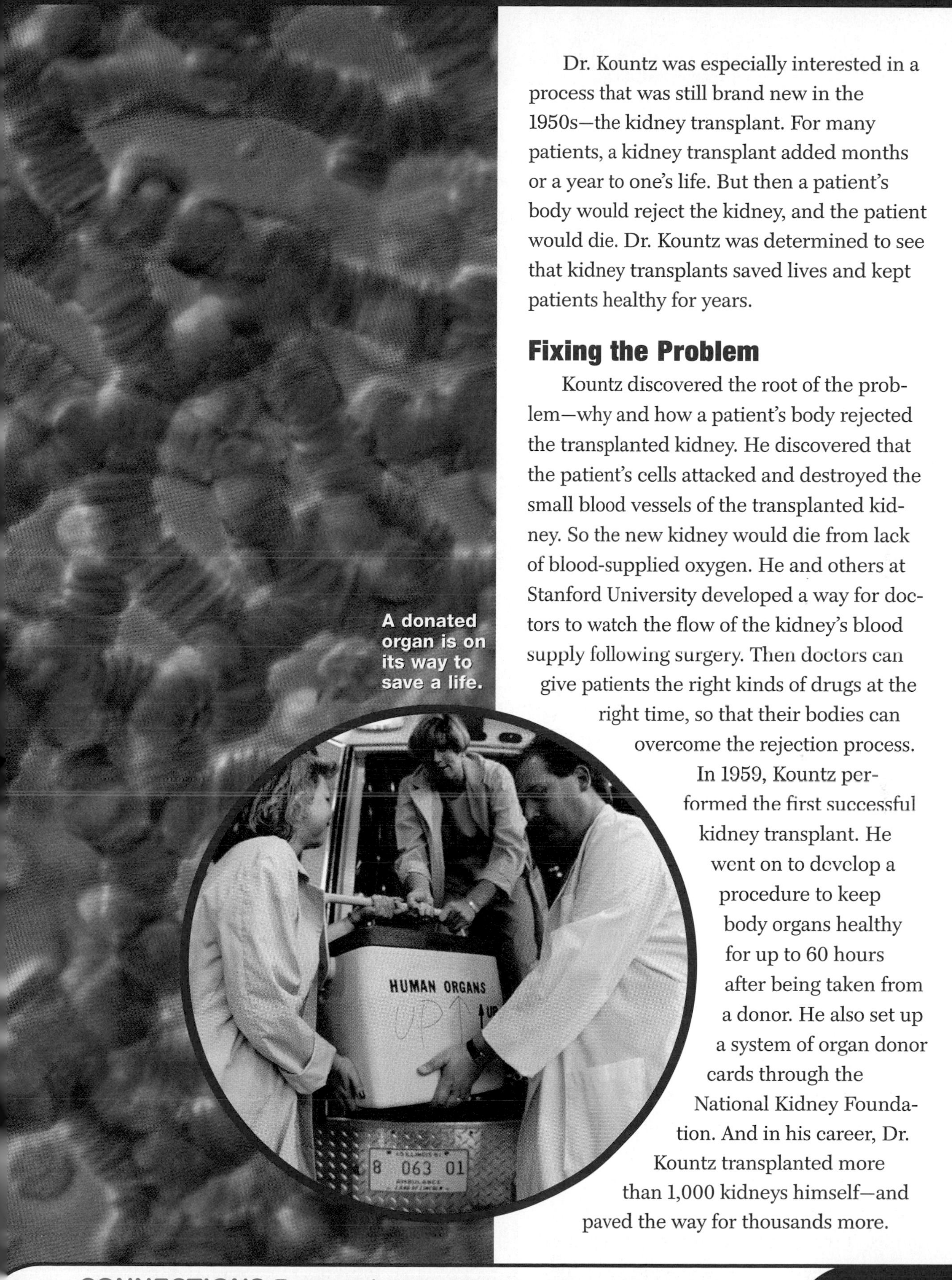

A donated organ is on its way to save a life.

CONNECTIONS Research What kinds of medical breakthroughs has the last century brought? Locate an article that explains either a recent advance in medicine or the work that doctors and medical researchers are doing. Share your findings with your class.

SCIENCE *Online*
For more information, visit science.glencoe.com

Chapter 6 Study Guide

Reviewing Main Ideas

Section 1 The Respiratory System

1. The respiratory system brings oxygen into the body and removes carbon dioxide.
2. Inhaled air passes through the nasal cavity, pharynx, larynx, trachea, bronchi, and into the alveoli of the lungs. *Why does the trachea, shown in the illustration, have cartilage but the esophagus does not?*

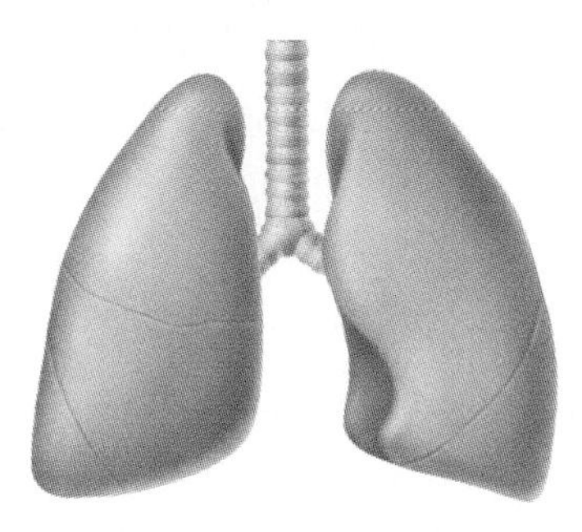

3. Breathing is the movement of the chest that brings air into the lungs and removes waste gases. The chemical reaction in the cells that needs oxygen to release energy from glucose is called cellular respiration.
4. The exchange of oxygen and carbon dioxide happens by the process of diffusion. In the lungs, oxygen diffuses into the capillaries from the alveoli. Carbon dioxide diffuses from the capillaries into the alveoli. In the body tissues, oxygen diffuses from the capillaries into the cells. Carbon dioxide diffuses from the cells into the capillaries.
5. Smoking causes many problems throughout the respiratory system, including chronic bronchitis, emphysema, and lung cancer. *What other body system is affected severely by smoking?*

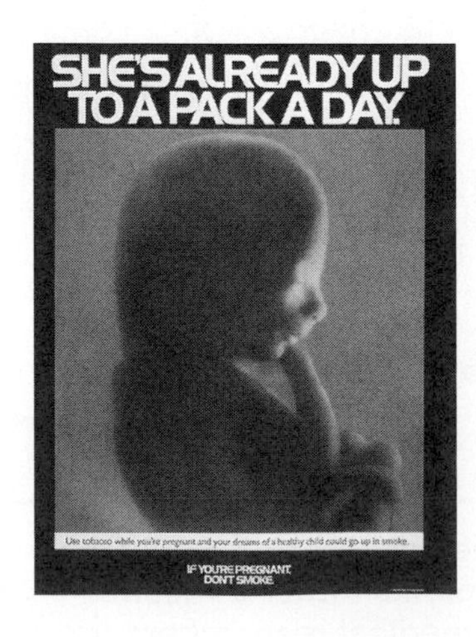

Section 2 The Excretory System

1. The kidneys are the major organs of the urinary system. They filter wastes from all of the blood in the body. *How does a kidney, shown in the photo regulate fluid levels in the body?*

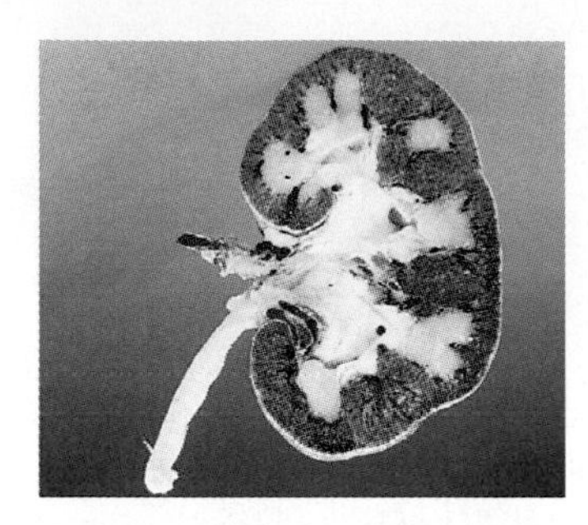

2. The kidney is a two-stage filtration system. The first filtration occurs when water, sugar, salt, and wastes from the blood pass into the cuplike part of the nephron. The capillaries surrounding the tubule part of the nephron perform the second filtration. In this filtration, most of the water, sugar, and salt are reabsorbed and returned to the blood.
3. The urinary system is part of the excretory system. The skin, lungs, liver, and large intestine are also excretory organs.
4. Urine can be tested for signs of urinary tract disease and other diseases.
5. A person who has only one kidney still can live normally. When kidneys fail to work, an artificial kidney can be used to filter the blood in a process called dialysis.

After You Read

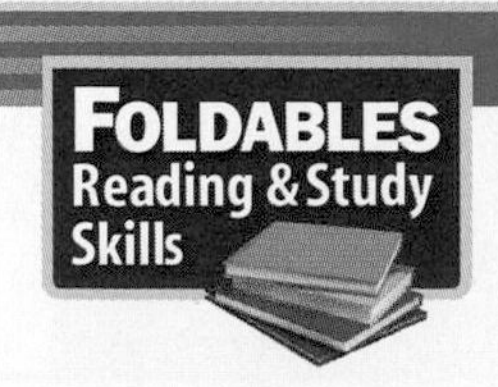

Now that you've read the chapter, write the answer to the *Want* question under the *Learned* tab of your Foldable.

Chapter 6 Study Guide

Visualizing Main Ideas

Complete the following table on the respiratory and excretory systems.

Human Body Systems		
	Respiratory System	Excretory System
Major Organs		
Wastes Eliminated		
Disorders		

Vocabulary Review

Vocabulary Words

a. alveoli
b. asthma
c. bladder
d. bronchi
e. diaphragm
f. emphysema
g. kidney
h. larynx
i. nephron
j. pharynx
k. trachea
l. ureter
m. urethra
n. urinary system
o. urine

Study Tip

Listening is a learning tool, too. Try recording a reading of your notes on tape and replaying it for yourself a few times a week.

Using Vocabulary

For each set of vocabulary words below, explain the relationship that exists.

1. alveoli, bronchi
2. bladder, urine
3. larynx, pharynx
4. ureter, urethra
5. alveoli, emphysema
6. nephron, kidney
7. urethra, bladder
8. asthma, bronchi
9. kidney, urine
10. diaphragm, alveoli

Chapter 6 Assessment

Checking Concepts

Choose the word or phrase that best answers the question.

1. When you inhale, which of the following contracts and moves down?
A) bronchioles **C)** nephrons
B) diaphragm **D)** kidneys

2. Air is moistened, filtered, and warmed in which of the following structures?
A) larynx **C)** nasal cavity
B) pharynx **D)** trachea

3. Exchange of gases occurs between capillaries and which of the following structures?
A) alveoli **C)** bronchioles
B) bronchi **D)** trachea

4. Which of the following is a lung disorder that can occur as an allergic reaction?
A) asthma **C)** emphysema
B) atherosclerosis **D)** cancer

5. When you exhale, which way does the rib cage move?
A) up **C)** out
B) down **D)** stays the same

6. Which of the following conditions does smoking worsen?
A) arthritis **C)** excretion
B) respiration **D)** emphysema

7. Urine is held temporarily in which of the following structures?
A) kidneys **C)** ureter
B) bladder **D)** urethra

8. What are the filtering units of the kidneys?
A) nephrons **C)** neurons
B) ureters **D)** alveoli

9. Approximately 1 L of water is lost per day through which of the following?
A) sweat **C)** urine
B) lungs **D)** large intestine

10. Which of the following substances is not reabsorbed by blood after it passes through the kidneys?
A) salt **C)** wastes
B) sugar **D)** water

Thinking Critically

11. Explain why certain foods, such as peanuts, can cause choking in small children.

12. Why is it an advantage to have lungs with many smaller air sacs instead of having just two large sacs, like balloons?

13. Explain the damage to cilia, alveoli, and lungs from smoking.

14. What happens to the blood if the kidneys stop working?

15. Small, solid particles called kidney stones can form in the kidneys. Explain why it is often painful when a kidney stone passes into the ureter.

Developing Skills

16. Interpreting Data Study the data below. How much of each substance is reabsorbed into the blood in the kidneys? What substance is excreted completely in the urine?

Materials Filtered by the Kidneys

Substance Filtered in Urine	Amount Moving Through Kidney	Amount Excreted
Water	125 L	1 L
Salt	350 g	10 g
Urea	1 g	1 g
Glucose	50 g	0 g

17. Recognizing Cause and Effect Discuss how lack of oxygen is related to lack of energy.

18. Making and Using Graphs Make a circle graph of total lung capacity using the following data:

- volume of air in a normal inhalation or exhalation = 500 mL
- volume of additional air that can be inhaled forcefully after a normal inhalation = 3,000 mL
- volume of additional air that can be exhaled forcefully after a normal expiration = 1,100 mL
- volume of air still left in the lungs after all the air that can be exhaled has been forcefully exhaled = 1,200 mL

19. Forming Hypotheses Make a hypothesis about the number of breaths a person might take per minute in each of these situations: asleep, exercising, and on top of Mount Everest. Give a reason for each hypothesis.

20. Concept Mapping Make an events chain concept map showing how urine forms in the kidneys. Begin with, "In the nephron …"

Performance Assessment

21. Questionnaire and Interview Prepare a questionnaire that can be used to interview a health specialist who works with lung cancer patients. Include questions on reasons for choosing the career, new methods of treatment, and the most encouraging or discouraging part of the job.

Technology

Go to the Glencoe Science Web site at **science.glencoe.com** or use the **Glencoe Science CD-ROM** for additional chapter assessment.

Test Practice

For one week, research scientists collected and accurately measured the amount of body water lost and gained per day for four different patients. They placed their results in the following table.

Body Water Gained (+) and Lost (–)

Person	Day 1 (L)	Day 2 (L)	Day 3 (L)	Day 4 (L)
Mr. Stoler	+0.05	+0.15	−0.35	+0.12
Mr. Jemma	−0.01	0.00	−0.20	−0.01
Mr. Lowe	0.00	+0.10	−0.28	+0.01
Mr. Cheng	−0.50	−0.50	−0.55	−0.32

Study the table and answer the following questions.

1. According to this information, which patient may be suffering from dehydration or an excessive amount of body water loss?
A) Mr. Stoler
B) Ms. Jemma
C) Mr. Lowe
D) Mr. Cheng

2. According to the table, it was probably very hot in each patient's hospital room during _____.
F) day one
G) day two
H) day three
J) day four

CHAPTER 7

Animal Behavior

Eye contact is made, dirt flies, and the silence is shattered. Massive horns clash as two bighorn sheep butt heads. Nearby, a spider spins a web to catch its food. Overhead, the honking of a V-shaped string of geese echoes through the valley. Do organisms learn these actions or do they occur automatically? In this chapter, you will examine the unique behaviors of animals. Also, you'll read about different types of behavior and learn about animal communication.

What do you think?

Science Journal Look at the picture below with a classmate. Discuss what you think this might be or what is happening. Here's a hint: *This instinctive reaction is triggered by their parent's arrival.* Write your answer or best guess in your Science Journal.

One way you communicate is by speaking. Other animals communicate without the use of sound. For example, a gull chick pecks at its parent's beak to get food. Try the activity below to see if you can communicate without speaking.

Observe how humans communicate without using sound

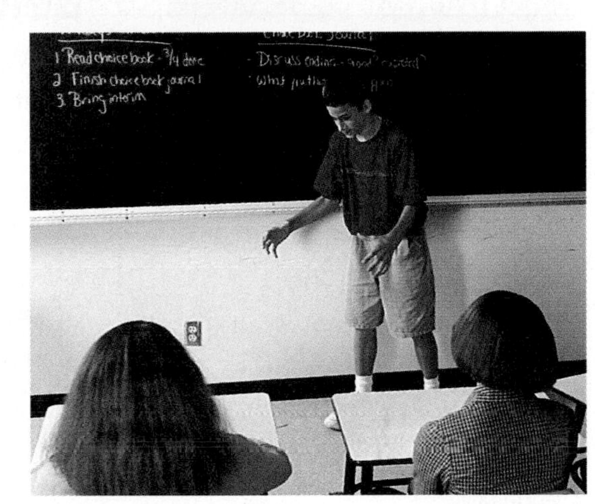

1. Form groups of students. Have one person choose an object and describe that object using gestures.
2. The other students observe and try to identify the object that is being described.
3. Each student in the group should choose an object and describe it without speaking while the others observe and identify the object.

Observe

In your Science Journal, describe how you and the other students were able to communicate without speaking to one another.

Before You Read

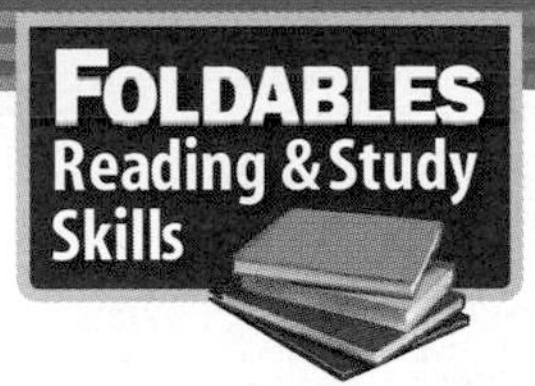

Making a Compare and Contrast Study Fold As you study behaviors, make the following Foldable to help find the similarities and differences between the behaviors of two animals.

1. Place a sheet of paper in front of you so the short side is at the top. Fold the paper in half from the left to the right side. Fold top to bottom but do not crease. Then unfold.
2. Label *Observed Behaviors of Animal 1* and *Observed Behaviors of Animal 2* across the front of the paper, as shown.
3. Through one thickness of paper, cut along the middle fold line to form two tabs, as shown.
4. Before you read the chapter, choose two animals to compare.
5. As you read the chapter, list the behaviors you learn about Animal 1 and Animal 2 under the tabs.

Types of Behavior

As You Read

What You'll Learn

- **Identify** the differences between innate and learned behavior.
- **Explain** how reflexes and instincts help organisms survive.
- **Identify** examples of imprinting and conditioning.

Vocabulary

behavior
innate behavior
reflex
instinct
imprinting
conditioning
insight

Why It's Important

Innate behavior helps you survive on your own.

Behavior

When you come home from school, does your dog run to meet you? Your dog barks and wags its tail as you scratch behind its ears. Sitting at your feet, it watches every move you make. Why do dogs do these things? In nature, dogs are pack animals that generally follow a leader. They have been living with people for about 12,000 years. Domesticated dogs treat people as part of their own pack, as shown in **Figure 1B.**

Animals are different from one another in their behavior. They are born with certain behaviors, and they learn others. **Behavior** is the way an organism interacts with other organisms and its environment. Anything in the environment that causes a reaction is called a stimulus. A stimulus can be external, such as a rival male entering another male's territory, or internal, such as hunger or thirst. You are the stimulus that causes your dog to bark and wag its tail. Your dog's reaction to you is a response.

Figure 1
Dogs are pack animals by nature. A This pack of wild dogs must work together to survive. B This domesticated dog has accepted a human as its leader.

A

B

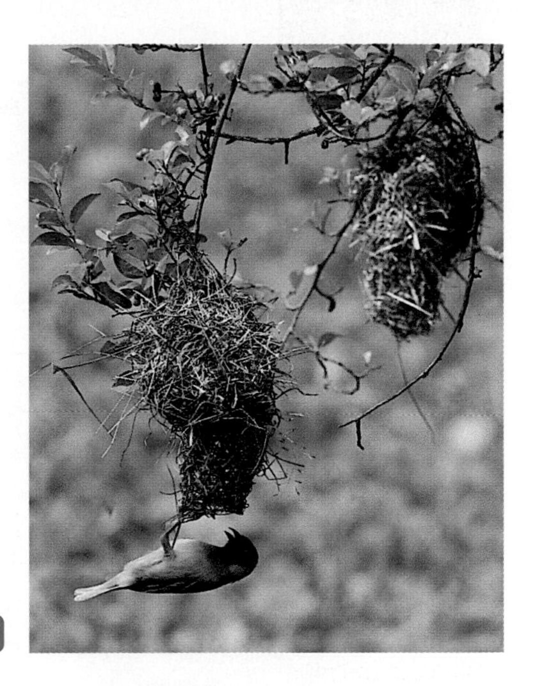

Figure 2
Bird nests come in different sizes and shapes. A Cliff swallows build nests out of mud. B Hummingbirds build delicate cup-shaped nests on branches of trees. C This male weaverbird is knotting the ends of leaves together to secure the nest.

Innate Behavior

A behavior that an organism is born with is called an **innate behavior.** These types of behaviors are inherited. They don't have to be learned.

Innate behavior patterns occur the first time an animal responds to a particular internal or external stimulus. For birds like the swallows in **Figure 2A** and the hummingbird in **Figure 2B** building a nest is innate behavior. When it's time for the female weaverbird to lay eggs, the male weaverbird builds an elaborate nest, as shown in **Figure 2C.** Although a young male's first attempt may be messy, the nest is constructed correctly.

The behavior of animals that have short life spans is mostly innate behavior. Most insects do not learn from their parents. In many cases, the parents have died or moved on by the time the young hatch. Yet every insect reacts innately to its environment. A moth will fly toward a light, and a cockroach will run away from it. They don't learn this behavior. Innate behavior allows animals to respond instantly. This quick response often means the difference between life and death.

Reflexes The simplest innate behaviors are reflex actions. A **reflex** is an automatic response that does not involve a message from the brain. Sneezing, shivering, yawning, jerking your hand away from a hot surface, and blinking your eyes when something is thrown toward you are all reflex actions.

In humans a reflex message passes almost instantly from a sense organ along the nerve to the spinal cord and back to the muscles. The message does not go to the brain. You are aware of the reaction only after it has happened. Your body reacts on its own. A reflex is not the result of conscious thinking.

Health INTEGRATION

A tap on a tendon in your knee causes your leg to stretch. This is known as the knee-jerk reflex. Abnormalities in this reflex tell doctors of a possible problem in the central nervous system. Research other types of reflexes and write a report about them in your Science Journal.

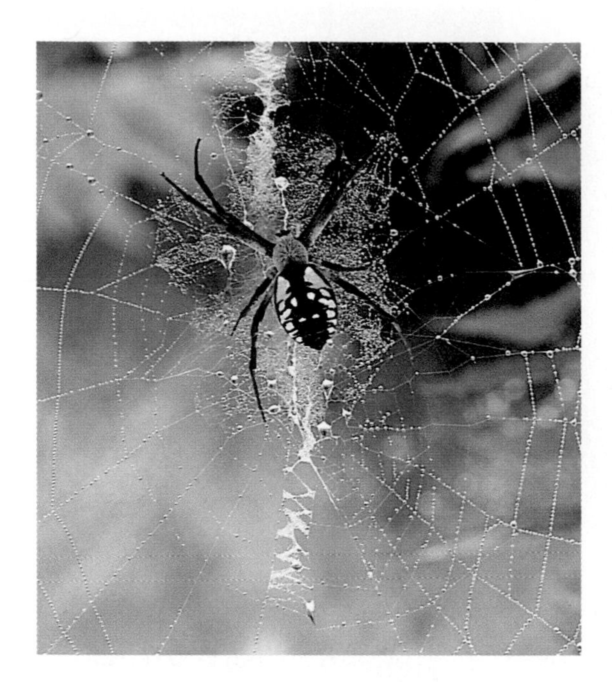

Figure 3
Spiders, like this orb weaver spider, know how to spin webs as soon as they hatch.

Instincts An **instinct** is a complex pattern of innate behavior. Spinning a web like the one in **Figure 3** is complicated, yet spiders spin webs correctly on the first try. Unlike reflexes, instinctive behaviors can take weeks to complete. Instinctive behavior begins when the animal recognizes a stimulus and continues until all parts of the behavior have been performed.

✔ **Reading Check** ***What is the difference between a reflex and an instinct?***

Learned Behavior

All animals have innate and learned behaviors. Learned behavior develops during an animal's lifetime. Animals with more complex brains exhibit more behaviors that are the result of learning. However, the behavior of insects, spiders, and other arthropods is mostly instinctive behavior. Fish, reptiles, amphibians, birds, and mammals all learn. Learning is the result of experience or practice.

Learning is important for animals because it allows them to respond to changing situations. In changing environments, animals that have the ability to learn a new behavior are more likely to survive. This is especially important for animals with long life spans. The longer an animal lives, the more likely it is that the environment in which it lives will change.

Learning also can modify instincts. For example, grouse and quail chicks, shown in **Figure 4,** leave their nests the day they hatch. They can run and find food, but they can't fly. When something moves above them, they instantly crouch and keep perfectly still until the danger has passed. They will crouch without moving even if the falling object is only a leaf. Older birds have learned that leaves will not harm them, but they freeze when a hawk moves overhead.

Figure 4
As they grow older, these quail chicks will learn which organisms to avoid. *Why is it important for young quail to react the same toward all organisms?*

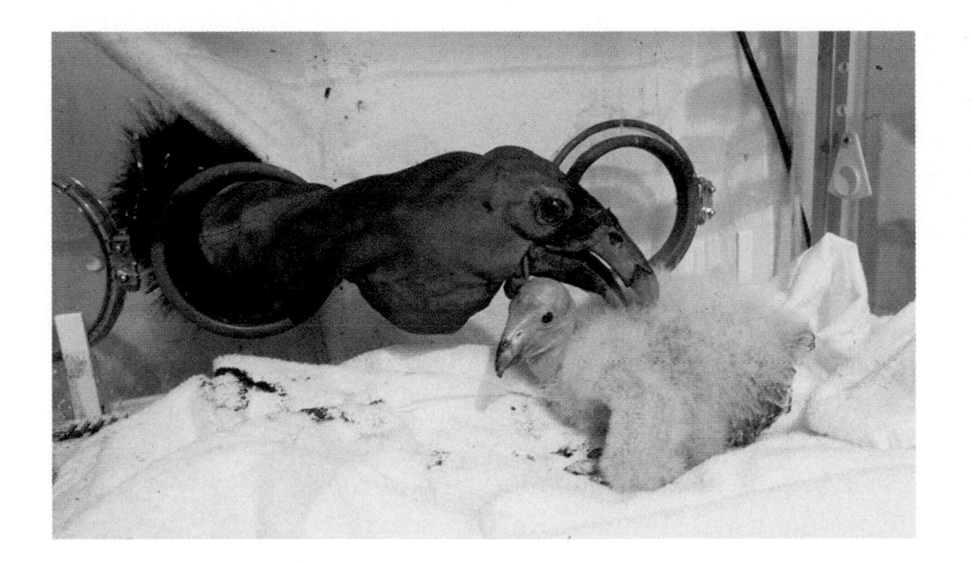

Figure 5
When feeding chicks in captivity, puppets of adult condors are used so the chicks don't associate humans with food.

Research Visit the Glencoe Science Web site at **science.glencoe.com** for the latest information about raising condors to be released into the wild. Communicate to your class what you learn.

Imprinting Learned behavior includes imprinting, trial and error, conditioning, and insight. Have you ever seen young ducks following their mother? This is an important behavior because the adult bird has had more experience in finding food, escaping predators, and getting along in the world. **Imprinting** occurs when an animal forms a social attachment, like the condor in **Figure 5,** to another organism within a specific time period after birth or hatching.

Konrad Lorenz, an Austrian naturalist, developed the concept of imprinting. Working with geese, he discovered that a gosling follows the first moving object it sees after hatching. The moving object, whatever it is, is imprinted as its parent. This behavior works well when the first moving object a gosling sees is an adult female goose. But goslings hatched in an incubator might see a human first and imprint on him or her. Animals that become imprinted toward animals of another species have difficulty recognizing members of their own species.

Figure 6
Were you able to tie your shoes on the first attempt? *What other things do you do every day that required learning?*

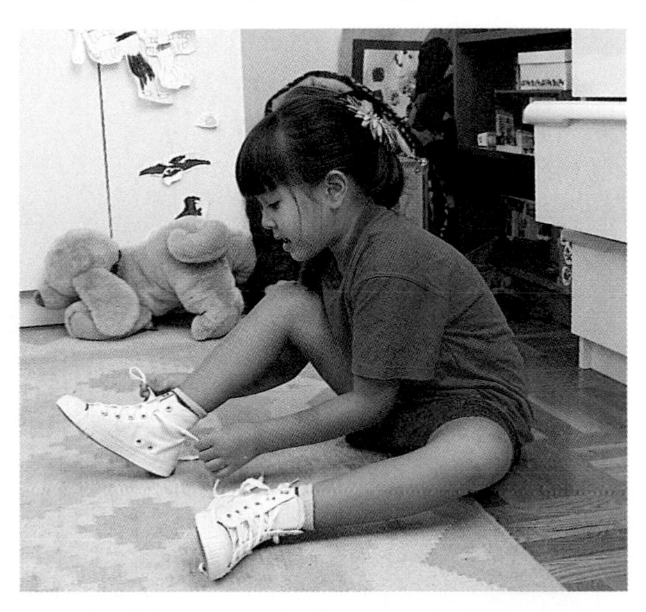

Trial and Error Can you remember when you learned to ride a bicycle? You probably fell many times before you learned how to balance on the bicycle. After a while you could ride without having to think about it. You have many skills that you have learned through trial and error such as feeding yourself and tying your shoes, as shown in **Figure 6.**

Behavior that is modified by experience is called trial-and-error learning. Many animals learn by trial and error. When baby chicks first try feeding themselves, they peck at many stones before they get any food. As a result of trial and error, they learn to peck only at food particles.

Mini LAB

Observing Conditioning

Procedure

1. Obtain several **photos of different foods and landscapes** from your teacher.
2. Show each picture to a classmate for 20 s.
3. Record how each photo made your partner feel.

Analysis

1. How did your partner feel after looking at the photos of food?
2. What effect did the landscape pictures have on your partner?
3. Infer how advertising might condition consumers to buy specific food products.

Conditioning Do you have an aquarium in your school or home? If you put your hand above the tank, the fish probably will swim to the top of the tank expecting to be fed. They have learned that a hand shape above them means food. What would happen if you tapped on the glass right before you fed them? After a while the fish probably will swim to the top of the tank if you just tap on the glass. Because they are used to being fed after you tap on the glass, they associate the tap with food.

Animals often learn new behaviors by conditioning. In **conditioning,** behavior is modified so that a response to one stimulus becomes associated with a different stimulus. There are two types of conditioning. One type introduces a new stimulus before the usual stimulus. Russian scientist Ivan P. Pavlov performed experiments with this type of conditioning. He knew that the sight and smell of food made hungry dogs secrete saliva. Pavlov added another stimulus. He rang a bell before he fed the dogs. The dogs began to connect the sound of the bell with food. Then Pavlov rang the bell without giving the dogs food. They salivated when the bell was rung even though he did not show them food. The dogs, like the one in **Figure 7,** were conditioned to respond to the bell.

In the second type of conditioning, the new stimulus is given after the affected behavior. Getting an allowance for doing chores is an example of this type of conditioning. You do your chores because you want to receive your allowance. You have been conditioned to perform an activity that you may not have done if you had not been offered a reward.

Reading Check *How does conditioning modify behavior?*

Figure 7
In Pavlov's experiment, a dog was conditioned to salivate when a bell was rung. It associated the bell with food.

Insight How does learned behavior help an animal deal with a new situation? Suppose you have a new math problem to solve. Do you begin by acting as though you've never seen it before, or do you use what you have learned previously in math to solve the problem? If you use what you have learned, then you have used a kind of learned behavior called insight. **Insight** is a form of reasoning that allows animals to use past experiences to solve new problems. In experiments with chimpanzees, as shown in **Figure 8,** bananas were placed out of the chimpanzees' reach. Instead of giving up, they piled up boxes found in the room, climbed them, and reached the bananas. At some time in their lives, the chimpanzees must have solved a similar problem. The chimpanzees demonstrated insight during the experiment. Much of adult human learning is based on insight. When you were a baby, you learned by trial and error. As you grow older, you will rely more on insight.

Figure 8
This illustration shows how chimpanzees may use insight to solve problems.

Section  Assessment

1. How is innate behavior different from learned behavior?
2. Compare a reflex with an instinct.
3. What is the difference between an internal and external stimulus?
4. Compare imprinting and conditioning.
5. **Think Critically** Use what you know about conditioning to explain how the term *mouthwatering food* might have come about.

Skill Builder Activities

6. **Researching Information** How are dogs trained to sniff out certain substances? **For more help, refer to the Science Skill Handbook.**
7. **Using an Electronic Spreadsheet** Make a spreadsheet of the behaviors in this section. Sort the behaviors according to whether they are innate or learned behaviors. Then identify the type of innate or learned behavior. **For more help, refer to the Technology Skill Handbook.**

SECTION 2

Behavioral Interactions

As You Read

What You'll Learn

- **Explain** why behavioral adaptations are important.
- **Describe** how courtship behavior increases reproductive success.
- **Explain** the importance of social behavior and cyclic behavior.

Vocabulary

social behavior
society
aggression
courtship behavior
pheromone
cyclic behavior
hibernation
migration

Why It's Important

Organisms must be able to communicate with each other to survive.

Instinctive Behavior Patterns

Complex interactions of innate behaviors between organisms result in many types of animal behavior. For example, courtship and mating within most animal groups are instinctive ritual behaviors that help animals recognize possible mates. Animals also protect themselves and their food sources by defending their territories. Instinctive behavior, just like natural hair color, is inherited.

Social Behavior

Animals often live in groups. One reason, shown in **Figure 9,** is that large numbers provide safety. A lion is less likely to attack a herd of zebras than a lone zebra. Sometimes animals in large groups help keep each other warm. Also, migrating animal groups are less likely to get lost than animals that travel alone.

Interactions among organisms of the same species are examples of **social behavior.** Social behaviors include courtship and mating, caring for the young, claiming territories, protecting each other, and getting food. These inherited behaviors provide advantages that promote survival of the species.

Reading Check *Why is social behavior important?*

Figure 9
When several zebras are close together their stripes make it difficult for predators to pick out one individual.

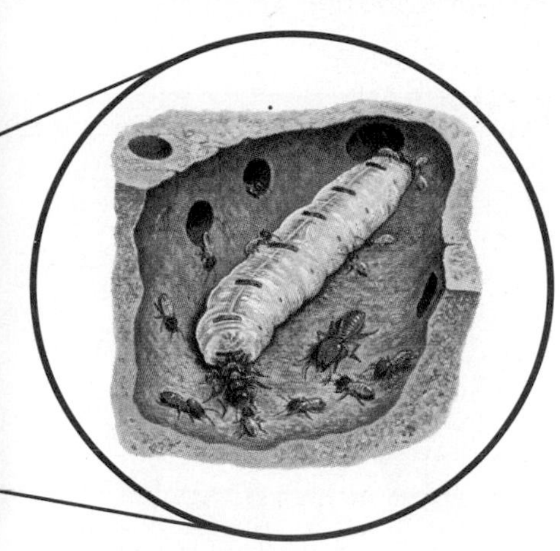

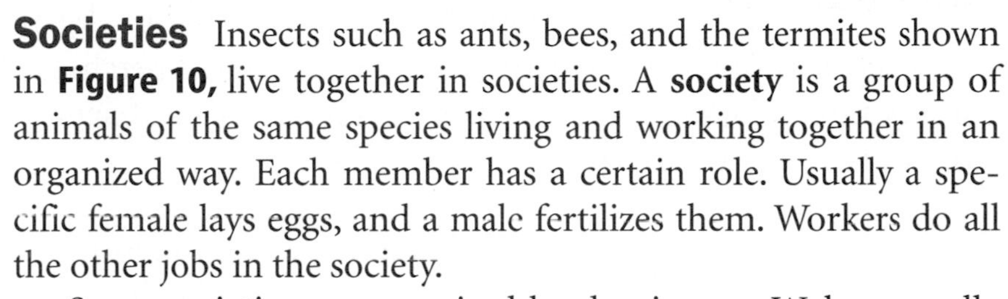

Figure 10
Termites built this large mound in Australia. The mound has a network of tunnels and chambers for the queen to deposit eggs into.

Societies Insects such as ants, bees, and the termites shown in **Figure 10,** live together in societies. A **society** is a group of animals of the same species living and working together in an organized way. Each member has a certain role. Usually a specific female lays eggs, and a male fertilizes them. Workers do all the other jobs in the society.

Some societies are organized by dominance. Wolves usually live together in packs. A wolf pack has a dominant female. The top female controls the mating of the other females. If plenty of food is available, she mates and then allows the others to do so. If food is scarce, she allows less mating. During such times, she is usually the only one to mate.

Territorial Behavior

Many animals set up territories for feeding, mating, and raising young. A territory is an area that an animal defends from other members of the same species. Ownership of a territory occurs in different ways. Songbirds sing, sea lions bellow, and squirrels chatter to claim territories. Other animals leave scent marks. Some animals, like the tiger in **Figure 11,** patrol an area and attack other animals of the same species who enter their territory. Why do animals defend their territories? Territories contain food, shelter, and potential mates. If an animal has a territory, it will be able to mate and produce offspring. Defending territories is an instinctive behavior. It improves the survival rate of an animal's offspring.

Figure 11
A tiger's territory may include several miles. It will confront any other tiger who enters it.

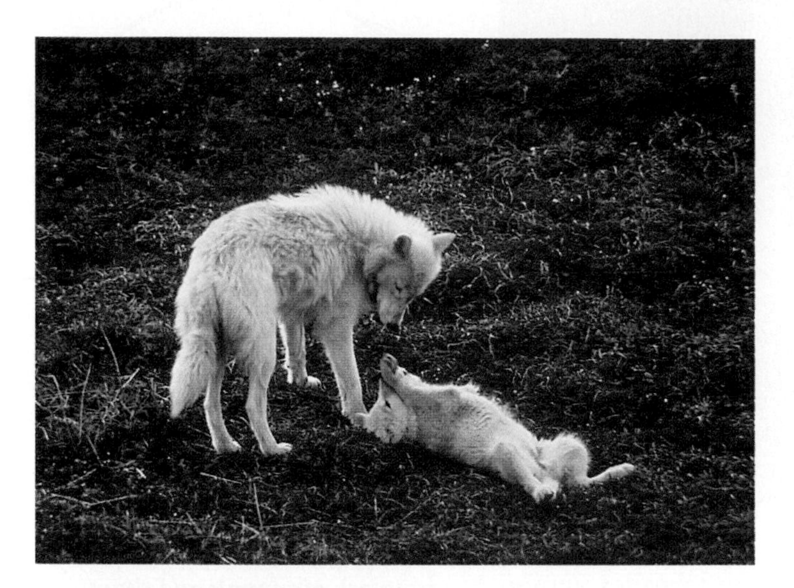

Figure 12
Young wolves roll over and make themselves as small as possible to show their submission to adult wolves.

Aggression Have you ever watched as one dog approached another dog that was eating a bone? What happened to the appearance of the dog with the bone? Did its hair on its back stick up? Did it curl its lips and make growling noises? This behavior is aggression. **Aggression** is a forceful behavior used to dominate or control another animal. Fighting and threatening are aggressive behaviors animals use to defend their territories, protect their young, or to get food.

Many animals demonstrate aggression. Some birds let their wings droop below their tail feathers. It may take another bird's perch and thrust its head forward in a pecking motion as a sign of aggression. Cats lay their ears flat, arch their backs, and hiss.

Submission Animals of the same species seldom fight to the death. Teeth, beaks, claws, and horns are used for killing prey or for defending against members of a different species.

To avoid being attacked and injured by an individual of its own species, an animal shows submission. Postures that make an animal appear smaller often are used to communicate surrender. In some animal groups, one individual is usually dominant. Members of the group show submissive behavior toward the dominant individual. This stops further aggressive behavior by the dominant animal. Young animals also display submissive behaviors toward parents or dominant animals, as shown in **Figure 12.**

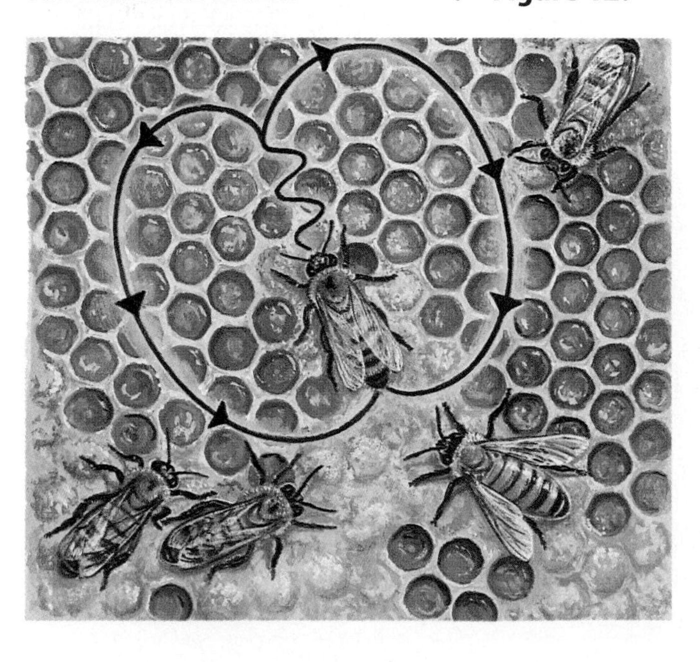

Figure 13
During the waggle dance, if the source is far from the hive, the dance takes the form of a figure eight. The angle of the waggle is equal to the angle from the hive between the Sun and nectar source.

Communication

In all social behavior, communication is important. Communication is an action by a sender that influences the behavior of a receiver. How do you communicate with the people around you? You may talk, make noises, or gesture like you did in this chapter's Explore Activity. Honeybees perform a dance, as shown in **Figure 13,** to communicate to other bees in the hive where a food source is. Animals in a group communicate with sounds, scents, and actions. Alarm calls, chemicals, speech, courtship behavior, and aggression are forms of communication.

Figure 14
This male Emperor of Germany bird of paradise attracts mates by posturing and fanning its tail.

Courtship Behavior A male bird of paradise, shown in **Figure 14,** spreads its tail feathers and struts. A male sage grouse fans its tail, fluffs its feathers, and blows up its two red air sacs. These are examples of behavior that animals perform before mating. This type of behavior is called **courtship behavior.** Courtship behaviors allow male and female members of a species to recognize each other. These behaviors also stimulate males and females so they are ready to mate at the same time. This helps ensure reproductive success.

In most species the males are more colorful and perform courtship displays to attract a mate. Some courtship behaviors allow males and females to find each other across distances.

Chemical Communication Ants are sometimes seen moving single file toward a piece of food. Male dogs frequently urinate on objects and plants. Both behaviors are based on chemical communication. The ants have laid down chemical trails that others of their species can follow. The dog is letting other dogs know he has been there. In these behaviors, the animals are using a chemical called a pheromone to communicate. A **pheromone** (FER uh mohn) is a chemical that is produced by one animal to influence the behavior of another animal of the same species. They are powerful chemicals needed only in small amounts. They remain in the environment so that the sender and the receiver can communicate without being in the same place at the same time. They can advertise the presence of an animal to predators, as well as to the intended receiver of the message.

Males and females use pheromones to establish territories, warn of danger, and attract mates. Certain ants, mice, and snails release alarm pheromones when injured or threatened.

Demonstrating Chemical Communication

Procedure 

1. Obtain a sample of perfume or air freshener.
2. Spray it into the air to leave a scent trail as you move around the house or apartment to a hiding place.
3. Have someone try to discover where you are by following the scent of the substance.

Analysis

1. What was the difference between the first and last room you were in?
2. Would this be an efficient way for humans to communicate? Explain.

Figure 15
Many animals use sound to communicate.

A Frogs often croak loud enough to be heard far away.

B Pileated woodpecker calls often can be heard above everything else in the forest.

C Howler monkeys got their name because of the sounds they make.

Sound Communication Male crickets rub one forewing against the other forewing. This produces chirping sounds that attract females. Each cricket species produces several calls that are different from other cricket species. These calls are used by researchers to identify different species. Male mosquitoes have hairs on their antennae that sense buzzing sounds produced by females of their same species. The tiny hairs vibrate only to the frequency emitted by a female of the same species.

Vertebrates use a number of different forms of sound communication. Rabbits thump the ground, gorillas pound their chests, beavers slap the water with their flat tails, and frogs, like the one in **Figure 15,** croak. Do you think that sound communication in noisy environments is useful? Seabirds that live where waves pound the shore rather than in some quieter place must rely on visual signals, not sound, for communication.

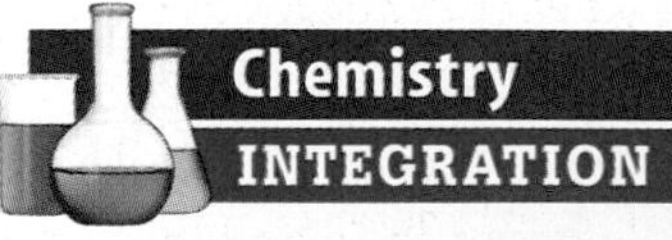

Chemistry INTEGRATION

The light produced by fireflies is a particle of visible light that radiates when chemicals produce a high-energy state and then return to their normal state. Hypothesize how this helps fireflies survive. Write your hypothesis in your Science Journal.

Light Communication Certain kinds of flies, marine organisms, and beetles have a special form of communication called bioluminescence. Bioluminescence, shown in **Figure 16,** is the ability of certain living things to give off light. This light is produced through a series of chemical reactions in the organism's body. Probably the most familiar bioluminescent organisms in North America are fireflies. They are not flies, but beetles. The flash of light is produced on the underside of the last abdominal segments and is used to locate a prospective mate. Each species has its own characteristic flashing. Males fly close to the ground and emit flashes of light. Females must flash an answer at exactly the correct time to attract males.

NATIONAL GEOGRAPHIC **VISUALIZING BIOLUMINESCENCE**

Figure 16

Many marine organisms use bioluminescence as a form of communication. This visible light is produced by a chemical reaction and often confuses predators or attracts mates. Each organism on this page is shown in its normal and bioluminescent state.

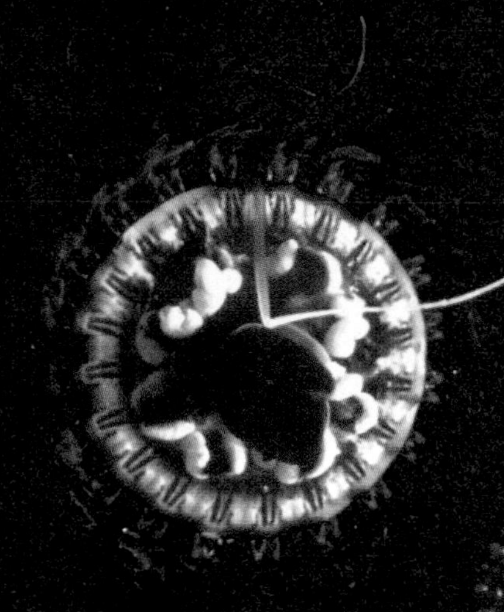

▲ JELLYFISH This jellyfish lights up like a neon sign when it is threatened.

▼ KRILL The blue dots shown below this krill are all that are visible when krill bioluminesce. The krill may use bioluminescence to confuse predators.

▲ DEEP-SEA SEA STAR The sea star uses light to warn predators of its unpleasant taste.

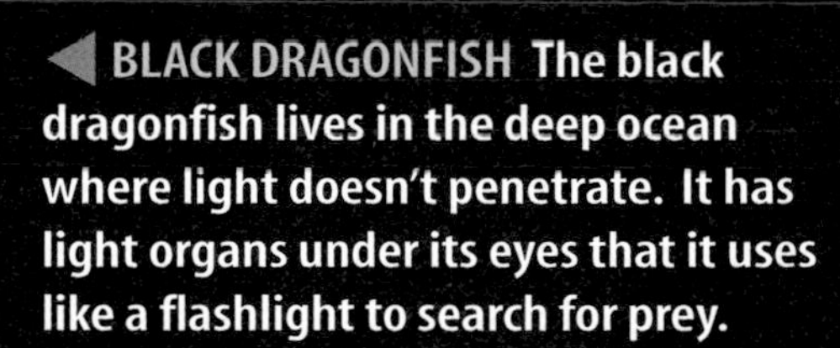

◀ BLACK DRAGONFISH The black dragonfish lives in the deep ocean where light doesn't penetrate. It has light organs under its eyes that it uses like a flashlight to search for prey.

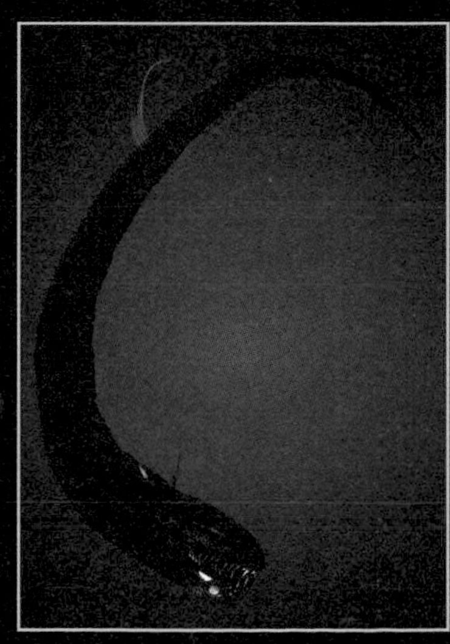

Activity Model and Invent

Animal Habitats

Zoos, animal parks, and aquariums are safe places for wild animals. Years ago, captive animals were kept in small cages or behind glass windows. Almost no attempt was made to provide natural habitats for the animals. People who came to see the animals could not observe the animal's normal behavior. Now, most captive animals are kept in exhibit areas that closely resemble their natural habitats. These areas provide suitable environments for the animals so that they can interact with members of their same species and have healthier, longer lives.

Recognize the Problem

What types of environments are best suited for raising animals in captivity?

Thinking Critically

How can the habitats provided at an animal park affect the behavior of animals?

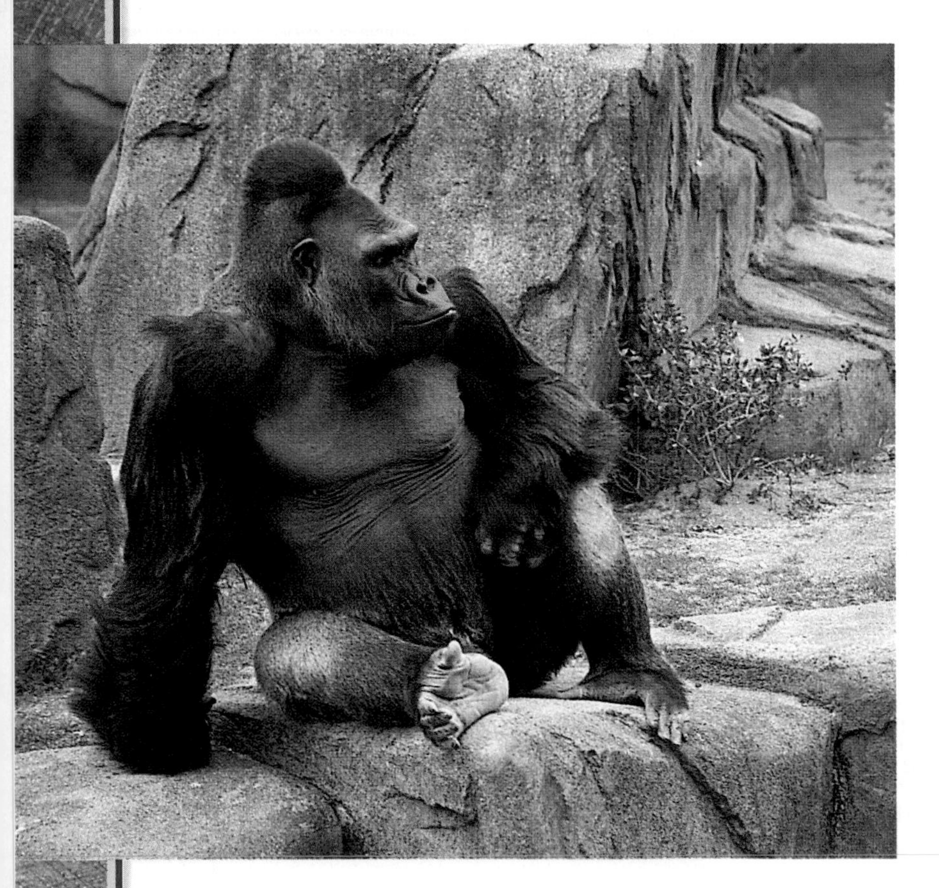

Goals

- **Research** the natural habitat and basic needs of one animal.
- **Design** and model an appropriate zoo, animal park, or aquarium environment for this animal. Working cooperatively with your classmates, design an entire zoo or animal park.

Possible Materials

poster board
markers or colored pencils
materials that can be used to make a scale model

Data Source

SCIENCE*Online* Go to the Glencoe Science Web site at **science.glencoe.com** for more information about existing zoos, animal parks, and aquariums.

Planning the Model

1. Choose an animal to research. Find out where this animal is found in nature. What does it eat? What are its natural predators? Does it exhibit unique territorial, courtship, or other types of behavior? How is this animal adapted to its natural environment?
2. **Design** a model of a proposed habitat in which this animal can live successfully. Don't forget to include all of the things, such as shelter, food, and water, that your animal will need to survive. Will there be any other organisms in the habitat?

Check the Model Plans

1. **Research** how zoos, animal parks, or aquariums provide habitats for animals. Information may be obtained by viewing the Glencoe Science Web site and contacting scientists who work at zoos, animal parks, and aquariums.
2. **Present** your design to your class in the form of a poster, slide show, or video. Compare your proposed habitat with that of the animal's natural environment. Make sure you include a picture of your animal in its natural environment.

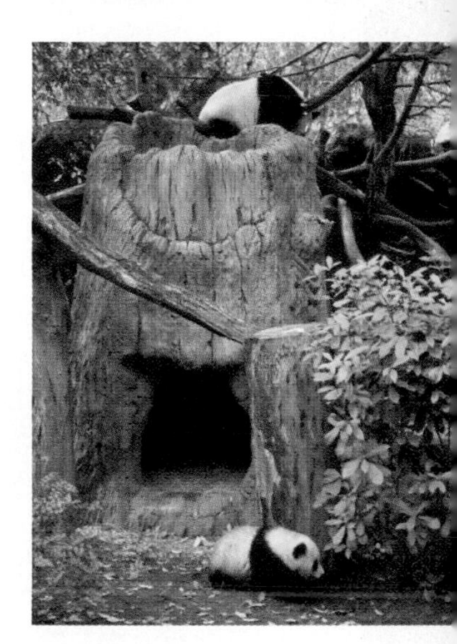

Making the Model

1. Using all of the information you have gathered, create a model exhibit area for your animal.
2. Indicate what other plants and animals may be present in the exhibit area.

Analyzing and Applying Results

1. **Decide** whether all of the animals studied in this activity can coexist in the same zoo or wildlife preserve.
2. **Predict** which animals could be grouped together in exhibit areas.
3. **Determine** how large your zoo or wildlife preserve needs to be. Which animals require a large habitat?
4. Using the information provided by the rest of your classmates, design an entire zoo or aquarium that could include the majority of animals studied.
5. **Analyze** problems that might exist in your design. Suggest some ways you might want to improve your design.

Give an oral presentation to another class on the importance of providing natural habitats for captive animals. **For more help, refer to the Science Skill Handbook.**

Going to the Dogs

A simple and surprising stroll showed that dogs really are humans' best friends

You've probably seen visually impaired people walking with their trusted and gentle four-legged guides—or "seeing-eye" dogs. The specially trained dogs serve as eyes for people who can't see, making it possible for them to lead independent lives. But what you probably didn't know is that about 80 years ago, a doctor and his patient discovered this canine ability entirely by accident!

Many people were killed or injured during World War I. Near the end of that war, Dr. Gerhard Stalling and his dog strolled with a patient—a German soldier who had been blinded—around hospital grounds in Germany.

German shepherds make excellent guide dogs.

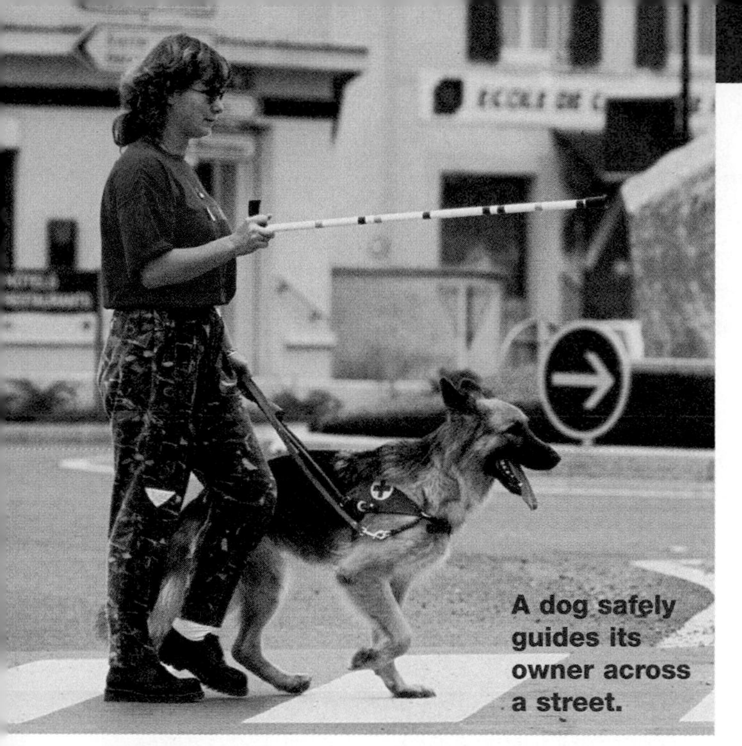

A dog safely guides its owner across a street.

While they were walking, the doctor was briefly called away. The dog and the soldier stayed outside. A few moments later, when the doctor returned, the dog and the soldier were gone! Searching the paths frantically, Dr. Stalling made an astonishing discovery. His pet had led the soldier safely around the hospital grounds. And together the two strolled peacefully back toward the doctor.

School for Dogs

Inspired by what his dog could do, Dr. Stalling set up the first school in the world dedicated to training dogs as guides. Dorothy Eustis, an American woman working as a dog trainer for the International Red Cross in Switzerland, traveled to Stalling's school about ten years later. A report of her visit and study of the way Stalling trained dogs appeared in a New York City newspaper in 1927.

Hearing the story, Morris Frank, a visually impaired American, became determined to get himself a guide dog. He wrote to Dorothy Eustis and asked that she train a dog for him. She accepted his request on one condition. She wanted Frank to join her in Switzerland for the training process. Frank and his guide dog Buddy returned to New Jersey in 1928. Within a year, Frank set up a training facility in New Jersey, "The Seeing Eye, Inc."

German shepherds, golden retrievers, and Labrador retrievers seem to make the best guide dogs. They learn hand gestures and simple commands to lead visually impaired people across streets and safely around obstacles. This is what scientists call "learned behavior." Animals gain learned behavior through experience. Learning happens gradually and in steps. In fact, scientists say that learning is a somewhat permanent change in behavior due to experience. But, a guide dog not only learns to respond to special commands, it must also know when *not* to obey. If its human owner urges the dog to cross the street and the dog sees that a car is approaching and refuses, the dog has learned to disobey the command. This trait, called "intelligent disobedience," ensures the safety of the owner and the dog—a sure sign that dogs are still humans' best friends.

This girl gets to help train a future guide dog for The Seeing Eye, Inc.

CONNECTIONS **Write** Lead a blindfolded partner around the classroom. Help your partner avoid obstacles. Then trade places. Write in your Science Journal about your experience leading and being led.

SCIENCE Online
For more information, visit science.glencoe.com

Chapter 7 Study Guide

Reviewing Main Ideas

Section 1 Types of Behavior

1. Behavior that an animal has when it's born is innate behavior. Other animal behaviors are learned through experience. *In the figure below, what type of behavior is the dog exhibiting?*

2. Reflexes are simple innate behaviors. An instinct is a complex pattern of innate behavior.
3. Learned behavior includes imprinting, in which an animal forms a social attachment immediately after birth.
4. Behavior modified by experience is learning by trial and error.
5. Conditioning occurs when the response to one stimulus becomes associated with another. Insight uses past experiences to solve new problems.

Section 2 Behavioral Interactions

1. Behavioral adaptations such as defense of territory, courtship behavior, and social behavior help species of animals survive and reproduce.
2. Courtship behaviors allow males and females to recognize each other and prepare to mate.

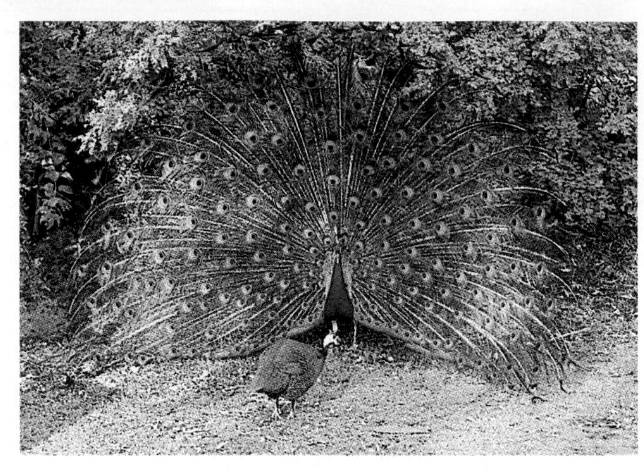

3. Interactions among members of the same species are social behaviors. *What type of social behavior is this male peacock displaying?*
4. Communication among organisms occurs in several ways including chemical, sound, and light. *How will other ants, like the one shown, be able to locate food that is far from their nest?*

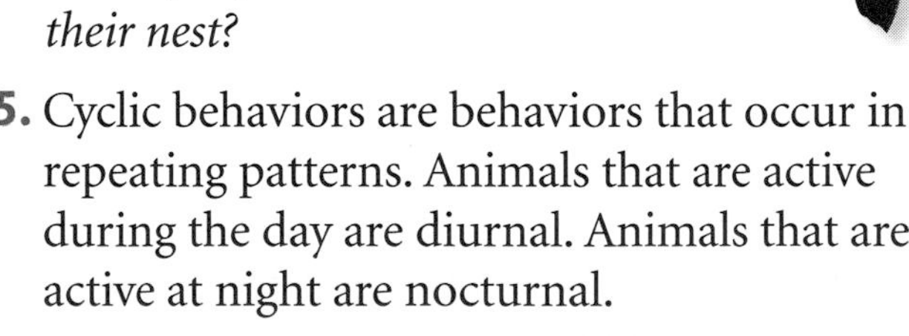

5. Cyclic behaviors are behaviors that occur in repeating patterns. Animals that are active during the day are diurnal. Animals that are active at night are nocturnal.

After You Read

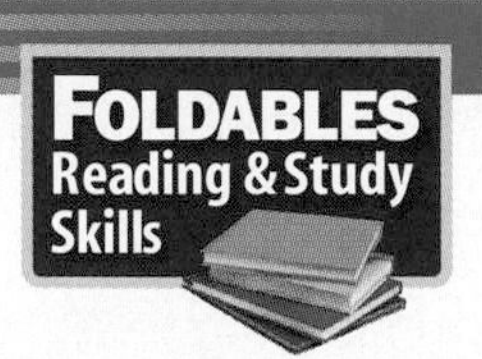

Compare and contrast the behaviors of Animal 1 and Animal 2 listed in your foldable. How many of the behaviors you listed were innate? Learned?

Chapter 7 Study Guide

Visualizing Main Ideas

Complete the following concept map on types of behavior.

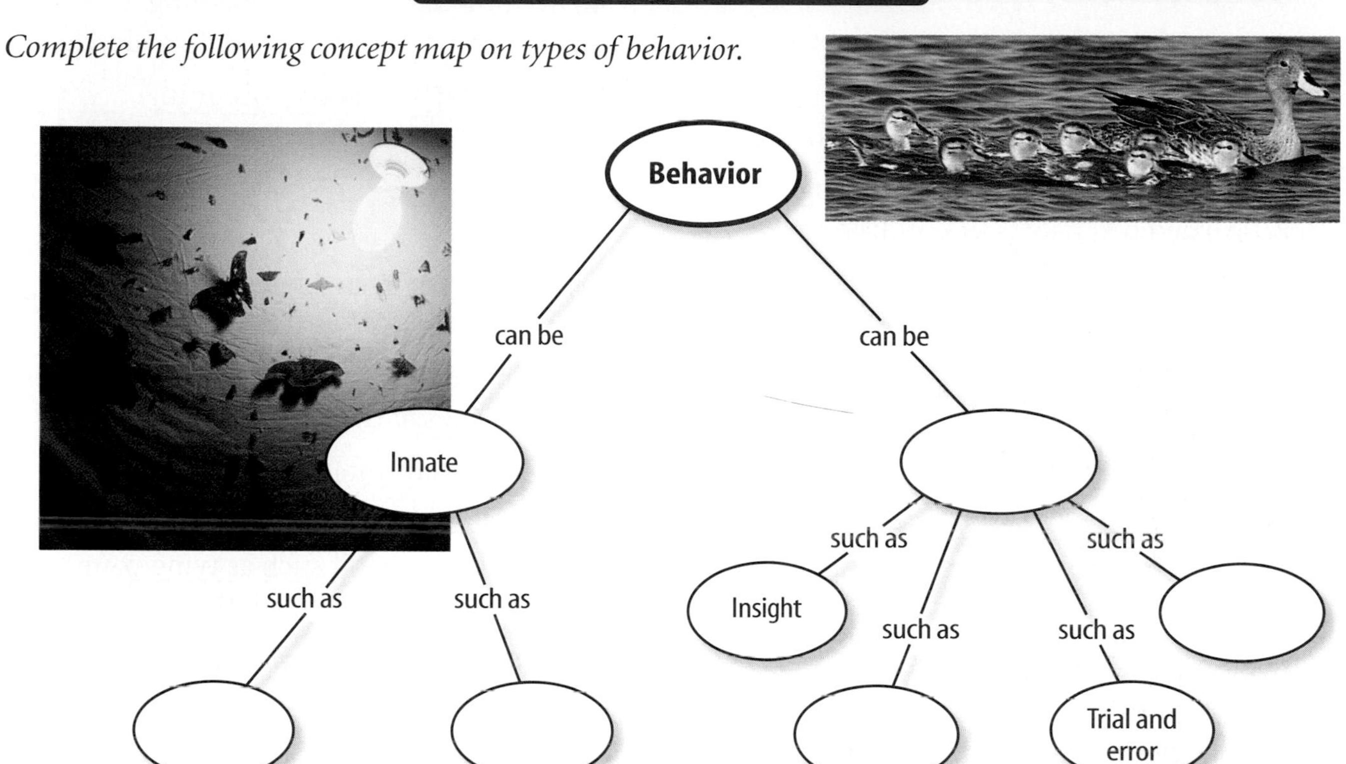

Vocabulary Review

Vocabulary Words

a. aggression
b. behavior
c. conditioning
d. courtship behavior
e. cyclic behavior
f. hibernation
g. imprinting
h. innate behavior
i. insight
j. instinct
k. migration
l. pheromone
m. reflex
n. social behavior
o. society

Study Tip

Take good notes, even during lab. Lab experiments reinforce key concepts, and looking back on these notes can help you better understand what happened and why.

Using Vocabulary

Explain the differences between the vocabulary words given below. Then explain how the words are related.

1. conditioning, imprinting
2. innate behavior, social behavior
3. insight, instinct
4. social behavior, society
5. instinct, reflex
6. hibernation, migration
7. courtship behavior, pheromone
8. cyclic behavior, migration
9. aggression, social behavior
10. behavior, reflex

Chapter 7 Assessment

Checking Concepts

Choose the word or phrase that best answers the question.

1. What is an instinct an example of?
A) innate behavior **C)** imprinting
B) learned behavior **D)** conditioning

2. What is a spider spinning a web an example of?
A) conditioning **C)** learned behavior
B) imprinting **D)** an instinct

3. Which animals depend least on instinct and most on learning?
A) birds **C)** mammals
B) fish **D)** amphibians

4. What is an area that an animal defends from other members of the same species called?
A) society **C)** migration
B) territory **D)** aggression

5. What is a forceful act used to dominate or control?
A) courtship **C)** aggression
B) reflex **D)** hibernation

6. Which of the following is NOT an example of courtship behavior?
A) fluffing feathers
B) taking over a perch
C) singing songs
D) releasing pheromones

7. What is an organized group of animals doing specific jobs called?
A) community **C)** society
B) territory **D)** circadian rhythm

8. What is the response of inactivity and slowed metabolism that occurs during cold conditions?
A) hibernation **C)** migration
B) imprinting **D)** circadian rhythm

9. Which of the following is a reflex?
A) writing **C)** sneezing
B) talking **D)** riding a bicycle

10. What are behaviors that occur in repeated patterns called?
A) cyclic **C)** reflex
B) imprinting **D)** society

Thinking Critically

11. Explain the type of behavior involved when the bell rings at the end of class.

12. Discuss the advantages and disadvantages of migration as a means of survival.

13. Explain how a habit such as tying your shoes, is different from a reflex.

14. Use one example to explain how behavior increases an animal's chance for survival.

15. Hens lay more eggs in the spring when the number of daylight hours increases. How can farmers use this knowledge of behavior to their advantage?

Developing Skills

16. Testing a Hypothesis Design an experiment to test a hypothesis about a specific response to a stimulus from an animal.

17. Recording Observations Make observations of a dog, cat, or bird for a week. Record what you see. How did the animal communicate with other animals and with you?

18. **Forming a Hypothesis** Make a hypothesis about how frogs communicate with each other. How could you test your hypothesis?

19. **Classifying** Make a list of 25 things that you do regularly. Classify each as an innate or learned behavior. Which behaviors do you have more of?

20. **Concept Mapping** Complete the following concept map about communication. Use these words: *light, sound, chirping, bioluminescence,* and *buzzing.*

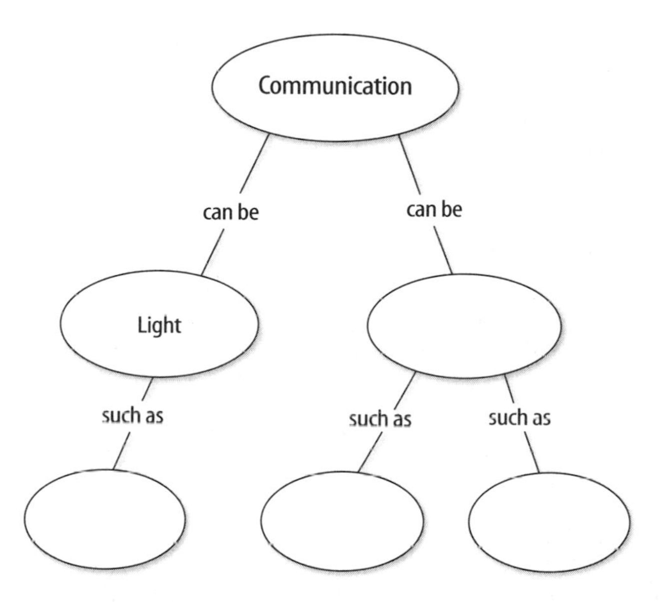

Performance Assessment

21. **Poster** Draw a map showing the migration route of monarch butterflies, gray whales, or blackpoll warblers.

TECHNOLOGY

Go to the Glencoe Science Web site at **science.glencoe.com** or use the **Glencoe Science CD-ROM** for additional chapter assessment.

Test Practice

A biologist is given illustrations of different behaviors. The different types of behaviors are listed below.

Types of Behavior

Behavior	Example
1	
2	
3	
4	

Study the table and answer the following questions.

1. A reflex is an automatic response to a stimulus. Which one of the behaviors in the table is an example of a reflex?

   A) one
   B) two
   C) three
   D) four

2. Trial and error is a type of learned behavior that is modified by experience. Which of the behaviors in the table is an example of a trial-and-error behavior?

   F) one
   G) two
   H) three
   J) four

Standardized Test Practice

Reading Comprehension

Read the passage. Then read each question that follows the passage. Decide which is the best answer to each question.

Enzymes in Humans

A catalyst is a substance that makes a chemical reaction happen faster than it would happen by itself. Interestingly, it affects the rate of the reaction without permanently entering into the reaction. More than 2,000 catalysts are necessary for the human body to function well. These catalysts are called enzymes.

Enzymes are a kind of protein. How an enzyme works depends on what shape it has. A special place on an enzyme attaches to chemicals. This site is called the active site. Enzyme activity can be compared to a lock and a key. Only the correct chemicals, or keys, will fit into the enzyme, or lock. The enzyme brings chemicals together so they can react. This is how enzymes speed up reactions—by making the reactants come together in a more direct way than if they were left to just bump into each other <u>randomly</u>. One enzyme can be used over and over to activate the same reaction. Some enzymes can help reactions go in either direction. In order for an enzyme to work properly, the temperature and pH must be within a certain range. Enough energy and enough reactants also must be present.

Two main types of enzymes are metabolic and digestive. Metabolic enzymes catalyze the reactions within cells. They help phosphorus turn into bone, iron attach to red blood cells, and wounds to heal. Digestive enzymes help with the breakdown of foods, allowing nutrients to be absorbed into the bloodstream and used by the body.

Enzymes are essential for many reactions within the human body. Amylase is an enzyme found in saliva. It starts digesting your food before you even swallow!

An enzyme called carbonic anhydrase helps remove carbon dioxide from your cells. You breathe the carbon dioxide out and replace it with oxygen. Carbonic anhydrase enzyme makes the reaction 107 times faster than if it had to happen on its own! You can see how people depend on enzyme catalysts to maintain health.

Enzymes can be found in all living things. Enzymes also have been used in industry for nearly 100 years. Some of the products that depend on the action of enzymes are leather, alcohol, medications, baking products, detergents, and even fruit juice!

Test-Taking Tip Make sure that you understand what you are reading as you read a passage. If you are confused by something, stop and read the information again.

1. Based on the information in the passage, it can be concluded that _____.

A) enzymes are found only in humans
B) amylase is an enzyme that removes oxygen
C) humans have more than 2,000 enzymes
D) another word for the locks found in doors is enzyme

2. Enzymes are important to human health because they _____.

F) are used to help open locks
G) are used to make leather and fruit juice
H) control the reactions in human bodies
J) are found in all living things

Standardized Test Practice

Reasoning and Skills

Read each question and select the best answer.

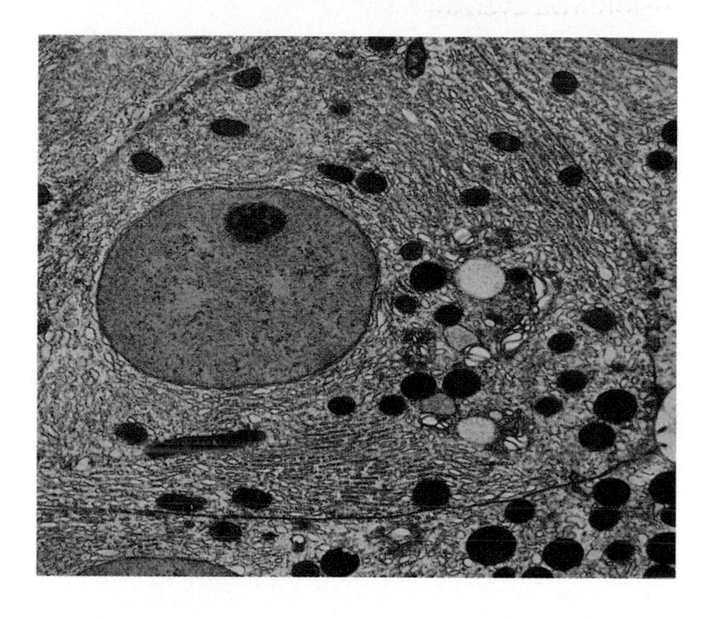

1. The object shown here is a(n) ______________ because it contains structures surrounded by membranes.

A) prokaryotic cell
B) mitochondrion
C) Golgi body
D) eukaryotic cell

Test-Taking Tip Think about the way cells are classified into groups.

2. Which of the following chemical compounds does not influence the growth of plants?

F) cytokinins
G) gibberellins
H) auxins
J) pheromones

Test-Taking Tip Think about the roles of hormones in plants and animals.

3. Campers dig a hole to bury their food scraps. Before they leave, they refill the hole with soil and place several large rocks on top. Later, a pair of raccoons explores the campground area, sniffing the ground. Eventually, the animals dig around and under the rocks to get to the food scraps. This is an example of how animals _____.

A) can be affected by pheromones.
B) use insight.
C) show cyclic behavior.
D) prepare for hibernation.

Test-Taking Tip Think about the definition of each type of animal behavior listed.

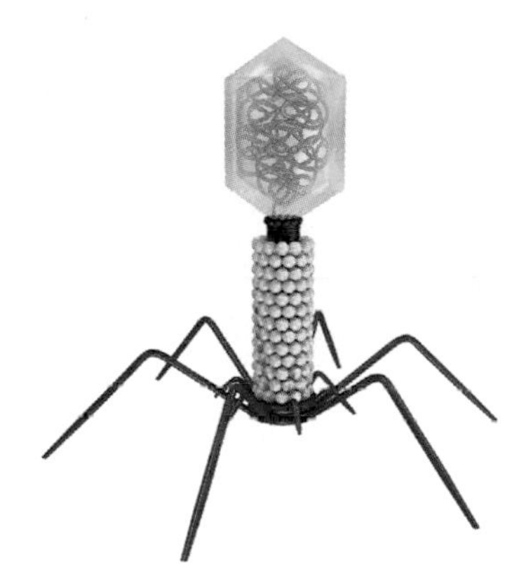

Consider this question carefully before writing your answer on a separate sheet of paper.

4. The virus particle pictured above is not inside a host cell. Give at least one good reason why it could be considered a living organism and one good reason why it could not.

Test-Taking Tip Compare the characteristics of a virus with the characteristics of living organisms.

Reproduction and Heredity

How Are Cargo Ships & Cancer Cells Connected?

Below, a present-day cargo ship glides through a harbor in New Jersey. In 1943, during World War II, another cargo ship floated in an Italian harbor. The ship carried a certain type of chemical. When a bomb struck the ship, the chemical accidentally was released. Later, when doctors examined the sailors who were exposed to the chemical, they noticed that the sailors had low numbers of white blood cells. The chemical had interfered with the genetic material in certain cells, preventing the cells from reproducing. Since cancer cells (such as the ones at lower left) are cells that reproduce without control, scientists wondered whether this chemical could be used to fight cancer. A compound related to the chemical became the first drug developed to fight cancer. Since then, many other cancer-fighting drugs have been developed.

SCIENCE CONNECTION

CANCER AND CELL REPRODUCTION Cancer is characterized by the uncontrolled reproduction of cells. Scientists have learned that normal cells can become cancer cells when the genes that control cell reproduction are changed by chemicals, radiation, or viruses. Find out which types of cancer are most common in the United States, and investigate what steps people can take to help reduce the chances of getting cancer. Create a brochure that presents your findings.

CHAPTER 8

Cell Reproduction

How does a cut on your skin heal? Why doesn't a baby chicken grow up to look like a duck? Why do turtles, like the one in the photo to the right, and most other animals need to have two parents, when a sweet potato plant can be grown from just one potato? In this chapter, you will find answers to these questions as you learn about cell reproduction. You also will learn what genetic material is and how it functions.

What do you think?

Science Journal Look at the picture below with a classmate. Discuss what you think this might be. Here is a hint: *These structures contain important information for cells.* Write your answer or best guess in your Science Journal.

Most flower and vegetable seeds sprout and grow into entire plants in just a few weeks. Although all of the cells in a seed have information and instructions to produce a new plant, only some of the cells in the seed use the information. Where are these cells in seeds? Do the following activity to find out.

Infer about seed growth 

1. Carefully split open two bean seeds that have soaked in water overnight.
2. Observe both halves and record your observations.
3. Wrap all four halves in a moist paper towel. Then put them into a self-sealing, plastic bag and seal the bag. Wash your hands.
4. Make observations for a few days.

Observe

In your Science Journal, describe what you observe. Hypothesize about which cells in seeds use information about how plants grow.

Before You Read

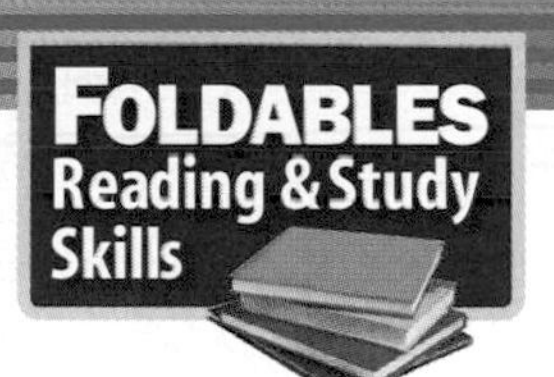

Making an Organizational Study Fold When information is grouped into clear categories, it is easier to make sense of what you are learning. Make the following Foldable to help you organize information about cell reproduction.

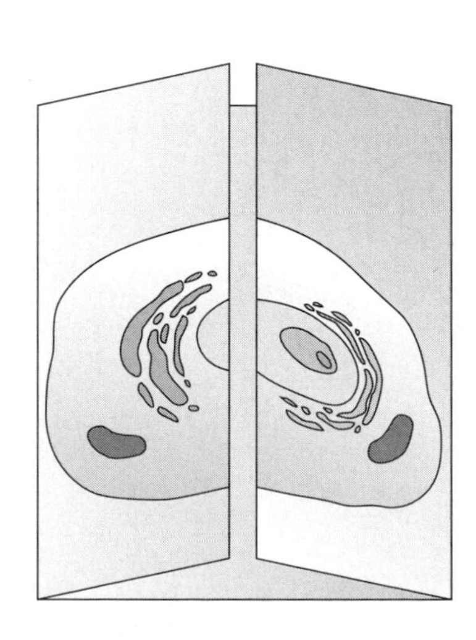

1. Place a sheet of paper in front of you so the long side is at the top. Fold the paper in half from the left side to the right side and then unfold.
2. Fold in each side to the center line to divide the paper into fourths.
3. Use a pencil to draw a cell on the front of your Foldable as shown.
4. As you read the chapter, use a pen to illustrate how the cell divides into two cells. Under the flaps, list how cells divide. In the middle section, list why cells divide.

SECTION

Cell Division and Mitosis

As You Read

What **You'll Learn**

- **Explain** why mitosis is important.
- **Examine** the steps of mitosis.
- **Compare** mitosis in plant and animal cells.
- **List** two examples of asexual reproduction.

Vocabulary

mitosis
chromosome
asexual reproduction

Why **It's Important**

Your growth, like that of many organisms, depends on cell division.

Why is cell division important?

What do you, an octopus, and an oak tree have in common? You share many characteristics, but an important one is that you are all made of cells—trillions of cells. Where did all of those cells come from? As amazing as it might seem, many organisms start as just one cell. That cell divides and becomes two, two become four, four become eight, and so on. Many-celled organisms, including you, grow because cell division increases the total number of cells in an organism. Even after growth stops, cell division is still important. Every day, billions of red blood cells in your body wear out and are replaced. During the few seconds it takes you to read this sentence, your bone marrow produced about six million red blood cells. Cell division is important to one-celled organisms, too—it's how they reproduce themselves, as shown in **Figure 1B.** Cell division isn't as simple as just cutting the cell in half, so how do cells divide?

The Cell Cycle

A living organism has a life cycle. A life cycle begins with the organism's formation, is followed by growth and development, and finally ends in death. Right now, you are in a stage of your life cycle called adolescence, which is a period of active growth and development. Individual cells also have life cycles.

Figure 1
All organisms use cell division.

A **Many-celled organisms, such as this octopus, grow by increasing the numbers of their cells.**

B **Like this amoeba, a one-celled organism reaches a certain size and then reproduces.**

Figure 2
Interphase is the longest part of the cell cycle. *When do chromosomes duplicate?*

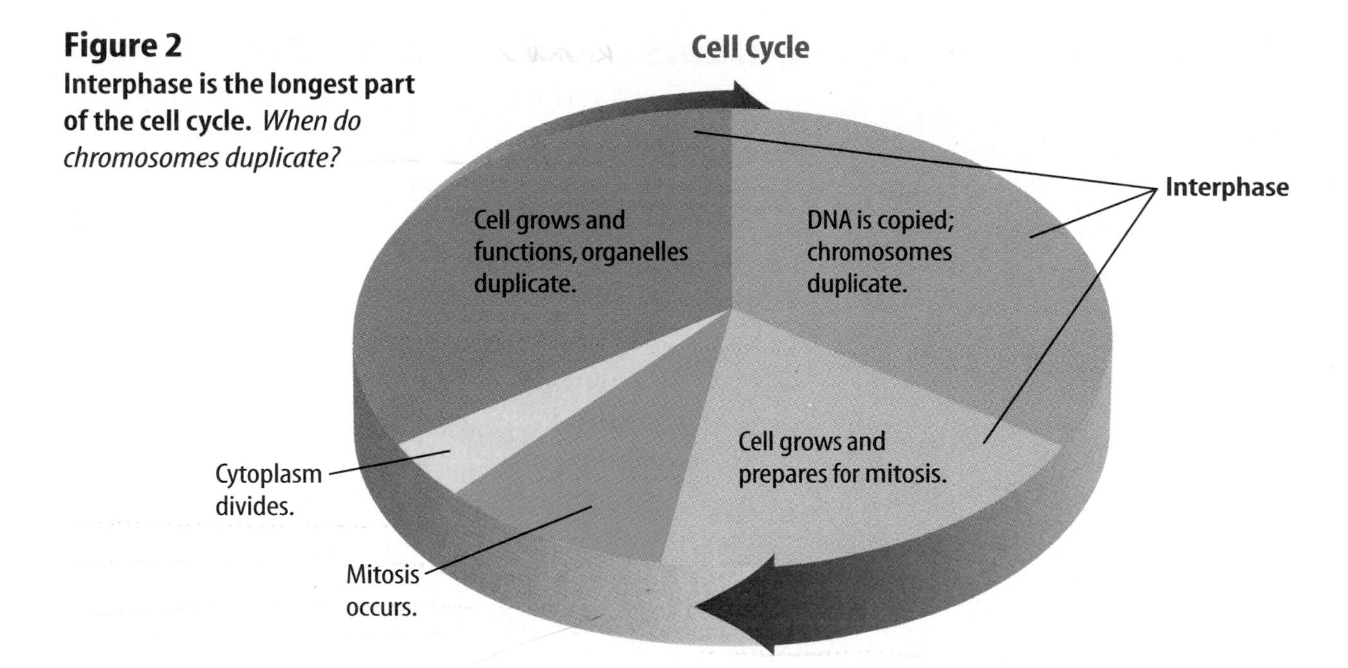

Length of Cycle The cell cycle, as shown in **Figure 2,** is a series of events that takes place from one cell division to the next. The time it takes to complete a cell cycle is not the same in all cells. For example, the cycle for cells in some bean plants takes about 19 h to complete. Cells in animal embryos divide rapidly and can complete their cycles is less than 20 min. In some human cells, the cell cycle takes about 16 h. Cells in humans that are needed for repair, growth, or replacement, like skin and bone cells, constantly repeat the cycle.

Interphase Most of the life of any eukaryotic cell—a cell with a nucleus—is spent in a period of growth and development called interphase. Cells in your body that no longer divide, such as nerve and muscle cells, are always in interphase. An actively dividing cell, such as a skin cell, copies its hereditary material and prepares for cell division during interphase.

Why is it important for a cell to copy its hereditary information before dividing? Imagine that you have a part in a play and the director has one complete copy of the script. If the director gave only one page to each person in the play, no one would have the entire script. Instead the director makes a complete, separate copy of the script for each member of the cast so that each one can learn his or her part. Before a cell divides, a copy of the hereditary material must be made so that each of the two new cells will get a complete copy. Just as the actors in the play need the entire script, each cell needs a complete set of hereditary material to carry out life functions.

After interphase, cell division begins. The nucleus divides, and then the cytoplasm separates to form two new cells.

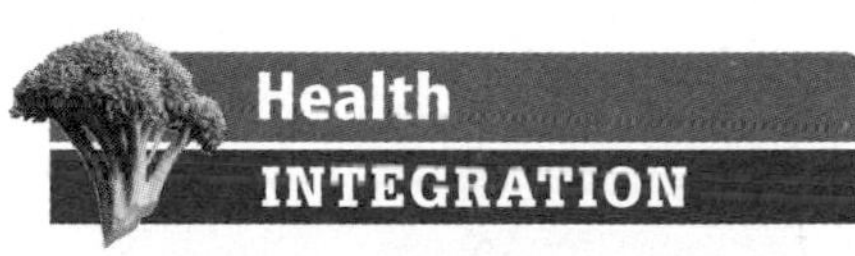

In most cells, the cell cycle is well controlled. However, cancerous cells have uncontrolled cell division. Some cancerous cells form a mass of cells called a tumor. Find out why some tumors are harmful to an organism. Write what you find out in your Science Journal.

Research Nerve cells in adults usually do not undergo mitosis. Visit the Glencoe Science Web site at **science.glencoe.com** for more information about nerve cell regeneration. Communicate to your class what you learn.

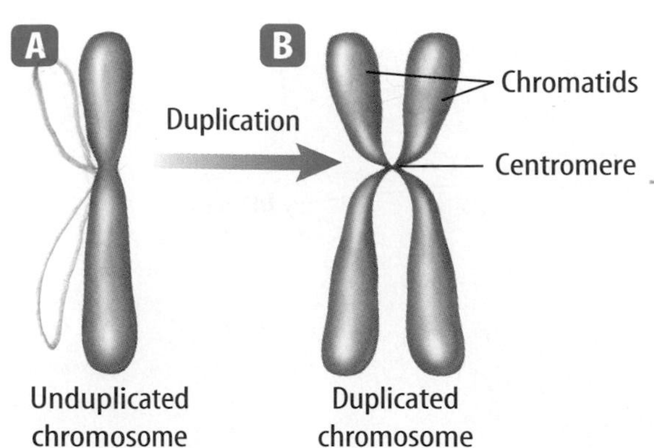

Figure 3
DNA is copied during interphase. A An unduplicated chromosome has one strand of DNA. B A duplicated chromosome has two identical DNA strands, called chromatids, that are held together at a region called the centromere.

Mitosis

Mitosis (mi TOH sus) is the process in which the nucleus divides to form two identical nuclei. Each new nucleus also is identical to the original nucleus. Mitosis is described as a series of phases, or steps. The steps of mitosis in order are named prophase, metaphase, anaphase, and telophase.

Steps of Mitosis When any nucleus divides, the chromosomes (KROH muh sohmz) play the important part. A **chromosome** is a structure in the nucleus that contains hereditary material. During interphase, each chromosome duplicates. When the nucleus is ready to divide, each duplicated chromosome coils tightly into two thickened, identical strands called chromatids, as shown in **Figure 3.**

Reading Check *How are chromosomes and chromatids related?*

During prophase, the pairs of chromatids are fully visible when viewed under a microscope. The nucleolus and the nuclear membrane disintegrate. Two small structures called centrioles (SEN tree olz) move to opposite ends of the cell. Between the centrioles, threadlike spindle fibers begin to stretch across the cell. Plant cells also form spindle fibers during mitosis but do not have centrioles.

In metaphase, the pairs of chromatids line up across the center of the cell. The centromere of each pair usually becomes attached to two spindle fibers—one from each side of the cell.

In anaphase, each centromere divides and the spindle fibers shorten. Each pair of chromatids separates, and chromatids begin to move to opposite ends of the cell. The separated chromatids are now called chromosomes. In the final step, telophase, spindle fibers start to disappear, the chromosomes start to uncoil, and a new nucleus forms.

Figure 4
The cell plate shown in this plant cell appears when the cytoplasm is being divided.

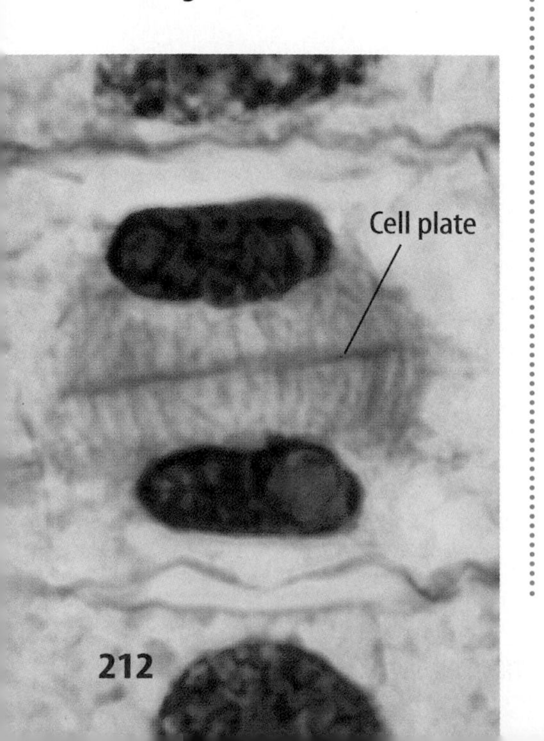

Division of the Cytoplasm For most cells, after the nucleus has divided, the cytoplasm separates and two new cells are formed. In animal cells, the cell membrane pinches in the middle, like a balloon with a string tightened around it, and the cytoplasm divides. In plant cells, the appearance of a cell plate, as shown in **Figure 4,** tells you that the cytoplasm is being divided. New cell walls form along the cell plate, and new cell membranes develop inside the cell walls. Following division of the cytoplasm, most new cells begin the period of growth, or interphase, again. Review cell division for an animal cell using the illustrations in **Figure 5.**

Figure 5
Cell division for an animal cell is shown here. Each micrograph shown in this figure is magnified 600 times.

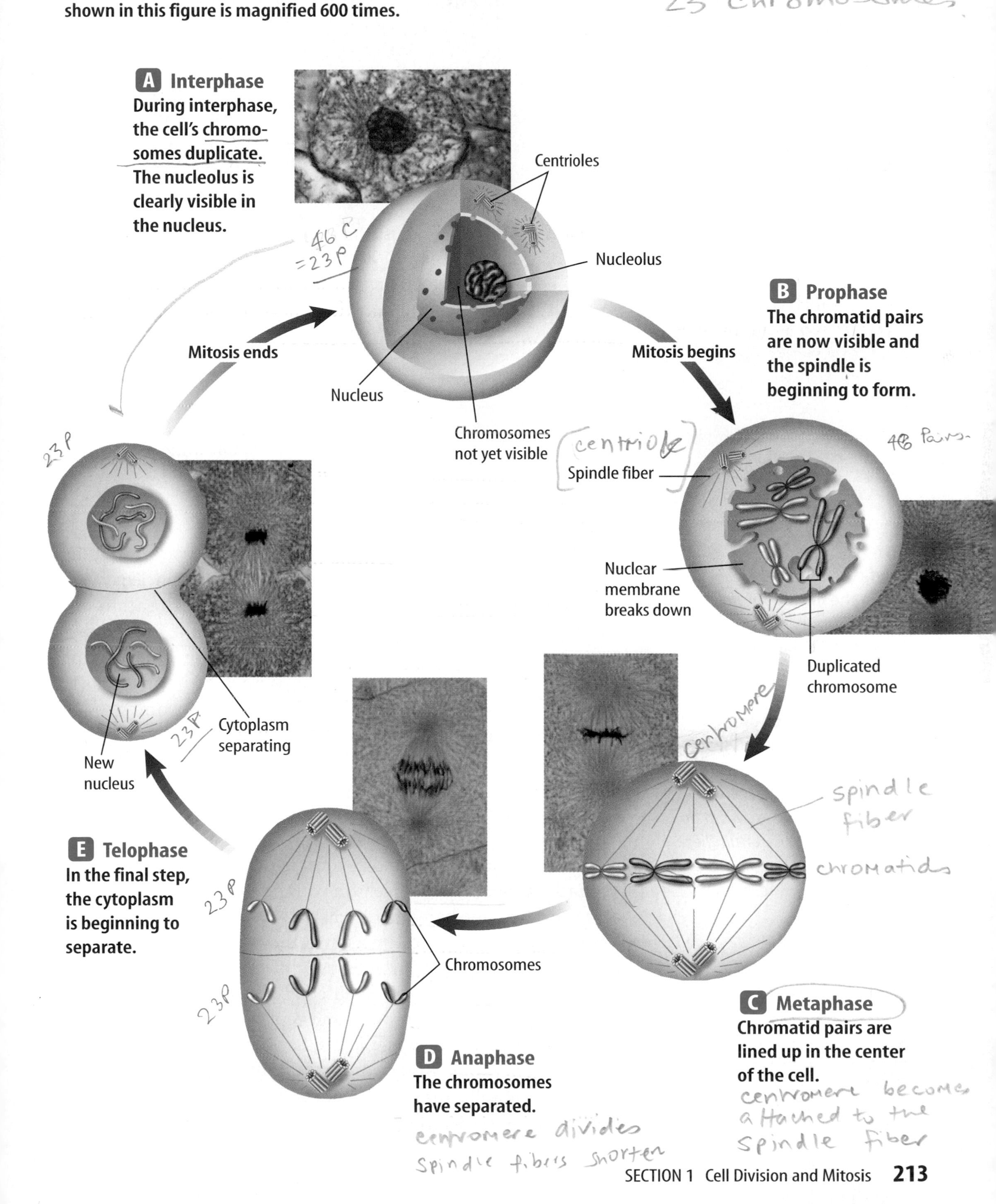

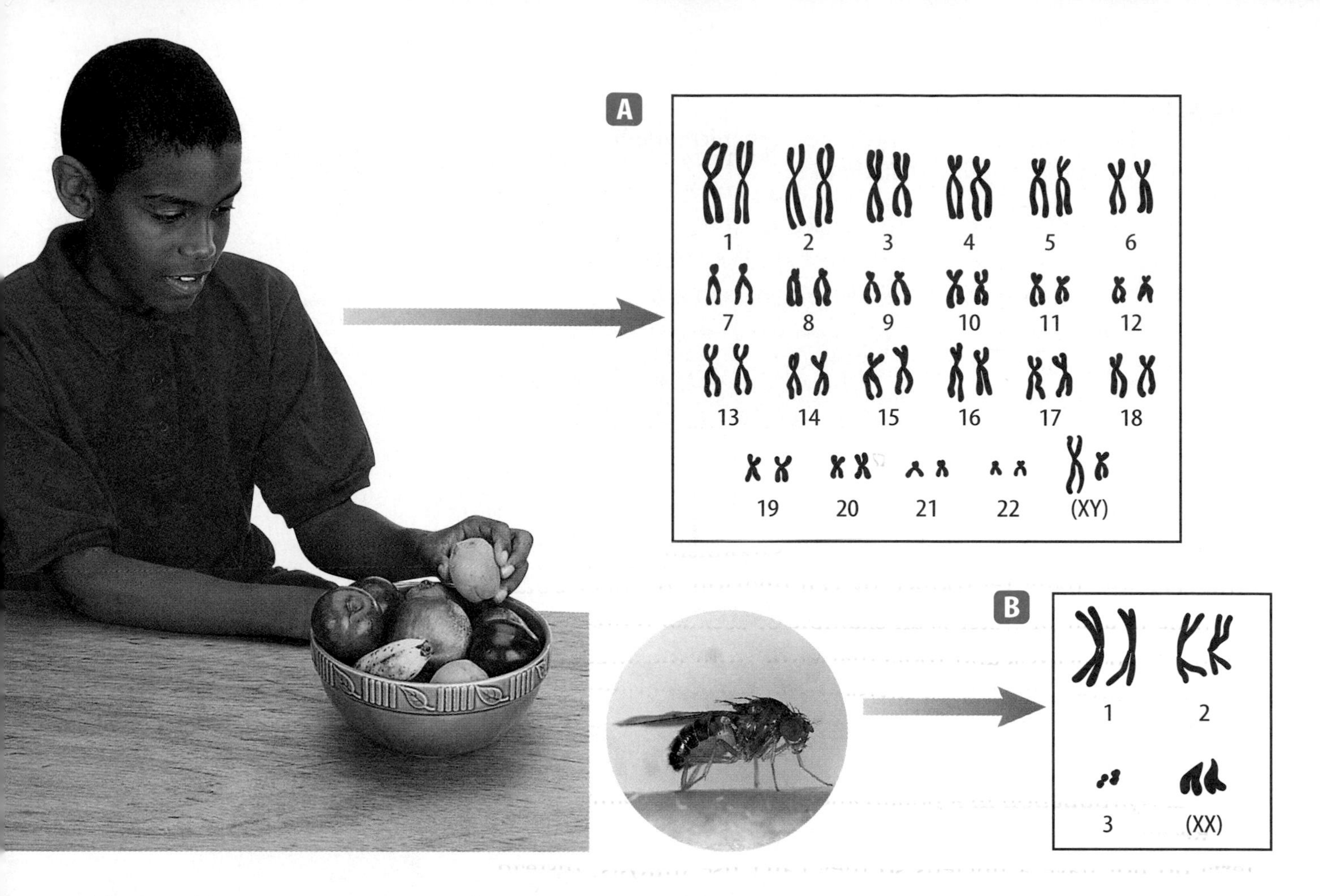

Figure 6
Pairs of chromosomes are found in the nucleus of most cells. All chromosomes shown here are in their duplicated form. A Most human cells have 23 pairs of chromosomes including one pair of chromosomes that help determine sex such as the XY pair above. B Most fruit fly cells have four pairs of chromosomes. *What do you think the XX pair in fruit flies helps determine?*

Results of Mitosis You should remember two important things about mitosis. First, it is the division of a nucleus. Second, it produces two new nuclei that are identical to each other and the original nucleus. Each new nucleus has the same number and type of chromosomes. Every cell in your body, except sex cells, has a nucleus with 46 chromosomes—23 pairs. This is because you began as one cell with 46 chromosomes in its nucleus. Skin cells, produced to replace or repair your skin, have the same 46 chromosomes as the original single cell you developed from. Each cell in a fruit fly has eight chromosomes, so each new cell produced by mitosis has a copy of those eight chromosomes. **Figure 6** shows the chromosomes found in most human cells and those found in most fruit fly cells.

Each of the trillions of cells in your body, except sex cells, has a copy of the same hereditary material. Even though all actors in a play have copies of the same script, they do not learn the same lines. Likewise, all of your cells use different parts of the same hereditary material to become different types of cells.

Cell division allows growth and replaces worn out or damaged cells. You are much larger and have more cells than a baby mainly because of cell division. If you cut yourself, the wound heals because cell division replaces damaged cells. Another way some organisms use cell division is to produce new organisms.

Asexual Reproduction

Reproduction is the process by which an organism produces others of its same kind. Among living organisms, there are two types of reproduction—sexual and asexual. Sexual reproduction usually requires two organisms. In **asexual reproduction,** a new organism (sometimes more than one) is produced from one organism. The new organism will have hereditary material identical to the hereditary material of the parent organism.

✔ Reading Check *How many organisms are needed for asexual reproduction?*

Cellular Asexual Reproduction Organisms with eukaryotic cells asexually reproduce by cell division. A sweet potato growing in a jar of water is an example of asexual reproduction. All the stems, leaves, and roots that grow from the sweet potato have been produced by cell division and have the same hereditary material. New strawberry plants can be reproduced asexually from horizontal stems called runners. **Figure 7** shows asexual reproduction in a potato and a strawberry plant.

Recall that mitosis is the division of a nucleus. However, bacteria do not have a nucleus so they can't use mitosis. Instead, bacteria reproduce asexually by fission. During fission, an organism whose cells do not contain a nucleus copies its genetic material and then divides into two identical organisms.

Mini LAB

Modeling Mitosis

Procedure

1. Make models of cell division using **materials supplied by your teacher.**
2. Use four chromosomes in your model.
3. When finished, arrange the models in the order in which mitosis occurs.

Analysis

1. In which steps is the nucleus visible?
2. How many cells does a dividing cell form?

Figure 7
Many plants can reproduce asexually.

A **A new potato plant can grow from each sprout on this potato.**

B *How does the genetic material in the small strawberry plant compare to the genetic material in the large strawberry plant?*

Figure 8
Some organisms use cell division for budding and regeneration.

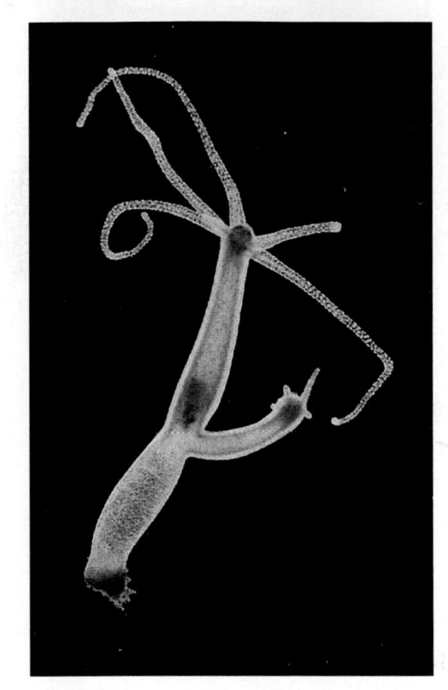

A Hydra, a freshwater animal, can reproduce asexually by budding. The bud is a small exact copy of the adult.

B Some sea stars reproduce asexually by shedding arms. Each arm can grow into a new sea star.

Budding and Regeneration Look at **Figure 8A.** A new organism is growing from the body of the parent organism. This organism, called a hydra, is reproducing by budding. Budding is a type of asexual reproduction made possible because of cell division. When the bud on the adult becomes large enough, it breaks away to live on its own.

Could you grow a new finger? Some organisms can regrow damaged or lost body parts, as shown in **Figure 8B.** Regeneration is the process that uses cell division to regrow body parts. Sponges, planaria, sea stars, and some other organisms can use regeneration for asexual reproduction. If these organisms break into pieces, a whole new organism will grow from each piece. Because sea stars eat oysters, oyster farmers dislike them. What would happen if an oyster farmer collected sea stars, cut them into pieces, and threw them back into the ocean?

Section  Assessment

1. What is mitosis and how does it differ in plants and animals?
2. Give two examples of asexual reproduction in many-celled organisms.
3. What happens to chromosomes before mitosis begins?
4. After a cell undergoes mitosis, how are the two new cells alike?
5. **Think Critically** Why is it important for the nuclear membrane to disintegrate during mitosis?

Skill Builder Activities

6. **Testing a Hypothesis** A piece of leaf, stem, or root can grow into a new plant. Hypothesize how you would use one of these plant parts to grow a new plant. Test your idea. **For more help, refer to the Science Skill Handbook.**
7. **Solving One-Step Equations** If a cell undergoes cell division every 5 min, how many cells will there be after 1 h? Calculate and record the answer in your Science Journal. **For more help, refer to the Math Skill Handbook.**

Activity

Mitosis in Plant Cells

Reproduction of most cells in plants and animals uses mitosis and cell division. In this activity, you will study mitosis in plant cells by examining prepared slides of onion root-tip cells.

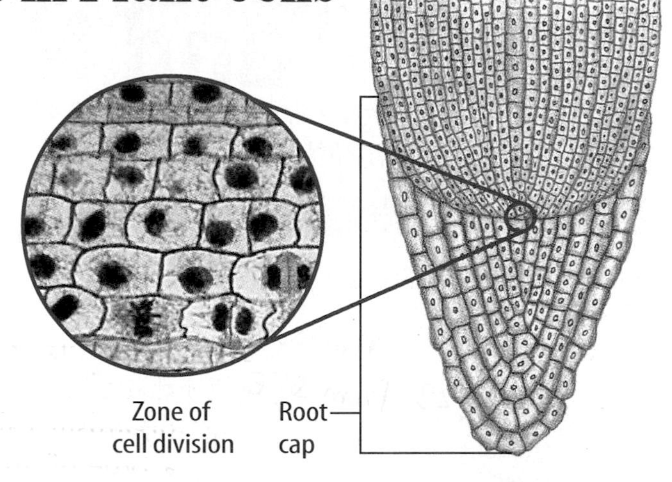

What You'll Investigate

How can plant cells in different stages of mitosis be distinguished from each other?

Materials

prepared slide of an onion root tip
microscope

Goals

- **Compare** cells in different stages of mitosis and observe the location of their chromosomes.
- **Observe** what stage of mitosis is most common in onion root tips.

Safety Precautions  

Procedure

1. Copy the data table in your Science Journal.
2. **Obtain** a prepared slide of cells from an onion root tip.
3. Set your microscope on low power and examine the onion root tip. Move the slide until you can see the tip of the root. You will see several large, round cells. These cells are called the root cap. Move your slide until you see the cells in the area just behind the root cap. Turn the nosepiece to high power.
4. Find one area of cells where you can see the most stages of mitosis. Count how many cells you see in each stage and record your data in the table.
5. Turn the microscope back to low power. Remove the onion root-tip slide.

Conclude and Apply

1. **Compare** the cells in the region behind the root cap to those in the root cap.
2. **Calculate** the percent of cells found in each stage of mitosis. Infer which stage of mitosis takes the longest period of time.

Number of Root-Tip Cells Observed

Stage of Mitosis	Number of Cells Observed	Percent of Cells Observed
Prophase		
Metaphase		
Anaphase		
Telophase		
Total		

Communicating Your Data

Write a story as if you were a cell in an onion root tip. Describe what changes occur as you go through mitosis. Use some of your drawings to illustrate the story. Share your story with your class. **For more help, refer to the Science Skill Handbook.**

SECTION

Sexual Reproduction and Meiosis

As You Read

***What* You'll Learn**

- **Describe** the stages of meiosis and how sex cells are produced.
- **Explain** why meiosis is needed for sexual reproduction.
- **Name** the cells that are involved in fertilization.
- **Explain** how fertilization occurs in sexual reproduction.

Vocabulary

sexual reproduction
sperm
egg
fertilization
zygote
diploid
haploid
meiosis

***Why* It's Important**

Because of meiosis and sexual reproduction, no one is exactly like you.

Sexual Reproduction

Sexual reproduction is another way that a new organism can be produced. During **sexual reproduction,** two sex cells, sometimes called an egg and a sperm, come together. Sex cells, like those in **Figure 9,** are formed from cells in reproductive organs. **Sperm** are formed in the male reproductive organs. **Eggs** are formed in the female reproductive organs. The joining of an egg and a sperm is called **fertilization,** and the cell that forms is called a **zygote** (ZI goht). Generally, the egg and the sperm come from two different organisms of the same species. Following fertilization, cell division begins. A new organism with a unique identity develops.

Diploid Cells Your body forms two types of cells—body cells and sex cells. Body cells far outnumber sex cells. Your brain, skin, bones, and other tissues and organs are formed from body cells. A typical human body cell has 46 chromosomes. Each chromosome has a mate that is similar to it in size and shape and has similar DNA. Human body cells have 23 pairs of chromosomes. When cells have pairs of similar chromosomes, they are said to be **diploid** (DIH ployd).

Figure 9
A human sperm or egg contains 23 chromosomes. The chromosomes are shown in their duplicated form.

Human egg and many sperm

Magnification: 790×

Haploid Cells Because sex cells do not have pairs of chromosomes, they are said to be **haploid** (HA ployd). They have only half the number of chromosomes as body cells. *Haploid* means "single form." Human sex cells have only 23 chromosomes—one from each of the 23 pairs of similar chromosomes. Compare the chromosomes found in a sex cell, as shown in **Figure 9,** to the full set of human chromosomes seen in **Figure 6A.**

✔ **Reading Check** ***How many chromosomes are usually in each human sperm?***

Meiosis and Sex Cells

A process called **meiosis** (mi OH sus) produces haploid sex cells. What would happen in sexual reproduction if two diploid cells combined? The offspring would have twice as many chromosomes as its parent. Although plants with twice the number of chromosomes as the parent plants are often produced, most animals do not survive with a double number of chromosomes. Meiosis ensures that the offspring will have the same diploid number as its parent, as shown in **Figure 10.** After two haploid sex cells combine, a diploid zygote is produced that develops into a new diploid organism.

During meiosis, two divisions of the nucleus occur. These divisions are called meiosis I and meiosis II. The steps of each division have names like those in mitosis and are numbered for the division in which they occur.

The human egg releases a chemical into the surrounding fluid that attracts sperm. Usually, only one sperm fertilizes the egg. After the sperm nucleus enters the egg, the cell membrane of the egg changes in a way that prevents other sperm from entering. What adaptation in this process guarantees that the zygote will be diploid? Write a paragraph describing your ideas in your Science Journal.

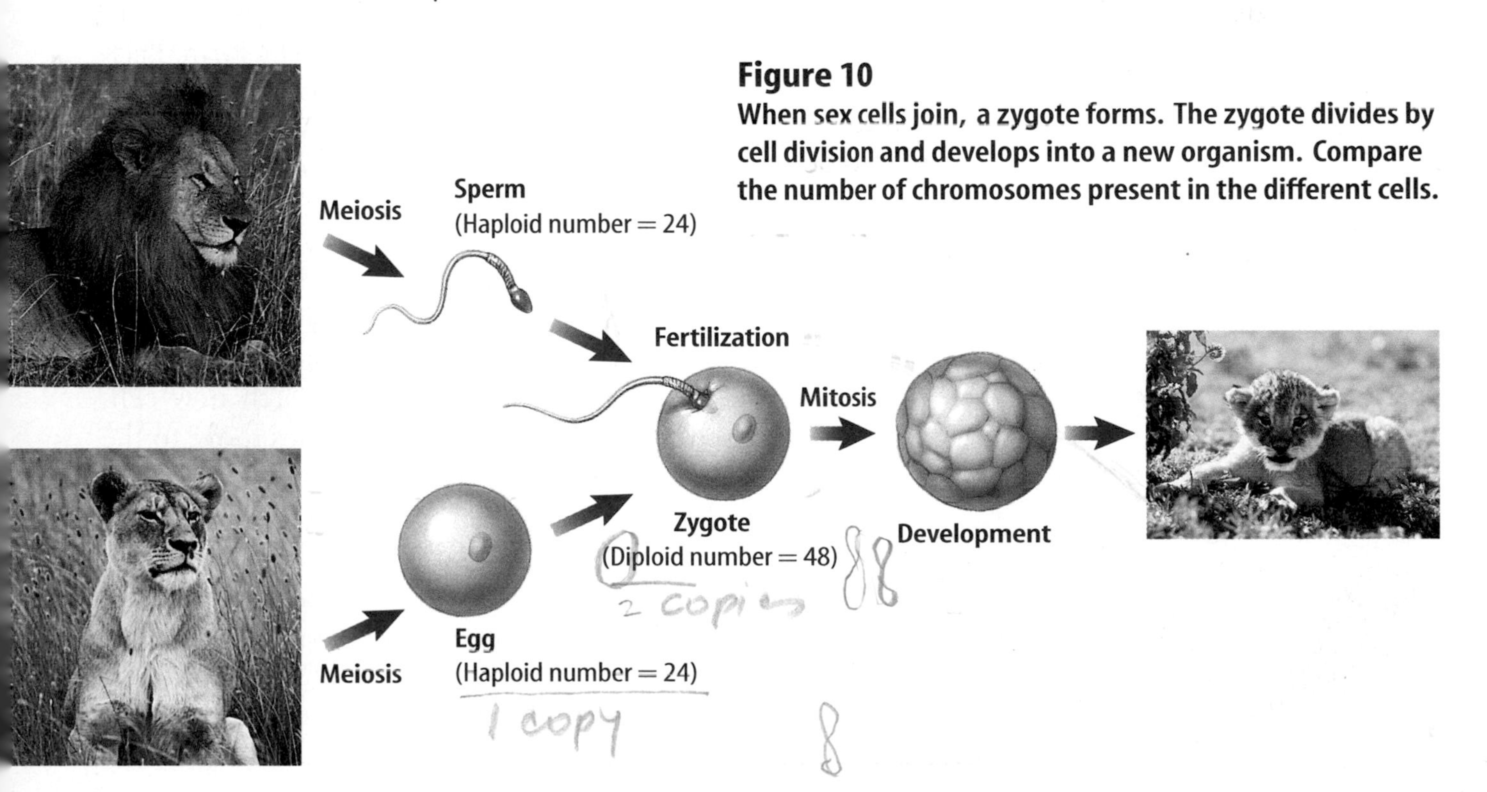

Figure 10
When sex cells join, a zygote forms. The zygote divides by cell division and develops into a new organism. Compare the number of chromosomes present in the different cells.

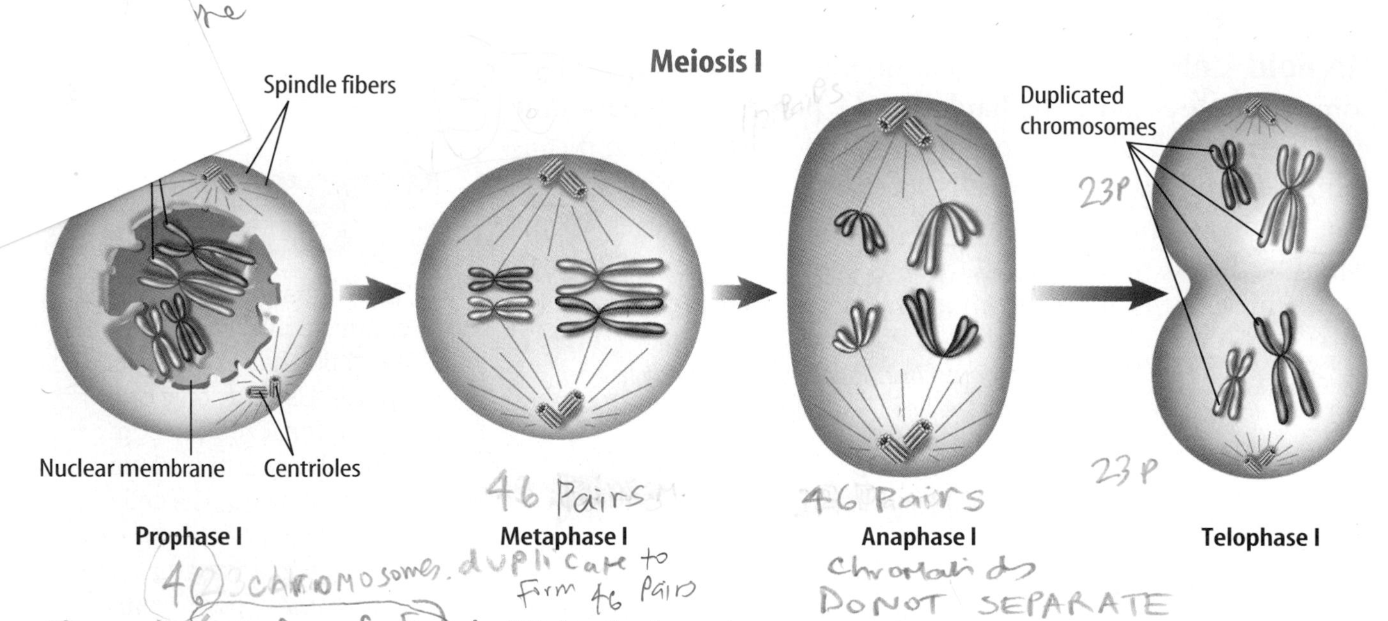

Figure 11 Meiosis has two divisions of the nucleus—meiosis I and meiosis II. *How many sex cells are finally formed after both divisions are completed?*

Meiosis I Before meiosis begins, each chromosome is duplicated, just as in mitosis. When the cell is ready for meiosis, each duplicated chromosome is visible under the microscope as two chromatids. As shown in **Figure 11,** the events of prophase I are similar to those of prophase in mitosis. In meiosis, each duplicated chromosome comes near its similar duplicated mate. In mitosis they do not come near each other.

In metaphase I, the pairs of duplicated chromosomes line up in the center of the cell. The centromere of each chromatid pair becomes attached to one spindle fiber so, the chromatids do not separate in anaphase I. The two pairs of chromatids of each similar pair move away from each other to opposite ends of the cell. Each duplicated chromosome still has two chromatids. Then, in telophase I, the cytoplasm divides, and two new cells form. Each new cell has one duplicated chromosome from each similar pair.

Reading Check *What happens to duplicated chromosomes during anaphase I?*

Meiosis II The two cells formed during meiosis I now begin meiosis II. The chromatids of each duplicated chromosome will be separated during this division. In prophase II, the duplicated chromosomes and spindle fibers reappear in each new cell. Then in metaphase II, the duplicated chromosomes move to the center of the cell. Unlike what occurs in metaphase I, each centromere now attaches to two spindle fibers instead of one. The centromere divides during anaphase II, and the chromatids separate and move to opposite ends of the cell. Each chromatid now is an individual chromosome. As telophase II begins, the spindle fibers disappear, and a nuclear membrane forms around the chromosomes at each end of the cell. When meiosis II is finished, the cytoplasm divides.

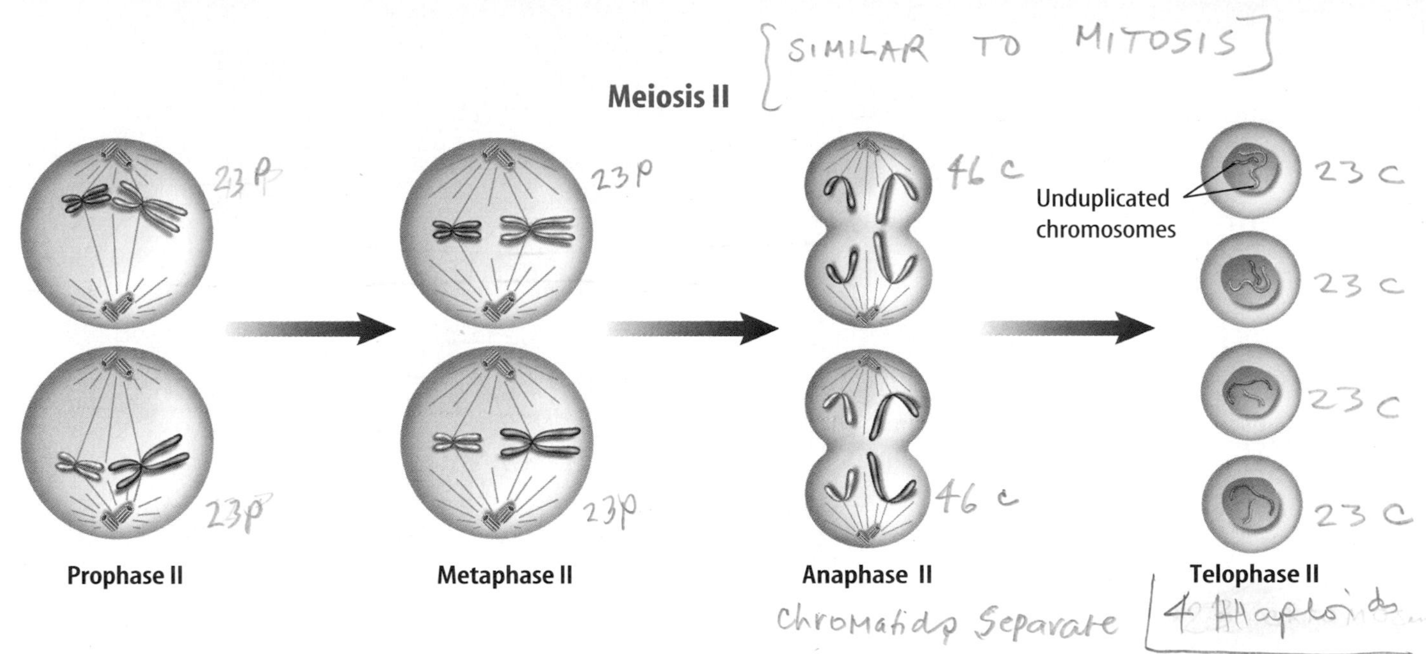

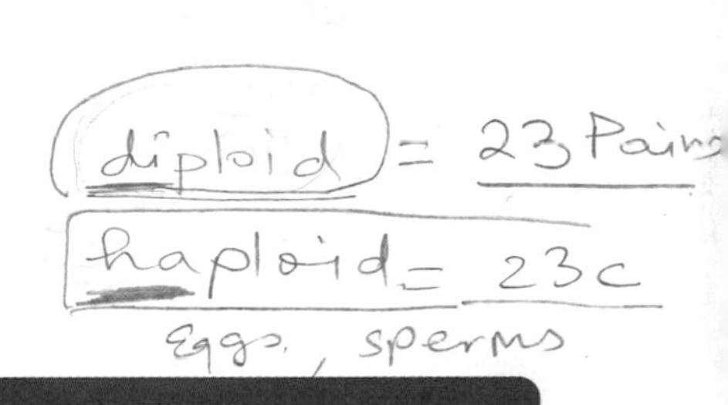

Summary of Meiosis Two cells form during meiosis I. In meiosis II, both of these cells form two cells. The two divisions of the nucleus result in four sex cells. Each has one-half the number of chromosomes in its nucleus that was in the original nucleus. From a human cell with 46 paired chromosomes, meiosis produces four sex cells each with 23 unpaired chromosomes.

Problem-Solving Activity

How can chromosome numbers be predicted?

Offspring get half of their chromosomes from one parent and half from the other. What happens if each parent has a different diploid number of chromosomes?

Identifying the Problem

A zebra and a donkey can mate to produce a zonkey. Zebras have a diploid number of 46. Donkeys have a diploid number of 62.

Solving the Problem

1. How many chromosomes would the zonkey receive from each parent?
2. What is the chromosome number of the zonkey?
3. What would happen when meiosis occurs in the zonkey's reproductive organs?
4. Predict why zonkeys are usually sterile.

Donkey
62 Chromosomes

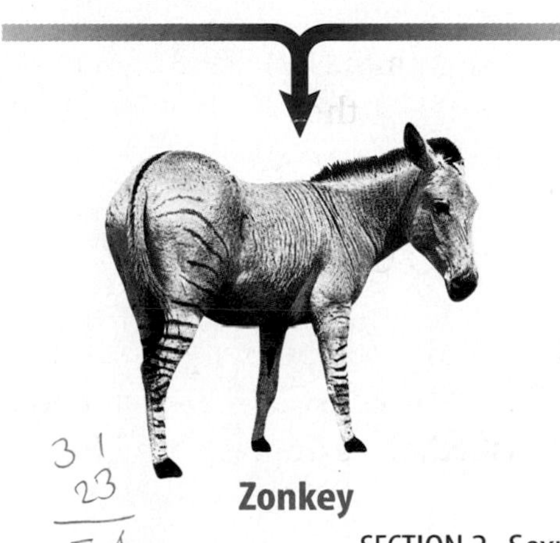

Zonkey

Zebra
46 Chromosomes

VISUALIZING POLYPLOIDY IN PLANTS

Figure 12

You received a haploid (n) set of chromosomes from each of your parents, making you a diploid (2n) organism. In nature, however, many plants are polyploid—they have three (3n), four (4n), or more sets of chromosomes. We depend on some of these plants for food.

▲ TRIPLOID Bright yellow bananas typically come from triploid (3n) banana plants. Plants with an odd number of chromosome sets usually cannot reproduce sexually and have very small seeds or none at all.

▲ TETRAPLOID Polyploidy occurs naturally in many plants—including peanuts and daylilies—due to mistakes in mitosis or meiosis.

▼ HEXAPLOID Modern cultivated strains of oats have six sets of chromosomes, making them hexaploid (6n) plants.

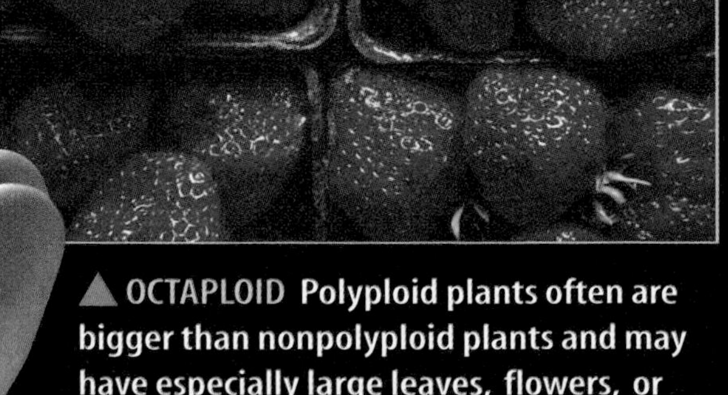

▲ OCTAPLOID Polyploid plants often are bigger than nonpolyploid plants and may have especially large leaves, flowers, or fruits. Strawberries are an example of octaploid (8n) plants.

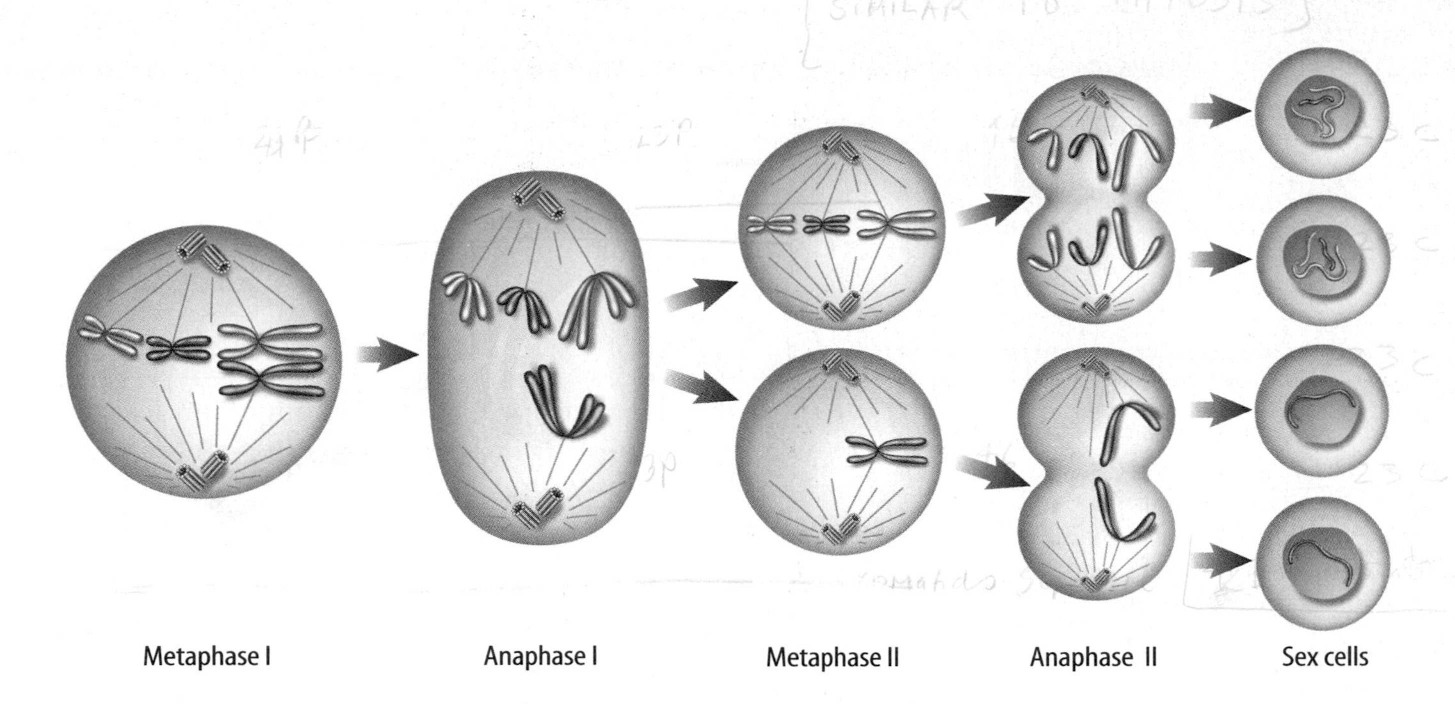

Figure 13
This diploid cell has four chromosomes. During anaphase I, one pair of duplicated chromosomes did not separate. *How many chromosomes does each sex cell usually have?*

Mistakes in Meiosis Meiosis occurs many times in reproductive organs. Although mistakes in plants, as shown in **Figure 12,** are common, mistakes are less common in animals. These mistakes can produce sex cells with too many or too few chromosomes, as shown in **Figure 13.** Sometimes, zygotes produced from these sex cells die. If the zygote lives, every cell in the organism that grows from that zygote usually will have the wrong number of chromosomes. Organisms with the wrong number of chromosomes may not grow normally.

Section  Assessment

1. Compare and contrast sexual and asexual reproduction.
2. What is a zygote, and how is it formed?
3. Give two examples of sex cells. Where are sex cells formed?
4. Compare what happens to chromosomes during anaphase I and anaphase II.
5. **Think Critically** Plants grown from runners and leaf cuttings have the same traits as the parent plant. Plants grown from seeds can vary from the parent plants in many ways. Suggest an explanation for why this can happen.

Skill Builder Activities

6. **Making and Using Tables** Make a table to compare mitosis and meiosis in humans. Vertical headings should include: *What Type of Cell (Body or Sex), Beginning Cell (Haploid or Diploid), Number of Cells Produced, End-Product Cell (Haploid or Diploid),* and *Number of Chromosomes in Cells Produced.* **For more help, refer to the Science Skill Handbook.**
7. **Communicating** Write a poem, song, or another memory device to help you remember the steps and outcome of meiosis. **For more help, refer to the Science Skill Handbook.**

DNA

As You Read

What You'll Learn

- **Identify** the parts of a DNA molecule and its structure.
- **Explain** how DNA copies itself.
- **Describe** the structure and function of each kind of RNA.

Vocabulary

DNA
gene
RNA
mutation

Why It's Important

DNA helps determine nearly everything your body is and does.

What is DNA?

Why was the alphabet one of the first things you learned when you started school? Letters are a code that you need to know before you learn to read. A cell also uses a code that is stored in its hereditary material. The code is a chemical called deoxyribonucleic (dee AHK sih ri boh noo klay ihk) acid, or **DNA.** It contains information for an organism's growth and function. **Figure 14** shows how DNA is stored in cells that have a nucleus. When a cell divides, the DNA code is copied and passed to the new cells. In this way, new cells receive the same coded information that was in the original cell. Every cell that has ever been formed in your body or in any other organism contains DNA.

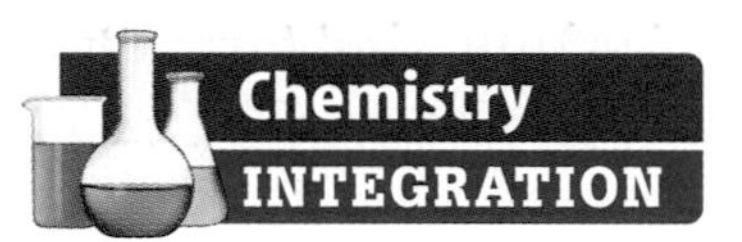

Discovering DNA Since the mid-1800s, scientists have known that the nuclei of cells contain large molecules called nucleic acids. By 1950, chemists had learned what the nucleic acid DNA was made of, but they didn't understand how the parts of DNA were arranged.

Figure 14
DNA is part of the chromosomes found in a cell's nucleus.

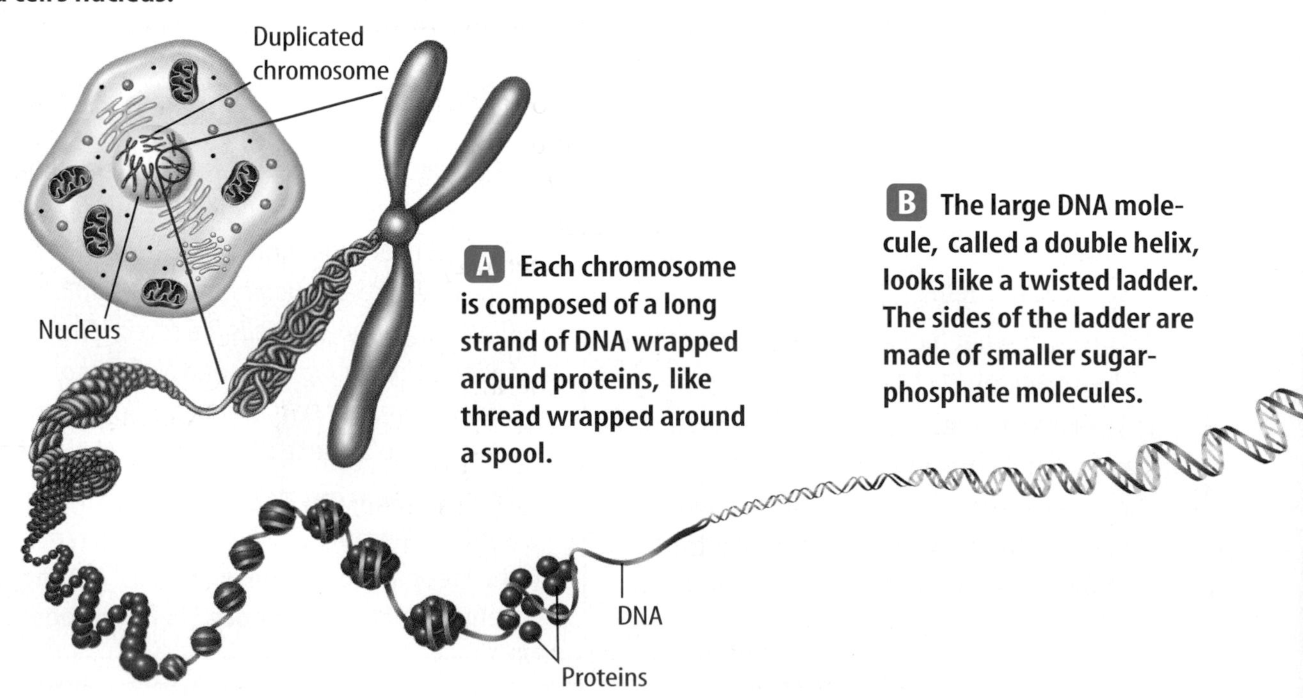

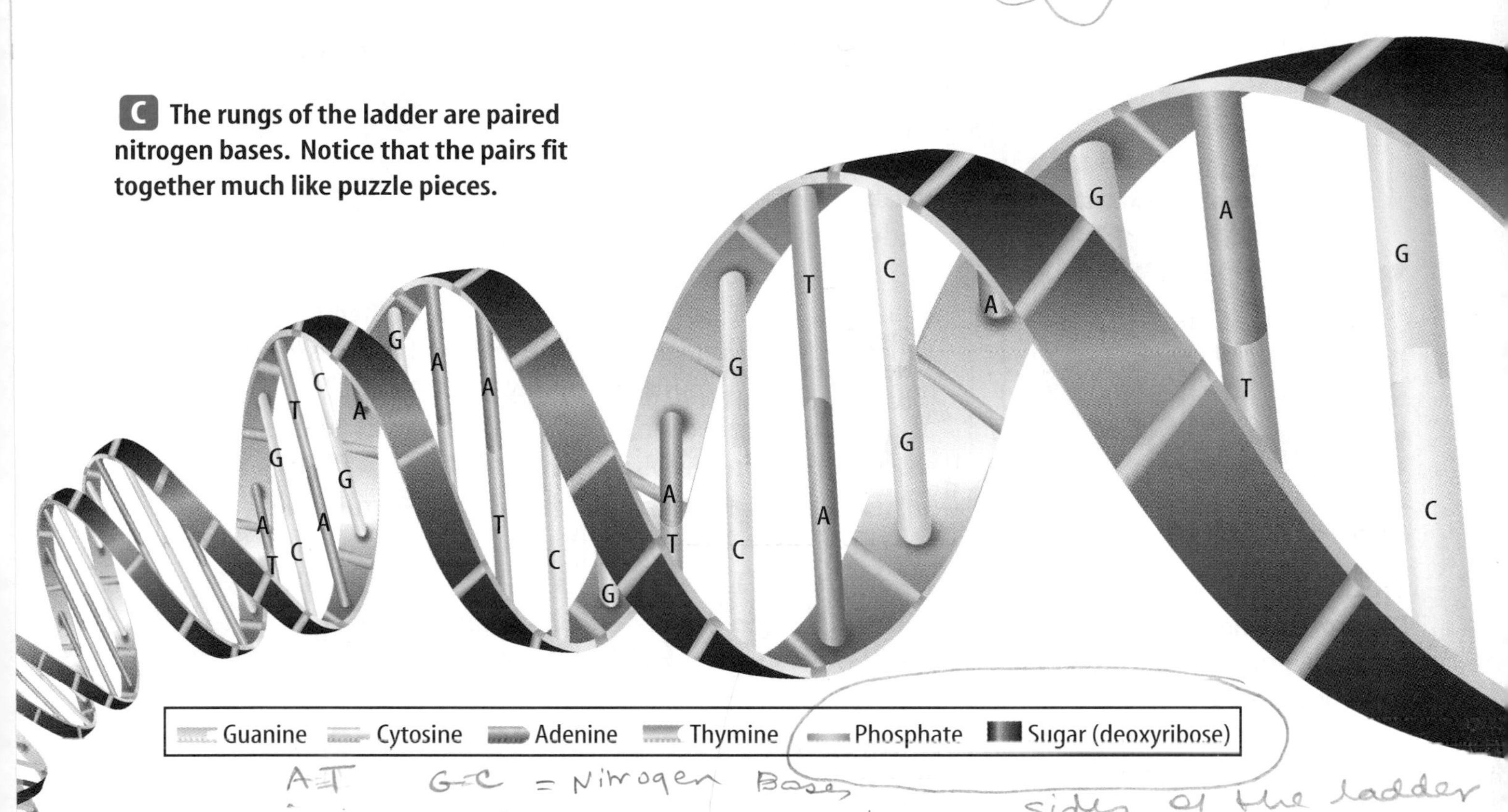

C The rungs of the ladder are paired nitrogen bases. Notice that the pairs fit together much like puzzle pieces.

DNA's Structure In 1952, scientist Rosalind Franklin discovered that DNA is two chains of molecules in a spiral form. By using an X-ray technique, Dr. Franklin showed that the large spiral was probably made up of two spirals. As it turned out, the structure of DNA is similar to a twisted ladder. In 1953, using the work of Franklin and others, scientists James Watson and Francis Crick made a model of a DNA molecule.

A DNA Model What does DNA look like? According to the Watson and Crick DNA model, each side of the ladder is made up of sugar-phosphate molecules. Each molecule consists of the sugar called deoxyribose (dee AHK sih ri bohs) and a phosphate group. The rungs of the ladder are made up of other molecules called nitrogen bases. Four kinds of nitrogen bases are found in DNA—adenine (AD un een), guanine (GWAHN een), cytosine (SITE uh seen), and thymine (THI meen). The bases are represented by the letters A, G, C, and T. The amount of cytosine in cells always equals the amount of guanine, and the amount of adenine always equals the amount of thymine. This led to the hypothesis that these bases occur as pairs in DNA. **Figure 14** shows that adenine always pairs with thymine, and guanine always pairs with cytosine. Like interlocking pieces of a puzzle, each base bonds only with its correct partner.

Reading Check *What are the nitrogen base pairs in a DNA molecule?*

Modeling DNA Replication

Procedure

1. Suppose you have a segment of DNA that is six nitrogen base pairs in length. On **paper,** using the letters A, T, C, and G, write a combination of six pairs remembering that A and T are always a pair and C and G are always a pair.
2. Duplicate your segment of DNA. On paper, diagram how this happens and show the new DNA segments.

Analysis

Compare the order of bases of the original DNA to the new DNA molecules.

Controlling Genes You might think that because most cells in an organism have exactly the same chromosomes and the same genes, they would make the same proteins, but they don't. In many-celled organisms like you, each cell uses only some of the thousands of genes that it has to make proteins. Just as each actor uses only the lines from the script for his or her role, each cell uses only the genes that direct the making of proteins that it needs. For example, muscle proteins are made in muscle cells, as represented in **Figure 18,** but not in nerve cells.

Cells must be able to control genes by turning some genes off and turning other genes on. They do this in many different ways. Sometimes the DNA is twisted so tightly that no RNA can be made. Other times, chemicals bind to the DNA so that it cannot be used. If the incorrect proteins are produced, the organism cannot function properly.

Figure 18
Each cell in the body produces only the proteins that are necessary to do its job.

Mutations

Sometimes mistakes happen when DNA is being copied. Imagine that the copy of the script the director gave you was missing three pages. You use your copy to learn your lines. When you begin rehearsing for the play, everyone is ready for one of the scenes except for you. What happened? You check your copy of the script against the original and find that three of the pages are missing. Because your script is different from the others, you cannot perform your part correctly.

If DNA is not copied exactly, the proteins made from the instructions might not be made correctly. These mistakes, called **mutations,** are any permanent change in the DNA sequence of a gene or chromosome of a cell. Some mutations include cells that receive an entire extra chromosome or are missing a chromosome. Outside factors such as X rays, sunlight, and some chemicals have been known to cause mutations.

When are mutations likely to occur?

Figure 19
Because of a defect on chromosome 2, the mutant fruit fly has short wings and cannot fly. *Could this defect be transferred to the mutant's offspring? Explain.*

Results of a Mutation Genes control the traits you inherit. Without correctly coded proteins, an organism can't grow, repair, or maintain itself. A change in a gene or chromosome can change the traits of an organism, as illustrated in **Figure 19.**

If the mutation occurs in a body cell, it might or might not be life threatening to the organism. However, if a mutation occurs in a sex cell, then all the cells that are formed from that sex cell will have that mutation. Mutations add variety to a species when the organism reproduces. Many mutations are harmful to organisms, often causing their death. Some mutations do not appear to have any effect on the organism, and some can even be beneficial. For example, a mutation to a plant might cause it to produce a chemical that certain insects avoid. If these insects normally eat the plant, the mutation will help the plant survive.

Research Visit the Glencoe Science Web site at **science.glencoe.com** for more information about what genes are present on the chromosomes of a fruit fly. Make a poster that shows one of the chromosomes and some of the genes found on that chromosome.

Section  Assessment

1. How does DNA make a copy of itself?
2. How are the codes for proteins carried from the nucleus to the ribosomes?
3. A single strand of DNA has the bases AGTAAC. Using letters, show a matching DNA strand from this pattern.
4. How is tRNA used when cells build proteins?
5. **Think Critically** You begin as one cell. Compare the DNA in one of your brain cells to the DNA in one of your heart cells.

Skill Builder Activities

6. **Concept Mapping** Using a network tree concept map, show how DNA and RNA are alike and how they are different. **For more help, refer to the Science Skill Handbook.**
7. **Using a Word Processor** Use a word processor to make an outline of the events that led up to the discovery of DNA. Use library resources to find this information. **For more help, refer to the Technology Skill Handbook.**

Activity Use the Internet

Mutations

Mutations can result in dominant or recessive genes. A recessive characteristic can appear only if an organism has two recessive genes for that characteristic. However, a dominant characteristic can appear if an organism has one or two dominant genes for that characteristic. Why do some mutations result in more common traits while others do not?

Fantail Pigeon

Recognize the Problem

How can a mutation become a common trait?

Form a Hypothesis

Form a hypothesis about how a mutation can become a common trait.

Goals

- **Observe** traits of various animals.
- **Research** how mutations become traits.
- Gather data about mutations.
- Make a frequency table of your findings and communicate them to other students.

Data Source

SCIENCE*Online* Go to the Glencoe Science Web site at **science.glencoe.com** for more information on common genetic traits in different animals, recessive and dominant genes, and data from other students.

White tiger

Test Your Hypothesis

Plan

1. **Observe** common traits in various animals, such as household pets or animals you might see in a zoo.
2. **Learn** what genes carry these traits in each animal.
3. **Research** the traits to discover which ones are results of mutations. Are all mutations dominant? Are any of these mutations beneficial?

Do

1. Make sure your teacher approves your plan before you start.
2. Visit the Glencoe Science Web site for links to different sites about mutations and genetics.
3. **Decide** if a mutation is beneficial, harmful, or neither. Record your data in your Science Journal.

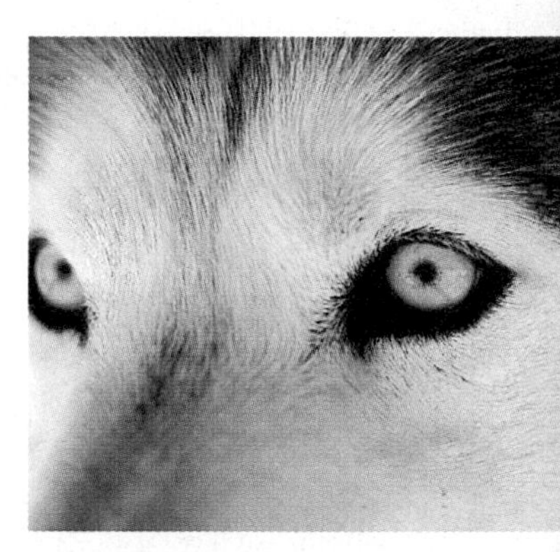

Siberian Husky's eyes

Analyze Your Data

1. **Record** in your Science Journal a list of traits that are results of mutations.
2. **Describe** an animal, such as a pet or an animal you've seen in the zoo. Point out which traits are known to be the result of a mutation.
3. Make a chart that compares recessive mutations to dominant mutations. Which are more common?
4. Share your data with other students by posting it on the Glencoe Science Web site.

Draw Conclusions

1. **Compare** your findings to those of your classmates and other data on the Glencoe Science Web site. What were some of the traits your classmates found that you did not? Which were the most common?
2. Look at your chart of mutations. Are all mutations beneficial? When might a mutation be harmful to an organism?
3. How would your data be affected if you had performed this activity when one of these common mutations first appeared? Do you think you would see more or less animals with this trait?
4. Mutations occur every day but we only see a few of them. Infer how many mutations over millions of years can lead to a new species.

Communicating Your Data

SCIENCE Online Find this *Use the Internet* activity on the Glencoe Science Web site at **science.glencoe.com. Post** your data in the table provided. Combine your data with that of other students and make a chart that shows all of the data.

Oops! Accidents in SCIENCE

SOMETIMES GREAT DISCOVERIES HAPPEN BY ACCIDENT!

A Tangled

How did a scientist get chromosomes to separate?

Thanks to chromosomes, each of us is unique!

Viewed under a microscope, chromosomes in cells sometimes look a lot like tangled spaghetti. That's why during the early 1900s, scientists had such a hard time figuring out how many chromosomes are in each human cell.

Tale

Imagine then, how Dr. Tao-Chiuh Hsu (dow shew•SEW) must have felt when he looked into a microscope and saw "beautifully scattered chromosomes." The problem was, Hsu didn't know what he had done to separate the chromosomes into countable strands.

"I tried to study those slides and set up some more cultures to repeat the miracle," Hsu explained. "But nothing happened."

For three months, Hsu toiled in the lab, changing every variable he could think of to make the chromosomes separate again. In April 1952, he reduced the amount of salt and increased the amount of water in the solution used to prepare the cells for study, and his efforts were finally rewarded. Hsu quickly realized that the chromosomes separated because of osmosis.

Osmosis is the movement of water molecules through cell membranes. This movement occurs in predictable ways. The water molecules move from areas with higher concentrations of water to areas with lower concentrations of water. In Hsu's case, the solution had a higher concentration of water than the cell did. So water moved from the solution into the cell and the cell swelled until it finally exploded. The chromosomes suddenly were visible as separate strands.

What made the cells swell the first time? Apparently, a lab technician had mixed the solution incorrectly. "Since nearly four months had elapsed, there was no way to trace who actually had prepared that particular [solution]," Hsu noted. "Therefore, this heroine must remain anonymous."

The Real Count

Although Hsu's view of the chromosomes was fairly clear, he mistakenly estimated the number of chromosomes in a human cell. He put the count at 48, which was the number that most scientists of the day accepted. By 1956, however, other scientists improved upon Hsu's techniques and concluded that there are 46 chromosomes in a human cell. Because chromosomes contain the genes that determine each person's characteristics, this discovery helped scientists better understand genetic diseases and disorders. Scientists also have a better idea of why every person, including you, is unique.

These chromosomes are magnified 500 times.

CONNECTIONS **Research** Until the 1950s, scientists believed that there were 48 chromosomes in a human cell. Research the developments that led scientists to the conclusion that the human cell has 46 chromosomes. Use the Glencoe Science Web site to get started.

SCIENCE Online
For more information, visit science.glencoe.com

Chapter 8 Study Guide

Reviewing Main Ideas

Section 1 Cell Division and Mitosis

1. The life cycle of a cell has two parts—growth and development and cell division. Cell division includes mitosis and the division of the cytoplasm.
2. In mitosis, the nucleus divides to form two identical nuclei. Mitosis occurs in four continuous steps, or phases—prophase, metaphase, anaphase, and telophase.
3. Cell division in animal cells and plant cells is similar, but plant cells do not have centrioles and animal cells do not form cell walls.
4. Organisms use cell division to grow, to replace cells, and for asexual reproduction. Asexual reproduction produces organisms with DNA identical to the parent's DNA. Fission, budding, and regeneration can be used for asexual reproduction. *How would cell division help heal this broken bone?*

Section 2 Sexual Reproduction and Meiosis

1. Sexual reproduction results when a male sex cell enters the female sex cell. This event is called fertilization, and the cell that forms is called the zygote.
2. Before fertilization, meiosis occurs in the reproductive organs, producing four haploid sex cells from one diploid cell.
3. During meiosis, two divisions of the nucleus occur.
4. Meiosis ensures that offspring produced by fertilization have the same number of chromosomes as their parents. *If the diploid number of a frog is 26, how many chromosomes does this tadpole have?*

Section 3 DNA

1. DNA—the genetic material of all organisms—is a large molecule made up of two twisted strands of sugar-phosphate molecules and nitrogen bases.
2. All cells contain DNA. The section of DNA on a chromosome that directs the making of a specific protein is a gene.
3. DNA can copy itself and is the pattern from which RNA is made. Messenger RNA, ribosomal RNA, and transfer RNA are used to make proteins.
4. Sometimes changes in DNA occur. Permanent changes in DNA are called mutations. *Why does this fruit fly have four wings instead of the normal two?*

After You Read

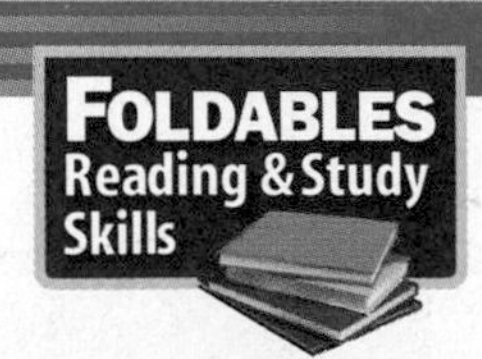

To help you review cell reproduction, use the Organizational Study Fold about the cell you made at the beginning of the chapter.

Visualizing Main Ideas

Think of four ways that organisms can use mitosis and fill out the spider diagram below.

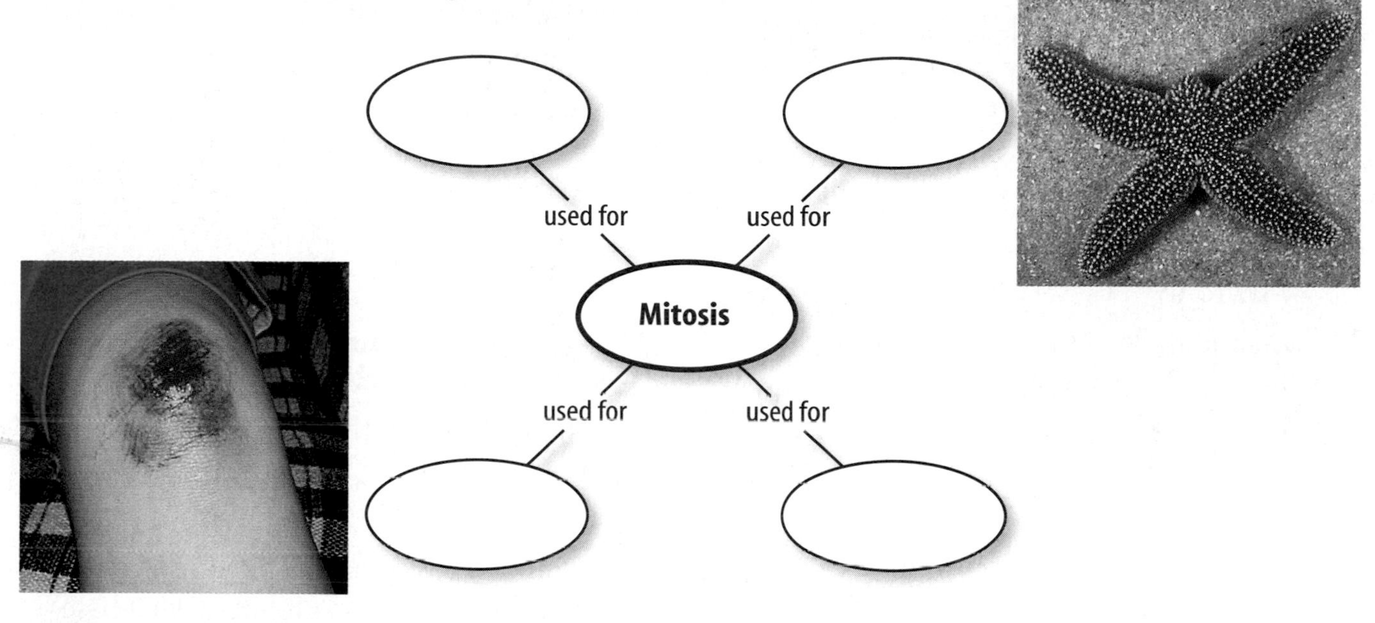

Vocabulary Review

Vocabulary Words

- **a.** asexual reproduction
- **b.** chromosome
- **c.** diploid
- **d.** DNA
- **e.** egg
- **f.** fertilization
- **g.** gene
- **h.** haploid
- **i.** meiosis
- **j.** mitosis
- **k.** mutation
- **l.** RNA
- **m.** sexual reproduction
- **n.** sperm
- **o.** zygote

Study Tip

Be a teacher—organize a group of friends and instruct each person to review a section of the chapter for the group. Teaching helps you remember and understand information thoroughly.

Using Vocabulary

Replace each underlined word in the following statements with the correct vocabulary word.

1. Muscle and skin cells are sex cells.
2. Digestion produces two identical cells.
3. An example of a nucleic acid is sugar.
4. A cell is the code for a protein.
5. A diploid sperm is formed during meiosis.
6. Budding is a type of meiosis.
7. A ribosome is a structure in the nucleus that contains hereditary material.
8. Respiration produces four sex cells.
9. As a result of fission, a new organism develops that has its own unique identity.
10. An error made during the copying of DNA is called a protein.

Chapter 8 Assessment

Checking Concepts

Choose the word or phrase that best answers the question.

1. Which of the following is a double spiral molecule with pairs of nitrogen bases?
   A) RNA **C)** protein
   B) amino acid **D)** DNA

2. What is in RNA but NOT in DNA?
   A) thymine **C)** adenine
   B) thyroid **D)** uracil

3. If a diploid tomato cell has 24 chromosomes, how many chromosomes will the tomato's sex cells have?
   A) 6 **C)** 24
   B) 12 **D)** 48

4. During a cell's life cycle, when do chromosomes duplicate?
   A) anaphase **C)** interphase
   B) metaphase **D)** telophase

5. When do chromatids separate during mitosis?
   A) anaphase **C)** metaphase
   B) prophase **D)** telophase

6. How many chromosomes are in the original cell compared to those in the new cells formed by cell division?
   A) the same amount **C)** twice as many
   B) half as many **D)** four times as many

7. What can budding, fission, and regeneration be used for?
   A) mutations
   B) sexual reproduction
   C) cell cycles
   D) asexual reproduction

8. What is any permanent change in a gene or a chromosome called?
   A) fission **C)** replication
   B) reproduction **D)** mutation

9. What does meiosis produce?
   A) cells with the diploid chromosome number
   B) cells with identical chromosomes
   C) sex cells
   D) a zygote

10. What type of nucleic acid carries the codes for making proteins from the nucleus to the ribosome?
    A) DNA **C)** protein
    B) RNA **D)** genes

Thinking Critically

11. If the sequence of bases on one side of DNA is ATCCGTC, what is the sequence on its other side?

12. A strand of RNA made using the DNA pattern ATCCGTC would have what base sequence? Look at **Figure 14** for a hint.

13. Will a mutation in a human skin cell be passed on to the person's offspring? Explain.

14. What occurs in mitosis that gives the new cells identical DNA?

15. How could a zygote end up with an extra chromosome?

Developing Skills

16. **Classifying** Copy and complete this table about DNA and RNA.

DNA and RNA

	DNA	RNA
Number of Strands		
Type of Sugar		
Letter Names of Bases		
Where Found		

17. Concept Mapping Complete the events chain concept map of DNA synthesis.

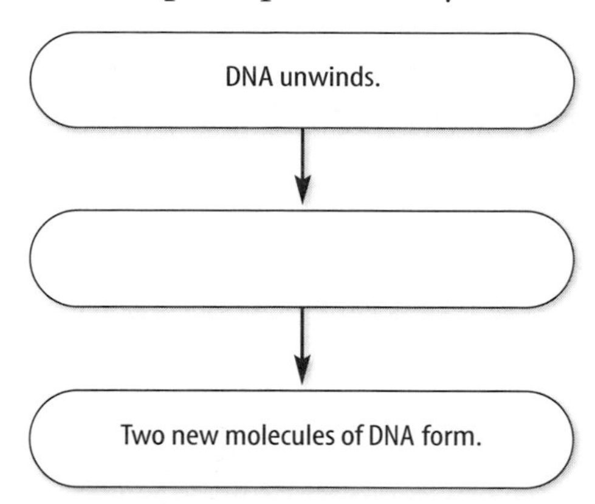

18. Comparing and Contrasting Meiosis is two divisions of a reproductive cell's nucleus. It occurs in a continuous series of steps. Compare and contrast the steps of meiosis I to the steps of meiosis II.

19. Forming Hypotheses Make a hypothesis about the effect of an incorrect mitotic division on the new cells produced.

20. Concept Mapping Make an events chain concept map of what occurs from interphase in the parent cell to the formation of the zygote. Tell whether the chromosome's number at each stage is haploid or diploid.

Performance Assessment

21. Flash Cards Make a set of 11 flash cards with drawings of a cell that show the different stages of meiosis. Shuffle your cards and then put them in the correct order. Give them to another student in the class to try.

TECHNOLOGY

Go to the Glencoe Science Web site at **science.glencoe.com** or use the **Glencoe Science CD-ROM** for additional chapter assessment.

Test Practice

A scientist studied the reproduction of human skin cells. The scientist examined several skin cells using a microscope. The table below summarizes what she learned.

Skin Cells

Cell	Phase of Division	Characteristic
1	Anaphase	Chromosome separation
2	Telophase	Cytoplasm division
3	Prophase	Visible chromosomes
4	Metaphase	Chromosomes line up

Use the information in the table to answer the following questions.

1. What process is taking place in all of the cells?

A) cell division
B) fertilization
C) cytoplasm division
D) chromosome separation

2. Which is the correct order of the stages, from first to last, in the cell division of a skin cell?

F) 3, 4, 1, 2
G) 1, 3, 2, 4
H) 1, 2, 4, 3
J) 2, 1, 3, 4

3. Since the process described in the table produces two new identical cells, before it begins the chromosomes in the cell must _____ .

A) divide in half
B) find a mate
C) duplicate
D) disintegrate

CHAPTER

Plant Reproduction

Saplings and other plants grow among the remains of trees that were destroyed by fire. Where did these new plants come from? Did they grow from seeds that survived the fire? Perhaps they grew from plant roots and stems that survived underground. In either case, these plants are the result of plant reproduction. In this chapter, you will learn how different groups of plants reproduce and how plants can be dispersed from place to place.

What do you think?

Science Journal Look at the picture below with a classmate. Discuss what this might be. Here's a hint: *In many plants these are colorful and have pleasant aromas.* Write your answer or best guess in your Science Journal.

You might know that most plants grow from seeds. Seeds are usually found in the fruits of plants. When you eat watermelon, it can contain many small seeds. Do all fruits contain seeds? Do this activity to find out.

Predict where seeds are found

1. Obtain two grapes from your teacher. Each grape should be from a different plant.
2. Split each grape in half and examine the insides of each grape. **WARNING:** *Do not eat the grapes.*

Observe

Were seeds found in both grapes? Hypothesize how new grape plants could be grown if no seeds are produced. In your Science Journal list three other fruits you know of that do not contain seeds.

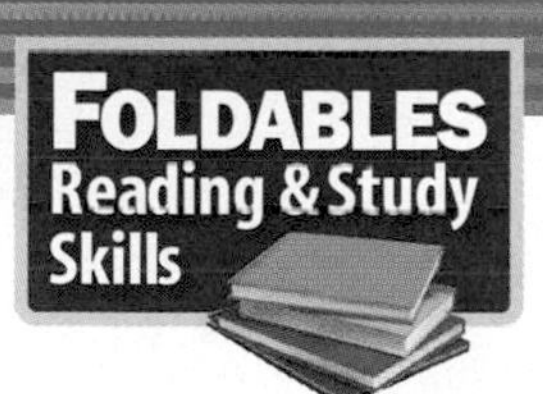

Before You Read

Making a Venn Diagram Study Fold **Make the following Foldable to compare and contrast sexual and asexual characteristics of a plant.**

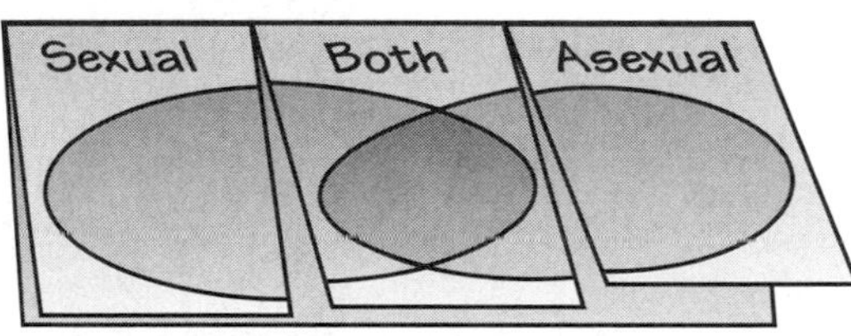

1. Place a sheet of paper in the front of you so the long side is at the top. Fold the paper in half from top to bottom.
2. Fold both sides in. Unfold the paper so three sections show.
3. Through the top thickness of the paper, cut along each of the fold lines to the top fold, forming three tabs. Label each tab *Sexual, Both,* and *Asexual* as shown.
4. Before you read the chapter, draw circles across the front of the page, as shown.
5. As you read the chapter write information about sexual and asexual reproduction under the left and right tabs.

SECTION 1

Introduction to Plant Reproduction

As You Read

What You'll Learn

- **Distinguish** between the two types of plant reproduction.
- **Describe** the two stages in a plant's life cycle.

Vocabulary

spore
gametophyte stage
sporophyte stage

Why It's Important

You can grow new plants without using seeds.

Types of Reproduction

Do people and plants have anything in common? You don't have leaves or roots, and a plant doesn't have a heart or a brain. Despite these differences, you are alike in many ways—you need water, oxygen, energy, and food to grow. Like humans, plants also can reproduce and make similar copies of themselves. Although humans have only one type of reproduction, most plants can reproduce in two different ways, as shown in **Figure 1.**

Sexual reproduction in plants and animals requires the production of sex cells—usually called sperm and eggs—in reproductive organs. The offspring produced by sexual reproduction are genetically different from either parent organism.

A second type of reproduction is called asexual reproduction. This type of reproduction does not require the production of sex cells. During asexual reproduction, one organism produces offspring that are genetically identical to it. Most plants have this type of reproduction, but humans and most other animals don't.

Figure 1
Many plants reproduce sexually with flowers that contain male and female parts.

A In crocus flowers, bees and other insects help get the sperm to the egg.

B Other plants can reproduce asexually. A cutting from this impatiens plant can be placed in water and will grow new roots. This new plant can then be planted in soil.

Figure 2
Asexual reproduction in plants takes many forms.

A **The eyes on these potatoes have begun to sprout. If a potato is cut into pieces, each piece that contains an eye can be planted and will grow into a new potato plant.**

B **The grass plants spread by reproducing asexually.**

Asexual Plant Reproduction Do you like to eat oranges and grapes that have seeds, or do you like seedless fruit? If these plants do not produce seeds, how do growers get new plants? Growers can produce new plants by asexual reproduction because many plant cells have the ability to grow into a variety of cell types. New plants can be grown from just a few cells in the laboratory. Under the right conditions, an entire plant can grow from one leaf or just a portion of the stem or root. When growers use these methods to start new plants, they must make sure that the leaf, stem, or root cuttings have plenty of water and anything else that they need to survive.

Asexual reproduction has been used to produce plants for centuries. The white potatoes shown in **Figure 2A** were probably produced asexually. Many plants, such as lawn grasses shown in **Figure 2B,** can spread and cover wide areas because their stems grow underground and produce new grass plants asexually along the length of the stem.

Sexual Plant Reproduction Although plants and animals have sexual reproduction, there are differences in the way that it occurs. An important event in sexual reproduction is fertilization. Fertilization occurs when a sperm and egg combine to produce the first cell of the new organism, the zygote. How do the sperm and egg get together in plants? In some plants, water or wind help bring the sperm to the egg. For other plants, animals such as insects help bring the egg and sperm together.

Reading Check *How does fertilization occur in plants?*

Mini LAB

Observing Asexual Reproduction

Procedure

1. Using a pair of **scissors,** cut a stem with at least two pairs of leaves from a **coleus or another houseplant.**
2. Carefully remove the bottom pair of leaves.
3. Place the cut end of the stem into a **cup that is half-filled with water** for two weeks. Wash your hands.
4. Remove the new plant from the water and plant it in a small **container** of **soil.**

Analysis

1. Draw and label your results in your **Science Journal.**
2. Predict how the new plant and the plant from which it was taken are genetically related.

Figure 3
Some plants can fertilize themselves. Others require two different plants before fertilization can occur.

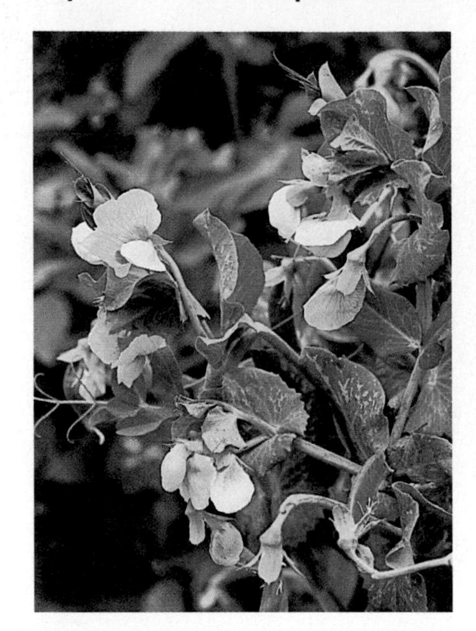

A **Flowers of pea plants contain male and female structures, and each flower can fertilize itself.**

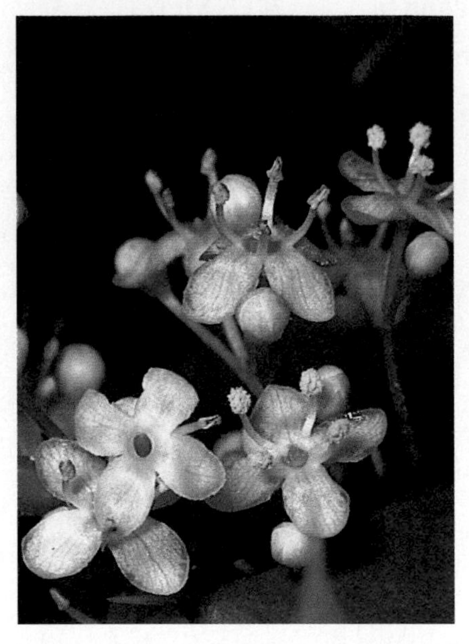

B **These holly flowers contain only male reproductive structures, so they can't fertilize themselves.**

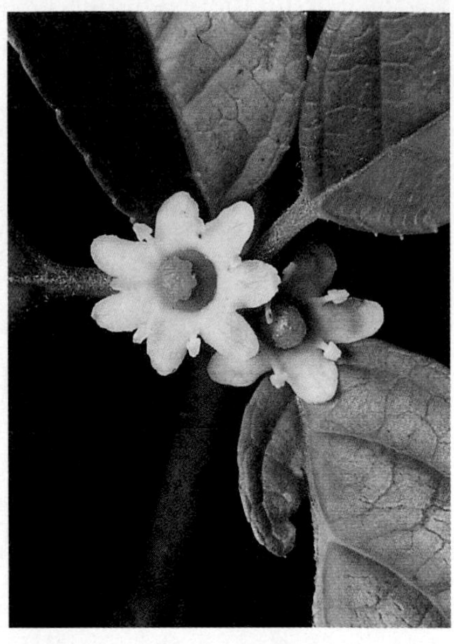

C **Compare the flowers of this female holly plant to those of the male plant.**

Reproductive Organs A plant's female reproductive organs produce eggs and male reproductive organs produce sperm. Depending on the species, these reproductive organs can be on the same plant or on separate plants, as shown in **Figure 3.** If a plant has both organs, it usually can reproduce by itself. However, some plants that have both sex organs still must exchange sex cells with other plants of the same type to reproduce.

In some plant species, the male and female reproductive organs are on separate plants. For example, holly plants are either female or male. For fertilization to occur, holly plants with flowers that have different sex organs must be near each other. In that case, after the eggs in female holly flowers are fertilized, berries can form.

Another difference between you and a plant is how and when plants produce sperm and eggs. You will begin to understand this difference as you examine the life cycle of a plant.

Research Visit the Glencoe Science Web site at **science.glencoe.com** to find out more about male and female plants. In your Science Journal, list four plants that have male and female repoductive structures on separate plants.

Plant Life Cycles

All organisms have life cycles. Your life cycle started when a sperm and an egg came together to produce the zygote that would grow and develop into the person you are today. A plant also has a life cycle. It can start when an egg and a sperm come together, eventually producing a mature plant.

Two Stages During your life cycle, all structures in your body are formed by cell division and made up of diploid cells—cells with a full set of chromosomes. However, sex cells form by meiosis and are haploid—they have half a set of chromosomes.

Plants have a two-stage life cycle, as shown in **Figure 4.** The two stages are the gametophyte (guh MEE tuh fite) stage and the sporophyte (SPOHR uh fite) stage.

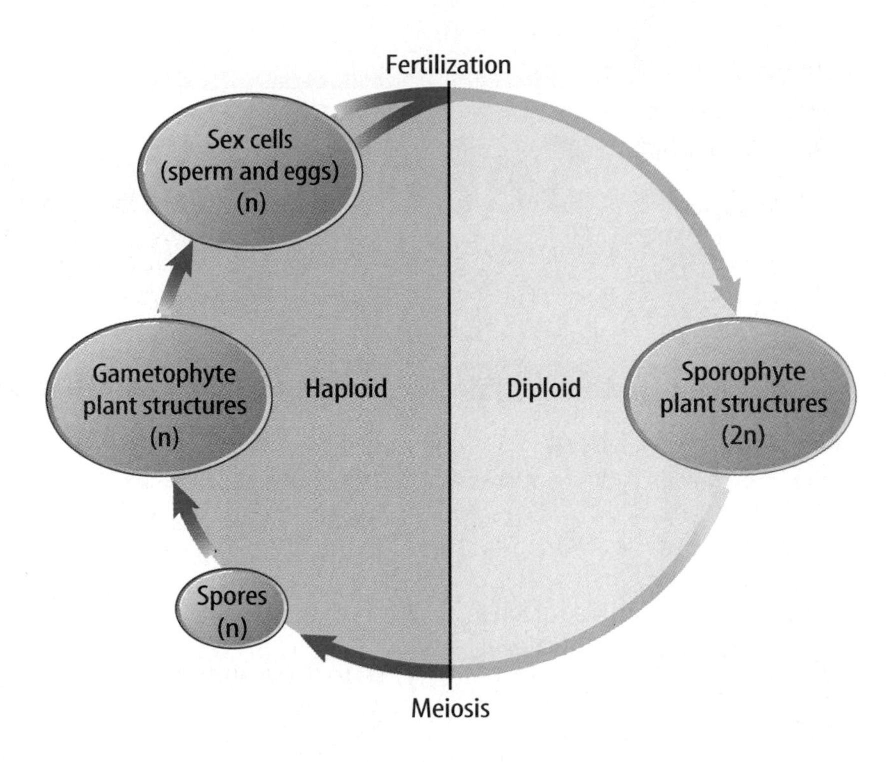

Figure 4
Plants produce diploid and haploid plant structures.

Gametophyte Stage When cells in reproductive organs undergo meiosis and produce haploid cells called **spores,** the **gametophyte stage** begins. Some plants release spores into their surroundings. Spores divide by cell division to form plant structures or an entire new plant. The cells in these structures or plants are haploid. Some of these cells undergo cell division and form sex cells.

Sporophyte Stage Fertilization—the joining of haploid sex cells—begins the **sporophyte stage.** Cells formed in this stage have the diploid number of chromosomes. Meiosis occurs in some of these plant structures to form spores, and the cycle begins again.

Section  Assessment

1. Name two types of plant reproduction.
2. Compare and contrast the gametophyte stage and the sporophyte stage.
3. Describe how plants can be grown using asexual reproduction.
4. Explain how sexual reproduction is different in plants and animals.
5. **Think Critically** You see a plant that you like and want to grow an identical one. What type of plant reproduction would you use? Why?

Skill Builder Activities

6. **Drawing Conclusions** You use a microscope to observe the nuclei of several cells from a plant. Each one has only half the number of chromosomes you would expect. What do you conclude about this stage of its life cycle? **For more help, refer to the Science Skill Handbook.**
7. **Communicating** In your Science Journal write your own analogy about the diploid and haploid stages of a plant life cycle. **For more help, refer to the Science Skill Handbook.**

SECTION 2

Seedless Reproduction

As You Read

What You'll Learn

- **Examine** the life cycles of a moss and a fern.
- **Explain** why spores are important to seedless plants.
- **Identify** some special structures used by ferns for reproduction.

Vocabulary

frond
rhizome
sori
prothallus

Why It's Important

Seedless plants have adaptations for reproduction on land.

The Importance of Spores

If you want to grow plants like ferns and moss plants, you can't go to a garden store and buy a package of seeds—they don't produce seeds. You could, however, grow them from spores. These plants produce haploid spores at the end of their sporophyte stage in structures called spore cases. When the spore case breaks open, the spores are released and spread by wind or water. The spores, shown in **Figure 5,** can grow into plants that will produce sex cells.

Seedless plants include all nonvascular plants and some vascular plants. Nonvascular plants do not have structures that transport water and substances throughout the plant. Instead, water and substances simply move from cell to cell. Vascular plants have tubelike cells that transport water and substances throughout the plant.

Nonvascular Seedless Plants

If you walked in a damp, shaded forest, you probably would see mosses covering the ground or growing on a log. Mosses, liverworts, and hornworts are all nonvascular plants.

The sporophyte stage of most nonvascular plants is so small that it can be easily overlooked. Moss plants have a life cycle that is typical of how sexual reproduction occurs in this plant group.

Figure 5
Spores come in a variety of shapes. All spores are small and have a waterproof coating. Some, like the horsetail spores, have winglike structures that uncoil and allow them to be blown easily by the wind.

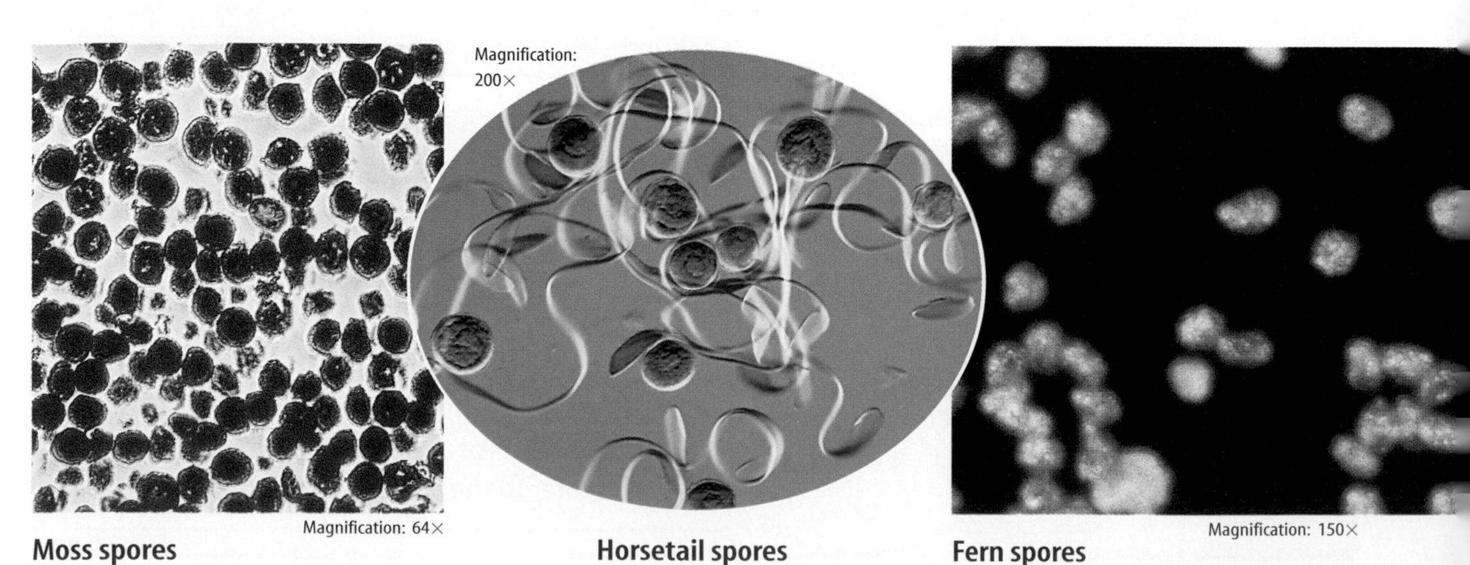

Moss spores **Horsetail spores** **Fern spores**

The Moss Life Cycle You recognize mosses as green, low-growing masses of plants. This is the gametophyte stage, which produces the sex cells. But the next time you see some moss growing, get down and look at it closely. If you see any brownish stalks growing up from the tip of the gametophyte plants, you are looking at the sporophyte stage. The sporophyte stage does not carry on photosynthesis. It depends on the gametophyte for nutrients and water. On the tip of the stalk is a tiny capsule. Inside the capsule millions of spores have been produced. When environmental conditions are just right, the capsule opens and the spores either fall to the ground or are blown away by the wind. New moss gametophytes can grow from each spore and the cycle begins again, as shown in **Figure 6.**

Figure 6
The life cycle of a moss alternates between gametophyte and sporophyte stages. *What is produced by the gametophyte stage?*

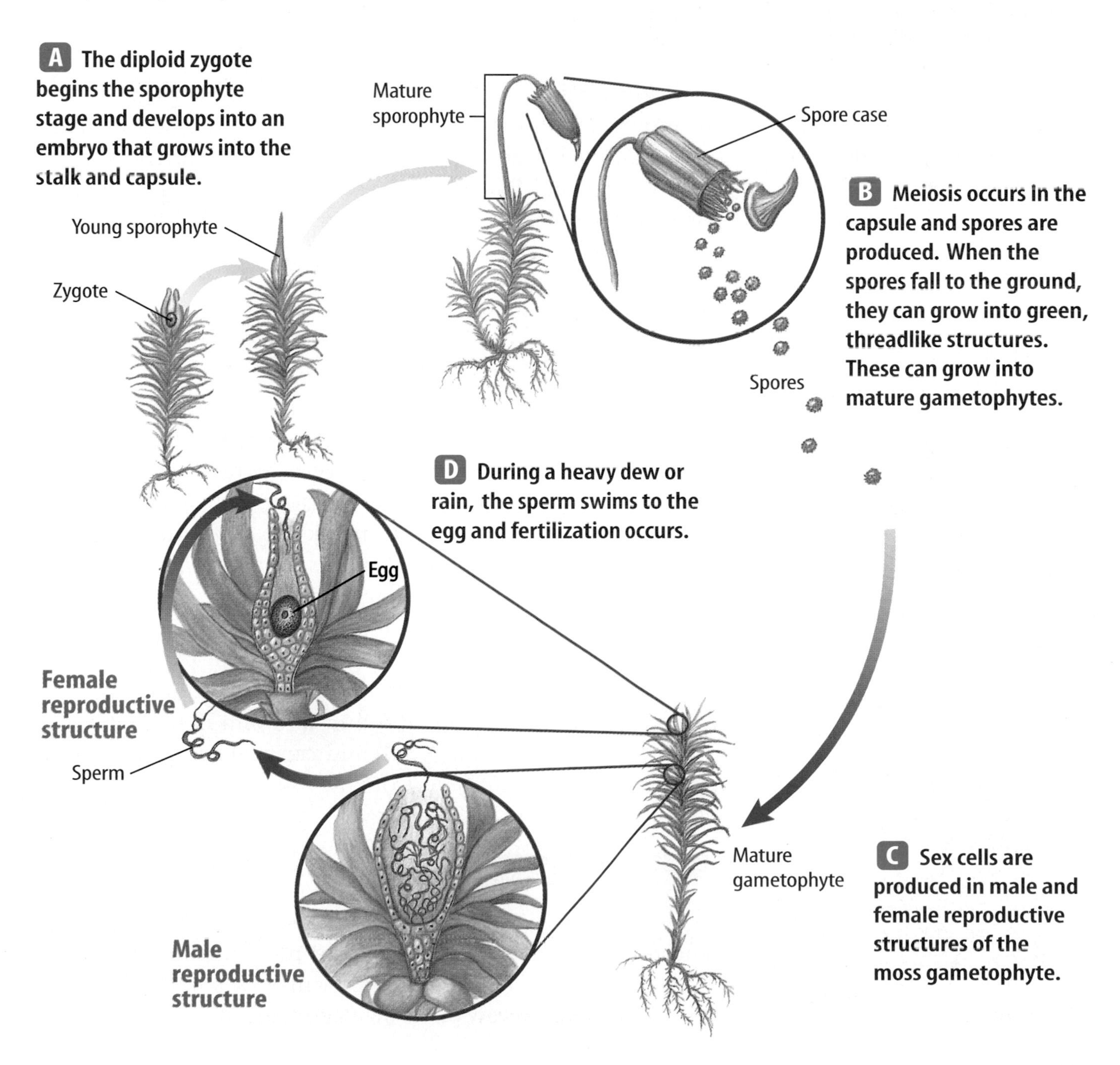

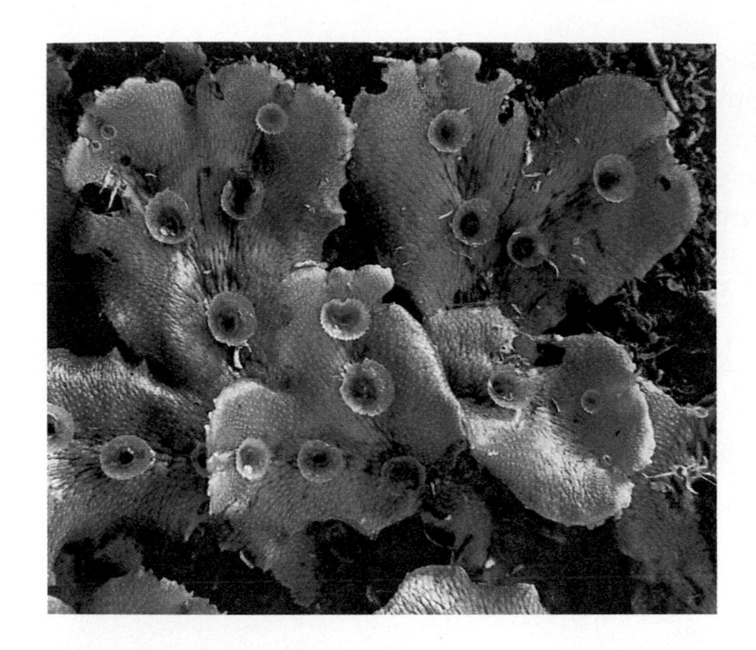

Figure 7
Small balls of cells grow in cup-like structures on the surface of the liverwort.

Nonvascular Plants and Asexual Reproduction Nonvascular plants also can reproduce asexually. For example, if a piece of a moss gametophyte plant breaks off, it can grow into a new plant. Liverworts can form small balls of cells on the surface of the gametophyte plant, as shown in **Figure 7.** These are carried away by water and grow into new gametophyte plants if they settle in a damp environment.

Vascular Seedless Plants

Millions of years ago most plants on Earth were vascular seedless plants. Today they are not as widespread.

Most vascular seedless plants are ferns. Other plants in this group include horsetails and club mosses. All of these plants have vascular tissue to transport water from their roots to the rest of the plant. Unlike the nonvascular plants, the gametophyte of vascular seedless plants is the part that is small and often overlooked.

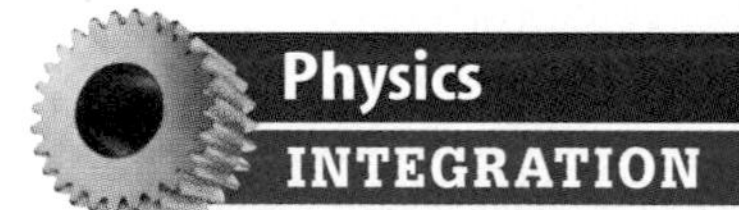

Catapults have been used by humans for thousands of years to launch objects. The spore cases of ferns act like tiny catapults as they eject their spores. In your Science Journal list tools, toys, and other objects that use catapult technology to work.

The Fern Life Cycle The fern plants that you see in nature or as houseplants are fern sporophyte plants. Fern leaves are called **fronds.** They grow from an underground stem called a **rhizome.** Roots that anchor the plant and absorb water and nutrients also grow from the rhizome. Fern sporophytes make their own food by photosynthesis. Fern spores are produced in structures called **sori** (singular, *sorus*), usually located on the underside of the fronds. Sori can look like crusty rust-, brown-, or dark-colored bumps. Sometimes they are mistaken for a disease or for something growing on the fronds.

If a fern spore lands on damp soil or rocks, it can grow into a small, green, heart-shaped gametophyte plant called a **prothallus** (proh THA lus). A prothallus is hard to see because most of them are only about 5 mm to 6 mm in diameter. The prothallus contains chlorophyll and can make its own food. It absorbs water and nutrients from the soil. The life cycle of a fern is shown in **Figure 8.**

Reading Check *What is the gametophyte plant of a fern called?*

Ferns may reproduce asexually, also. Fern rhizomes grow and form branches. New fronds and roots develop from each branch. The new rhizome branch can be separated from the main plant. It can grow on its own and form more fern plants.

Figure 8
A fern's life cycle and a moss's are similar. However, the fern sporophyte and gametophyte are photosynthetic and can grow on their own.

A Meiosis takes place inside each spore case to produce thousands of spores.

Spore case

Spore

B Spores are ejected and fall to the ground. Each can grow into a prothallus, which is the gametophyte plant.

Spore grows to form prothallus

Female reproductive structure

Egg

C The prothallus contains the male and female reproductive structures where sex cells form.

Male reproductive structure

Sperm

D Water is needed for the sperm to swim to the egg. Fertilization occurs and a zygote is produced.

Zygote

E The zygote is the beginning of the sporophyte stage and grows into the familiar fern plant.

Young sporophyte growing on gametophyte

Section 2 Assessment

1. Describe the life cycle of mosses.
2. Explain the stages in the life cycle of a fern.
3. Compare and contrast the gametophyte plant of the moss with the gametophyte plant of the fern.
4. List several ways that seedless plants reproduce asexually.
5. **Think Critically** Why might some seedless plants reproduce only asexually during dry times of the year?

Skill Builder Activities

6. **Concept Mapping** Use an events-chain concept map to show the events in the life cycle of either a moss or fern. **For more help, refer to the Science Skill Handbook.**
7. **Solving One-Step Equations** Moss spores are usually no more than 0.1 mm in diameter. About how many spores would it take to equal the diameter of a penny? **For more help, refer to the Math Skill Handbook.**

Activity

Comparing Seedless Plants

All seedless plants have specialized structures that produce spores. Although these sporophyte structures have a similar function, they look different. The gametophyte plants also are different from each other. Do this activity and observe the similarities and differences among three groups of seedless plants.

What You'll Investigate

How are the gametophyte stages and the sporophyte stages of liverworts, mosses, and ferns similar and different?

Materials

live mosses, liverworts, and ferns
with gametophytes and sporophytes
hand lens
forceps
dropper
microscope slides and coverslips (2)
microscope
dissecting needle
pencil with eraser

Goals

- **Describe** the sporophyte and gametophyte forms of liverworts, mosses, and ferns.
- **Identify** the spore-producing structures of liverworts, mosses, and ferns.

Safety Precautions

Procedure

1. Obtain a gametophyte of each plant. With a hand lens, observe the rhizoids, leafy parts, and stemlike parts, if any are present.
2. Obtain a sporophyte of each plant and use a hand lens to observe it.

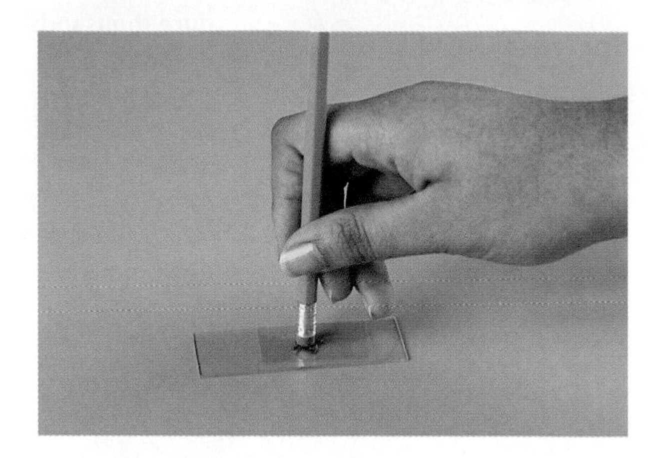

3. Locate the spore structure on the moss plant. Remove it and place it in a drop of water on the slide. Place a coverslip over it. Use the eraser of a pencil to gently push on the coverslip to release the spores. **WARNING:** *Do not break the coverslip.* Observe the spores under low and high power.
4. Make labeled drawings of all observations in your Science Journal.
5. Repeat steps 3 and 4 using a fern.

Conclude and Apply

1. For each plant, compare the gametophyte's appearance to the sporophyte's appearance.
2. **List** structure(s) common to all three plants.
3. **Hypothesize** about why each plant produces a large number of spores.

Communicating Your Data

Prepare a bulletin board that shows differences between the sporophyte and gametophyte stages of liverworts, mosses, and ferns. **For more help, refer to the Science Skill Handbook.**

SECTION 3

Seed Reproduction

The Importance of Pollen and Seeds

All the plants described so far have been seedless plants. However, the fruits and vegetables that you eat come from seed plants. Oak, maple, and other shade trees are also seed plants. All flowers are produced by seed plants. In fact, most of the plants on Earth are seed plants. How do you think they became such a successful group? Reproduction that involves pollen and seeds is part of the answer.

Pollen In seed plants, some spores develop into small structures called pollen grains. A **pollen grain,** as shown in **Figure 9,** has a water-resistant covering and contains gametophyte parts that can produce the sperm. The sperm of seed plants do not need to swim to the female part of the plant. Instead, they are carried as part of the pollen grain by gravity, wind, water currents, or animals. The transfer of pollen grains to the female part of the plant is called **pollination.**

After the pollen grain reaches the female part of a plant, sperm and a pollen tube are produced. The sperm moves through the pollen tube, then fertilization can occur.

As You Read

What **You'll Learn**

- **Examine** the life cycles of typical gymnosperms and angiosperms.
- **Describe** the structure and function of the flower.
- **Discuss** methods of seed dispersal in seed plants.

Vocabulary

pollen grain, pollination, ovule, stamen, pistil, ovary, germination

Why **It's Important**

Seeds from cones and flowers produce most plants on Earth.

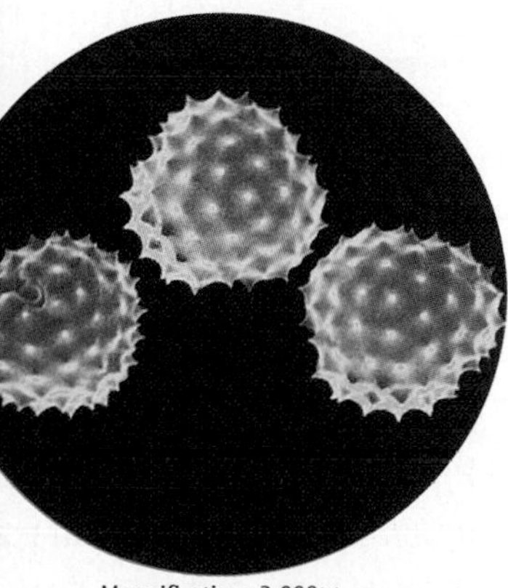

Magnification: 3,000×

Figure 9
The waterproof covering of a pollen grain is unique and can be used to identify the plant that it came from. This pollen from a ragweed plant is a common cause of hay fever.

Figure 10
Seeds have three main parts—a seed coat, stored food, and an embryo. This pine seed also has a wing.
What is the function of the wing?

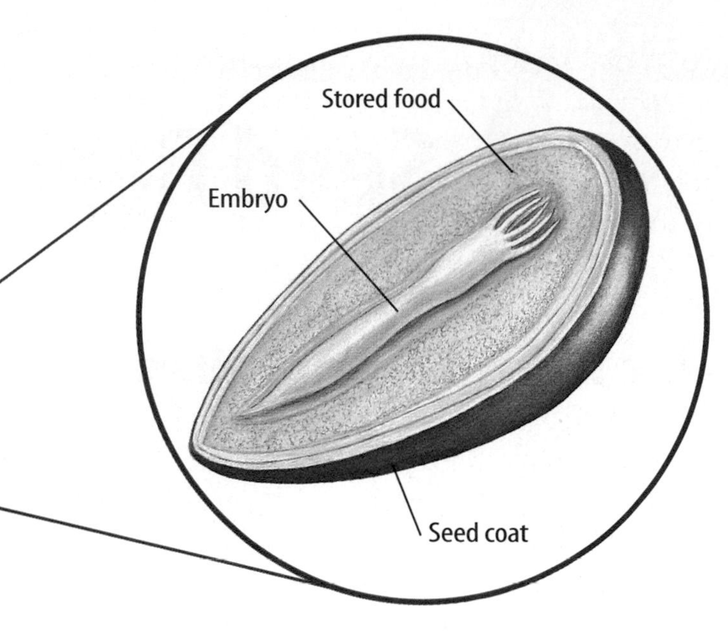

Seeds Following fertilization, the female part can develop into a seed. A seed consists of an embryo, stored food, and a protective seed coat, as shown in **Figure 10.** The embryo has structures that eventually will produce the plant's stem, leaves, and roots. In the seed, the embryo grows to a certain stage and then stops until the seed is planted. The stored food provides energy that is needed when the plant embryo begins to grow into a plant. Because the seed contains an embryo and stored food, a new plant can develop more rapidly from a seed than from a spore.

✔ Reading Check *What are the three parts of a seed?*

Gymnosperms (JIHM nuh spurmz) and angiosperms are seed plants. One difference between the two groups is the way seeds develop. In gymnosperms, seeds usually develop in cones—in angiosperms, seeds develop in flowers.

SCIENCE Online

Research Seed banks conserve seeds of many useful and endangered plants. Visit the Glencoe Science Web site at **science.glencoe.com** to find out more about seed banks. In your Science Journal list three organizations that manage seed banks.

Gymnosperm Reproduction

If you have collected pine cones or used them in a craft project, you probably noticed that many shapes and sizes of cones exist. You probably also noticed that some cones contain seeds. Cones are the reproductive structures of gymnosperms. Each gymnosperm species has a different cone.

Gymnosperm plants include pines, firs, cedars, cycads, and ginkgoes. The pine is a familiar gymnosperm. Production of seeds in pines is typical of most gymnosperms.

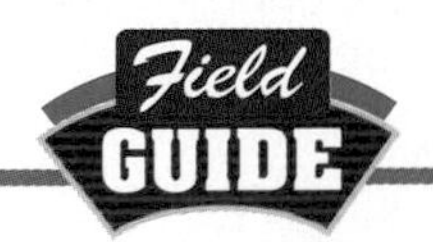

Do all gymnosperm plants produce the same type of cones? To find out more about cones, see the **Cones Field Guide** at the back of this book.

Cones A pine tree or shrub is a sporophyte plant that produces male cones and female cones as shown in **Figure 11.** Male and female gametophyte structures are produced in the cones but you'd need a magnifying glass to see these structures clearly.

A mature female cone consists of a spiral of woody scales on a short stem. At the base of each scale are two ovules. The egg is produced in the **ovule.** Pollen grains are produced in the smaller male cones. In the spring, clouds of pollen are released from the male cones. Anything near pine trees might be covered with the yellow, dustlike pollen.

Figure 11
Seed formation in pines, as in most gymnosperms, involves male and female cones.

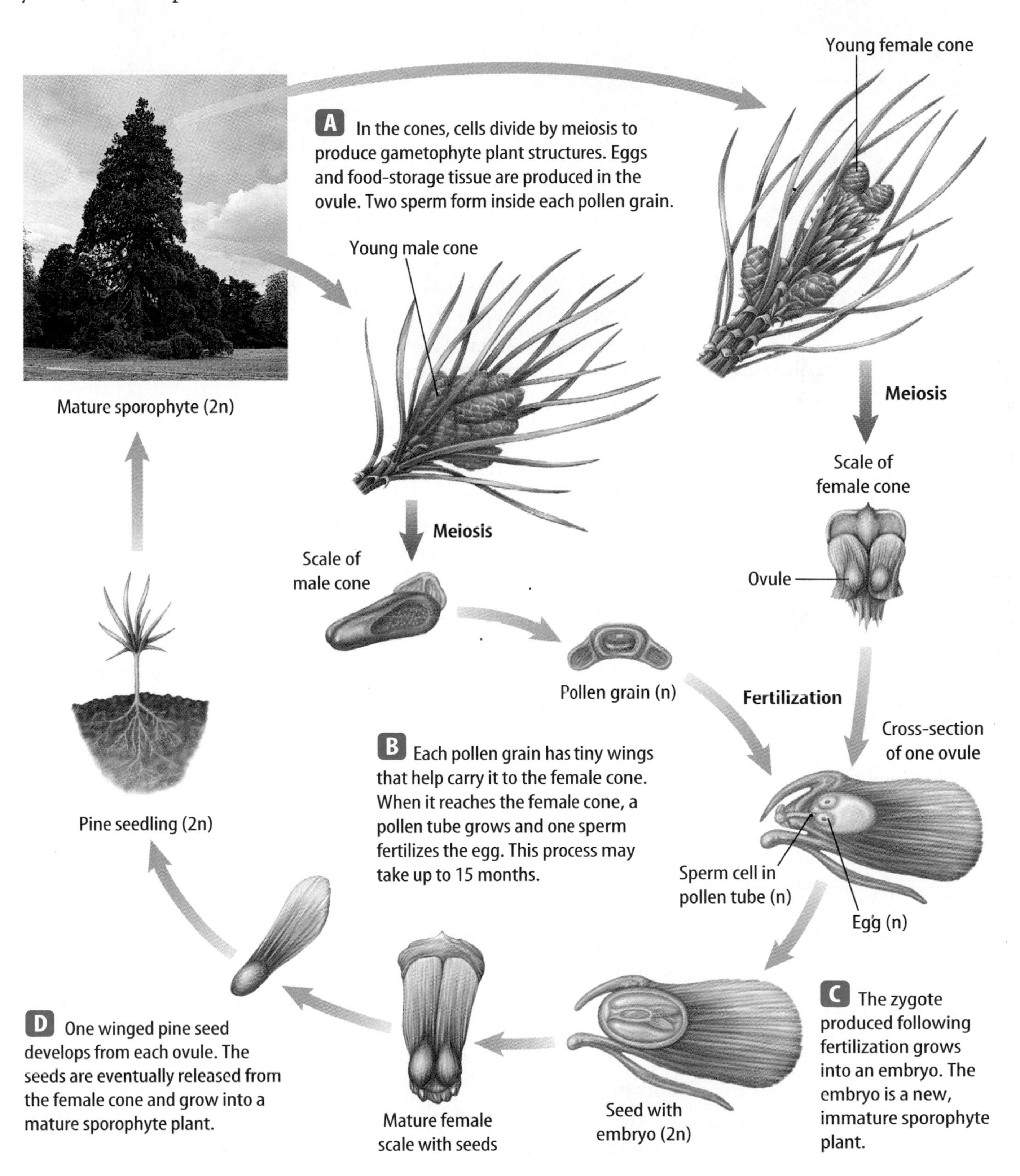

Figure 12
Seed development can take more than one year in pines. The female cone looks different at various stages of the seed-production process.

Cone at pollination   Cone at the end of the first year   Mature, second-year cone

Gymnosperm Seeds Wind usually carries the pollen from male cones to female cones. However, most of the pollen falls on other plants, the ground, and bodies of water. To be useful, the pollen has to be blown between the scales of a female cone. There it can be trapped in the sticky fluid secreted by the ovule. If the pollen grain and the female cone are the same species, fertilization and the formation of a seed can take place.

If you are near a pine tree when the female cones release their seeds, you might hear a crackling noise as the cones' scales open. It can take a long time for seeds to be released from a female pine cone. From the moment a pollen grain falls on the female cone until the seeds are released, can take two or three years, as shown in **Figure 12.** In the right environment, each seed can grow into a new pine sporophyte.

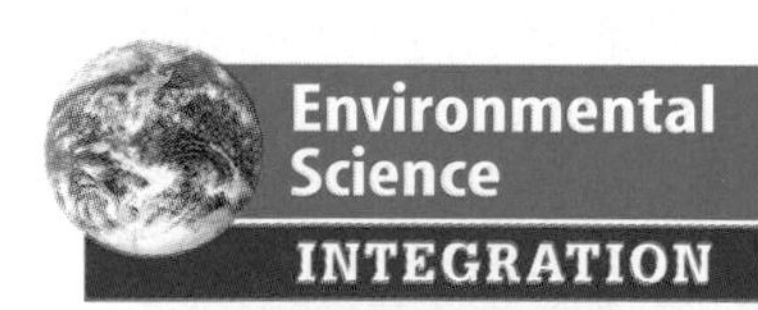

Some gymnosperm seeds will not germinate until the heat of a fire causes the cones to open and release the seeds. Without fires, these plants cannot reproduce. In your Science Journal, explain why some forest fires could be good for the environment.

Angiosperm Reproduction

You might not know it, but you are already familiar with angiosperms. If you had cereal for breakfast or bread in a sandwich for lunch, you ate parts of angiosperms. Flowers that you send or receive for special occasions are from angiosperms. Most of the seed plants on Earth today are angiosperms.

All angiosperms have flowers. The sporophyte plant produces the flowers. Flowers are important because they contain the reproductive organs that contain gametophyte structures that produce sperm or eggs for sexual reproduction.

The Flower When you think of a flower, you probably imagine something with a pleasant aroma and colorful petals. Although many such flowers do exist, some flowers are drab and have no aroma, like the flowers of the maple tree shown in **Figure 13.** Why do you think such variety among flowers exists?

Most flowers have four main parts—petals, sepals, stamen, and pistil—as shown in **Figure 14.** Generally, the colorful parts of a flower are the petals. Outside the petals are usually leaflike parts called sepals. Sepals form the outside of the flower bud. Sometimes petals and sepals are the same color.

Inside the flower are the reproductive organs of the plant. The **stamen** is the male reproductive organ. Pollen is produced in the stamen. The **pistil** is the female reproductive organ. The **ovary** is the swollen base of the pistil where ovules are found. Not all flowers have every one of the four parts. Remember the holly plants you learned about at the beginning of the chapter? What flower part would be missing on a flower from a male holly plant?

Figure 13
Maple trees produce clusters of flowers early in the spring. *How are these flowers different from those of the crocus seen earlier?*

Reading Check *Where are ovules found in the flower?*

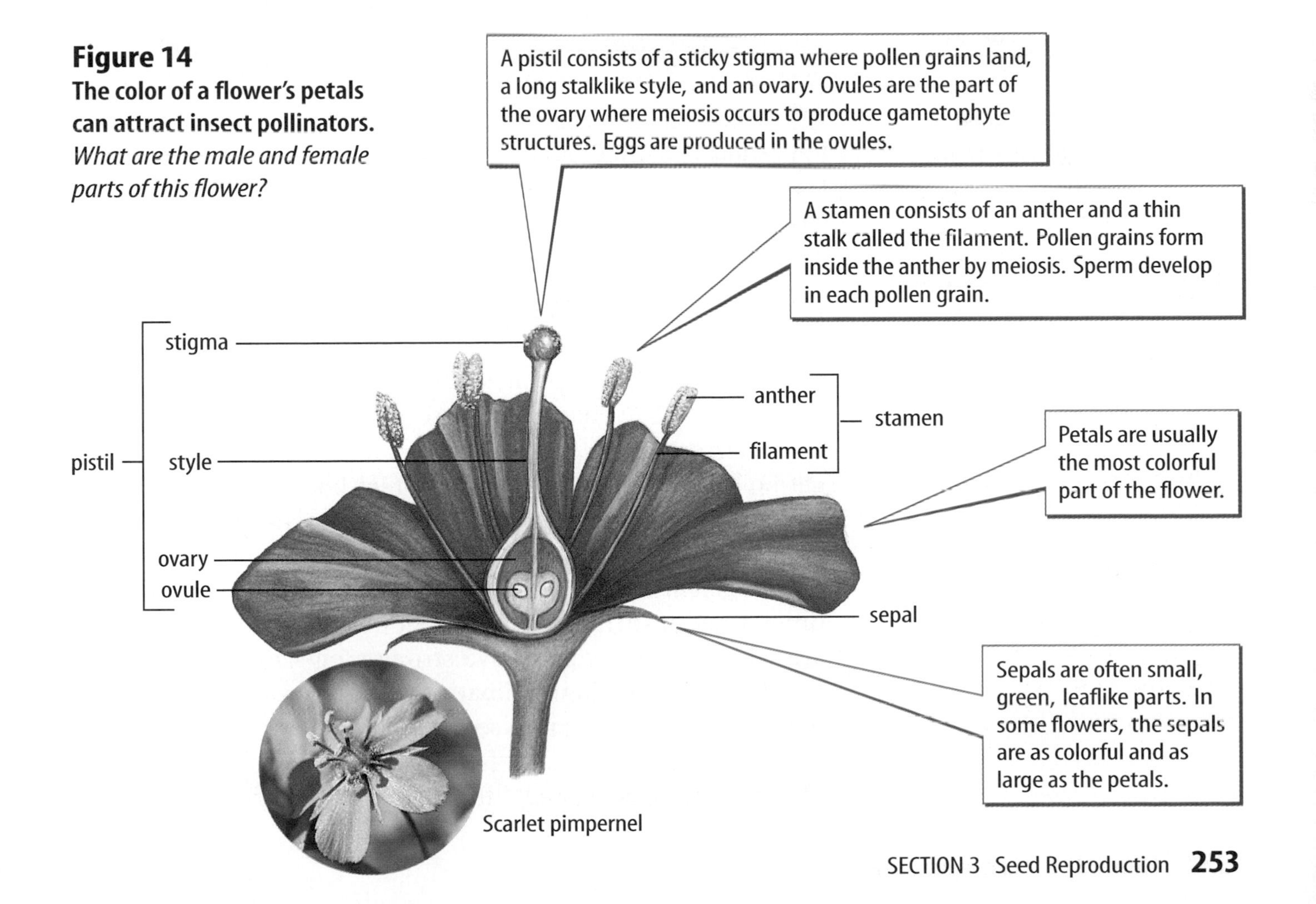

Figure 14
The color of a flower's petals can attract insect pollinators. *What are the male and female parts of this flower?*

Figure 15
Looking at flowers will give you a clue about how each one is pollinated.

A **Honeybees are important pollinators. They are attracted to brightly colored flowers, especially blue and yellow flowers.**

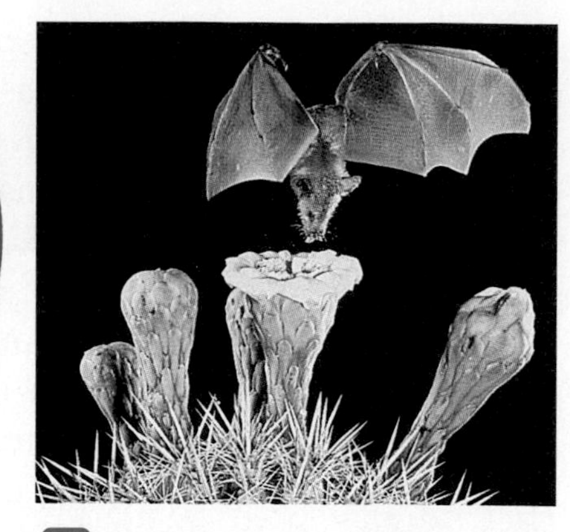

B **Flowers that are pollinated at night, like this cactus flower being pollinated by a bat, are usually white.**

C **Flowers that are pollinated by hummingbirds usually are brightly colored, especially bright red and yellow.**

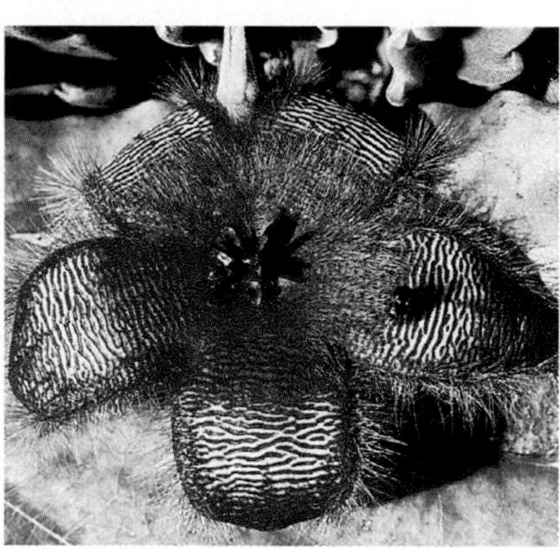

D **Flowers that are pollinated by flies usually are dull red or brown. They often have a strong odor like rotten meat.**

E **The flower of this wheat plant does not have a strong odor and is not brightly colored. Wind, not an animal, is the pollinator of wheat and most other grasses.**

Importance of Flowers The appearance of a plant's flowers can tell you something about the life of the plant. Large flowers with brightly colored petals often attract insects and other animals, as shown in **Figure 15.** These animals might eat the flower, its nectar, or pollen. As they move about the flower, the animals get pollen on their wings, legs, or other body parts. Later, these animals spread the flower's pollen to other plants that they visit. Other flowers depend on wind, rain, or gravity to spread their pollen. Their petals can be small or absent. Flowers that open only at night, such as the cactus flower in **Figure 15B,** usually are white or yellow and have strong scents to attract animal pollinators. Following pollination and fertilization, the ovules of flowers can develop into seeds.

✔ **Reading Check** *How do animals spread pollen?*

Angiosperm Seeds The development of angiosperm seeds is shown in **Figure 16.** Pollen grains reach the stigma in a variety of ways. Pollen is carried by wind, rain, or animals such as insects, birds, and mammals. A flower is pollinated when pollen grains land on the sticky stigma. A pollen tube grows from the pollen grain down through the style. The pollen tube enters the ovary and reaches an ovule. The sperm then travels down the pollen tube and fertilizes the egg in the ovule. A zygote forms and grows into the plant embryo.

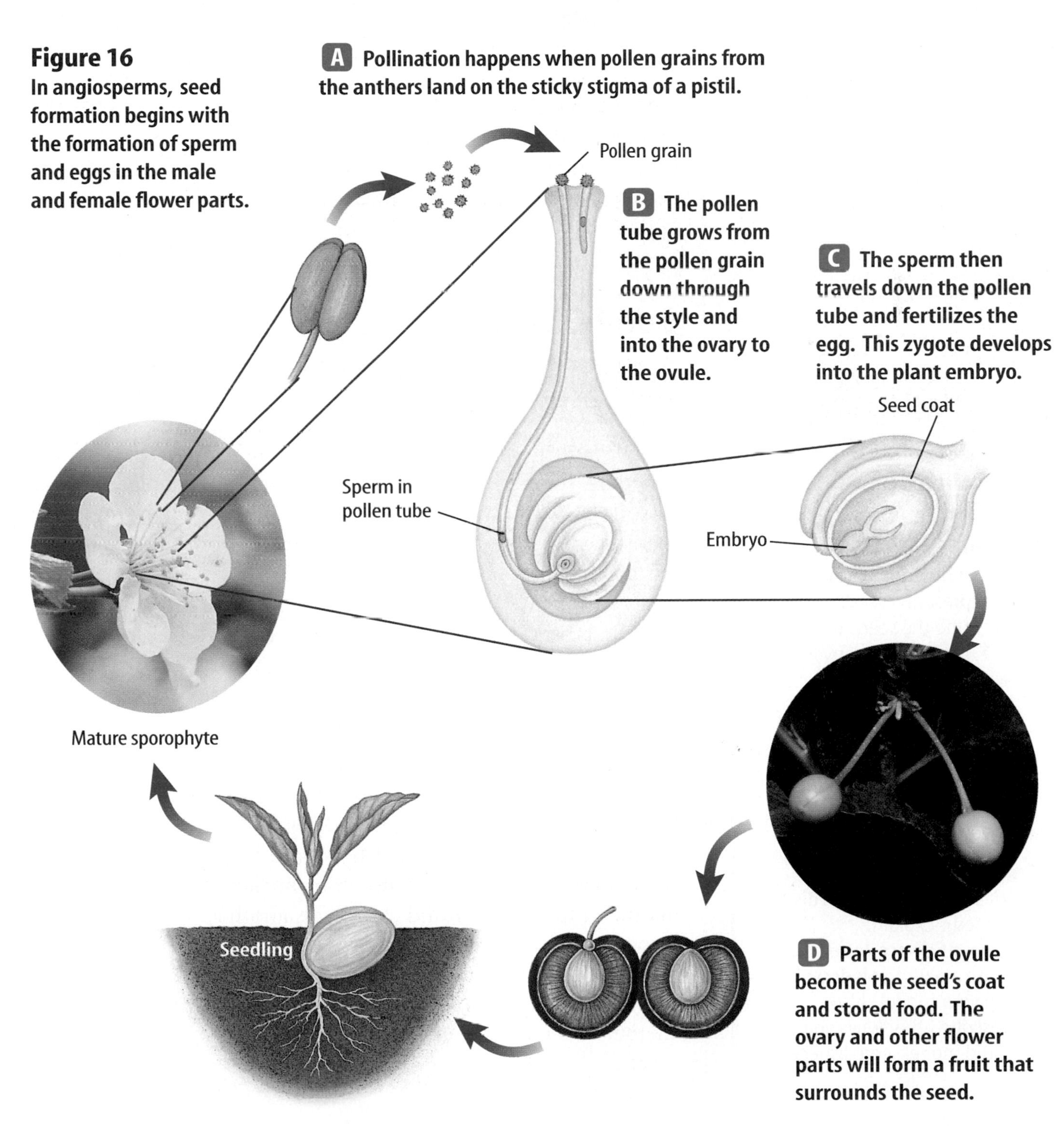

Figure 16
In angiosperms, seed formation begins with the formation of sperm and eggs in the male and female flower parts.

Figure 17
Seeds of land plants are capable of surviving unfavorable environmental conditions.
1. Immature plant
2. Cotyledon(s)
3. Seed coat
4. Endosperm

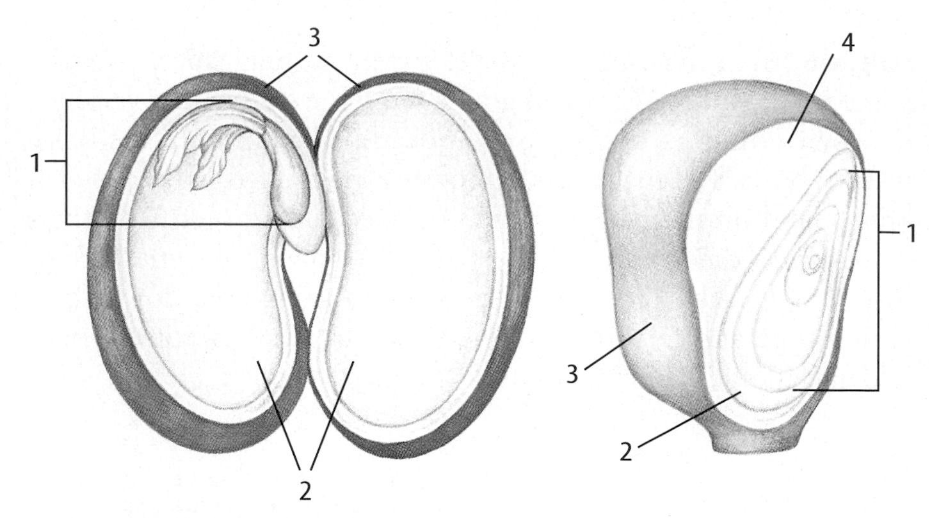

Seed Development Parts of the ovule develop into the stored food and the seed coat that surround the embryo, and a seed is formed, as shown in **Figure 17.** In the seeds of some plants, like beans and peanuts, the food is stored in structures called cotyledons. The seeds of other plants, like corn and wheat, have food stored in a tissue called endosperm.

Seed Dispersal

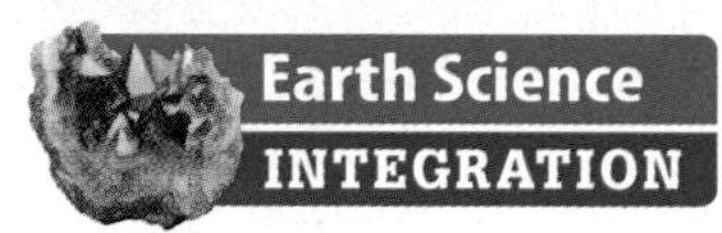

Sometimes, plants just seem to appear. They probably grew from a seed, but where did the seed come from? Plants have many ways of dispersing their seeds, as shown in **Figure 18.** Most seeds grow only when they are placed on or in soil. Do you know how seeds naturally get to the soil? For many seeds, gravity is the answer. They fall onto the soil from the parent plant on which they grew. However, in nature some seeds can be spread great distances from the parent plant.

Wind dispersal usually occurs because a seed has an attached structure that moves it with air currents. Sometimes, small seeds become airborne when released by the plant.

✔ Reading Check *How can wind be used to disperse seeds?*

Animals can disperse many seeds. Some seeds are eaten with fruits, pass through an animal's digestive system, and are dispersed as the animal moves from place to place. Seeds can be carried great distances and stored or buried by animals. Hitchhiking on fur, feathers, and clothing is another way that animals disperse seeds.

Water also disperses seeds. Raindrops can knock seeds out of a dry fruit. Some fruits and seeds float on flowing water or ocean currents. When you touch the seedpod of an impatiens flower, it explodes. The tiny seeds are ejected and spread some distance from the plant.

TRY AT HOME
Mini LAB

Modeling Seed Dispersal

Procedure

1. Find a **button** you can use to represent a seed.
2. Examine the seeds pictured in **Figure 18** and invent a way that your button seed could be dispersed by wind, water, on the fur of an animal, or by humans.
3. Bring your button seed to class and demonstrate how it could be dispersed.

Analysis

1. Was your button seed dispersed? Explain.
2. In your **Science Journal,** write a paragraph describing your model. Also describe other ways you could model seed dispersal.

NATIONAL GEOGRAPHIC

VISUALIZING SEED DISPERSAL

Figure 18

Plants have many adaptations for dispersing seeds, often enlisting the aid of wind, water, or animals.

▲ Equipped with tiny hooks, burrs cling tightly to fur and feathers.

▲ Pressure builds within the seedpods of this jewelweed plant until the pod bursts, flinging seeds far and wide.

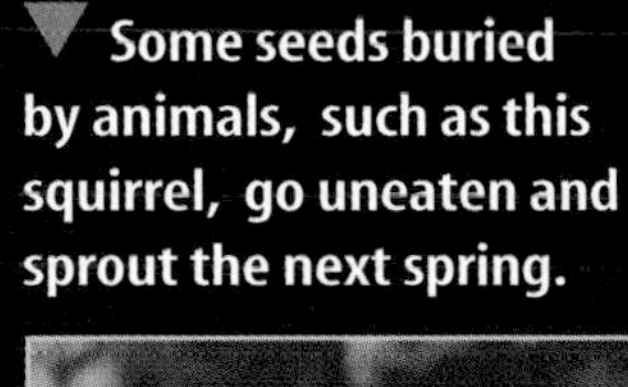

▼ Some seeds buried by animals, such as this squirrel, go uneaten and sprout the next spring.

▼ Dandelion seeds are easily dislodged and sail away on a puff of wind.

▲ Encased in a thick, buoyant husk, a coconut may be carried hundreds of kilometers by ocean currents.

▶ Blackberry seeds eaten by this white-footed mouse will pass through its digestive tract and be deposited in a new location.

Germination A series of events that results in the growth of a plant from a seed is called **germination.** When dispersed from the plant, some seeds germinate in just a few days and other seeds take weeks or months to grow. Some seeds can stay in a resting stage for hundreds of years. In 1982, seeds of the East Indian lotus sprouted after 466 years.

Seeds will not germinate until environmental conditions are right. Temperature, the presence or absence of light, availability of water, and amount of oxygen present can affect germination. Sometimes the seed must pass through an animal's digestive system before it will germinate. Germination begins when seed tissues absorb water. This causes the seed to swell and the seed coat to break open.

Math Skills Activity

Calculating the Number of Seeds That Will Germinate

Example Problem

The label on a packet of carrot seeds says that it contains about 200 seeds. It also claims that 95 percent of the seeds will germinate. How many seeds should germinate if the packet is correct?

1. *This is what you know:*
   quantity = 200
   percentage = 95
2. *This is what you need to find:*
   95 percent of 200
3. *This is the equation you need to use:*
   $$\frac{95}{100} = \frac{x}{200}$$
4. *Solve the equation for* x:
   $$x = \frac{95 \times 200}{100}$$
   Check your answer by dividing by 200 then multiplying by 100. Do you get the original percentage of 95?

Practice Problem

The label on a packet of 50 corn kernels claims that 98 percent will germinate. How many kernels will germinate if the packet is correct?

For more help, refer to the Math Skill Handbook.

Figure 19
Although germination in all seeds is similar, some differences exist.

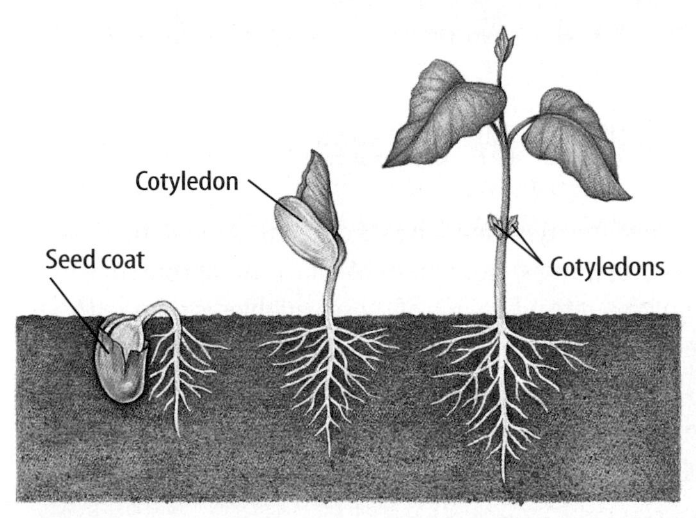

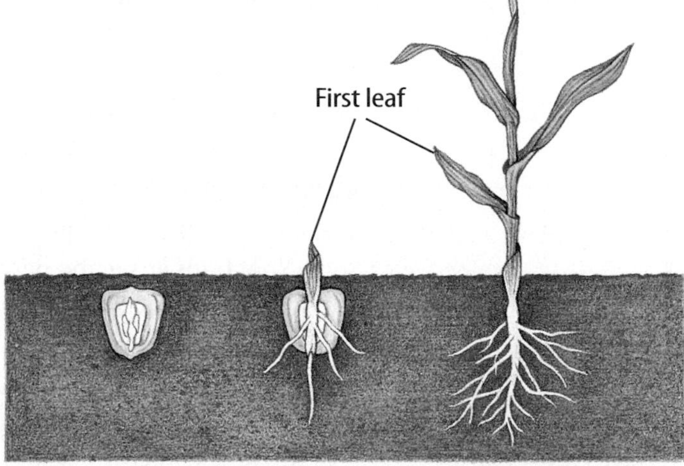

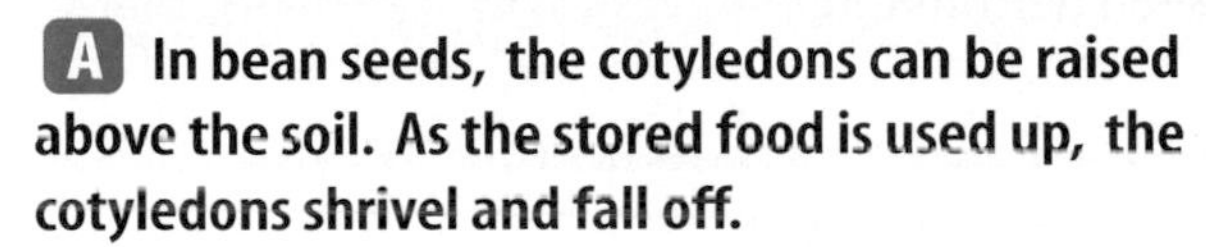
A In bean seeds, the cotyledons can be raised above the soil. As the stored food is used up, the cotyledons shrivel and fall off.

B In corn, the stored food in the endosperm remains in the soil and is gradually used up as the young plant grows.

Next, a series of chemical reactions occurs that releases energy from the stored food in the cotyledons or endosperm for growth. Eventually, a root grows from the seed, followed by a stem and leaves as shown in **Figure 19.** After the plant emerges from the soil, photosynthesis can begin. Photosynthesis provides food as the plant continues to grow.

Section  Assessment

1. Compare and contrast life cycles of angiosperms and gymnosperms.
2. Diagram a flower that has all four parts and label them.
3. List three methods of seed dispersal in plants.
4. Describe the three parts of a seed and give the function of each.
5. **Think Critically** Some conifers have female cones on the top half of the tree and male cones on the bottom half. Why would this arrangement of cones on a tree be important?

Skill Builder Activities

6. **Researching Information** Find out what conditions are needed for seed germination of three different garden plants, such as corn, peas, and beans. How long does each type of seed take to germinate? **For more help, refer to the Science Skill Handbook.**
7. **Communicating** Observe live specimens of several different types of flowers. In your Science Journal, describe their structures. Include numbers of petals, sepals, stamens, and pistil. **For more help, refer to the Science Skill Handbook.**

Activity *Design Your Own Experiment*

Germination Rate of Seeds

Many environmental factors affect the germination rate of seeds. Among these are soil temperature, air temperature, moisture content of soil, and salt content of soil. What happens to the germination rate when one of these variables is changed? Can you determine a way to predict the best conditions for seed germination?

Recognize the Problem

How do environmental factors affect seed germination?

Form a Hypothesis

Based on your knowledge of seed germination, state a hypothesis about how environmental factors affect germination rates.

Possible Materials

seeds
water
salt
potting soil
plant trays or plastic cups
**seedling warming cables*
thermometer
graduated cylinder
beakers
**Alternate materials*

Goals

- **Design** an experiment to test the effect of an environmental factor on seed germination rate.
- **Compare** germination rates under different conditions.

Safety Precautions

Some kinds of seeds are poisonous. Do not place any seeds in your mouth. Be careful when using any electrical equipment to avoid shock hazards.

Test Your Hypothesis

Plan

1. As a group, agree upon and write your hypothesis and decide how you will test it. Identify which results will confirm the hypothesis.
2. **List** the steps you need to take to test your hypothesis. Be specific, and describe exactly what you will do at each step. List your materials.
3. **Prepare** a data table in your Science Journal to record your observations.
4. Reread your entire experiment to make sure that all of the steps are in a logical order.
5. **Identify** all constants, variables, and controls of the experiment.

Do

1. Make sure your teacher approves your plan and your data table before you proceed.
2. Use the same type and amount of soil in each tray.
3. While the experiment is going on, record your observations accurately and complete the data table in your Science Journal.

Analyze Your Data

1. **Compare** the germination rate in the two groups of seeds.
2. **Compare** your results with those of other groups.
3. Did changing the variable affect germination rates? Explain.
4. Make a bar graph of your experimental results.

Draw Conclusions

1. **Interpret** your graph to estimate the conditions that give the best germination rate.
2. What things affect the germination rate?

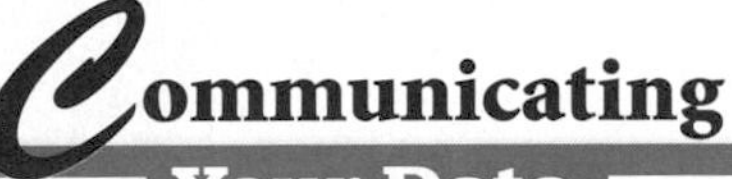

Write a short article for a local newspaper telling about this experiment. Give some ideas about when and how to plant seeds in the garden and the conditions needed for germination.

TIME SCIENCE AND Society

SCIENCE ISSUES THAT AFFECT YOU!

Genetic

What would happen if you crossed a cactus with a rose? Well, you'd either get an extra spiky flower, or a bush that didn't need to be watered very often. Until recently, this sort of mix was the stuff science fiction was made of. But now, with the help of genetic engineering, it may be possible.

Genetic engineering is a way of taking genes from one species and giving them to another. One purpose of genetic engineering is to transfer an organism's traits. For example, scientists have changed grass by adding to it the gene from another grass species. This gene makes the grass grow so slowly, it doesn't have to be mowed very often.

How is this done? It all starts with genes—sections of DNA found in the cells of all living things. Genes produce certain characteristics, or traits, in an organism, like the color of a flower or whether a person has blond hair or black hair. Scientists have found a way to exchange genes and their traits among bacteria, viruses, plants, animals, and even humans. In 1983, the first plant was genetically modified, or changed. Since then, many crops in the U.S. have been modified in this way, including soybeans, potatoes, and tomatoes.

One common genetically engineered crop is corn. To modify it, scientists took a gene from a particular bacterium. The gene "instructed" the bacterium to produce a natural toxin that killed certain insects. This gene was placed into another bacterium, which was placed into a corn plant. This bacterial "taxi" carried the gene into the plant's DNA, giving it the new trait. The seeds from the genetically modified corn produced crops that resisted harmful insects.

Clifton Poodry: Biologist

Clifton Poodry was born in Buffalo, New York, on the Tonawanda Seneca Indian reservation. On the reservation, Native Americans took a great deal of pride in free thinking. "This way of thinking helps if you want to do scientific research," said Poodry. "As a scientist, one thing you get to do is pursue a question that is of interest to you, even if it is not of interest to someone else."

In graduate school, Poodry studied how cells were organized and how they develop structures in the human body. He was especially interested in learning how an organism develops from one cell to an organism with many different cells. He based his work on the fruit fly.

But equally exciting to Poodry was the fact that in college he shared the labs with men and women from many different ethnic backgrounds—all of whom were interested in research and science. That excitement, and the idea of many people coming together to share knowledge, led him to become a university professor. Today, he works to increase the number of minority students who go into scientific research. He heads the Division of Minority Opportunities in Research, a part of the National Institutes of Health.

Engineering

In addition to making plants resist insects, genetic engineering can make plants grow bigger and faster. Genetic engineering also has produced herbicide-resistant plants. This allows farmers to produce more crops with less chemicals. Scientists predict that genetic engineering will soon produce crops that are more nutritious and that can resist cold, heat, or even drought. This will help farmers increase their harvests and make more food available.

However, not everyone thinks genetic engineering is so great. Since it is a relatively new process, some people are worried about the long-term risks. One concern is that people might be allergic to modified foods and not realize it until it's too late. Other people say that genetic engineering is unnatural. Also, farmers must purchase the patented genetically modified seeds each growing season from the companies that make them, rather than saving and replanting the seeds from their current crops.

Genetically modified "super" tomatoes and "super" corn can resist heat, cold, drought, and insects.

People in favor of genetic engineering reply that there are always risks with new technology, but proper precautions are being taken. Each new plant is tested and then approved by U.S. governmental agencies. And they say that most "natural" crops aren't really natural. They are really hybrid plants bred by agriculturists, and they couldn't survive on their own.

As genetic engineering continues, so does the debate.

CONNECTIONS Debate Research the pros and cons of genetic engineering on the Glencoe Science Web site and in your school's media center. Decide whether you are for or against genetic engineering. Debate your conclusions with your classmates.

SCIENCE *Online* For more information, visit science.glencoe.com

Chapter 9 Study Guide

Reviewing Main Ideas

Section 1 Introduction to Plant Reproduction

1. Plants reproduce sexually and asexually. Sexual reproduction involves the formation of sex cells and fertilization.
2. Asexual reproduction does not involve sex cells and produces plants genetically identical to the parent plant. *How do fern plants produced from the same rhizome compare genetically?*

3. Plant life cycles include a gametophyte and a sporophyte stage. The gametophyte stage begins with meiosis. The sporophyte stage begins when the egg is fertilized by a sperm.
4. In some plant life cycles, the sporophyte and gametophyte stages are separate and not dependent on each other. In other plant life cycles, they are part of the same organism.

Section 2 Seedless Reproduction

1. For liverworts and mosses, the gametophyte stage is the familiar plant form. The sporophyte stage produces spores.
2. In ferns, the sporophyte stage, not the gametophyte stage, is the familiar plant form.
3. Seedless plants, like mosses and ferns, use sexual reproduction to produce spores. *Why do seedless plants such as these produce so many small spores?*

Section 3 Seed Reproduction

1. In seed plants the male reproductive organs produce pollen grains that eventually contain sperm. Eggs are produced in the ovules of the female reproductive organs.
2. The male and female reproductive organs of gymnosperms are called cones. Wind usually moves pollen from the male cone to the female cone for pollination.
3. The reproductive organs of angiosperms are in a flower. The male reproductive organ is the stamen, and the female reproductive organ is the pistil. Gravity, wind, rain, and animals can pollinate a flower. *How would these flowers become pollinated?*

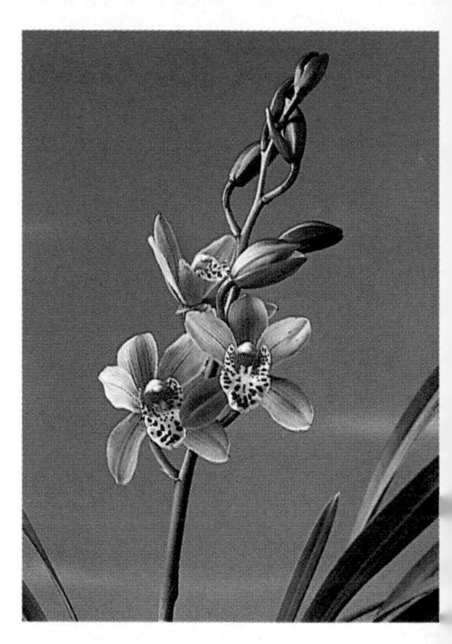

4. Seeds of gymnosperms and angiosperms are dispersed in many ways. Wind, water, and animals spread seeds. Some plants can eject their seeds.
5. Germination is a series of events that results in the growth of a plant from a seed.

After You Read

FOLDABLES Reading & Study Skills

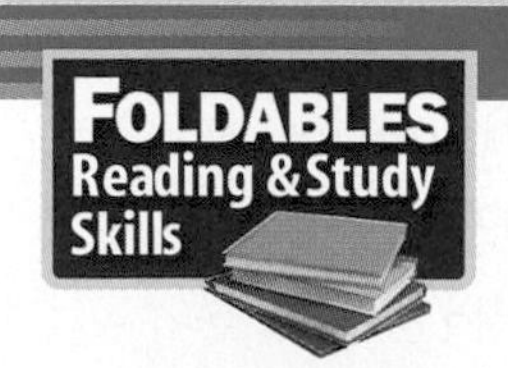

On the front of your Venn Diagram Study Fold where the circles overlap, write common characteristics of sexual and asexual reproduction.

Visualizing Main Ideas

Complete the following table that compares reproduction in different plant groups.

Plant Reproduction

Plant Group	Seeds?	Pollen?	Cones?	Flowers?
Mosses				
Ferns				
Gymnosperms				
Angiosperms				

Vocabulary Review

Vocabulary Words

a. frond
b. gametophyte stage
c. germination
d. ovary
e. ovule
f. pistil
g. pollen grain
h. pollination
i. prothallus
j. rhizome
k. sori
l. spore
m. sporophyte stage
n. stamen

Study Tip

Read the chapter before you go over it in class. Being familiar with the material before your teacher explains it gives you better understanding and an opportunity to ask questions.

Using Vocabulary

Replace the underlined word or phrase with the correct vocabulary word(s).

1. A sori is the leaf of a fern.
2. In seed plants, the anther contains the egg.
3. The plant structures in the sporophyte stage are made up of haploid cells.
4. The green, leafy moss plant is part of the prothallus in the moss life cycle.
5. Two parts of a sporophyte fern are stamen and pistil.
6. The female reproductive organ of the flower is the rhizome.
7. The ovule is the swollen base of the pistil.

Chapter 9 Assessment

Checking Concepts

Choose the word or phrase that best answers the question.

1. How are colorful flowers usually pollinated?
   A) insects
   B) wind
   C) clothing
   D) gravity
2. What type of reproduction produces plants that are genetically identical?
   A) asexual
   B) sexual
   C) spore
   D) flower
3. Which of the following terms describes the cells in the gametophyte stage?
   A) haploid
   B) prokaryote
   C) diploid
   D) missing a nucleus
4. What structures do ferns form when they reproduce sexually?
   A) spores
   B) anthers
   C) seeds
   D) flowers
5. What contains food for the plant embryo?
   A) endosperm
   B) pollen grain
   C) stigma
   D) root
6. What disperses most dandelion seeds?
   A) rain
   B) animals
   C) wind
   D) insects
7. What is the series of events that results in a plant growing from a seed?
   A) pollination
   B) prothallus
   C) germination
   D) fertilization
8. In plants, meiosis is used to produce what before fertilization?
   A) prothallus
   B) seeds
   C) flowers
   D) spores
9. Ovules and pollen grains take part in what process?
   A) germination
   B) asexual reproduction
   C) seed dispersal
   D) sexual reproduction
10. What part of the flower receives the pollen grain from the anther?
    A) sepal
    B) petal
    C) stamen
    D) stigma

Thinking Critically

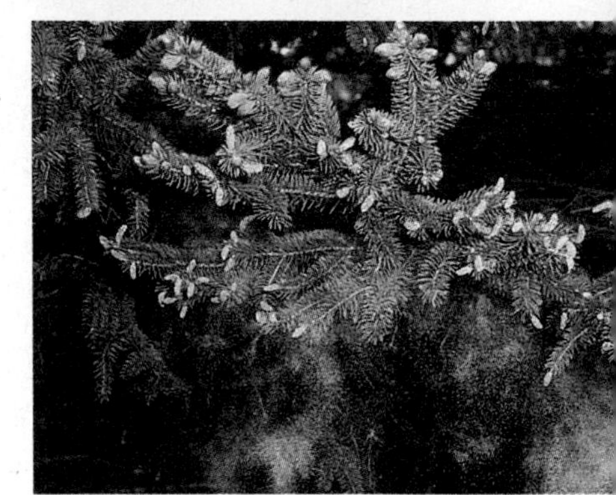

11. Explain why male cones produce so many pollen grains.
12. Could a seed without an embryo germinate? Explain your answer.
13. Discuss the importance of water in the sexual reproduction of nonvascular plants and ferns.
14. In mosses, why is the sporophyte stage dependent on the gametophyte stage?
15. What features of flowers ensure pollination?

Developing Skills

16. **Making and Using Graphs** Make a bar graph for the following data table about onion seeds. Put days on the horizontal axis and temperature on the vertical axis.

Onion Seed Data

Temperature (°C)	10	15	20	25	30	35
Days to Germinate	13	7	5	4	4	13

17. **Comparing and Contrasting** Describe the differences and similarities between the fern sporophyte and gametophyte stages.

18. Predicting Observe pictures of flowers or actual flowers and predict how they are pollinated. Explain your prediction.

19. Interpreting Scientific Illustrations Using **Figure 16,** sequence these events.
pollen is trapped on the stigma
pollen tube reaches the ovule
fertilization
pollen released from the anther
pollen tube forms through the style
a seed forms

20. Concept Mapping Complete this concept map of a typical plant life cycle.

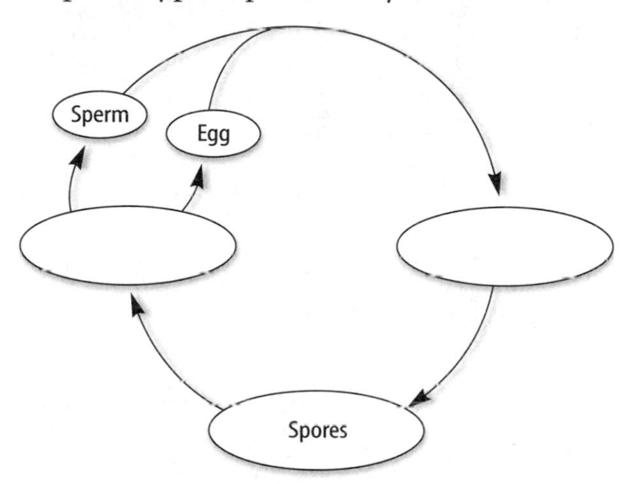

Performance Assessment

21. Seed Mosaic Collect several different types of seeds and use them to make a mosaic picture of a flower.

22. Newspaper Story Write a newspaper story to tell people about the importance of gravity, water, wind, insects, and other animals in plant life cycles.

Technology

Go to the Glencoe Science Web site at **science.glencoe.com** or use the **Glencoe Science CD-ROM** for additional chapter assessment.

Test Practice

Four groups containing 15 plants each were set up to determine how many flowers were pollinated in a 24-h period. Each plant had five flowers. A botanist recorded the data in the following table.

Pollination Data

Plant Group (15 plants per group)	Number of Bees	Number of Birds	Number of Flowers Pollinated after 24 hrs.
1	5	1	16
2	10	1	35
3	15	1	49
4	20	1	73

Study the table and answer the following questions.

1. Which hypothesis probably is being tested in this experiment?

A) A greater number of bees increases the rate of pollination.
B) Birds increase the chance of pollination.
C) A combination of birds and bees gives the best chance of successful pollination.
D) Increasing the number of plants in a group results in increased pollination.

2. The pollination that is taking place in this experiment is part of which process?

F) asexual reproduction
G) sexual reproduction
H) seed dispersal
J) germination

CHAPTER 10

Regulation and Reproduction

The control room blinks with monitors and panels of dials and buttons. Not much is going to get past this complex monitoring system. Your body also is designed with a system that monitors and controls the actions of many of your body's functions. In this chapter, you'll learn about this system—the endocrine system. You'll also study the human reproductive system and the stages of growth.

What do you think?

Science Journal Look at the picture below with a classmate. Discuss what this might be. Here's a hint: *This object could be considered a small beginning.* Write your answer or best guess in your Science Journal.

Your body has systems that work together to control your body's activities. One of these systems sends chemical messages through your blood to certain tissues, which, in turn, respond. You may feel the results of this system's action, but you cannot see them. Do the activity below to see how a chemical signal can be sent.

Model a chemical message

1. Cut a 10-cm-tall Y shape from filter paper and place it on a plastic, ceramic, or glass plate.
2. Sprinkle baking soda on one arm of the Y and salt on the other arm.
3. Using a dropper, place five or six drops of vinegar halfway up the leg of the Y.

Observe

Describe in your Science Journal how the chemical moves along the paper and the reaction(s) it causes.

Before You Read

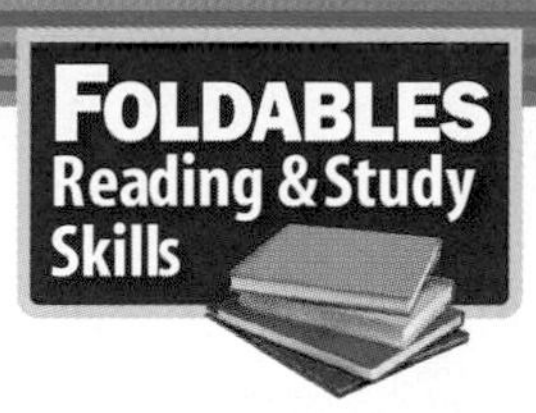

Making a Sequence Study Fold

Make the following Foldable to help you predict what might occur next in the sequence of life.

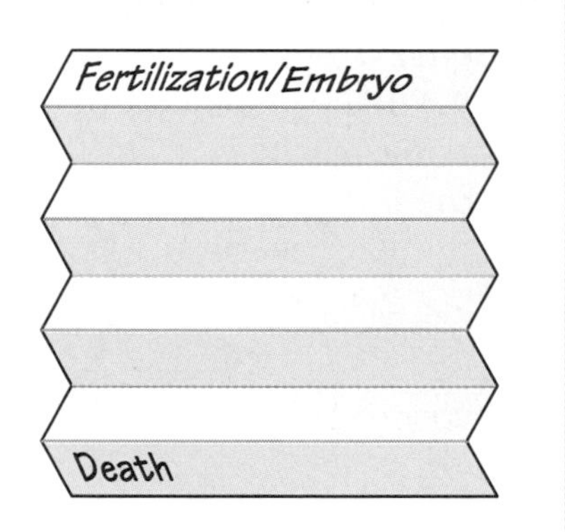

1. Place a sheet of paper in front of you so the short side is at the top. Fold the paper in half from top to bottom. Then fold it in half again top to bottom two more times. Unfold all the folds.
2. Using the fold lines as a guide, refold the paper into a fan. Unfold all the folds again.
3. Before you read the chapter list as many stages of life as you can on your foldable, beginning with *Fertilization/Embryo* and ending with *Death.* As you read the chapter add to your list.

SECTION 1

The Endocrine System

As You Read

What You'll Learn

- **Define** how hormones function.
- **Identify** different endocrine glands and the effects of the hormones they produce.
- **Describe** how a feedback system works in your body.

Vocabulary
hormone

Why It's Important

The endocrine system uses chemicals to control many systems in your body.

Functions of the Endocrine System

You go through the dark hallways of a haunted house. You can't see a thing. Your heart is pounding. Suddenly, a monster steps out in front of you. You scream and jump backwards. Your body is prepared to defend itself or get away. Preparing the body for fight or flight in times of emergency, as shown in **Figure 1,** is one of the functions of the body's control systems.

Chemical Messengers Your body is made up of systems that are controlled, or regulated, to work together. The nervous system and the endocrine (EN duh krun) system are the control systems of your body. The nervous system sends messages to and from the brain throughout the body. The endocrine system uses **hormones** (HOR mohnz)—chemicals that are made in tissues called glands found throughout your body. Hormones from endocrine glands are released directly into your bloodstream. They affect specific tissues called target tissues, usually located in the body far from the hormone-producing gland. The body doesn't react as quickly to messages from the endocrine system as it does to those of the nervous system.

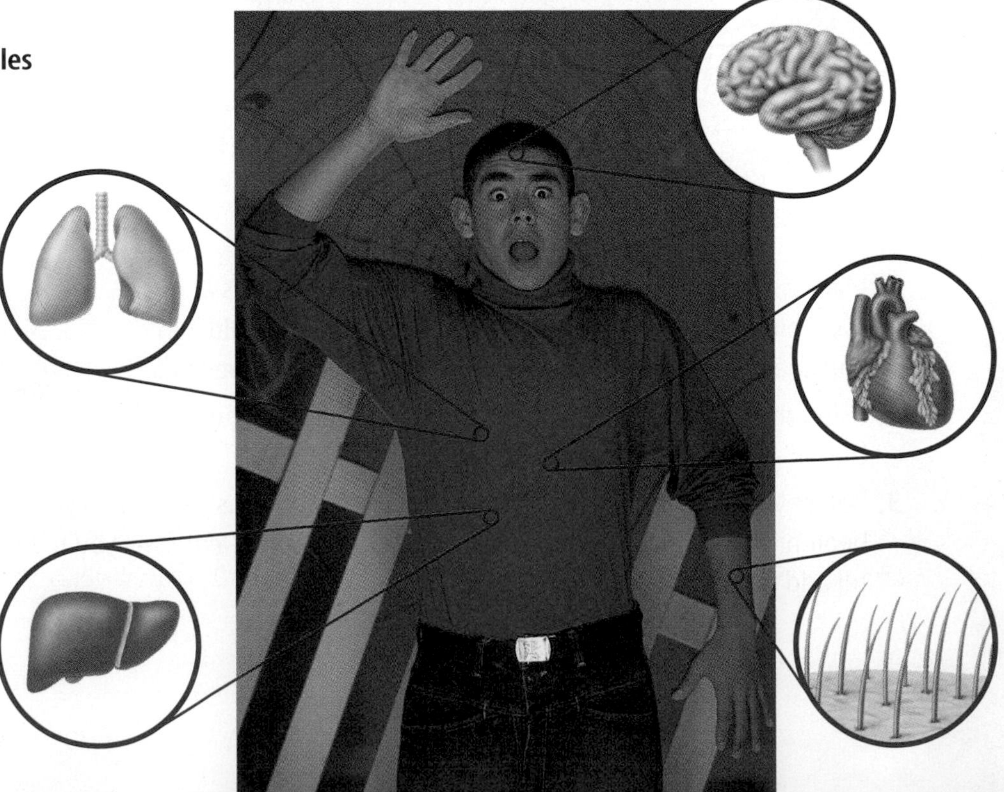

Figure 1
Your endocrine system enables many parts of your body to respond with an immediate reaction to a fearful situation.

Endocrine Glands

Unlike some glands, such as your mouth's saliva glands, that release their products through small tubes called ducts, endocrine glands are ductless. Hormones from endocrine glands pour directly into the blood to reach target tissues. Hormones regulate certain cellular activities. **Figure 2** on the next page describes some of your major endocrine glands and how they function to regulate your body.

What is the function of hormones?

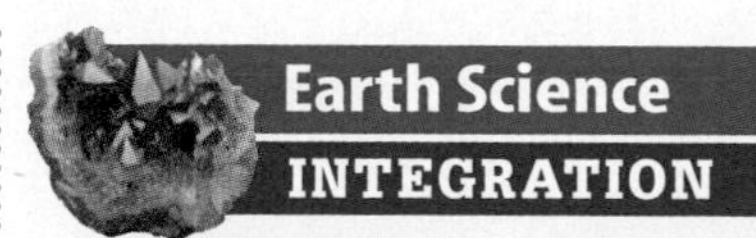

Without the element iodine, the thyroid gland cannot function properly. Iodine is found in seawater, soil, and rocks. How does your body take in iodine? Write your answer in your Science Journal.

Math Skills Activity

Calculating Blood Sugar Percentage

Example Problem

Calculate how much higher the blood sugar (glucose) level of a diabetic is before breakfast when compared to a nondiabetic before breakfast. Express this number as a percentage of the nondiabetic sugar level before breakfast.

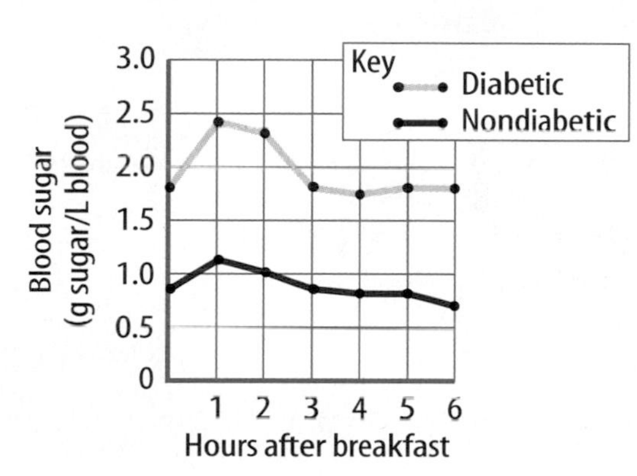

Solution

1 *This is what you know:*

blood sugar of a nondiabetic person at 0 h = 0.85 g sugar/L blood
blood sugar of a diabetic person at 0 h = 1.8 g sugar/L blood

2 *This is what you must do first:*

Find the difference between the two values. 1.8 g/L − 0.85 g/L = 0.95 g/L

3 *This is the equation you need to use:*

$$\frac{\text{difference between values}}{\text{nondiabetic value}} \times 100\% = \text{percent difference}$$

4 *Substitute in the known values:*

$$\frac{0.95}{0.85} \times 100\% = 111\%$$

At 0 h before breakfast, a diabetic's blood sugar is 111 percent higher than that of a nondiabetic.

Practice Problem

Express as a percentage how much higher the blood sugar value is for a diabetic person compared to a nondiabetic person 1 h, 3 h, and 6 h after breakfast.

For more help, refer to the Math Skill Handbook.

NATIONAL GEOGRAPHIC

VISUALIZING THE ENDOCRINE SYSTEM

Figure 2

Your endocrine system is involved in regulating and coordinating many body functions, from growth and development to reproduction. This complex system consists of many diverse glands and organs, including the nine shown here. Endocrine glands produce chemical messenger molecules, called hormones, that circulate in the bloodstream. Hormones exert their influence only on the specific target cells to which they bind.

PINEAL GLAND Shaped like a tiny pine cone, the pineal gland lies deep in the brain. It produces melatonin, a hormone that may function as a sort of body clock by regulating wake/sleep patterns.

PITUITARY GLAND A pea-size structure attached to the hypothalamus of the brain, the pituitary gland produces hormones that affect a wide range of body activities, from growth to reproduction.

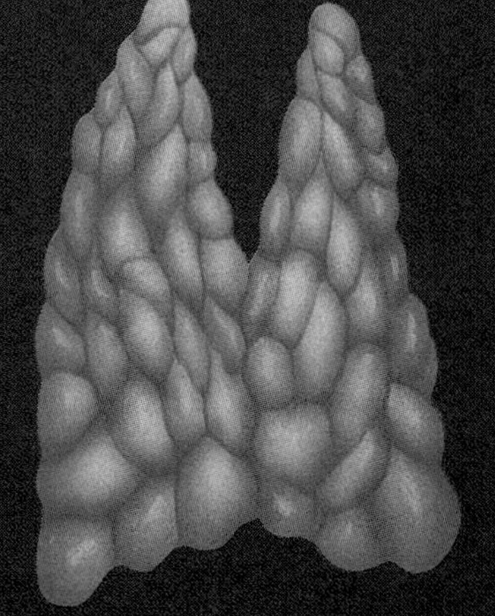

THYMUS The thymus is located in the upper chest, just behind the sternum. Hormones produced by this organ stimulate the production of certain infection-fighting cells.

TESTES These paired male reproductive organs primarily produce testosterone, a hormone that controls the development and maintenance of male sexual traits. Testosterone also plays an important role in the production of sperm.

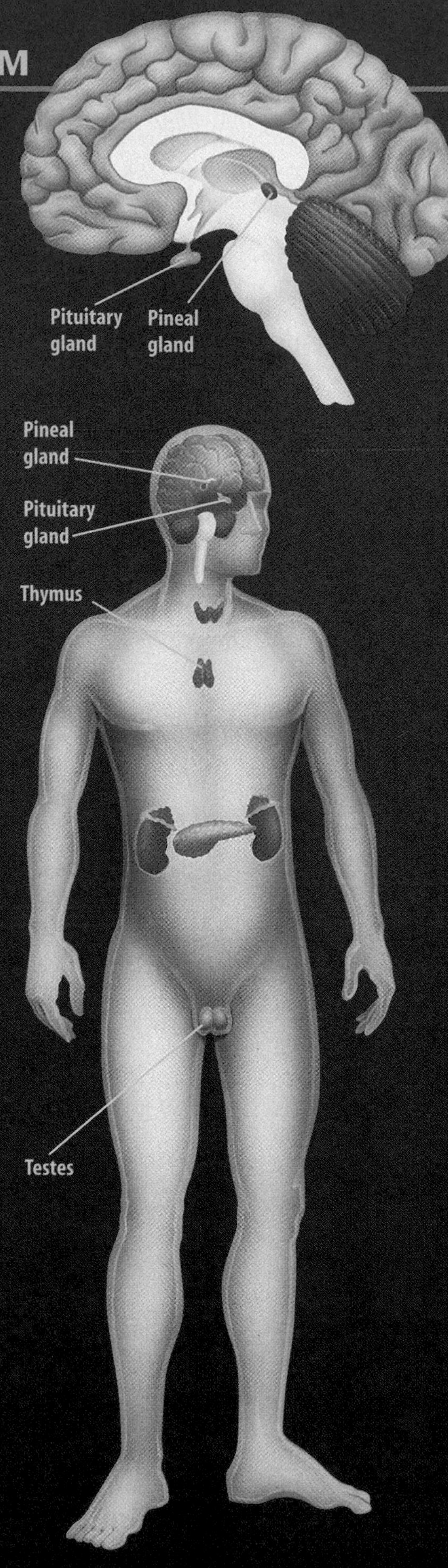

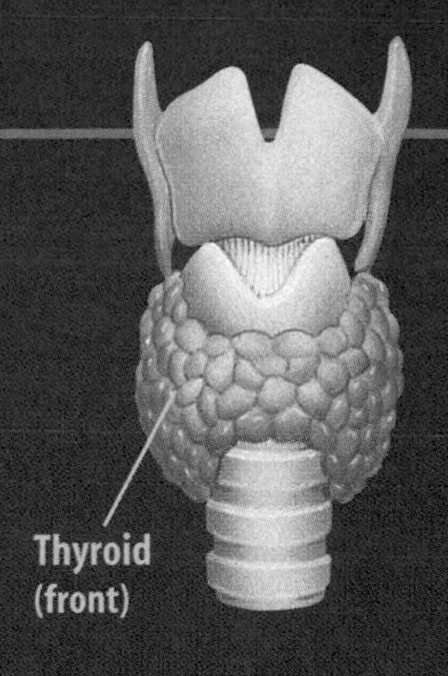

THYROID GLAND Located below the larynx, the bi-lobed thyroid gland is richly supplied with blood vessels. It produces hormones that regulate metabolic rate, control the uptake of calcium by bones, and promote normal nervous system development.

PARATHYROID GLANDS Attached to the back surface of the thyroid are tiny parathyroids, which help regulate calcium levels in the body. Calcium is important for bone growth and maintenance, as well as for muscle contraction and nerve impulse transmission.

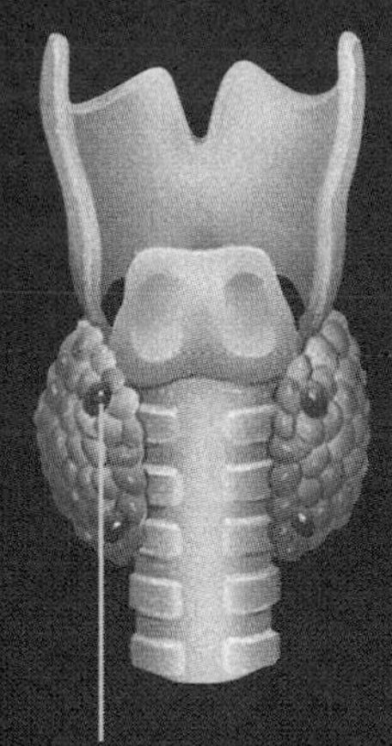

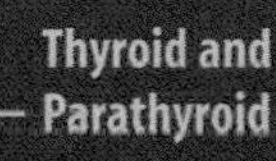

Adrenal gland

Kidney

Pancreas

Ovaries

ADRENAL GLANDS On top of each of your kidneys is an adrenal gland. This complex endocrine gland produces a variety of hormones. Some play a critical role in helping your body adapt to physical and emotional stress. Others help stabilize blood sugar levels.

PANCREAS Scattered throughout the pancreas are millions of tiny clusters of endocrine tissue called the islets of Langerhans. Cells that make up the islets produce hormones that help control sugar levels in the bloodstream.

OVARIES Found deep in the pelvic cavity, ovaries produce female sex hormones known as estrogen and progesterone. These hormones regulate the female reproductive cycle and are responsible for producing and maintaining female sex characteristics.

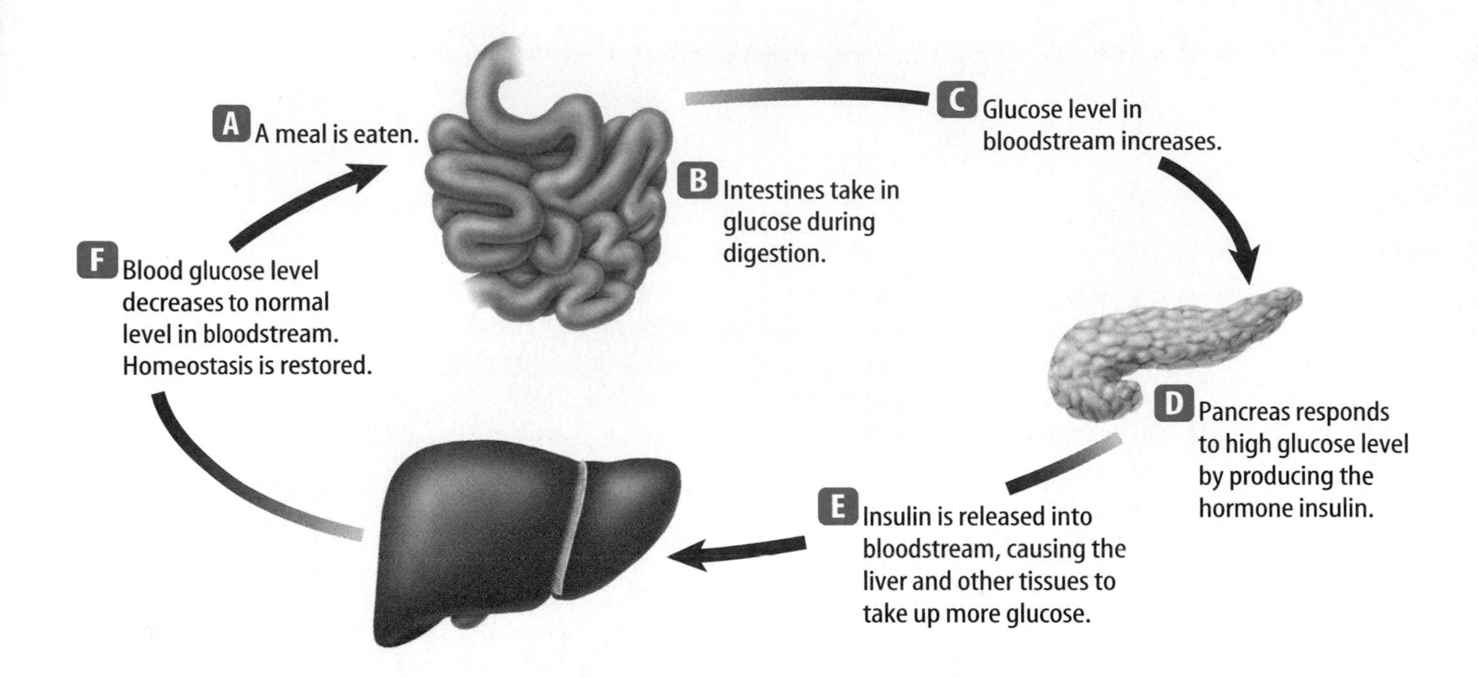

Figure 3
Many internal body conditions, such as hormone level, blood sugar level, and body temperature, are controlled by negative-feedback systems. Using a negative-feedback system, the pancreas controls the level of glucose in your bloodstream.

A Negative-Feedback System

To control the amount of hormones that are in your body, the endocrine system sends chemical messages back and forth within itself. This is called a negative-feedback system. It works much the way a thermostat works. When the temperature in a room drops below a set level, the thermostat signals the furnace to turn on. Once the furnace has raised the temperature in the room to the set level, the thermostat signals the furnace to shut off. It will continue to stay off until the thermostat signals that the temperature has dropped again. **Figure 3** shows how a negative-feedback system controls the level of glucose in your bloodstream.

Section 1 Assessment

1. Compare and contrast the human body's two control systems.
2. What is the function of hormones?
3. Choose one endocrine gland and explain how it works.
4. What is a negative-feedback system?
5. **Think Critically** Glucose is required for cellular respiration, the process that releases energy within cells. How would lack of insulin affect this process?

Skill Builder Activities

6. **Predicting** Predict why the circulatory system is a good mechanism for delivering hormones throughout the body. **For more help, refer to the Science Skill Handbook.**
7. **Researching Information** Research recent treatments for growth disorders involving the pituitary gland. Write a brief paragraph of your results in your Science Journal. **For more help, refer to the Science Skill Handbook.**

The Reproductive System

Reproduction and the Endocrine System

Reproduction is the process that continues life on Earth. Most human body systems, such as the digestive system and the nervous system, are the same in males and females, but this is not true for the reproductive system. Males and females each have structures specialized for their roles in reproduction. Although structurally different, both the male and female reproductive systems are adapted to allow for a series of events that can lead to the birth of a baby.

Hormones are the key to how the human reproductive system functions, as shown in **Figure 4.** Sex hormones are necessary for the development of sexual characteristics, such as breast development in females and facial hair growth in males. Hormones from the pituitary gland also begin the production of eggs in females and sperm in males. Eggs and sperm transfer hereditary information from one generation to the next.

As You Read

***What* You'll Learn**

- **Identify** the function of the reproductive system.
- **Compare and contrast** the major structures of the male and female reproductive systems.
- **Sequence** the stages of the menstrual cycle.

Vocabulary

testes
sperm
semen
ovary
ovulation
uterus
vagina
menstrual cycle
menstruation

***Why* It's Important**

The reproductive system helps ensure that life continues on Earth.

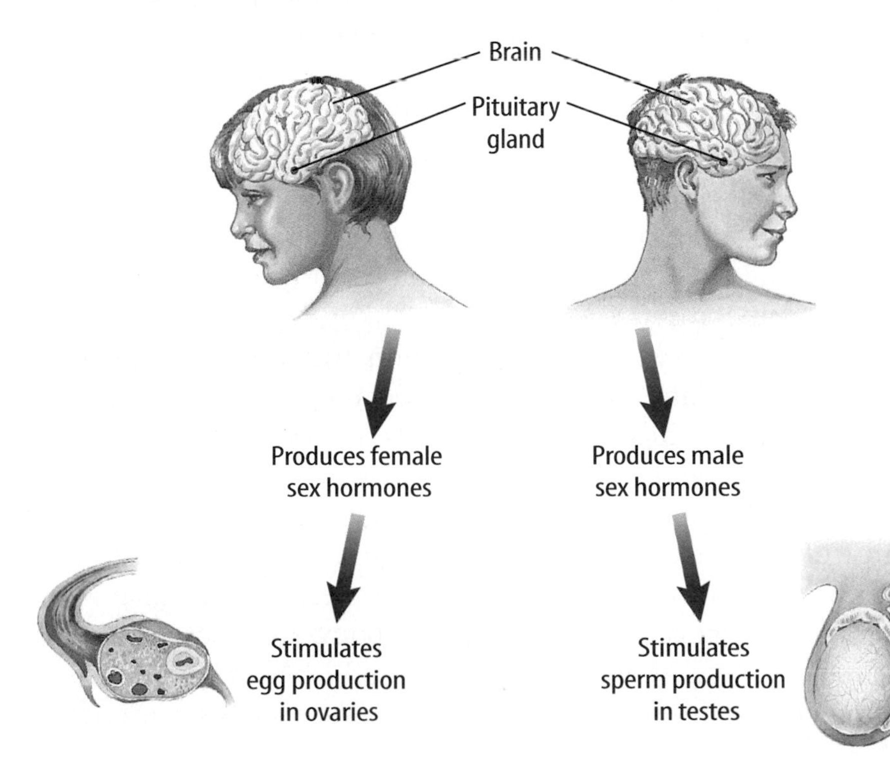

Figure 4
The pituitary gland produces hormones that control the male and female reproductive systems.

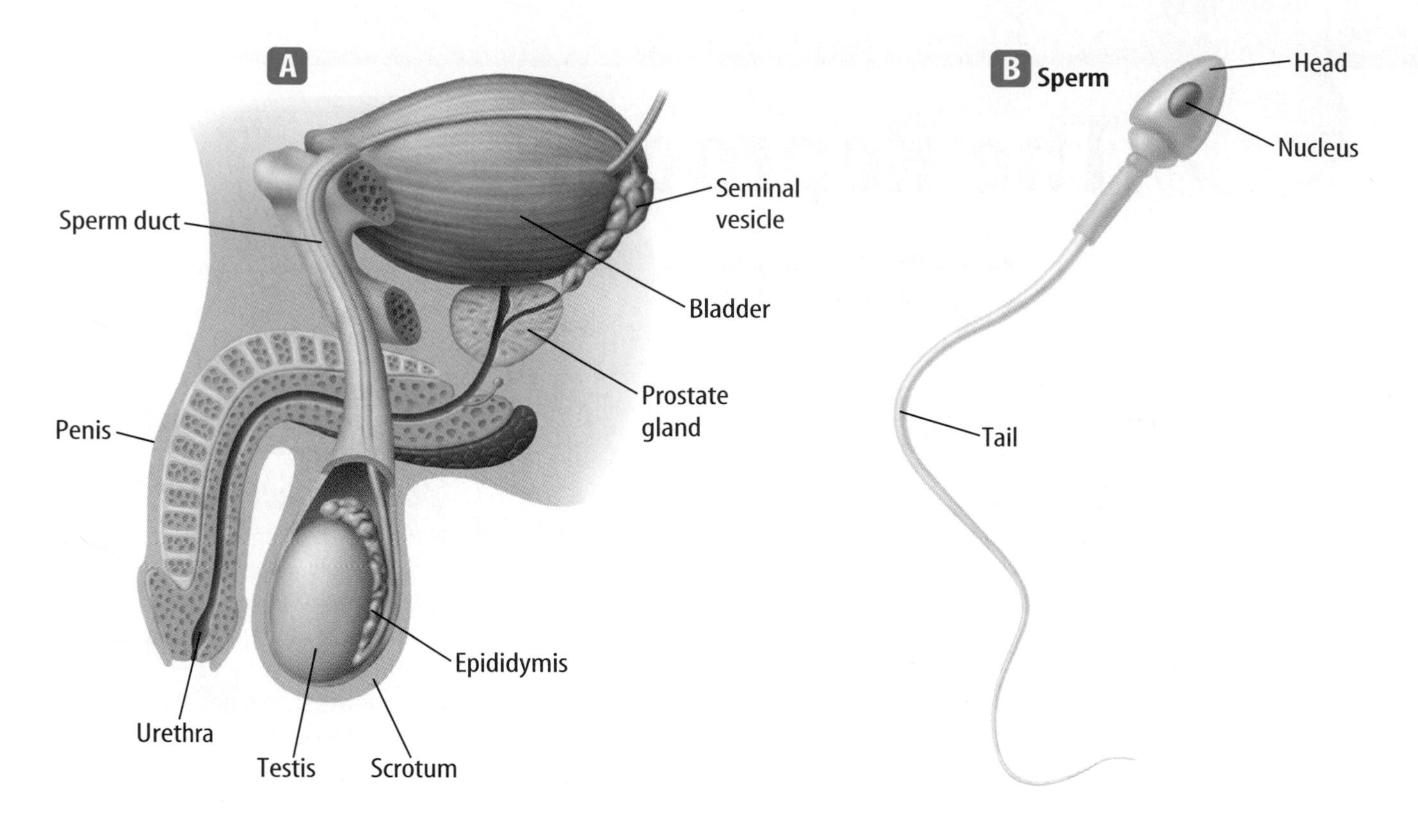

Figure 5
The structures of A the male reproductive system are shown with B a close-up of the sperm, which is produced in the testis. Sperm are produced throughout the life of a male.

The Male Reproductive System

The male reproductive system is made up of external and internal organs. The external organs of the male reproductive system are the penis and scrotum, shown in **Figure 5.** The scrotum contains two organs called testes (TES teez). As males mature sexually, the **testes** begin to produce testosterone, the male hormone, and **sperm,** which are male reproductive cells.

Sperm Each sperm cell has a head and tail. The head contains hereditary information, and the tail moves the sperm. Because the scrotum is located outside the body cavity, the testes, where sperm are produced, are kept at a lower temperature than the rest of the body. Sperm are produced in greater numbers at lower temperatures.

Many organs help in the production, transportation, and storage of sperm. After sperm are produced, they travel from the testes through sperm ducts that circle the bladder. Behind the bladder, a gland called the seminal vesicle provides sperm with a fluid. This fluid supplies the sperm with an energy source and helps them move. This mixture of sperm and fluid is called **semen** (SEE mun). Semen leaves the body through the urethra, which is the same tube that carries urine from the body. However, semen and urine never mix. A muscle at the back of the bladder contracts to prevent urine from entering the urethra as sperm leave the body.

The Female Reproductive System

Unlike male reproductive organs, most of the reproductive organs of the female are inside the body. The **ovaries**—the female sex organs—are located in the lower part of the body cavity. Each of the two ovaries is about the size and shape of an almond. **Figure 6** shows the different organs of the female reproductive system.

Research Visit the Glencoe Science Web site at **science.glencoe.com** for information about ovarian cysts. Make a small pamphlet explaining what cysts are and how they can be treated.

The Egg When a female is born, she already has all of the cells in her ovaries that eventually will develop into eggs—the female reproductive cells. At puberty, eggs start to develop in her ovaries because of specific sex hormones.

About once a month, an egg is released from an ovary in a hormone-controlled process called **ovulation** (ahv yuh LAY shun). The two ovaries release eggs on alternating months. One month, an egg is released from an ovary. The next month, the other ovary releases an egg and so on. After the egg is released, it enters the oviduct. Sometimes a sperm fertilizes the egg. If fertilization takes place, it usually happens in an oviduct. Short, hairlike structures called cilia help sweep the egg through the oviduct toward the uterus (YEWT uh rus).

✔ Reading Check *When are eggs released by the ovaries?*

The **uterus** is a hollow, pear-shaped, muscular organ with thick walls in which a fertilized egg develops. The lower end of the uterus, the cervix, narrows and is connected to the outside of the body by a muscular tube called the **vagina** (vuh JI nuh). The vagina also is called the birth canal because during birth, a baby travels through this tube from the uterus to the outside of the mother's body.

Figure 6
The structures of the female reproductive system are shown from the A side of the body and from the B front. *Where in the female reproductive system do the eggs develop?*

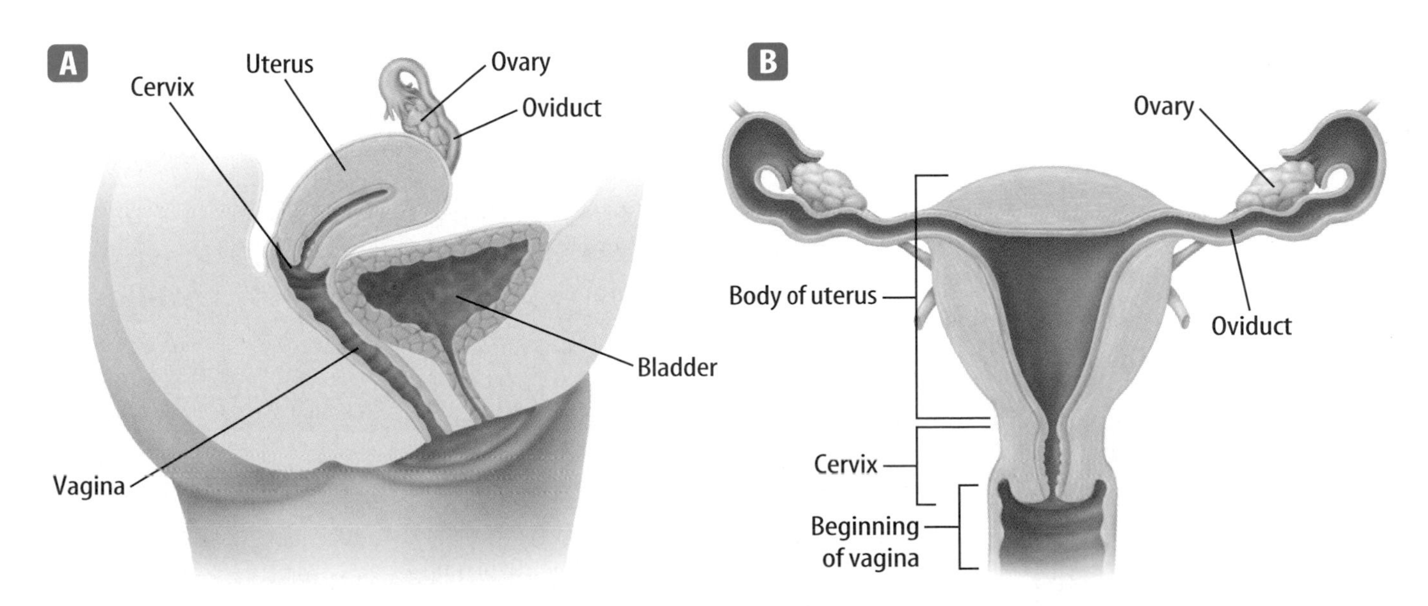

Graphing Hormone Levels

Procedure

Make a line graph of this table.

Hormone Changes	
Day	**Level of Hormone**
1	12
5	14
9	15
13	70
17	13
21	12
25	8

Analysis

1. On what day is the highest level of hormone present?
2. What event takes place around the time of the highest hormone level?

The Menstrual Cycle

How is the female body prepared for having a baby? The **menstrual cycle** is the monthly cycle of changes in the female reproductive system. Before and after an egg is released from an ovary, the uterus undergoes changes. The menstrual cycle of a human female averages 28 days. However, the cycle can vary in some individuals from 20 to 40 days. Changes include the maturing of an egg, the production of female sex hormones, and the preparation of the uterus to receive a fertilized egg.

✔ Reading Check *What is the menstrual cycle?*

Endocrine Control Hormones control the entire menstrual cycle. The pituitary gland responds to chemical messages from the hypothalamus by releasing several hormones. These hormones start the development of eggs in the ovary. They also start the production of other hormones in the ovary, including estrogen (ES truh jun) and progesterone (proh JES tuh rohn). The interaction of all these hormones results in the physical processes of the menstrual cycle.

Phase One As shown in **Figure 7,** the first day of phase 1 starts when menstrual flow begins. Menstrual flow consists of blood and tissue cells released from the thickened lining of the uterus. This flow usually continues for four to six days and is called **menstruation** (men STRAY shun).

Figure 7
The three phases of the menstrual cycle make up the monthly changes in the female reproductive system.

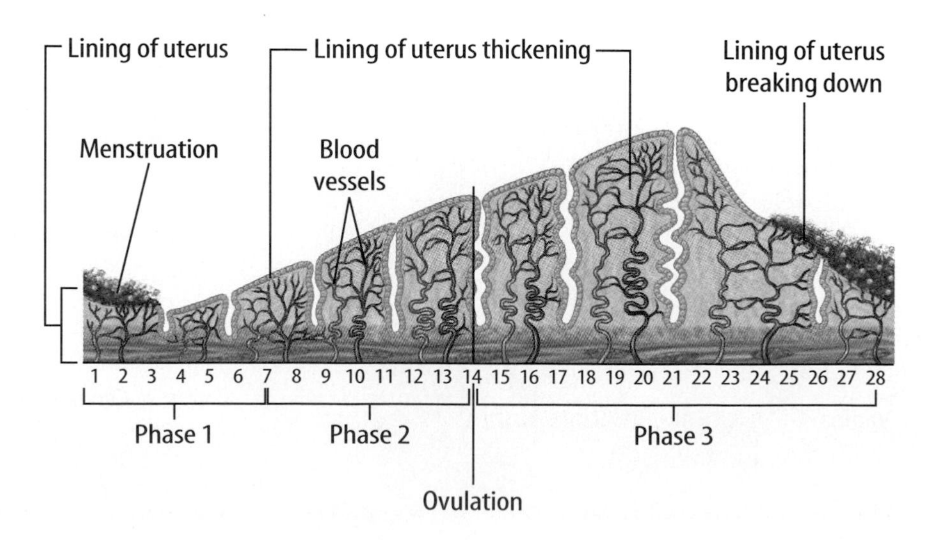

Phase Two Hormones cause the lining of the uterus to thicken in phase 2. Hormones also control the development of an egg in the ovary. Ovulation occurs about 14 days before menstruation begins. Once the egg is released, it must be fertilized within 24 h or it usually begins to break down. Because sperm can survive in a female's body for up to three days, fertilization can occur soon after ovulation.

Figure 8
This older woman enjoys exercising with her granddaughter.

Phase Three Hormones produced by the ovaries continue to cause an increase in the thickness of the uterine lining during phase 3. If a fertilized egg does arrive, the uterus is ready to support and nourish the developing embryo. If the egg is not fertilized, the lining of the uterus breaks down as the hormone levels decrease. Menstruation begins and the cycle repeats itself.

Menopause For most females the first menstrual period happens between ages nine years and 13 years and continues until 45 years of age to 60 years of age. Then, a gradual reduction of menstruation takes place as hormone production by the ovaries begins to shut down. Menopause occurs when both ovulation and menstrual periods end. It can take several years for the completion of menopause. As **Figure 8** indicates, menopause does not inhibit a woman's ability to enjoy an active life.

Section  Assessment

1. What is the major function of male and female reproductive systems in humans?
2. Explain the movement of sperm through the male reproductive system.
3. Compare and contrast the major organs and structures of the male and female reproductive systems.
4. Using diagrams and captions, sequence the stages of the menstrual cycle in a human female.
5. **Think Critically** Adolescent females often require additional amounts of iron in their diet. Explain.

Skill Builder Activities

6. **Concept Mapping** Make an events chain concept map to sequence the movement of an egg through the female reproductive system. **For more help, refer to the Science Skill Handbook.**
7. **Solving One-Step Equations** Usually, one egg is released each month during a female's reproductive years. If menstruation begins at 12 years of age and ends at 50 years of age, calculate the possible number of eggs her body can release during her reproductive years. **For more help, refer to the Math Skill Handbook.**

Activity

Interpreting Diagrams

Starting in adolescence, hormones cause the development of eggs in the ovary and changes in the uterus. These changes prepare the uterus to accept a fertilized egg that can attach itself in the wall of the uterus. What happens to an unfertilized egg?

What You'll Investigate

What changes occur to the uterus during a female's monthly menstrual cycle?

Materials

paper and pencil

Goals

- **Observe** the stages of the menstrual cycle in the diagram.
- **Relate** the process of ovulation to the cycle.

Procedure

1. The diagrams below show what is explained in this chapter on the menstrual cycle.
2. Use the information in this chapter and the diagrams below to complete a data table.
3. On approximately what day in a 28-day cycle is the egg released from the ovary?

Menstruation Cycle

Days	Condition of Uterus	What Happens
1–6		
7–12		
13–14		
15–18		

Conclude and Apply

1. How many days does the average menstrual cycle last?
2. On what days does the lining of the uterus build up?
3. **Infer** why this process is called a cycle.
4. **Calculate** how many days before menstruation ovulation usually occurs.

Communicating Your Data

Compare your data table with those of other students in your class. **For more help, refer to the Science Skill Handbook.**

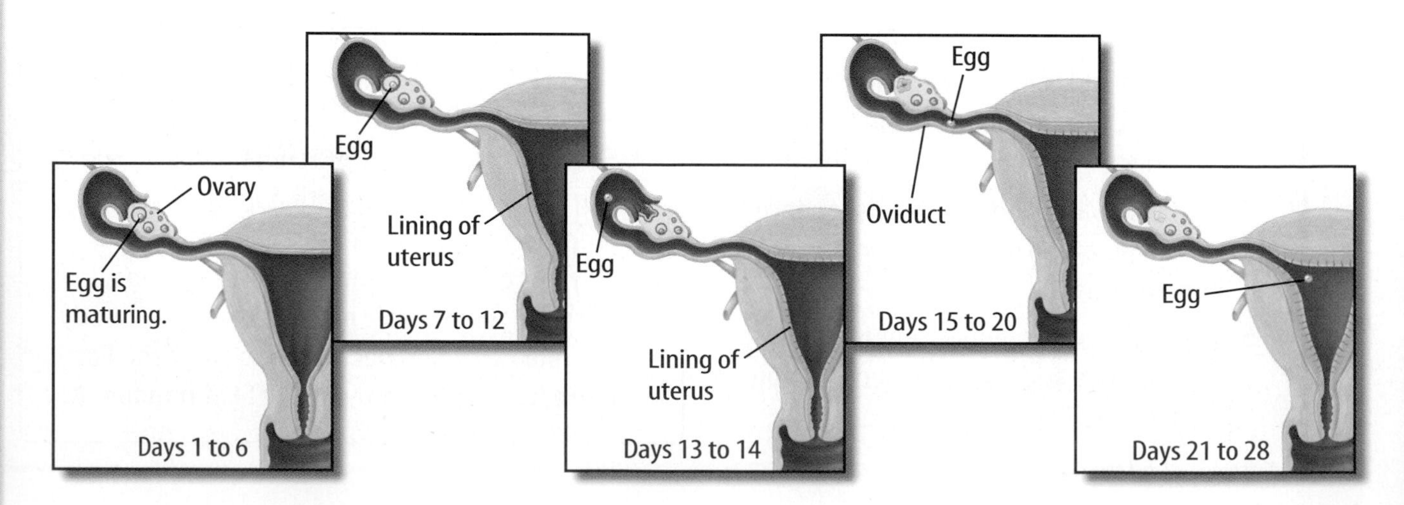

Human Life Stages

The Function of the Reproductive System

Before the invention of powerful microscopes, some people imagined an egg or a sperm to be a tiny person that grew inside a female. In the latter part of the 1700s, experiments using amphibians showed that contact between an egg and sperm is necessary for the development of life. With the development of the cell theory in the 1800s, scientists recognized that a human develops from an egg that has been fertilized by a sperm. The uniting of a sperm and an egg is known as fertilization. Fertilization, as shown in **Figure 9,** usually takes place in the oviduct.

As You Read

***What* You'll Learn**

- **Describe** the fertilization of a human egg.
- **List** the major events in the development of an embryo and fetus.
- **Describe** the developmental stages of infancy, childhood, adolescence, and adulthood.

Vocabulary

pregnancy, embryo, amniotic sac, fetus, fetal stress

***Why* It's Important**

Fertilization begins the entire process of human growth and development.

Fertilization

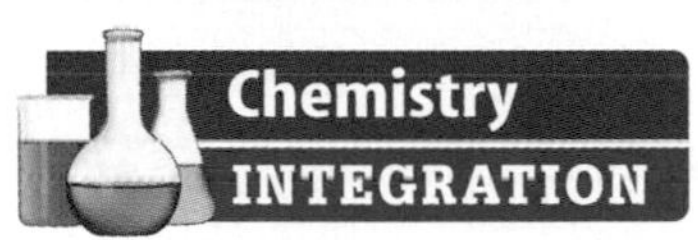

Although 200 millon to 300 million sperm can be deposited in the vagina, only several thousand reach an egg in the oviduct. As they enter the female, the sperm come into contact with chemical secretions in the vagina. It appears that this contact causes a change in the membrane of the sperm. The sperm then become capable of fertilizing the egg. The one sperm that makes successful contact with the egg releases an enzyme from the saclike structure on its head. Enzymes help speed up chemical reactions that have a direct effect on the protective membranes on the egg's surface. The structure of the egg's membrane is disrupted, and the sperm head can enter the egg.

Zygote Formation Once a sperm has entered the egg, changes in the electric charge of the egg's membrane prevent other sperm from entering the egg. At this point, the nucleus of the successful sperm joins with the nucleus of the egg. This joining of nuclei creates a fertilized cell called the zygote. It begins to undergo many cell divisions.

Figure 9
After the sperm releases enzymes that disrupt the egg's membrane, it penetrates the egg.

Figure 10
The development of fraternal and identical twins is different.

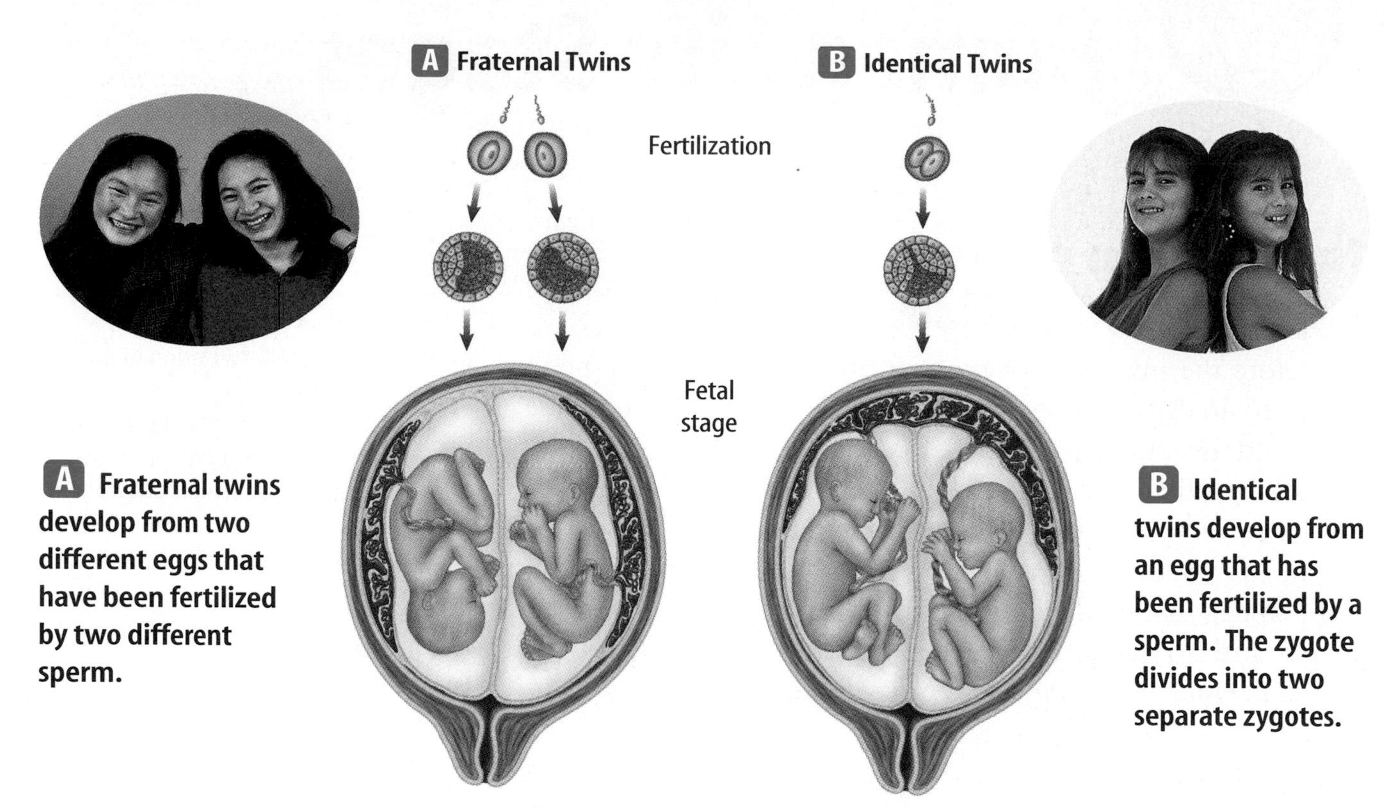

A **Fraternal twins develop from two different eggs that have been fertilized by two different sperm.**

B **Identical twins develop from an egg that has been fertilized by a sperm. The zygote divides into two separate zygotes.**

Multiple Births

Sometimes two eggs leave the ovary at the same time. If both eggs are fertilized and both develop, fraternal twins are born. Fraternal twins, as shown in **Figure 10A,** can be two girls, two boys, or a boy and a girl. Because fraternal twins come from two eggs, they only resemble each other.

Because identical twin zygotes develop from the same egg and sperm, as explained in **Figure 10B,** they have the same hereditary information. These identical zygotes develop into identical twins, which are either two girls or two boys. Multiple births also can occur when three or more eggs are produced at one time or when the zygote separates into three or more parts.

Development Before Birth

After fertilization, the zygote moves along the oviduct to the uterus. During this time, the zygote is dividing and forming into a ball of cells. After about seven days, the zygote attaches to the wall of the uterus, which has been thickening in preparation to receive a zygote, as shown in **Figure 11.** If attached to the wall of the uterus, the zygote will develop into a baby in about nine months. This period of development from fertilized egg to birth is known as **pregnancy.**

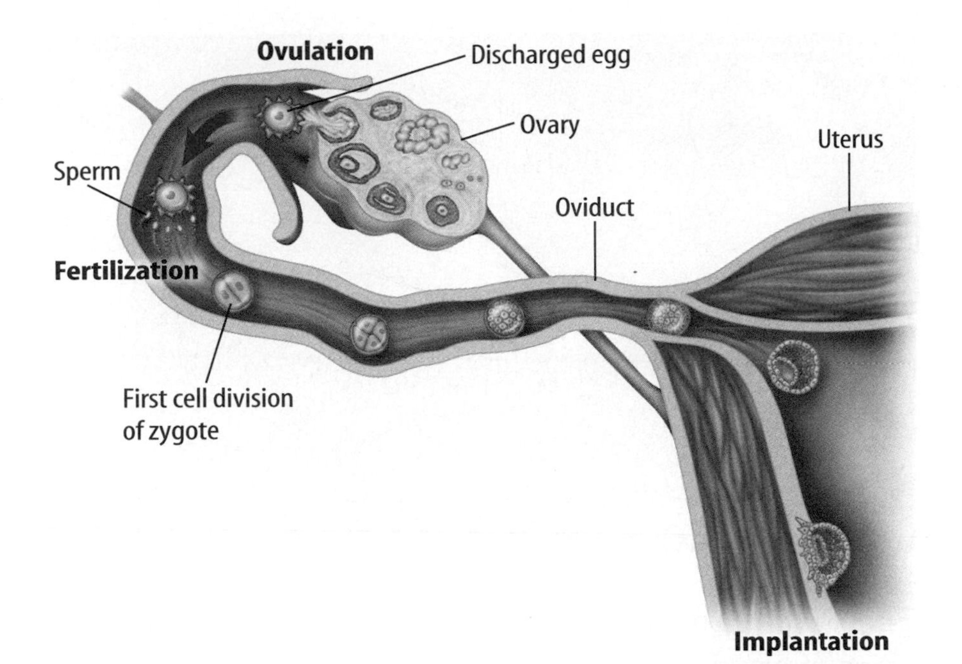

Figure 11
After a few days of rapid cell division, the zygote, now a ball of cells, reaches the lining of the uterus, where it attaches itself to the lining for development.

The Embryo After the zygote attaches to the wall of the uterus, it is known as an **embryo,** illustrated in **Figure 12.** It receives nutrients from fluids in the uterus until the placenta (pluh SENT uh) develops from tissues of the uterus and the embryo. An umbilical cord develops that connects the embryo to the placenta. In the placenta, materials diffuse between the mother's blood and the embryo's blood, but their bloods do not mix. Blood vessels in the umbilical cord carry nutrients and oxygen from the mother's blood through the placenta to the embryo. Other substances in the mother's blood can move into the embryo, including drugs, toxins, and disease organisms. Wastes from the embryo are carried in other blood vessels in the umbilical cord through the placenta to the mother's blood.

Reading Check ***Why must a pregnant woman avoid alcohol, tobacco, and harmful drugs?***

Pregnancy in humans lasts about 38 to 39 weeks. During the third week, a thin membrane called the **amniotic** (am nee AH tihk) **sac** begins to form around the embryo. The amniotic sac is filled with a clear liquid called amniotic fluid, which acts as a cushion for the embryo and stores nutrients and wastes.

During the first two months of development, the embryo's major organs form and the heart structure begins to beat. At five weeks, the embryo has a head with eyes, nose, and mouth features. During the sixth and seventh weeks, fingers and toes develop.

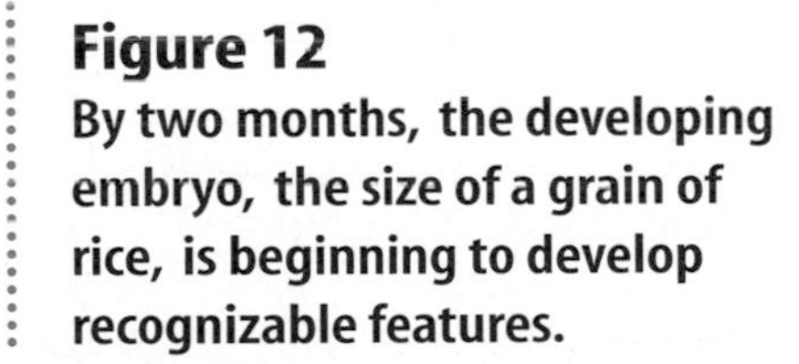

Figure 12
By two months, the developing embryo, the size of a grain of rice, is beginning to develop recognizable features.

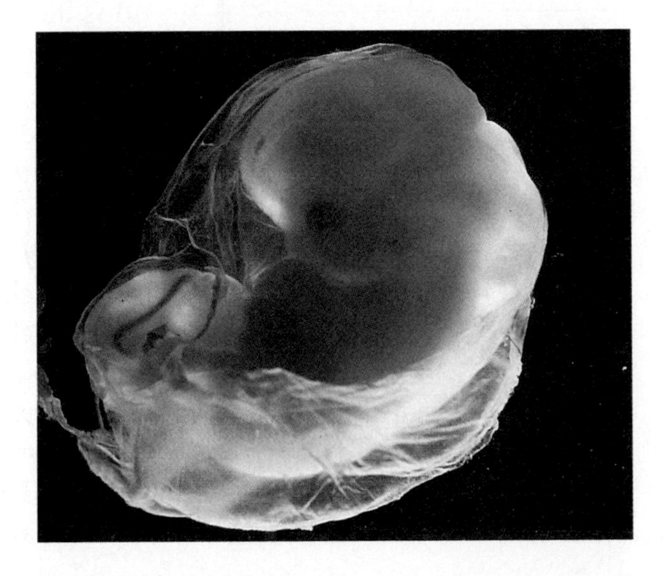

Figure 13
A fetus at about 16 weeks is approximately 15 cm long and weighs 140 g.

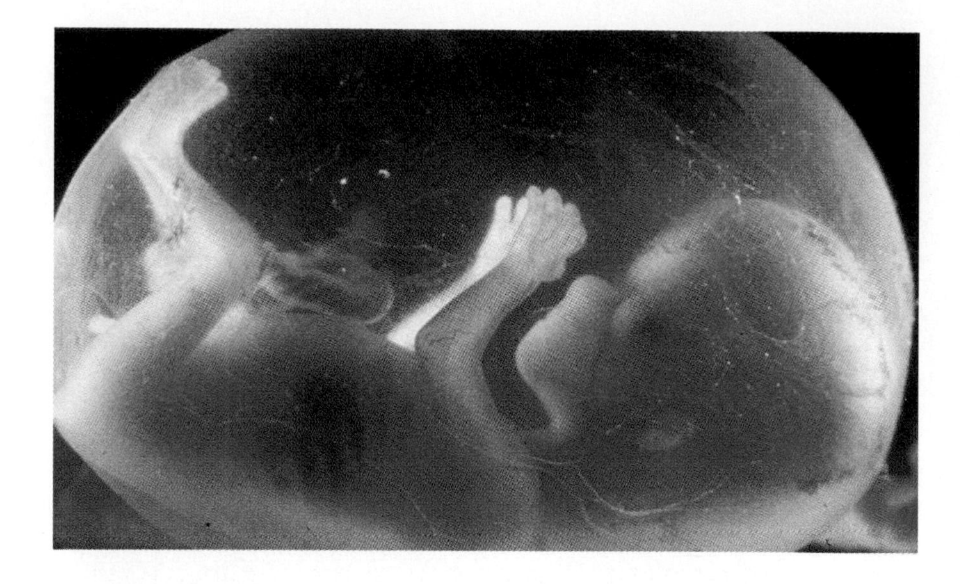

The Fetus After the first two months of pregnancy, the developing embryo is called a **fetus,** shown in **Figure 13.** At this time, body organs are present. Around the third month, the fetus is 8 cm to 9 cm long. The mother may feel the fetus move. The fetus can even suck its thumb. By the fourth month, an ultrasound test can determine the sex of the fetus. The fetus is 30 cm to 38 cm in length by the end of the seventh month of pregnancy. Fatty tissue builds up under the skin, and the fetus looks less wrinkled. By the ninth month, the fetus usually has shifted to a head-down position within the uterus, a position beneficial for delivery. The head usually is in contact with the opening of the uterus to the vagina. The fetus is about 50 cm in length and weighs from 2.5 kg to 3.5 kg.

TRY AT HOME
Mini LAB

Interpreting Fetal Development

Procedure
Make a bar graph of the following data.

Fetal Development

End of Month	Length (cm)
3	8
4	15
5	25
6	30
7	35
8	40
9	51

Analysis
1. During which month does the greatest increase in length occur?
2. On average, how many centimeters does the baby grow per month?

The Birthing Process

The process of childbirth, as shown in **Figure 14,** begins with labor, the muscular contractions of the uterus. As the contractions increase in strength and number, the amniotic sac usually breaks and releases its fluid. Over a period of hours, the contractions cause the opening of the uterus to widen. More powerful and more frequent contractions push the baby out through the vagina into its new environment.

Delivery Often a mother is given assistance by a doctor during the delivery of the baby. As the baby emerges from the birth canal, a check is made to determine if the umbilical cord is wrapped around the baby's neck or any body part. When the head is free, any fluid in the baby's nose and mouth is removed by suction. After the head and shoulders appear, contractions force the baby out completely. Up to an hour after delivery, contractions occur that push the placenta out of the mother's body.

Cesarean Section Sometimes a baby must be delivered before labor begins or before it is completed. At other times, a baby cannot be delivered through the birth canal because the mother's pelvis might be too small or the baby might be in the wrong birthing position. In cases like these, surgery called a cesarean (suh SEER ee uhn) section is performed. An incision is made through the mother's abdominal wall, then through the wall of the uterus. The baby is delivered through this opening.

Reading Check *What is a cesarean section?*

After Birth When the baby is born, it is attached to the umbilical cord. The person assisting with the birth clamps the cord in two places and cuts it between the clamps. The baby does not feel any pain from this procedure. The baby might cry, which is the result of air being forced into its lungs. The scar that forms where the cord was attached is called the navel.

SCIENCE Online

Research Visit the Glencoe Science Web site at **science.glencoe.com** for more information about cesarean section delivery. Communicate what you learn to your class.

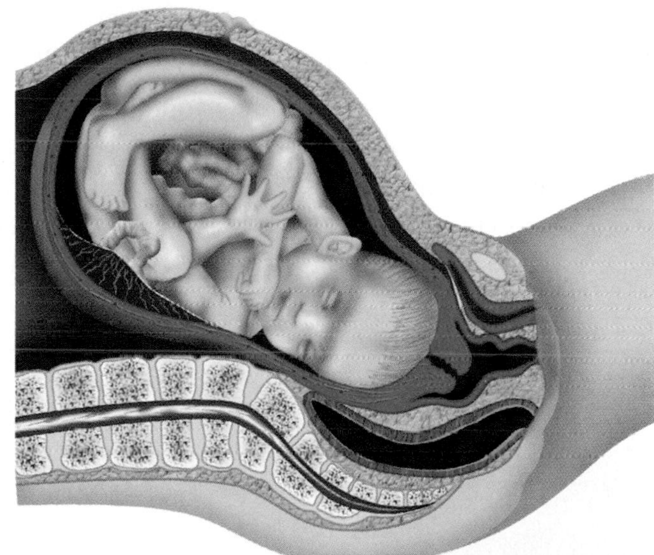

A The fetus moves into the opening of the birth canal, and the uterus begins to widen.

Figure 14
Childbirth begins with labor. The opening to the uterus widens, and the baby passes through.

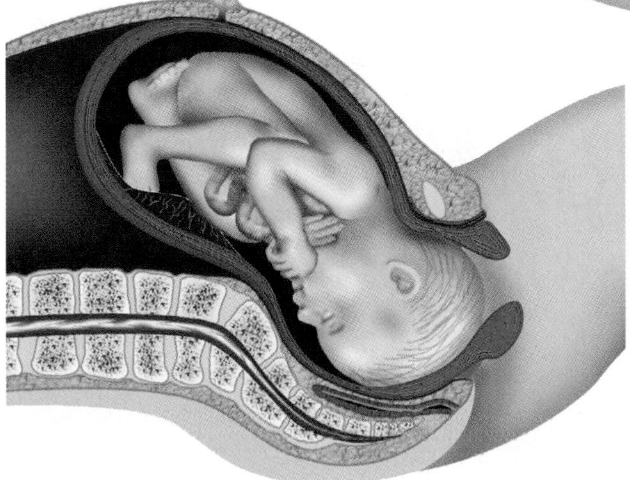

B The base of the uterus is completely dilated.

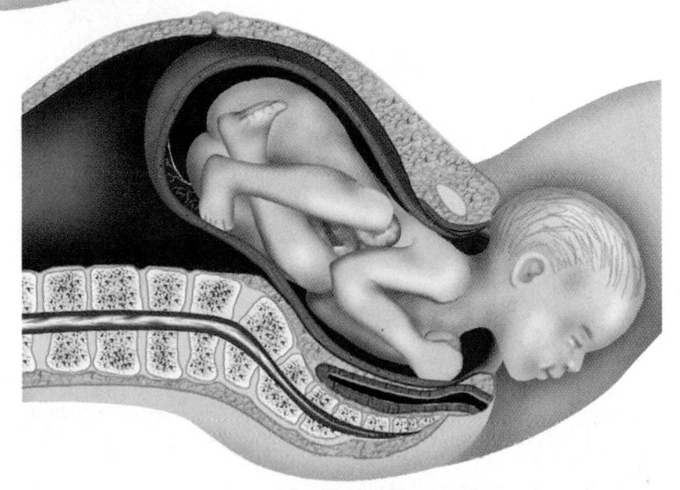

C The fetus is pushed out through the birth canal.

Stages After Birth

Defined stages of development occur after birth, based on the major developments that take place during those specific years. Infancy lasts from birth to around 18 months of age. Childhood extends from the end of infancy to sexual maturity, or puberty. The years of adolescence vary, but they usually are considered to be the teen years. Adulthood covers the years of age from the early 20s until life ends, with older adulthood considered to be over 60. The age spans of these different stages are not set, and scientists differ in their opinions regarding them.

Infancy What type of environment must the infant adjust to after birth? The experiences the fetus goes through during birth cause **fetal stress.** The fetus has emerged from an environment that was dark, watery, a constant temperature, and nearly soundless. In addition, the fetus might have been forced through the constricted birth canal. However, in a short period of time, the infant's body becomes adapted to its new world.

The first four weeks after birth are known as the neonatal(nee oh NAY tul) period. The term *neonatal* means "newborn." During this time, the baby's body begins to function normally. Unlike the newborn of some other animals, human babies, shown in **Figure 15A,** depend on other humans for their survival. In contrast, many other animals, such as horses like those shown in **Figure 15B,** begin walking a few hours after they are born.

Figure 15
Human babies are more dependent upon their caregivers than many other mammals are.

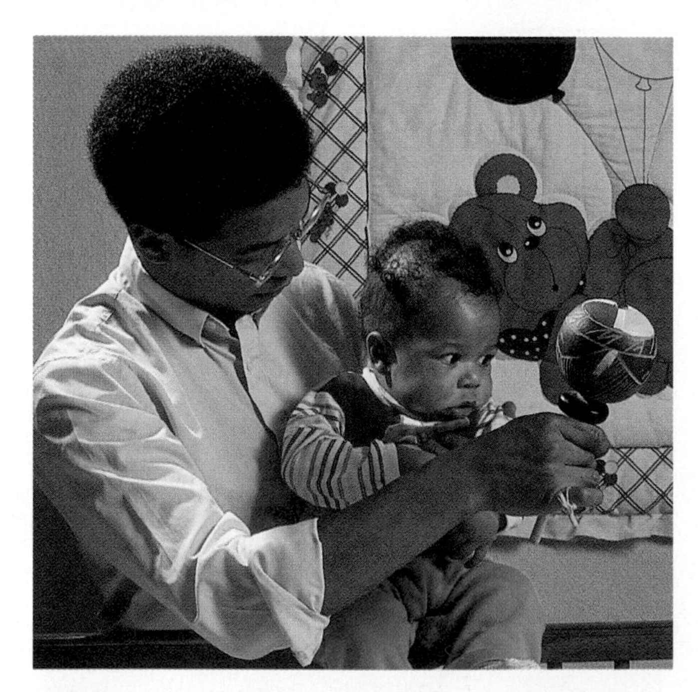

A Infants and toddlers are completely dependent upon caregivers for all their needs.

B Other young mammals are more self-sufficient. This colt is able to stand within an hour after birth.

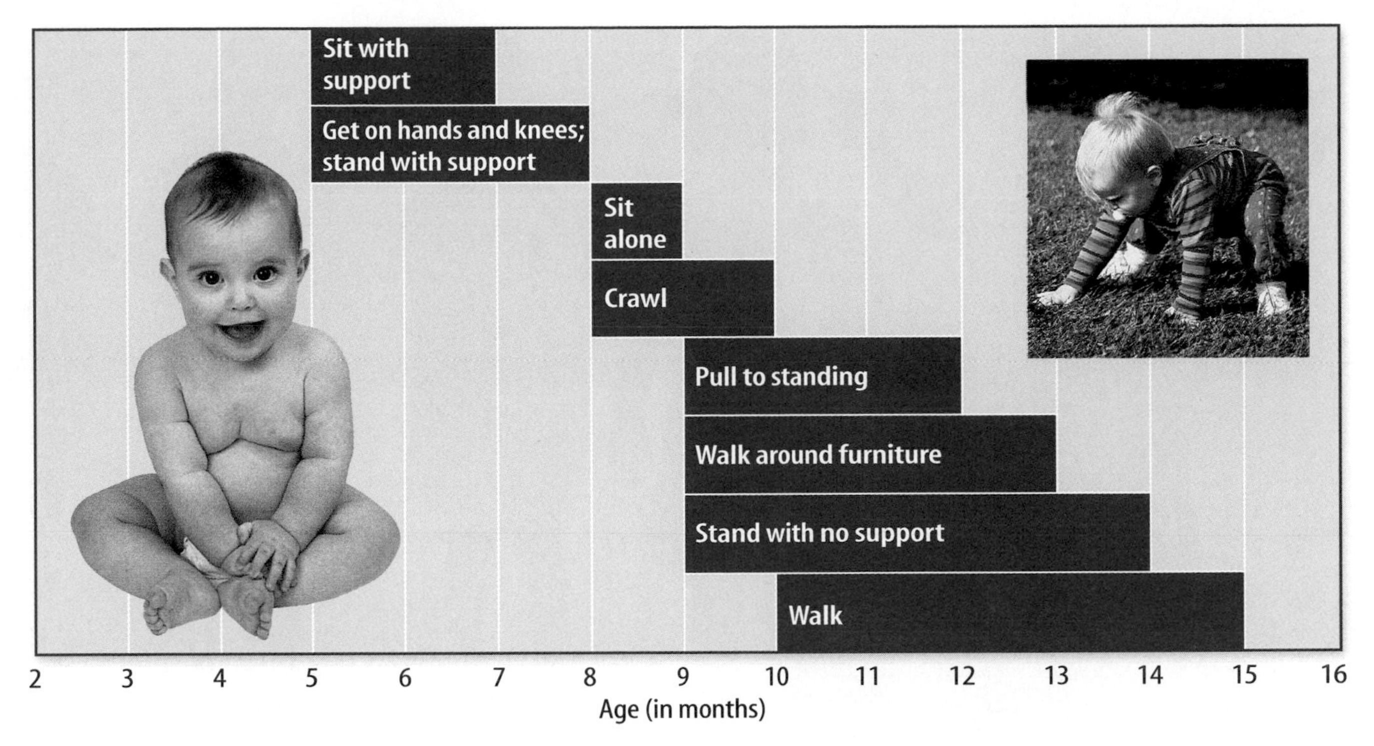

Figure 16
Infants show rapid development in their nervous and muscular systems through 18 months of age.

During these first 18 months, infants show increased physical coordination, mental development, and rapid growth. Many infants will triple their weight in the first year. **Figure 16** shows the extremely rapid development of the nervous and muscular systems during this stage, which enables infants to start interacting with the world around them.

Childhood After infancy is childhood, which lasts until about puberty, or sexual maturity. Sexual maturity occurs around 12 years of age. Overall, growth during early childhood is rather rapid, although the physical growth rate for height and weight is not as rapid as it is in infancy. Between two and three years of age, the child learns to control his or her bladder and bowels. At age two to three, most children can speak in simple sentences. Around age four, the child is able to get dressed and undressed with some help. By age five, many children can read a limited number of words. By age six, children usually have lost their chubby baby appearance, as seen in **Figure 17.** However, muscular coordination and mental abilities continue to develop. Throughout this stage, children develop their abilities to speak, read, write, and reason. These ages of development are only guidelines because each child develops at a different rate.

Figure 17
Children grow and develop at different rates, like these kindergartners.

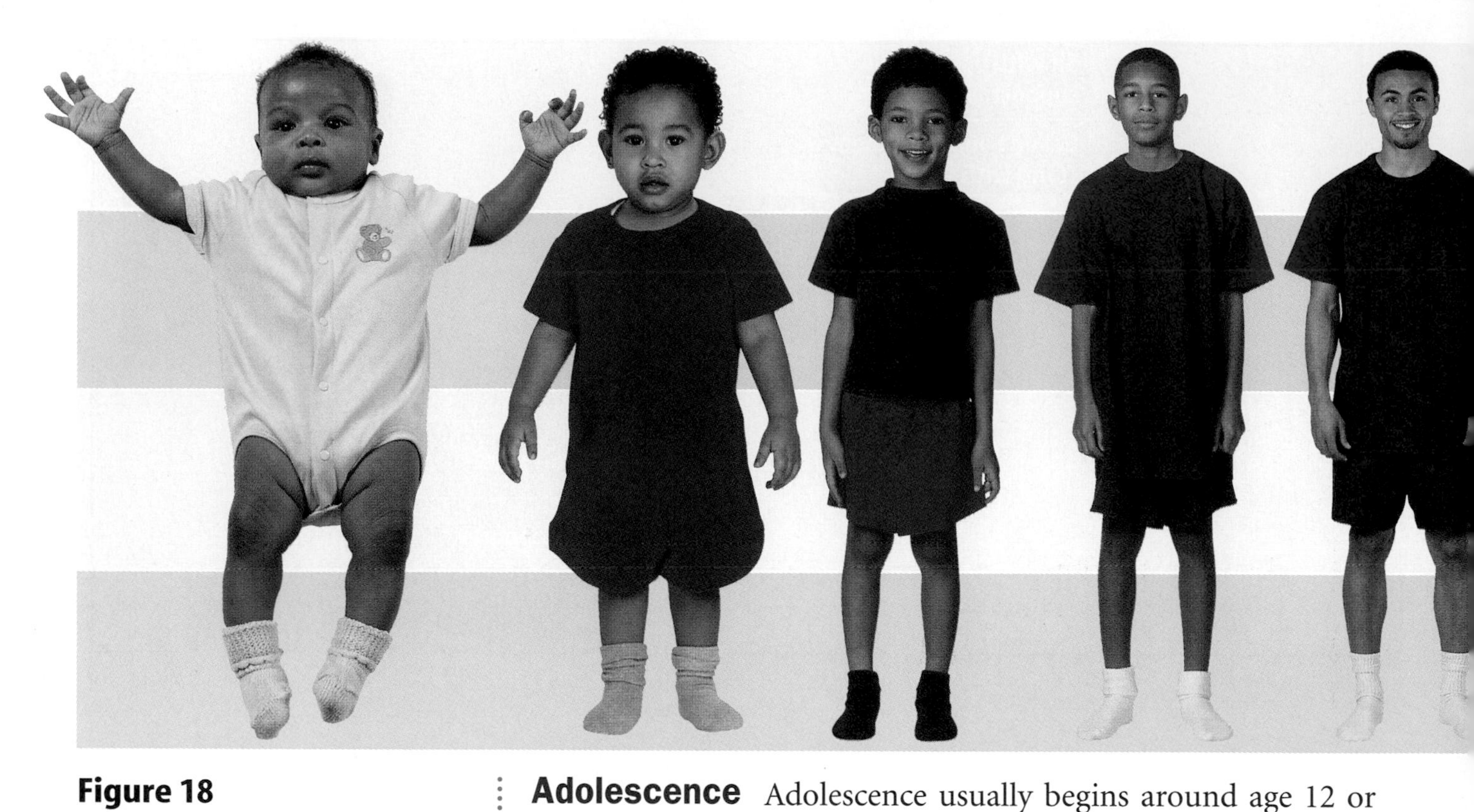

Figure 18
The proportions of body parts change over time as the body develops. *Describe how the head changes proportion.*

Adolescence Adolescence usually begins around age 12 or 13. A part of adolescence is puberty—the time of development when a person becomes physically able to reproduce. For girls, puberty occurs between ages nine and 13. For boys, puberty occurs between ages 13 and 16. During puberty, hormones produced by the pituitary gland cause changes in the body. These hormones produce reproductive cells and sex hormones. Secondary sex characteristics also develop. In females, the breasts develop, pubic and underarm hair appears, and fatty tissue is added to the buttocks and thighs. In males, the hormones cause a deepened voice, an increase in muscle size, and the growth of facial, pubic, and underarm hair.

Adolescence usually is when the final growth spurt occurs. Because the time when hormones begin working varies among individuals and between males and females, growth rates differ. Girls often begin their final growth phase at about age 11 and end around age 16. Boys usually start their growth spurt at age 13 and end around 18 years of age.

Physics INTEGRATION

During adolescence, the body parts do not all grow at the same rate. The legs grow longer before the upper body lengthens. This changes the body's center of gravity, the point at which the body maintains its balance. This is one cause of teenager clumsiness. In your Science Journal, write a paragraph about how this might affect playing sports.

Adulthood The final stage of development, adulthood, begins with the end of adolescence and continues through old age. This is when the growth of the muscular and skeletal system stops. **Figure 18** shows how body proportions change as you age.

People from age 45 to age 60 are sometimes considered middle-aged adults. During these years, physical strength begins to decline. Blood circulation and respiration become less efficient. Bones become more brittle, and the skin becomes wrinkled.

Older Adulthood People over the age of 60 may experience an overall decline in their physical body systems. The cells that make up these systems no longer function as well as they did at a younger age. Connective tissues lose their elasticity, causing muscles and joints to be less flexible. Bones become thinner and more brittle. Hearing and vision are less sensitive. The lungs and heart work less efficiently. However, exercise and eating well over a lifetime can help extend the health of one's body systems. Many healthy older adults enjoy full lives and embrace challenges, as shown in **Figure 19.**

Figure 19
Astronaut and Senator John Glenn traveled into space twice. In 1962, at age 40, he was the first U.S. citizen to orbit Earth. He was part of the space shuttle crew in 1998 at age 77. Senator Glenn has helped change people's views of what many older adults are capable of doing.

What physical changes occur during late adulthood?

Human Life Spans Seventy-five years is the average life span—from birth to death—of humans, although an increasing number of people live much longer. However, body systems break down with age, resulting in eventual death. Death can occur earlier than old age for many reasons, including diseases, accidents, and bad health choices.

Section 3 Assessment

1. What happens when an egg is fertilized in a female?
2. What happens to an embryo during the first two months of pregnancy?
3. Describe the major events that occur during childbirth.
4. What stage of development are you in? What physical changes have occurred or will occur during this stage of human development?
5. **Think Critically** Why is it hard to compare the growth and development of different adolescents?

Skill Builder Activities

6. **Making Models** Use references to construct a time line that highlights the major events in the various stages of development from the embryo to adulthood. **For more help, refer to the Science Skill Handbook.**
7. **Using an Electronic Spreadsheet** Using your text and other resources, make a spreadsheet for the stages of human development from a zygote to a fetus. Title one column *Zygote,* another *Embryo,* and a third *Fetus.* Complete the spreadsheet. **For more help, refer to the Technology Skill Handbook.**

Activity

Changing Body Proportions

The ancient Greeks believed the perfect body was completely balanced. Arms and legs should not be too long or short. A person's head should not be too large or small. The extra large muscles of a body builder would have been ugly to the Greeks. How do you think they viewed the bodies of infants and children? Infants and young children have much different body proportions than adults, and teenagers often go through growth spurts that quickly change their body proportions. How do body proportions differ among people?

What You'll Investigate

How do the body proportions differ between adolescent males and females?

Materials

tape measure
erasable pencil
graph paper

Goals

- **Measure** specific body proportions of adolescents.
- **Infer** how body proportions differ between adolescent males and females.

Procedure

1. Copy the data table in your Science Journal and record the gender of each person that you measure.
2. Measure each person's head circumference by starting in the middle of the forehead and wrapping the tape measure around the head. Record these measurements.
3. Measure each person's arm length from the top of the shoulder to the tip of the middle finger while the arm is held straight out to the side of the body. Record these measurements.
4. Ask each person to remove his or her shoes and stand next to a wall. Mark their height with an erasable pencil and measure their height from the floor to the mark. Record these measurements in the data table.
5. **Combine** your data with that of your classmates. Find the averages of head circumference, arm length, and height. Then, find these averages for males and females.
6. Make a bar graph of your calculations in step 5. Plot the measurements on the y-axis and plot all of the averages along the x-axis.
7. **Calculate** the proportion of average head circumference to average height for everyone in your class by dividing the average head circumference by the average height. Repeat this calculation for males and females.

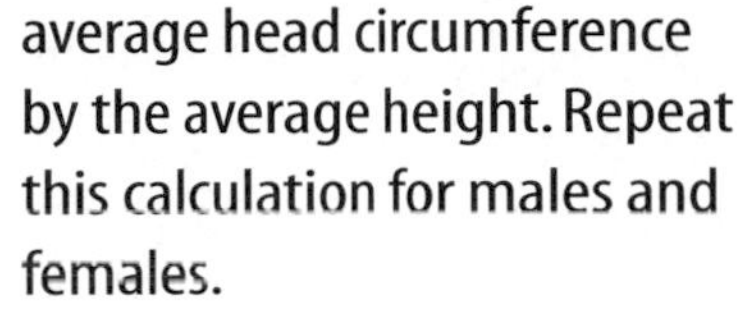

8. **Calculate** the proportion of average arm length to average height for everyone in your class by dividing the average arm length by the average height. Repeat this calculation for males and females.

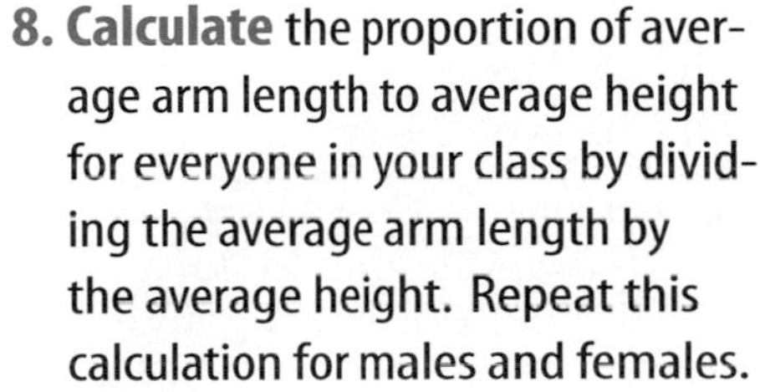

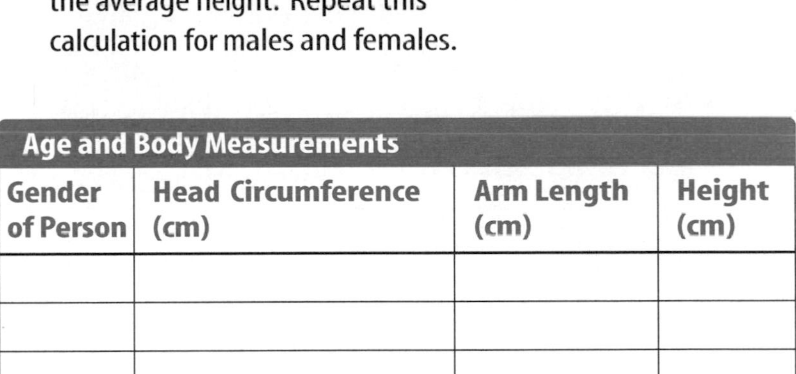

Age and Body Measurements			
Gender of Person	Head Circumference (cm)	Arm Length (cm)	Height (cm)

Conclude and Apply

1. Do adolescent males or females have larger head circumferences or longer arms? Which group has the larger proportion of head circumference or arm length to height?
2. Does this activity support the information in this chapter about the differences between growth rates of adolescent males and females? Explain.

Communicating Your Data

On poster board, **construct** data tables showing your results and those of your classmates. Discuss with your classmates why these results might be different.

Science Stats

Facts About Infants

Did you know...

...Humans and chimpanzees share about 99 percent of their genes. Although humans look different than chimps, reproduction is similar and gestation is the same—about nine months. Youngsters of both species lose their baby teeth at about six years of age.

Female kangaroo and joey

...Unlike humans and most other mammals, the newborn kangaroo develops in its mother's pouch longer than in her uterus. About one month after fertilization, the kangaroo is born and moves from its mother's uterus into her pouch. It spends seven to ten months in the pouch, drinking milk and growing.

...The blue whale calf is the biggest newborn in the world. At birth, it measures about 7 m and weighs about 2,700 kg. The average human newborn measures about 50 cm and weighs about 3.3 kg. In its first year of life, a blue whale calf gains about 90 kg every day. Compare that to an average human baby who gains about 10 kg during his or her entire first year.

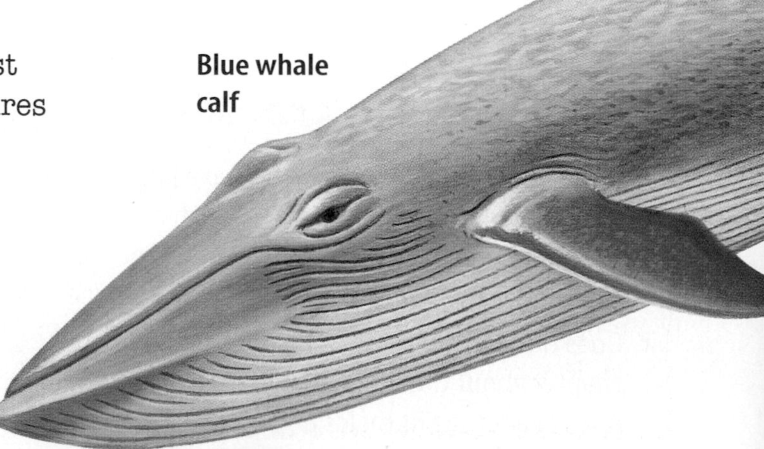
Blue whale calf

Mammal Facts				
Mammal	**Average Gestation**	**Average Birth Weight**	**Average Adult Weight**	**Average Life Span (years)**
African Elephant	22 months	136 kg	4,989.5 kg	35
Blue Whale	12 months	1,800 kg	135,000 kg	60
Human	**9 months**	**3.3 kg**	**59–76 kg**	**76***
Brown Bear	7 months	0.23–0.5 kg	350 kg	22.5
Cat	2 months	99 g	2.7–7 kg	13.5
Kangaroo	1 month	0.75–1.0 g	45 kg	5
Golden Hamster	2.5 weeks	0.3 g	112 g	2

* In the United States

1-day-old to 7-day-old mice

...Of about 4,000 species of mammals, only three lay eggs. These species are the platypus, the short-beaked echidna (ih KIHD nuh), and the long-beaked echidna. No other mammals lay eggs.

Echidna

... House mice can have up to ten litters per year, each one containing up to seven mice. Mice have this many offspring because so few of them survive.

Do the Math

1. Look at the data table above. Make a generalization about a mammal's birth weight and the length of its gestation period.
2. Make a bar graph that compares the length of the human gestation period to that of two other mammals.
3. Assume that a female of each mammal listed in the table above is pregnant once during her life. Which mammal is pregnant for the greatest proportion of her life?

Go Further

Do research to find out which species of vertebrate animals has the longest life span and which has the shortest. Present your findings in a table that also shows the life span of humans.

Chapter 10 Study Guide

Reviewing Main Ideas

Section 1 The Endocrine System

1. Endocrine glands secrete hormones directly into the bloodstream.
2. Hormones affect specific tissues throughout the body. *How can a gland near your head control chemical activities in other parts of your body?*

3. A change in the body causes an endocrine gland to function. When homeostasis is reached, the endocrine gland receives a signal to slow or stop its production.

Section 2 The Reproductive System

1. The reproductive system allows new organisms to be formed.
2. The testes produce sperm that leave the male through the penis.
3. The female ovary produces an egg. If fertilized, it becomes a zygote and later develops into a fetus within the uterus. *How are the structures of the egg and sperm suited for their functions?*

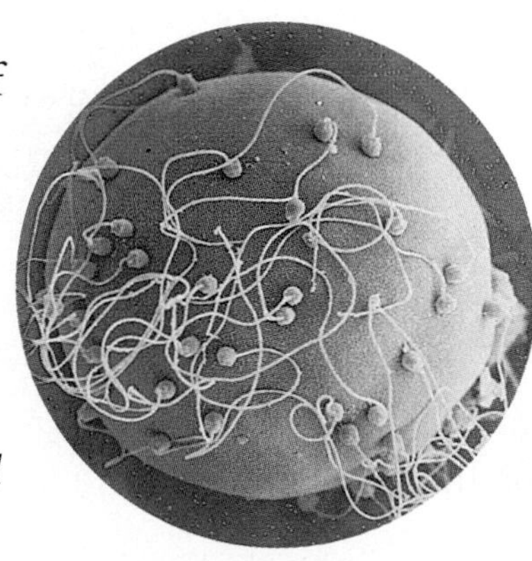

4. When an egg is not fertilized, the built-up lining of the uterus is shed in a process called menstruation. This process begins 14 days after ovulation.

Section 3 Human Life Stages

1. After fertilization, the zygote undergoes developmental changes to become an embryo, then a fetus. Twins occur when two eggs are fertilized or when a zygote divides after fertilization.
2. Birth begins with labor—muscular contractions of the uterus. The amniotic sac breaks. Then, usually after several hours, the contractions force the baby out of the mother's body. *Why is the first stage in the birthing process called labor?*

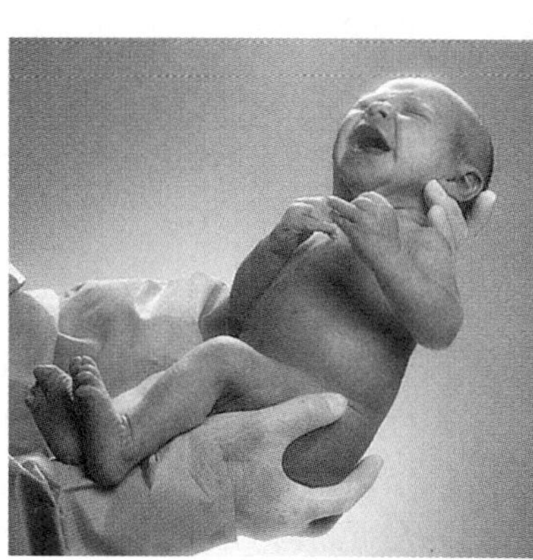

3. Infancy is the stage of development from birth to 18 months of age. It is a period of rapid growth of mental and physical skills. Childhood, which lasts until age 12, is marked by development of muscular coordination and mental abilities.
4. Adolescence is the stage of development when a person becomes physically able to reproduce. The final stage of development is adulthood. Physical development is complete and body systems become less efficient. Death occurs at the end of life.

FOLDABLES Reading & Study Skills

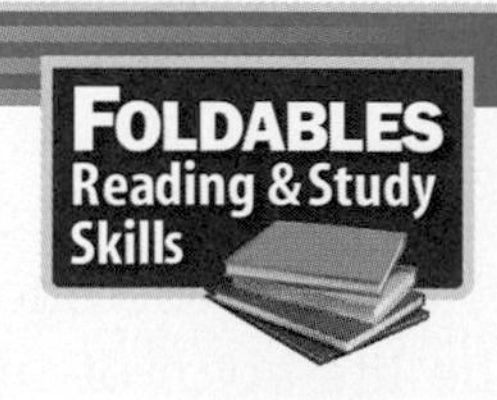

After You Read

Using the information on your Foldable as an outline, explain each stage of human life.

Visualizing Main Ideas

Complete the following table on life stages.

Human Development

Stages of Life	Age Range	Physical Development
Infant		sits, stands, words spoken
		walks, speaks, writes, reads
Adolescent		
		end of muscular and skeletal growth

Vocabulary Review

Vocabulary Words

- **a.** amniotic sac
- **b.** embryo
- **c.** fetal stress
- **d.** fetus
- **e.** hormone
- **f.** menstrual cycle
- **g.** menstruation
- **h.** ovary
- **i.** ovulation
- **j.** pregnancy
- **k.** semen
- **l.** sperm
- **m.** testes
- **n.** uterus
- **o.** vagina

Study Tip

Use lists to help you memorize facts. For example, when trying to memorize the stages of human development, write them down several times on a piece of paper until you know them.

Using Vocabulary

Replace the underlined words with the correct vocabulary word(s).

1. <u>Testes</u> is a mixture of sperm and fluid.
2. The time of the development until the birth of a baby is known as <u>menstruation</u>.
3. During the first two months of pregnancy, the unborn child is known as <u>fetal stress</u>.
4. The <u>vagina</u> is a hollow, pear-shaped muscular organ.
5. The <u>ovary</u> is the membrane that protects the unborn child.
6. After two months of pregnancy, the unborn child is known as a(n) <u>embryo</u>.
7. The <u>testes</u> is the organ that produces eggs.

Chapter 10 Assessment

Checking Concepts

Choose the word or phrase that best answers the question.

1. What are the chemicals produced by the endocrine system?
   A) enzymes
   B) target tissues
   C) hormones
   D) saliva
2. Which gland produces melatonin?
   A) adrenal
   B) thyroid
   C) pancreas
   D) pineal
3. Where does the embryo develop?
   A) oviduct
   B) ovary
   C) uterus
   D) vagina
4. What is the monthly process that releases an egg called?
   A) fertilization
   B) ovulation
   C) menstruation
   D) puberty
5. What is the union of an egg and a sperm?
   A) fertilization
   B) ovulation
   C) menstruation
   D) puberty
6. Where is the egg usually fertilized?
   A) oviduct
   B) uterus
   C) vagina
   D) ovary
7. When does puberty occur?
   A) childhood
   B) adulthood
   C) adolescence
   D) infancy
8. Which sex characteristics are common to males and females?
   A) breasts
   B) pubic hair
   C) increased fat
   D) increased muscles
9. During which period does growth stop?
   A) childhood
   B) adulthood
   C) adolescence
   D) infancy
10. During what stage of development does the amniotic sac form?
    A) zygote
    B) embryo
    C) fetus
    D) newborn

Thinking Critically

11. List the effects that adrenal gland hormones can have on your body as you prepare to run a race.
12. Explain the similar functions of the ovaries and testes.
13. Identify the structure in the following diagram in which each process occurs: ovulation, fertilization, and implantation.

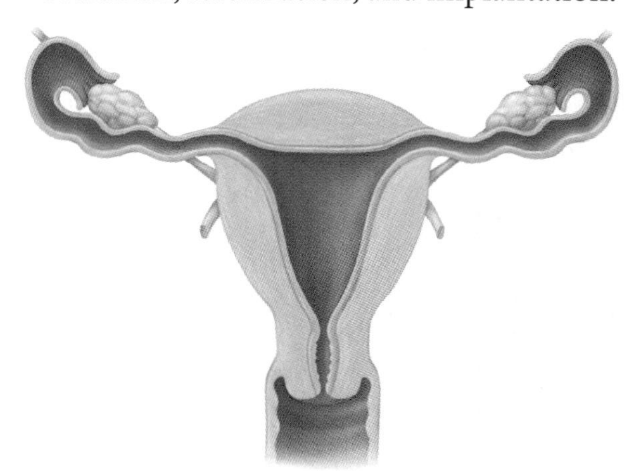

14. How is your endocrine system like the thermostat in your house?
15. Are quadruplets always identical or always fraternal, or can they be either? Explain.

Developing Skills

16. **Predicting** During the ninth month of pregnancy, the fetus develops a white, greasy coating. Predict what the function of this coating might be.
17. **Forming Hypotheses** Make a hypothesis about the effect of raising identical twins apart from each other.
18. **Classifying** Classify each of the following structures as female or male and internal or external: ovary, penis, scrotum, testes, uterus, and vagina.

19. Concept Mapping Complete the following concept map of egg release and implantation using the appropriate scientific words.

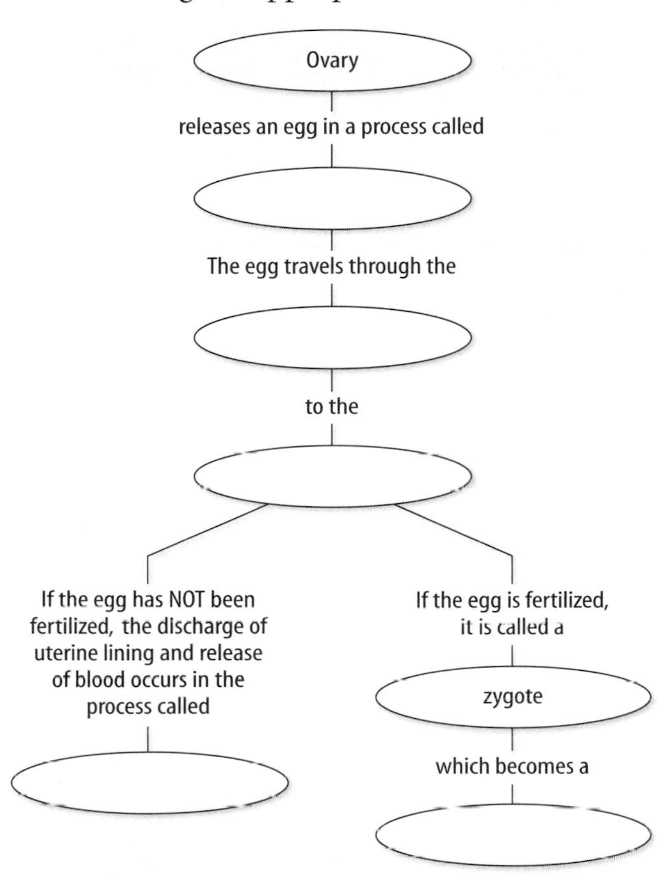

Performance Assessment

20. Letter Find newspaper or magazine articles on the effects of smoking on the health of the developing embryo and newborn. Write a letter to the editor about why a mother's smoking is damaging her unborn baby's health.

Technology

Go to the Glencoe Science Web site at **science.glencoe.com** or use the **Glencoe Science CD-ROM** for additional chapter assessment.

Test Practice

In health class, Angela decided to do a report about the cases of syphilis in the United States. She brought the following graph to accompany her report, which shows syphilis rates by year between 1970 and 1997.

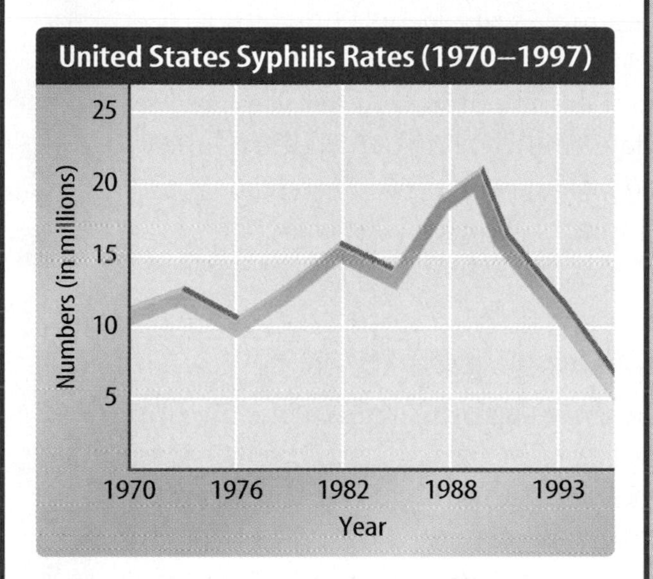

Study the graph and answer the following questions.

1. According to the information in the graph, when did an epidemic of syphilis occur in the United States?

A) 1970–1972
B) 1982–1984
C) 1988–1990
D) 1992–1994

2. A reasonable hypothesis for the information in the graph is that the number of people infected with syphilis _____.

F) is increasing
G) is decreasing
H) has remained the same
J) is related to gender

CHAPTER

Heredity

Wherever you go, look around you. You don't have the same skin color, the same kind of hair, or the same height as everyone else. Why do you resemble some people but do not look like others at all? In this chapter, you'll find out how differences are determined, and you will learn how to predict when certain traits might appear. You also will learn what causes some hereditary disorders.

What do you think?

Science Journal Look at the picture below with a classmate. Discuss what you think this might be or what is happening. Here's a hint: *The secret to why you look the way you do is found in this picture.* Write your answer or best guess in your Science Journal.

You and your best friend enjoy the same sports, like the same food, and even have similar haircuts. But, there are noticeable differences between your appearances. Most of these differences are controlled by the genes you inherited from your parents. In the following activity, you will observe one of these differences.

Observe dimples on faces

1. Notice the two students in the photographs. One student has dimples when she smiles, and the other student doesn't have dimples.
2. Ask your classmates to smile naturally. In your Science Journal, record the name of each classmate and whether each one has dimples.

Observe

In your Science Journal, calculate the percentage of students who have dimples. Are facial dimples a common feature among your classmates?

Before You Read

FOLDABLES Reading & Study Skills

Making a Classify Study Fold As you read this chapter about heredity, you can use the following Foldable to help you classify characteristics. When you classify, you organize objects or events into groups based on their common features.

1. Place a sheet of paper in front of you so the short side is at the top. Fold both sides in to divide the paper into thirds. Unfold the paper so three columns show.
2. Fold the paper in half from top to bottom. Then fold it in half again two more times. Unfold all the folds.
3. Trace over all the fold lines and label the columns you created as shown: *Personal Characteristics*, *Inherited*, and *Not Inherited*. List personal characteristics down the left-hand column, as shown.
4. Before you read the chapter, predict which characteristics are inherited or not inherited. As you read the chapter, check and change the table.

SECTION

Genetics

As You Read

What You'll Learn

- **Explain** how traits are inherited.
- **Identify** Mendel's role in the history of genetics.
- **Use** a Punnett square to predict the results of crosses.
- **Compare and contrast** the difference between an individual's genotype and phenotype.

Vocabulary

heredity
allele
genetics
hybrid
dominant
recessive
Punnett square
genotype
phenotype
homozygous
heterozygous

Why It's Important

Heredity and genetics help explain why people are different.

Inheriting Traits

Do you look more like one parent or grandparent? Do you have your father's eyes? What about Aunt Isabella's cheekbones? Eye color, nose shape, and many other physical features are some of the traits that are inherited from parents, as **Figure 1** shows. An organism is a collection of traits, all inherited from its parents. **Heredity** (huh REH duh tee) is the passing of traits from parent to offspring. What controls these traits?

What is genetics? Generally, genes on chromosomes control an organism's form and function. The different forms of a trait that a gene may have are called **alleles** (uh LEELZ). When a pair of chromosomes separates during meiosis (mi OH sus), alleles for each trait also separate into different sex cells. As a result, every sex cell has one allele for each trait, as shown in **Figure 2.** The allele in one sex cell may control one form of the trait, such as having facial dimples. The allele in the other sex cell may control a different form of the trait—not having dimples. The study of how traits are inherited through the interactions of alleles is the science of **genetics** (juh NET ihks).

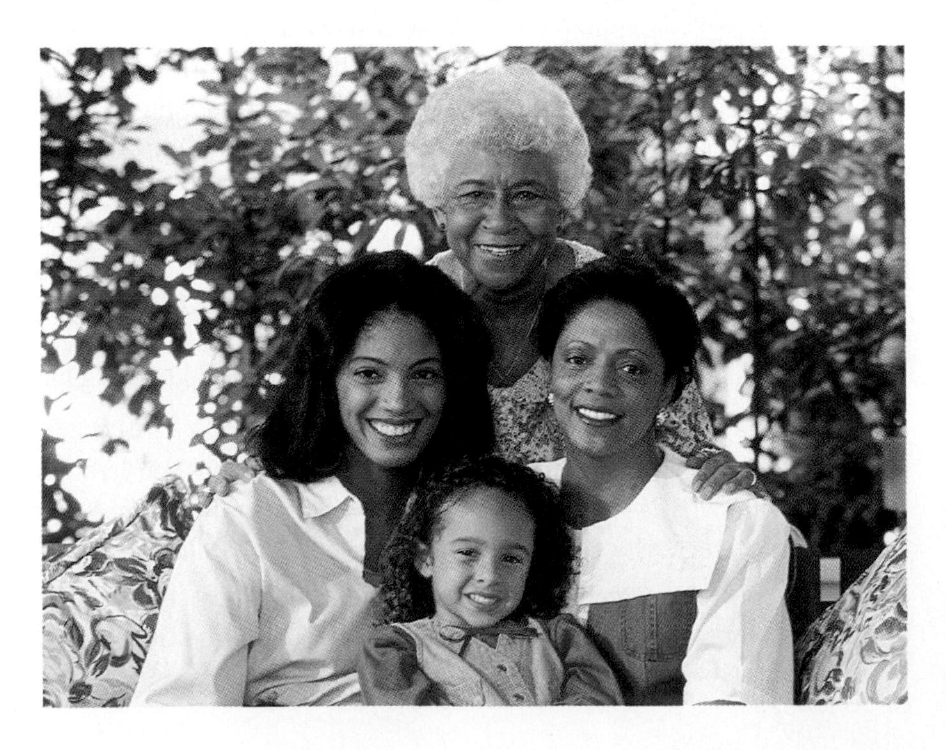

Figure 1
Note the strong family resemblance among these four generations.

Figure 2
An allele is one form of a gene. Alleles separate into separate sex cells during meiosis. In this example, the alleles that control the trait for dimples include *D,* the presence of dimples, and *d,* the absence of dimples.

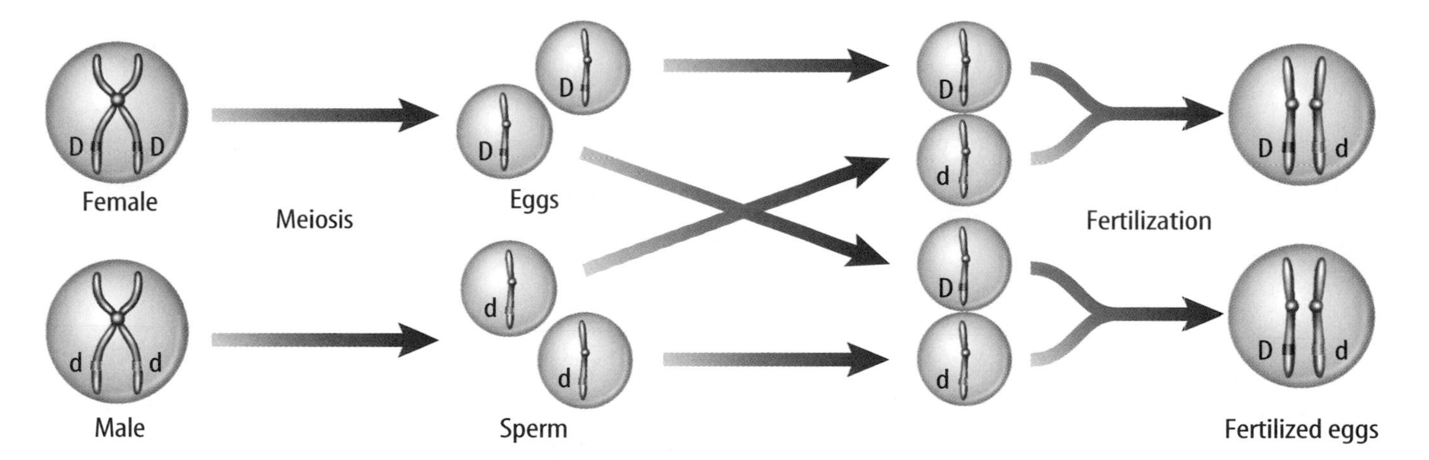

A **The alleles that control a trait are located on each duplicated chromosome.**

B **During meiosis, duplicated chromosomes separate.**

C **During fertilization, each parent donates one chromosome. This results in two alleles for the trait of dimples in the new individual formed.**

Mendel—The Father of Genetics

Did you know that an experiment with pea plants helped scientists understand why your eyes are the color that they are? Gregor Mendel was an Austrian monk who studied mathematics and science but became a gardener in a monastery. His interest in plants began as a boy in his father's orchard where he could predict the possible types of flowers and fruits that would result from crossbreeding two plants. Curiosity about the connection between the color of a pea flower and the type of seed that same plant produced inspired him to begin experimenting with garden peas in 1856. Mendel made careful use of scientific methods, which resulted in the first recorded study of how traits pass from one generation to the next. After eight years, Mendel presented his results with pea plants to scientists.

Before Mendel, scientists mostly relied on observation and description, and often studied many traits at one time. Mendel was the first to trace one trait through several generations. He was also the first to use the mathematics of probability to explain heredity. The use of math in plant science was a new concept and not widely accepted then. Mendel's work was forgotten for a long time. In 1900, three plant scientists, working separately, reached the same conclusions as Mendel. Each plant scientist had discovered Mendel's writings while doing his own research. Since then, Mendel has been known as the father of genetics.

SCIENCE *Online*
Research Visit the Glencoe Science Web site at **science.glencoe.com** for more information about early genetics experiments. Write a paragraph in your Science Journal about a scientist, other than Gregor Mendel, who studied genetics.

Table 1 Traits Compared by Mendel							
Traits	Shape of Seeds	Color of Seeds	Color of Pods	Shape of Pods	Plant Height	Position of Flowers	Flower Color
Dominant Trait	Round	Yellow	Green	Full	Tall	At leaf junctions	Purple
Recessive Trait	Wrinkled	Green	Yellow	Flat, constricted	Short	At tips of branches	White

Genetics in a Garden

Each time Mendel studied a trait, he crossed two plants with different expressions of the trait and found that the new plants all looked like one of the two parents. He called these new plants **hybrids** (HI brudz) because they received different genetic information, or different alleles, for a trait from each parent. The results of these studies made Mendel even more curious about how traits are inherited.

Garden peas are easy to breed for pure traits. An organism that always produces the same traits generation after generation is called a purebred. For example, tall plants that always produce seeds that produce tall plants are purebred for the trait of tall height. **Table 1** shows other pea plant traits that Mendel studied.

✔ **Reading Check** *Why might farmers plant purebred crop seeds?*

Dominant and Recessive Factors In nature, insects randomly pollinate plants as they move from flower to flower. In his experiments, Mendel used pollen from the flowers of purebred tall plants to pollinate by hand the flowers of purebred short plants. This process is called cross-pollination. He found that tall plants crossed with short plants produced seeds that produced all tall plants. Whatever caused the plants to be short had disappeared. Mendel called the tall form the **dominant** (DAHM uh nunt) factor because it dominated, or covered up, the short form. He called the form that seemed to disappear the **recessive** (rih SES ihv) factor. Today, these are called dominant alleles and recessive alleles. What happened to the recessive form? **Figure 3** answers this question.

Comparing Common Traits

Procedure

1. Safely survey as many **dogs** in your neighborhood as you can for the presence of a solid color or spotted coat, short or long hair, and floppy ears or ears that stand up straight.
2. Make a data table that lists each of the traits. Record your data in the data table.

Analysis

1. Compare the number of dogs that have one form of a trait with those that have the other form. How do those two groups compare?
2. What can you conclude about the variations you noticed in the dogs?

NATIONAL GEOGRAPHIC **VISUALIZING MENDEL'S EXPERIMENTS**

Figure 3

Gregor Mendel discovered that the experiments he carried out on garden plants provided an understanding of heredity. For eight years he crossed plants that had different characteristics and recorded how those characteristics were passed from generation to generation. One such characteristic, or trait, was the color of pea pods. The results of Mendel's experiment on pea pod color are shown below.

Parents

1st Generation

Ⓐ One of the so-called "parent plants" in Mendel's experiment had pods that were green, a dominant trait. The other parent plant had pods that were yellow, a recessive trait.

Ⓑ Mendel discovered that the two "parents" produced a generation of plants with green pods. The recessive color—yellow—did not appear in any of the pods.

2nd Generation

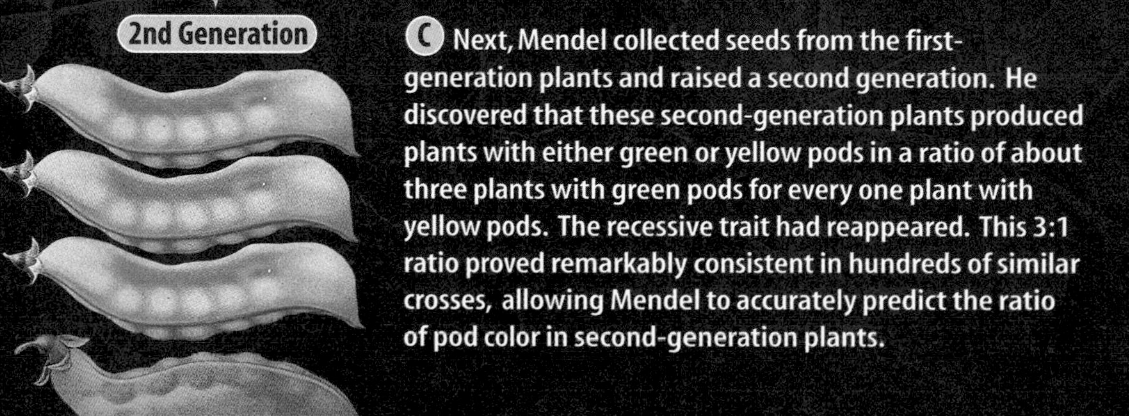

Ⓒ Next, Mendel collected seeds from the first-generation plants and raised a second generation. He discovered that these second-generation plants produced plants with either green or yellow pods in a ratio of about three plants with green pods for every one plant with yellow pods. The recessive trait had reappeared. This 3:1 ratio proved remarkably consistent in hundreds of similar crosses, allowing Mendel to accurately predict the ratio of pod color in second-generation plants.

Using Probability to Make Predictions If you and your sister can't agree on what movie to see, you could solve the problem by tossing a coin. When you toss a coin, you're dealing with probabilities. Probability is a branch of mathematics that helps you predict the chance that something will happen. If your sister chooses tails while the coin is in the air, what is the probability that the coin will land tail-side up? Because a coin has two sides, there are two possible outcomes, heads or tails. One outcome is tails. Therefore, the probability of one side of a coin showing is one out of two, or 50 percent.

Mendel also dealt with probabilities. One of the things that made his predictions accurate was that he worked with large numbers of plants. He studied almost 30,000 pea plants over a period of eight years. By doing so, Mendel increased his chances of seeing a repeatable pattern. Valid scientific conclusions need to be based on results that can be duplicated.

Figure 4
This snapdragon's phenotype is red. *Can you tell what the flower's genotype for color is? Explain your answer.*

Punnett Squares Suppose you wanted to know what colors of pea plant flowers you would get if you pollinated white flowers on one pea plant with pollen from purple flowers on a different plant. How could you predict what the offspring would look like without making the cross? A handy tool used to predict results in Mendelian genetics is the **Punnett** (PUN ut) **square.** In a Punnett square, letters represent dominant and recessive alleles. An uppercase letter stands for a dominant allele. A lowercase letter stands for a recessive allele. The letters are a form of code. They show the **genotype** (JEE nuh tipe), or genetic makeup, of an organism. Once you understand what the letters mean, you can tell a lot about the inheritance of a trait in an organism.

The way an organism looks and behaves as a result of its genotype is its **phenotype** (FEE nuh tipe), as shown in **Figure 4.** If you have brown hair, then the phenotype for your hair color is brown.

Alleles Determine Traits Most cells in your body have two alleles for every trait. These alleles are located on chromosomes within the nucleus of cells. An organism with two alleles that are the same is called **homozygous** (hoh muh ZI gus). For Mendel's peas, this would be written as *TT* (homozygous for the tall-dominant trait) or *tt* (homozygous for the short-recessive trait). An organism that has two different alleles for a trait is called **heterozygous** (het uh roh ZI gus). The hybrid plants Mendel produced were all heterozygous for height, *Tt.*

What is the difference between homozygous and heterozygous organisms?

Making a Punnett Square In a Punnett square for predicting one trait, the letters representing the two alleles from one parent are written along the top of the grid, one letter per section. Those of the second parent are placed down the side of the grid, one letter per section. Each square of the grid is filled in with one allele donated by each parent. The letters that you use to fill in each of the squares represent the genotypes of possible offspring that the parents could produce.

Math Skills Activity

Calculating Probability Using a Punnett Square

You can determine the probability of certain traits by using a Punnett square. Letters are used to represent the two alleles from each parent and are combined to determine the possible genotypes of the offspring.

Example Problem

One dog carries heterozygous, black-fur traits (Bb), and its mate carries homogeneous, blond-fur traits (bb). Calculate the probability of the puppy having black fur.

Solution

1 *This is what you know:*
dominant allele is represented by *B*
recessive allele is represented by *b*

2 *This is what you need to find:*
the probability of a puppy's fur color being black using a Punnett square

3 *This is the diagram you need to use:*

Black dog

Blond dog		B	b
	b		
	b		

4 *Complete the Punnett square by taking each letter in each column and combining it with each letter from each row in the corresponding square.*

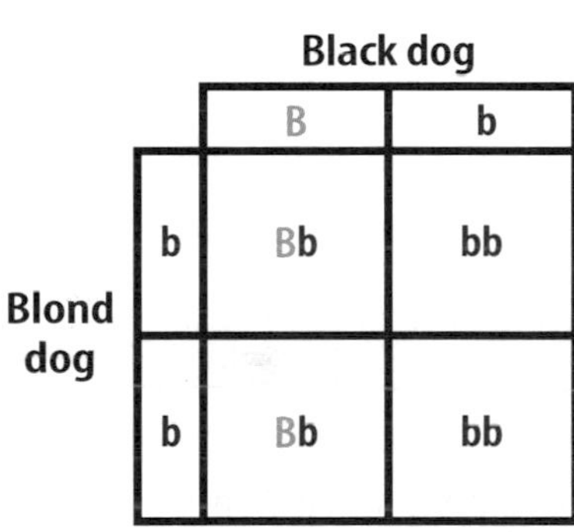

Genotypes of offspring:
2Bb, 2bb
Phenotypes of offspring:
2 black, 2 blond

5 *Find the needed probability. There are two Bb genotypes and four possible outcomes.*

$$P(\text{black fur}) = \frac{\text{number of ways to get black fur}}{\text{number of possible outcomes}}$$

$$= \frac{2}{4} = \frac{1}{2} = 50\%$$

Practice Problem

Use a Punnett square to determine the probability of each of the offspring's genotype and phenotype when two heterozygous, tall-dominant traits (Tt) are crossed with each other.

For more help, refer to the Math Skill Handbook.

SECTION 2

Genetics Since Mendel

As You Read

What You'll Learn

- **Explain** how traits are inherited by incomplete dominance.
- **Compare** multiple alleles and polygenic inheritance, and give examples of each.
- **Describe** two human genetic disorders and how they are inherited.
- **Explain** how sex-linked traits are passed to offspring.

Vocabulary

incomplete dominance
polygenic inheritance
sex-linked gene

Why It's Important

Most of your inherited traits involve more complex patterns of inheritance than Mendel discovered.

Incomplete Dominance

Not even in science do things remain the same. After Mendel's work was rediscovered in 1900, scientists repeated his experiments. For some plants, such as peas, Mendel's results proved true. However, when different plants were crossed, the results were sometimes different. One scientist crossed purebred red four-o'clock plants with purebred white four-o'clock plants. He expected to get all red flowers, but they were pink. Neither allele for flower color seemed dominant. Had the colors become blended like paint colors? He crossed the pink-flowered plants with each other, and red, pink, and white flowers were produced. The red and white alleles had not become blended. Instead, when the allele for white flowers and the allele for red flowers combined, the result was an intermediate phenotype—a pink flower. When the offspring of two homozygous parents show an intermediate phenotype, this inheritance is called **incomplete dominance.** Other examples of incomplete dominance include the feather color of some chicken breeds and the coat color of some horse breeds, as shown in **Figure 5.**

Figure 5
When A a chestnut horse is bred with B a cremello horse, all offspring will be C palomino. The Punnett square shown in D can be used to predict this result. ***How does the color of the palomino horse in C show that the coat color of horses may be inherited by incomplete dominance?***

A

B

Multiple Alleles Mendel studied traits in peas that were controlled by just two alleles. However, many traits are controlled by more than two alleles. A trait that is controlled by more than two alleles is said to be controlled by multiple alleles. Traits controlled by multiple alleles produce more than three phenotypes of that trait.

Imagine that only three types of coins are made—nickels, dimes, and quarters. If every person can have only two coins, six different combinations are possible. In this problem, the coins represent alleles of a trait. The sum of each two-coin combination represents the phenotype. Can you name the six different phenotypes possible with two coins?

Blood type in humans is an example of multiple alleles that produce only four phenotypes. The alleles for blood types are called A, B, and O. The O allele is recessive to both the A and B alleles. When a person inherits one A allele and one B allele for blood type, both are expressed—phenotype AB. A person with phenotype A blood has the genetic makeup, or genotype—AA or AO. Someone with phenotype B blood has the genotype BB or BO. Finally, a person with phenotype O blood has the genotype OO.

✔ **Reading Check** ***What are the six different blood type genotypes?***

Research Visit the Glencoe Science Web site at **science.glencoe.com** for information on the importance of blood types in blood transfusions. In your Science Journal, draw a chart showing which blood types can be used safely during transfusions.

C

D

Chestnut horse (CC)

Cremello horse (C'C')	C	C
C'	CC'	CC'
C'	CC'	CC'

Genotypes: All CC'
Phenotypes: All palomino horses

Mini LAB

Interpreting Polygenic Inheritance

Procedure

1. Measure the hand spans of your classmates.
2. Using a **ruler,** measure from the tip of the thumb to the tip of the little finger when the hand is stretched out. Read the measurement to the nearest centimeter.
3. Record the name and hand-span measurement of each person in a data table.

Analysis

1. What range of hand spans did you find?
2. Are hand spans inherited as a simple Mendelian pattern or as a polygenic or incomplete dominance pattern? Explain.

Polygenic Inheritance

Eye color is an example of a trait that is produced by a combination of many genes. **Polygenic** (pahl ih JEHN ihk) **inheritance** occurs when a group of gene pairs acts together to produce a trait. The effects of many alleles produces a wide variety of phenotypes. For this reason, it may be hard to classify all the different shades of eye color.

Your height and the color of your eyes and skin are just some of the many human traits controlled by polygenic inheritance. It is estimated that three to six gene pairs control your skin color. Even more gene pairs might control the color of your hair and eyes. The environment also plays an important role in the expression of traits controlled by polygenic inheritance. Polygenic inheritance is common and includes such traits as grain color in wheat and milk production in cows. Egg production in chickens is also a polygenic trait.

Impact of the Environment Your environment plays a role in how some of your genes are expressed or whether they are expressed at all, as shown in **Figure 6.** Environmental influences can be internal or external. For example, most male birds are more brightly colored than females. Chemicals in their bodies determine whether the gene for brightly colored feathers is expressed.

Although genes determine many of your traits, you might be able to influence their expression by the decisions you make. Some people have genes that make them at risk for developing certain cancers. Whether they get cancer might depend on external environmental factors. For instance, if some people at risk for skin cancer limit their exposure to the Sun and take care of their skin, they might never develop cancer.

Reading Check *What environmental factors might affect the size of leaves on a tree?*

Figure 6
Himalayan rabbits have alleles for dark-colored fur. However, this allele is able to express itself only at lower temperatures. Only the areas located farthest from the rabbit's main body heat (ears, nose, feet, tail) have dark-colored fur.

Human Genes and Mutations

Sometimes a gene undergoes a change that results in a trait that is expressed differently. Occasionally errors occur in the DNA when it is copied inside of a cell. Such changes and errors are called mutations. Not all mutations are harmful. They might be helpful or have no effect on an organism.

Certain chemicals are known to produce mutations in plants or animals, including humans. X rays and radioactive substances are other causes of some mutations. Mutations are changes in genes.

Chromosome Disorders In addition to individual mutations, problems can occur if the incorrect number of chromosomes is inherited. Every organism has a specific number of chromosomes. However, mistakes in the process of meiosis can result in a new organism with more or fewer chromosomes than normal. A change in the total number of human chromosomes is usually fatal to the unborn embryo or fetus, or the baby may die soon after birth.

Look at the human chromosomes in **Figure 7.** If three copies of chromosome 21 are produced in the fertilized human egg, Down's syndrome results. Individuals with Down's syndrome can be short, exhibit learning disabilities, and have heart problems. Such individuals can lead normal lives if they have no severe health complications.

Figure 7
Humans usually have 23 pairs of chromosomes. Notice that three copies of chromosome 21 are present in this photo, rather than the usual two chromosomes. This change in chromosome number results in Down's syndrome. Chris Burke, a well-known actor, has this syndrome.

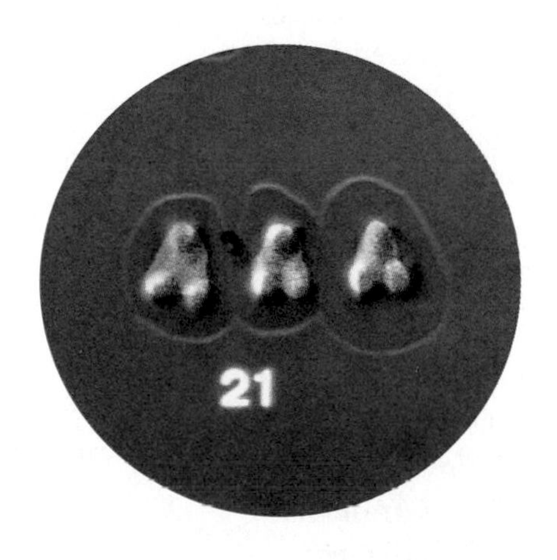

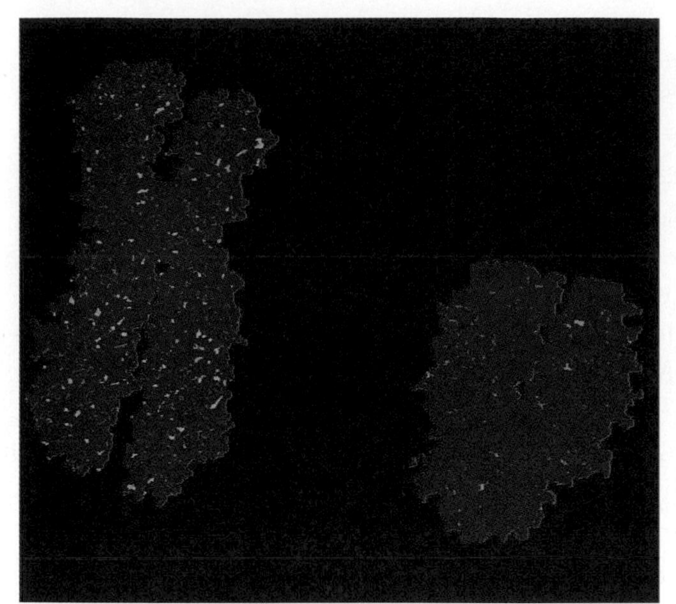

Magnification: 10,000×

Figure 8
Sex in many organisms is determined by X and Y chromosomes.
How do the X (left) and Y (right) chromosomes differ from one another in shape and size?

Recessive Genetic Disorders

Many human genetic disorders, such as cystic fibrosis, are caused by recessive genes. Some recessive genes are the result of a mutation within the gene. Many of these alleles are rare. Such genetic disorders occur when both parents have a recessive allele responsible for this disorder. Because the parents are heterozygous, they don't show any symptoms. However, if each parent passes the recessive allele to the child, the child inherits both recessive alleles and will have a recessive genetic disorder.

How is cystic fibrosis inherited?

Cystic fibrosis is a homozygous recessive disorder. It is the most common genetic disorder that can lead to death among Caucasian Americans. In most people, a thin fluid is produced that lubricates the lungs and intestinal tract. People with cystic fibrosis produce thick mucus instead of this thin fluid. The thick mucus builds up in the lungs and makes it hard to breathe. This buildup often results in repeated bacterial respiratory infections. The thick mucus also reduces or prevents the flow of substances necessary for digesting food. Physical therapy, special diets, and new drug therapies have increased the life spans of patients with cystic fibrosis.

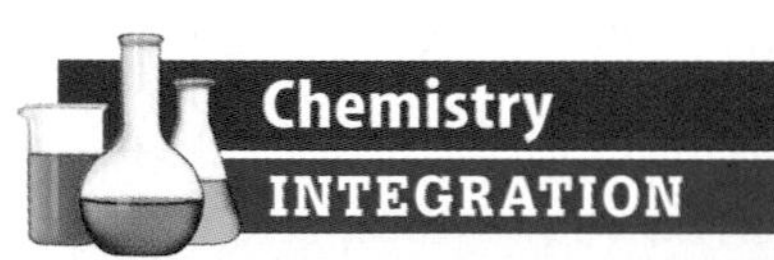

People with PKU, a recessive disorder, cannot produce the enzyme needed for the breakdown of a substance found in some artificially sweetened drinks. Soft-drink cans must be labeled to ensure that individuals with this disorder do not unknowingly consume the substance. Explain in your Science Journal how a person can be born with PKU if neither parent has this recessive disorder.

Sex Determination

What determines the sex of an individual? Much information on sex inheritance came from studies of fruit flies. Fruit flies have only four pairs of chromosomes. Because the chromosomes are large and few in number, they are easy to study. Scientists identified one pair that contains genes that determine the sex of the organism. They labeled the pair XX in females and XY in males. Geneticists use these labels when studying organisms, including humans. You can see human X and Y chromosomes in **Figure 8.**

Each egg produced by a female normally contains one X chromosome. Males produce sperm that normally have either an X or a Y chromosome. When a sperm with an X chromosome fertilizes an egg, the offspring is a female, XX. A male offspring, XY, is the result of a Y-containing sperm fertilizing an egg. What pair of sex chromosomes is in each of your cells? Sometimes chromosomes do not separate during meiosis. When this occurs, an individual can inherit an abnormal number of sex chromosomes.

Sex-Linked Disorders

Some inherited conditions are linked with the X and Y chromosomes. An allele inherited on a sex chromosome is called a **sex-linked gene.** Color blindness is a sex-linked disorder in which people cannot distinguish between certain colors, particularly red and green. This trait is a recessive allele on the X chromosome. Because males have only one X chromosome, a male with this allele on his X chromosome is color-blind. However, a color-blind female occurs only when both of her X chromosomes have the allele for this trait.

The allele for the distinct patches of three different colors found in calico cats is recessive and carried on the X chromosome. As shown in **Figure 9,** calico cats have inherited two X chromosomes with this recessive allele—one from both parents.

Female carrier of calico gene (X^cX)

Male carrier of calico gene (X^cY)

	X^c	X
X^c	X^cX^c	XX^c
Y	X^cY	XY

Genotypes: X^cX^c, X^cX, X^cY, XY
Phenotypes: One calico female, one carrier female, one carrier male, one normal male

Figure 9
Calico cat fur is a homozygous recessive sex-linked trait. Female cats that are heterozygous are not calico but are only carriers. Two recessive alleles must be present for this allele to be expressed. *Why aren't all the females calico?*

Pedigrees Trace Traits

How can you trace a trait through a family? A pedigree is a visual tool for following a trait through generations of a family. Males are represented by squares and females by circles. A completely filled circle or square shows that the trait is seen in that person. Half-colored circles or squares indicate carriers. A carrier is heterozygous for the trait and it is not seen. People represented by empty circles or squares do not have the trait and are not carriers. The pedigree in **Figure 10** shows how the trait for color blindness is carried through a family.

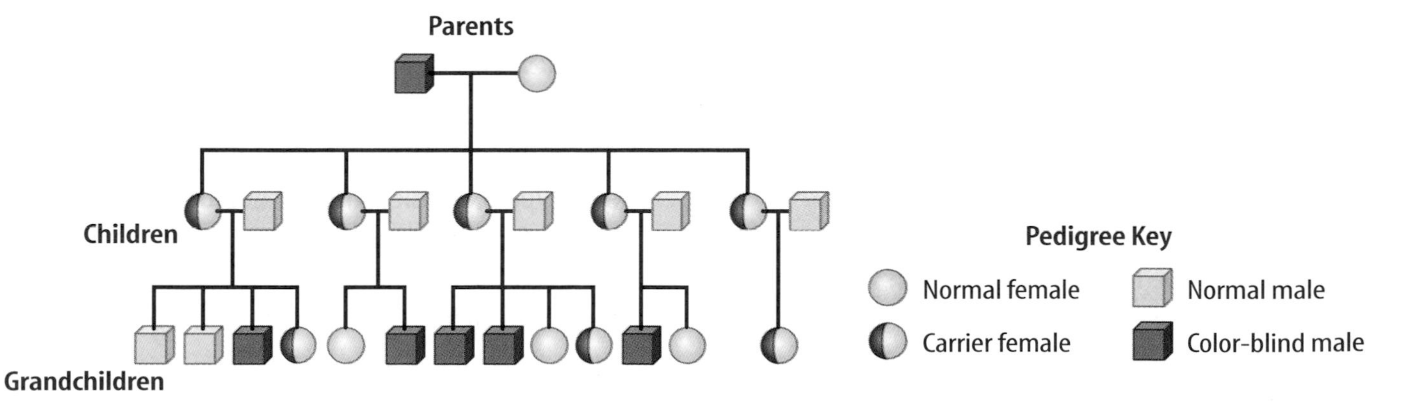

Figure 10
The symbols in this pedigree's key mean the same thing on all pedigree charts. The grandfather in this family was color-blind and married to a woman who was not a carrier of the color-blind allele. *Why are no women in this family color-blind?*

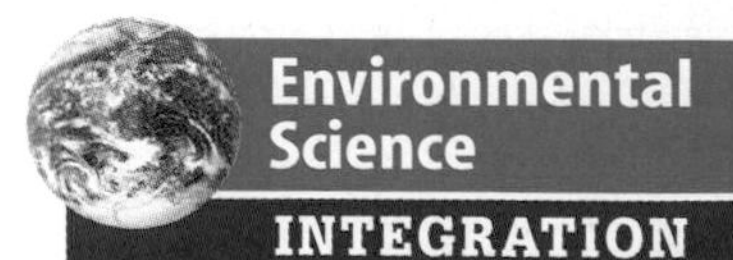

Crop plants are now being genetically engineered to produce chemicals that kill specific pests that feed on them. Some of the pollen from pesticide-resistant canola crops is capable of spreading up to 8 km from the plant, while corn and potato pollen can spread up to 1 km. What might be the effects of pollen landing on other plants?

Recombinant DNA Making recombinant DNA is one method of genetic engineering. Recombinant DNA is made by inserting a useful segment of DNA from one organism into a bacterium, as illustrated in **Figure 12.** Large quantities of human insulin are made by some genetically-engineered organisms. People with Type 1 diabetes need this insulin because their pancreases produce too little or no insulin. Other uses include the production of growth hormone to treat dwarfism and chemicals to treat cancer.

Gene Therapy Gene therapy is a kind of genetic engineering. In gene therapy, a normal allele is placed in a virus, as shown in **Figure 13.** The virus then delivers the normal allele when it infects its target cell. The normal allele replaces the defective one. Scientists are conducting experiments that use this method to test ways of controlling cystic fibrosis and some kinds of cancer. More than 2,000 people already have taken part in gene therapy experiments. Gene therapy might be a method of curing several other genetic disorders in the future.

Figure 13
Gene therapy involves placing a normal allele in a cell that has a mutation. When the normal allele begins to function, a genetic disorder such as cystic fibrosis (CF) may be corrected.

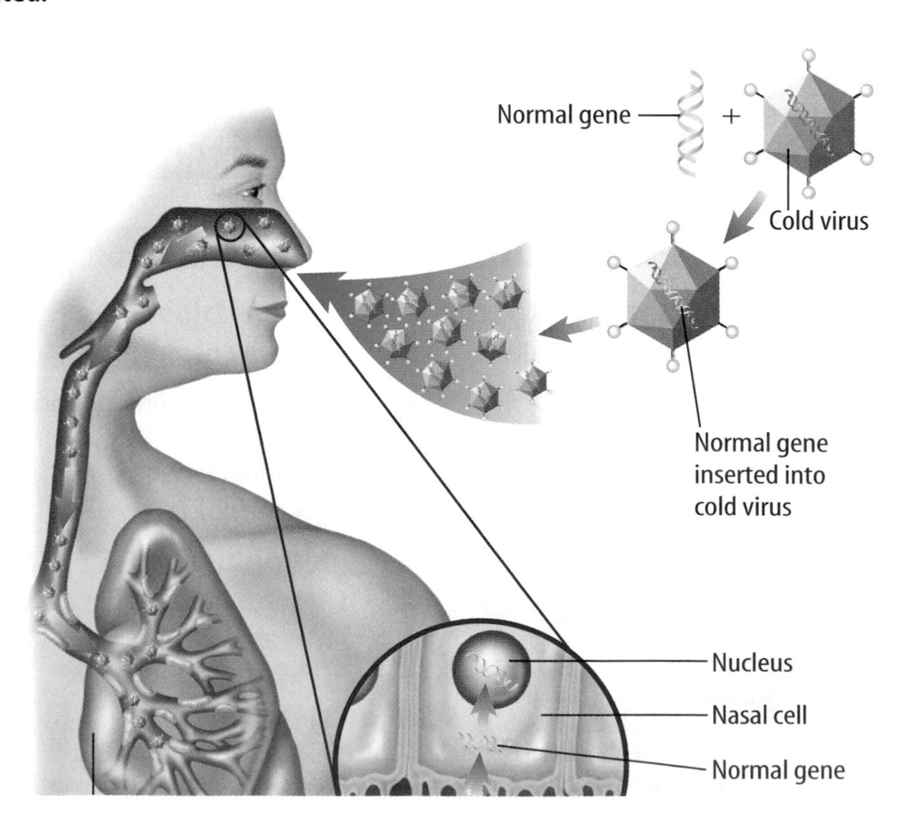

Genetically Engineered Plants For thousands of years people have improved the plants they use for food and clothing even without the knowledge of genotypes. Until recently, these improvements were the results of selecting plants with the most desired traits to breed for the next generation. This process is called selective breeding. Recent advances in genetics have not replaced selective breeding. Although a plant can be bred for a particular phenotype, the genotype and pedigree of the plants also are considered.

Figure 14
Genetically engineered produce is sometimes labeled. This allows consumers to make informed choices about their foods.

Genetic engineering can produce improvements in crop plants, such as corn, wheat, and rice. One type of genetic engineering involves finding the genes that produce desired traits in one plant and then inserting those genes into a different plant. Scientists recently have made genetically engineered tomatoes with a gene that allows tomatoes to be picked green and transported great distances before they ripen completely. Ripe, firm tomatoes are then available in the local market. In the future, additional food crops may be genetically engineered so that they are not desirable food for insects.

Reading Check *What other types of traits would be considered desirable in plants?*

Because some people might prefer foods that are not changed genetically, some stores label such produce, as shown in **Figure 14.** The long-term effects of consuming genetically engineered plants are unknown.

Section  Assessment

1. Give examples of areas in which advances in genetics are important.
2. Compare and contrast the technologies of using recombinant DNA and gene therapy.
3. What are some benefits of genetically engineered crops?
4. How does selective breeding differ from genetic engineering?
5. **Think Critically** Why might some people be opposed to genetically engineered plants?

Skill Builder Activities

6. **Concept Mapping** Make an events chain concept map of the steps used in making recombinant DNA. **For more help, refer to the Science Skill Handbook.**
7. **Using a Word Processor** Use a computer word processing program to write predictions about how advances in genetics might affect your life in the next ten years. **For more help, refer to the Technology Skill Handbook.**

Activity Design Your Own Experiment

Tests for Color Blindness

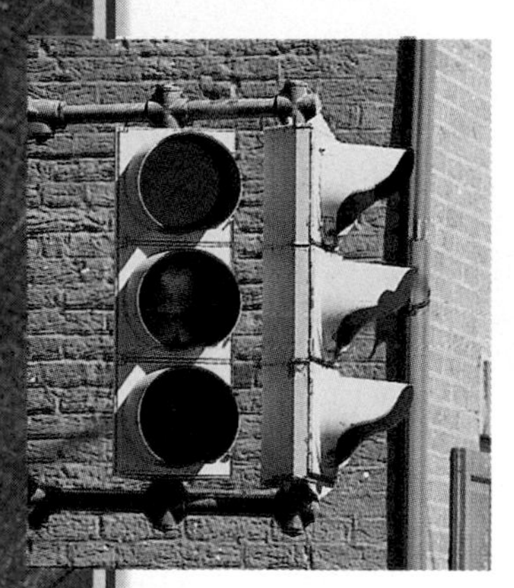

What do color-blind people see? People who have inherited color blindness can see most colors, but they have difficulty telling the difference between two specific colors. You have three genes that help you see color. One gene lets you see red, another blue, and the third gene allows you to see green. In the most common type of color blindness, red-green color blindness, the green gene does not work properly. What percentage of people are color-blind?

Recognize the Problem

What percentages of males and females in your school are color-blind?

Form a Hypothesis

Based on your reading and your own experiences, form a hypothesis about how common color blindness is among males and females.

Goals

- **Design** an experiment that tests for a specific type of color blindness in males and females.
- **Calculate** the percentage of males and females with the disorder.

Possible Materials

white paper or poster board
colored markers: red, orange, yellow, bright green, dark green, blue
computer and color printer
**Alternate materials*

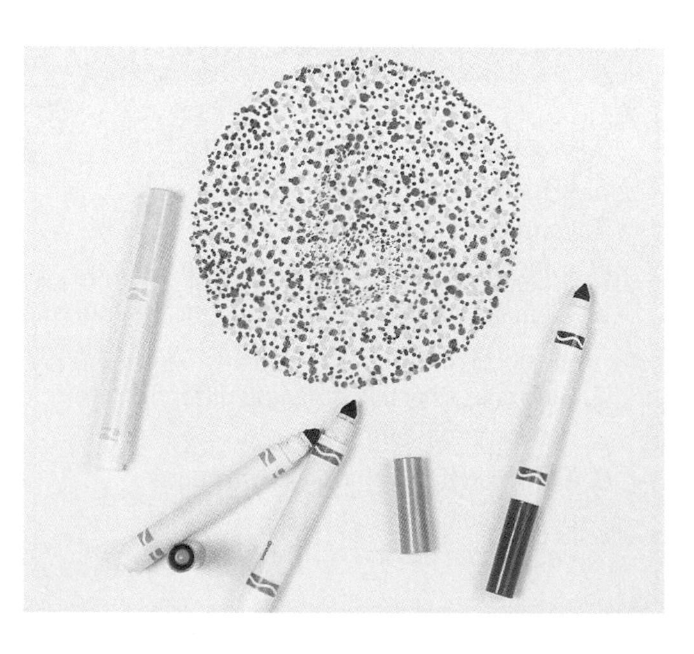

To a person with red-green color blindness, bright green appears tan in color, and dark green looks like brown. The color red also looks brown, making it difficult to tell the difference between green and red. A person without red-green color blindness will see a "6" in this test, while a red-green color-blind person will not see this number.

Test Your Hypothesis

Plan

1. Decide what type of color blindness you will test for—the common green-red color blindness or the more rare green-blue color blindness.
2. **List** the materials you will need and describe how you will create test pictures. Tests for color blindness use many circles of red, orange, and yellow as a background, with circles of dark and light green to make a picture or number. List the steps you will take to test your hypothesis.
3. Prepare a data table in your Science Journal to record your test results.
4. **Examine** your experiment to make sure all steps are in logical order.
5. **Identify** which pictures you will use as a control and which pictures you will use as variables.

Do

1. Make sure your teacher approves your plan before you start.
2. **Draw** the pictures that you will use to test for color blindness.
3. Carry out your experiment as planned and record your results in your data table.

Analyze Your Data

1. **Calculate** the percentage of males and females that tested positive for color blindness.
2. **Compare** the frequency of color blindness in males with the frequency of color blindness in females.

Draw Conclusions

1. Did the results support your hypothesis? Explain.
2. Use your results to explain why color blindness is called a sex-linked disorder.
3. **Infer** how common the color-blind disorder is in the general population.
4. **Predict** your results if you were to test a larger number of people.

Communicating Your Data

Using a word processor, **write** a short article for the advice column of a fashion magazine about how a color-blind person can avoid wearing outfits with clashing colors. **For more help, refer to the Technology Skill Handbook.**

Science Stats

The Human Genome

Did you know...

. . . The human genome is not very different from the genome of mice. As shown to the right, many of the genes that are found on mouse chromosome 17 are similar to genes on human chromosomes. Humans may be more closely related to other organisms than previously thought.

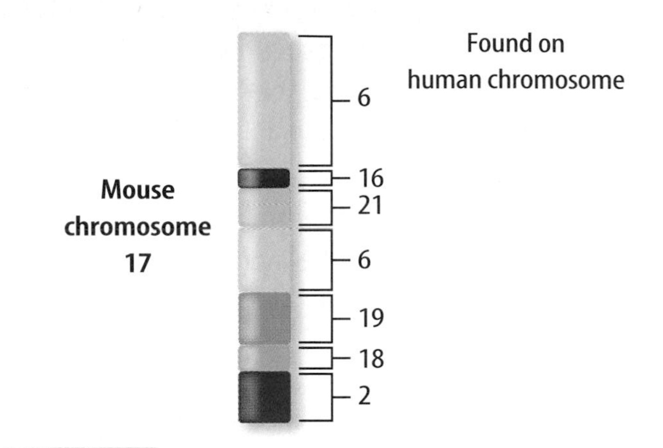

Human hair

DNA

. . . The strands of DNA in the human genome, if unwound and connected end to end, would be more than 1.5 m long—but only about 130 trillionths of a centimeter wide. Even an average human hair is as much as 200,000 times wider than that.

Centriole

Nucleus

Mitochondrion

Endoplasmic reticulum

Cytoplasm

. . . The biggest advance in genetics in years took place in February 2001. Scientists successfully mapped the human genome. There are 30,000 to 40,000 genes in the human genome. Genes are in the nucleus of each of the several trillion cells in your body.

. . . It would take about nine and one-half years to read aloud without stopping the 3 billion bits of instructions (called base pairs) in your genome.

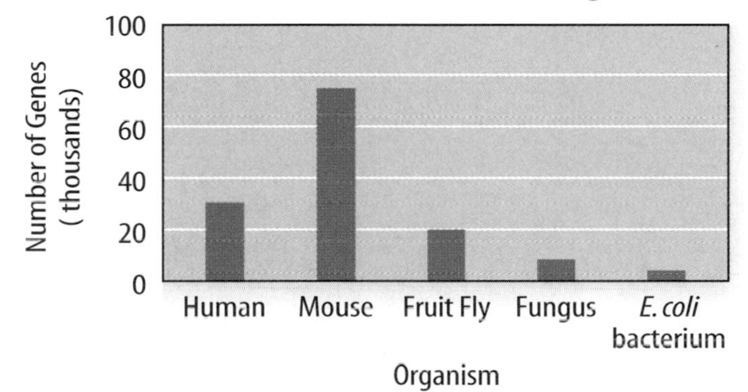

. . . Not all the DNA in your genes contains useful information. About 90 percent of it is "junk" DNA—meaningless sequences located in and between genes.

Do the Math

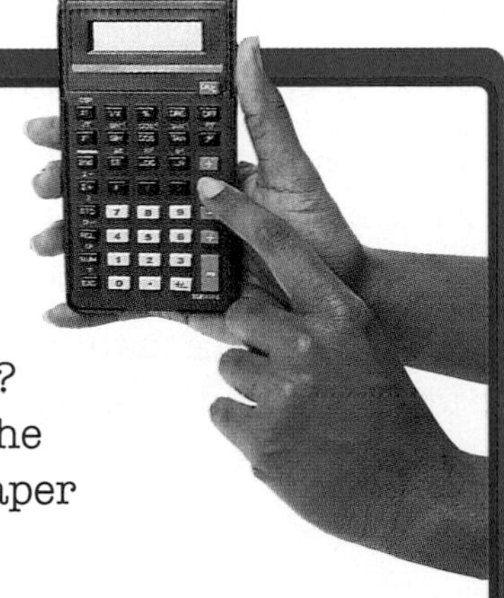

1. If one million base pairs of DNA take up 1 megabyte of storage space on a computer, how many gigabytes (1,024 megabytes) would the whole genome fill?
2. Consult the above graph. How many more genes are in the human genome than the genome of the fruit fly?
3. If you wrote the genetic information for each gene in the human genome on a separate sheet of 0.2-mm-thick paper and stacked the sheets, how tall would the stack be?

Go Further

By decoding the human genome scientists hope to identify the location of disease-causing genes. Research a genetic disease and share your results with your class.

Chapter 11 Study Guide

Reviewing Main Ideas

Section 1 Genetics

1. Genetics is the study of how traits are inherited. Gregor Mendel determined the basic laws of genetics.
2. Traits are controlled by alleles on chromosomes in the nuclei of cells.
3. Some alleles can be dominant and others can be recessive in action.
4. When a pair of chromosomes separates during meiosis, the different alleles for a trait move into separate sex cells. Mendel found that traits followed the laws of probability and that he could predict the outcome of genetic crosses. *How can a Punnett square help predict inheritance of traits?*

	F	f
F	FF	Ff
F	FF	Ff

Section 2 Genetics Since Mendel

1. Inheritance patterns studied since Mendel include incomplete dominance, multiple alleles, and polygenic inheritance.
2. These inheritance patterns allow a greater variety of phenotypes to be produced than would result from Mendelian inheritance.
3. Some disorders are the results of inheritance and can be harmful, even deadly, to those affected.
4. Pedigree charts help reveal patterns of the inheritance of a trait in a family. Pedigrees show that sex-linked traits are expressed more often in males than in females.

Section 3 Advances in Genetics

1. Genetic engineering uses biological and chemical methods to add or remove genes in an organism's DNA.
2. Recombinant DNA is one way genetic engineering can be performed using bacteria to make useful chemicals, including hormones.
3. Gene therapy shows promise for correcting many human genetic disorders by inserting normal alleles into cells.
4. Breakthroughs in the field of genetic engineering are allowing scientists to do many things, such as producing plants that are resistant to disease. *What types of crops might benefit from advances in genetic engineering? Give examples.*

After You Read

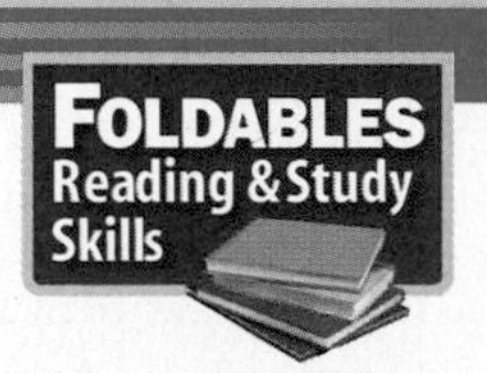

How many characteristics listed in your Classify Study Fold are inherited from your parents? How many are not inherited? Why are some not inherited?

Visualizing Main Ideas

Examine the following pedigree for diabetes and explain the inheritance pattern.

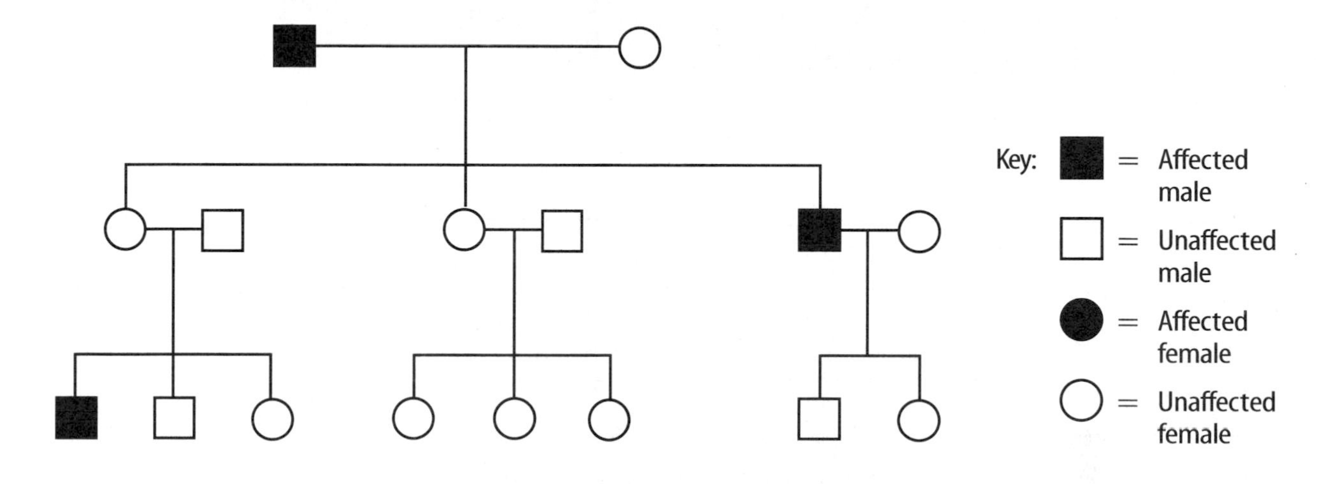

Vocabulary Review

Vocabulary Words

- **a.** allele
- **b.** dominant
- **c.** genetic engineering
- **d.** genetics
- **e.** genotype
- **f.** heredity
- **g.** heterozygous
- **h.** homozygous
- **i.** hybrid
- **j.** incomplete dominance
- **k.** phenotype
- **l.** polygenic inheritance
- **m.** Punnett square
- **n.** recessive
- **o.** sex-linked gene

Using Vocabulary

Make the sentences to the right true by replacing the underlined word with the correct vocabulary word.

Study Tip

When you encounter new vocabulary, write it down in a sentence. This will help you understand, remember, and use new vocabulary words.

1. Alternate forms of a gene are called genetics.

2. The outward appearance of a trait is a genotype.

3. Human height, eye color, and skin color are all traits controlled by sex-linked genes.

4. An allele that produces a trait in the heterozygous condition is recessive.

5. Polygenic inheritance is the branch of biology that deals with the study of heredity.

6. The actual combination of alleles of an organism is its phenotype.

7. Hybrid is moving fragments of DNA from one organism and inserting them into another organism.

8. A phenotype is a helpful device for predicting the proportions of possible genotypes.

9. Genetics is the passing of traits from parents to offspring.

10. Red-green color blindness and hemophilia are two human genetic disorders that are caused by a genotype.

Chapter 11 Assessment

Checking Concepts

Choose the word or phrase that best answers the question.

1. Which of the following are located in the nuclei on chromosomes?
A) genes
B) pedigrees
C) carbohydrates
D) zygotes

2. Which of the following describes the allele that causes color blindness?
A) dominant
B) carried on the Y chromosome
C) carried on the X chromosome
D) present only in males

3. What is it called when the presence of two different alleles results in an intermediate phenotype?
A) incomplete dominance
B) polygenic inheritance
C) multiple alleles
D) sex-linked genes

4. What separates during meiosis?
A) proteins
B) phenotypes
C) alleles
D) pedigrees

5. What controls traits in organisms?
A) cell membrane
B) cell wall
C) genes
D) Punnett squares

6. Which of the following is a use for a Punnett square?
A) to dominate the outcome of a cross
B) to predict the outcome of a cross
C) to assure the outcome of a cross
D) to number the outcome of a cross

7. What term describes the inheritance of cystic fibrosis?
A) polygenic inheritance
B) multiple alleles
C) incomplete dominance
D) recessive genes

8. What type of inheritance is eye color?
A) polygenic inheritance
B) multiple alleles
C) incomplete dominance
D) recessive genes

9. What chromosome(s) did the father contribute if a normal female is produced?
A) X
B) XX
C) Y
D) XY

10. What type of inheritance is blood type?
A) polygenic inheritance
B) multiple alleles
C) incomplete dominance
D) recessive genes

Thinking Critically

11. Explain the relationship among DNA, genes, alleles, and chromosomes.

12. Explain how the parents and offspring represented in this Punnett square have the same phenotype.

	F	f
F	FF	Ff
F	FF	Ff

13. Explain why two rabbits with the same genes might not be colored the same if one is raised in Maine and one is raised in Texas.

14. Why would a person who receives genetic therapy for a disorder still be able to pass the disorder to his or her children?

Developing Skills

15. Predicting Two organisms were found to have different genotypes but the same phenotype. Predict what these phenotypes might be. Explain.

Chapter 11 Assessment

16. **Classifying** Classify the inheritance pattern for each of the following:
    a. many different phenotypes produced by one pair of alleles;
    b. many phenotypes produced by more than one pair of alleles; two phenotypes from two alleles; three phenotypes from two alleles.

17. **Comparing and Contrasting** Compare and contrast Mendelian inheritance with incomplete dominance.

18. **Interpreting Scientific Illustrations** What were the genotypes of the parents that produced the following Punnett square?

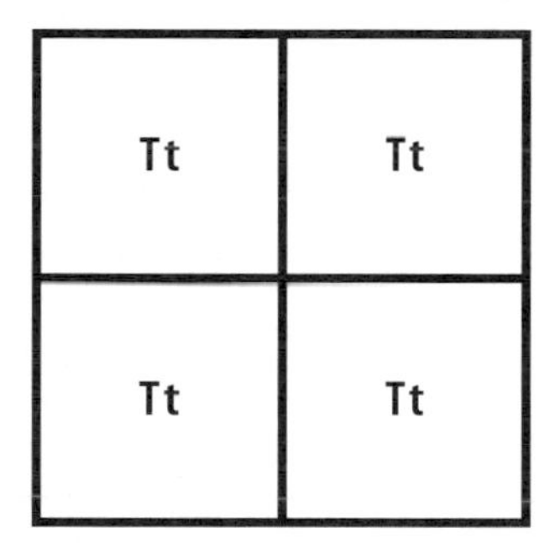

Performance Assessment

19. **Newspaper Article** Write a newspaper article to announce a new, genetically engineered plant. Include the method of developing the plant, the characteristic changed, and the terms that you would expect to see. Read your article to the class.

Technology

Go to the Glencoe Science Web site at **science.glencoe.com** or use the **Glencoe Science CD-ROM** for additional chapter assessment.

Test Practice

A scientist is studying pea plants. The scientist made this Punnett square to predict the color traits of the offspring of two parent pea plants.

	Parent (Yy)	
Parent (Yy)	**Y**	**y**
Y	YY	Yy
y	Yy	??

Study the Punnett square and answer the following questions.

1. Which of these genotypes will complete this Punnett square?
   A) YY
   B) Yy
   C) yy
   D) Yx

2. In peas, the color yellow (Y) is dominant to the color green (y). According to this Punnett square, most of the offspring of the two yellow pea plants probably will be _____.
   F) orange
   G) green
   H) yellow
   J) red

Standardized Test Practice

Reading Comprehension

Read the passage. Then read each question that follows the passage. Decide which is the best answer to each question.

Genetic Engineering and Your Food

In recent years, scientists have made tremendous advances in the study of DNA. DNA is the material found in each cell that determines a cell's type, activities, and development. The DNA of a cell contains all the instructions that a cell inherits, including traits that help that organism survive. Scientists now can change an organism's DNA. This is called genetic engineering.

In the past, people were able to affect the DNA of organisms through breeding. The miniature poodle is a great example of this. People wanted smaller poodles, so they selected small poodles from the poodle population and bred them to produce small offspring. Eventually, a new variety of poodle, called the miniature poodle, developed. It has slightly different DNA than larger poodles.

Today, however, scientists can take small pieces of DNA from one organism and add them to another organism's DNA. Scientists have done this a lot with plants such as corn. They wanted corn to have two different traits—resistance to a weed-killing chemical often used by farmers and the ability to make a bug-killing substance called Cry1A(b). To develop a variety of corn with both of those traits, scientists took DNA from bacteria that display these traits and added it to the DNA of corn. The result was a new variety of corn that is not harmed by the weed-killing chemicals and that naturally produces Cry1A(b).

Because scientists have identified and learned about different DNA over the last few decades, they now know which traits result from different DNA. As a result, scientists are able to genetically engineer such things as new varieties of corn. Scientists also are exploring many other ideas for using genetic engineering to develop new crop varieties. Soon, these new varieties of crops will be able to be used all over the world to help feed millions of people.

Test-Taking Tip Consider how the actions of scientists have changed over the years from breeding to genetic engineering.

This genetically engineered corn contains DNA from bacteria.

1. Scientists were able to create new varieties of corn by _____.
   - **A)** using less Cry1A(b)
   - **B)** successfully breeding dogs
   - **C)** asking farmers what they needed
   - **D)** using DNA from bacteria

2. According to the passage, which of the following must have happened first?
   - **F)** the breeding of miniature poodles
   - **G)** the transfer of small sections of DNA from bacteria to other organisms
   - **H)** the discovery that DNA determines a cell's traits
   - **J)** the development of a genetically engineered variety of corn

Standardized Test Practice

Reasoning and Skills

Read each question and choose the best answer.

1. The genetic makeup of an organism is called its genotype. Two plants are bred and all of their offspring have a Tt genotype. The genotype of one parent is TT. Which of the following is the most likely explanation for the Tt genotype of the offspring?

A) The other parent plant's genotype is tt.
B) The offspring's phenotype is controlled by the T allele.
C) The offspring are called heterozygotes.
D) The T and t are two alleles of the same gene.

Test-Taking Tip If a question contains a lot of information about genotypes, draw a Punnett square to keep the information organized.

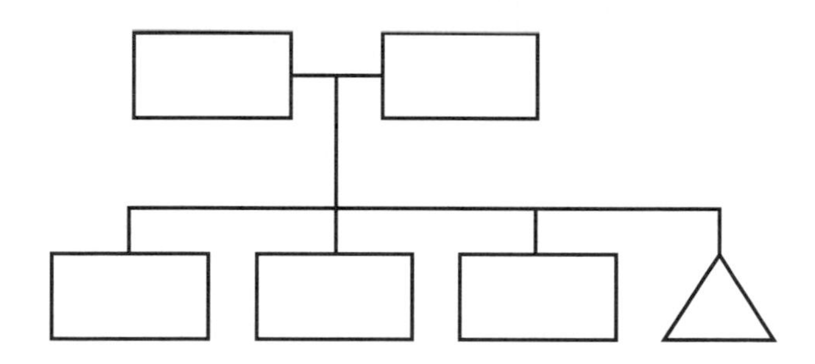

2. When an organism has a dominant and a recessive allele of a gene, only the dominant allele controls the organism's phenotype. Phenotypes can be represented by using shapes. According to the diagram above, the rectangle results from what kind of allele?

F) recessive
G) mutant
H) dominant
J) incomplete dominant

Test-Taking Tip As you read the question, remember the difference between dominant and recessive alleles.

3. Some alleles have incomplete dominance, which means they make phenotypes that are intermediate between the two alleles. The drawing above shows a cross between a shaded plant and a white plant that carry alleles with incomplete dominance for flower color. Which of the following depicts what the offsprings' flowers might look like?

Test-Taking Tip Cover up the answer choices, reread the question, and imagine what the correct answer should be. Then, read each answer choice.

Consider this question carefully before writing your answer on a separate sheet of paper.

4. The U.S. Department of Health funds many genetics research projects. How could this research help medical doctors?

Test-Taking Tip Use scratch paper to list as many different ways that genetics relates to disease and medicine as you can think of. Then, write about one or two of these.

UNIT 4 Ecology

How Are Cotton & Cookies Connected?

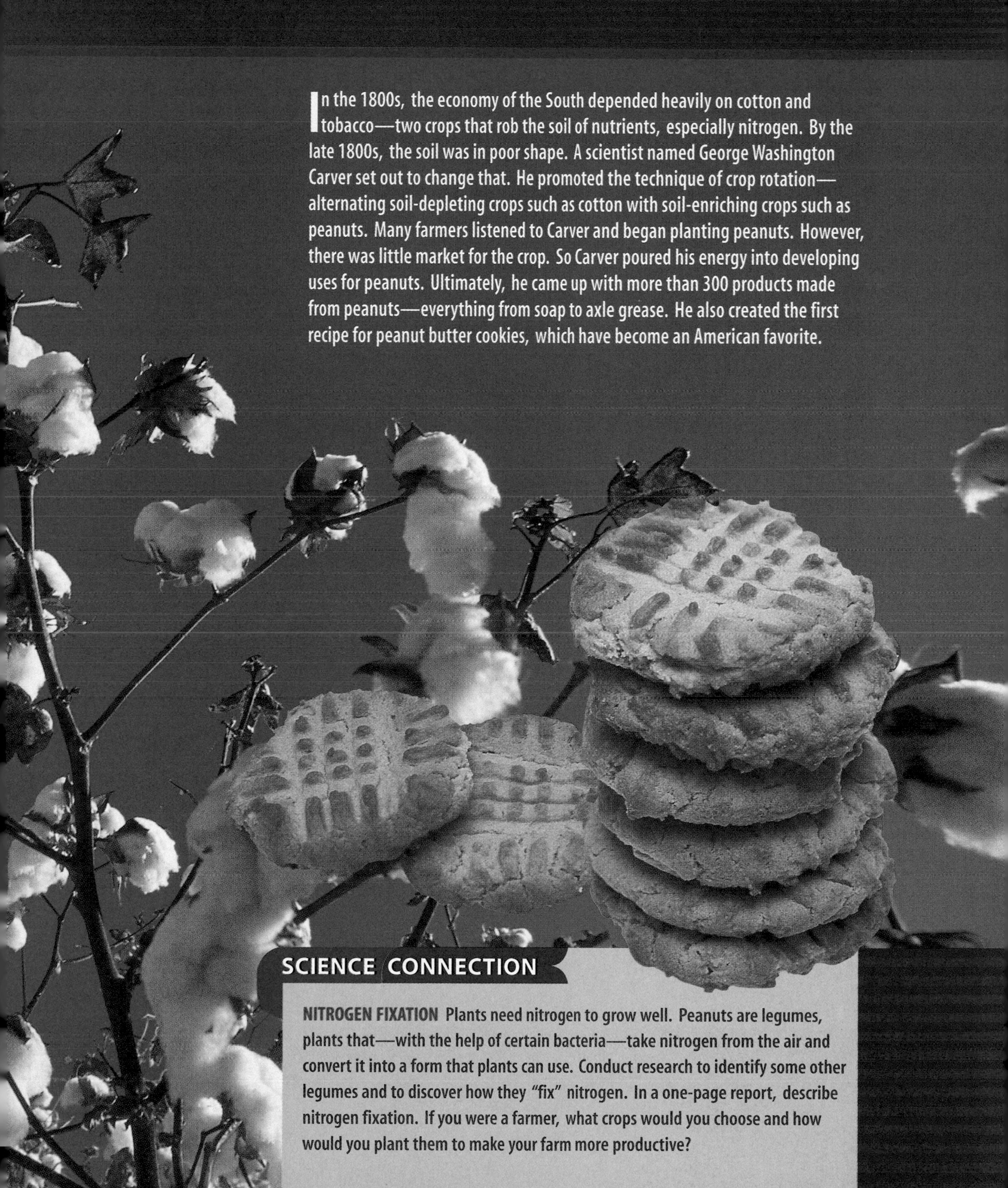

In the 1800s, the economy of the South depended heavily on cotton and tobacco—two crops that rob the soil of nutrients, especially nitrogen. By the late 1800s, the soil was in poor shape. A scientist named George Washington Carver set out to change that. He promoted the technique of crop rotation—alternating soil-depleting crops such as cotton with soil-enriching crops such as peanuts. Many farmers listened to Carver and began planting peanuts. However, there was little market for the crop. So Carver poured his energy into developing uses for peanuts. Ultimately, he came up with more than 300 products made from peanuts—everything from soap to axle grease. He also created the first recipe for peanut butter cookies, which have become an American favorite.

SCIENCE CONNECTION

NITROGEN FIXATION Plants need nitrogen to grow well. Peanuts are legumes, plants that—with the help of certain bacteria—take nitrogen from the air and convert it into a form that plants can use. Conduct research to identify some other legumes and to discover how they "fix" nitrogen. In a one-page report, describe nitrogen fixation. If you were a farmer, what crops would you choose and how would you plant them to make your farm more productive?

CHAPTER 12 Interactions of Life

Why would a powerful rhinoceros allow birds to perch on its back? Why aren't these birds safely perched in a tree? How do they find food? You don't have to go to Africa to see birds on the back of a rhino. You can see these animals at zoos or wildlife parks. In this chapter, you will learn how living organisms interact with each other and their surroundings. You also will learn about the roles each organism plays in the flow of energy through the environment.

What do you think?

Science Journal Look at the picture below with a classmate. Discuss what you think this might be or what is happening. Here's a hint: *It's a city within a city.* Write your answer or best guess in your Science Journal.

In your lifetime, you probably have taken thousands of footsteps on grassy lawns or playing fields. If you take a close look at the grass, you'll see that each blade is attached to roots in the soil. How do the grass plants obtain everything they need to live and grow? What other kinds of organisms live in the grass? The following activity will give you a chance to take a closer look at the life in a lawn.

Examine sod from a lawn

1. Examine a section of sod from a lawn.
2. How do the roots of the grass plants hold the soil?
3. Do you see signs of other living things besides grass?

Observe

In your Science Journal, answer the above questions and describe any organisms that are present in your section of sod. Explain how these organisms might affect the growth of grass plants. Draw a picture of your section of sod.

Before You Read

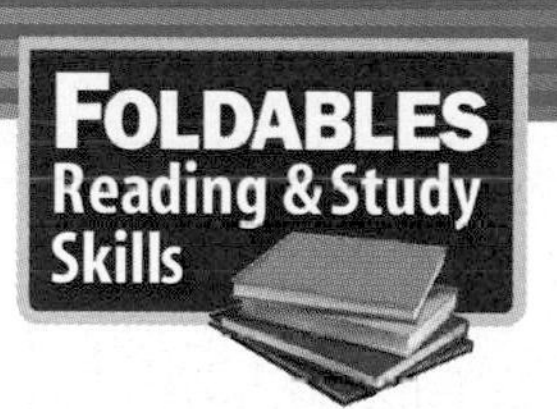

Making a Concept Map Study Fold The following Foldable will help you organize information by diagramming ideas about your favorite wild animal.

1. Place a sheet of paper in front of you with the short side at the top. Fold the paper in half from the left side to the right side.
2. Fold from top to bottom to divide the paper into thirds, then open up the three folds.
3. Through the top thickness of paper, cut along each of the fold lines to the side fold, forming three tabs.
4. Label *Organism, Population,* and *Community* across the front of the paper, as shown. Write the name of your favorite wild animal under the *Organism* tab.
5. Before you read the chapter, write what you know about your favorite animal under the top tab. As you read the chapter, write how this animal is part of a population and a community under the middle and bottom tabs.

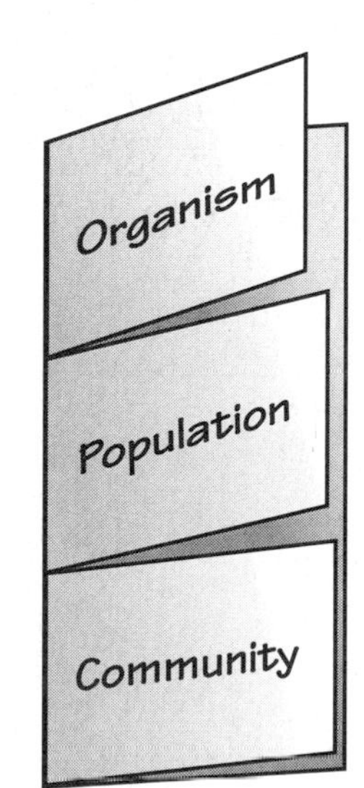

SECTION 1

Living Earth

As You Read

***What* You'll Learn**

- **Identify** places where life is found on Earth.
- **Define** ecology.
- **Observe** how the environment influences life.

Vocabulary

biosphere
ecosystem
ecology
population
community
habitat

***Why* It's Important**

All living things on Earth depend on each other for survival.

The Biosphere

What makes Earth different from other planets in the solar system? One difference is Earth's abundance of living organisms. The part of Earth that supports life is the **biosphere** (BI uh sfihr). The biosphere includes the top portion of Earth's crust, all the waters that cover Earth's surface, and the atmosphere that surrounds Earth.

Reading Check *What three things make up the biosphere?*

As **Figure 1** shows, the biosphere is made up of different environments that are home to different kinds of organisms. For example, desert environments receive little rain. Cactus plants, coyotes, and lizards are included in the life of the desert. Tropical rain forest environments receive plenty of rain and warm weather. Parrots, monkeys, and tens of thousands of other organisms live in the rain forest. Coral reefs form in warm, shallow ocean waters. Arctic regions near the north pole are covered with ice and snow. Polar bears, seals, and walruses live in the arctic.

Figure 1
Earth's biosphere consists of many environments, including ocean waters, polar regions, and deserts.

Desert

Arctic

Coral reef

Life on Earth In our solar system, Earth is the third planet from the Sun. The amount of energy that reaches Earth from the Sun helps make the temperature just right for life. Mercury, the planet closest to the Sun, is too hot during the day and too cold at night to make life possible there. Venus, the second planet from the Sun, has a thick, carbon dioxide atmosphere and high temperatures. It is unlikely that life could survive there. Mars, the fourth planet, is much colder than Earth because it is farther from the Sun and has a thinner atmosphere. It might support microscopic life, but none has been found. The planets beyond Mars probably do not receive enough heat and light from the Sun to have the right conditions for life.

Ecosystems

On a visit to Yellowstone National Park in Wyoming, you might see a prairie scene like the one shown in **Figure 2.** Bison graze on prairie grass. Cowbirds follow the bison, catching grasshoppers that jump away from the bisons' hooves. This scene is part of an ecosystem. An **ecosystem** consists of all the organisms living in an area and the nonliving features of their environment. Bison, grass, birds, and insects are living organisms of this prairie ecosystem. Water, temperature, sunlight, soil, and air are nonliving features of this prairie ecosystem. **Ecology** is the study of interactions that occur among organisms and their environment. Ecologists are scientists who study these interactions.

✔ Reading Check *What is an ecosystem?*

Figure 2
Ecosystems are made up of living organisms and the nonliving features of their environment. In this prairie ecosystem, cowbirds eat insects and bison graze on grass. *What other kinds of organisms might live in this ecosystem?*

Populations

Suppose you meet an ecologist who studies how a herd of bison moves from place to place and how the female bison in the herd care for their young. This ecologist is studying the members of a population. A **population** is made up of all the organisms in an ecosystem that belong to the same species. For example, all the bison in a prairie ecosystem are one population. All the cowbirds in this ecosystem make up a different population. The grasshoppers make up yet another population.

Ecologists often study how populations interact. For example, an ecologist might try to answer questions about several prairie species. How does grazing by bison affect the growth of prairie grass? How does grazing influence the insects that live in the grass and the birds that eat those insects? This ecologist is studying a community. A **community** refers to all the populations in an ecosystem. The prairie community is made of populations of bison, grasshoppers, cowbirds, and all other species in the prairie ecosystem. An arctic community might include populations of fish, seals that eat fish, and polar bears that hunt and eat seals. **Figure 3** shows how organisms, populations, communities, and ecosystems are related.

SCIENCE Online

Research Visit the Glencoe Science Web site at **science.glencoe.com** and find out the estimated human population size for the world today. In your Science Journal, create a graph that shows the population change between the year 2000 and this year.

Figure 3
The living world is arranged in several levels of organization.

Figure 4
The trees of the forest provide a habitat for woodpeckers and other birds. This salamander's habitat is the moist forest floor.

Habitats

Each organism in an ecosystem needs a place to live. The place in which an organism lives is called its **habitat.** The animals shown in **Figure 4** live in a forest ecosystem. Trees are the woodpecker's habitat. These birds use their strong beaks to pry insects from tree bark or break open acorns and nuts. Woodpeckers usually nest in holes in dead trees. The salamander's habitat is the forest floor, beneath fallen leaves and twigs. Salamanders avoid sunlight and seek damp, dark places. This animal eats small worms, insects, and slugs. An organism's habitat provides the kinds of food and shelter, the temperature, and the amount of moisture the organism needs to survive.

Section  Assessment

1. What is the biosphere?
2. What is ecology?
3. How are the terms *habitat* and *biosphere* related to each other?
4. What is the major difference between a community and a population? Give one example of each.
5. **Think Critically** Does the amount of rain that falls in an area determine which kinds of organisms can live there? Why or why not?

Skill Builder Activities

6. **Forming Hypotheses** Make a hypothesis about how one nonliving feature of an ecosystem would affect the growth of dandelions in that ecosystem. **For more help, refer to the Science Skill Handbook.**
7. **Communicating** Pretend you are a nonhuman organism in the wild. Describe what you are and list living and nonliving features of the environment that affect you. **For more help, refer to the Science Skill Handbook.**

SECTION

Populations

As You Read

What You'll Learn

- **Identify** methods for estimating population sizes.
- **Explain** how competition limits population growth.
- **List** factors that influence changes in population size.

Vocabulary
limiting factor
carrying capacity

Why It's Important

Competition caused by population growth affects many organisms, including humans.

Competition

Some pet shops sell lizards, snakes, and other reptiles. Crickets are raised as a food supply for pet reptiles. In the wild, crickets come out at night and feed on plant material. During the day, they hide in dark areas, beneath leaves or under buildings. Pet shop workers who raise crickets make sure that the insects have plenty of food, water, and hiding places. As the cricket population grows, the workers increase the crickets' food supply and the number of hiding places. To avoid crowding, some of the crickets could be moved into larger containers.

Food and Space Organisms living in the wild do not always have enough food or living space. The Gila woodpecker, shown in **Figure 5,** lives in the Sonoran Desert of Arizona and Mexico. This bird makes its nest in a hole that it drills in a saguaro (suh GWAR oh) cactus. If an area has too many Gila woodpeckers or too few saguaros, the woodpeckers must compete with each other for nesting spots. Competition occurs when two or more organisms seek the same resource at the same time.

Growth Limits Competition limits population size. If the amount of available nesting space is limited, some woodpeckers will not be able to raise young. Gila woodpeckers eat cactus fruit, berries, and insects. If food becomes scarce, some woodpeckers might not survive to reproduce. Competition for food, living space, or other resources can prevent population growth.

In nature, the most intense competition is usually among individuals of the same species, because they need the same kinds of food and shelter. Competition also takes place among individuals of different species. For example, after a Gila woodpecker has abandoned its nesting hole, owls, flycatchers, snakes, and lizards compete for the shelter of the empty hole.

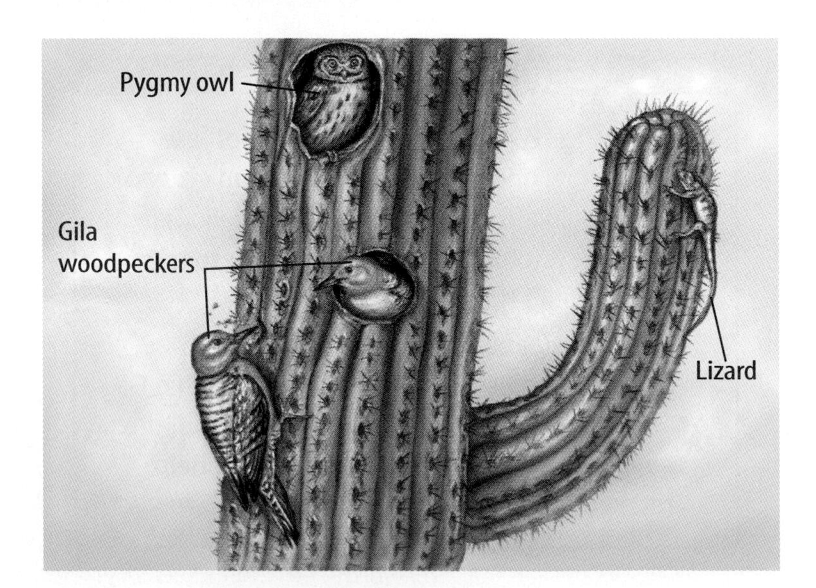

Figure 5
Gila woodpeckers make nesting holes in the saguaro cactus. Many animals compete for the shelter these holes provide.

Population Size

Ecologists often need to measure the size of a population. This information can indicate whether or not a population is healthy and growing. Population counts can help identify populations that could be in danger of disappearing.

Some populations are easy to measure. If you were raising crickets, you could measure the size of your cricket population simply by counting all the crickets in the container. What if you wanted to compare the cricket populations in two different containers? You would calculate the number of crickets per square meter (m^2) of your container. The size of a population that occupies a specific area is called population density. **Figure 6** shows human population density in different places in the world.

✔ Reading Check *What is population density?*

Measuring Populations Counting crickets can be tricky. They look alike, move a lot, and hide. The same cricket could be counted more than once, and others could be completely missed. Ecologists have similar problems when measuring wildlife populations. One of the methods they use is called trap-mark-release. Suppose you want to count wild rabbits. Rabbits live underground and come out at dawn and dusk to eat. Ecologists set traps that capture rabbits without injuring them. Each captured rabbit is marked and released. Later, another sample of rabbits is captured. Some of these rabbits will have marks, but many will not. By comparing the number of marked and unmarked rabbits in the second sample, ecologists can estimate the population size.

TRY AT HOME Mini LAB

Observing Seedling Competition

Procedure

1. Fill **two plant pots** with **moist potting soil.**
2. Plant **radish seeds** in one pot, following the spacing instructions on the seed packet. Label this pot "Recommended Spacing."
3. Plant radish seeds in the second pot, spaced half the recommended distance apart. Label this pot "Densely Populated." Wash your hands.
4. Keep the soil moist. When the seeds sprout, move them to a well-lit area.
5. Measure the height of the seedlings every two days for two weeks. Record the data in your **Science Journal.**

Analysis

1. Which plants grew faster?
2. Which plants looked healthiest after two weeks?
3. How did competition influence the plants?

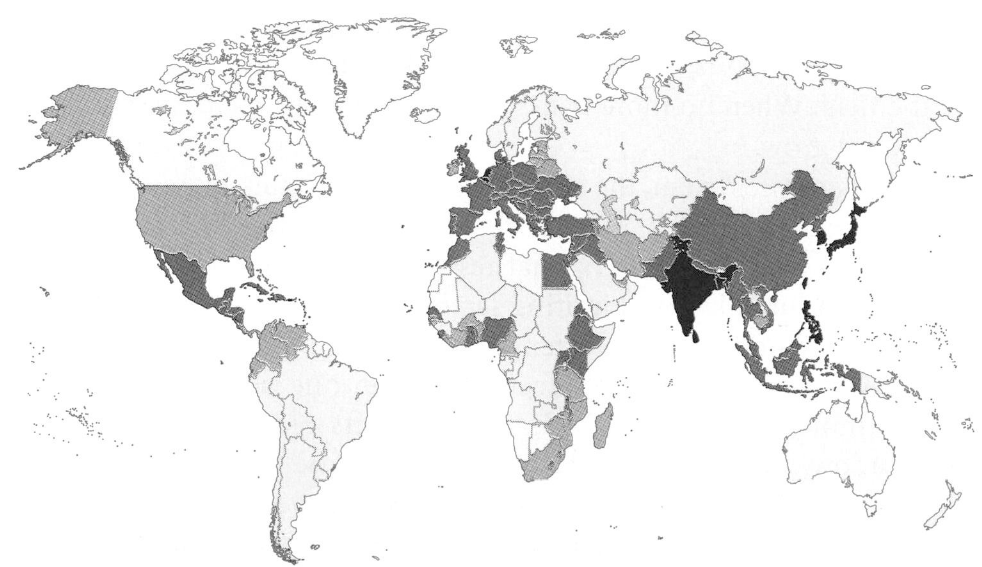

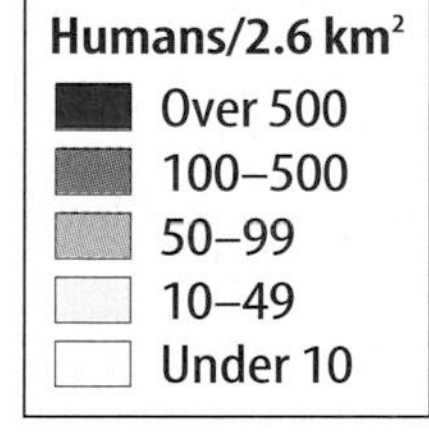

Figure 6
This map shows human population density. *Which countries have the highest population density?*

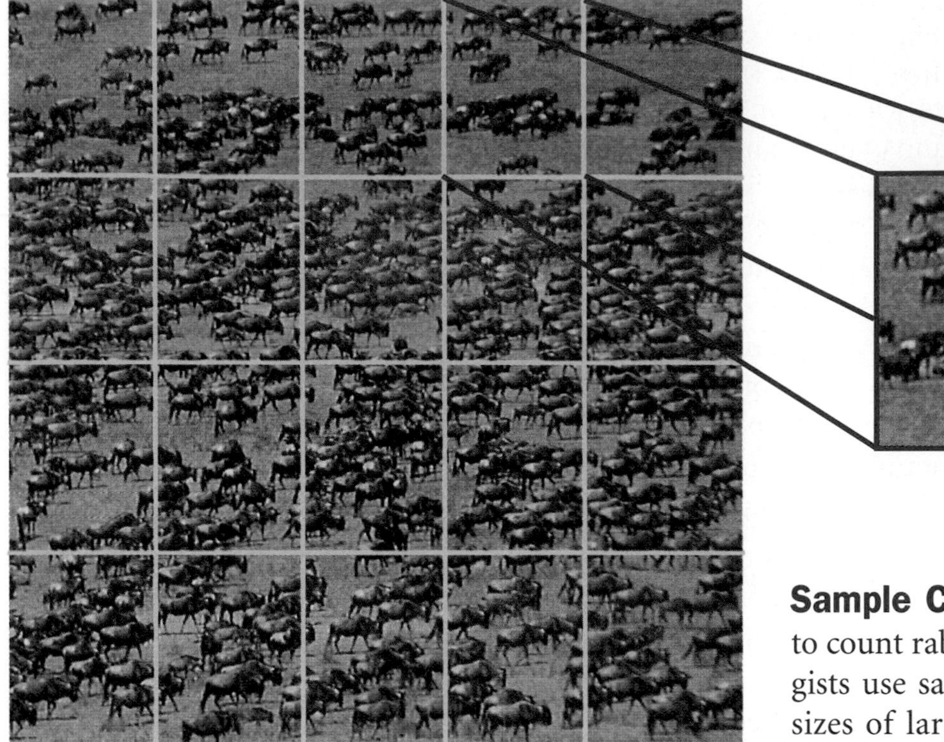

Figure 7
Ecologists can estimate population size by making a sample count. Wildebeests graze on the grassy plains of Africa. *How could you use the enlarged square to estimate the number of wildebeests in the entire photograph?*

Sample Counts What if you wanted to count rabbits over a large area? Ecologists use sample counts to estimate the sizes of large populations. To estimate the number of rabbits in a 100-acre area, for example, you could count the rabbits in one acre and multiply by 100 to estimate the population size. **Figure 7** shows another approach to sample counting.

Limiting Factors One grass plant can produce hundreds of seeds. Imagine those seeds drifting onto a vacant field. Many of the seeds sprout and grow into grass plants that produce hundreds more seeds. Soon the field is covered with grass. Can this grass population keep growing forever? Suppose the seeds of wildflowers or trees drift onto the field. If those seeds sprout, trees and flowers would compete with grasses for sunlight, soil, and water. Even if the grasses did not have to compete with other plants, they might eventually use up all the space in the field. When no more living space is available, the population cannot grow.

In any ecosystem, the availability of food, water, living space, mates, nesting sites, and other resources is often limited. A **limiting factor** is anything that restricts the number of individuals in a population. Limiting factors include living and nonliving features of the ecosystem.

A limiting factor can affect more than one population in a community. Suppose a lack of rain limits plant growth in a meadow. Fewer plants produce fewer seeds. For seed-eating mice, this reduction in the food supply could become a limiting factor. A smaller mouse population could, in turn, become a limiting factor for the hawks and owls that feed on mice.

Carrying Capacity A population of robins lives in a grove of trees in a park. Over several years, the number of robins increases and nesting space becomes scarce. Nesting space is a limiting factor that prevents the robin population from getting any larger. This ecosystem has reached its carrying capacity for robins. **Carrying capacity** is the largest number of individuals of one species that an ecosystem can support over time. If a population begins to exceed the environment's carrying capacity, some individuals will not have enough resources. They could die or be forced to move elsewhere, like the deer shown in **Figure 8.**

Figure 8
These deer might have moved into a residential area because a nearby forest's carrying capacity for deer has been reached.

✔ Reading Check ***How are limiting factors related to carrying capacity?***

Problem-Solving Activity

Do you have too many crickets?

You've decided to raise crickets to sell to pet stores. A friend says you should not allow the cricket population density to go over 210 crickets/m^2. Use what you've learned in this section to measure the population density in your cricket tanks.

Identifying the Problem

The table on the right lists the areas and populations of your three cricket tanks. How can you determine if too many crickets are in one tank? If a tank contains too many crickets, what could you do? Explain why too many crickets in a tank might be a problem.

Cricket Population

Tank	Area (m^2)	Number of Crickets
1	0.80	200
2	0.80	150
3	1.5	315

Solving the Problem

1. Do any of the tanks contain too many crickets? Could you make the population density of the three tanks equal by moving crickets from one tank to another? If so, which tank would you move crickets into?
2. The population density of wild crickets living in a field is 2.4 crickets/m^2. If the field has an area of 250 m^2, what is the approximate size of the cricket population? Why would the population density of crickets in a field be lower than the population density of crickets in a tank?

Data Update For an online update of recent data on human birthrates and death rates around the world, visit the Glencoe Science Web site at **science.glencoe.com** and select the appropriate chapter.

Biotic Potential What would happen if no limiting factors restricted the growth of a population? Think about a population that has an unlimited supply of food, water, and living space. The climate is favorable. Population growth is not limited by diseases, predators, or competition with other species. Under ideal conditions like these, the population would continue to grow.

The highest rate of reproduction under ideal conditions is a population's biotic potential. The larger the number of offspring that are produced by parent organisms, the higher the biotic potential of the species will be. Compare an avocado tree to a tangerine tree. Assume that each tree produces the same number of fruits. Each avocado fruit contains one large seed. Each tangerine fruit contains a dozen seeds or more. Because the tangerine tree produces more seeds per fruit, it has a higher biotic potential than the avocado tree.

Changes in Populations

Birthrates and death rates also influence the size of a population and its rate of growth. A population gets larger when the number of individuals born is greater than the number of individuals that die. When the number of deaths is greater than the number of births, populations get smaller. Take the squirrels living in New York City's Central Park as an example. In one year, if 900 squirrels are born and 800 die, the population increases by 100. If 400 squirrels are born and 500 die, the population decreases by 100.

The same is true for human populations. **Table 1** shows birthrates, death rates, and population changes for several countries around the world. In countries with faster population growth, birthrates are much higher than death rates. In countries with slower population growth, birthrates are only slightly higher than death rates. In Germany, where the population is getting smaller, the birthrate is lower than the death rate.

Table 1 Population Growth

	Birthrate*	Death Rate*	Population Increase (percent)
Rapid-Growth Countries			
Jordan	38.8	5.5	3.3
Uganda	50.8	21.8	2.9
Zimbabwe	34.3	9.4	5.2
Slow-Growth Countries			
Germany	9.4	10.8	−1.5
Sweden	10.8	10.6	0.1
United States	14.8	8.8	0.6

*Number per 1,000 people

Figure 9
The mangrove seeds sprout while they are still attached to the parent tree. Some sprouted seeds drop into the mud below the parent tree and continue to grow. Others drop into the water and can be carried away by tides and ocean currents. When they wash ashore, they might start a new population of mangroves or add to an existing mangrove population.

Moving Around Most animals can move easily from place to place, and these movements can affect population size. For example, a male mountain sheep might wander many miles in search of a mate. After he finds a mate, their offspring might establish a completely new population of mountain sheep far from the male's original population.

Many bird species move from one place to another during their annual migrations. During the summer, populations of Baltimore orioles are found throughout eastern North America. During the winter, these populations disappear because the birds migrate to Central America. They spend the winter there, where the climate is mild and food supplies are plentiful. When summer approaches, the orioles migrate back to North America.

Even plants and microscopic organisms can move from place to place, carried by wind, water, or animals. The tiny spores of mushrooms, mosses, and ferns float through the air. The seeds of dandelions, maple trees, and other plants have feathery or winglike growths that allow them to be carried by wind. Spine-covered seeds hitch rides by clinging to animal fur or people's clothing. Many kinds of seeds can be transported by river and ocean currents. Mangrove trees growing along Florida's Gulf Coast, shown in **Figure 9,** provide an example of how water moves seeds.

Mini LAB

Comparing Biotic Potential

Procedure

1. Remove all the seeds from a **whole fruit.** Do not put fruit or seeds in your mouth.
2. Count the total number of seeds in the fruit. Wash your hands, then record these data in your Science Journal.
3. Compare your seed totals with those of classmates who examined other types of fruit.

Analysis

1. Which type of fruit had the most seeds? Which had the fewest seeds?
2. What is an advantage of producing many seeds? Can you think of a possible disadvantage?
3. To estimate the total number of seeds produced by a tomato plant, what would you need to know?

NATIONAL GEOGRAPHIC

VISUALIZING POPULATION GROWTH

Figure 10

When a species enters an ecosystem that has abundant food, water, and other resources, its population can flourish. Beginning with a few organisms, the population increases until the number of organisms and available resources are in balance. At that point, population growth slows or stops. A graph of these changes over time produces an S-curve, as shown here for coyotes.

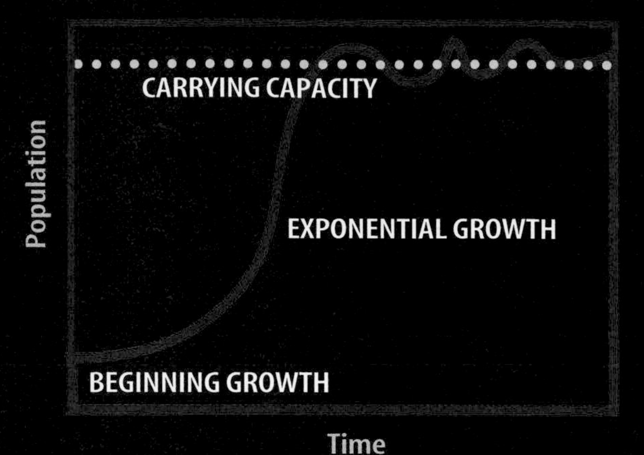

BEGINNING GROWTH During the first few years, population growth is slow, because there are few adults to produce young. As the population grows, so does the number of breeding adults.

EXPONENTIAL GROWTH As the number of adults in the population grows, so does the number of births. The coyote population undergoes exponential growth, quickly increasing in size.

CARRYING CAPACITY As resources become less plentiful, the birthrate declines and the death rate may rise. Population growth slows. The coyote population has reached the environmental carrying capacity—the maximum number of coyotes that the environment can sustain.

Exponential Growth Imagine what might happen if a pair of coyotes moves into a valley where no other coyotes live. Food and water are abundant, and there are plenty of areas where female coyotes can build dens for their young. This population grows quickly in a pattern called exponential growth. Exponential growth means that the larger a population becomes, the faster it grows.

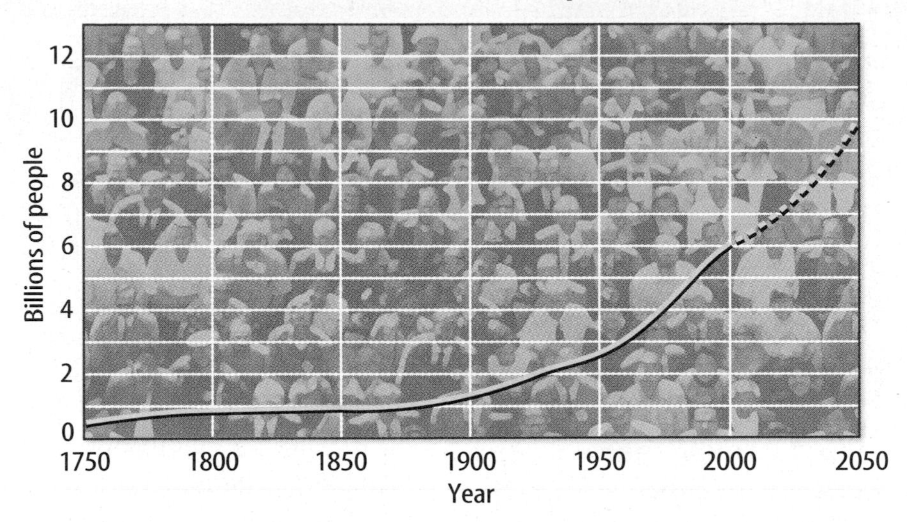

Figure 11
The size of the human population is increasing by about 1.6 percent per year. *What factors affect human population growth?*

After several years, the population becomes so large that the coyotes begin to compete for food and den sites. Population growth slows, and the number of coyotes remains fairly constant and reaches equilibrium. This ecosystem has reached its carrying capacity for coyotes. A graph that describes each stage in this pattern of population growth is shown in **Figure 10.** As you can see in **Figure 11,** Earth's human population shows exponential growth. In the year 2000, Earth's human population exceeded 6 billion. By the year 2050, it is estimated that Earth's human population could reach 10 billion.

Section  Assessment

1. How can an ecologist predict the size of a population without counting every organism in the population?
2. Why does competition between individuals of the same species tend to be greater than competition between individuals of different species?
3. How do birthrates and death rates influence the size of a population?
4. How does carrying capacity influence the number of organisms in an ecosystem?
5. **Think Critically** Why does the supply of food and water in an ecosystem usually affect population size more than other limiting factors?

Skill Builder Activities

6. **Making and Using Tables** Construct a table using the following data on changes in the size of a deer population in Arizona. In 1910 there were 6 deer; in 1915, 36 deer; in 1920, 143 deer; in 1925, 86 deer; and in 1935, 26 deer. Propose a hypothesis to explain what might have caused these changes. **For more help, refer to the Science Skill Handbook.**
7. **Solving One-Step Equations** A vacant lot that measures 12 m × 12 m contains 46 dandelion plants, 212 grass plants, and 14 bindweed plants. What is the population density, per square meter, of each species? **For more help, refer to the Math Skill Handbook.**

SECTION

Interactions Within Communities

As You Read

What You'll Learn

- **Describe** how organisms obtain energy for life.
- **Explain** how organisms interact.
- **Recognize** that every organism occupies a niche.

Vocabulary

producer
consumer
symbiosis
mutualism
commensalism
parasitism
niche

Why It's Important

How organisms obtain food and meet other needs is critical for their survival.

Obtaining Energy

Just as a car engine needs a constant supply of gasoline, living organisms need a constant supply of energy. The energy that fuels most life on Earth comes from the Sun. Some organisms use the Sun's energy to create energy-rich molecules through the process of photosynthesis. The energy-rich molecules, usually sugars, serve as food. They are made up of different combinations of carbon, hydrogen, and oxygen atoms. Energy is stored in the chemical bonds that hold the atoms of these molecules together. When the molecules break apart—for example, during digestion—the energy in the chemical bonds is released to fuel life processes.

Producers Organisms that use an outside energy source like the Sun to make energy-rich molecules are called **producers.** Most producers contain chlorophyll (KLOR uh fihl), a chemical that is required for photosynthesis. As shown in **Figure 12,** green plants are producers. Some producers do not contain chlorophyll and do not use energy from the Sun. Instead, they make energy-rich molecules through a process called chemosynthesis (kee moh SIHN thuh sus). These organisms can be found near volcanic vents on the ocean floor. Inorganic molecules in the water provide the energy source for chemosynthesis.

Figure 12
Green plants, including the grasses that surround this pond, are producers. The pond also contains many other producers, including microscopic organisms like A *Euglena* and B simple plantlike organisms called algae.

Consumers

Herbivores

Carnivores

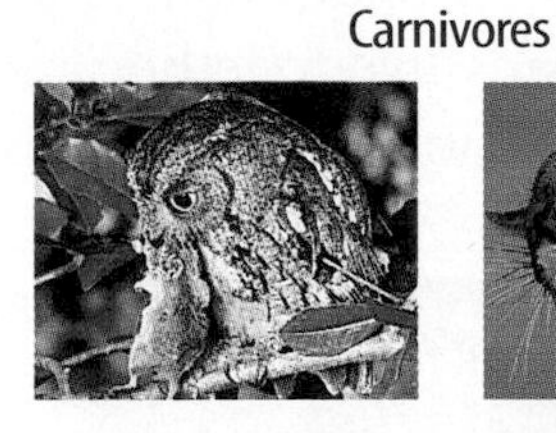

Omnivores

Decomposers

Figure 13
Four categories of consumers are shown. *What kind of consumer is a cactus wren? A mushroom?*

Consumers Organisms that cannot make their own energy-rich molecules are called **consumers.** Consumers obtain energy by eating other organisms. **Figure 13** shows the four general categories of consumers. Herbivores are the vegetarians of the world. They include rabbits, deer, and other plant eaters. Carnivores are animals that eat other animals. Frogs and spiders are carnivores that eat insects. Omnivores, including pigs and humans, eat mostly plants and animals. Decomposers, including fungi, bacteria, and earthworms, consume wastes and dead organisms. Decomposers help recycle once-living matter by breaking it down into simple, energy-rich substances. These substances might serve as food for decomposers, be absorbed by plant roots, or be consumed by other organisms.

How are producers different from consumers?

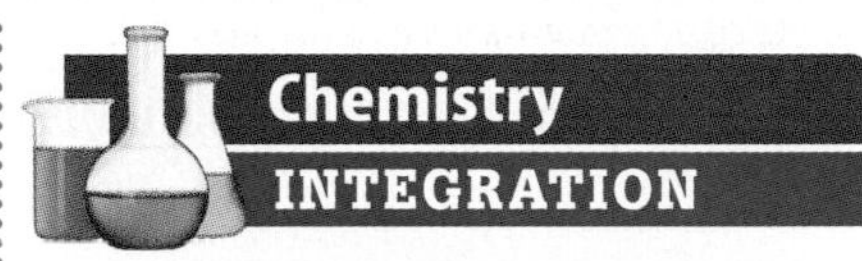

Glucose is a nutrient molecule produced during photosynthesis. Look up the chemical structure of glucose and draw it in your Science Journal.

Food Chains Ecology includes the study of how organisms depend on each other for food. A food chain is a simple model of the feeding relationships in an ecosystem. For example, shrubs are food for deer, and deer are food for mountain lions, as illustrated in **Figure 14.** What food chain would include you?

Figure 14
Food chains illustrate how consumers obtain energy from other organisms in an ecosystem.

Symbiotic Relationships

Figure 15
Many examples of symbiotic relationships exist in nature.

Not all relationships among organisms involve food. Many organisms live together and share resources in other ways. Any close relationship between species is called **symbiosis.**

A Lichens are a result of mutualism.

Mutualism You may have noticed crusty lichens growing on fences, trees, or rocks. Lichens, like those shown in **Figure 15A,** are made up of an alga or a cyanobacterium that lives within the tissues of a fungus. Through photosynthesis, the cyanobacterium or alga supplies energy to itself and the fungus. The fungus provides a protected space in which the cyanobacterium or alga can live. Both organisms benefit from this association. A symbiotic relationship in which both species benefit is called **mutualism** (MYEW chuh wuh lih zum).

B Clown fish and sea anemones have a commensal relationship.

Commensalism If you've ever visited a marine aquarium, you might have seen the ocean organisms shown in **Figure 15B.** The creature with gently waving, tubelike tentacles is a sea anemone. The tentacles contain a mild poison. Anemones use their tentacles to capture shrimp, fish, and other small animals to eat. The striped clown fish can swim among the tentacles without being harmed. The anemone's tentacles protect the clown fish from predators. In this relationship, the clown fish benefits but the sea anemone is not helped or hurt. A symbiotic relationship in which one organism benefits and the other is not affected is called **commensalism** (kuh MEN suh lih zum).

Magnification: 128×

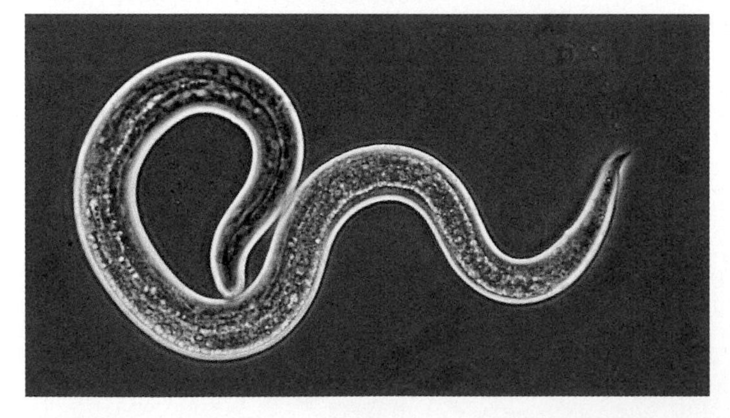

C Some roundworms are parasites that rob nutrients from their hosts.

Parasitism Pet cats or dogs sometimes have to be treated for worms. Roundworms, like the one shown in **Figure 15C,** are common in puppies. This roundworm attaches itself to the inside of the puppy's intestine and feeds on nutrients in the puppy's blood. The puppy may have abdominal pain, bloating, and diarrhea. If the infection is severe, the puppy might die. A symbiotic relationship in which one organism benefits but the other is harmed is called **parasitism** (PER uh suh tih zum).

Niches

One habitat might contain hundreds or even thousands of species. Look at the rotting log habitat shown in **Figure 16.** A rotting log in a forest can be home to many species of insects, including termites that eat decaying wood and ants that feed on the termites. Other species that live on or under the rotting log include millipedes, centipedes, spiders, and worms. You might think that competition for resources would make it impossible for so many species to live in the same habitat. However, each species has different requirements for its survival. As a result, each species has its own niche (NIHCH). A **niche** refers to how an organism survives, how it obtains food and shelter, how it finds a mate and cares for its young, and how it avoids danger.

Reading Check *Why does each species have its own niche?*

Special adaptations that improve survival are often part of an organism's niche. Milkweed plants contain a poison that prevents many insects from feeding on them. Monarch butterfly caterpillars have an adaptation that allows them to eat milkweed. Monarchs can take advantage of a food resource that other species cannot use. Milkweed poison also helps protect monarchs from predators. When the caterpillars eat milkweed, they become slightly poisonous. Birds avoid eating monarchs because they learn that the caterpillars and adult butterflies have an awful taste and can make them sick.

Health INTEGRATION

The poison in milkweed is similar to the drug digitalis. Small amounts of digitalis are used to treat heart ailments in humans, but it is poisonous in large doses. Look up digitalis and explain in your Science Journal how it affects the human body.

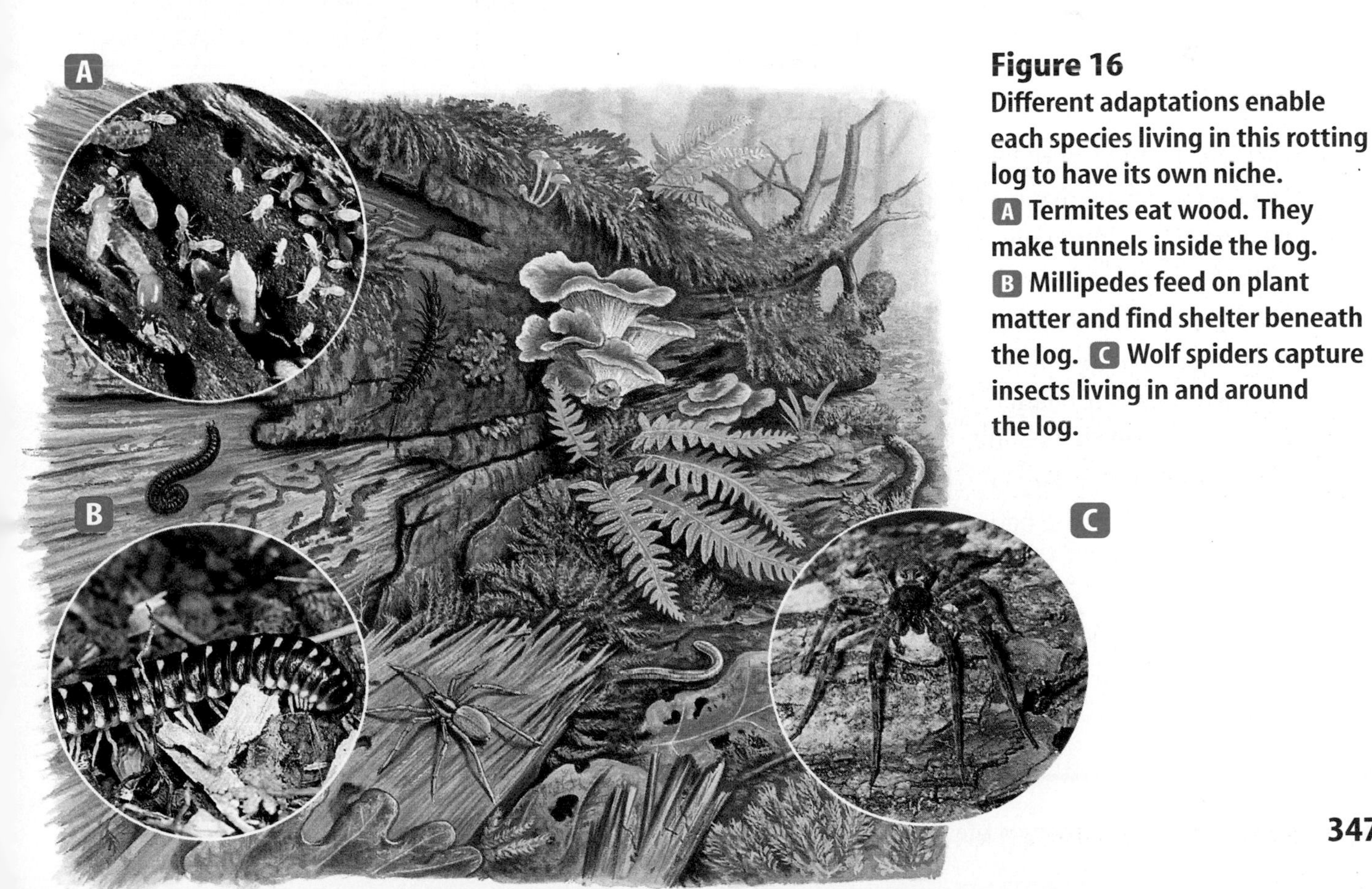

Figure 16
Different adaptations enable each species living in this rotting log to have its own niche.
A Termites eat wood. They make tunnels inside the log.
B Millipedes feed on plant matter and find shelter beneath the log. C Wolf spiders capture insects living in and around the log.

Predator and Prey When you think of survival in the wild, you might imagine an antelope running away from a lion. An organism's niche includes how it avoids being eaten and how it finds or captures its food. Predators, like the one shown in **Figure 17,** are consumers that capture and eat other consumers. The prey is the organism that is captured by the predator. The presence of predators usually increases the number of different species that can live in an ecosystem. Predators limit the size of prey populations. As a result, food and other resources are less likely to become scarce, and competition between species is reduced.

Figure 17
The alligator is a predator. The turtle is its prey.

Cooperation Individual organisms often cooperate in ways that improve survival. For example, a white-tailed deer that detects the presence of wolves or coyotes will alert the other deer in the herd. Many insects, such as ants and honeybees, live in social groups. Different individuals perform different tasks required for the survival of the entire nest. Soldier ants protect workers that go out of the nest to gather food. Worker ants feed and care for ant larvae that hatch from eggs laid by the queen. These cooperative actions improve survival and are a part of the species' niche.

Section 3 Assessment

1. Explain why all consumers ultimately depend on producers for food.
2. Draw a food chain that models the feeding relationships of three species in a community. Choose a food chain other than the one shown in **Figure 14.**
3. Make up two imaginary organisms that have a mutualistic relationship. Give them names and explain how they benefit from the association.
4. What is the difference between a habitat and a niche?
5. **Think Critically** A parasite can obtain food only from a host organism. Most parasites weaken but do not kill their hosts. Why?

Skill Builder Activities

6. **Manipulating Variables and Controls** You are sure that Animal A benefits from a relationship with Plant B, but you are not sure if Plant B benefits, is harmed, or is unaffected by the relationship. Design an experiment to compare how well Plant B grows on its own and when Animal A is present. **For more help, refer to the Science Skill Handbook.**
7. **Using Graphics Software** Use graphics software to make three different food chains. Represent each organism with a shape that resembles it. For example, you could use a leaf shape to represent a plant. Label each shape. **For more help, refer to the Technology Skill Handbook.**

Activity

Feeding Habits of Planaria

You probably have watched minnows darting about in a stream. It is not as easy to observe organisms that live at the bottom of a stream, beneath rocks, logs, and dead leaves. Countless stream organisms, including insect larvae, worms, and microscopic organisms, live out of your view. One such organism is a type of flatworm called a planarian. In this activity, you will find out about the eating habits of planarians.

What You'll Investigate

What food items do planarians prefer to eat?

Materials

small bowl
planarians (several)
lettuce leaf
raw liver or meat
guppies (several)
pond or stream water
magnifying lens

Goals

- **Observe** the food preference of planarians.
- **Infer** what planarians eat in the wild.

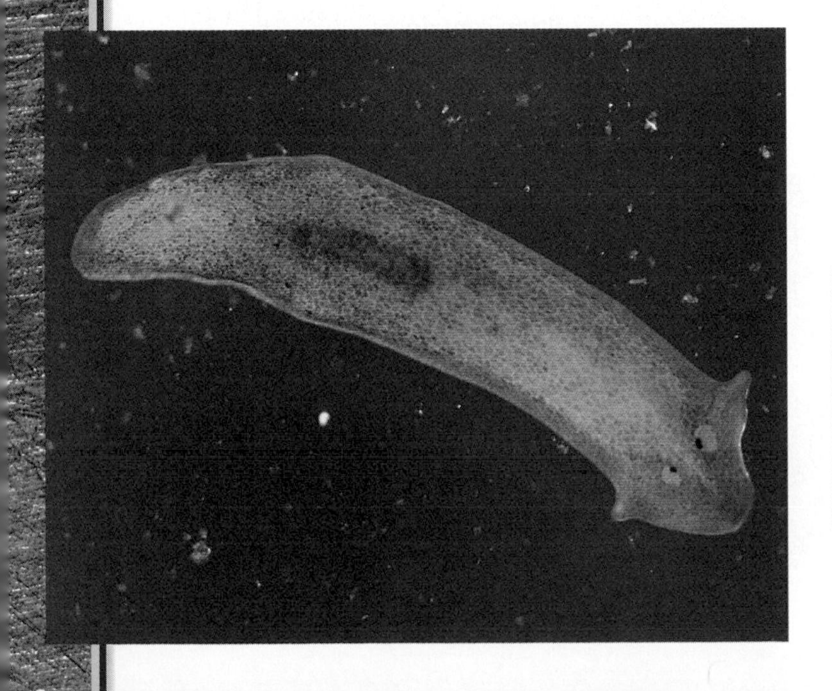

Safety Precautions

Procedure

1. Fill the bowl with stream water.
2. Place a lettuce leaf, piece of raw liver, and several guppies in the bowl. Add the planarians. Wash your hands.
3. **Observe** what happens inside the bowl for at least 20 minutes. Do not disturb the bowl or its contents. Use a magnifying lens to look at the planarians.
4. **Record** all of your observations in your Science Journal.

Conclude and Apply

1. Which food did the planarians prefer?
2. **Infer** what planarians might eat when in their natural environment.
3. Based on your observations during this activity, what is a planarian's niche in a stream ecosystem?
4. **Predict** where in a stream you might find planarians. Use references to find out whether your prediction is correct.

Communicating Your Data

Share your results with other students in your class. Plan an adult-supervised trip with several classmates to a local stream to search for planarians in their native habitat. **For more help, refer to the Science Skill Handbook.**

Activity Design Your Own Experiment

Population Growth in Fruit Flies

Populations can grow at an exponential rate only if the environment provides the right amount of food, shelter, air, moisture, heat, living space, and other factors. You probably have seen fruit flies hovering near ripe bananas or other fruit. Fruit flies are fast-growing organisms often raised in science laboratories. The flies are kept in culture tubes and fed a diet of specially prepared food flakes. Can you improve on this standard growing method to achieve faster population growth?

Recognize the Problem

Will a change in one environmental factor affect the growth of a fruit fly population?

Form a Hypothesis

Based on your reading about fruit flies, state a hypothesis about how changing one environmental factor will affect the rate of growth of a fruit fly population.

Goals

- **Identify** the environmental factors needed by a population of fruit flies.
- **Design** an experiment to investigate how a change in one environmental factor affects in any way the size of a fruit fly population.
- **Observe** and **measure** changes in population size.

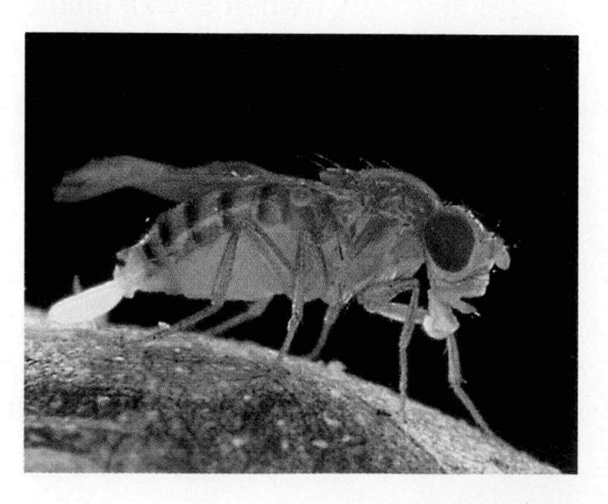

Possible Materials

fruit flies
standard fruit fly culture kit
food items (banana, orange peel, or other fruit)
water
heating or cooling source
culture containers
cloth, plastic, or other tops for culture containers
hand lens

Safety Precautions

Test Your Hypothesis

Plan

1. As a group, decide on one environmental factor to investigate. Agree on a hypothesis about how a change in this factor will affect population growth. Decide how you will test your hypothesis, and identify the experimental results that would support your hypothesis.
2. **List** the steps you will need to take to test your hypothesis. Describe exactly what you will do. List your materials.
3. **Determine** the method you will use to measure changes in the size of your fruit fly populations.
4. Prepare a data table in your Science Journal to record weekly measurements of your fruit fly populations.
5. Read the entire experiment and make sure all of the steps are in a logical order.
6. **Research** the standard method used to raise fruit flies in the laboratory. Use this method as the control in your experiment.
7. **Identify** all constants, variables, and controls in your experiment.

Do

1. Make sure your teacher approves your plan before you start.
2. Carry out your experiment.
3. **Measure** the growth of your fruit fly populations weekly and record the data in your data table.

Analyze Your Data

1. What were the constants in your experiment? The variables?
2. **Compare** changes in the size of your control population with changes in your experimental population. Which population grew faster?
3. Using the information in your data table, make a line graph that shows how the sizes of your two fruit fly populations changed over time. Use a different colored pencil for each population's line on the graph.

Draw Conclusions

1. Did the results support your hypothesis? Explain.
2. **Compare** the growth of your control and experimental populations. Did either population reach exponential growth? How do you know?

Compare the results of your experiment with those of other students in your class. **For more help, refer to the Science Skill Handbook.**

TIME **SCIENCE** AND **HISTORY**

SCIENCE CAN CHANGE THE COURSE OF HISTORY!

YOU CAN COUNT

The Census gives a snapshot of the people of the United States

The doorbell rings and you hear someone at the door say to your mom, "I'm working for the U.S. Census Bureau, doing follow-up interviews. Do you have a few minutes to answer some questions?" What does this person—and the U.S. government—want to know about your family?

Counting people is important to the United States and to many other countries around the world. It helps governments determine the distribution of people in the various regions of a nation. To obtain this information, the government takes a census—a count of how many people are living in their country on a particular day at a particular time, and in a particular place. A census is a snapshot of a country's population. The time at which the count occurs is called the "census moment." Some countries close their borders for a day or two so everyone will "sit still" for the census camera at the census moment, as was done in Nigeria in 1991.

Counting on the Count

When the United States government was formed, its founders set up the House of Representatives based on population. Areas with more people had more government representatives, and areas with fewer people had fewer representatives. In 1787, the requirement for a census became part of the Constitution. A census must be taken every ten years so the proper number of representatives for each state can be calculated.

Over the years, the U.S. Census Bureau has added questions to obtain more information than just a population count. In 1810, questions about manufacturing were added. In 1850, as more immigrants began coming to the United States, a new question about where people were born was added. In 1880, census takers asked people whether or not they were married. And in 1950, the first electronic computers were used to add up the census results.

Next, read on to find out more about the census.

ON IT

Growing by the Numbers

Chances are you just blinked your eyes. While you did it, three people were added to the world's population. There, you blinked again—that's another three people! It may seem impossible, but that's how quickly the world's population is growing. It adds up to 184 people every minute, 11,040 every hour, 264,960 every day, and 97 million every year! On October 12, 1999, the official number of people on the planet reached a record 6 billion.

The Short Form

Before 1970, United States census data was collected by field workers. They went door to door to count the number of people living in each household. Since then, the census has been done mostly by mail. People are sent a form they must fill out. The form asks for the number of people living at an address and their names, races, ages, and relationships. Answers to these and other questions are confidential. Census workers visit some homes to check on the accuracy of the information. The census helps the government to figure out how the population is aging. Census data are also important in deciding how to distribute government services and funding.

The 2000 Snapshot

One of the findings of the 2000 Census is that the U.S. population is becoming more equally spread out across age groups. By analyzing the data from the census, officials estimate that by 2020 the population of children, middle-aged people, and senior citizens will be about equal. It's predicted also that there will be more people who are over 100 years old than ever before.

Martha F. Riche researches population changes in the United States. She was also a director of the Census Bureau. Riche thinks that the more equal distribution in age will lead to challenges for the nation. How will we meet the demands of more people who are living longer? Will we need to build more hospitals to care for them? Will more children mean a need to build more schools? Federal, state, and local governments will be using the results of the 2000 Census for years to come as they plan our future.

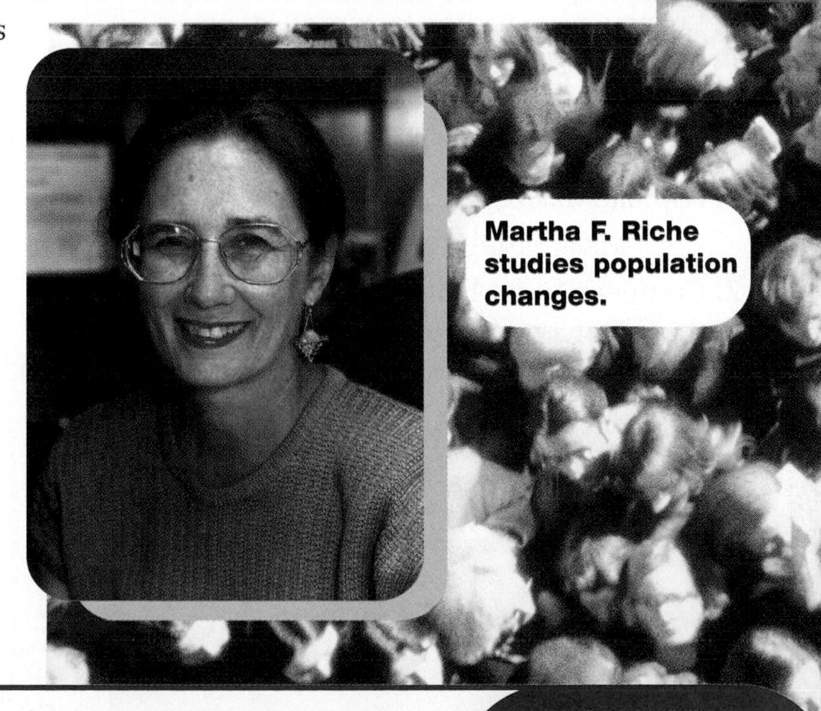

Martha F. Riche studies population changes.

CONNECTIONS Census **Develop a school census. What questions will you ask? (Don't ask questions that are too personal.) Who will ask them? How will you make sure you counted everyone? Using the results, can you make any predictions about your school's future or its current students?**

SCIENCE *Online*
For more information, visit science.glencoe.com

Chapter 12 Study Guide

Reviewing Main Ideas

Section 1 Living Earth

1. Ecology is the study of interactions that take place in the biosphere. *Is ice-covered Antarctica a part of Earth's biosphere? Why or why not?*

2. Populations are made up of all organisms of the same species living in an area.
3. Communities are made up of all the populations of different species of organisms living in one ecosystem.
4. Living and nonliving factors affect an organism's ability to survive in its habitat.

Section 2 Populations

1. Population size can be estimated by counting a sample of a total population.
2. Competition for limiting factors can restrict the size of a population. *What limiting factors might influence the size of a rabbit population?*

3. Population growth is affected by birthrate, death rate, and the movement of individuals into or out of a community.
4. Exponential population growth can occur in environments that provide a species with plenty of food, shelter, and other resources.

Section 3 Interactions Within Communities

1. All life requires energy.
2. Most producers use the Sun's energy to make food in the form of energy-rich molecules. Consumers obtain their food by eating other organisms.
3. Mutualism, commensalism, and parasitism are the three kinds of symbiosis.
4. Every species has its own niche, which includes adaptations for survival. *What adaptations are involved in the relationship between the milkweed plant and the caterpillar of the monarch butterfly?*

After You Read

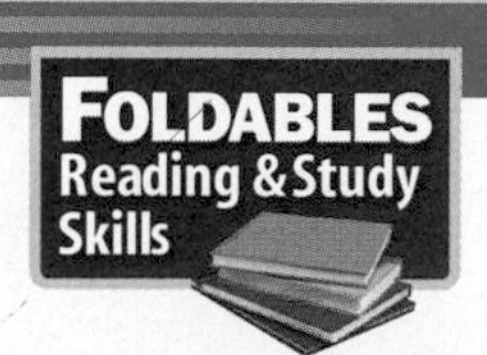

Under the population tab of your Concept Map Study Fold, write what would happen if there were an increase in the population of your animal.

Visualizing Main Ideas

Complete the following concept map on communities.

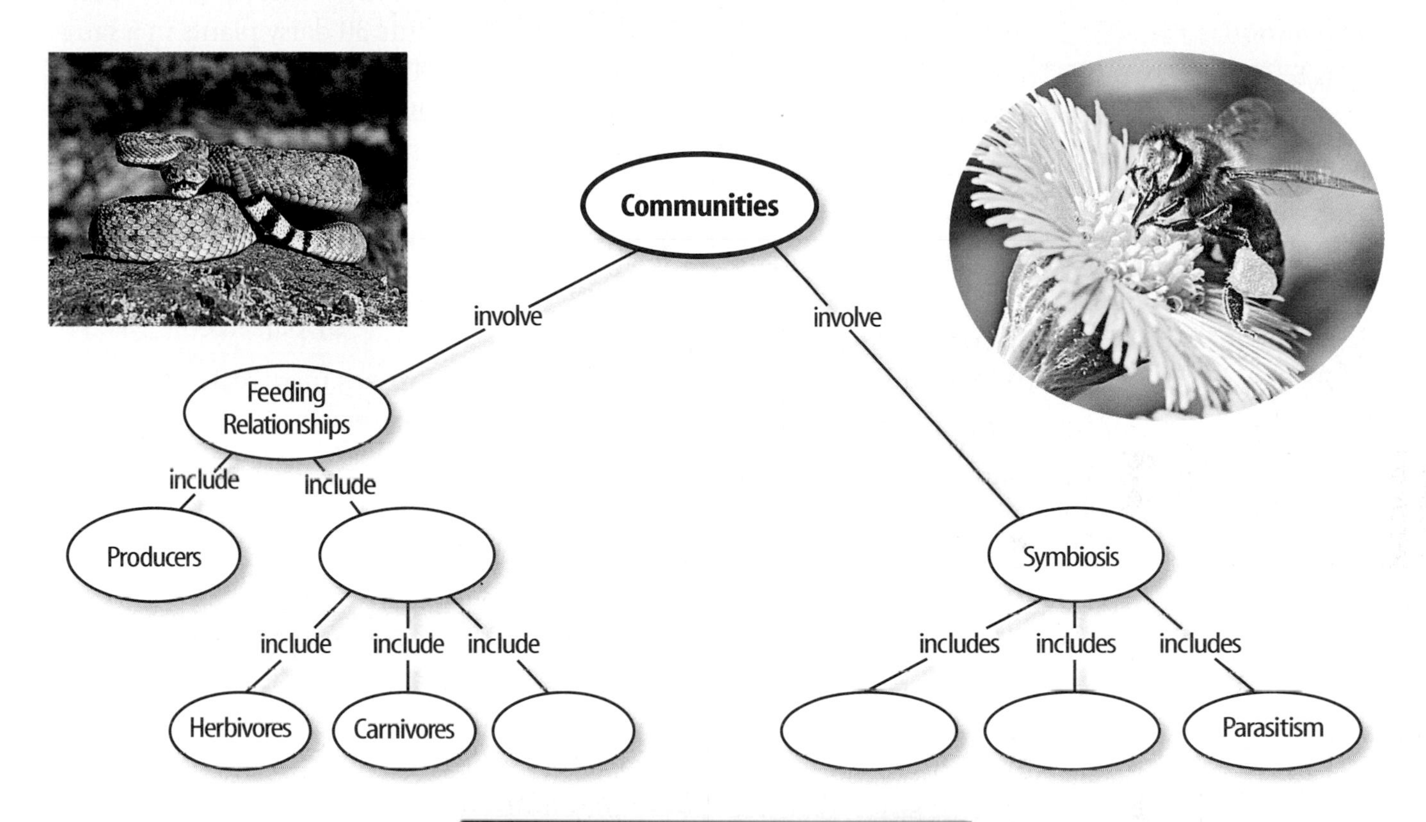

Vocabulary Review

Vocabulary Words

a. biosphere
b. carrying capacity
c. commensalism
d. community
e. consumer
f. ecology
g. ecosystem
h. habitat
i. limiting factor
j. mutualism
k. niche
l. parasitism
m. population
n. producer
o. symbiosis

THE PRINCETON REVIEW Study Tip

Get together with a friend to study. Quiz each other about specific topics from your textbook and class material to prepare for a test.

Using Vocabulary

Explain the difference between the vocabulary words in each of the following sets.

1. niche, habitat
2. mutualism, commensalism
3. limiting factor, carrying capacity
4. biosphere, ecosystem
5. producer, consumer
6. population, ecosystem
7. community, population
8. parasitism, symbiosis
9. ecosystem, ecology
10. parasitism, commensalism

Chapter 12 Assessment

Checking Concepts

Choose the word or phrase that best answers the question.

1. Which of the following is a living factor in the environment?
A) animals
B) air
C) sunlight
D) soil

2. What is made up of all the populations in an area?
A) niches
B) habitats
C) community
D) ecosystem

3. What does the number of individuals in a population that occupies an area of a specific size describe?
A) clumping
B) size
C) spacing
D) density

4. Which of the following animals is an example of an herbivore?
A) wolf
B) moss
C) tree
D) rabbit

5. What term best describes a symbiotic relationship in which one species is helped and the other is harmed?
A) mutualism
B) parasitism
C) commensalism
D) consumerism

6. Which of the following conditions tends to increase the size of a population?
A) births exceed deaths
B) population size exceeds the carrying capacity
C) movements out of an area exceed movements into the area
D) severe drought

7. Which of the following is most likely to be a limiting factor in a population of fish living in the shallow water of a large lake?
A) sunlight
B) water
C) food
D) soil

8. An ecologist wants to know the size of a population of wild daisy plants growing in a meadow. The meadow measures 1,000 m^2. The ecologist counts 30 daisy plants in a sample area that is 100 m^2. What is the estimated population of daisies in the entire meadow?
A) 3
B) 30
C) 300
D) 3,000

9. Which of these organisms is a producer?
A) mole
B) owl
C) whale
D) oak tree

10. Which pair of words is incorrect?
A) black bear—carnivore
B) grasshopper—herbivore
C) pig—omnivore
D) lion—carnivore

Thinking Critically

11. Why does a parasite have a harmful effect on the organism it infects?

12. What factors affect carrying capacity?

13. Describe your own habitat and niche.

14. The female cowbird lays eggs in the nest of another bird. The other birds care for and feed the cowbird chicks when they hatch. Which type of symbiosis is this?

15. Explain how several different niches can exist in the same habitat.

Developing Skills

16. Making Models Place the following organisms in the correct sequence to model a food chain: grass, snake, mouse, and hawk.

17. Predicting Dandelion seeds can float great distances on the wind with the help of white, featherlike attachments. Predict how a dandelion seed's ability to be carried on the wind helps reduce competition among dandelion plants.

18. Classifying Classify the following relationships as parasitism, commensalism, or mutualism: a shark and a remora fish that cleans and eats parasites from the shark's gills; head lice and a human; a spiny sea urchin and a tiny fish that hides from predators by floating among the sea urchin's spines.

19. Comparing and Contrasting Compare and contrast the diets of omnivores and herbivores. Give examples of each.

20. Making and Using Tables Complete the following table.

Types of Symbiosis

Organism A	Organism B	Relationship
Gains	Doesn't gain or lose	
Gains		Mutualism
Gains	Loses	

Performance Assessment

21. Poster Use photographs from old magazines to create a poster that shows at least three different food chains. Display your poster for your classmates.

Technology

Go to the Glencoe Science Web site at **science.glencoe.com** or use the **Glencoe Science CD-ROM** for additional chapter assessment.

Test Practice

A food web shows how organisms in a particular ecosystem depend on each other for food. The food web below shows how the plants and animals in a grassland ecosystem obtain energy from each other.

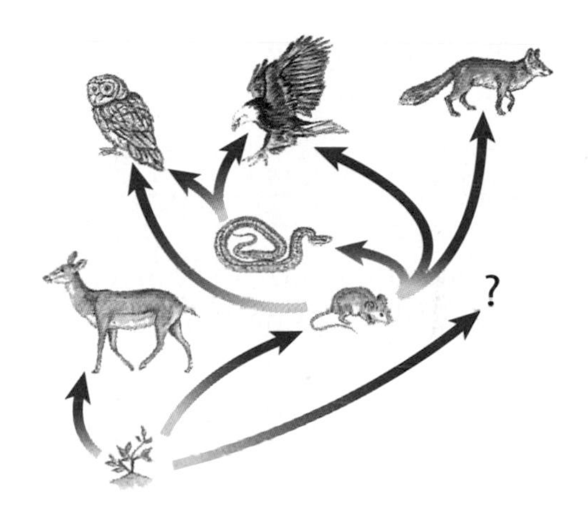

Study the picture and answer the following questions.

1. Other organisms also live in this habitat. Which of the following organisms could fill in the blank space in this food web?
A) tree
B) bison
C) alligator
D) hawk

2. Suppose all the snakes were removed from this ecosystem. Which of the following statements represents the most reasonable prediction of what could happen in this ecosystem?
F) The plants would die.
G) The owls would start eating foxes.
H) There would be no more predators to eat the mice.
J) The eagles would start eating more mice.

CHAPTER

13 The Nonliving Environment

Could you write a story about what would happen if the Sun stopped shining? Most life on Earth depends on the Sun's energy. In this chapter, you'll learn about how organisms called producers use energy to make food and how other organisms called consumers take in that food. You'll also read about cycles in nature such as the water, carbon, and nitrogen cycles, and many other nonliving factors that affect your life.

What do you think?

Science Journal Look at the picture below with a classmate. Discuss what this might be. Here's a hint: *It's a factory that relies on sunlight for its energy supply.* Write your answer or best guess in your Science Journal.

Do you live in a dry, sandy region covered with cactus plants or desert scrub? Is your home in the mountains? Does snow fall during the winter? Perhaps you live near the coast, where flowers bloom year-round. Earth has many ecosystems. In this chapter, you'll learn why the nonliving factors in each ecosystem are different. The following activity will get you started.

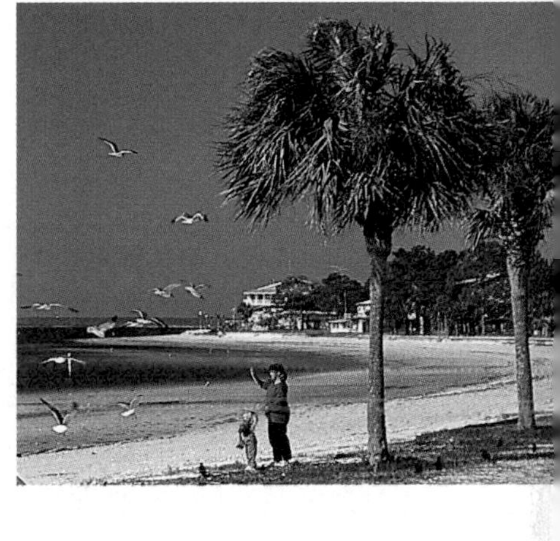

Compare climate differences

1. Locate your city or town on a globe or world map. Find your latitude. Latitude shows your distance from the equator and is expressed in degrees, minutes, and seconds.
2. Locate another city with the same latitude as your city but on a different continent.
3. Locate a third city with latitude close to the equator.
4. Using references, compare average annual precipitation and average high and low temperatures for all three cities.

Observe

In your Science Journal, hypothesize how latitude affects average temperatures and rainfall.

Before You Read

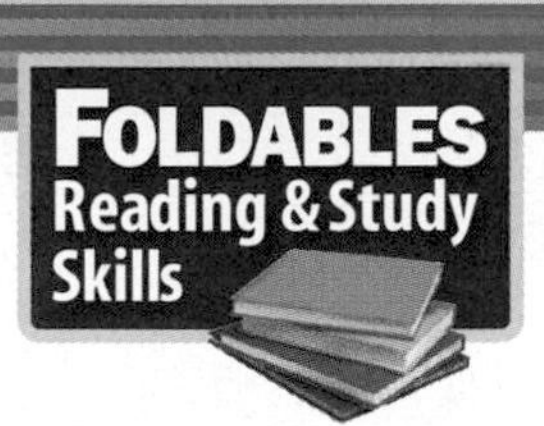

Making a Cause and Effect Study Fold Make the following Foldable to help you understand the cause and effect relationship of the nonliving environment.

Nonliving
Water
Soil
Wind
Temperature
Elevation

1. Place a sheet of paper in front of you so the long side is at the top. Fold the left and right sides in to divide the paper into thirds. Then fold it in half from left to right. Unfold all the folds.
2. Using the fold lines as a guide, refold the paper into a fan. Unfold all the folds again.
3. Before you read the chapter, draw a picture of a familiar ecosystem on one side of the paper. On the other side, label the folds *Nonliving, Water, Soil, Wind, Temperature,* and *Elevation* as shown.
4. As you read the chapter, write on the folds how each nonliving factor affects the environment you drew.

SECTION 1

Abiotic Factors

As You Read

What You'll Learn

- **Identify** common abiotic factors in most ecosystems.
- **List** the components of air that are needed for life.
- **Explain** how climate influences life in an ecosystem.

Vocabulary

biotic
abiotic
atmosphere
soil
climate

Why It's Important

Knowing how organisms depend on the nonliving world can help humans maintain a healthy environment.

Environmental Factors

Living organisms depend on one another for food and shelter. The leaves of plants provide food and a home for grasshoppers, caterpillars, and other insects. Many birds depend on insects for food. Dead plants and animals decay and become part of the soil. The features of the environment that are alive, or were once alive, are called **biotic** (bi AH tihk) factors. The term *biotic* means "living."

Biotic factors are not the only things in an environment that are important to life. Most plants cannot grow without sunlight, air, water, and soil. Animals cannot survive without air, water, or the warmth that sunlight provides. The nonliving, physical features of the environment are called **abiotic** (ay bi AH tihk) factors. The prefix *a* means "not." The term *abiotic* means "not living." Abiotic factors include air, water, soil, sunlight, temperature, and climate. The abiotic factors in an environment often determine which kinds of organisms can live there. For example, water is an important abiotic factor in the environment, as shown in **Figure 1.**

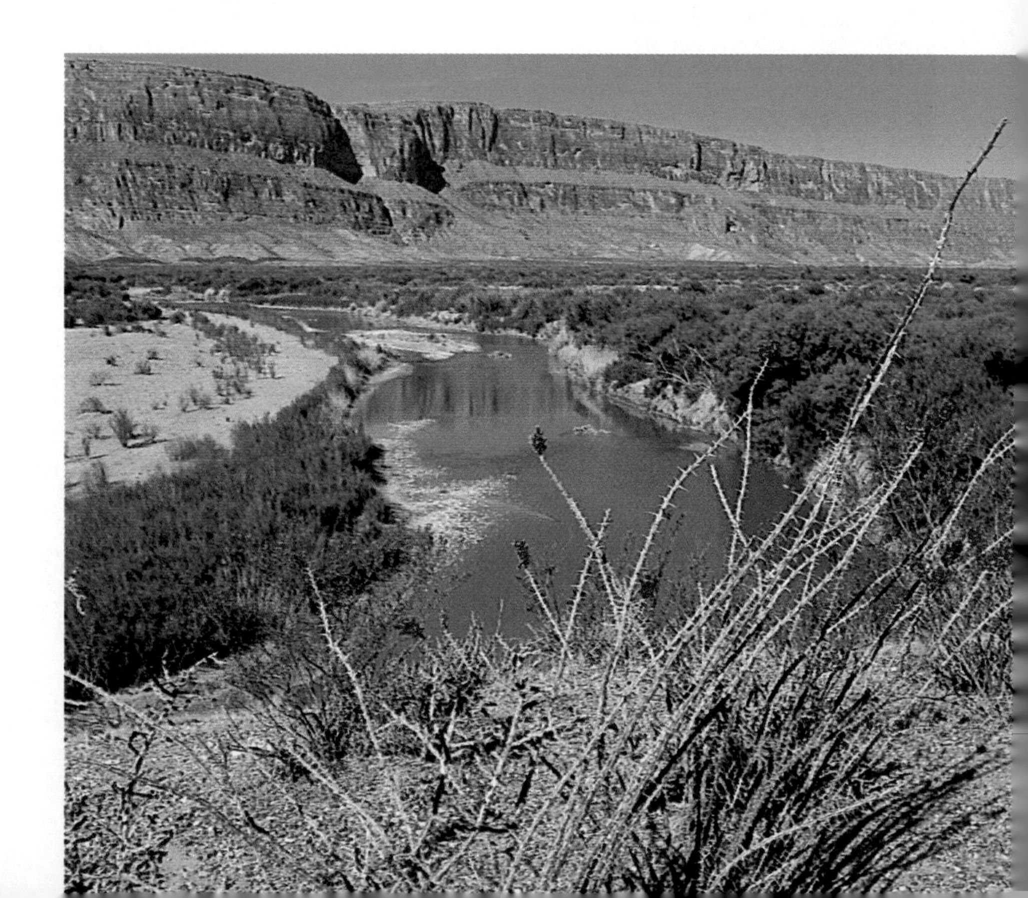

Figure 1
Abiotic factors—air, water, soil, sunlight, temperature, and climate—influence all life on Earth.

Air

Air is invisible and plentiful, so it is easily overlooked as an abiotic factor of the environment. The air that surrounds Earth is called the **atmosphere.** Air contains 78 percent nitrogen, 21 percent oxygen, 0.94 percent argon, 0.03 percent carbon dioxide, and trace amounts of other gases. Some of these gases provide substances that support life.

Carbon dioxide (CO_2) is required for photosynthesis. Photosynthesis—a series of chemical reactions—uses CO_2, water, and energy from sunlight to produce sugar molecules. Organisms like plants that can use photosynthesis are called producers because they produce their own food. During photosynthesis, oxygen is released into the atmosphere.

When a candle burns, oxygen from the air chemically combines with the molecules of candle wax. Chemical energy stored in the wax is converted and released as heat and light energy. In a similar way, cells use oxygen to release the chemical energy stored in sugar molecules. This process is called respiration. Through respiration, cells obtain the energy needed for all life processes. Air-breathing animals aren't the only organisms that need oxygen. Plants, some bacteria, algae, fish, and most other organisms also need oxygen for respiration.

Figure 2
Water is an important abiotic factor in deserts and rain forests.

A Life in deserts is limited to species that can survive for long periods without water.

B Thousands of species can live in lush rain forests where rain falls almost every day.

Water

Water is essential to life on Earth. It is a major ingredient of the fluid inside the cells of all organisms. In fact, most organisms are 50 percent to 95 percent water. Respiration, digestion, photosynthesis, and many other important life processes can take place only in the presence of water. As **Figure 2** shows, environments that have plenty of water usually support a greater diversity of and a larger number of organisms than environments that have little water.

Determining Soil Makeup

Procedure

1. Collect 2 cups of **soil.** Remove large pieces of debris and break up clods.
2. Put the soil in a **quart jar or similar container that has a lid.**
3. Fill the container with **water** and add 1 teaspoon of **dishwashing liquid.**
4. Put the lid on tightly and shake the container.
5. After 1 min, measure and record the depth of sand that settled on the bottom.
6. After 2 h, measure and record the depth of silt that settles on top of the sand.
7. After 24 h, measure and record the depth of the layer between the silt and the floating organic matter.

Analysis

1. Clay particles are so small that they can remain suspended in water. Where is the clay in your sample?
2. Is sand, silt, or clay the greatest part of your soil sample?

Soil

Soil is a mixture of mineral and rock particles, the remains of dead organisms, water, and air. It is the topmost layer of Earth's crust, and it supports plant growth. Soil is formed, in part, of rock that has been broken down into tiny particles.

Soil is considered an abiotic factor because most of it is made up of nonliving rock and mineral particles. However, soil also contains living organisms and the decaying remains of dead organisms. Soil life includes bacteria, fungi, insects, and worms. The decaying matter found in soil is called humus. Soils contain different combinations of sand, clay, and humus. The type of soil present in a region has an important influence on the kinds of plant life that grow there.

Sunlight

All life requires energy, and sunlight is the energy source for almost all life on Earth. During photosynthesis, producers convert light energy into chemical energy that is stored in sugar molecules. Consumers are organisms that cannot make their own food. Energy is passed to consumers when they eat producers or other consumers. As shown in **Figure 3,** photosynthesis cannot take place if light is never available.

A

B

Figure 3
Photosynthesis requires light. A Little sunlight reaches the shady forest floor, so plant growth beneath trees is limited. B Sunlight does not reach into deep lake or ocean waters. Photosynthesis can take place only in shallow water or near the water's surface. *How do fish that live at the bottom of the deep ocean obtain energy?*

Figure 4
Temperature is an abiotic factor that can affect an organism's survival.

A The penguin has a thick layer of fat to hold in heat and keep the bird from freezing. These emperor penguins huddle together for added warmth.

B The Arabian camel stores fat only in its hump. This way, the camel loses heat from other parts of its body, which helps it stay cool in the hot desert.

Temperature

Sunlight supplies life on Earth with light energy for photosynthesis and heat energy for warmth. Most organisms can survive only if their body temperatures stay within the range of 0°C to 50°C. Water freezes at 0°C. The penguins in **Figure 4** are adapted for survival in the freezing Antarctic. Camels can survive the hot temperatures of the Arabian Desert because their bodies are adapted for staying cool. The temperature of a region depends in part on the amount of sunlight it receives. The amount of sunlight depends on the land's latitude and elevation.

What does sunlight provide for life on Earth?

Latitude In this chapter's Explore Activity, you discovered that temperature is affected by latitude. You found that cities located at latitudes farther from the equator tend to have colder temperatures than cities at latitudes nearer to the equator. As **Figure 5** shows, polar regions receive less of the Sun's energy than equatorial regions. Near the equator, sunlight strikes Earth directly. Near the poles, sunlight strikes Earth at an angle, which spreads the energy over a larger area.

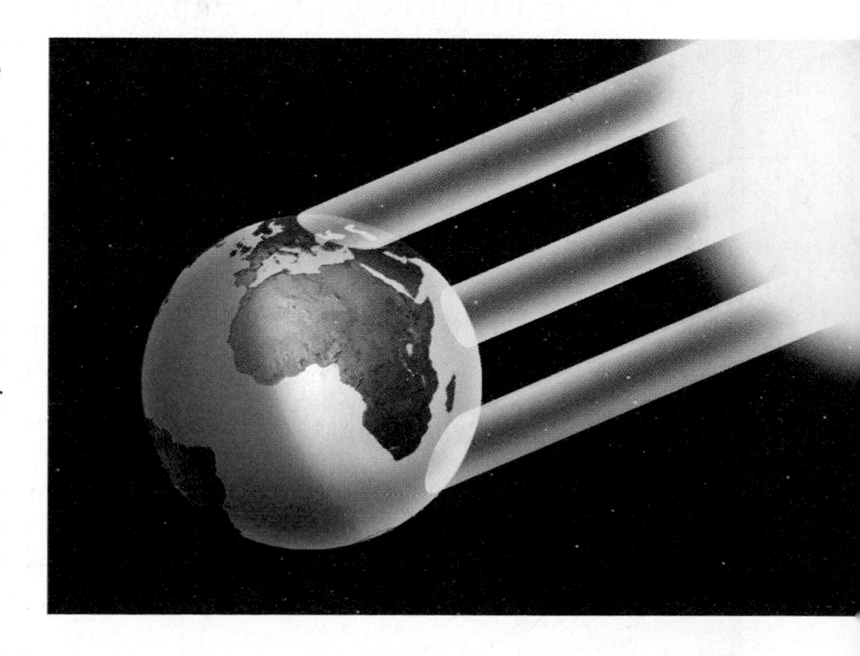

Figure 5
Because Earth is curved, latitudes farther from the equator are colder than latitudes near the equator.

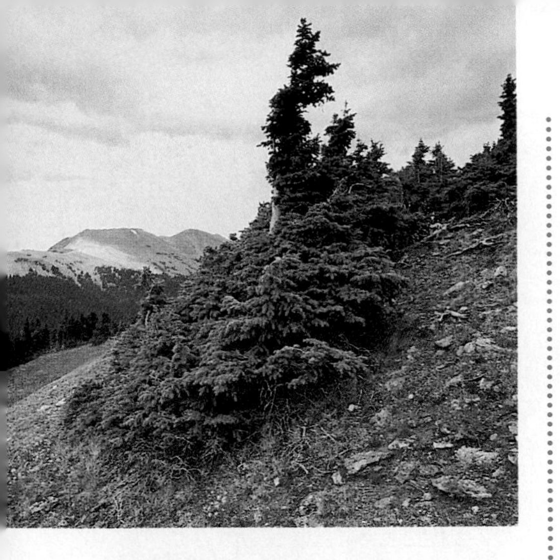

Figure 6
The stunted growth of these trees is a result of abiotic factors.

Elevation If you have climbed or driven up a mountain, you probably noticed that the temperature got cooler as you went higher. A region's elevation, or distance above sea level, affects its temperature. Earth's atmosphere acts as insulation that traps the Sun's heat. At higher elevations, the atmosphere is thinner than it is at lower elevations. Air becomes warmer when sunlight heats the air molecules. Because there are fewer air molecules at higher elevations, air temperatures there tend to be cooler.

Figure 6 shows how elevation affects other abiotic conditions, including soil and wind. At higher elevations, trees are shorter and the ground is rocky. Above the timberline—the elevation beyond which trees do not grow—plant life is limited to low-growing plants. The tops of some mountains are so cold that no plants can survive. Some mountain peaks are covered with snow year-round.

Math Skills Activity

Graphing Temperature Versus Elevation

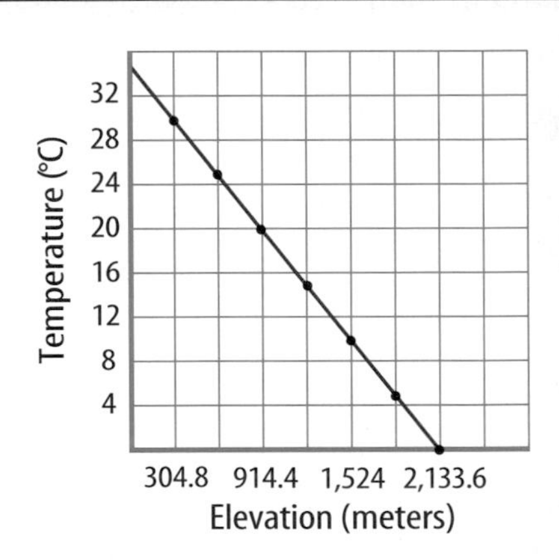

Example Problem

You climb a mountain and record the temperature every 1,000 m of elevation. The temperature is 30°C at 304.8 m, 25°C at 609.6 m, 20°C at 914.4 m, 15°C at 1,219.2 m, and 5°C at 1,828.8 m. Make a graph of the data. Use your graph to predict the temperature at an altitude of 2,133.6 m.

Solution

1 *This is what you know:*
The data can be written as ordered pairs (elevation, temperature). The ordered pairs for these data are (304.8, 30), (609.6, 25), (914.4, 20), (1,219.2, 15), (1,828.8, 5).

2 *This is what you want to find:*
Predict the temperature at an elevation of 2,133.6 m.

3 *This is what you need to do:*
Graph the data by plotting elevation on the x-axis and temperature on the y-axis. Draw a line to connect the data points on your graph.

4 *Predict the temperature at 2,133.6 m:*
Extend the graph line to predict the temperature at 2,133.6 m.

Practice Problem

Temperatures on another mountain are 33°C at sea level, 31°C at 125 m, 29°C at 250 m, and 26°C at 425 m. Graph the data and predict the temperature at 550 m.

For more help, refer to the Math Skill Handbook.

Climate

In Fairbanks, Alaska, winter temperatures may be as low as −52°C, and more than a meter of snow might fall in one month. In Key West, Florida, snow never falls and winter temperatures rarely dip below 5°C. These two cities have different climates. **Climate** refers to an area's average weather conditions over time, including temperature, rainfall or other precipitation, and wind.

For the majority of living things, temperature and precipitation are the two most important components of climate. The average temperature and rainfall in an area influence the type of life found there. Suppose a region has an average temperature of 25°C and receives an average of less than 25 cm of rain every year. It is likely to be the home of cactus plants and other desert life. A region with similar temperatures that receives more than 300 cm of rain every year is probably a tropical rain forest.

Wind Heat energy from the Sun not only determines temperature, but also is responsible for the wind. The air is made up of molecules of gas. As the temperature increases, the molecules spread farther apart. As a result, warm air is lighter than cold air. Colder air sinks below warmer air and pushes it upward, as shown in **Figure 7.** These motions create air currents that are called wind.

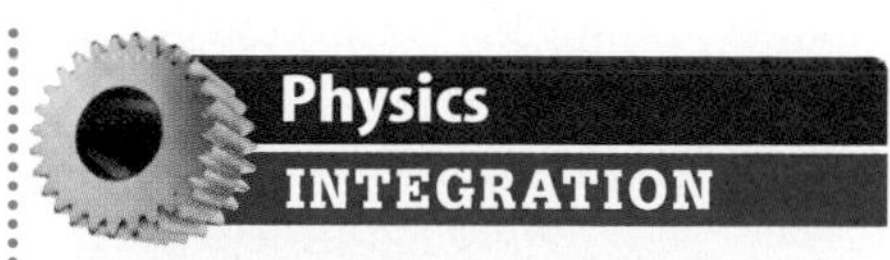

Gravity pulls the gases of the atmosphere toward Earth's surface. Also, the weight of the air at the top of the atmosphere presses down on the air below it. In your Science Journal, explain why air at sea level is thicker than air at the top of a mountain.

Data Update Visit the Glencoe Science Web site at **science.glencoe.com** to look up recent weather data for your area. In your Science Journal, describe how these weather conditions affect plants or animals that live in your area.

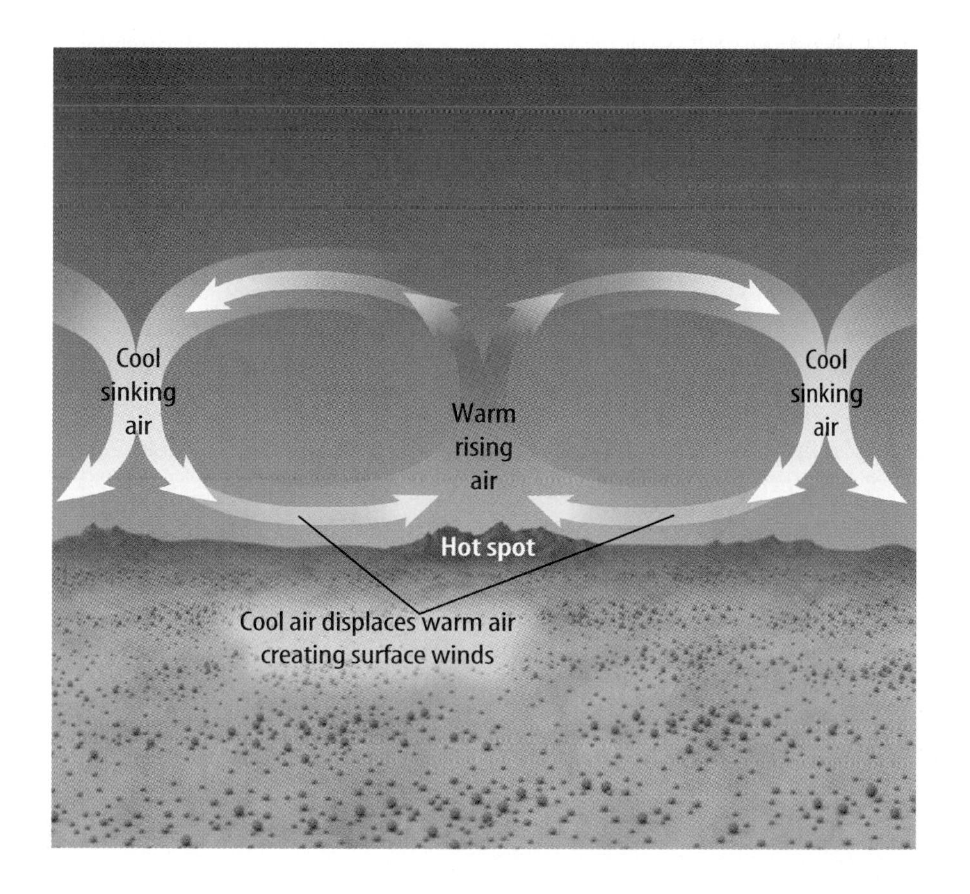

Figure 7
Winds are created when sunlight heats some portions of Earth's surface more than others. In areas that receive more heat, the air becomes warmer. Cold air sinks beneath the warm air, forcing the warm air upward.

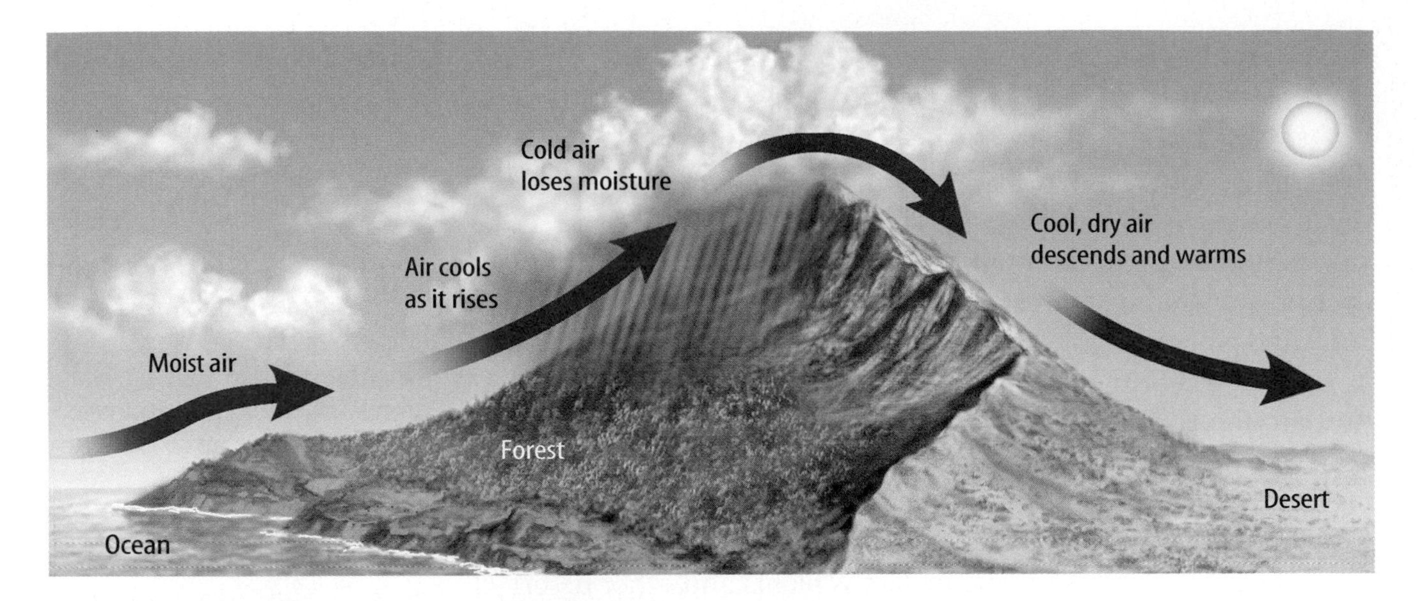

Figure 8
In Washington State, the western side of the Cascade Mountains receives an average of 101 cm of rain each year. The eastern side of the Cascades is in a rain shadow that receives only about 25 cm of rain per year.

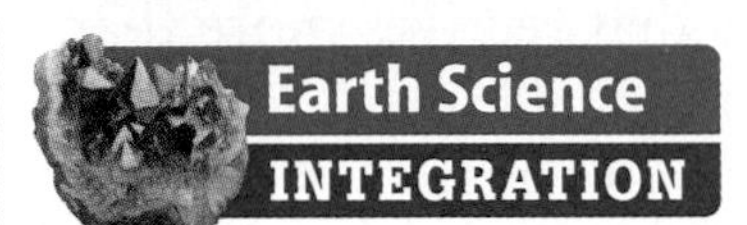

The Rain Shadow Effect The presence of mountains can affect rainfall patterns. As **Figure 8** shows, wind blowing toward one side of a mountain is forced upward by the mountain's shape. As the air nears the top of the mountain, it cools. When air cools, the moisture it contains falls as rain or snow. By the time the cool air crosses over the top of the mountain, it has lost most of its moisture. The other side of the mountain range receives much less precipitation. It is not uncommon to find lush forests on one side of a mountain range and desert on the other side.

Section

Assessment

1. What is the difference between biotic and abiotic factors?
2. What substances in the air are required for life on Earth?
3. Why is soil considered an abiotic factor and a biotic factor?
4. Why is climate an important abiotic factor?
5. **Think Critically** On day 1 of a hiking trip, you walk in shade under tall trees. On day 2, the trees are shorter and farther apart. On day 3, you see small plants but no trees. On day 4, you see snow. What abiotic factors might contribute to these changes?

Skill Builder Activities

6. **Identifying and Manipulating Variables and Controls** Describe an experiment to find out how much water different types of dry soil can hold. **For more help, refer to the Science Skill Handbook.**
7. **Using an Electronic Spreadsheet** Obtain two months of temperature and precipitation data for two cities in your state. Enter the data in a spreadsheet and calculate average daily temperature and rainfall. Use your calculations to compare the two climates. **For more help, refer to the Technology Skill Handbook.**

Activity

Humus Farm

Soil contains abiotic factors, including rock particles and minerals. Soil also contains biotic factors, such as bacteria, molds, fungi, worms, insects, and decayed organisms. The crumbly, dark brown soil found in gardens or forests contains a high percentage of humus. Humus is formed primarily from the decayed remains of plants, animals, and animal droppings. It adds essential nutrients to the soil, including nitrogen. In this activity, you will cultivate your own humus.

What You'll Investigate

How does humus form?

Materials

widemouth jar
soil
grass clippings or green leaves
water
marker
metric ruler
graduated cylinder

Goals

- **Observe** the formation of humus.
- **Observe** biotic factors in the soil.
- **Infer** how humus forms naturally.

Safety Precautions

Wash your hands thoroughly after handling soil, grass clippings, or leaves.

Humus Formation

Date	Observations

Procedure

1. Copy the data table below into your Science Journal.
2. Place 4 cm of soil in the jar. Pour 30 mL of water into the jar to moisten the soil.
3. Place 2 cm of grass clippings or green leaves on top of the soil in the jar.
4. Use a marker to mark the height of the grass clippings or green leaves in the jar.
5. Put the jar in a sunny place. Every other day, add 30 mL of water to it. In your Science Journal, write a prediction of what you think will happen in your jar.
6. **Observe** your jar every other day for four weeks. Record your observations in your data table.

Conclude and Apply

1. **Describe** what happened during your investigation.
2. **Infer** how molds and bacteria help the process of humus formation.
3. **Infer** how humus forms on forest floors or in grasslands.

Communicating Your Data

Compare your humus farm with those of your classmates. With several classmates, write a recipe for creating the richest humus. Ask your teacher to post your recipe in the classroom. **For more help, refer to the Science Skill Handbook.**

SECTION

Cycles in Nature

As You Read

What You'll Learn

- **Explain** the importance of Earth's water cycle.
- **Diagram** the carbon cycle.
- **Recognize** the role of nitrogen in life on Earth.

Vocabulary

evaporation
condensation
water cycle
nitrogen fixation
nitrogen cycle
carbon cycle

Why It's Important

The recycling of matter on Earth demonstrates natural processes.

The Cycles of Matter

Imagine an aquarium tank containing water, fish, snails, plants, algae, and bacteria. The tank is sealed so that only light can enter. Food, water, and air cannot be added. Will the organisms in this environment survive? Through photosynthesis, plants and algae produce their own food. They also supply oxygen to the tank. Fish and snails take in oxygen and eat plants and algae. Wastes from fish and snails fertilize plants and algae. Organisms that die are decomposed by the bacteria. The organisms in this closed environment can survive because the materials are recycled. A constant supply of light energy is the only requirement. Earth's biosphere also contains a fixed amount of water, carbon, nitrogen, oxygen, and other materials required for life. These materials cycle through the environment and are reused by different organisms.

Water Cycle

If you leave a glass of water on a sunny windowsill, the water will disappear. It evaporates. **Evaporation** takes place when liquid water changes into water vapor, which is a gas, and enters the atmosphere, as shown in **Figure 9.** Water evaporates from the surfaces of lakes, streams, puddles, and oceans. Water vapor enters the atmosphere from plant leaves in a process known as transpiration (trans puh RAY shun). Animals release water vapor into the air when they exhale. Water also returns to the environment from animal wastes.

Figure 9
Water vapor is a gas that is present in the atmosphere.

A Water evaporates after a summer rain.

B Water also evaporates from the ocean.

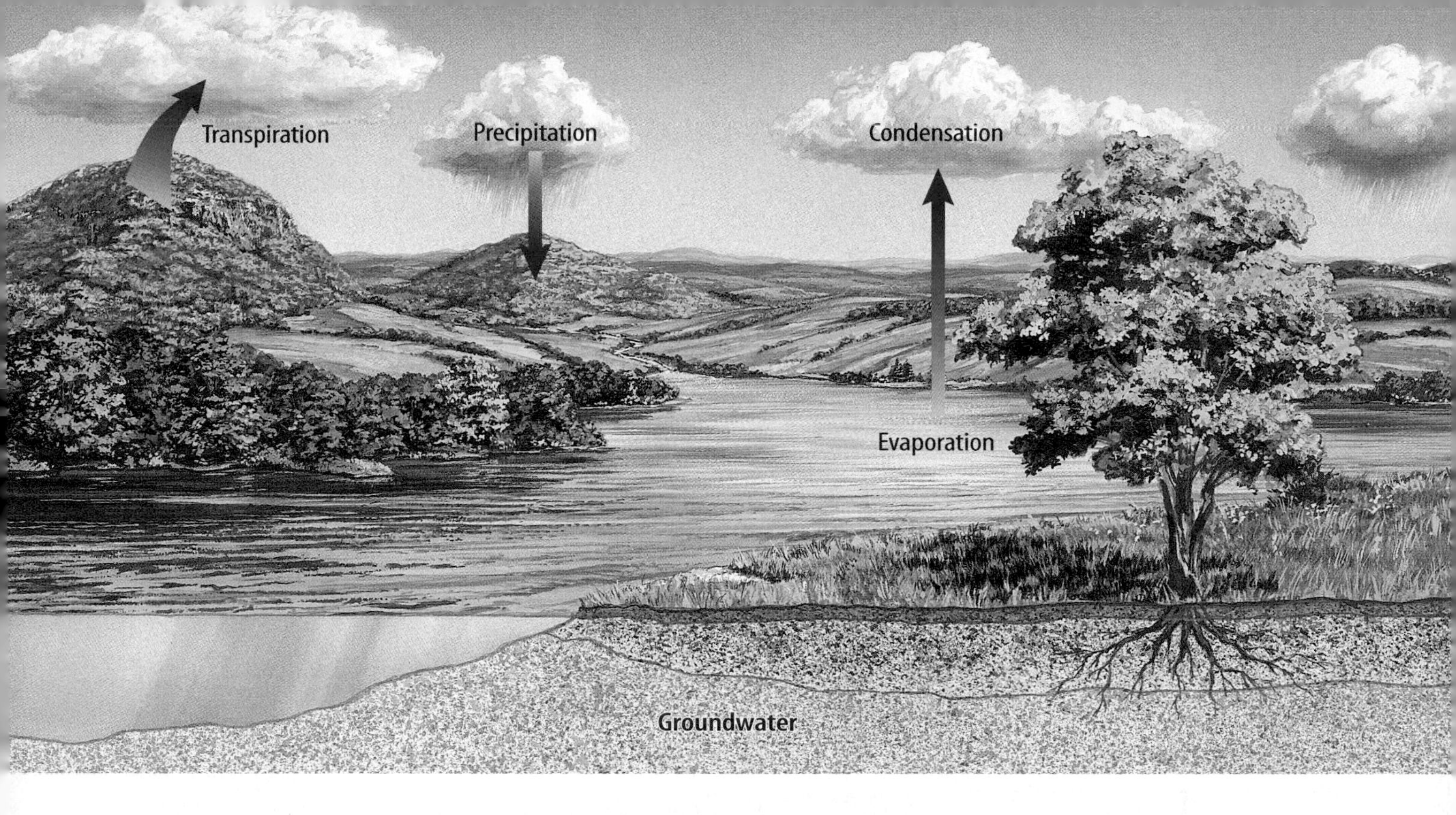

Figure 10
The water cycle involves evaporation, condensation, and precipitation. Water molecules can follow several pathways through the water cycle. *How many water cycle pathways can you identify from this diagram?*

Condensation Water vapor that has been released into the atmosphere eventually comes into contact with colder air. The temperature of the water vapor drops. Over time, the water vapor cools enough to change back into liquid water. The process of changing from a gas to a liquid is called **condensation.** Water vapor condenses on particles of dust in the air, forming tiny droplets. At first, the droplets clump together to form clouds. When they become large and heavy enough, they fall to the ground as rain or other precipitation. As the diagram in **Figure 10** shows, the **water cycle** is a model that describes how water moves from the surface of Earth to the atmosphere and back to the surface again.

Water Use **Table 1** gives data on the amount of water people take from reservoirs, rivers, and lakes for use in households, businesses, agriculture, and power production. These actions can reduce the amount of water that evaporates into the atmosphere. They also can influence how much water returns to the atmosphere by limiting the amount of water available to plants and animals.

Table 1 U.S. Estimated Water Use in 1990

Water Use	Millions of Gallons per Day	Percent of Total
Homes and Businesses	39,100	11.5
Industry and Mining	27,800	8.2
Farms and Ranches	141,000	41.5
Electricity Production	131,800	38.6

Nitrogen Cycle

The element nitrogen is important to all living things. Nitrogen is a necessary ingredient of proteins. Proteins are required for the life processes that take place in the cells of all organisms. Nitrogen is also an essential part of the DNA of all organisms. Although nitrogen is the most plentiful gas in the atmosphere, most organisms cannot use nitrogen directly from the air. Plants need nitrogen that has been combined with other elements to form nitrogen compounds. Through a process called **nitrogen fixation,** some types of soil bacteria can form the nitrogen compounds that plants need. Plants absorb these nitrogen compounds through their roots. Animals obtain the nitrogen they need by eating plants or other animals. When dead organisms decay, the nitrogen in their bodies returns to the soil or to the atmosphere. This transfer of nitrogen from the atmosphere to the soil, to living organisms, and back to the atmosphere is called the **nitrogen cycle,** shown in **Figure 11.**

✔ **Reading Check** *What is nitrogen fixation?*

Figure 11
During the nitrogen cycle, nitrogen gas from the atmosphere is converted to a soil compound that plants can use.

Figure 12
Nitrogen fixation is important to plant growth.

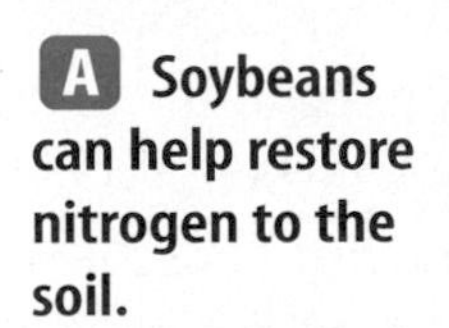

A **Soybeans can help restore nitrogen to the soil.**

B **The swollen nodules on the roots of the soybean plants contain colonies of nitrogen-fixing bacteria.**

C **The bacteria depend on the plant for food. The plant depends on the bacteria to form the nitrogen compounds the plant needs.**

Magnification: 1,000×

Soil Nitrogen Human activities can affect the part of the nitrogen cycle that takes place in the soil. If a farmer grows a crop, such as corn or wheat, most of the plant material is taken away when the crop is harvested. The plants are not left in the field to decay and return their nitrogen compounds to the soil. If these nitrogen compounds are not replaced, the soil could become infertile. You might have noticed that adding fertilizer to soil can make plants grow greener, bushier, or taller. Most fertilizers contain the kinds of nitrogen compounds that plants need for growth. Fertilizers can be used to replace soil nitrogen in crop fields, lawns, and gardens. Compost and animal manure also contain nitrogen compounds that plants can use. They also can be added to soil to improve fertility.

Another method farmers use to replace soil nitrogen is to grow nitrogen-fixing crops. Most nitrogen-fixing bacteria live on or in the roots of certain plants. Some plants, such as peas, clover, and beans including the soybeans shown in **Figure 12,** have roots with swollen nodules that contain nitrogen-fixing bacteria. These bacteria supply nitrogen compounds to the soybean plants and add nitrogen compounds to the soil.

Mini LAB

Comparing Fertilizers

Procedure

1. Examine the three numbers (e.g., 5-10-5) on the **labels of three brands of houseplant fertilizer.** The numbers indicate the percentages of nitrogen, phosphorus, and potassium, respectively, that the product contains.
2. Compare the prices of the three brands of fertilizer.
3. Compare the amount of each brand needed to fertilize a typical houseplant.

Analysis

1. Which brand has the highest percentage of nitrogen?
2. Which brand is the most expensive source of nitrogen? The least expensive?

NATIONAL GEOGRAPHIC

VISUALIZING THE CARBON CYCLE

Figure 13

Carbon—in the form of different kinds of carbon-containing molecules—moves through an endless cycle. The diagram below shows several stages of the carbon cycle. It begins when plants and algae remove carbon from the environment during photosynthesis. This carbon returns to the atmosphere via several carbon-cycle pathways.

Ⓐ Air contains carbon in the form of carbon dioxide gas. Plants and algae use carbon dioxide to make sugars, which are energy-rich, carbon-containing compounds.

Ⓑ Organisms break down sugar molecules made by plants and algae to obtain energy for life and growth. Carbon dioxide is released as a waste.

Ⓒ Burning fossil fuels and wood releases carbon dioxide into the atmosphere.

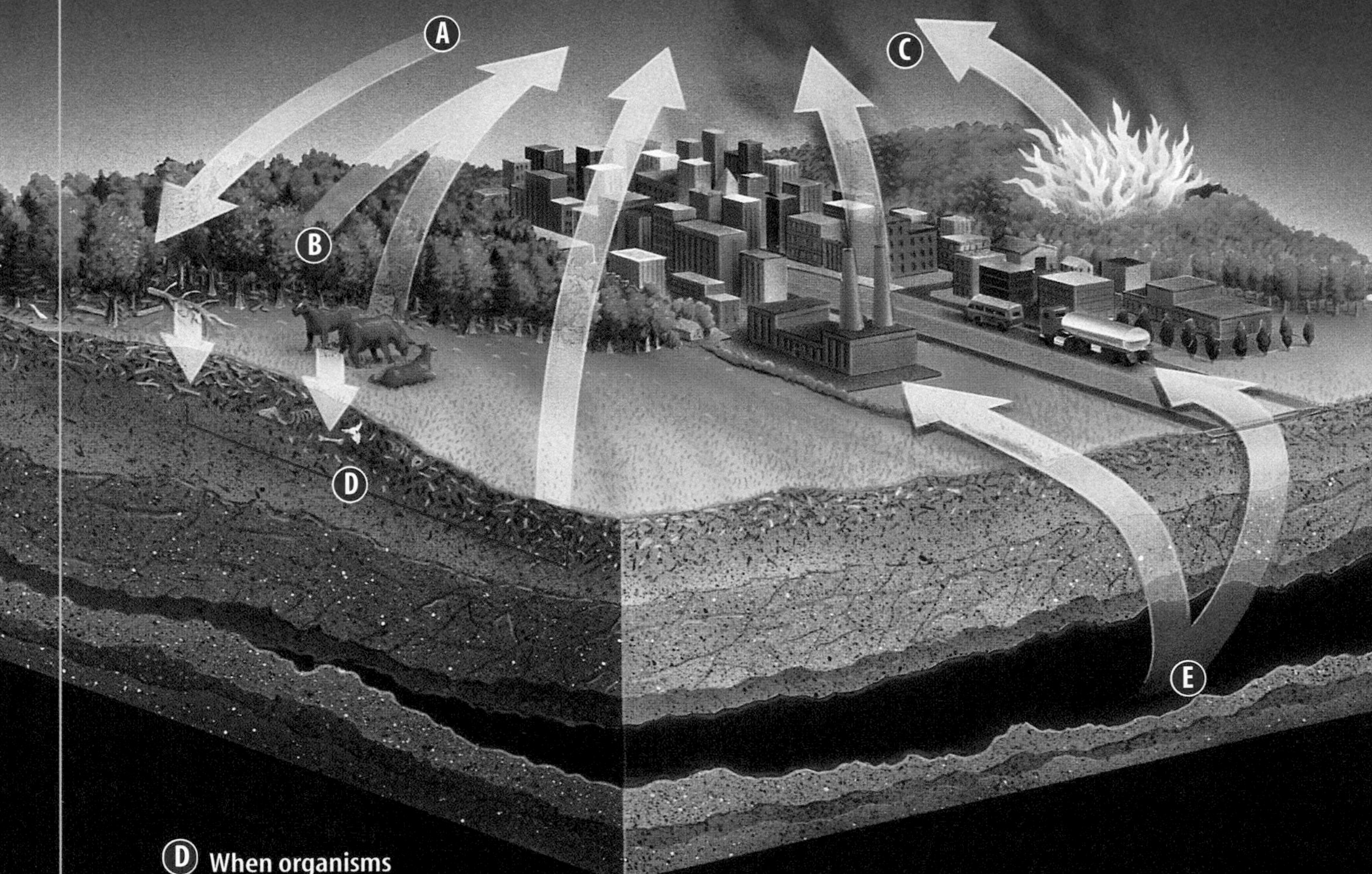

Ⓓ When organisms die, their carbon-containing molecules become part of the soil. The molecules are broken down by fungi, bacteria, and other decomposers. During this decay process, carbon dioxide is released into the air.

Ⓔ Under certain conditions, the remains of some dead organisms may gradually be changed into fossil fuels such as coal, gas, and oil. These carbon compounds are energy rich.

The Carbon Cycle

Carbon atoms are found in the molecules that make up living organisms. Carbon is an important part of soil humus, which is formed when dead organisms decay, and it is found in the atmosphere as carbon dioxide gas (CO_2). The **carbon cycle** describes how carbon molecules move between the living and nonliving world, as shown in **Figure 13.**

The carbon cycle begins when producers remove CO_2 from the air during photosynthesis. They use CO_2, water, and sunlight to produce energy-rich sugar molecules. Energy is released from these molecules during respiration—the chemical process that provides energy for cells. Respiration uses oxygen and releases CO_2. Photosynthesis uses CO_2 and releases oxygen. These two processes help recycle carbon on Earth.

✔ Reading Check *How does carbon dioxide enter the atmosphere?*

Human activities also release CO_2 into the atmosphere. Fossil fuels such as gasoline, coal, and heating oil are the remains of organisms that lived millions of years ago. These fuels are made of energy-rich, carbon-based molecules. When people burn these fuels, CO_2 is released into the atmosphere as a waste product. People also use wood for building and for fuel. Trees that are harvested for these purposes no longer remove CO_2 from the atmosphere during photosynthesis. The amount of CO_2 in the atmosphere is increasing. Extra CO_2 could trap more heat from the Sun and cause average temperatures on Earth to rise.

SCIENCE *Online*

Research Visit the Glencoe Science Web site at **science.glencoe.com** for the chemical equations that describe photosynthesis and respiration. In your Science Journal, write these equations and use them to explain how respiration is the reverse of photosynthesis.

Section  Assessment

1. Describe the water cycle.
2. Explain how respiration can be considered the reverse of photosynthesis.
3. How might burning fossil fuels affect the composition of gases in the atmosphere?
4. Why do plants, animals, and other organisms need nitrogen?
5. **Think Critically** Most chemical fertilizers contain nitrogen, phosphorus, and potassium. Why don't they contain carbon? How do plants obtain carbon?

Skill Builder Activities

6. **Identifying and Manipulating Variables and Controls** Describe an experiment that would determine whether extra carbon dioxide enhances the growth of tomato plants. **For more help, refer to the Science Skill Handbook.**
7. **Communicating** Pretend you are a carbon molecule. Write a fictional account of your travels from the atmosphere, through at least two organisms, and back to the atmosphere. **For more help, refer to the Science Skill Handbook.**

SECTION

Energy Flow

As You Read

What You'll Learn

- **Explain** how organisms produce energy-rich compounds.
- **Describe** how energy flows through ecosystems.
- **Recognize** how much energy is available at different levels in a food chain.

Vocabulary

chemosynthesis
energy pyramid
food web

Why It's Important

All living things, including people, need a constant supply of energy.

Converting Energy

All living things are made of matter, and all living things need energy. Matter and energy move through the natural world in different ways. Matter can be recycled over and over again. The recycling of matter requires energy. Energy is not recycled, but it is converted from one form to another. The conversion of energy is important to all life on Earth.

Photosynthesis During photosynthesis, producers convert light energy into the chemical energy in sugar molecules. Some of these sugar molecules are broken down as energy is needed. Others are used to build complex carbohydrate molecules that become part of the producer's body. Fats and proteins also contain stored energy.

Chemosynthesis Not all producers rely on light for energy. During the 1970s, scientists exploring the ocean floor were amazed to find communities teeming with life. These communities were at a depth of almost 3.2 km and living in total darkness. They were found near powerful hydrothermal vents like the one shown in **Figure 14.**

Figure 14
A Chemicals in the water that flows from hydrothermal vents provide bacteria with a source of energy. B The bacterial producers use this energy to make nutrients through the process of chemosynthesis. Consumers, such as tubeworms, feed on the bacteria.

Magnification: 38,000×

Hydrothermal Vents A hydrothermal vent is a deep crack in the ocean floor through which the heat of molten magma can escape. The water from hydrothermal vents is extremely hot from contact with molten rock that lies deep in Earth's crust.

Because no sunlight reaches these deep ocean regions, plants or algae cannot grow there. How do the organisms living in this community obtain energy? Scientists learned that the hot water contains nutrients such as sulfur molecules that bacteria use to produce their own food. The production of energy-rich nutrient molecules from chemicals is called **chemosynthesis** (kee moh SIN thuh sus). Consumers living in the hydrothermal vent communities rely on chemosynthetic bacteria for nutrients and energy. Chemosynthesis and photosynthesis allow producers to make their own energy-rich molecules.

✔ **Reading Check** *What is chemosynthesis?*

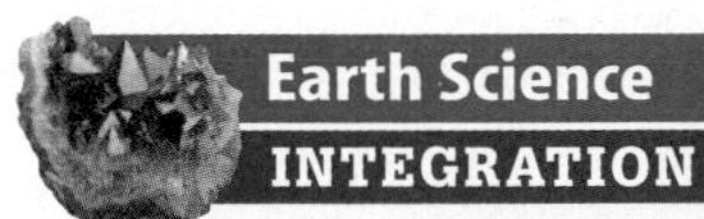

The first hydrothermal vent community discovered was found along the Galápagos rift zone. A rift zone forms where two plates of Earth's crust are spreading apart. In your Science Journal, describe the energy source that heats the water in the hydrothermal vents of the Galápagos rift zone.

Energy Transfer

Energy can be converted from one form to another. It also can be transferred from one organism to another. Consumers cannot make their own food. Instead, they obtain energy by eating producers or other consumers. This way, energy stored in the molecules of one organism is transferred to another organism. At the same time, the matter that makes up those molecules is transferred from one organism to another. Throughout nature, energy and matter move from organism to organism when one organism becomes food for another organism.

Food Chains A food chain is a way of showing how matter and energy pass from one organism to another. Producers—plants, algae, and other organisms that are capable of photosynthesis or chemosynthesis—are always the first step in a food chain. Animals that consume producers such as herbivores are the second step. Carnivores and omnivores—animals that eat other consumers—are the third and higher steps of food chains. One example of a food chain is shown in **Figure 15.**

Figure 15
In this food chain, grasses are producers, marmots are herbivores that eat the grasses, and grizzly bears are consumers that eat marmots. The arrows show the direction in which matter and energy flow.

Food Webs A forest community includes many feeding relationships. These relationships can be too complex to show with a food chain. For example, grizzly bears eat many different organisms, including berries, insects, chipmunks, and fish. Berries are eaten by bears, birds, insects, and other animals. A bear carcass might be eaten by wolves, birds, or insects. A **food web** is a model that shows all the possible feeding relationships among the organisms in a community. A food web is made up of many different food chains, as shown in **Figure 16.**

Energy Pyramids

Food chains usually have at least three links, but rarely more than five. This limit exists because the amount of available energy is reduced as you move from one level to the next in a food chain. Imagine a grass plant that absorbs energy from the Sun. The plant uses some of this energy to grow and produce seeds. Some of the energy is stored in the seeds.

Figure 16
Compared to a food chain, a food web provides a more complete model of the feeding relationships in a community.

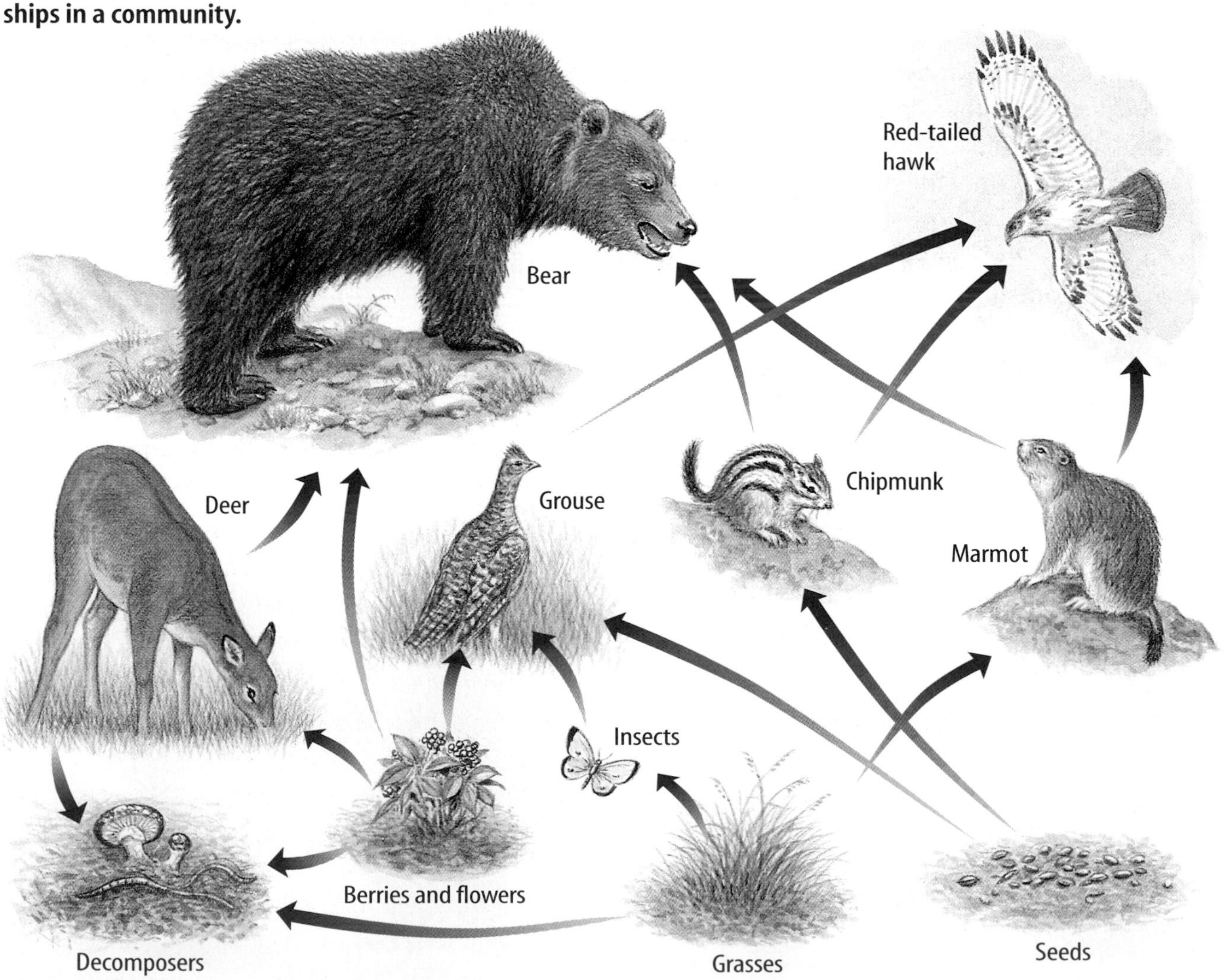

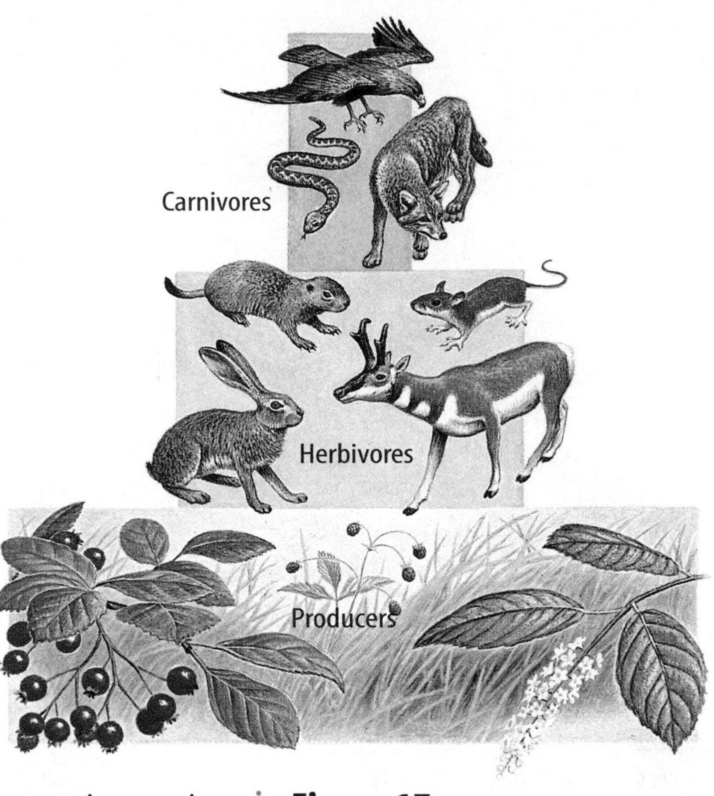

Available Energy When a mouse eats grass seeds, energy stored in the seeds is transferred to the mouse. However, most of the energy the plant absorbed from the Sun was used for the plant's growth. Much less energy is stored in the seeds eaten by the mouse. The mouse uses much of the energy remaining in the seeds for its own life processes, including respiration, digestion, and growth. A hawk that eats the mouse obtains even less energy.

The same thing happens at every feeding level of a food chain. The amount of available energy is reduced from one feeding level to another. An **energy pyramid,** like the one in **Figure 17,** shows the amount of energy available at each feeding level in an ecosystem. The bottom layer of the pyramid, which represents all of the producers, is the first feeding level. It is the largest level because it contains the most energy and the largest number of organisms. As you move up the pyramid, each level becomes smaller. Only about ten percent of the energy available at each feeding level of an energy pyramid is transferred to the next higher level.

Figure 17
This energy pyramid shows that each feeding level contains less energy than the level below it. *What would happen if the hawks and snakes outnumbered the rabbits and mice in this ecosystem?*

✔ Reading Check *Why does the first feeding level of an energy pyramid contain the most energy?*

Section  Assessment

1. Compare and contrast photosynthesis and chemosynthesis.
2. Explain how your three favorite foods provide you with energy from the Sun.
3. What is the difference between a food web and an energy pyramid?
4. Why is there a limit to the number of links in a food chain?
5. **Think Critically** Use your knowledge of food chains and the energy pyramid to explain why the number of mice in a grassland ecosystem is greater than the number of hawks.

Skill Builder Activities

6. **Classifying** Classify each species as photosynthetic or chemosynthetic: *Red hattus* uses red light to make its food; *Selen dion* makes food if the element selenium is present. **For more help, refer to the Science Skill Handbook.**
7. **Solving One-Step Equations** A forest has 24,055,000 kilocalories (kcals) of producers, 2,515,000 kcals of herbivores, and 235,000 kcals of carnivores. How much energy is lost between producers and herbivores? Between herbivores and carnivores? **For more help, refer to the Math Skill Handbook.**

Activity

Where does the mass of a plant come from?

An enormous oak tree starts out as a tiny acorn. The acorn sprouts in dark, moist soil. Roots grow down through the soil. Its stem and leaves grow up toward the light and air. Year after year, the tree grows taller, its trunk grows thicker, and its roots grow deeper. It becomes a towering oak that produces thousands of acorns of its own. An oak tree has much more mass than an acorn. Where does this mass come from? The soil? The air? In this activity, you'll find out by conducting an experiment with radish plants.

What You'll Investigate

Does all of the matter in a radish plant come from the soil?

Goals

- **Measure** the mass of soil before and after radish plants have been grown in it.
- **Measure** the mass of radish plants grown in the soil.
- **Analyze** the data to determine whether the mass gained by the plants equals the mass lost by the soil.

Materials

8-oz plastic or paper cup
potting soil to fill cup
scale or balance
radish seeds (4)
water
paper towels

Safety Precautions 

Procedure

1. Copy the data table into your Science Journal.
2. Fill the cup with dry soil.
3. Find the mass of the cup of soil and record this value in your data table.
4. Moisten the soil in the cup. Plant four radish seeds 2 cm deep in the soil. Space the seeds an equal distance apart. Wash your hands.
5. Add water to keep the soil barely moist as the seeds sprout and grow.
6. When the plants have developed four to six true leaves, usually after two to three weeks, carefully remove the plants from the soil. Gently brush the soil off the roots. Make sure all the soil remains in the cup.
7. Spread the plants out on a paper towel. Place the plants and the cup of soil in a warm area to dry out.
8. When the plants are dry, measure their mass and record this value in your data table. Write this number with a plus sign in the Gain or Loss column.
9. When the soil is dry, find the mass of the cup of soil. Record this value in your data table. Subtract the End mass from the Start mass and record this number with a minus sign in the Gain or Loss column.

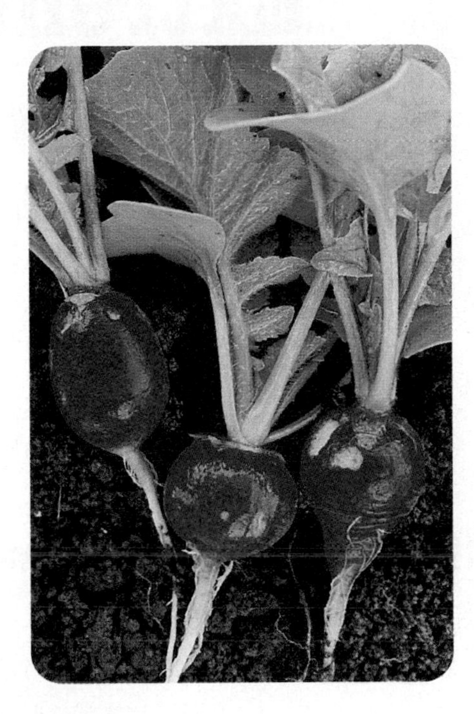

Mass of Soil and Radish Plants

	Start	End	Gain (+) or Loss (−)
Mass of dry soil and cup			
Mass of dried radish plants	0 g		

Conclude and Apply

1. In the early 1600s, a Belgian scientist named J. B. van Helmont conducted this experiment with a willow tree. What is the advantage of using radishes instead of a tree?
2. How much mass was gained or lost by the soil? By the radish plants?
3. Did the mass of the plants come completely from the soil? How do you know?
4. If all of the mass gained by the plants did not come from the soil, where could it have come from?

Compare your conclusions with those of other students in your class. **For more help, refer to the Science Skill Handbook.**

Science Stats

Extreme Climates

Did you know...

... The greatest snowfall in one year occurred at Mount Baker in Washington State. Approximately 2,896 cm of snow fell on Mount Baker during the 1998-99, 12-month snowfall season. That's enough snow to bury an eight-story building.

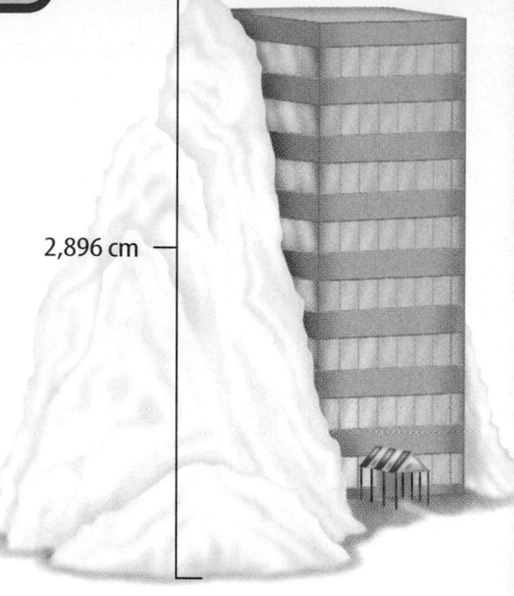

... The hottest climate in the United States is found in Death Valley, California. In July 1913, Death Valley reached approximately 57°C. This is the hottest officially recognized temperature on Earth. As a comparison, a comfortable room temperature is about 20°C.

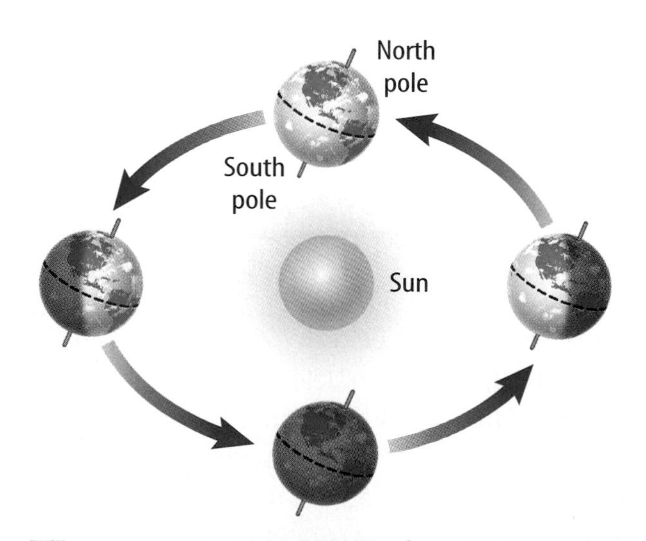

... The record for the lowest temperature was set in Antarctica in 1983. The temperature was a frigid −89°C. As a comparison, the temperature of your freezer at home is about −15°C.

... The south pole receives sunshine for less than 50 percent of the days in a year. Because Earth is tilted, the south pole is pointed away from the Sun for about half the year and receives very little during that time.

... The fastest tornado winds have been measured at a speed of about 512 km/h. That's faster than the blades of some helicopters, which can rotate at about 450 km/h.

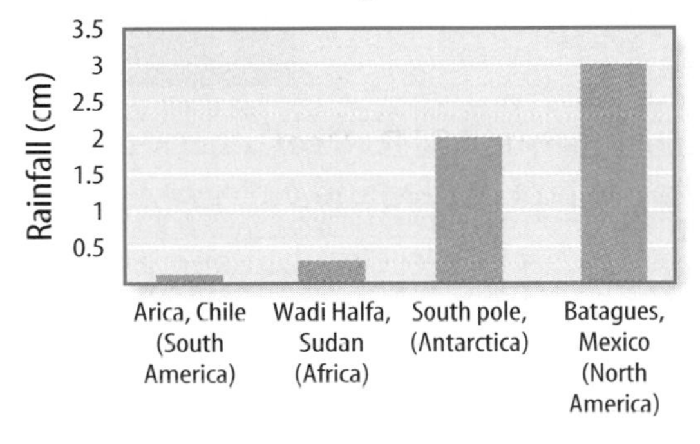

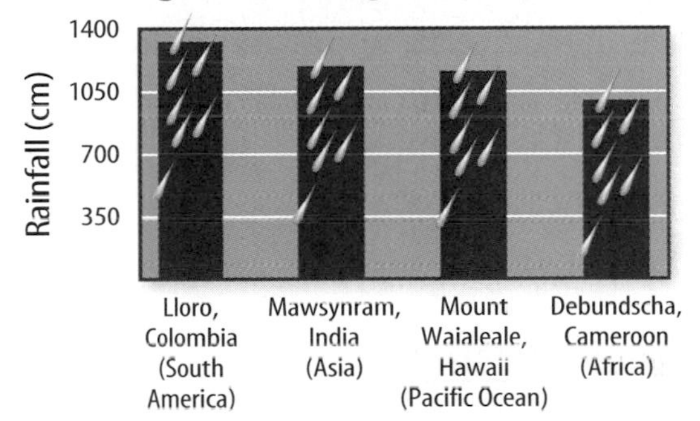

Do the Math

1. Look at the graph above. How many years of average south pole precipitation would it take to equal a single year of average precipitation in Lloro, Colombia?
2. What is the difference in degrees Celsius between the world record low temperature and the world record high temperature?
3. What was the average monthly snowfall at Mount Baker during the 1998-99 snowfall season?

Go Further

Go to **science.glencoe.com** and find out the average monthly rainfall in a tropical rain forest. Make a line graph to show how the amount of precipitation changes during the 12 months of the year.

Reviewing Main Ideas

Section 1 Abiotic Factors

1. Abiotic factors include air, water, soil, sunlight, temperature, and climate. *What abiotic factors are required for this squirrel's survival? Explain.*
2. The availability of water and light influences where life exists on Earth.
3. Soil and climate have an important influence on the types of organisms that can survive in different environments.
4. High latitudes and elevations generally have lower average temperatures.

Section 2 Cycles in Nature

1. Matter is limited on Earth and is recycled through the environment. *How do green plants help recycle oxygen?*

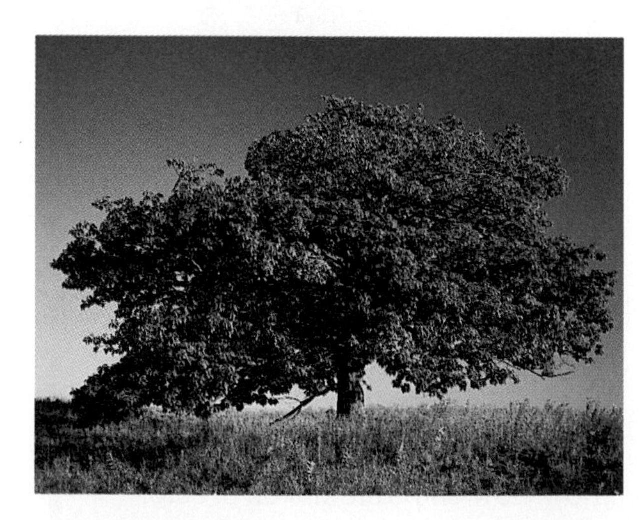

2. The water cycle involves evaporation, condensation, and precipitation.
3. The carbon cycle involves photosynthesis and respiration.
4. Nitrogen in the form of soil compounds enters plants, which are then consumed by other organisms.

Section 3 Energy Flow

1. Producers make energy-rich molecules through photosynthesis or chemosynthesis. *How do seaweeds in shallow water obtain energy?*
2. When organisms feed on other organisms, they obtain matter and energy.
3. Matter can be recycled, but energy cannot.
4. Food webs are models of the complex feeding relationships in communities.
5. Available energy decreases as you go to higher feeding levels in an energy pyramid. *What happens to most of the energy in an apple that you eat?*

After You Read

FOLDABLES Reading & Study Skills

Find a student who drew a different ecosystem on his or her Cause and Effect Study Fold. Then, compare and contrast the information on your two Foldables.

Visualizing Main Ideas

This diagram shows photosynthesis in a leaf. Fill in the blank lines with the terms light, carbon dioxide, *and* oxygen.

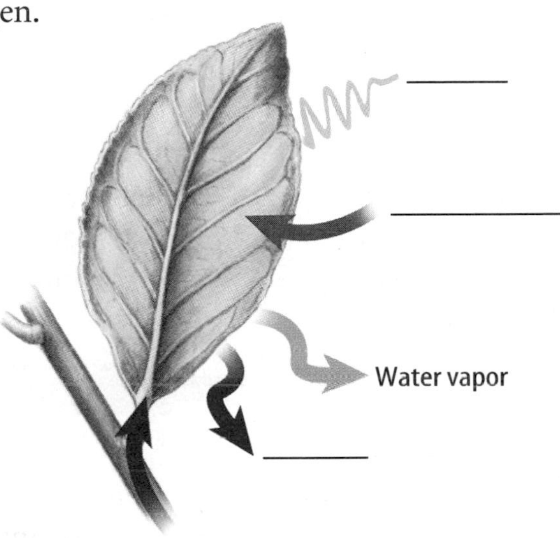

Vocabulary Review

Vocabulary Words

a. abiotic
b. atmosphere
c. biotic
d. carbon cycle
e. chemosynthesis
f. climate
g. condensation
h. energy pyramid
i. evaporation
j. food web
k. nitrogen cycle
l. nitrogen fixation
m. soil
n. water cycle

Using Vocabulary

Which vocabulary word best corresponds to each of the following events?

1. A liquid changes to a gas.
2. Some types of bacteria form nitrogen compounds in the soil.
3. Decaying plants add nitrogen to the soil.
4. Chemical energy is used to make energy-rich molecules.
5. Decaying plants add carbon to the soil.
6. A gas changes to a liquid.
7. Water flows downhill into a stream. The stream flows into a lake, and water evaporates from the lake.
8. Burning coal and exhaust from automobiles release carbon into the air.

Study Tip

Write out the full questions and answers to end-of-chapter quizzes, not just the answers. This will help you form complete responses to important questions.

Chapter 13 Assessment

Checking Concepts

Choose the word or phrase that best answers the question.

1. Which of the following is an abiotic factor?
   A) penguins **C)** soil bacteria
   B) rain **D)** redwood trees

2. Which group makes up the largest level of an energy pyramid?
   A) herbivores **C)** decomposers
   B) producers **D)** carnivores

3. You climb up the western slope of the Cascade Mountains and down the eastern side. Which of the following weather changes do you observe?
   A) Warm and wet changes to cold and wet, then cold and dry, then warm and dry.
   B) Cold and wet changes to warm and wet, then warm and dry, then cold and dry.
   C) Warm and wet changes to cold and wet, then cold and dry, then warm and wet.
   D) Warm and dry changes to cold and dry, then warm and dry, then cold and dry.

4. Which of the following applies to latitudes farther from the equator?
   A) higher elevations
   B) higher temperatures
   C) higher precipitation levels
   D) lower temperatures

5. Water vapor forming droplets that form clouds directly involves which process?
   A) condensation **C)** evaporation
   B) respiration **D)** transpiration

6. Which one of the following components of air is least necessary for life on Earth?
   A) argon **C)** carbon dioxide
   B) nitrogen **D)** oxygen

7. What do plants make that requires nitrogen?
   A) sugars **C)** fats
   B) proteins **D)** carbohydrates

8. Which of the following processes removes carbon dioxide from the air?
   A) condensation **C)** burning
   B) photosynthesis **D)** respiration

9. Earth receives a constant supply of which of the following items?
   A) light energy **C)** nitrogen
   B) carbon **D)** water

10. Which of these is an energy source for chemosynthesis?
    A) sunlight **C)** sulfur molecules
    B) moonlight **D)** carnivores

Thinking Critically

11. A country has many starving people. Should they grow vegetables and corn to eat, or should they grow corn to feed cattle so they can eat beef? Explain.

12. Why is a food web a better model than a food chain?

13. Do bacteria need nitrogen? Why or why not?

14. It is often easier to walk through an old, mature forest of tall trees than through a young forest that is full of small trees. Why?

15. The Inyo Mountains are located in central California. Explain why giant sequoia trees grow on the west side of the mountains and Death Valley, a desert, is on the east side.

Developing Skills

16. **Classifying** Classify each of the following environmental concerns according to the cycle it affects—carbon, nitrogen, or water.
    a. algal blooms caused by excess fertilizer
    b. acid rain damage to pine trees
    c. the unnatural warming of Earth

17. Recognizing Cause and Effect A lake in Kenya has been taken over by a floating weed. What could you do to determine if nitrogen fertilizer runoff from farms is causing the problem?

18. Making and Using Graphs Abiotic factors, such as climate, cause populations to move from place to place. Make a bar graph of the following migration distances.

Mighty Migrators

Species	Distance (km)
Desert locust	4,800
Caribou	800
Green turtle	1,900
Arctic tern	35,000
Gray whale	19,000

19. Forming Hypotheses For each hectare of land, ecologists found 10,000 kcals of producers, 10,000 kcals of herbivores, and 2,000 kcals of carnivores. Suggest a reason why producer and herbivore levels are equal.

20. Concept Mapping Draw a food web of these organisms: *caterpillars and rabbits eat grasses, raccoons eat rabbits and mice, mice eat grass seeds,* and *birds eat caterpillars.*

Performance Assessment

21. Poster Use magazine photographs to make a visual representation of the water cycle.

TECHNOLOGY

Go to the Glencoe Science Web site at **science.glencoe.com** or use the **Glencoe Science CD-ROM** for additional chapter assessment.

Test Practice

Food chains model how energy is transferred from one organism to another organism in the environment. The diagram below shows a food chain that includes aquatic plants and animals and a land animal.

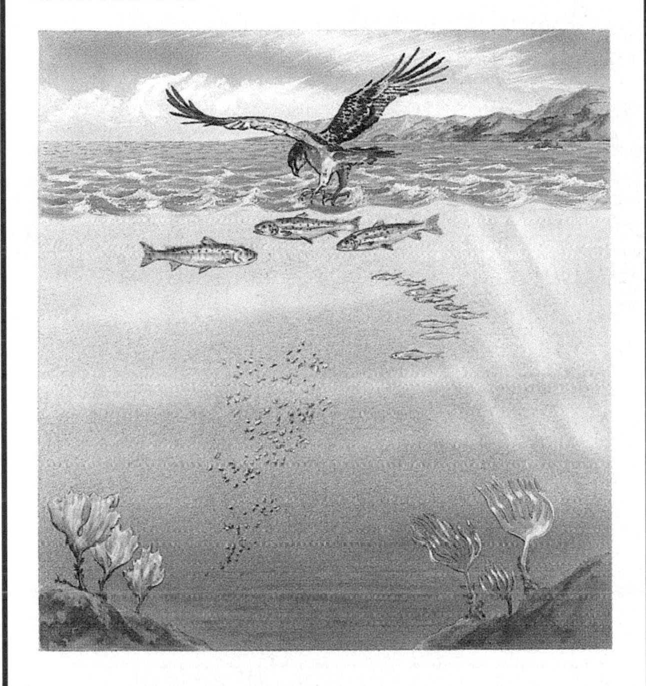

Study the diagram and answer the following questions.

1. Which of the following statements is true based on the order of the food chain shown above?

A) Algae eat plankton.
B) Plankton eat salmon.
C) Herring eat plankton.
D) Salmon eat algae.

2. If the supply of salmon were suddenly depleted, the numbers of which of the following might also be depleted?

F) algae
G) plankton
H) herring
J) eagles

Standardized Test Practice

Reading Comprehension

Read the passage. Then read each question that follows the passage. Decide which is the best answer to each question.

Interactions in Ecosystems

Fearing for their safety and the safety of their livestock, early settlers of northern Wisconsin killed the native timberwolves. Timberwolves are a natural predator of white-tailed deer. Over time the deer population increased in size. The available vegetation could not support the deer population. Even though emergency feeding stations were set up, thousands of deer died of starvation. The deer population now is kept down by controlled hunting seasons. In some areas wolves have been reintroduced.

An ecosystem consists of organisms, from many different species, living together and connected by the flow of energy, nutrients and matter. Organisms in an ecosystem can be classified as either producers or consumers. Most producers use the Sun's radiant energy and convert it into chemical energy through photosynthesis. Consumers take in and use this chemical energy. Herbivores eat producers, carnivores eat other consumers, and omnivores eat both producers and consumers. As organisms die, decomposers take in and use the energy in the dead organisms. In doing so, they release nutrients into the soil and carbon dioxide into the air that are used by producers again.

The loss of one species from an ecosystem may lead to the overpopulation or extinction of other species. This loss degrades the ecosystem upon which humans and other organisms depend for clean air, water, and food.

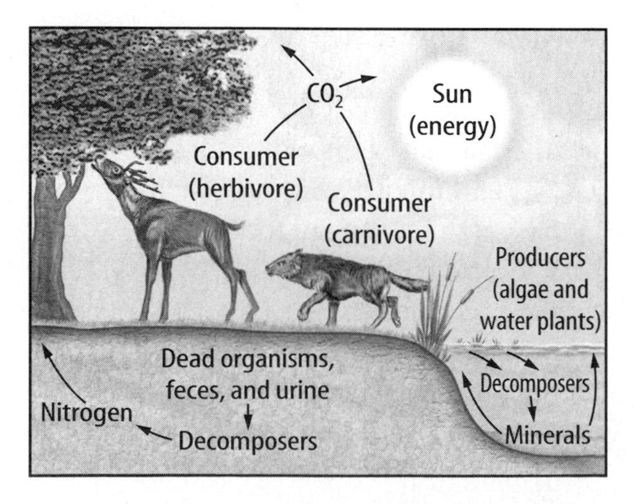

The major biotic components of an ecosystem

Test-Taking Tip Use the figure to help you visualize the ecosystem that is being described in the passage.

1. Food chains are a way of showing how energy, nutrients and matter flow through an ecosystem. Which of the following is a food chain of the ecosystem described in the passage?

A) carnivore, producer, herbivore
B) producer, herbivore, carnivore
C) carnivore, producer, decomposer
D) decomposer, carnivore, herbivore

2. Predators are consumers that capture and eat other consumers. The presence of a predator limits the size of the prey population. This means that food and other resources are less likely to become scarce. What is the predator in this passage?

F) vegetation
G) deer
H) Sun
J) timberwolf

Standardized Test Practice

Reasoning and Skills

Read each question and choose the best answer.

1. Within an ecosystem there are many populations as well as abiotic factors. Groups of populations that interact within a specific area of an ecosystem are referred to as which of the following?

A) a habitat
B) a community
C) a species
D) an atmosphere

Test-Taking Tip Think about the levels of an ecosystem and how they relate to each other.

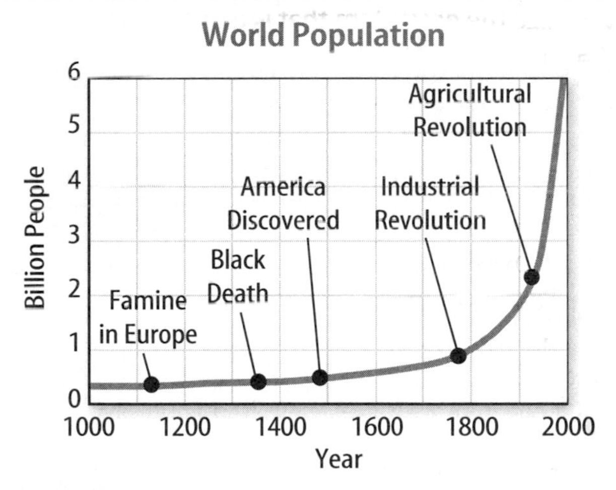

2. Refer to the Population Growth graph. In which of the following years were the birth rate and the death rate nearly equal?

F) 1,800
G) 2,000
H) 1,200
J) 1,600

Test-Taking Tip Consider what you know about the effects of birth and death rates on population size.

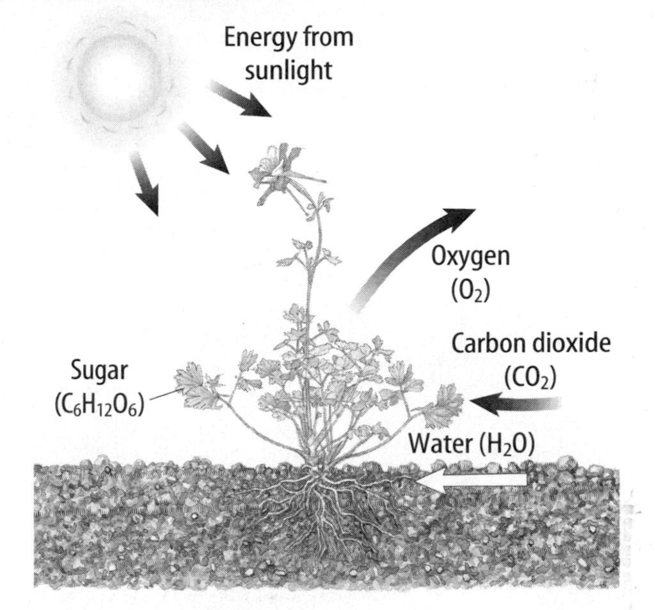

3. The conversion of energy is important to all life on Earth. Some producers use sunlight as an energy source, converting it into chemical energy through photosynthesis. Other producers that live where sunlight does not reach them, can use which of the following as an energy source?

A) water
B) air
C) soil
D) chemicals

Test-Taking Tip Read about converting energy before answering the question.

Consider this question carefully before writing your answer on a separate sheet of paper.

4. Consider what you have learned about ecosystems. Compare and contrast biotic and abiotic factors of the environment. List some abiotic factors and describe why each is important to life.

Test-Taking Tip Make a concept map of the environmental factors of an ecosystem to help answer the question.

UNIT 5 Earth and the Solar System

How Are Volcanoes & Fish Connected?

It's hard to know exactly what happened four and a half billion years ago, when Earth was very young. But it's likely that Earth was much more volcanically active than it is today. Along with lava and ash, volcanoes emit gases—including water vapor. Some scientists think that ancient volcanoes spewed tremendous amounts of water vapor into the early atmosphere. When the water vapor cooled, it would have condensed to form liquid water. Then the water would have fallen to the surface and collected in low areas, creating the oceans. Scientists hypothesize that roughly three and a half billion to four billion years ago, the first living things developed in the oceans. According to this hypothesis, these early life-forms gradually gave rise to more and more complex organisms—including the multitudes of fish that swim through the world's waters.

SCIENCE CONNECTION

VOLCANOES AND COMETS Not all scientists agree with the hypothesis that Earth's oceans were formed primarily by emissions from ancient volcanoes. For example, some researchers suggest that the water may have come largely from comets. Divide the class into two teams. Have one team investigate the volcano hypothesis, while the other team researches the comet hypothesis. Then hold a class debate, with each team presenting evidence in support of its hypothesis.

Plate Tectonics

Characterized by volcanoes and scenic vistas, the East African Rift Valley marks a place where Earth's crust is being pulled apart. If the pulling continues over millions of years, Africa will separate into two landmasses. In this chapter, you'll learn about Rift Valleys and other features explained by the theory of plate tectonics. You'll also learn about the fossil, climate, and rock clues that indicate that Earth's continents have drifted over time.

What do you think?

Science Journal Look at the picture below with a classmate. Discuss what you think this might be or what is happening. Here's a hint: *A river runs through this dog leg.* Write your answer or best guess in your Science Journal.

Can you imagine a giant landmass that broke into many separate continents and Earth scientists working to reconstruct Earth's past? Do this activity to learn about clues that can be used to reassemble a supercontinent.

Reassemble an image 

1. Collect interesting photographs from an old magazine.
2. You and a partner each select one photo, but don't show them to each other. Then each of you cut your photos into pieces no smaller than about 5 cm or 6 cm.
3. Trade your cut-up photo for your partner's.
4. Observe the pieces, and reassemble the photograph your partner has cut up.

Observe

In your Science Journal, describe the characteristics of the cut-up photograph that helped you put the image back together. Think of other examples in which characteristics of objects are used to match them up with other objects.

Before You Read

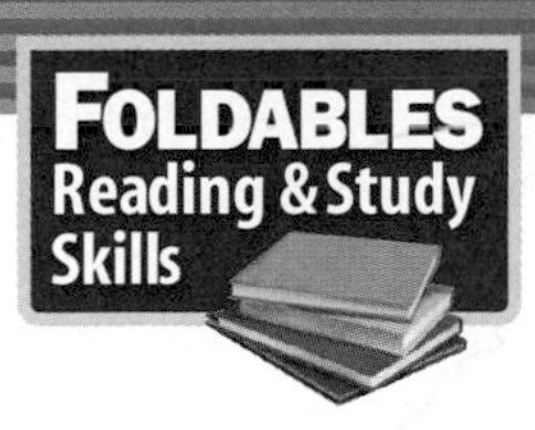

Making a Know-Want-Learn Study Fold **It would be helpful to identify what you already know and what you want to know. Make the following Foldable to help you focus on reading about plate tectonics.**

1. Place a sheet of paper in front of you so the long side is at the top. Fold the paper in half from top to bottom.
2. Fold both sides in to divide the paper into thirds. Unfold the paper so three sections show.
3. Through the top thickness of paper, cut along each of the fold lines to the topfold, forming three tabs. Label the tabs *Know, Want,* and *Learn,* as shown.
4. Before you read the chapter, write what you know about plate tectonics under the left tab and what you want to know under the middle tab.
5. As you read the chapter, write what you learn about plate tectonics under the right tab.

SECTION

Continental Drift

As You Read

What You'll Learn

- **Describe** the hypothesis of continental drift.
- **Identify** evidence supporting continental drift.

Vocabulary

continental drift
Pangaea

Why It's Important

The hypothesis of continental drift led to plate tectonics—a theory that explains many processes in Earth.

Evidence for Continental Drift

If you look at a map of Earth's surface, you can see that the edges of some continents look as though they could fit together like a puzzle. Other people also have noticed this fact. For example, Dutch mapmaker Abraham Ortelius noted the fit between the coastlines of South America and Africa more than 400 years ago.

Pangaea German meteorologist Alfred Wegener (VEG nur) thought that the fit of the continents wasn't just a coincidence. He suggested that all the continents were joined together at some time in the past. In a 1912 lecture, he proposed the hypothesis of continental drift. According to the hypothesis of **continental drift,** continents have moved slowly to their current locations. Wegener suggested that all continents once were connected as one large landmass, shown in **Figure 1,** that broke apart about 200 million years ago. He called this large landmass **Pangaea** (pan JEE uh), which means "all land."

Reading Check *Who proposed continental drift?*

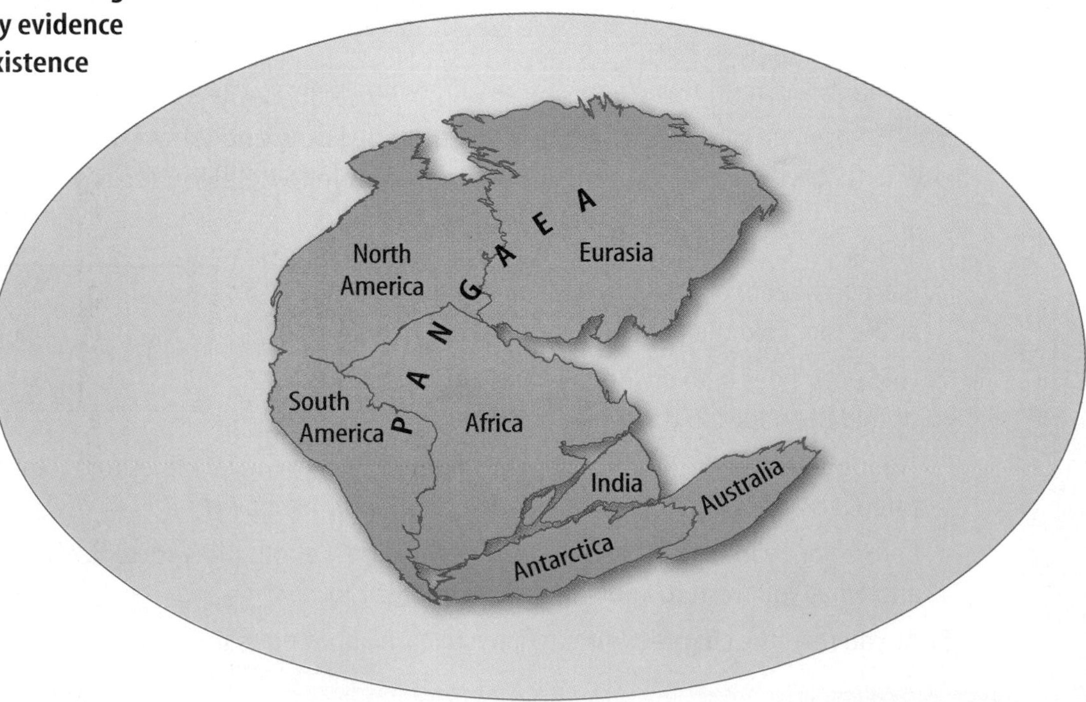

Figure 1
This illustration represents how the continents once were joined to form Pangaea. This fitting together of continents according to shape is not the only evidence supporting the past existence of Pangaea.

A Controversial Idea Wegener's ideas about continental drift were controversial. It wasn't until long after Wegener's death in 1930 that his basic hypothesis was accepted. The evidence Wegener presented hadn't been enough to convince many people during his lifetime. He was unable to explain exactly how the continents drifted apart. He proposed that the continents plowed through the ocean floor, driven by the spin of Earth. Physicists and geologists of the time strongly disagreed with Wegener's explanation. They pointed out that continental drift would not be necessary to explain many of Wegener's observations. Other important observations that came later eventually supported Wegener's earlier evidence.

Research Visit the Glencoe Science Web site at **science.glencoe.com** for more information about the continental drift hypothesis. Communicate to your class what you learn.

Fossil Clues Besides the puzzlelike fit of the continents, fossils provided support for continental drift. Fossils of the reptile *Mesosaurus* have been found in South America and Africa, as shown in **Figure 2.** This swimming reptile lived in freshwater and on land. How could fossils of *Mesosaurus* be found on land areas separated by a large ocean of salt water? It probably couldn't swim between the continents. Wegener hypothesized that this reptile lived on both continents when they were joined.

Reading Check *How do* Mesosaurus *fossils support the past existence of Pangaea?*

Figure 2
Fossil remains of plants and animals that lived in Pangaea have been found on more than one continent. *How do the locations of* Glossopteris, Mesosaurus, Kannemeyerid, Labyrinthodont, *and other fossils support Wegener's hypothesis of continental drift?*

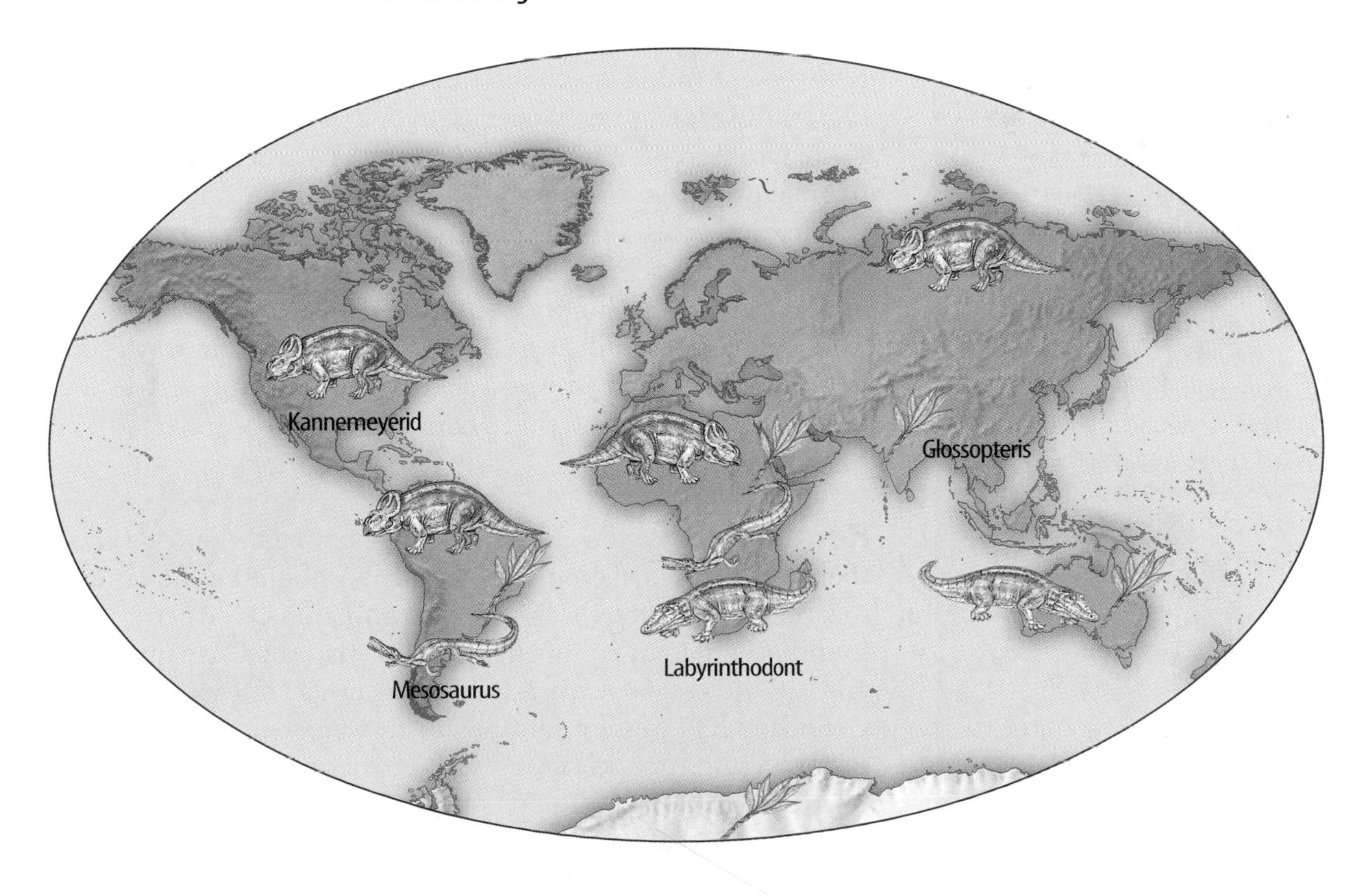

Figure 3
This fossil plant, *Glossopteris,* grew in a temperate climate.

Interpreting Fossil Data

Procedure

1. Build a three-layer landmass using **clay or modeling dough.**
2. Mold the clay into mountain ranges.
3. Place similar **"fossils"** into the clay at various locations around the landmass.
4. Form five continents from the one landmass. Also, form two smaller landmasses out of different clay with different mountain ranges and fossils.
5. Place the five continents and two smaller landmasses around the room.
6. Have someone who did not make or place the landmasses make a model that shows how they once were positioned.
7. Return the clay to its container so it can be used again.

Analysis

What clues were useful in reconstructing the original landmass?

A Widespread Plant Another fossil that supports the hypothesis of continental drift is *Glossopteris* (glahs AHP tur us). **Figure 3** shows this fossil plant, which has been found in Africa, Australia, India, South America, and Antarctica. The presence of *Glossopteris* in so many areas also supported Wegener's idea that all of these regions once were connected and had similar climates.

Climate Clues Wegener used continental drift to explain evidence of changing climates. For example, fossils of warm-weather plants were found on the island of Spitsbergen in the Arctic Ocean. To explain this, Wegener hypothesized that Spitsbergen drifted from tropical regions to the arctic. Wegener also used continental drift to explain evidence of glaciers found in temperate and tropical areas. Glacial deposits and rock surfaces scoured and polished by glaciers are found in South America, Africa, India, and Australia. This shows that parts of these continents were covered with glaciers in the past. How could you explain why glacial deposits are found in areas where no glaciers exist today? Wegener thought that these continents were connected and partly covered with ice near Earth's south pole long ago.

Rock Clues If the continents were connected at one time, then rocks that make up the continents should be the same in locations where they were joined. Similar rock structures are found on different continents. Parts of the Appalachian Mountains of the eastern United States are similar to those found in Greenland and western Europe. If you were to study rocks from eastern South America and western Africa, you would find other rock structures that also are similar. Rock clues like these support the idea that the continents were connected in the past.

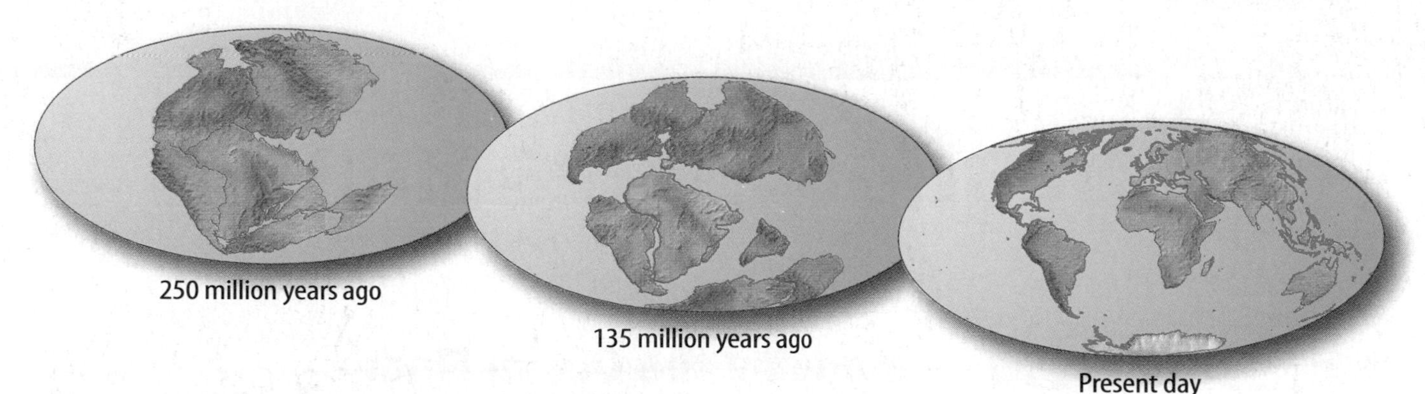

How could continents drift?

Although Wegener provided evidence to support his hypothesis of continental drift, he couldn't explain how, when, or why these changes, shown in **Figure 4,** took place. The idea suggested that lower-density, continental material somehow had to plow through higher-density, ocean-floor material. The force behind this plowing was thought to be the spin of Earth on its axis—a notion that was quickly rejected by physicists. Because other scientists could not provide explanations either, Wegener's idea of continental drift was initially rejected. The idea was so radically different at that time that most people closed their minds to it.

Rock, fossil, and climate clues were the main types of evidence for continental drift. After Wegener's death, more clues were found, largely because of advances in technology, and new ideas that related to continental drift were developed. You'll learn about one of these new ideas, seafloor spreading, in the next section. Seafloor spreading helped provide an explanation of how the continents could move.

Figure 4
These computer models show the probable course the continents have taken. On the far left is their position 250 million years ago. In the middle is their position 135 million years ago. At right is their current position.

Section  Assessment

1. Why were Wegener's ideas about continental drift initially rejected?
2. How did Wegener use climate clues to support his hypothesis of continental drift?
3. What rock clues were used to support the hypothesis of continental drift?
4. In what ways do fossils help support the hypothesis of continental drift?
5. **Think Critically** Why would you expect to see similar rocks and rock structures on two landmasses that were connected at one time?

Skill Builder Activities

6. **Comparing and Contrasting** Compare and contrast the locations of fossils of the temperate plant *Glossopteris,* as shown in **Figure 2,** with the climate that exists at each location today. **For more help, refer to the Science Skill Handbook.**
7. **Communicating** Imagine that you are Alfred Wegener in the year 1912. In your Science Journal, write a letter to another scientist explaining your idea about continental drift. Try to convince this scientist that your hypothesis is correct. **For more help, refer to the Science Skill Handbook.**

SECTION

Seafloor Spreading

As You Read

What You'll Learn

- **Explain** seafloor spreading.
- **Recognize** how age and magnetic clues support seafloor spreading.

Vocabulary
seafloor spreading

Why It's Important

Seafloor spreading helps explain how continents moved apart.

Mapping the Ocean Floor

If you were to lower a rope from a boat until it reached the seafloor, you could record the depth of the ocean at that particular point. In how many different locations would you have to do this to create an accurate map of the seafloor? This is exactly how it was done until World War I, when the use of sound waves was introduced to detect submarines. During the 1940s and 1950s, scientists began using sound waves on moving ships to map large areas of the ocean floor in detail. Sound waves echo off the ocean bottom—the longer the sound waves take to return to the ship, the deeper the water is.

Using sound waves, researchers discovered an underwater system of ridges, or mountains, and valleys like those found on the continents. In the Atlantic, the Pacific, and in other oceans around the world, a system of ridges, called the mid-ocean ridges, is present. These underwater mountain ranges, shown in **Figure 5,** stretch along the center of much of Earth's ocean floor. This discovery raised the curiosity of many scientists. What formed these mid-ocean ridges?

Reading Check *How were mid-ocean ridges discovered?*

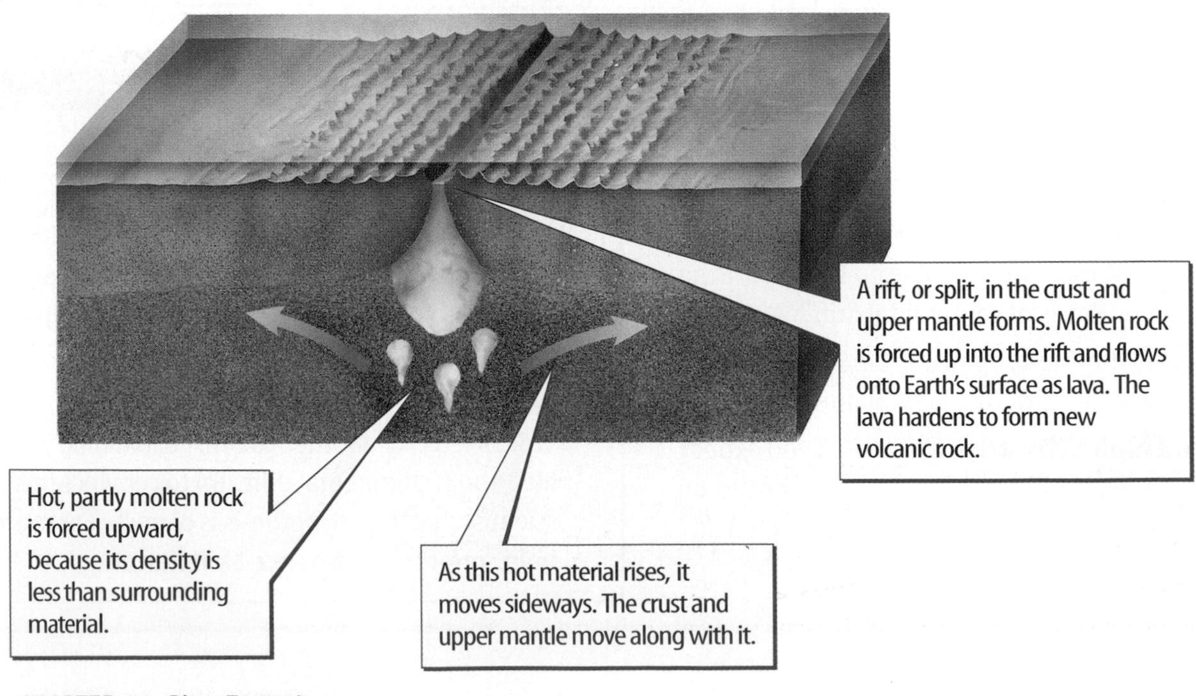

Figure 5
As the seafloor spreads apart at a mid-ocean ridge, new seafloor is created. The older seafloor moves away from the ridge in opposite directions.

The Seafloor Moves In the early 1960s, Princeton University scientist Harry Hess suggested an explanation. His now-famous theory is known as **seafloor spreading.** Hess proposed that hot, less dense material below Earth's crust rises toward the surface at the mid-ocean ridges. Then, it flows sideways, carrying the seafloor away from the ridge in both directions, as seen in **Figure 5.**

As the seafloor spreads apart, magma moves upward and flows from the cracks. It becomes solid as it cools and forms new seafloor. As new seafloor moves away from the mid-ocean ridge, it cools, contracts, and becomes denser. This denser, colder seafloor sinks, helping to form the ridge. The theory of seafloor spreading was later supported by the following observations.

How does new seafloor form at mid-ocean ridges?

Figure 6
Many new discoveries have been made on the seafloor. These giant tube worms inhabit areas near hot water vents along mid-ocean ridges.

Evidence for Spreading In 1968, scientists aboard the research ship *Glomar Challenger* began gathering information about the rocks on the seafloor. *Glomar Challenger* was equipped with a drilling rig that allowed scientists to drill into the seafloor to obtain rock samples. They made a remarkable discovery as they studied the ages of the rocks. Scientists found that the youngest rocks are located at the mid-ocean ridges. The ages of the rocks become increasingly older in samples obtained farther from the ridges, adding to the evidence for seafloor spreading.

Using submersibles along mid-ocean ridges, new seafloor features and life-forms also were discovered there, as shown in **Figure 6.** As molten material rises along the ridges, it brings heat and chemicals that support exotic life-forms in deep, ocean water. Among these are giant clams, mussels, and tube worms.

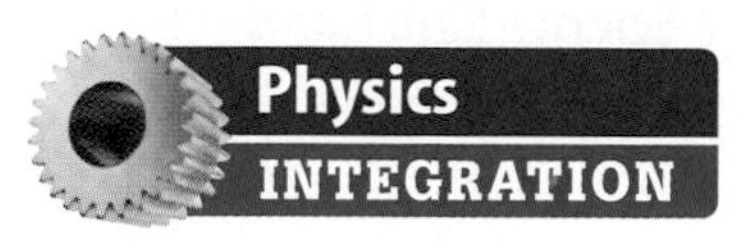

Magnetic Clues Earth's magnetic field has a north and a south pole. Magnetic lines, or directions, of force leave Earth near the south pole and enter Earth near the north pole. During a magnetic reversal, the lines of magnetic force run the opposite way. Scientists have determined that Earth's magnetic field has reversed itself many times in the past. These reversals occur over intervals of thousands or even millions of years. The reversals are recorded in rocks forming along mid-ocean ridges.

Find out what the Curie point is and describe in your Science Journal what happens to iron-bearing minerals when they are heated to the Curie point. Explain how this is important to studies of seafloor spreading.

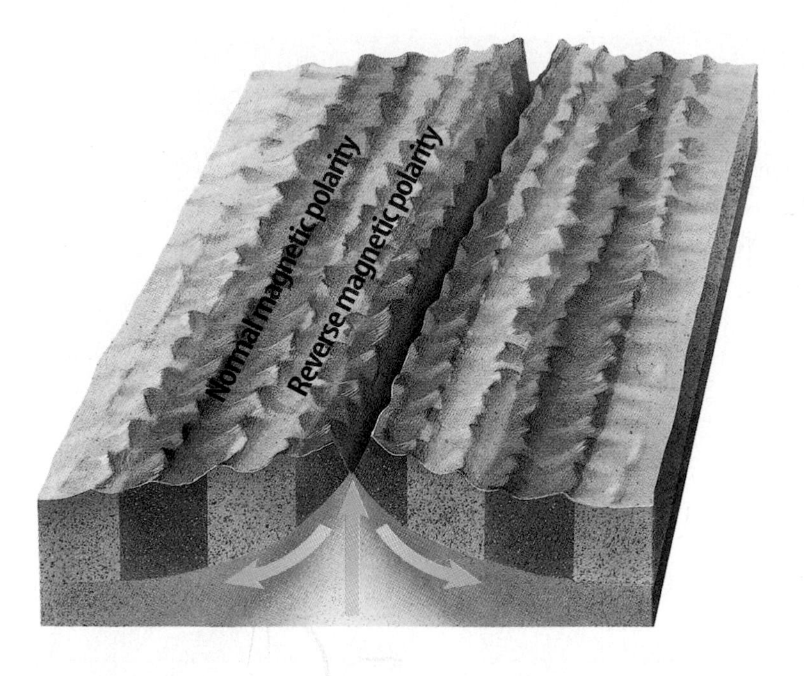

Figure 7
Changes in Earth's magnetic field are preserved in rock that forms on both sides of mid-ocean ridges. *Why is this considered to be evidence of seafloor spreading?*

Magnetic Time Scale Iron-bearing minerals, such as magnetite, that are found in the rocks of the seafloor can record Earth's magnetic field direction when they form. Whenever Earth's magnetic field reverses, newly forming iron minerals will record the magnetic reversal.

Using a sensing device called a magnetometer (mag nuh TAH muh tur) to detect magnetic fields, scientists found that rocks on the ocean floor show many periods of magnetic reversal. The magnetic alignment in the rocks reverses back and forth over time in strips parallel to the mid-ocean ridges, as shown in **Figure 7.** A strong magnetic reading is recorded when the polarity of a rock is the same as the polarity of Earth's magnetic field today. Because of this, normal polarities in rocks show up as large peaks. This discovery provided strong support that seafloor spreading was indeed occurring. The magnetic reversals showed that new rock was being formed at the mid-ocean ridges. This helped explain how the crust could move—something that the continental drift hypothesis could not do.

Section  Assessment

1. What properties of iron-bearing minerals on the seafloor support the theory of seafloor spreading?
2. How do the ages of the rocks on the ocean floor support the theory of seafloor spreading?
3. How did Harry Hess's hypothesis explain seafloor movement?
4. Why does some partly molten material rise toward Earth's surface?
5. **Think Critically** The ideas of Hess, Wegener, and others emphasize that Earth is a dynamic planet. How is seafloor spreading different from continental drift?

Skill Builder Activities

6. **Concept Mapping** Make a concept map that includes evidence for seafloor spreading using the following phrases: *ages increase away from ridge, pattern of magnetic field reversals, mid-ocean ridge, pattern of ages,* and *reverses back and forth.* **For more help, refer to the Science Skill Handbook.**
7. **Solving One-Step Equations** North America is moving about 1.25 cm per year away from a ridge in the middle of the Atlantic Ocean. Using this rate, how much farther apart will North America and the ridge be in 200 million years? **For more help, refer to the Math Skill Handbook.**

Activity

Seafloor Spreading Rates

How did scientists use their knowledge of seafloor spreading and magnetic field reversals to reconstruct Pangaea? Try this activity to see how you can determine where a continent may have been located in the past.

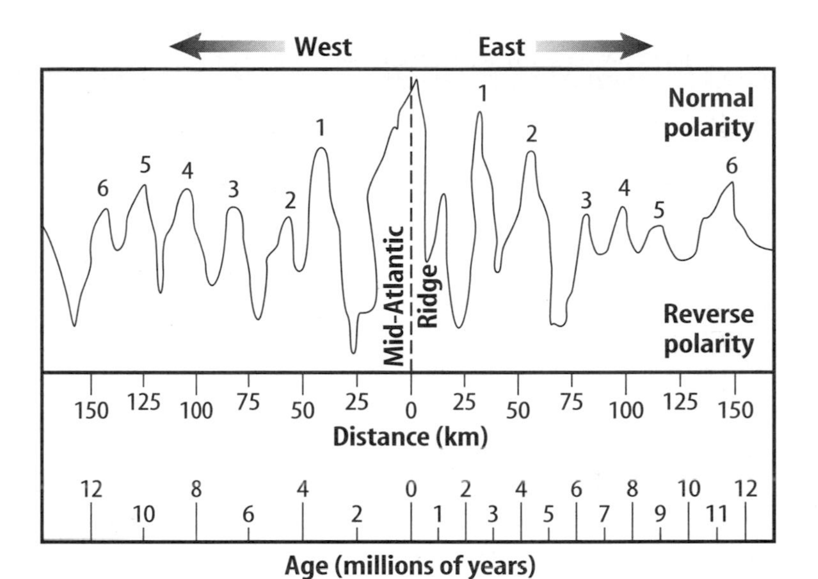

What You'll Investigate

Can you use clues, such as magnetic field reversals on Earth, to help reconstruct Pangaea?

Materials

metric ruler
pencil

Goals

- **Interpret** data about magnetic field reversals. Use these magnetic clues to reconstruct Pangaea.

Procedure

1. Study the magnetic field graph above. You will be working only with normal polarity readings, which are the peaks above the baseline in the top half of the graph.
2. Place the long edge of a ruler vertically on the graph. Slide the ruler so that it lines up with the center of peak 1 west of the Mid-Atlantic Ridge.
3. **Determine** and record the distance and age that line up with the center of peak 1 west. Repeat this process for peak 1 east of the ridge.
4. **Calculate** the average age and distance for this pair of peaks.
5. Repeat steps 2 through 4 for the remaining pairs of normal-polarity peaks.
6. **Calculate** the rate of movement in cm per year for the six pairs of peaks. Use the formula rate = distance/time. Convert kilometers to centimeters. For example, to calculate a rate using normal-polarity peak 5, west of the ridge:

$$\text{rate} = \frac{125\text{ km}}{10\text{ million years}} = \frac{12.5\text{ km}}{\text{million years}} = \frac{1{,}250{,}000\text{ cm}}{1{,}000{,}000\text{ years}} = 1.25\text{ cm/year}$$

Conclude and Apply

1. **Compare** the age of igneous rock found near the mid-ocean ridge with that of igneous rock found farther away from the ridge.
2. If the distance from a point on the coast of Africa to the Mid-Atlantic Ridge is approximately 2,400 km, calculate how long ago that point in Africa was at or near the Mid-Atlantic Ridge.
3. How could you use this method to reconstruct Pangaea?

SECTION 3

Theory of Plate Tectonics

As You Read

What You'll Learn

- **Compare and contrast** different types of plate boundaries.
- **Explain** how heat inside Earth causes plate tectonics.
- **Recognize** features caused by plate tectonics.

Vocabulary

plate tectonics
plate
lithosphere
asthenosphere
convection current

Why It's Important

Plate tectonics explains how many of Earth's features form.

Plate Tectonics

The idea of seafloor spreading showed that more than just continents were moving, as Wegener had thought. It was now clear to scientists that sections of the seafloor and continents move in relation to one another.

Plate Movements In the 1960s, scientists developed a new theory that combined continental drift and seafloor spreading. According to the theory of **plate tectonics,** Earth's crust and part of the upper mantle are broken into sections. These sections, called **plates,** move on a plasticlike layer of the mantle. The plates can be thought of as rafts that float and move on this layer.

Composition of Earth's Plates Plates are made of the crust and a part of the upper mantle, as shown in **Figure 8.** These two parts combined are the **lithosphere** (LIH thuh sfihr). This rigid layer is about 100 km thick and generally is less dense than material underneath. The plasticlike layer below the lithosphere is called the **asthenosphere** (as THE nuh sfihr). The rigid plates of the lithosphere float and move around on the asthenosphere.

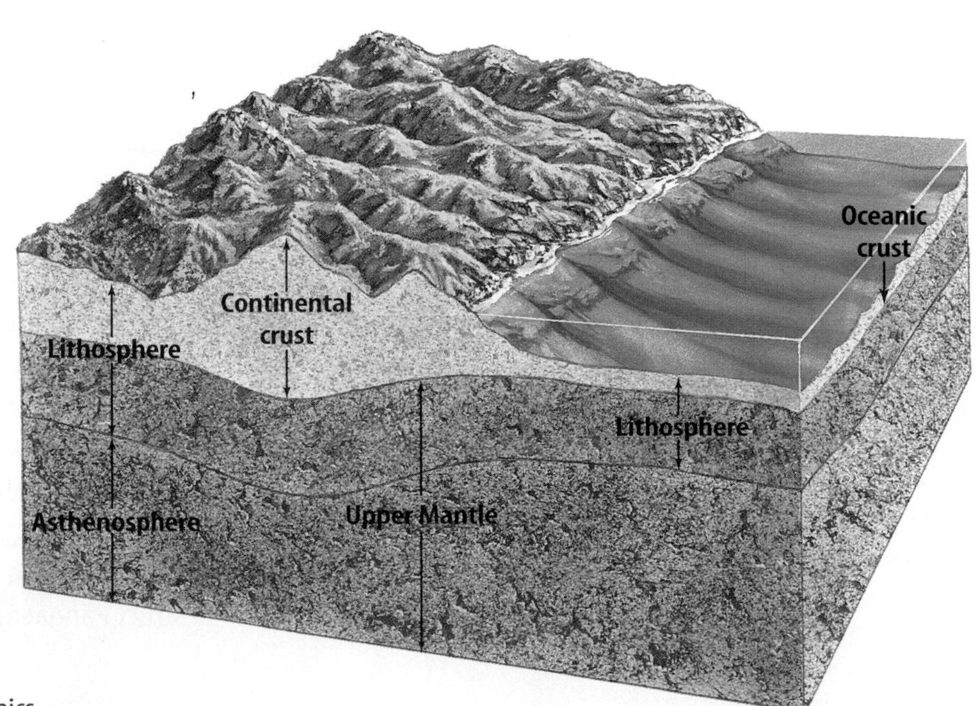

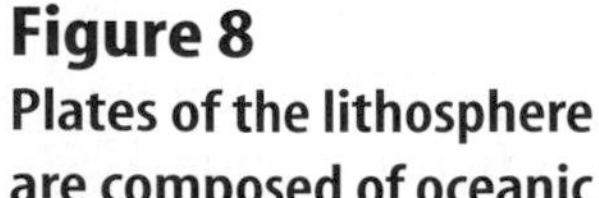

Figure 8
Plates of the lithosphere are composed of oceanic crust, continental crust, and rigid upper mantle.

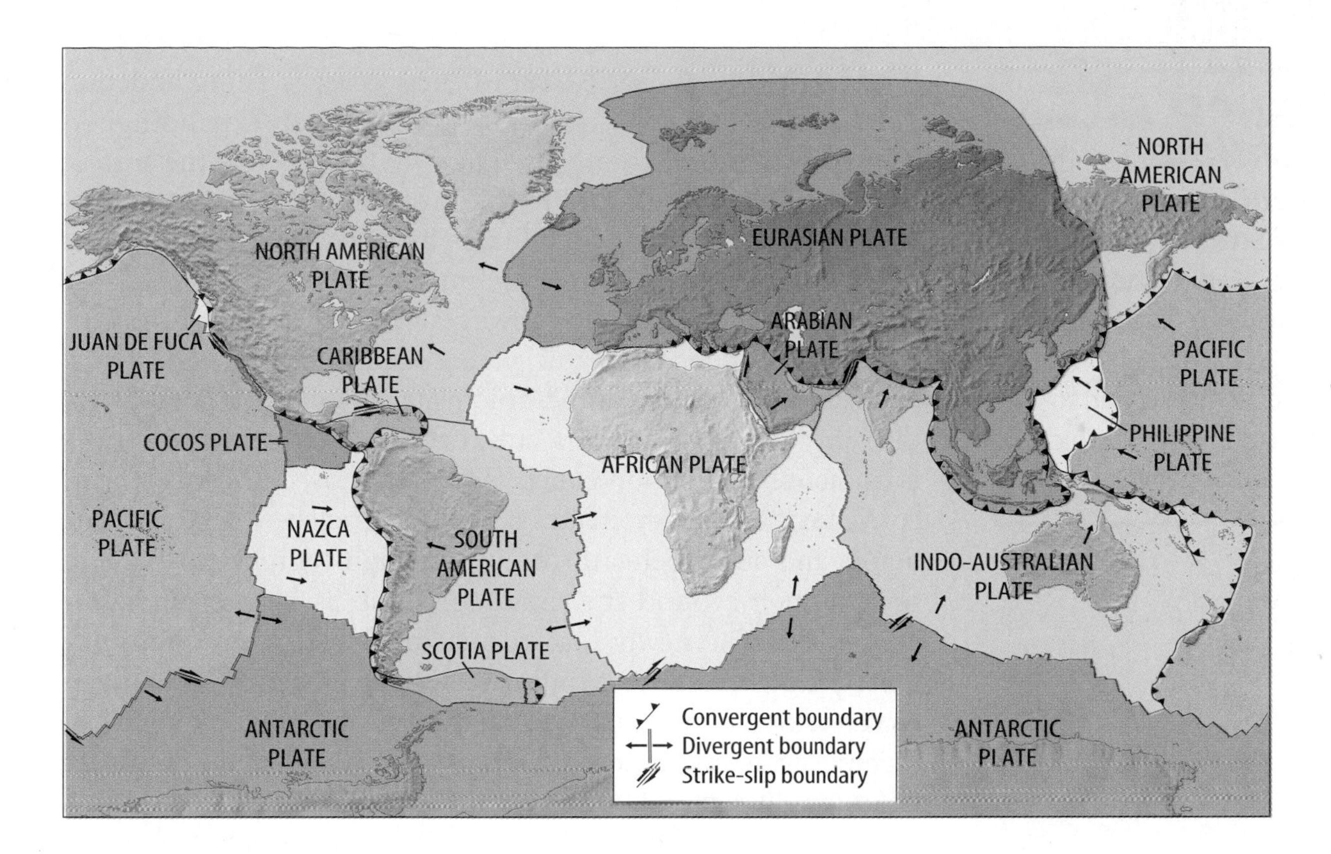

Plate Boundaries

When plates move, they can interact in several ways. They can move toward each other and converge, or collide. They also can pull apart or slide alongside one another. When the plates interact, the result of their movement is seen at the plate boundaries, as in **Figure 9.**

Figure 9
This diagram shows the major plates of the lithosphere, their direction of movement, and the type of boundary between them.
Based on what is shown in this figure, what is happening where the Nazca Plate meets the Pacific Plate?

✔ Reading Check *What are the general ways that plates interact?*

Movement along any plate boundary means that changes must happen at other boundaries. What is happening to the Atlantic Ocean floor between the North American and African Plates? Compare this with what is happening along the western margin of South America.

Plates Moving Apart The boundary between two plates that are moving apart is called a divergent boundary. You learned about divergent boundaries when you read about seafloor spreading. In the Atlantic Ocean, the North American Plate is moving away from the Eurasian and the African Plates, as shown in **Figure 9.** That divergent boundary is called the Mid-Atlantic Ridge. The Great Rift Valley in eastern Africa might become a divergent plate boundary. There, a valley has formed where a continental plate is being pulled apart. **Figure 10** shows a side view of what a rift valley might look like and illustrates how the hot material rises up where plates separate.

SCIENCE *Online*

Research Visit the Glencoe Science Web site at **science.glencoe.com** for recent news or magazine articles about earthquakes and volcanic activity related to plate tectonics. Communicate to your class what you learned.

Plates Moving Together If new crust is being added at one location, why doesn't Earth's surface keep expanding? As new crust is added in one place, it disappears below the surface at another. The disappearance of crust can occur when seafloor cools, becomes denser, and sinks. This occurs where two plates move together at a convergent boundary.

When an oceanic plate converges with a less dense continental plate, the denser oceanic plate sinks under the continental plate. The area where an oceanic plate subducts, or goes down, into the mantle is called a subduction zone. Some volcanoes form above subduction zones. **Figure 10** shows how this type of convergent boundary creates a deep-sea trench where one plate bends and sinks beneath the other. High temperatures cause rock to melt around the subducting slab as it goes under the other plate. The newly formed magma is forced upward along these plate boundaries, forming volcanoes. The Andes mountain range of South America contains many volcanoes. They were formed at the convergent boundary of the Nazca and the South American Plates.

Problem-Solving Activity

How well do the continents fit together?

Recall the Explore Activity you performed at the beginning of this chapter. While you were trying to fit pieces of a cut-up photograph together, what clues did you use?

Identifying the Problem

Take a copy of a map of the world and cut out each continent. Lay them on a tabletop and try to fit them together, using techniques you used in the Explore Activity. You will find that the pieces of your Earth puzzle—the continents—do not fit together well. Yet, several of the areas on some continents fit together extremely well.

Take out another world map—one that shows the continental shelves as well as the continents. Copy it and cut out the continents, this time including the continental shelves.

Solving the Problem

1. Does including the continental shelves solve the problem of fitting the continents together?
2. Why should continental shelves be included with maps of the continents?

NATIONAL GEOGRAPHIC VISUALIZING PLATE BOUNDARIES

Figure 10

By diverging at some boundaries and converging at others, Earth's plates are continually—but gradually—reshaping the landscape around you. The Mid-Atlantic Ridge, for example, was formed when the North and South American Plates pulled apart from the Eurasian and African Plates (see globe). Some features that occur along plate boundaries—rift valleys, volcanoes, and mountain ranges—are shown on the right and below.

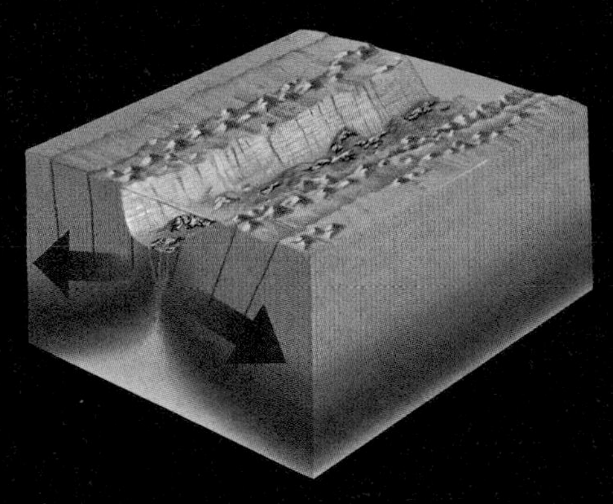

A RIFT VALLEY When continental plates pull apart, they can form rift valleys. The African continent is separating now along the East African Rift Valley.

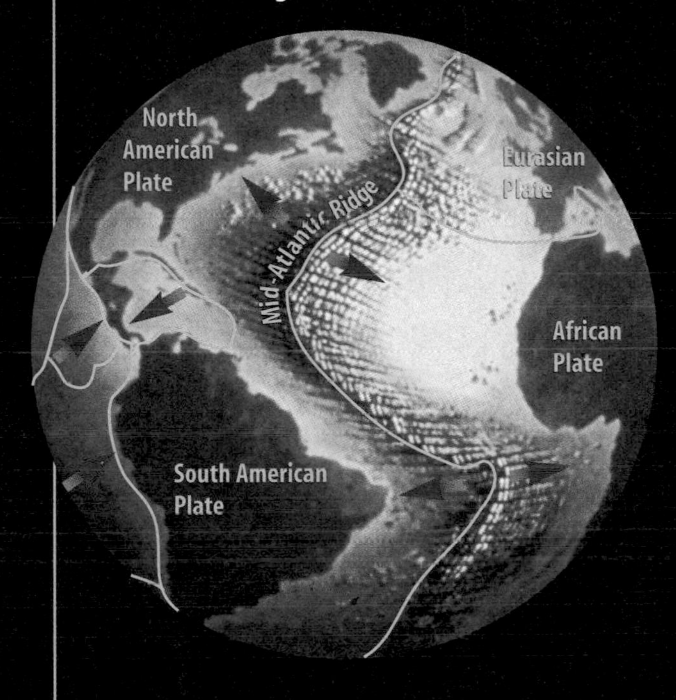

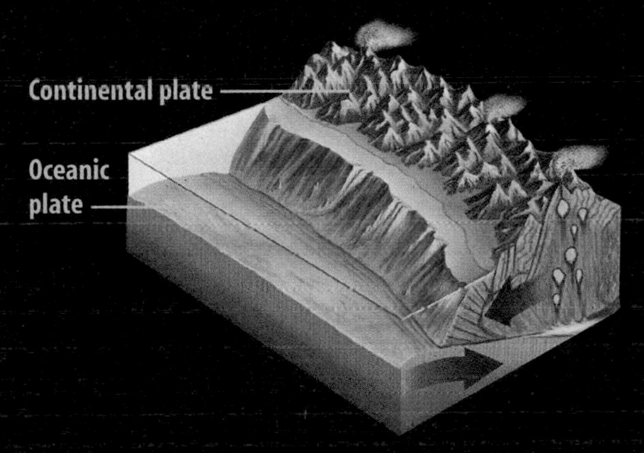

SUBDUCTION Where oceanic and continental plates collide, the oceanic plate plunges beneath the less dense continental plate. As the plate descends, molten rock (yellow) forms and rises toward the surface, creating volcanoes.

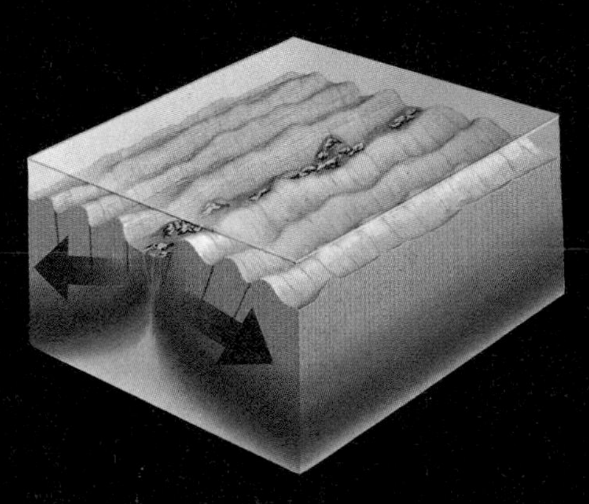

SEA-FLOOR SPREADING A mid-ocean ridge, like the Mid-Atlantic Ridge, forms where oceanic plates continue to separate. As rising magma (yellow) cools, it forms new oceanic crust.

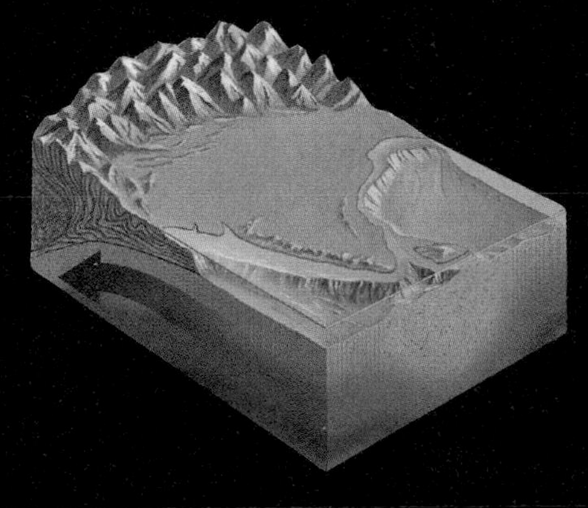

CONTINENTAL COLLISION Where two continental plates collide, they push up the crust to form mountain ranges such as the Himalaya.

Where Plates Collide A subduction zone also can form where two oceanic plates converge. In this case, the colder, older, denser oceanic plate bends and sinks down into the mantle. The Mariana Islands in the western Pacific are a chain of volcanic islands formed where two oceanic plates collide.

Usually, no subduction occurs when two continental plates collide, as shown in **Figure 10.** Because both of these plates are less dense than the material in the asthenosphere, the two plates collide and crumple up, forming mountain ranges. Earthquakes are common at these convergent boundaries. However, volcanoes do not form because there is no, or little, subduction. The Himalaya in Asia are forming where the Indo-Australian Plate collides with the Eurasian Plate.

Where Plates Slide Past Each Other The third type of plate boundary is called a transform boundary. Transform boundaries occur where two plates slide past one another. They move in opposite directions or in the same direction at different rates. When one plate slips past another suddenly, earthquakes occur. The Pacific Plate is sliding past the North American Plate, forming the famous San Andreas Fault in California, as seen in **Figure 11.** The San Andreas Fault is part of a transform plate boundary. It has been the site of many earthquakes.

Figure 11
The San Andreas Fault in California occurs along the transform plate boundary where the Pacific Plate is sliding past the North American Plate.

A Overall, the two plates are moving in roughly the same direction. *Why, then, do the red arrows show movement in opposite directions?*

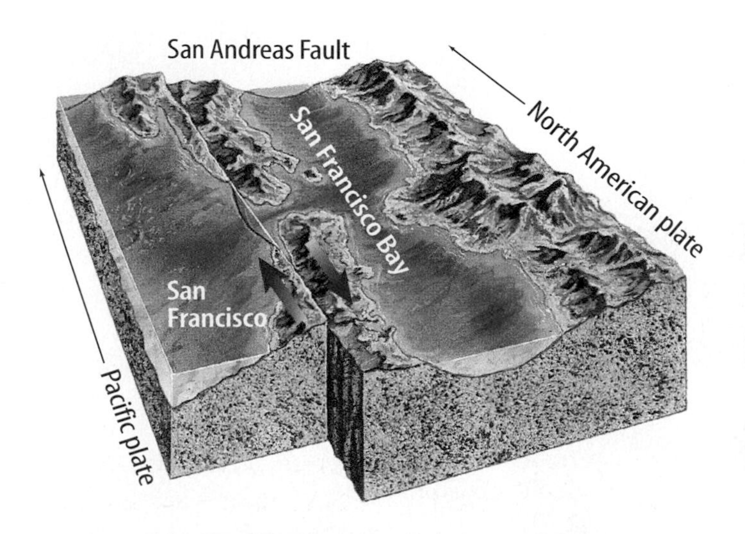

B This photograph shows an aerial view of the San Andreas Fault.

Causes of Plate Tectonics

Many new discoveries have been made about Earth's crust since Wegener's day, but one question still remains. What causes the plates to move? Scientists now think they have a good idea. They think that plates move by the same basic process that occurs when you heat soup.

Convection Inside Earth Soup that is cooking in a pan on the stove contains currents caused by an unequal distribution of heat in the pan. Hot, less dense soup is forced upward by the surrounding, cooler soup. As the hot soup reaches the surface, it cools and sinks back down into the pan. This entire cycle of heating, rising, cooling, and sinking is called a **convection current.** A version of this same process, occurring in the mantle, is thought to be the force behind plate tectonics. Scientists suggest that differences in density cause hot, plasticlike rock to be forced upward toward the surface.

Moving Mantle Material Wegener wasn't able to come up with an explanation for why plates move. Today, researchers who study the movement of heat in Earth's interior have proposed several possible explanations. All of the hypotheses use convection in one way or another. It is, therefore, the transfer of heat inside Earth that provides the energy to move plates and causes many of Earth's surface features. One hypothesis is shown in **Figure 12.** It relates plate motion directly to the movement of convection currents. According to this hypothesis, convection currents cause the movements of plates.

Modeling Convection Currents

Procedure

1. Pour **water** into a **clear, colorless casserole dish** until it is 5 cm from the top.
2. Center the dish on a **hot plate** and heat it. **WARNING:** *Wear* ***thermal mitts*** *to protect your hands.*
3. Add a few drops of **food coloring** to the water above the center of the hot plate.
4. Looking from the side of the dish, observe what happens in the water.
5. Illustrate your observations in your **Science Journal.**

Analysis

1. Determine whether any currents form in the water.
2. Infer what causes the currents to form.

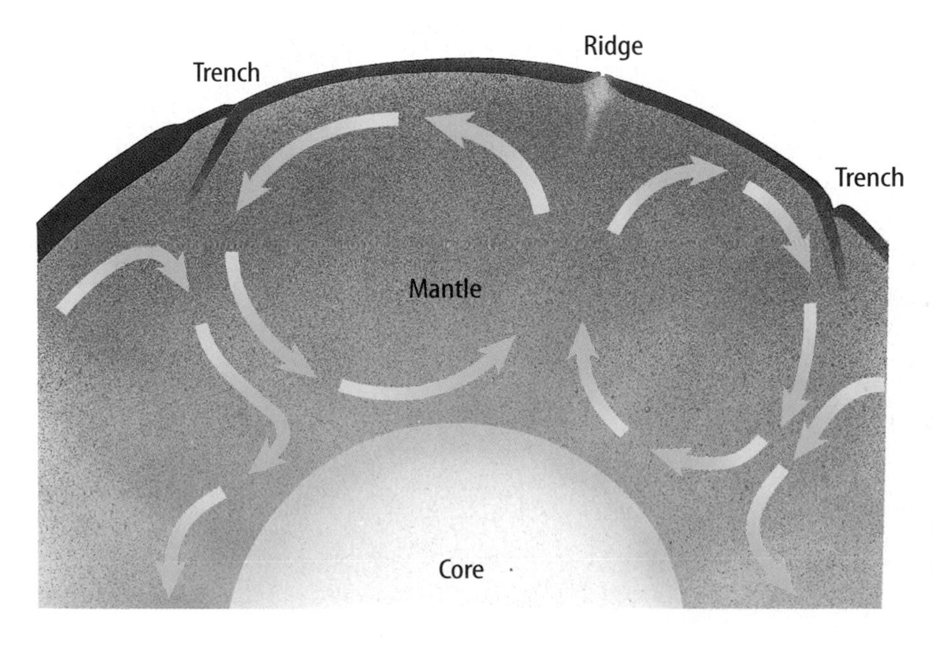

Figure 12
In one hypothesis, convection currents occur throughout the mantle. Such convection currents (see arrows) are the driving force of plate tectonics.

Features Caused by Plate Tectonics

Earth is a dynamic planet with a hot interior. This heat leads to convection, which powers the movement of plates. As the plates move, they interact. The interaction of plates produces forces that build mountains, create ocean basins, and cause volcanoes. When rocks in Earth's crust break and move, energy is released in the form of seismic waves. Humans feel this release as earthquakes. You can see some of the effects of plate tectonics in mountainous regions, where volcanoes erupt, or where landscapes have changed from past earthquake or volcanic activity.

What happens when seismic energy is released as rocks in Earth's crust break and move?

Normal Faults and Rift Valleys Tension forces, which are forces that pull apart, can stretch Earth's crust. This causes large blocks of crust to break and tilt or slide down the broken surfaces of crust. When rocks break and move along surfaces, a fault forms. Faults interrupt rock layers by moving them out of place. Entire mountain ranges can form in the process, called fault-block mountains, as shown in **Figure 13.** Generally, the faults that form from pull-apart forces are normal faults—faults in which the rock layers above the fault move down when compared with rock layers below the fault.

Rift valleys and mid-ocean ridges can form where Earth's crust separates. Examples of rift valleys are the Great Rift Valley in Africa, and the valleys that occur in the middle of mid-ocean ridges. Examples of mid-ocean ridges include the Mid-Atlantic Ridge and the East Pacific Rise.

Figure 13
Fault-block mountains can form when Earth's crust is stretched by tectonic forces. The arrows indicate the directions of moving blocks. *What type of force occurs when Earth's crust is pulled in opposite directions?*

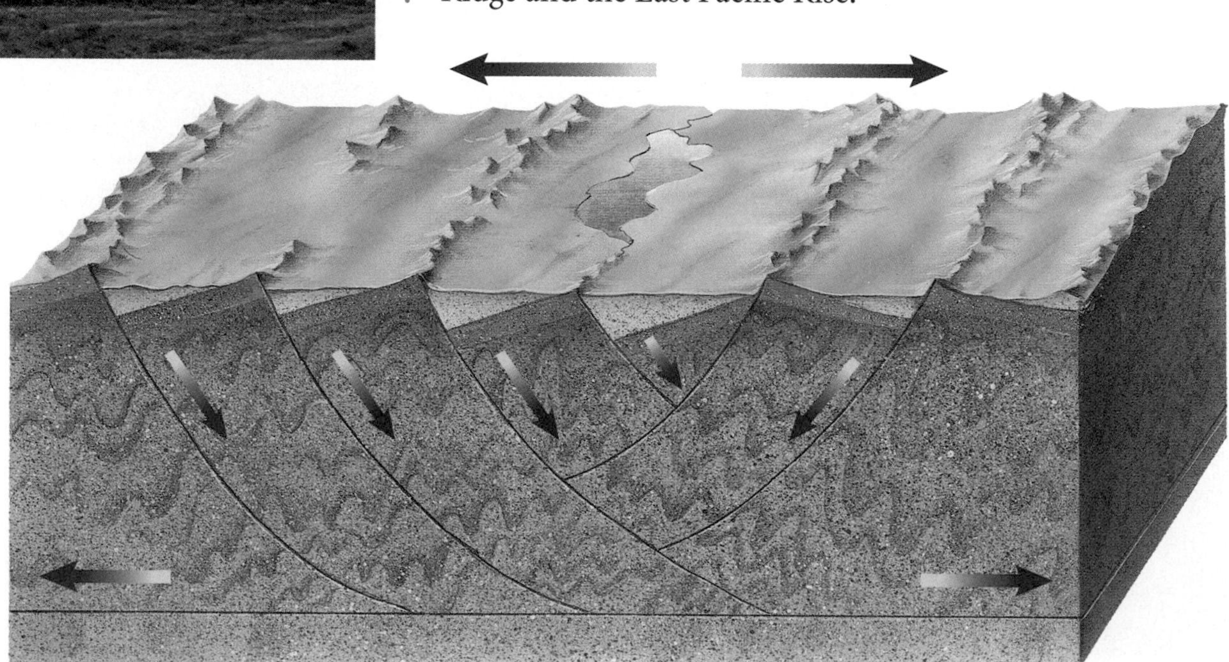

Mountains and Volcanoes Compression forces squeeze objects together. Where plates come together, compression forces produce several effects. As continental plates collide, the forces that are generated cause massive folding and faulting of rock layers into mountain ranges such as the Himalaya, shown in **Figure 14,** or the Appalachian Mountains. The type of faulting produced is generally reverse faulting. Along a reverse fault, the rock layers above the fault surface move up relative to the rock layers below the fault.

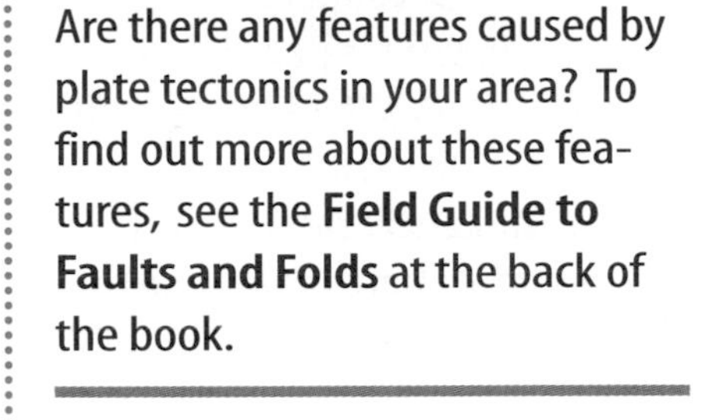

Are there any features caused by plate tectonics in your area? To find out more about these features, see the **Field Guide to Faults and Folds** at the back of the book.

Reading Check *What features occur where plates converge?*

As you learned earlier, when two oceanic plates converge, the denser plate is forced beneath the other plate. Curved chains of volcanic islands called island arcs form above the sinking plate. If an oceanic plate converges with a continental plate, the denser oceanic plate slides under the continental plate. Folding and faulting at the continental plate margin can thicken the continental crust to produce mountain ranges. Volcanoes also typically are formed at this type of convergent boundary.

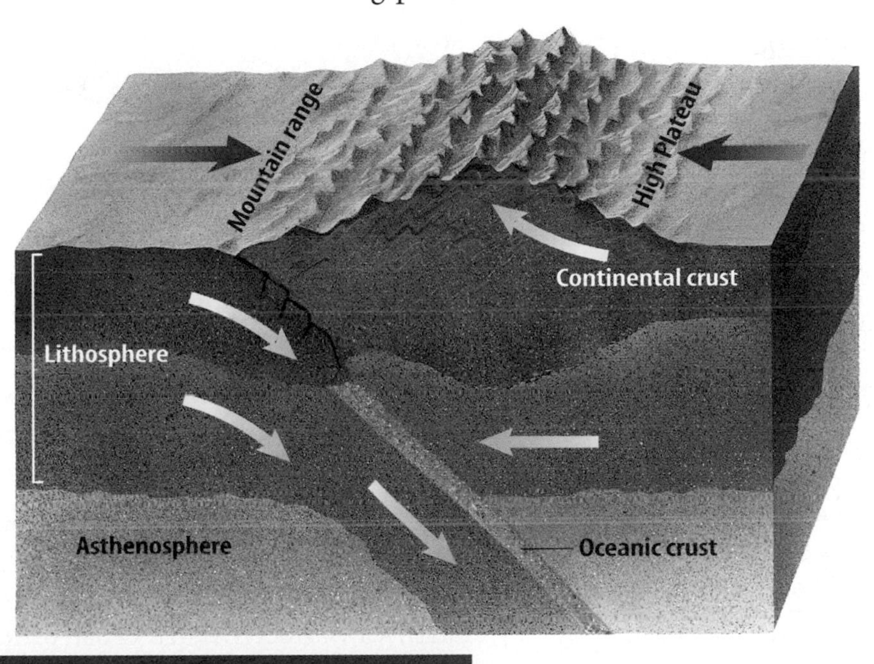

Figure 14
The Himalaya still are forming today as the Indo-Australian Plate collides with the Eurasian Plate.

Figure 15
Most of the movement along a strike-slip fault is parallel to Earth's surface. When movement occurs, human-built structures along a strike-slip fault are offset, as shown here in this road.

Strike-Slip Faults At transform boundaries, two plates slide past one another without converging or diverging. The plates stick and then slide, mostly in a horizontal direction, along large strike-slip faults. In a strike-slip fault, rocks on opposite sides of the fault move in opposite directions, or in the same direction at different rates. This type of fault movement is shown in **Figure 15.** One such example is the San Andreas Fault. When plates move suddenly, vibrations are generated inside Earth that are felt as an earthquake.

Earthquakes, volcanoes, and mountain ranges are evidence of plate motion. Plate tectonics explains how activity inside Earth can affect Earth's crust differently in different locations. You've seen how plates have moved since Pangaea separated. Is it possible to measure how far plates move each year?

Testing for Plate Tectonics

Until recently, the only tests scientists could use to check for plate movement were indirect. They could study the magnetic characteristics of rocks on the seafloor. They could study volcanoes and earthquakes. These methods supported the theory that the plates have moved and still are moving. However, they did not provide proof—only support—of the idea.

New methods had to be discovered to be able to measure the small amounts of movement of Earth's plates. One method, shown in **Figure 16,** uses lasers and a satellite. Now, scientists can measure exact movements of Earth's plates of as little as 1 cm per year.

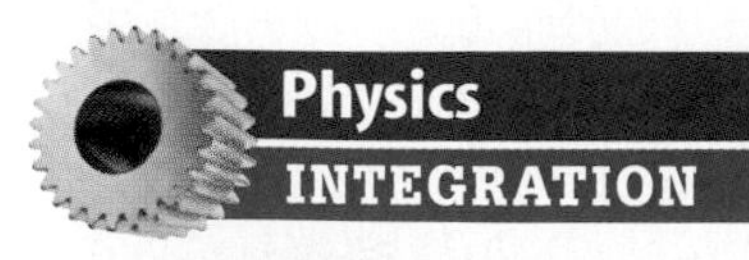

In which directions do forces act at convergent, divergent, and transform boundaries? Demonstrate these forces using wooden blocks or your hands.

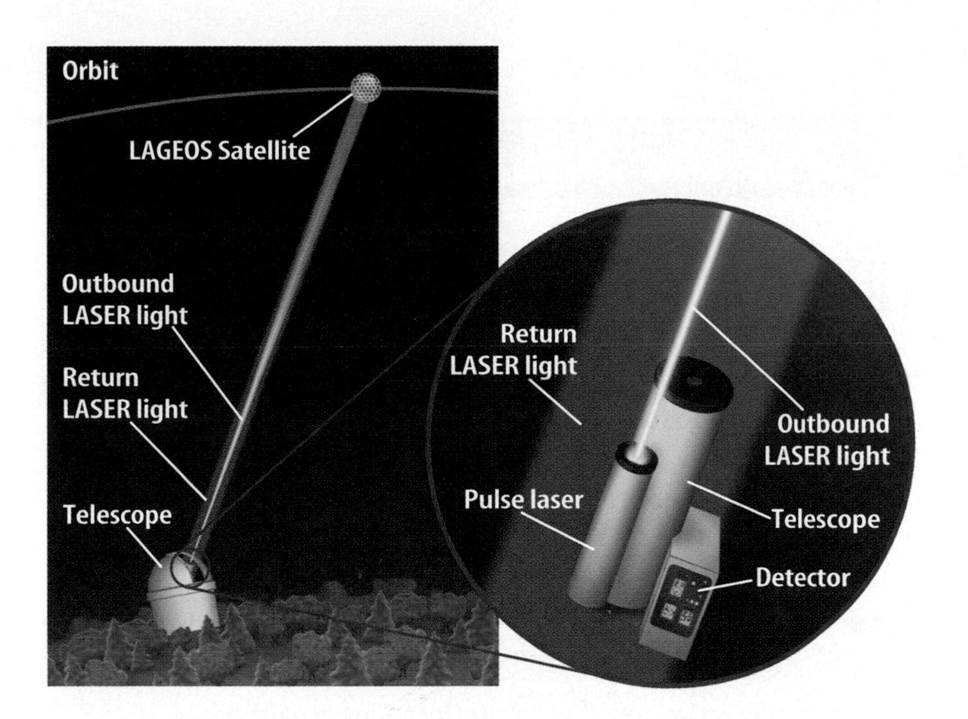

Figure 16
When using the Satellite Laser Ranging System, scientists on the ground aim laser pulses at a satellite. The pulses reflect off the satellite and are used to determine a precise location on the ground.

Current Data Satellite data show that Hawaii is moving toward Japan at a rate of about 8.3 cm per year. Maryland is moving away from England at a rate of 1.7 cm per year. Using such methods, scientists have observed that the plates move at rates ranging from about 1 cm to 12 cm per year.

Section 3 Assessment

1. What happens to plates at a transform plate boundary?
2. What occurs at plate boundaries that are associated with seafloor spreading?
3. Describe three types of plate boundaries where volcanic eruptions can occur.
4. How are convection currents related to plate tectonics?
5. **Think Critically** Using **Figure 9** and a world map, determine what natural disasters might occur in Iceland. Also determine what disasters might occur in Tibet. Explain why some Icelandic disasters are not expected to occur in Tibet.

Skill Builder Activities

6. **Predicting** Plate tectonic activity causes many events that can be dangerous to humans. One of these events is a seismic sea wave, or tsunami. Learn how scientists predict the arrival time of a tsunami in a coastal area. **For more help, refer to the Science Skill Handbook.**
7. **Using a Word Processor** Write three separate descriptions of the three basic types of plate boundaries—divergent boundaries, convergent boundaries, and transform boundaries. Then draw a sketch of an example of each boundary next to your description. **For more help, refer to the Technology Skill Handbook.**

Activity Use the Internet

Predicting Tectonic Activity

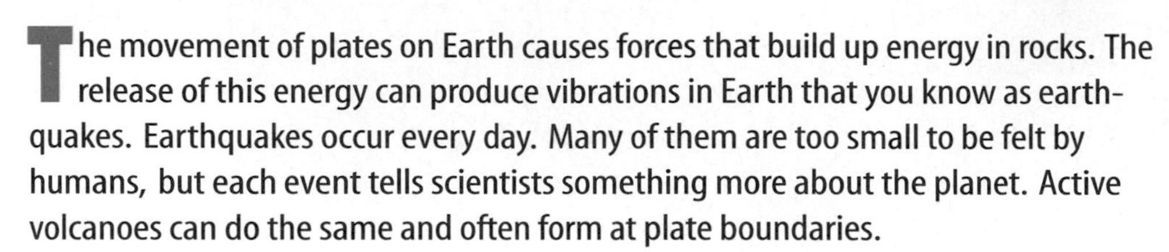

The movement of plates on Earth causes forces that build up energy in rocks. The release of this energy can produce vibrations in Earth that you know as earthquakes. Earthquakes occur every day. Many of them are too small to be felt by humans, but each event tells scientists something more about the planet. Active volcanoes can do the same and often form at plate boundaries.

Recognize the Problem

Can you predict tectonically active areas by plotting locations of earthquake epicenters and volcanic eruptions?

Form a Hypothesis

Think about where earthquakes and volcanoes have occurred in the past. Make a hypothesis about whether the locations of earthquake epicenters and active volcanoes can be used to predict tectonically active areas.

Goals

- **Research** the locations of earthquakes and volcanic eruptions around the world.
- **Plot** earthquake epicenters and the locations of volcanic eruptions obtained from the Glencoe Science Web site.
- **Predict** locations that are tectonically active based on a plot of the locations of earthquake epicenters and active volcanoes.

Data Sources

SCIENCE*Online* Go to the Glencoe Science Web site at **science.glencoe.com** for more information about earthquake and volcano sites, hints about earthquake and volcano sites, and data from other students.

Test Your Hypothesis

Plan

1. Make a data table in your Science Journal like the one shown.
2. Collect data for earthquake epicenters and volcanic eruptions for at least the past two weeks. Your data should include the longitude and latitude for each location. For help, refer to the data sources given on the opposite page.

Locations of Epicenters and Eruptions		
Earthquake Epicenter/ Volcanic Eruption	Longitude	Latitude

Do

1. Make sure your teacher approves your plan before you start.
2. **Plot** the locations of earthquake epicenters and volcanic eruptions on a map of the world. Use an overlay of tissue paper or plastic.
3. After you have collected the necessary data, predict where the tectonically active areas on Earth are.
4. **Compare and contrast** the areas that you predicted to be tectonically active with the plate boundary map shown in **Figure 9.**

Analyze Your Data

1. What areas on Earth do you predict to be the locations of tectonic activity?
2. How close did your prediction come to the actual location of tectonically active areas?

Draw Conclusions

1. How could you make your predictions closer to the locations of actual tectonic activity?
2. Would data from a longer period of time help? Explain.
3. What types of plate boundaries were close to your locations of earthquake epicenters? Volcanic eruptions?
4. **Explain** which types of plate boundaries produce volcanic eruptions. Be specific.

Communicating Your Data

SCIENCE Online Find this Internet activity on the Glencoe Science Web site at **science.glencoe.com.** **Post** your data in the table provided. **Compare** your data to those of other students. Combine your data with those of other students and **plot** these combined data on a map to **recognize** the relationship between plate boundaries, volcanic eruptions, and earthquake epicenters.

Science and Language Arts

Listening In

by Gordon Judge

Respond to the Reading

1. Who is narrating the poem?
2. Why might the narrator think he or she hasn't "moved for ages"?
3. Who or what is the narrator's "best mate"?

I'm just a bit of seafloor on this mighty solid sphere.
With no mind to be broadened, I'm quite content down here.
The mantle churns below me, and the sea's in turmoil, too;
But nothing much disturbs me, I'm rock solid through and through.

I do pick up occasional low-frequency vibrations –
(I think, although I can't be sure, they're sperm whales' conversations).
I know I shouldn't listen in, but what else can I do?
It seems they are all studying for degrees from the OU.

They've mentioned me in passing, as their minds begin improving:
I think I've heard them say "The theory says the sea-floor's moving…".
Well, that shook me, I can tell you; yes, it gave me quite a fright.
Yet I've not moved for ages, so I *know* it can't be right.

They call it "Plate Tectonics", this new theory in their noddle.
If they would only ask me, I could tell them it's all twaddle.
Apparently, I "oozed out from a mid-Atlantic split,
Solidified and cooled right down, then moved out bit by bit".

But, how can I be moving, when I know full well myself
That I'm quite firmly anchored to a continental shelf?
"Well, the continent is moving, too; you're *pushing* it, you see,"
I hear those OU whales intone, hydro-acoustically.

Now, my best mate's a sea floor in the mighty East Pacific.
He reckons life is balmy there: the summers are terrific!
He's heard the whale-talk, too, and found it pretty scary.
"Subduction" was the word he heard, which sounded rather hairy.

It was to be his fate, they claimed with undisguised great relish:
A hot and fiery end to things – it really would be hellish.
In fact, he'd end up underneath *my* continent, lengthwise,
So I would be the one to blame for my poor mate's demise.

Well, thank you very much, OU. You've upset my composure.
Next time you send your student whales to look at my exposure
I'll tell them it's a load of tosh: it's *they* who move, not me,
Those arty-smarty blobs of blubber, clogging up the sea!

Understanding Literature

Point of View Point of view refers to the perspective from which an author writes. This poem begins, "I'm just a bit of sea floor…." Right away, you know that the poem, or story, is being told from the point of view of the speaker, or the "first person." Not all first-person stories are told from the point of view of a person. The narrator in this poem is a geological feature, not a person. This point of view helps give the poem a fantastic or outlandish quality. It also gives a playful tone to the poem. What other effects does the first-person narration have on the story?

Science Connection Volcanoes can occur where two plates move toward each other. In the poem, the author gives several clues that a volcano will form. First, the narrator's "best mate" is a seafloor in the Pacific Ocean. When an oceanic plate and a continental plate collide, a volcano will form. The narrator also hears the word *subduction* spoken. Subduction zones occur when one plate sinks under another plate. Rocks melt in the zones where these plates converge, causing magma to move upward and form volcanic mountains. What other clues does the author give that a volcano will form?

Linking Science and Writing

Using Point of View Using the first-person point of view, write an account from the point of view of a living or nonliving thing. You could write an account of an object, such as a pencil, that you use or encounter every day. You also could write from the point of view of a living thing, such as a family pet. Be sure to use the personal pronoun "I" in your account.

Career Connection

Volcanologist

Ed Klimasauskas is a volcanologist at the Cascades Volcano Observatory in Washington State. His job is to study volcanoes in order to predict eruptions. Volcanologists' predictions can save lives: people can be evacuated from danger areas before an eruption occurs. Klimasauskas also educates the public about the hazards of volcanic eruptions and tells people who live near active volcanoes what they can do to be safe in case a volcano erupts. Volcanologists travel all over the world to study new sites.

SCIENCE*Online* To learn more about careers in volcanology, visit the Glencoe Science Web site at **science.glencoe.com.**

Chapter 14 Study Guide

Reviewing Main Ideas

Section 1 Continental Drift

1. Alfred Wegener suggested that the continents were joined together at some point in the past in a large landmass he called Pangaea. Wegener proposed that continents have moved slowly, over millions of years, to their current locations.
2. The puzzlelike fit of the continents, fossils, climatic evidence, and similar rock structures support Wegener's idea of continental drift. However, Wegener could not explain what process could cause the movement of the landmasses. *How do fossils support the hypothesis of continental drift?*

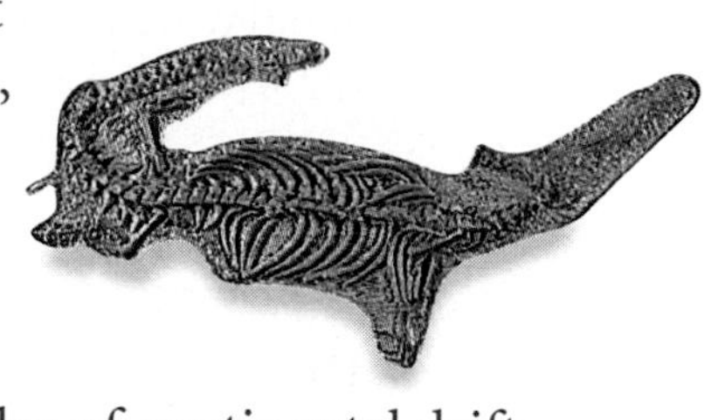

Section 2 Seafloor Spreading

1. Detailed mapping of the ocean floor in the 1950s showed underwater mountains and rift valleys.
2. In the 1960s, Harry Hess suggested seafloor spreading as an explanation for the formation of mid-ocean ridges. *How is magnetic evidence preserved in rocks forming along a mid-ocean ridge?*

Normal magnetic polarity
Reversed magnetic polarity
Mid-ocean ridge
Lithosphere

3. The theory of seafloor spreading is supported by magnetic evidence in rocks and by the ages of rocks on the ocean floor.

Section 3 Theory of Plate Tectonics

1. In the 1960s, scientists combined the ideas of continental drift and seafloor spreading to develop the theory of plate tectonics. The theory states that the surface of Earth is broken into sections called plates that move around on the asthenosphere.
2. Currents in Earth's mantle called convection currents transfer heat in Earth's interior. It is thought that this transfer of heat energy moves plates.
3. Earth is a dynamic planet. As the plates move, they interact, resulting in many of the features of Earth's surface. *How do converging plates form mountains?*

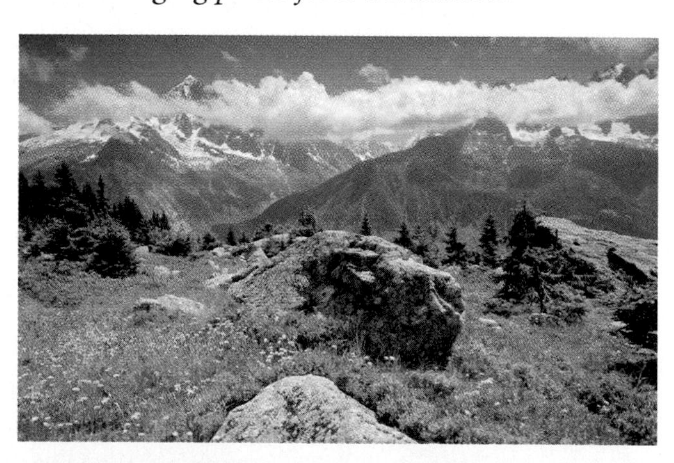

After You Read

FOLDABLES Reading & Study Skills

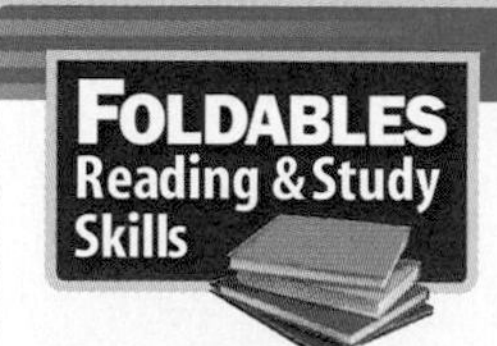

To help you review what you learned about plate tectonics, use the Foldable you made at the beginning of the chapter.

Chapter 14 Study Guide

Visualizing Main Ideas

Complete the concept map below about continental drift, seafloor spreading, and plate tectonics.

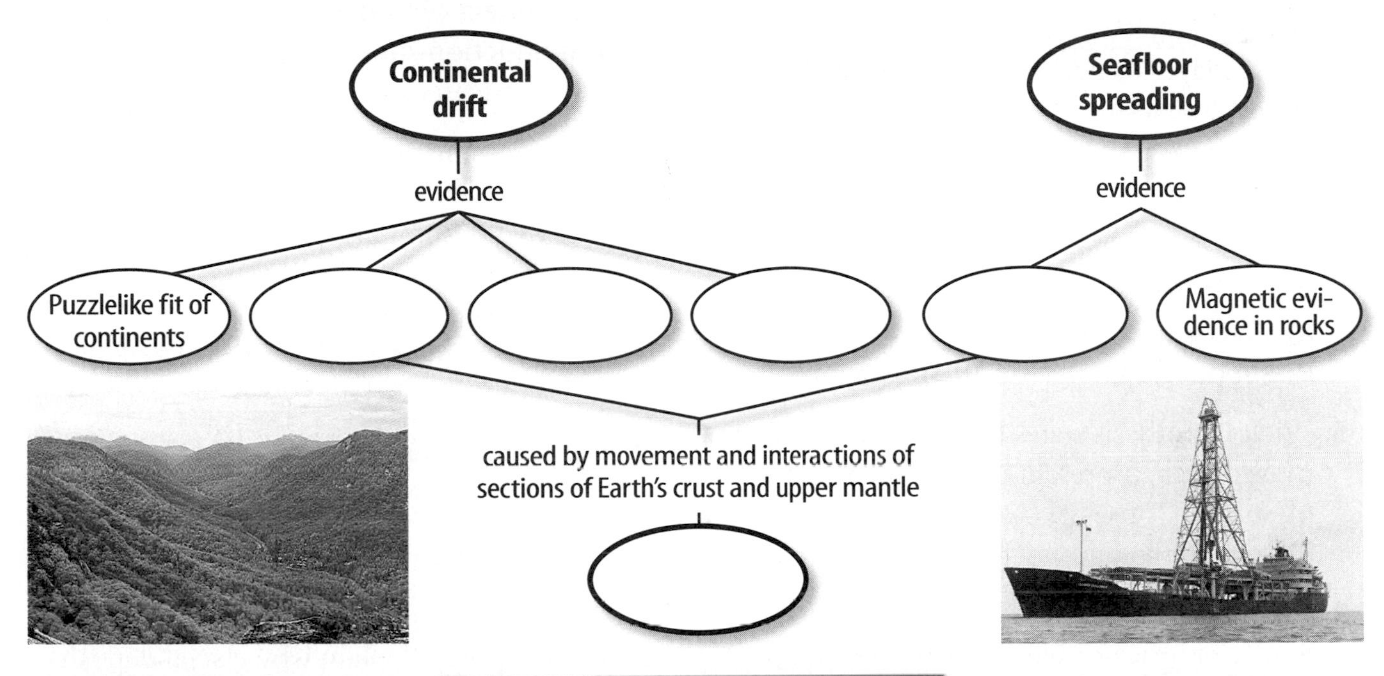

Vocabulary Review

Vocabulary Words

- **a.** asthenosphere
- **b.** continental drift
- **c.** convection current
- **d.** lithosphere
- **e.** Pangaea
- **f.** plate
- **g.** plate tectonics
- **h.** seafloor spreading

Using Vocabulary

Each phrase below describes a vocabulary term from the list. Write the term that matches the phrase describing it.

1. plasticlike layer below the lithosphere
2. idea that continents move slowly across Earth's surface
3. large, ancient landmass that consisted of all the continents on Earth
4. process that forms new seafloor as hot material is forced upward
5. driving force for plate movement
6. composed of oceanic or continental crust and upper mantle
7. explains locations of mountains, trenches, and volcanoes
8. piece of the lithosphere that moves over a plasticlike layer
9. theory proposed by Harry Hess that includes processes along mid-ocean ridges
10. forms as warm material rises and cold material sinks

Study Tip

Make a note of anything you don't understand so that you'll remember to ask your teacher about it.

Chapter 14 Assessment

Checking Concepts

Choose the word or phrase that best answers the question.

1. Which layer of Earth contains the asthenosphere?
A) crust
B) mantle
C) outer core
D) inner core

2. What type of plate boundary is the San Andreas Fault part of?
A) divergent
B) subduction
C) convergent
D) transform

3. What hypothesis states that continents slowly moved to their present positions on Earth?
A) subduction
B) seafloor spreading
C) continental drift
D) erosion

4. Which plate is subducting beneath the South American Plate to form the Andes mountain range?
A) North American
B) African
C) Indo-Australian
D) Nazca

5. Which of the following features indicates that many continents were once near Earth's south pole?
A) glacial deposits
B) mid-ocean ridges
C) volcanoes
D) earthquakes

6. What evidence in rocks supports the theory of seafloor spreading?
A) plate movement
B) magnetic reversals
C) subduction
D) convergence

7. Which type of plate boundary is the Mid-Atlantic Ridge a part of?
A) convergent
B) divergent
C) transform
D) lithosphere

8. What theory states that plates move around on the asthenosphere?
A) continental drift
B) seafloor spreading
C) subduction
D) plate tectonics

9. What forms when one plate slides past another plate?
A) transform boundary
B) divergent boundary
C) subduction zone
D) mid-ocean ridge

10. When oceanic plates collide, what volcanic landforms are made?
A) folded mountains
B) island arcs
C) strike-slip faults
D) mid-ocean ridges

Thinking Critically

11. Why do many earthquakes but few volcanic eruptions occur in the Himalaya?

12. Glacial deposits often form at high latitudes near the poles. Explain why glacial deposits have been found in Africa.

13. How is magnetism used to support the theory of seafloor spreading?

14. Explain why volcanoes do not form along the San Andreas Fault.

15. Explain why the fossil of an ocean fish found on two different continents would not be good evidence of continental drift.

Developing Skills

16. Forming Hypotheses Mount St. Helens in the Cascade Range is a volcano. Use **Figure 9** and a U.S. map to hypothesize how it might have formed.

17. Measuring in SI Movement along the African Rift Valley is about 2.1 cm per year. If plates continue to move apart at this rate, how much larger will the rift be (in meters) in 1,000 years? In 15,500 years?

18. Concept Mapping Make an events chain concept map that describes seafloor spreading along a divergent plate boundary. Choose from the following phrases: *magma cools to form new seafloor, convection currents circulate hot material along divergent boundary,* and *older seafloor is forced apart.*

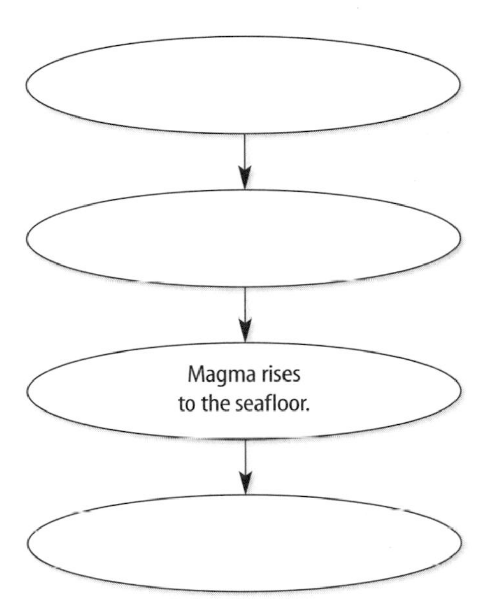

Performance Assessment

19. Observe and Infer In the MiniLab Modeling Convection Currents, you observed convection currents produced in water as it was heated. Repeat the experiment, placing sequins, pieces of wood, or pieces of rubber bands into the water. How do their movements support your observations and inferences from the MiniLab?

TECHNOLOGY

Go to the Glencoe Science Web site at **science.glencoe.com** or use the **Glencoe Science CD-ROM** for additional chapter assessment.

Test Practice

Ms. Fernandez was leading a class discussion on plate tectonics and Earth's interior.

Study the diagram below and then answer the following questions.

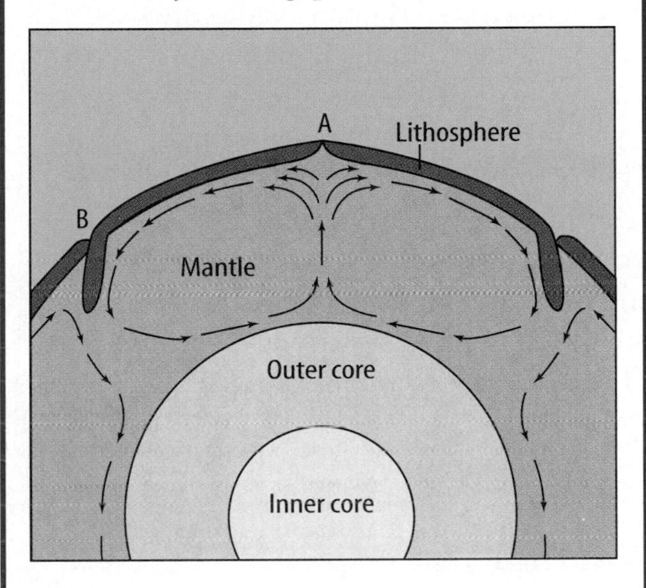

1. Suppose that the arrows in the diagram represent patterns of convection in Earth's mantle. Which type of plate boundary is most likely to form along the region labeled "A"?

A) transform
B) reverse
C) convergent
D) divergent

2. Which statement is true of the region marked "B" on the diagram?

F) Plates separate and slip past one another sideways.
G) Plates diverge and volcanoes form.
H) Plates converge and volcanoes form.
J) Plates collapse and form a strike-slip boundary.

CHAPTER

15 Earthquakes and Volcanoes

The ground shook violently and the bridge came tumbling down. The Loma Prieta earthquake of 1989 hit San Francisco hard. The Cypress freeway in Oakland was one of its many victims. You might be wondering what caused this disaster and what could have been done to prevent it. In this chapter, you'll begin to answer these questions as you read about earthquakes, volcanoes, and Earth's moving plates.

What do you think?

Science Journal Look at the picture below with a classmate. Discuss what this might be. Here's a hint: *It's a far-out view of something that starts very deep down.* Write your answer or best guess in your Science Journal.

One of the greatest dangers associated with an earthquake occurs when people are inside buildings during the event. If buildings were constructed so they would not fall down as easily when shaken by an earthquake, the number of deaths and the amount of destruction could be reduced. In the following activity, you will see how construction materials can be used to help strengthen a building.

Construct with strength

1. Using wooden blocks, construct a building with four walls. Place a piece of cardboard over the four walls as a ceiling.
2. Gently shake the table under your building. Describe what happens.
3. Reconstruct the building. Wrap large rubber bands around each section, or wall, of blocks. Then wrap large rubber bands around the entire building.
4. Gently shake the table again.

Observe

In your Science Journal, note any differences you observed as the two buildings were shaken. Hypothesize how the construction methods you used in this activity might be applied to the construction of real buildings.

Before You Read

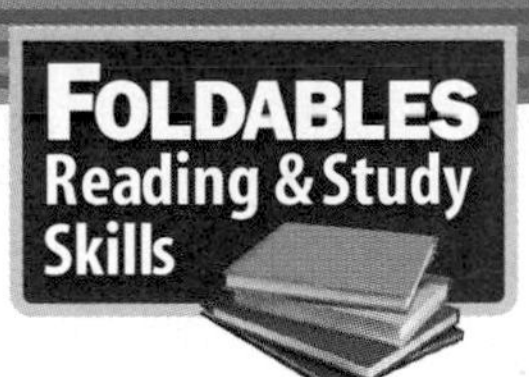

Making a Compare and Contrast Study Fold Make the following Foldable to help you see how earthquakes and volcanoes are similar and different.

1. Place a sheet of paper in front of you so the long side is at the top. Fold the paper in half from the left side to the right side and then unfold.
2. Fold each side in to the centerfold line to divide the paper into fourths.
3. Draw a volcano on one flap and label the flap *Volcanoes.* Draw an earthquake on the other flap and label it *Earthquakes.*
4. Before you read the chapter, write what you know about earthquakes and volcanoes on the back of each flap. As you read the chapter, add to your information.

SECTION

Earthquakes

As You Read

What You'll Learn

- **Explain** how earthquakes are caused by a buildup of strain in Earth's crust.
- **Compare and contrast** primary, secondary, and surface waves.
- **Recognize** earthquake hazards and how to prepare for them.

Vocabulary

earthquake
fault
seismic wave
focus
epicenter
seismograph
magnitude
tsunami
seismic safe

Why It's Important

Studying earthquakes will help you learn where they might occur and how you can prepare for their hazards.

What causes earthquakes?

If you've gone for a walk in the woods lately, maybe you picked up a stick along the way. If so, did you try to bend or break it? If you've ever bent a stick slowly, you might have noticed that it changes shape but usually springs back to normal form when you stop bending it. If you continue to bend the stick, you can do it for only so long before it changes permanently. When this elastic limit is passed, the stick may break, as shown in **Figure 1.** When the stick snaps, you can feel vibrations in the stick.

Elastic Rebound As hard as they seem, rocks act in much the same way when forces push or pull on them. If enough force is applied, rocks become strained, which means they change shape. They may even break, and the ends of the broken pieces may snap back. This snapping back is called elastic rebound.

Rocks usually change shape, or deform, slowly over long periods of time. As they are strained, potential energy builds up in them. This energy is released suddenly by the action of rocks breaking and moving. Such breaking, and the movement that follows, causes vibrations that move through rock or other earth materials. If they are large enough, these vibrations are felt as **earthquakes.**

 Reading Check *What is an earthquake?*

Figure 1
A stick can bend only so far before it breaks. A When a stick is bent, potential energy is stored in the stick. B The energy is released as vibrations when the stick breaks.

Figure 2
When rocks change shape by breaking, faults form. The type of fault formed depends on the type of stress exerted on the rock.

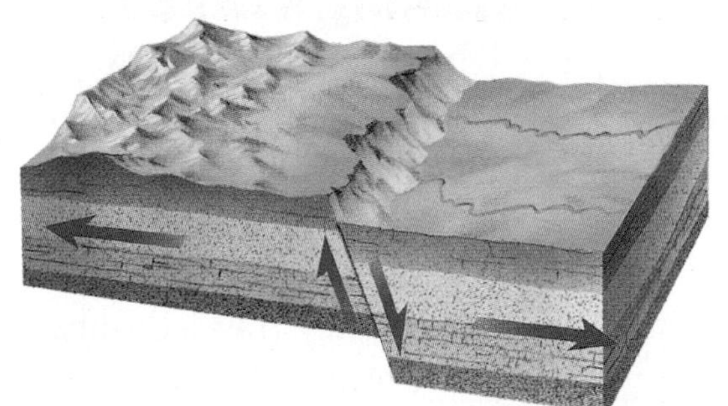

A When rocks are pulled apart, a normal fault may form.

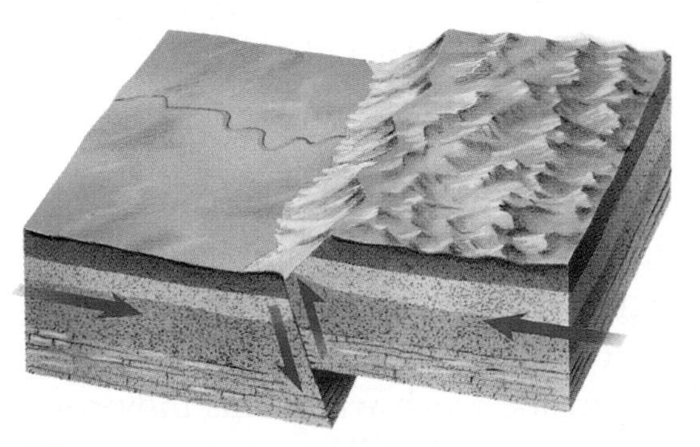

B When rocks are compressed, a reverse fault may form.

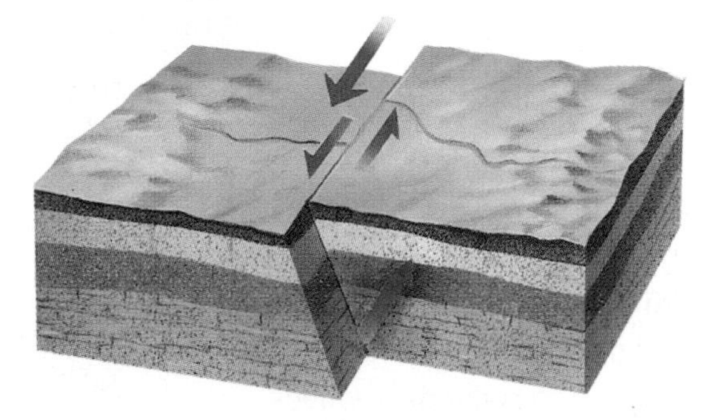

C When rocks are sheared, a strike-slip fault may form.

Types of Faults When a section of rock breaks, rocks on either side of the break might move because of elastic rebound. The surface of such a break is called a **fault.** Several types of faults exist. The type that forms depends on how forces were applied to the rocks.

When rocks are pulled apart under tension forces, normal faults form, as shown in **Figure 2A.** Along a normal fault, rock above the fault moves down compared to rock below the fault. Compression forces squeeze rocks together, like an accordion. Compression might cause rock above a fault to move up compared to rock below the fault. This movement forms reverse faults, as shown in **Figure 2B.** As illustrated in **Figure 2C,** rock experiencing shear forces can break to form a strike-slip fault. Shear forces cause rock on either side of a strike-slip fault to move past one another in opposite directions along Earth's surface. You could infer the motion of a strike-slip fault while walking along and observing an offset feature, such as a displaced fence line, on Earth's surface.

Where do the forces come from that cause rocks to deform by bending or breaking? Why do faults form and why do earthquakes occur in certain areas? As you'll learn later in this chapter, forces inside Earth are caused by the constant motion of plates, or sections, of Earth's crust and upper mantle.

Mini LAB

Observing Deformation

WARNING: *Do not taste or eat any lab materials. Wash hands when finished.*

Procedure

1. Remove the wrapper from three bars of **taffy.**
2. Hold a bar of taffy lengthwise between your hands and gently push on it from opposite directions.
3. Hold another bar of taffy and pull it in opposite directions.

Analysis

1. Which of the procedures that you performed on the taffy involved applying tension? Which involved applying compression?
2. Infer how to apply a shear stress to the third bar of taffy.

Making Waves

Do you recall the last time you shouted for a friend to save you a seat on the bus? When you called out, energy was transmitted through the air to your friend, who interpreted the familiar sound of your voice as belonging to you. These sound waves were released by your vocal cords and were affected by your tongue and mouth. They traveled outward through the air. Earthquakes also release waves. Earthquake waves are transmitted through materials in Earth and along Earth's surface. Earthquake waves are called **seismic waves.** In the two-page activity, you'll make waves similar to seismic waves by moving a coiled spring toy.

Earthquake Focus and Epicenter Movement along a fault releases strain energy. Strain energy is potential energy that builds up in rock when it is bent. When this potential energy is released, it moves outward from the fault in the form of seismic waves. The point inside Earth where this movement first occurs and energy is released is called the **focus** of an earthquake, as shown in **Figure 3.** The point on Earth's surface located directly above the earthquake focus is called the **epicenter** of the earthquake.

Reading Check *What is the focus of an earthquake?*

Figure 3
During an earthquake, several types of seismic waves form. Primary and secondary waves travel in all directions from the focus and can travel through Earth's interior. Surface waves travel at shallow depths and along Earth's surface. *Which seismic waves are the most destructive?*

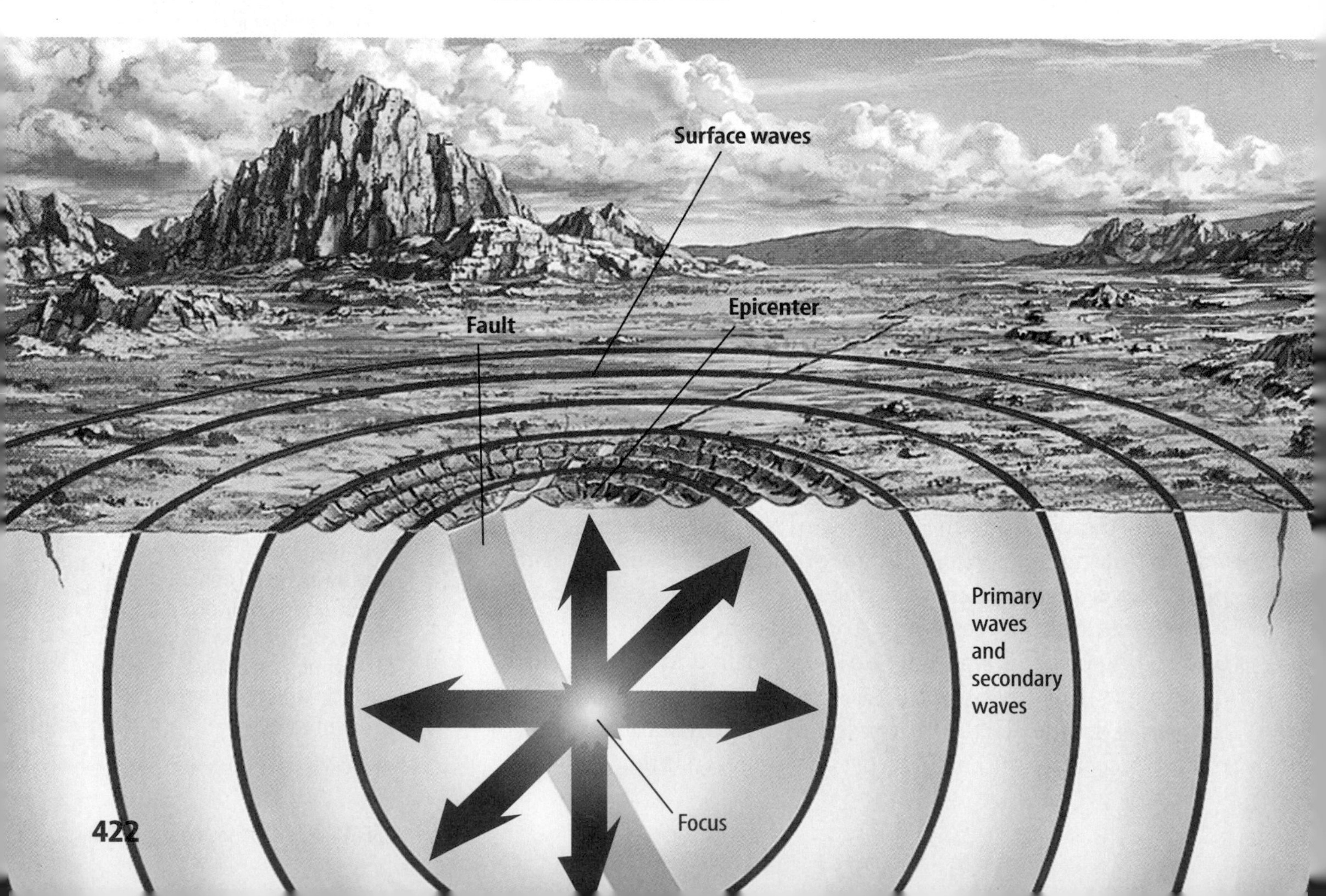

Seismic Waves After they are produced at the focus, seismic waves travel away from the focus in all directions, as illustrated in **Figure 3.** Some seismic waves travel throughout Earth's interior, and others travel along Earth's surface. The surface waves cause the most damage during an earthquake event.

Primary waves, also known as P-waves, travel the fastest through rock material by causing particles in the rock to move back and forth, or vibrate, in the same direction as the waves are moving. Secondary waves, known as S-waves, move through rock material by causing particles in the rock to vibrate at right angles to the direction in which the waves are moving. P- and S-waves travel through Earth's interior. Studying them has revealed much information about Earth's interior.

Surface waves are the slowest and largest of the seismic waves, and they cause most of the destruction during an earthquake. The movements of surface waves are complex. Some surface waves move along Earth's surface in a manner that moves rock and soil in a backward rolling motion. They have been observed moving across the land like waves of water. Some surface waves vibrate in a side-to-side, or swaying, motion parallel to Earth's surface. This motion can be particularly devastating to human-built structures.

Figure 4
Scientists study seismic waves using seismographs located around the world.

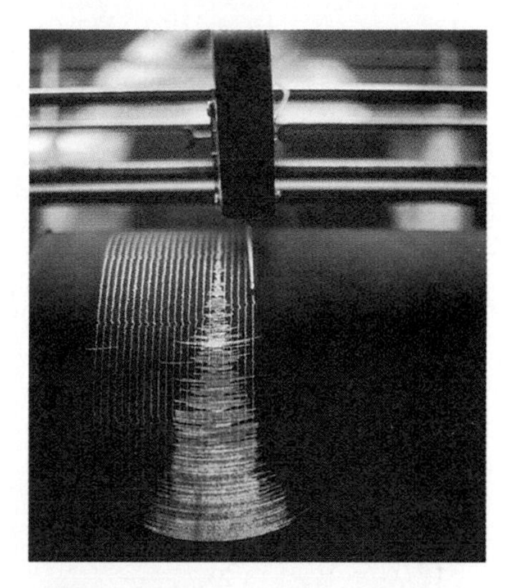

A **This seismograph records incoming seismic waves using a fixed mass.**

B **Some seismographs collect and store data on a computer.**

Learning from Earthquakes

On your way to lunch tomorrow, suppose you were to walk twice as fast as your friend does. What would happen to the distance between the two of you as you walked to the lunchroom? The distance between you and your friend would become greater the farther you walked, and you would arrive first. Using this same line of reasoning, scientists use the different speeds of seismic waves and their differing arrival times to calculate the distance to an earthquake epicenter.

Earthquake Measurements Seismologists are scientists who study earthquakes and seismic waves. The instrument they use to obtain a record of seismic waves from all over the world is called a **seismograph,** shown in **Figure 4A.**

One type of seismograph has a drum holding a roll of paper on a fixed frame. A pendulum with an attached pen is suspended from the frame. When seismic waves are received at the station, the drum vibrates but the pendulum remains at rest. The pen on the pendulum traces a record of the vibrations on the paper. The height of the lines traced on the paper is a measure of the energy released by the earthquake, also known as its **magnitude.**

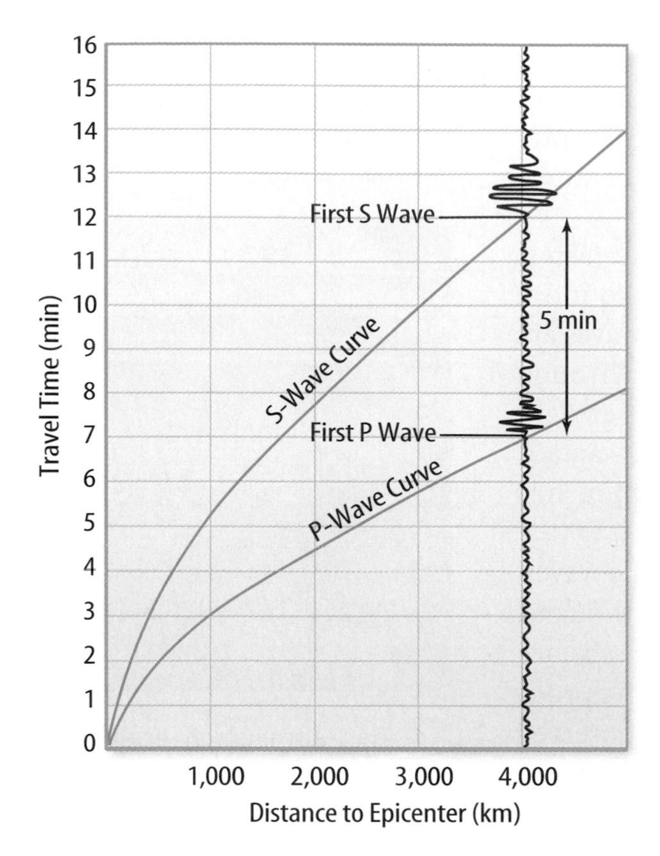

Figure 5
P- and S-waves travel at different speeds. These speeds are used to determine how close a seismograph station is to an earthquake.

Figure 6
After distances from at least three seismograph stations are determined, they are plotted as circles with radii equal to these distances on a map. The epicenter is the point at which the circles intersect.

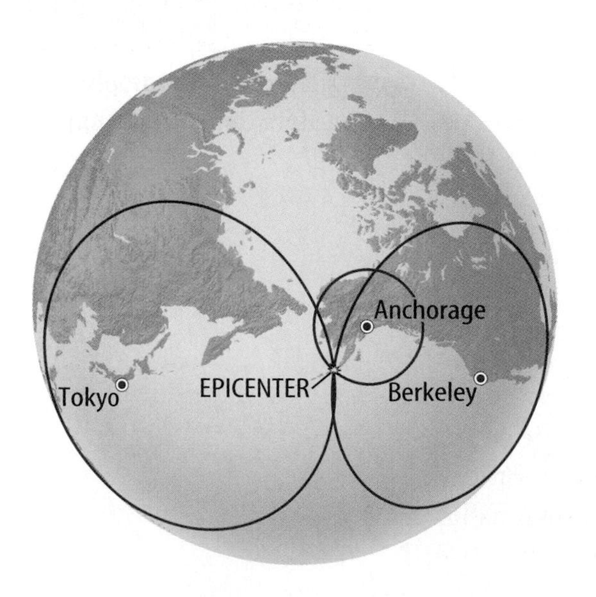

Epicenter Location When seismic-wave arrival times are recorded at a seismograph station, the distance from that station to the epicenter can be determined. The farther apart the arrival times for the different waves are, the farther away the earthquake epicenter is. This difference is shown by the graph in **Figure 5.** Using this information, scientists draw a circle with a radius equal to the distance from the earthquake for each of at least three seismograph stations, as illustrated in **Figure 6.** The point where the three circles meet is the location of the earthquake epicenter. Data from many stations normally are used to determine an epicenter location.

How strong are earthquakes?

As shown in **Table 1,** major earthquakes cause much loss of life. For example, on September 20, 1999, a major earthquake struck Taiwan, leaving more than 2,400 people dead, more than 8,700 injured, and at least 100,000 homeless. Sometimes earthquakes are felt and can cause destruction in areas hundreds of kilometers away from their epicenters. The Mexico City earthquake in 1985 is an example of this. The movement of the soft sediment underneath Mexico City caused extensive damage to this city, even though the epicenter was nearly 400 km away.

The Richter Scale Richter (RIHK tur) magnitude is based on measurements of amplitudes, or heights, of seismic waves as recorded on seismographs. Richter magnitude describes how much energy an earthquake releases. For each increase of 1.0 on the Richter scale, the amplitude of the highest recorded seismic wave increases by 10. However, about 32 times more energy is released for every increase of 1.0 on the scale. For example, an earthquake with a magnitude of 7.5 releases about 32 times more energy than one with a magnitude of 6.5, and the wave height for a 7.5-magnitude quake is ten times higher than for a quake with a magnitude of 6.5.

Earthquake Damage

Year	Location	Magnitude	Deaths
1989	Loma Prieta, CA	7.1	62
1990	Iran	7.7	50,000
1990	Luzon, Philippines	7.8	1,621
1993	Guam	8.1	none
1993	Marharashtra, India	6.4	30,000
1994	Northridge, CA	6.7	61
1995	Kobe, Japan	6.8	5,378
1997	Iran	7.3	1,500
1998	Afghanistan-Tajikistan border	5.9	2,323
1998	Afghanistan-Tajikistan border	6.1	4,000
1999	Colombia, South America	6.2	2,000
1999	Taiwan	7.7	2,400
2000	Indonesia	7.9	103
2001	India	7.7	20,000

Table 1 Strong Earthquakes

Another way to measure earthquakes is available. The modified Mercalli intensity scale measures the intensity of an earthquake. Intensity is a measure of the amount of structural and geologic damage done by an earthquake in a specific location. The range of intensities spans Roman numerals I through XII. The amount of damage done depends on several factors—the strength of the earthquake, the nature of the surface material, the design of structures, and the distance from the epicenter. An intensity-I earthquake would be felt only by a few people under ideal conditions, whereas an intensity-XII earthquake would cause major destruction to human-built structures and Earth's surface. The 1994 earthquake in Northridge, California was a Richter magnitude 6.7, and its intensity was listed at IX. An intensity-IX earthquake causes considerable damage to buildings and could cause cracks in the ground.

Tsunamis Most damage from an earthquake is caused by surface waves. Buildings can crack or fall down. Elevated bridges and highways can collapse. However, people living near the seashore must protect themselves against another hazard from earthquakes. When an earthquake occurs on the ocean floor, the sudden movement pushes against the water and powerful water waves are produced. These waves can travel outward from the earthquake thousands of kilometers in all directions.

When these seismic sea waves, or **tsunamis,** are far from shore, their energy is spread out over large distances and great water depths. The wave heights of tsunamis are less than a meter in deep water, and large ships can ride over them and not even know it. However, when tsunamis approach land, the waves slow down and their wave heights increase as they encounter the bottom of the seafloor. This creates huge tsunami waves that can be as much as 30 m in height. Just before a tsunami crashes to shore, the water near a shoreline may move rapidly out toward the sea. If this should happen, there is immediate danger that a tsunami is about to strike. **Figure 7** illustrates the behavior of a tsunami as it approaches the shore.

Research Visit the Glencoe Science Web site at **science.glencoe.com** for information on determining earthquake magnitudes.

NATIONAL GEOGRAPHIC **VISUALIZING TSUNAMIS**

Figure 7

The diagram below shows stages in the development of a tsunami. A tsunami is an ocean wave that is usually generated by an earthquake and is capable of inflicting great destruction.

▶ TSUNAMI ALERT The red dots on this map show the tide monitoring stations that make up part of the Tsunami Warning System for the Pacific Ocean. The map shows approximately how long it would take for tsunamis that originate at different places in the Pacific to reach Hawaii. Each ring represents two hours of travel time.

Displacement

Ⓐ The vibrations set off by a sudden movement along a fault in Earth's crust are transferred to the water's surface and spread across the ocean in a series of long waves.

Ⓑ The waves travel across the ocean at speeds ranging from about 500 to 950 kilometers per hour.

Ⓒ When a tsunami wave reaches shallow water, friction slows it down and causes it to roll up into a wall of water—sometimes 30 meters high—before it breaks against the shore.

Tsunami Warning System buoy

TSUNAMI

Earthquake Safety

You've just read about the destruction that earthquakes cause. Fortunately, there are ways to reduce the damage and the loss of life associated with earthquakes.

Learning the earthquake history of an area is one of the first things to do to protect yourself. If the area you are in has had earthquakes before, chances are it will again and you can prepare for that.

Is your home seismic safe? What could you do to make your home earthquake safe? As shown in **Figure 8,** it's a good idea to move all heavy objects to lower shelves so they can't fall on you. Make sure your gas hot-water heater and appliances are well secured. A new method of protecting against fire is to place sensors on your gas line that would shut off the gas when the vibrations of an earthquake are felt.

In the event of an earthquake, keep away from all windows and avoid anything that might fall on you. Watch for fallen power lines and other possible fire hazards. Collapsed buildings and piles of rubble can contain many sharp edges, so keep clear of these areas.

Seismic-Safe Structures If a building is considered **seismic safe,** it will be able to stand up against the vibrations caused by most earthquakes. Residents in earthquake-prone areas are constantly improving the way structures are built. Since 1971, stricter building codes have been enforced in California. Older buildings have been reinforced. Many high-rise office buildings now stand on huge steel-and-rubber supports that could enable them to ride out the vibrations of an earthquake. Underground water and gas pipes are replaced with pipes that will bend during an earthquake. This can help prevent broken gas lines and therefore reduce damage from fires.

Seismic-safe highways have cement pillars with spiral reinforcing rods placed within them. One structure that was severely damaged in the 1989 Loma Prieta, California earthquake was Interstate Highway 880. The collapsed highway was due to be renovated to make it seismic safe. It was built in the 1950s and did not have spiral reinforcing rods in its concrete columns. When the upper highway went in one direction, the lower one went in the opposite direction. The columns collapsed and the upper highway came down onto the lower one.

Figure 8
You can minimize your risk of getting hurt by preparing for an earthquake in advance.

A Placing heavy or breakable objects on lower shelves means they won't fall too far during an earthquake.

B Vibration sensors on gas lines shut off the supply of gas automatically during an earthquake. *What hazard can be prevented if the gas is turned off?*

Figure 9
One way to monitor changes along a fault is to detect any movement that occurs.

Predicting Earthquakes Imagine how many lives could be saved if only the time and location of a major earthquake could be predicted. Because most injuries from earthquakes occur when structures fall on top of people, it would help if people could be warned to move outside of buildings.

Researchers try to predict earthquakes by noting changes that precede them. That way, if such changes are observed again, an earthquake warning may be issued.

For example, movement along faults is monitored using laser-equipped, distance-measuring devices, such as the one shown in **Figure 9.** Changes in groundwater level or in electrical properties of rocks under stress have been measured by some scientists. Some people even study rock layers that have been affected by ancient earthquakes. Whether any of these studies will lead to the accurate and reliable prediction of earthquakes, no one knows. A major problem is that no single change in Earth occurs for all earthquakes. Each earthquake is unique.

Long-range forecasts predict whether an earthquake of a certain magnitude is likely to occur in a given area within 30 to 100 years. Forecasts of this nature are used to update building codes to make a given area more seismic safe.

Section  Assessment

1. What happens to rocks after their elastic limit is passed?
2. Which seismic wave arrives first at a seismograph station? Which arrives last? Which seismic waves cause most of the damage during an earthquake?
3. What improvements have been made to buildings and other structures to make them more seismic safe?
4. How can seismic waves be used to determine an earthquake's epicenter?
5. **Think Critically** Explain how a magnitude-8.0 earthquake could be classified as a low-intensity earthquake.

Skill Builder Activities

6. **Making and Using Tables** Use **Table 1** to research the earthquakes that struck Indonesia in 2000, Loma Prieta, California in 1989, and Iran in 1990. Although the three earthquakes were close in magnitude, explain why there was such a great difference in the number of deaths. **For more help, refer to the Science Skill Handbook.**
7. **Communicating** In your Science Journal, write a one-page description of how you would make your home or your classroom more seismic safe. **For more help, refer to the Science Skill Handbook.**

SECTION

Volcanoes

How do volcanoes form?

Much like air bubbles that are forced upward toward the bottom of an overturned bottle of denser syrup, molten rock, or magma, is forced upward toward Earth's surface by denser, cooler surrounding rock. Rising magma eventually can lead to an eruption, where magma, solids, and gas are spewed out to form cone-shaped mountains called **volcanoes.** As magma flows onto Earth's surface through a vent, or opening, it is called **lava.** Volcanoes have circular holes near their summits called craters. Lava and other volcanic materials can be expelled through a volcano's crater.

Some explosive eruptions throw lava and rock thousands of meters into the air. Bits of rock or solidified lava dropped from the air are called tephra. Tephra varies in size from volcanic ash to cinders to larger rocks called bombs or blocks.

Where Plates Collide Some volcanoes form because of collision of large plates of Earth's crust and upper mantle. This process has produced a string of volcanic islands, much like those illustrated in **Figure 10,** which includes Montserrat. These islands are forming as plates made up of oceanic crust and mantle collide. The older and denser oceanic plate subducts, or sinks beneath, the less dense plate, as shown in **Figure 10.** When one plate sinks under another plate, rock in and above the sinking plate melts, forming chambers of magma. This magma is the source for volcanic eruptions that have formed the Caribbean Islands.

As You Read

What You'll Learn

- **Explain** how volcanoes can affect people.
- **Describe** how types of materials are produced by volcanoes.
- **Compare** how three different volcano forms develop.

Vocabulary

volcano
lava
shield volcano
cinder cone volcano
composite volcano

Why It's Important

Volcanic eruptions can cause serious consequences for humans and other organisms.

Montserrat

Figure 10
A string of Caribbean Islands known as the Lesser Antilles form because of subduction. The island of Montserrat is among these.

A Volcanic ash blanketing an area can cause collapse of structures or—when mixed with precipitation—mudflows.

B Objects in the path of a pyroclastic flow are subject to complete destruction.

Figure 11
Several volcanic hazards are associated with explosive activity.

Modeling an Eruption

Procedure

1. Place **red-colored gelatin** into a **self-sealing plastic bag** until the bag is half-full.
2. Seal the bag and press the gelatin to the bottom of the bag.
3. Put a hole in the bottom of the bag with a **pin.**

Analysis

1. What parts of a volcano do the gelatin, the plastic bag, and the hole represent?
2. What force in nature did you mimic as you moved the gelatin to the bottom of the bag?
3. What factors in nature cause this force to increase and lead to an eruption?

Eruptions on a Caribbean Island Soufrière (soo free UR) Hills volcano on the island of Montserrat was considered dormant until recently. However, in 1995, Soufrière Hills volcano surprised its inhabitants with explosive activity. In July 1995, plumes of ash soared to heights of more than 10,000 m. This ash covered the capital city of Plymouth and many other villages, as shown in **Figure 11A.**

Every aspect of a once-calm tropical life changed when the volcano erupted. Glowing avalanches and hot, boiling mudflows destroyed villages and shut down the main harbor of the island and its airport. During activity on July 3, 1998, volcanic ash reached heights of more than 14,000 m. This ash settled over the entire island and was followed by mudflows brought on by heavy rains.

Pyroclastic flows are another hazard for inhabitants of Montserrat. They can occur anytime on any side of the volcano. Pyroclastic flows are massive avalanches of hot, glowing rock flowing on a cushion of intensely hot gases, as shown in **Figure 11B.** Speeds at which these flows travel can reach 200 km/h.

More than one half of Montserrat has been converted to a barren wasteland by the volcano. Virtually all of the farmland is now unusable, and most of the island's business and leisure centers are gone. Many of the inhabitants of the island have been evacuated to England, surrounding islands, or northern Montserrat, which is considered safe from volcanic activity.

Volcanic Risks According to the volcanic-risk map shown in **Figure 12,** inactive volcanic centers exist at Silver Hill, Centre Hill, and South Soufrière Hills. The active volcano, Soufrière Hills Volcano, is located just north of South Soufrière Hills. The risk map shows different zones of the island where inhabitants still are able to stay and locations from which they have been evacuated. Twenty people who had ignored evacuation orders were killed by pyroclastic flows from the June 25, 1997, event. These are the first and only deaths that have occurred since July 1995.

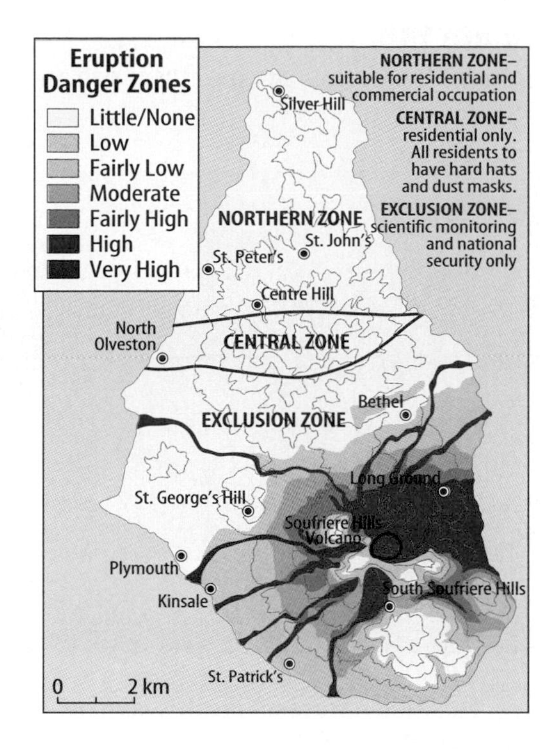

Figure 12
A volcanic risk map for Montserrat was prepared to warn inhabitants and visitors about unsafe areas on the island. *Do research and, using a computer, make your own volcanic risk map for a different volcano.*

Forms of Volcanoes

As you have learned, volcanoes can cause great destruction. However, volcanoes also add new rock to Earth's crust with each eruption. The way volcanoes add this new material to Earth's surface varies greatly. Different types of eruptions produce different types of volcanoes.

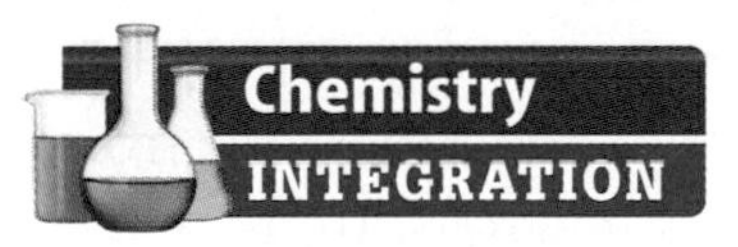

What determines how a volcano erupts? Some volcanic eruptions are violent, while during others lava flows out quietly around a vent. The composition of the magma plays a big part in determining the manner in which energy is released during a volcanic eruption. Lava that contains more silica, which is a compound consisting of silicon and oxygen, tends to be thicker and is more resistant to flow. Lava containing more iron and magnesium and less silica tends to flow easily. The amount of water vapor and other gases trapped in the lava also influences how lava erupts.

When you shake a bottle of carbonated soft drink before opening it, the pressure from the gas in the drink builds up and is released suddenly when the container is opened. Similarly, steam builds pressure in magma. This pressure is released as magma rises toward Earth's surface and eventually erupts. Sticky, silica-rich lava tends to trap water vapor and other gases.

Water is carried down from the surface of Earth into the mantle when one plate subducts beneath another, as in the case of the Lesser Antilles volcanoes. In hotter regions of Earth's interior, part of a descending plate and nearby rock will melt to form magma. The magma produced is more silica rich than the rock that melts to form the magma. Superheated steam produces tremendous pressure in such thick, silica-rich magmas. After enough pressure builds up, an eruption occurs. The type of lava and the gases contained in that lava determine the type of eruption that occurs.

Data Update For an online update of data on Soufrière Hills volcano on Montserrat, visit the Glencoe Science Web site at **science.glencoe.com** and select the appropriate chapter.

Figure 13
Volcanic landforms vary greatly in size and shape.

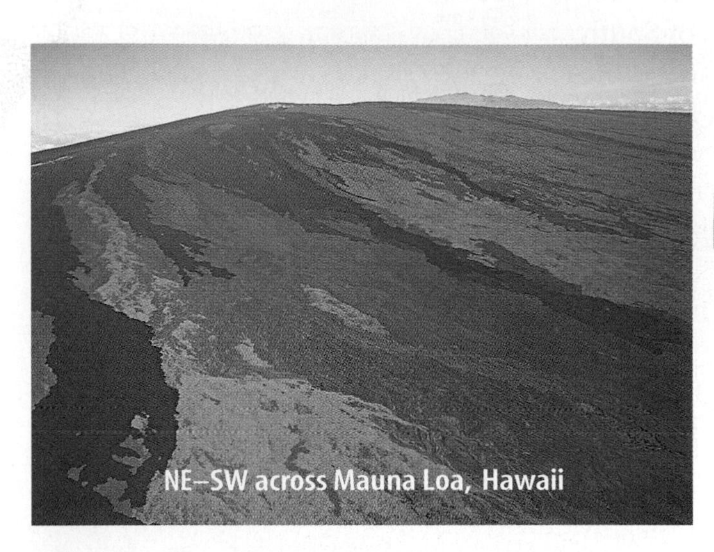

A **The fluid nature of basaltic lava has produced extensive flows at Mauna Loa, Hawaii—the largest active volcano on Earth.**

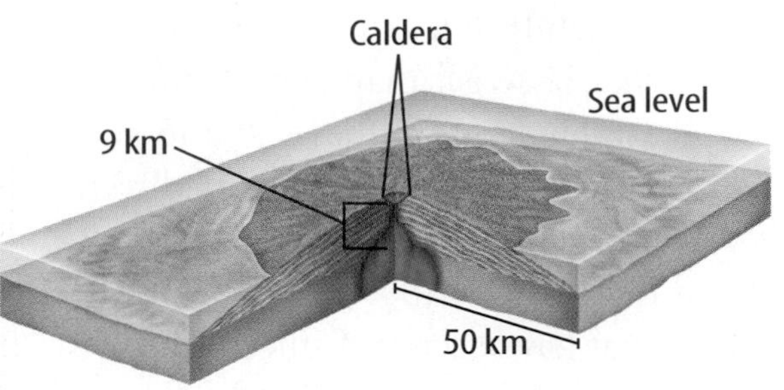

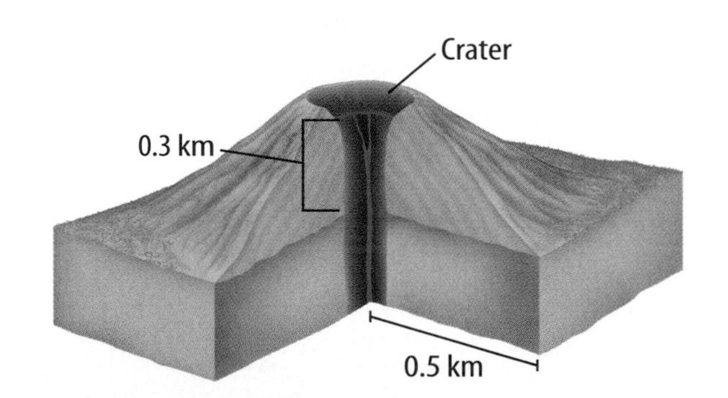

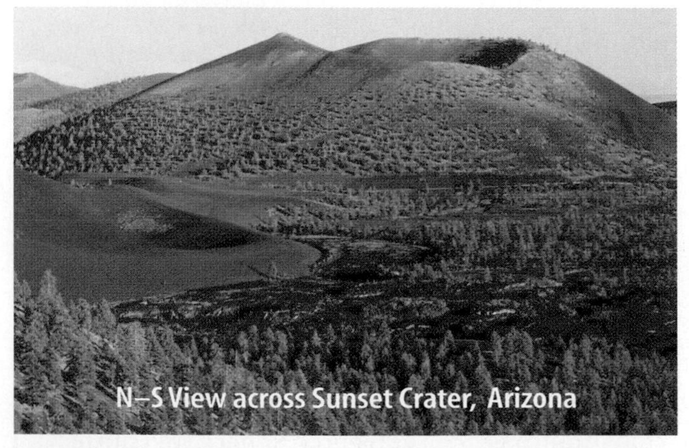

B **Sunset Crater is small and steep along its flanks—typical of a cinder cone. Compare the scale given for Sunset Crater with that shown in Figure 13A.**

Shield Volcanoes Basaltic lava, which is high in iron and magnesium and low in silica, flows in broad, flat layers. The buildup of basaltic layers forms a broad volcano with gently sloping sides called a **shield volcano.** Shield volcanoes, shown in **Figure 13A,** are the largest type of volcano to form. They form where magma is being forced up from extreme depths within Earth, or in areas where Earth's plates are moving apart. The separation of plates enables magma to be forced upward to Earth's surface.

✔ Reading Check ***What materials are shield volcanoes composed of?***

Cinder Cone Volcanoes Rising magma accumulates gases on its way to the surface. When the gas builds up enough pressure, it erupts. Moderate to violent eruptions throw volcanic ash, cinders, and lava high into the air. The lava cools quickly in midair and the particles of solidified lava, ash, and cinders fall back to Earth. This tephra forms a relatively small cone of volcanic material called a **cinder cone volcano.** Cinder cones are usually less than 300 m in height and often form in groups near other larger volcanoes. Because the eruption is powered by the high gas content, it usually doesn't last long. After the gas is released, the force behind the eruption is gone. Sunset Crater, an example of a cinder cone near Flagstaff, Arizona, is shown in **Figure 13B.**

Composite Volcanoes **Composite volcanoes** are steep-sided mountains composed of alternating layers of lava and tephra. They sometimes erupt violently, releasing large quantities of ash and gas. This forms a tephra layer of solid materials. Then a quieter eruption forms a lava layer.

Composite volcanoes form where one plate sinks beneath another. Soufrière Hills volcano is an example of a composite volcano. Another volcanic eruption from a composite volcano was the May 1980 eruption of Mount St. Helens in the state of Washington. It erupted explosively, spewing ash that fell on regions hundreds of kilometers away from the volcano. A composite volcano is shown in **Figure 13C.**

C Composite cones are intermediate in size and shape compared to shield volcanoes and cinder cone volcanoes.

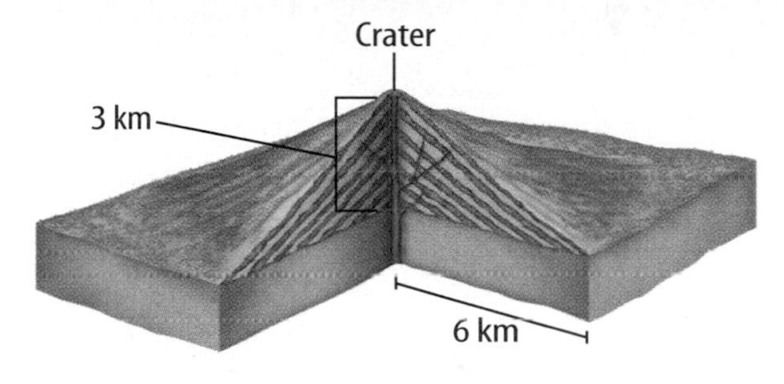

Fissure Eruptions Magma that is highly fluid can ooze from cracks or fissures in Earth's surface. This is the type of magma that usually is associated with fissure eruptions. The lava that erupts has a low viscosity, which means it can flow freely across the land to form flood basalts. Flood basalts that have been exposed to erosion for millions of years can become large, relatively flat landforms known as lava plateaus, as shown in **Figure 13D.** The Columbia River Plateau in the northwestern United States was formed about 15 million years ago when several fissures erupted and the flows built up layer upon layer.

D No modern example compares with the extensive flood basalts making up the Columbia River Plateau.

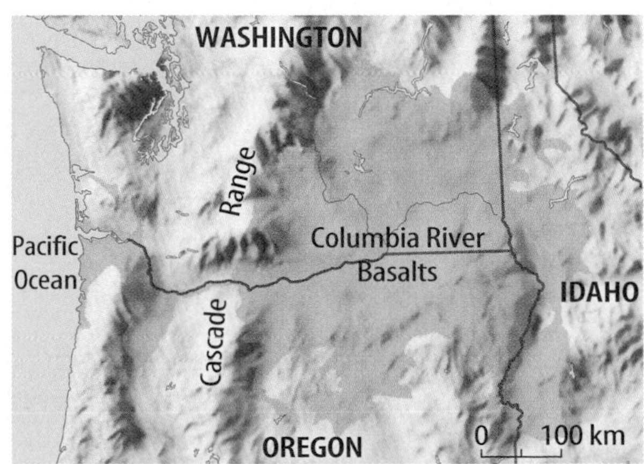

Table 2 Ten Selected Eruptions in History

Volcano (Year)	Type	Eruptive Force	Silica Content	Gas Content	Eruption Products
Tambora, Indonesia (1815)	composite	high	high	high	gas, cinders, ash
Krakatau, Indonesia (1883)	composite	high	high	high	gas, cinders, ash
Pelée, Martinique (1902)	composite	high	high	high	gas, ash
Katmai, Alaska (1912)	composite	high	high	high	lava, ash, gas
Paricutín, Mexico (1943)	cinder cone	moderate	high	low	gas, cinders, ash
Helgafell, Iceland (1973)	cinder cone	moderate	low	high	gas, ash
Mount St. Helens, Washington (1980)	composite	high	high	high	gas, ash
Kilauea Iki, Hawaii (1989)	shield	low	low	low	gas, lava
Pinatubo, Philippines (1991)	composite	high	high	high	gas, ash
Soufrière Hills, Montserrat (1995–)	composite	high	high	high	gas, ash, rocks

Large Eruptions The Columbia River Plateau covers about 200,000 km^2 and is up to 3 km thick in places. In addition, most of Earth's crust beneath the oceans is composed of basalt that was erupted from huge fissures where plates separate.

You have read about some variables that control the type of volcanic eruption that will occur. Examine **Table 2** for a summary of these important factors. In the next section, you'll learn that the type of magma produced is associated with properties of Earth's plates and how these plates interact.

Section  Assessment

1. Which types of lava eruptions cover the largest area on Earth's surface?
2. Describe the processes that have led to the formation of the Soufrière Hills volcano. How have the inhabitants of Montserrat been affected by this volcano?
3. Why does a cinder cone have such steep sides?
4. What types of materials are volcanoes like Mount St. Helens made of?
5. **Think Critically** How does silica-rich magma erupt?

Skill Builder Activities

6. **Predicting** Predict what type of eruption will occur if the magma inside a volcano is low in silica, high in iron and magnesium, and contains little gas. **For more help, refer to the Science Skill Handbook.**
7. **Solving One-Step Equations** Mauna Loa in Hawaii is a shield volcano that rises 9 km above the seafloor. Sunset Crater in Arizona rises to an elevation of 300 m. How many times higher is Mauna Loa than Sunset Crater? **For more help, refer to the Math Skill Handbook.**

Activity

Disruptive Eruptions

A volcano's structure can influence how it erupts. Some volcanoes have only one central vent, while others have numerous fissures that allow lava to escape. Materials in magma influence its viscosity, or how it flows. If magma is a thin fluid—not viscous—gases can escape easily. But if magma is thick—viscous—gases cannot escape as easily. This builds up pressure within a volcano.

What You'll Investigate

What determines the explosiveness of a volcanic eruption?

Materials

plastic film canisters
baking soda ($NaHCO_3$)
vinegar (CH_3COOH)
50-mL graduated cylinder
teaspoon

Goals

- **Infer** how a volcano's opening contributes to how explosive an eruption might be.
- **Hypothesize** how the viscosity of magma can influence an eruption.

Safety Precautions

This activity should be done outdoors. Goggles must be worn at all times. The caps of the film canisters fly off due to the chemical reaction that occurs inside them. Never put anything in your mouth while doing the experiment.

Procedure

1. Watch your teacher demonstrate this activity before attempting to do it yourself.
2. Add 15 ml of vinegar to a film canister.
3. Place 1 teaspoon of baking soda in the film canister's lid, using it as a type of plate.
4. Place the lid on top of the film canister, but do not cap it. The baking soda will fall into the vinegar. Move a safe distance away. Record your observations in your Science Journal.

5. Clean out your film canister and repeat the activity, but this time cap the canister quickly and tightly. Record your observations.

Conclude and Apply

1. Which of the two activities models a more explosive eruption?
2. Was the pressure greater inside the canister during the first or second activity? Why?
3. What do the bubbles have to do with the explosion? How do they influence the pressure in the container?
4. If the vinegar were a more viscous substance, how would the eruption be affected?

Communicating Your Data

Research three volcanic eruptions that have occurred in the past five years. Compare each eruption to one of the eruption styles you modeled in this activity. Communicate to your class what you learn.

SECTION

Earthquakes, Volcanoes, and Plate Tectonics

As You Read

What You'll Learn

- **Explain** how the locations of volcanoes and earthquake epicenters are related to tectonic plate boundaries.
- **Explain** how heat within Earth causes Earth's plates to move.

Vocabulary

rift
hot spot

Why It's Important

Most volcanoes and earthquakes are caused by the motion and interaction of Earth's plates.

Earth's Moving Plates

At the beginning of class, your teacher asks for volunteers to help set up the cafeteria for a special assembly. You and your classmates begin to move the tables carefully, like the students shown in **Figure 14.** As you move the tables, two or three of them crash into each other. Think about what could happen if the students moving those tables kept pushing on them. For a while one or two of the tables might keep another from moving. However, if enough force were used, the tables would slide past one another. One table might even slide up on top of another. It is because of this possibility that your teacher has asked that you move the tables carefully.

The movement of the tables and the possible collisions among them is like the movement of Earth's crust and uppermost mantle, called the lithosphere. Earth's lithosphere is broken into separate sections, or plates. When these plates move around, they collide, move apart, or slide past each other. The movement of these plates can cause vibrations known as earthquakes and can create conditions that cause volcanoes to form.

Figure 14
Like the tables pictured here, Earth's plates are in contact with one another and can slide beneath each other. The way Earth's plates interact at boundaries is an important control on the locations of earthquakes and volcanoes.

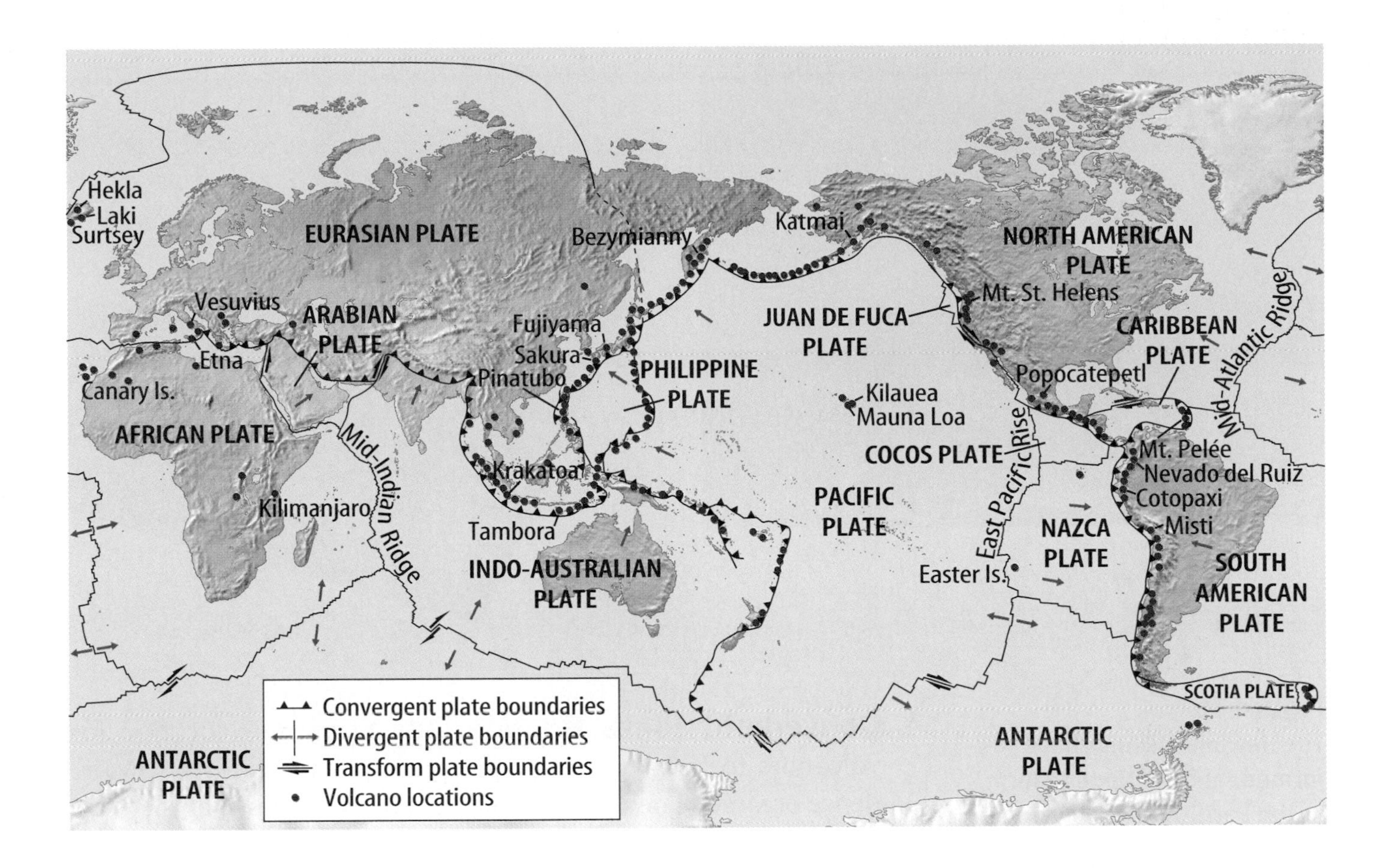

Figure 15
Earth's lithosphere is divided into about 13 major plates. Where plates collide, separate, and slip past one another at plate boundaries, interesting geological activity results.

Where Volcanoes Form

A plot of the location of plate boundaries and volcanoes on Earth shows that most volcanoes form along plate boundaries. Examine the map in **Figure 15.** Can you see how this indicates that plate tectonics and volcanic activity are related? Perhaps the energy involved in plate tectonics is causing magma to form deep under Earth's surface. You'll recall that the Soufrière Hills volcano formed where plates converge. Plate movement often explains why volcanoes form in certain areas.

Divergent Plate Boundaries Tectonic plates move apart at divergent plate boundaries. As the plates separate, long cracks called **rifts** form between them. Rifts contain fractures that serve as passageways for magma originating in the mantle. Rift zones account for most of the places where lava flows onto Earth's surface. Fissure eruptions often occur along rift zones. These eruptions form lava that cools and solidifies into basalt, the most abundant type of rock in Earth's crust.

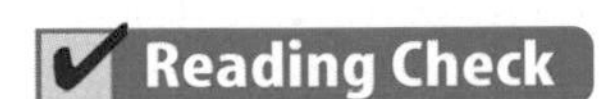

Where does magma along divergent boundaries originate?

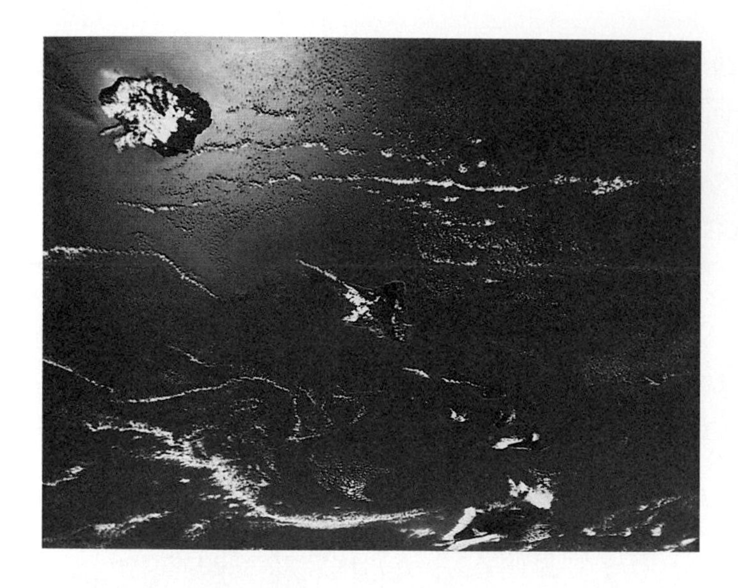

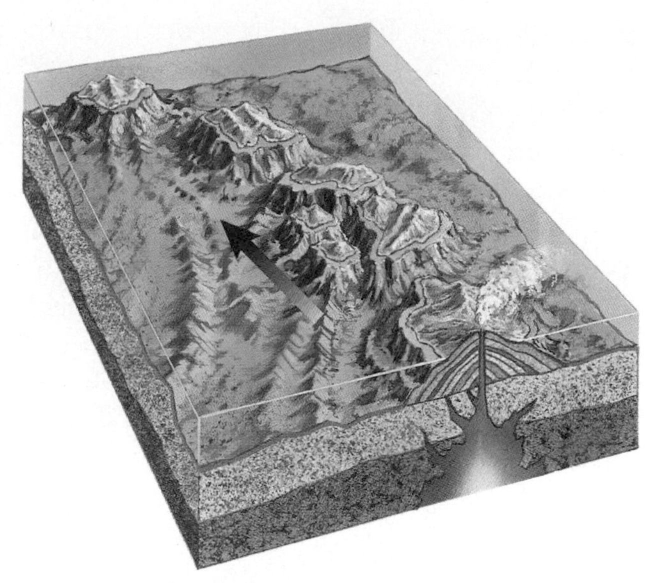

Figure 16
The Hawaiian Islands have formed, and continue to form, as the Pacific Plate moves over a hot spot. The arrow shows that the Pacific Plate is moving north-northwest.

Chemistry INTEGRATION

The melting point of a substance is the temperature at which a solid changes to a liquid. Depending on the substance, a change in pressure can raise or lower the melting point. Do research to find out how pressure affects the formation of magma in a mantle plume in a process called decompression melting.

Convergent Plate Boundaries A common location for volcanoes to form is along convergent plate boundaries. More dense oceanic plates sink beneath less dense plates that they collide with. This sets up conditions that form volcanoes.

When one plate sinks beneath another, basalt and sediment on an oceanic plate move down into the mantle. Water from the sediment and altered basalt lowers the melting point of the surrounding rock. Heat in the mantle causes part of the sinking plate and overlying mantle to melt. This melted material then is forced upward. Volcanoes have formed in this way all around the Pacific Ocean, where the Pacific Plate, among others, collides with several other plates. This belt of volcanoes surrounding the Pacific Ocean is called the Pacific Ring of Fire.

Hot Spots The Hawaiian Islands are volcanic islands that have not formed along a plate boundary. In fact, they are located well within the Pacific Plate. What process causes them to form? Large, rising bodies of magma, called **hot spots,** can force their way through Earth's mantle and crust, as shown in **Figure 16.** Scientists suggest that this is what is occurring at a hot spot that exists under the present location of Hawaii.

✔ Reading Check *What is a hot spot?*

Volcanoes on Earth usually form along rift zones, subduction zones (where one plate sinks beneath another), or over hot spots. At each of these locations, magma from deep within Earth rises toward the surface. Lava breaks through and flows out, where it piles up into layers or forms a volcanic cone.

Moving Plates Cause Earthquakes

Place two notebooks on your desk with the page edges facing each other. Then push them together slowly. The individual sheets of paper gradually will bend upward from the stress. If you continue to push on the notebooks, one will slip past the other suddenly. This sudden movement is like an earthquake.

Now imagine what would happen if tectonic plates were moving like the notebooks. What would happen if the plates collided and stopped moving? Forces generated by the locked-up plates would cause strain to build up. Both plates would begin to deform until the elastic limit was passed. The breaking and elastic rebound of the deformed material would produce vibrations felt as earthquakes.

Earthquakes often occur where tectonic plates come together at a convergent boundary, where tectonic plates move apart at a divergent boundary, and where tectonic plates grind past each other, called a transform boundary.

Earthquake Locations If you look at a map of earthquakes, you'll see that most occur in well-known belts. About 80 percent of them occur in the Pacific Ring of Fire—the same belt in which many of Earth's volcanoes occur. If you compare **Figure 17** with **Figure 15,** you will notice a definite relationship between earthquake epicenters and tectonic plate boundaries. Movement of the plates produces forces that generate the energy to cause earthquakes.

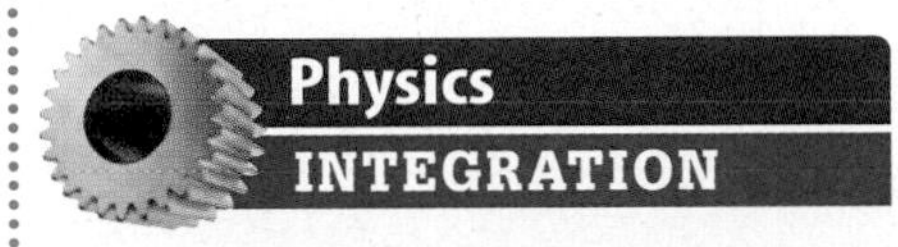

Friction is a force that opposes the motion of two objects in contact. Do research to find out the role of friction in plate movement and earthquakes.

Figure 17
Locations of earthquakes that have occurred between 1990 and 2000 are plotted below.

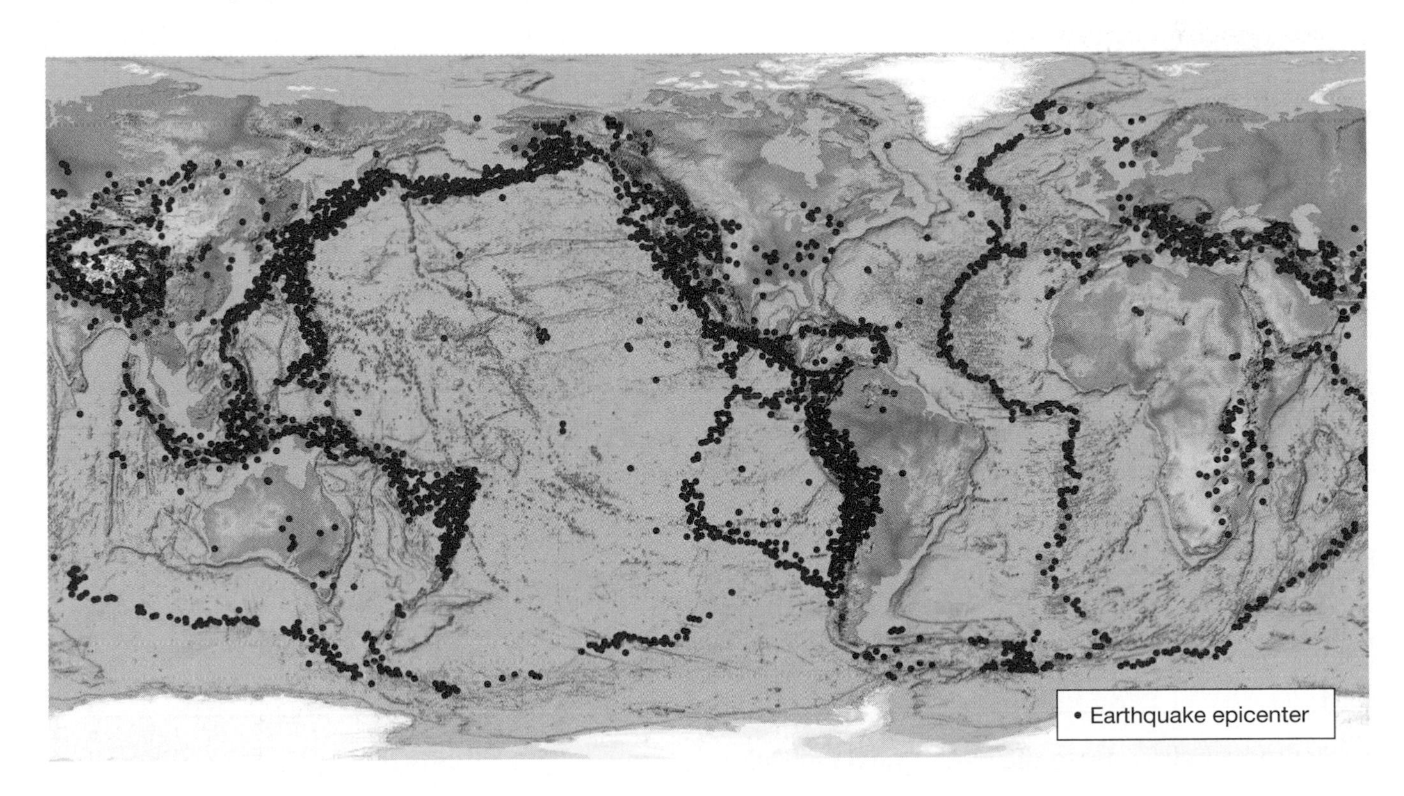

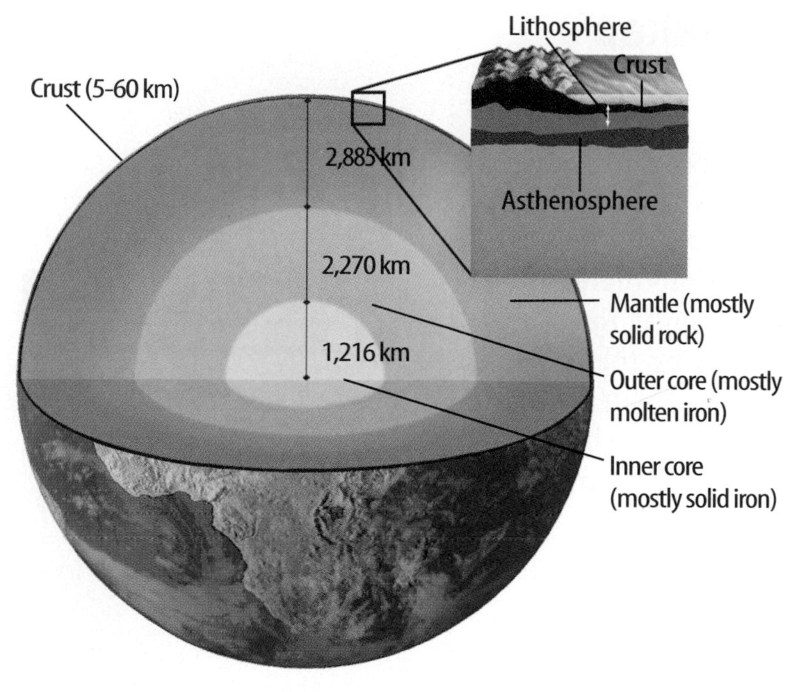

Figure 18
Seismic waves generated by earthquakes allow researchers to figure out the structure and composition of Earth's layers.

Earth's Plates and Interior

Researchers have learned much about Earth's interior and plate tectonics by studying seismic waves. The way in which seismic waves pass through a material depends on the properties of that material. Seismic wave speeds, and how they travel through different levels in the interior, have allowed scientists to map out the major layers of Earth, as shown in **Figure 18.**

For example, the asthenosphere was discovered when seismologists noted that seismic waves slowed when they reached the base of the lithosphere of Earth. This partially molten layer forms a warmer, softer layer over which the colder, brittle, rocky plates move.

Math Skills Activity

Calculating Time Traveled by Waves

Examine the table. What is the relationship between density of a region in Earth and the velocities of P-waves?

Density and Wave Velocity

Region	Density	P-Wave Velocity
Crust	2.8 g/cm^3	6 km/s
Upper mantle	3.3 g/cm^3	8 km/s

Example Problem

Calculate the time it would take P-waves to travel 100 km in the crust of Earth.

Solution

1. *This is what you know:* velocity: $v = 6$ km/s; distance: $d = 100$ km
2. *This is what you need to find:* time: t
3. *This is the equation you need to use:* $t = d/v$
4. *Solve the equation for* t *in seconds:* $t = (100 \text{ km})/(6 \text{ km/s}) = 16.7$ s

Practice Problem

Calculate the time it takes P-waves to travel 300 km in the upper mantle.

For more help, refer to the Math Skill Handbook.

What is driving Earth's plates? There are several hypotheses about where all the energy comes from to power the movement of Earth's plates.

In one case, mantle material deep inside Earth is heated by Earth's core. This hot, less dense rock material is forced toward the surface. The hotter, rising mantle material eventually cools. The cooler material then sinks into the mantle toward Earth's core, completing the convection current. Convection currents inside Earth, shown in **Figure 19,** provide the mechanism for plate motion, which then produces the conditions that cause volcanoes and earthquakes. Sometimes magma rises up directly within a plate. Volcanic activity in Yellowstone National Park is caused by a hot spot beneath the North American Plate. Such hot spots might be related to larger-scale convection in Earth's mantle.

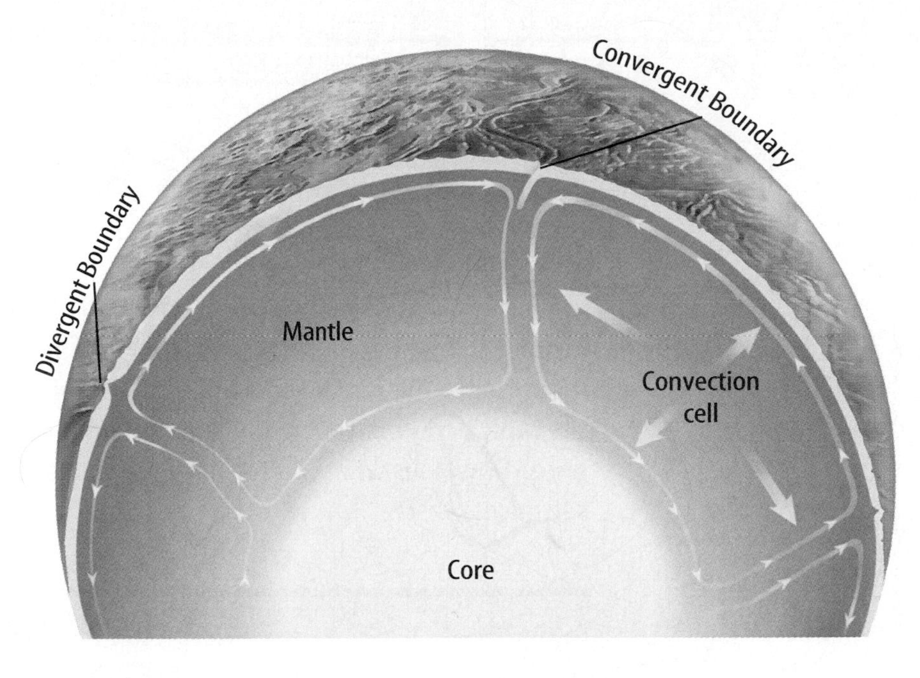

Figure 19
Convection of material in Earth's interior drives the motion of tectonic plates.

Section  Assessment

1. Along which type of tectonic plate boundary has the Soufrière Hills volcano formed?
2. At which type of tectonic boundary does rift-volcanism occur?
3. Explain how volcanoes in Hawaii form—even though they are located far from plate boundaries.
4. Why do most deep earthquakes occur at convergent boundaries?
5. **Think Critically** Subduction occurs where two oceanic plates or an oceanic and a continental plate converge. This causes water-rich sediment and altered rock to be forced down to great depths. Explain how water can help form a volcano.

Skill Builder Activities

6. **Forming Hypotheses** Write a hypothesis concerning the type of lava that will flow from a hot spot to form a volcano. While forming your hypothesis, consider that magma in a hot spot comes from deep inside Earth's mantle. Earth's mantle contains more iron and magnesium and less silica than the crust. **For more help, refer to the Science Skill Handbook.**
7. **Using an Electronic Spreadsheet** Research for listings of the major earthquakes over the past five years. Make a table that provides information for each of the earthquakes, including whether they are located at or near tectonic plate boundaries. **For more help, refer to the Technology Skill Handbook.**

Activity

Seismic Waves

If you and one of your friends hold a long piece of rope between you and move one end of the rope back and forth, you can send a wave through the length of the rope. Hold a ruler at the edge of a table securely with one end of it sticking out from the table's edge. If you bend the ruler slightly and then release it, what do you experience? How does what you see in the rope and what you feel in the ruler relate to seismic waves?

What You'll Investigate

How do seismic waves differ?

Materials

coiled spring toy
yarn or string
metric ruler

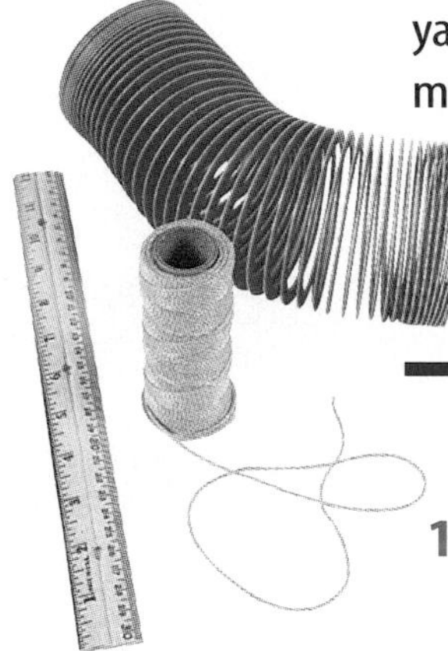

Goals

- **Demonstrate** the motion of primary, secondary, and surface waves.
- **Identify** how parts of the spring move in each of the waves.

Safety Precautions

Procedure

1. Copy the following data table in your Science Journal.
2. Tie a small piece of yarn or string to every tenth coil of the spring.
3. Place the spring on a smooth, flat surface. Stretch it so it is about 2 m long (1 m for shorter springs).
4. Hold your end of the spring firmly. Make a wave by having your partner snap the spring from side to side quickly.
5. **Record** your observations in your Science Journal and draw the wave you and your partner made in the data table.

Comparing Seismic Waves

Observation of Wave	Observation of Yarn or String	Drawing	Wave Type

6. Have your lab partner hold his or her end of the spring firmly. Make a wave by quickly pushing your end of the spring toward your partner and bringing it back to its original position.
7. **Record** your observations of the wave and of the yarn or string and draw the wave in the data table.
8. Have your lab partner hold his or her end of the spring firmly. Move the spring off of the table. Gently move your end of the spring side to side while at the same time moving it in a rolling motion, first up and away and then down and toward your partner.
9. **Record** your observations and draw the wave in the data table.

Conclude and Apply

1. Based on your observations, decide which of the waves that you and your partner have generated demonstrates a primary, or pressure, wave. Record in your data table and explain why you chose the wave you did.
2. Do the same for the secondary, or shear wave, and for the surface wave. Explain why you chose the wave you did.
3. Based on your observations of wave motion, which of the waves that you and your partner generated probably would cause the most damage during an earthquake? Explain your answers.
4. What was the purpose of the yarn or string?
5. **Compare and contrast** the motion of the yarn or string when primary and secondary waves travel through the spring. Which of these waves is a compression wave? Which is a transverse wave? Explain each answer.
6. Which wave most closely resembled wave motion in a body of water? How was it different? Explain.

Communicating Your Data

Compare your conclusions with those of other students in your class. **For more help, refer to the Science Skill Handbook.**

quake

The 1906 San Francisco earthquake taught people valuable lessons

"We found ourselves staggering and reeling. It was as if the earth was slipping gently from under our feet. Then came the sickening swaying of the earth that threw us flat upon our faces. We struggled in the street. We could not get on our feet. Then it seemed as though my head were split with the roar that crashed into my ears. Big buildings were crumbling as one might crush a biscuit in one's hand."

That's how survivor P. Barrett described the San Francisco earthquake of 1906. Duration of the quake on the morning of April 18—one minute. Yet, in that short time, Earth opened a gaping hole stretching more than 430 km. The tragic result was one of the worst natural disasters in U.S. history.

Fires caused by falling chimneys and fed by broken gas mains raged for three days. Despite the estimated 3,000 deaths and enormous devastation to San Francisco, the earthquake did have a positive effect. It led to major building changes that would help protect people and property from future quakes.

Before the 1906 earthquake, little was known about how or where earthquakes were likely to occur. And not much was known about the destructiveness of quakes. However, after the quake hit San Francisco, scientists and government workers set up a State Earthquake Investigation Commission.

The 1906 quake destroyed City Hall (above). Today, a rebuilt City Hall stands on the same site (right).

Scientists and engineers placed instruments in buildings and bridges, and nearby on the ground, to measure how structures respond to the motion of the ground during an earthquake. They found that structures built on soft, muddy clay shook worse than structures built on harder ground, like bedrock. They also discovered that buildings made from flexible materials, such as wood, are damaged less than those constructed of rigid materials, such as steel and brick. Rigid materials will snap, rather than sway, during an earthquake. These discoveries led to knowledge about quakes that is still used today to make buildings safer.

Thanks to these early findings, combined with recent developments in earthquake detection, California now has instruments tracking plate motions throughout the state. Computers analyze information from seismographs that have helped to map the San Andreas Fault—the area along which many California earthquakes take place. This information is helping scientists better understand how and when earthquakes might strike.

The 1906 quake also has led to building codes that require stronger construction materials for homes, offices, and bridges. Laws have been passed saying where hospitals, homes, and nuclear power plants can be built—away from soft ground and away from the San Andreas Fault.

Even today, scientists can't predict an earthquake. But thanks to what they learned from the 1906 quake—and others—people are safer today than ever before.

CONNECTIONS Write **Prepare a diary entry pretending to be a person who experienced the 1906 San Francisco earthquake. Possible events to include in your entry: What were you doing at 5:15 A.M.? What began to happen around you? What did you see and hear?**

SCIENCE *Online*

For more information, visit science.glencoe.com

Chapter 15 Study Guide

Reviewing Main Ideas

Section 1 Earthquakes

1. Earthquakes occur whenever rocks inside Earth pass their elastic limit, break, and experience elastic rebound.
2. The precise point where an earthquake occurs is the focus. The point on Earth's surface directly above an earthquake's focus is the epicenter. *What do scientists measure on seismograms to determine an epicenter?*

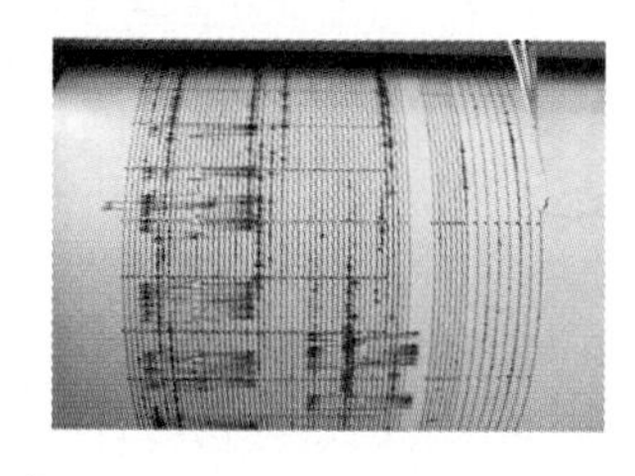

3. Seismic waves are vibrations inside Earth. P- and S-waves travel in all directions away from the earthquake focus and move throughout Earth, including the deep interior. Surface waves travel along the surface.
4. Earthquakes are measured by their magnitudes—the amount of energy they release—and by their intensity—the amount of damage they produce.

Section 2 Volcanoes

1. The Soufrière Hills volcano is a composite volcano formed by the convergence of two tectonic plates.
2. Material extruded from volcanoes varies greatly and includes lava, tephra, and gases. *What types of material erupted to form the volcano shown here?*

3. The way a volcano erupts is determined by the composition of the lava and the amount of water vapor and other gases in the lava.
4. Three different forms of volcanoes are shield volcanoes, cinder cone volcanoes, and composite volcanoes.

Section 3 Earthquakes, Volcanoes, and Plate Tectonics

1. The locations of volcanoes and earthquake epicenters are related to the locations of tectonic plate boundaries.
2. Volcanoes occur along rift zones and subduction zones, as well as at hot spots.
3. Most earthquakes occur at convergent, divergent, and transform plate boundaries, but some occur within the plates. *Which type of plate boundary is illustrated below? Be specific.*

After You Read

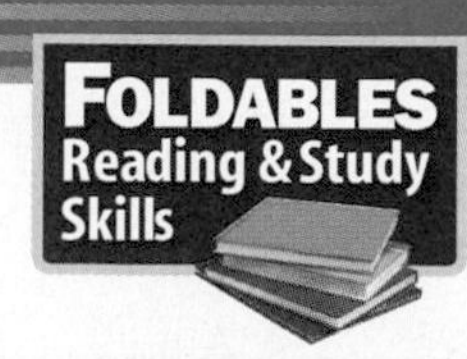

Record how earthquakes and volcanoes are similar and different on the center section of your Foldable.

Visualizing Main Ideas

Fill in the following table comparing characteristics of shield, composite, and cinder cone volcanoes.

Volcanoes

Characteristic	Shield Volcano	Cinder Cone Volcano	Composite Volcano
Relative size	large		
Nature of eruption			moderate to high eruptive force
Materials extruded	lava, gas	cinders, gas	
Composition of lava			variable
Ability of lava to flow		low	variable

Vocabulary Review

Vocabulary Words

a. cinder cone volcano
b. composite volcano
c. earthquake
d. epicenter
e. fault
f. focus
g. hot spot
h. lava
i. magnitude
j. rift
k. seismic safe
l. seismic wave
m. seismograph
n. shield volcano
o. tsunami
p. volcano

THE PRINCETON REVIEW Study Tip

Sit with a study partner and read aloud to each other from a chapter. Then discuss what you've been reading with each other.

Using Vocabulary

Explain the differences between the vocabulary words in each of the following sets.

1. fault, earthquake
2. seismic wave, seismic safe
3. shield volcano, composite volcano
4. focus, epicenter
5. seismic wave, seismograph
6. hot spot, focus
7. tsunami, seismic wave
8. epicenter, earthquake
9. focus, fault
10. cinder cone volcano, shield volcano

Chapter 15 Assessment

Checking Concepts

Choose the word or phrase that best answers the question.

1. Which type of plate boundary caused the formation of the Soufrière Hills volcano?
A) divergent **C)** rift
B) transform **D)** convergent

2. What is a cone-shaped mountain that is built from layers of lava?
A) volcano **C)** vent
B) lava flow **D)** crater

3. What are avalanches of hot, glowing rock flowing on a cushion of hot gases called?
A) lava flows **C)** mudflows
B) pyroclastic flows **D)** gas clouds

4. Which type of lava flows easily?
A) silica-rich lava **C)** basaltic lava
B) composite lava **D)** smooth lava

5. Which type of volcano is built from alternating layers of lava and tephra?
A) shield volcano
B) cinder cone volcano
C) lava dome
D) composite volcano

6. Which type of volcano forms a relatively small, steep-sided cone?
A) shield volcano
B) cinder cone volcano
C) lava dome
D) composite volcano

7. Which seismic wave moves through Earth at the fastest speed?
A) primary wave **C)** surface wave
B) secondary wave **D)** tsunami

8. Which of the following is a wave of water caused by an earthquake under the ocean?
A) primary wave **C)** surface wave
B) secondary wave **D)** tsunami

9. What is the point on Earth's surface directly above an earthquake's focus?
A) earthquake center
B) epicenter
C) wave center
D) focus

10. What is the cause of the volcanoes on Hawaii?
A) rift zone
B) hot spot
C) divergent plate boundary
D) convergent plate boundary

Thinking Critically

11. Why does the Soufrière Hills volcano erupt so explosively?

12. Compare and contrast composite and cinder cone volcanoes.

13. How can the composition of magma affect the way a volcano erupts?

14. Why do shield volcanoes have flatter slopes than composite volcanoes?

15. What factors determine an earthquake's intensity on the modified Mercalli scale?

Developing Skills

16. Making and Using Tables Use **Table 2** in Section 2 to answer the following questions. What general statement can be made about the eruptive force of composite volcanoes? What usually is erupted from shield volcanoes such as Kilauea Iki?

17. Comparing and Contrasting Compare and contrast magnitude and intensity.

18. Making Models Select one of the three forms of volcanoes and make a model, using appropriate materials.

19. Drawing Conclusions You are flying over an area that has just experienced an earthquake. You see that most of the buildings are damaged or destroyed and much of the surrounding countryside is disrupted. What level of intensity would you conclude for this earthquake?

20. Concept Mapping Complete this concept map on examples of features produced along plate boundaries. Use the following terms: *Mid-Atlantic Ridge, Soufrière Hills volcano, divergent, San Andreas Fault, convergent,* and *transform.*

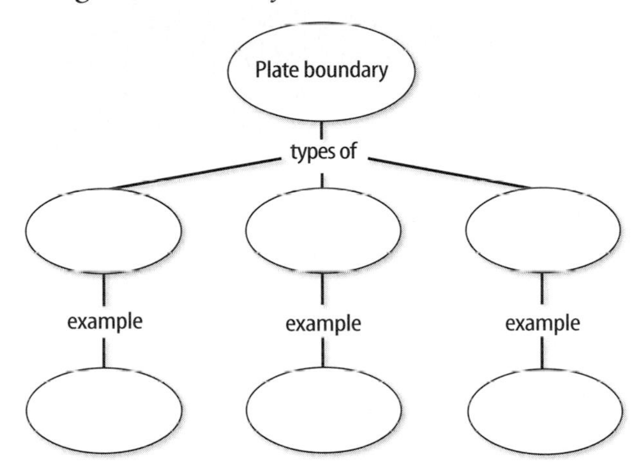

Performance Assessment

21. Oral Presentation Research the earthquake or volcano history of your state or community. Find out how long ago your area experienced earthquake- or volcano-related problems. Present your findings in a speech to your class.

TECHNOLOGY

Go to the Glencoe Science Web site at **science.glencoe.com** or use the **Glencoe Science CD-ROM** for additional chapter assessment.

Test Practice

The table below presents data about Earth's plate boundaries.

Plate Boundaries

Plate	Number of convergent boundaries	Number of divergent boundaries
African	1	4
Antarctic	1	2
Indo-Australian	4	2
Eurasian	4	0
North American	2	1
Pacific	6	1
South American	2	1

Study the table and answer the following questions.

1. Which plate has the most spreading boundaries?
A) African
B) Indo-Australian
C) Pacific
D) Antarctic

2. If composite volcanoes often form along convergent boundaries, which plate should be surrounded by the most composite volcanoes?
F) Pacific
G) Antarctic
H) Eurasian
J) Indo-Australian

Ocean Motion

Surfers in Hawaii experience firsthand the enormous power of moving water. It surprises people to learn that wind causes most waves, from small ripples to the giant waves of hurricanes, some more than 30 m high. Wind also creates surface currents. Other types of currents move through the ocean, too. In this chapter, you'll learn about the composition of ocean water, the interaction between the atmosphere and the oceans, and how waves, currents, and tides are created.

What do you think?

Science Journal Look at the picture below with a classmate. Discuss what you think this might be or what is happening. Here's a hint: *Daily fluctuations make this happen.* Write your answer or best guess in your Science Journal.

Surface currents are caused by wind, but wind cannot cause currents deep in the ocean. Instead, deep-water currents are created by differences in the density of ocean water. Several factors affect water density. One is temperature. Do the activity below to see how temperature differences create these kinds of currents.

Explore how currents work

1. In a bowl, mix ice and cold water to make ice water.
2. Fill a beaker with warm tap water.
3. Add a few drops of food coloring to the ice water and stir the mixture.
4. Use a dropper to place some of this ice water on top of the warm water.

Observe

In your Science Journal, describe what happened. Did adding cold water on top produce a current? Look up the word *convection* in a dictionary. Infer why the current you created is called a convection current.

Before You Read

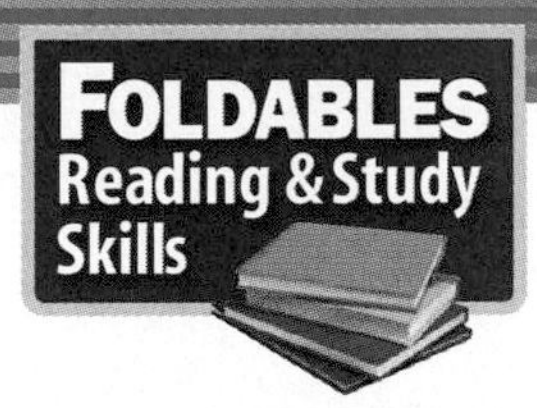

Making a Cause and Effect Study Fold **Make the following Foldable to help you understand the cause and effect relationship of ocean motion.**

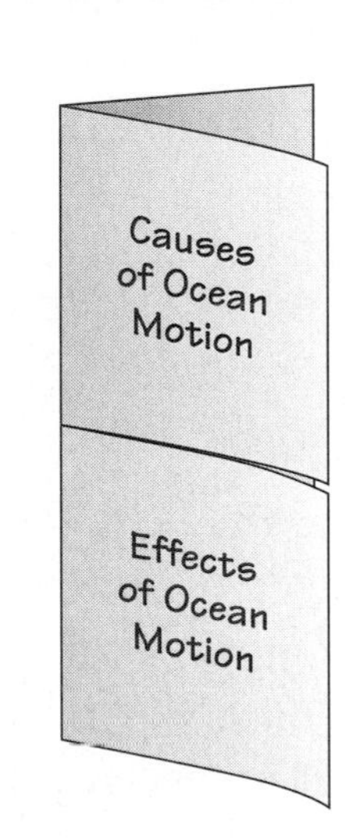

1. Place a sheet of paper in front of you so the long side is at the top. Fold the paper in half from the left side to the right side. Fold top to bottom and crease. Then unfold.
2. Through the top thickness of paper, cut along the middle fold line to form two tabs. Label the tabs *Causes of Ocean Motion* and *Effects of Ocean Motion.*
3. As you read the chapter, write what you learn about why the ocean moves and the types of ocean motion under the top tab.

SECTION 1 Ocean Water

As You Read

What You'll Learn

- **Identify** the origin of the water in Earth's oceans.
- **Explain** how dissolved salts and other substances get into seawater.
- **Describe** the composition of seawater.

Vocabulary

basin
salinity

Why It's Important

Oceans affect weather and provide food and natural resources.

Importance of Oceans

Imagine yourself lying on a beach and listening to the waves gently roll onto shore. A warm breeze blows off the water, making it seem as if you're in a tropical paradise. It's easy to appreciate the oceans under these circumstances, but the oceans affect your life in other ways, too.

Varied Resources Oceans are important for food, minerals, transportation, and weather. Today, as shown in **Figure 1,** a huge variety of different resources comes from the oceans of the world. Oceans also allow efficient transportation. Can you think of a product you own that was transported by a ship? Energy and mineral resources also are found in oceans. Oil wells often are drilled in shallow ocean water. Oceans affect weather and climate. Hurricanes develop in some tropical waters, and moist air masses move onto land from oceans. Ocean currents keep some places warm while creating cool, foggy days elsewhere.

Reading Check *What resources come from oceans?*

Figure 1
People depend on the oceans for many resources.

A Krill are tiny, shrimplike animals that live in the Antarctic Ocean. Some cultures use krill in noodles and rice cakes.

B Kelp is a fast-growing seaweed that is a source of algin, used in making ice cream, salad dressing, medicines, and cosmetics.

Origin of Oceans

During Earth's first billion years, its surface, shown in **Figure 2A,** was much more volcanically active than it is today. When volcanoes erupt, they spew lava and ash, and they give off water vapor, carbon dioxide, and other gases. Scientists hypothesize that about 4 billion years ago, this water vapor began to be stored in Earth's early atmosphere. Over millions of years, it cooled enough to condense into storm clouds. Torrential rains began to fall. Shown in **Figure 2B,** oceans were formed as this water filled low areas on Earth called **basins.** Today, 70 percent of Earth's surface is covered by ocean water.

Figure 2
Earth's oceans formed from water vapor.

A Water vapor was released into the atmosphere by volcanoes that also gave off other gases, such as carbon dioxide and nitrogen.

B Condensed water vapor formed storm clouds. Oceans formed when basins filled with water from torrential rains.

Composition of Oceans

Ocean water contains dissolved gases such as oxygen, carbon dioxide, and nitrogen. Oxygen is the gas that almost all organisms need for respiration. It enters the oceans in two ways—directly from the atmosphere and from organisms that photosynthesize. Carbon dioxide enters the ocean from the atmosphere and from organisms when they respire. The atmosphere is the only important source of nitrogen gas. Bacteria combine nitrogen and oxygen to create nitrates, which are important nutrients for plants.

If you've ever tasted ocean water, you know that it is salty. Ocean water contains many dissolved salts. Chloride, sodium, sulfate, magnesium, calcium, and potassium are some of the ions in seawater. An ion is a charged atom or group of atoms. Some of these ions come from rocks that are dissolved slowly by rivers and groundwater. These include calcium, magnesium, and sodium. Rivers carry these chemicals to the oceans. Erupting volcanoes add other ions, such as sulfate and chloride.

✔ Reading Check ***How do sodium and chloride ions get into seawater?***

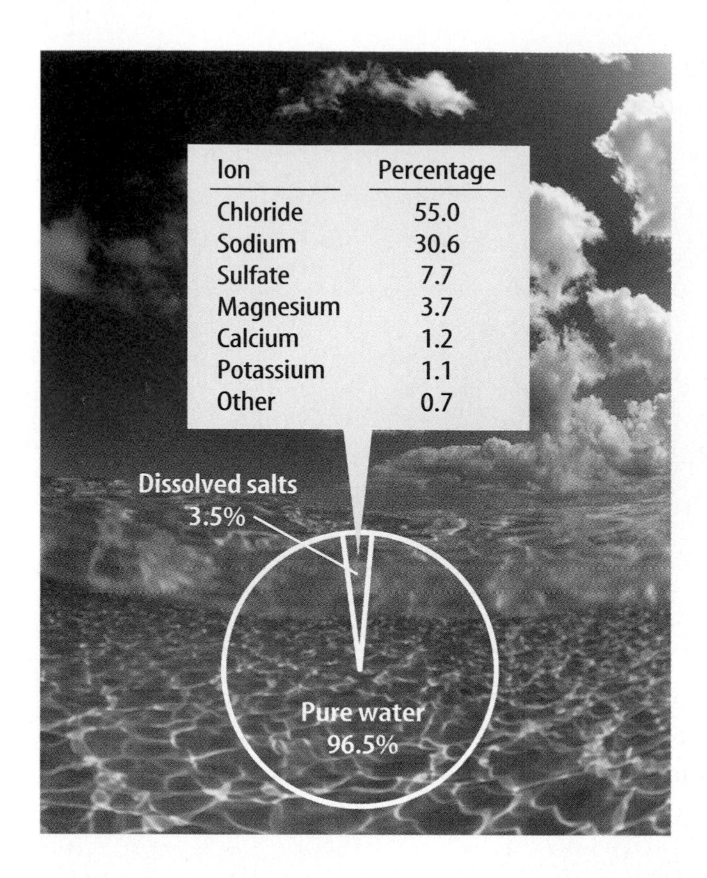

Ion	Percentage
Chloride	55.0
Sodium	30.6
Sulfate	7.7
Magnesium	3.7
Calcium	1.2
Potassium	1.1
Other	0.7

Figure 3
Ocean water contains about 3.5 percent dissolved salts. *If you evaporated 1,000 g of seawater, how many grams of salt would be left?*

Salts The most abundant elements in seawater are the hydrogen and oxygen that make up water. Ions of many other elements are found dissolved in seawater. When seawater is evaporated, these ions combine to form materials called salts. Sodium and chloride make up most of the ions in seawater. If seawater evaporates, the sodium and chloride ions combine to form a salt called halite. Halite is the common table salt you use to season food. It is this dissolved salt and similar ones that give ocean water its salty taste.

Salinity (say LIH nuh tee) is a measure of the amount of salts dissolved in seawater. It usually is measured in grams of dissolved salt per kilogram of seawater. One kilogram of ocean water contains about 35 g of dissolved salts, or 3.5 percent. The chart in **Figure 3** shows the most abundant ions in ocean water. The proportion and amount of dissolved salts in seawater remain nearly constant and have stayed about the same for hundreds of millions of years. This tells you that the composition of the oceans is in balance. Evidence that scientists have gathered indicates that Earth's oceans are not growing saltier.

Removal of Elements Although rivers, volcanoes, and the atmosphere constantly add material to the oceans, the oceans are considered to be in a steady state. This means that elements are added to the oceans at about the same rate that they are removed. They are removed when ocean water evaporates, leaving salt behind, and when organisms use the dissolved salts to make shells. Some marine animals remove calcium ions from the water to form bones. Other animals, such as oysters and clams, use the dissolved calcium to form shells. Some algae, called diatoms, have silica shells. Because many organisms use calcium and silicon, these elements are removed more quickly from seawater than elements such as chlorine or sodium.

Desalination Salt can be removed from ocean water by a process called desalination (dee sa luh NAY shun). If you have ever swum in the ocean, you know what happens when your skin dries. The white, flaky substance on your skin is salt. As seawater evaporates, salt is left behind. As demand for freshwater increases throughout the world, scientists are working on technology to remove salt to make seawater drinkable.

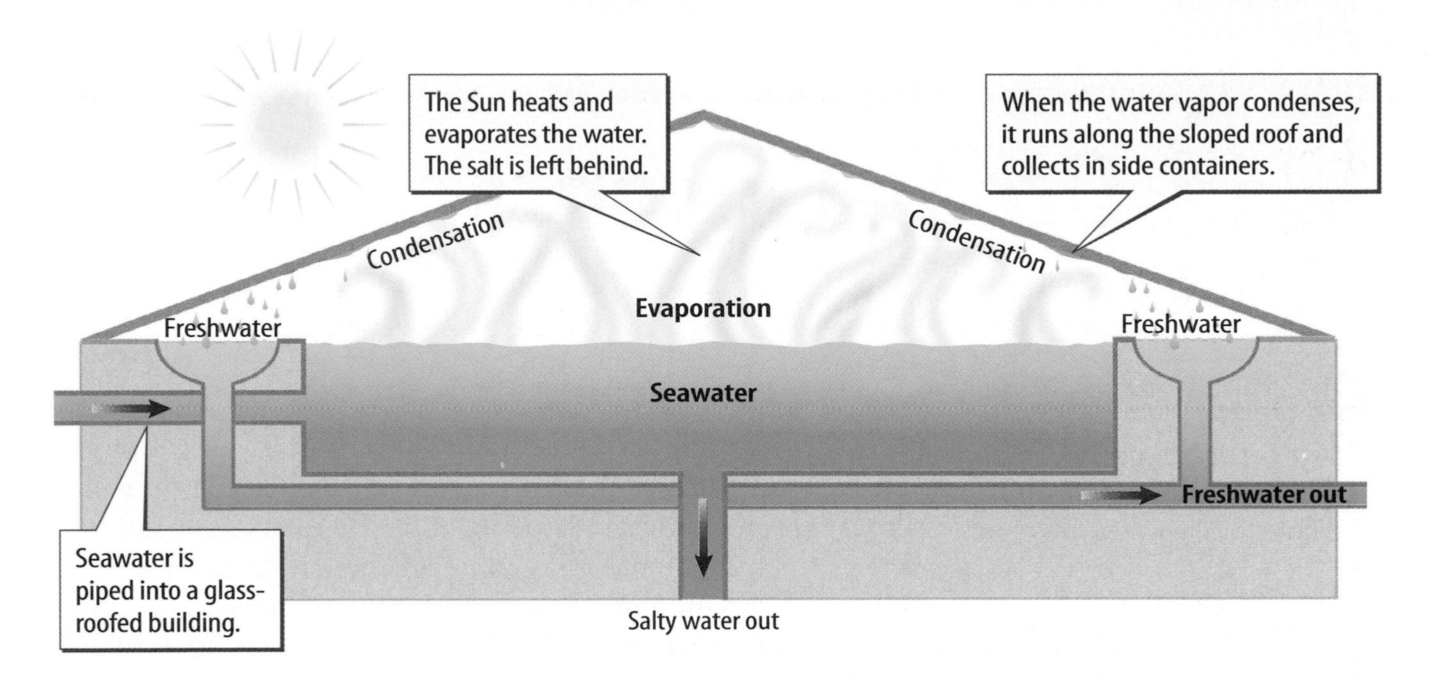

Figure 4
This desalination plant uses solar energy to produce freshwater.

Desalination Plants Some methods of desalination include evaporating seawater and collecting the freshwater as it condenses on a glass roof. **Figure 4** shows how a desalination plant that uses solar energy works. Other plants desalinate water by passing it through a membrane that removes the dissolved salts. Freshwater also can be obtained by melting frozen seawater. As seawater freezes, the ice crystals that form contain much less salt than the remaining water. The salty, unfrozen water then can be separated from the ice. The ice can be washed and melted to produce freshwater.

Section 1 Assessment

1. Describe at least five ways that Earth's oceans affect your life.
2. According to scientific hypothesis, how were Earth's oceans formed? When do scientists hypothesize they formed?
3. Where do the dissolved salts in ocean water come from?
4. How does oxygen get into oceans?
5. **Think Critically** Organisms in the oceans are important sources of food and medicine. What steps can humans take to ensure that these resources are available for future generations?

Skill Builder Activities

6. **Concept Mapping** Make a concept map that shows how sodium and chloride become dissolved in ocean water and what happens when the seawater evaporates. Use the terms *rivers, volcanoes, halite, source of, sodium, chloride,* and *combine to form*. **For more help, refer to the Science Skill Handbook.**
7. **Using Proportions** If the average salinity of seawater is 35 parts per thousand, how many grams of dissolved salts will 500 g of seawater contain? **For more help, refer to the Math Skill Handbook.**

SECTION 2

Ocean Currents

As You Read

***What* You'll Learn**

- **Explain** how winds and the Coriolis effect influence surface currents.
- **Discuss** the temperatures of coastal waters.
- **Describe** density currents.

Vocabulary

surface current
Coriolis effect
upwelling
density current

***Why* It's Important**

Ocean currents and the atmosphere transfer heat that creates the climate you live in.

Surface Currents

When you stir chocolate into a glass of milk, do you notice the milk swirling around in the glass in a circle? If so, you've observed something similar to an ocean current. Ocean currents are a mass movement, or flow, of ocean water. An ocean current is like a river within the ocean.

Surface currents move water horizontally—parallel to Earth's surface. These currents are powered by wind. The wind forces the ocean to move in huge, circular patterns. **Figure 5** shows these major surface currents. Notice that some currents are shown with red arrows and some are shown with blue arrows. Red arrows indicate warm currents. Blue arrows indicate cold currents. The currents on the ocean's surface are related to the general circulation of winds on Earth.

Surface currents move only the upper few hundred meters of seawater. Some seeds and plants are carried between continents by surface currents. Sailors take advantage of these currents along with winds to sail more efficiently from place to place.

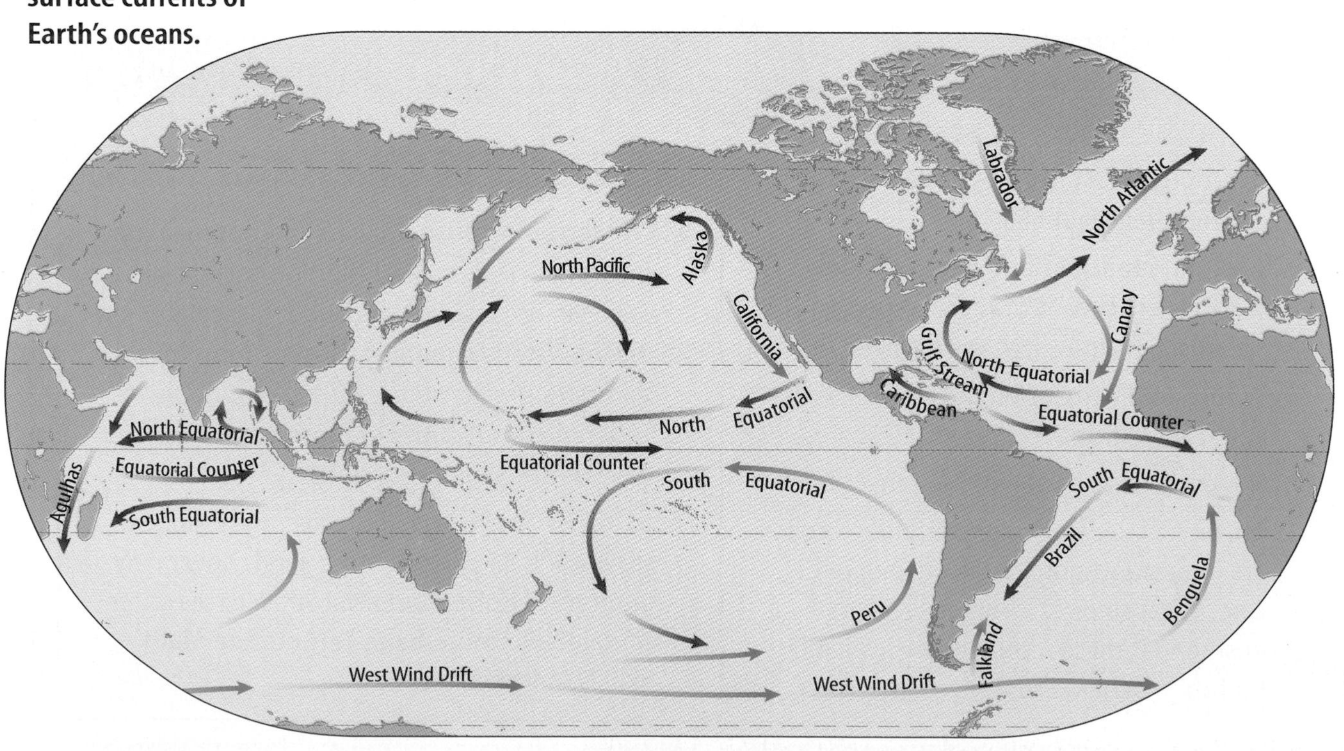

Figure 5
These are the major surface currents of Earth's oceans.

How Surface Currents Form Surface ocean currents and surface winds are affected by the Coriolis (kor ee OH lus) effect. The **Coriolis effect** is the shifting of winds and surface currents from their expected paths that is caused by Earth's rotation. Imagine that you try to draw a line straight out from the center of a disk to the edge of the disk. You probably could do that with no problem. But what would happen if the disk were slowly spinning like the one in **Figure 6A?** As the student tried to draw a straight line, the disk rotated and, as shown in **Figure 6B,** the line curved.

A similar thing happens to wind and surface currents. Because Earth rotates toward the east, winds appear to curve to the right in the northern hemisphere and to the left in the southern hemisphere. These surface winds can cause water to pile up in certain parts of the ocean. When gravity pulls water off the pile, the Coriolis effect turns the water. This causes surface water in the oceans to spiral around the piles of water. The Coriolis effect causes currents north of the equator to turn to the right. Currents south of the equator are turned to the left. Look again at the map of surface currents in **Figure 5** to see the results of the Coriolis effect.

Figure 6
A The student draws a line straight out from the center of the disk. B Because the disk was spinning, the line is curved.

The Gulf Stream Although satellites provide new information about ocean movements, much of what is known about surface currents comes from records that were kept by sailors of the nineteenth century. Sailors always have used surface currents to help them travel quickly. Sailing ships depend on some surface currents to carry them to the west and others to carry them east. During the American colonial era, ships floated on the 100-km-wide Gulf Stream current to go quickly from North America to England. Find the Gulf Stream current in the Atlantic Ocean on the map in **Figure 5.**

In the late 1700s, Deputy Postmaster General Benjamin Franklin received complaints about why it took longer to receive a letter from England than it did to send one there. Upon investigation, Franklin found that a Nantucket whaling captain's map furnished the answer. Going against the Gulf Stream delayed ships sailing west from England by up to 110 km per day.

Research Visit the Glencoe Science Web site at **science.glencoe.com** for more information about ocean currents. Communicate to your class what you learn.

Figure 7
Bottles and other floating objects that enter the ocean are used to gain information about surface currents.

Tracking Surface Currents Items that wash up on beaches, such as the bottle shown in **Figure 7,** provide information about ocean currents. Drift bottles containing messages and numbered cards are released from a variety of coastal locations. The bottles are carried by surface currents and might end up on a beach. The person who finds a bottle writes down the date and the location where the bottle was found. Then the card is sent back to the institution that launched the bottle. By doing this, valuable information is provided about the current that carried the bottle.

Warm and Cold Surface Currents Notice in **Figure 5** that currents on the west coasts of continents begin near the poles where the water is colder. The California Current that flows along the west coast of the United States is a cold surface current. East-coast currents originate near the equator where the water is warmer. Warm surface currents, such as the Gulf Stream, distribute heat from equatorial regions to other areas of Earth. **Figure 8** shows the warm water of the Gulf Stream in red and orange. Cooler water appears in blue and green.

As warm water flows away from the equator, heat is released to the atmosphere. The atmosphere is warmed. This transfer of heat influences climate.

Figure 8
Data about ocean temperature collected by a satellite were used to make this surface-temperature image of the Atlantic Ocean.

Figure 9
Winds push surface water away from the coast of Peru, causing upwelling. This process brings colder water to the surface.

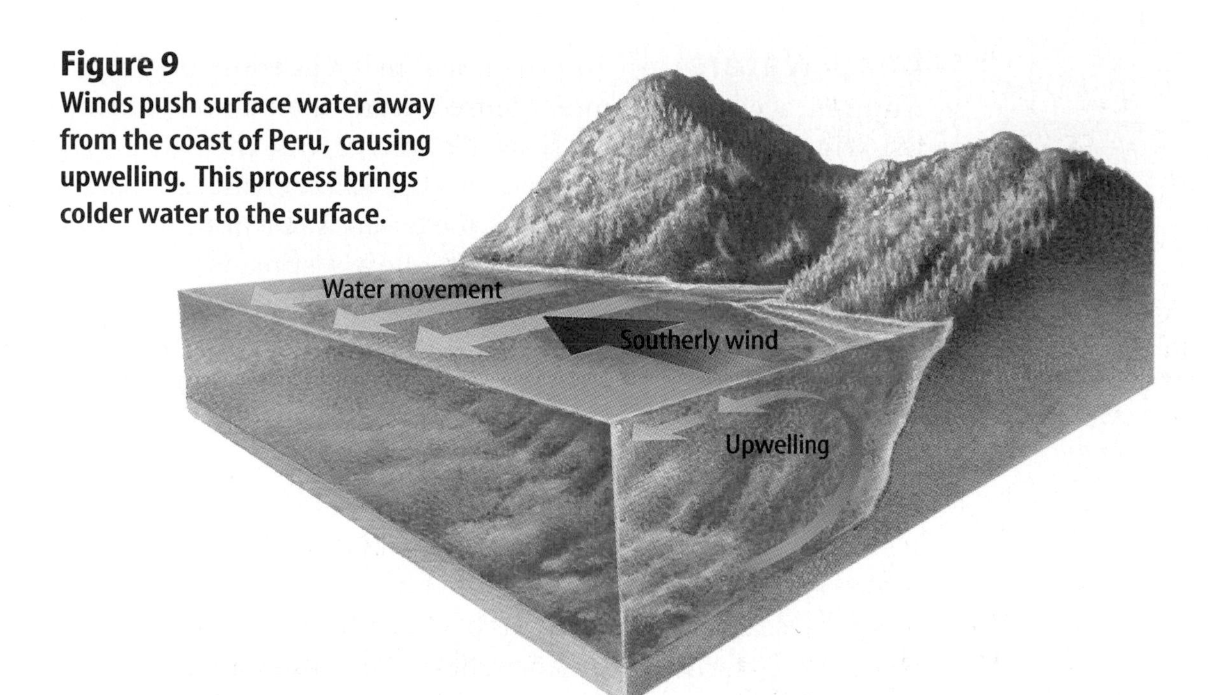

Upwelling

Upwelling is a circulation in the ocean that brings deep, cold water to the ocean surface. Along some coasts of continents, wind blowing parallel to the coast carries water away from the land because of the Coriolis effect, as shown in **Figure 9.** Cold, deep ocean water rises to the surface and replaces water that has moved away from shore. This water contains high concentrations of nutrients from organisms that died, sank to the bottom, and decayed. Nutrients promote plankton growth, which attracts fish. Areas of upwelling occur along the coasts of Oregon, Washington, and Peru and create important fishing grounds.

Density Currents

Deep in the ocean, waters circulate not because of wind but because of density differences. A **density current** forms when a mass of seawater becomes more dense than the surrounding water. Gravity causes more dense seawater to sink beneath less dense seawater. This deep, dense water then slowly spreads to the rest of the ocean.

The density of seawater can be increased if salinity increases, as you can see if you perform the MiniLAB on this page. It also can be increased by a decrease in temperature. In the Explore Activity, the cold water was more dense than the warm water in the beaker. The cold water sank to the bottom. This created a density current that moved the food coloring.

Changes in temperature and salinity work together to create density currents. Density currents circulate ocean water slowly—moving as little as a few meters per month.

Modeling a Density Current

Procedure

1. Fill a **clear plastic storage box** (shoe-box size) with room-temperature **water.**
2. Mix several spoonfuls of table **salt** into a **glass** of water at room temperature.
3. Add a few drops of **food coloring** to the saltwater solution. Pour the solution slowly into the freshwater in the large container.

Analysis

1. Describe what happened when you added salt water to freshwater.
2. How does this lab relate to density currents?

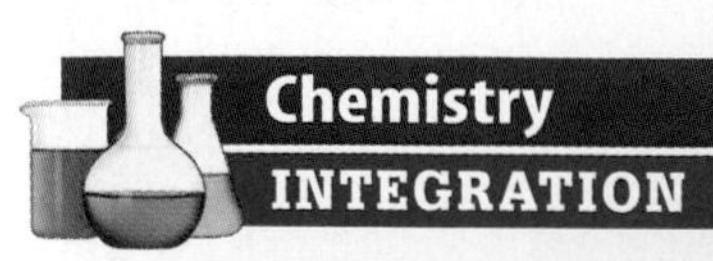

When salt dissolves in water, the freezing point of the mixture is lowered. The greater the number of particles dissolved in the water is, the more the freezing point is lowered. How does this help ocean water near the poles get colder than freshwater could?

Deep Waters An important density current begins in Antarctica where the most dense ocean water forms during the winter. As ice forms, seawater freezes, but the salt is left behind in the unfrozen water. This extra salt increases the salinity and, therefore, the density of the ocean water until it is very dense. This dense water sinks and slowly spreads along the ocean bottom toward the equator, forming a density current. In the Pacific Ocean, this water could take 1,000 years to reach the equator.

In the North Atlantic Ocean, cold, dense water forms around Norway, Greenland, and Labrador. These waters sink, forming North Atlantic Deep Water. In about the northern one third to one half of the Atlantic Ocean, North Atlantic Deep Water forms the bottom layer of ocean water. In the southern part of the Atlantic Ocean, it flows at depths of about 3,000 m, just above the denser water formed near Antarctica. The dense waters circulate more quickly in the Atlantic Ocean than in the Pacific Ocean. In the Atlantic, a density current could circulate in 275 years.

Math Skills Activity

Calculating Density

Example Problem

You have an aquarium full of freshwater in which you have dissolved salt. If the mass of the salt water is 123,000 g and its volume is 120,000 cm^3, what is the density of the salt water?

Solution

1. *This is what you know:* volume: $v = 120{,}000\ cm^3$; mass of salt water: $m = 123{,}000$ g
2. *This is what you need to find:* density of water: d
3. *This is the equation you need to use:* $d = m/v$
4. *Substitute the known values:* $d = 123{,}000\text{g} / 120{,}000\text{cm}^3 = 1.025\ \text{g/cm}^3$

Check your answer by multiplying your answer by the volume. Do you calculate the same mass of salt water that was given?

Practice Problems

1. Calculate the density of 78,000 cm^3 of salt water with a mass of 79,000 g.
2. If a sample of ocean water has a density of 1.03 g/cm^3 and a mass of 50,000 g, what is the volume of the water?

For more help, refer to the Math Skill Handbook.

Intermediate Waters A density current also occurs in the Mediterranean Sea, a nearly enclosed body of water. The warm temperatures and dry air in the region cause large amounts of water to evaporate from the surface of the sea. This evaporation increases the salinity and density of the water. This dense water from the Mediterranean flows through the narrow Straits of Gibraltar into the Atlantic Ocean at a depth of about 320 m. When it reaches the Atlantic, it flows to depths of 1,000 m to 2,000 m because it is more dense than the water in the upper parts of the North Atlantic Ocean. However, the water from the Mediterranean is less dense than the very cold, salty water flowing from the North Atlantic Ocean around Greenland, Norway, and Labrador. Therefore, as shown in **Figure 10,** the Mediterranean water forms a middle layer of water—the Mediterranean Intermediate Water.

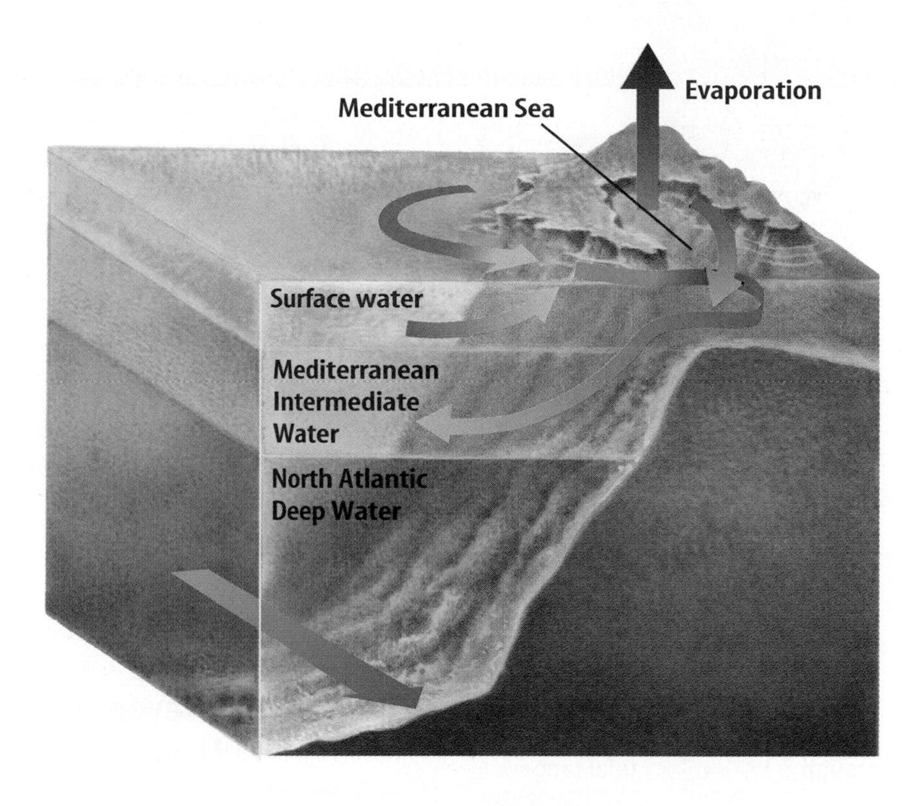

Figure 10
Dense layers of North Atlantic Deep Water form in the Greenland, Labrador, and Norwegian Seas. This water flows southward along the North Atlantic seafloor. Less dense water from the Mediterranean Sea forms Mediterranean Intermediate Water.

What causes the Mediterranean Intermediate Water to form?

Section 2 Assessment

1. What factors create surface currents?
2. What is the Coriolis effect?
3. How do density currents circulate water?
4. What is upwelling?
5. **Think Critically** The latitudes of San Diego, California, and Charleston, South Carolina, are exactly the same. However, the average yearly water temperature in the ocean off Charleston is much higher than the water temperature off San Diego. Explain why.

Skill Builder Activities

6. **Predicting** A river flows into the ocean. Predict what will happen to this layer of freshwater. Explain your prediction. **For more help, refer to the Science Skill Handbook.**
7. **Using an Electronic Spreadsheet** Make a spreadsheet that compares surface and density currents. Focus on characteristics such as wind, horizontal and vertical movement, temperature, and density. **For more help, refer to the Technology Skill Handbook.**

SECTION 3

Ocean Waves and Tides

As You Read

What You'll Learn

- **Describe** wave formation.
- **Distinguish** between the movement of water particles in a wave and the movement of the wave.
- **Explain** how ocean tides form.

Vocabulary

wave
crest
trough
breaker
tide
tidal range

Why It's Important

Waves and tides affect life and property in coastal areas.

Waves

If you've been to the seashore or seen a beach on TV, you've watched waves roll in. There is something hypnotic about ocean waves. They keep coming and coming, one after another. But what is an ocean wave? A **wave** is a rhythmic movement that carries energy through matter or space. In the ocean, waves like those in **Figure 11A** move through seawater.

Describing Waves Several terms are used to describe waves, as shown in **Figure 11B.** Notice that waves look like hills and valleys. The **crest** is the highest point of the wave. The **trough** (TRAWF) is the lowest point of the wave. Wavelength is the horizontal distance between the crests or between the troughs of two adjacent waves. Wave height is the vertical distance between crest and trough.

Half the distance of the wave height is called the amplitude (AM pluh tewd) of the wave. The amplitude squared is proportional to the amount of energy the wave carries. For example, a wave with twice the amplitude of the wave in **Figure 11** carries four times ($2 \times 2 = 4$) the energy. On a calm day, the amplitude of ocean waves is small. But during a storm, wave amplitude increases and the waves carry a lot more energy. Large waves can damage ships and coastal property.

Figure 11
Ocean waves carry energy through seawater.

A Identify the crests and troughs in this picture.

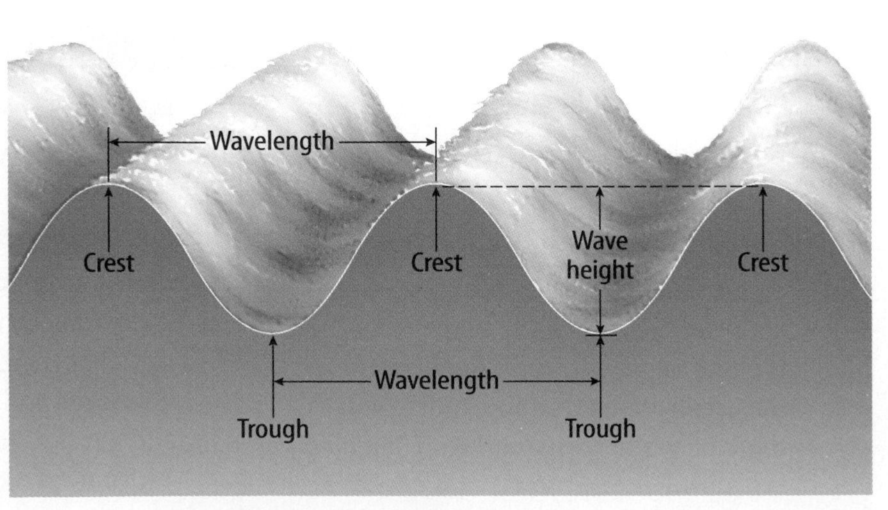

B The crest, trough, wavelength, and wave height describe a wave.

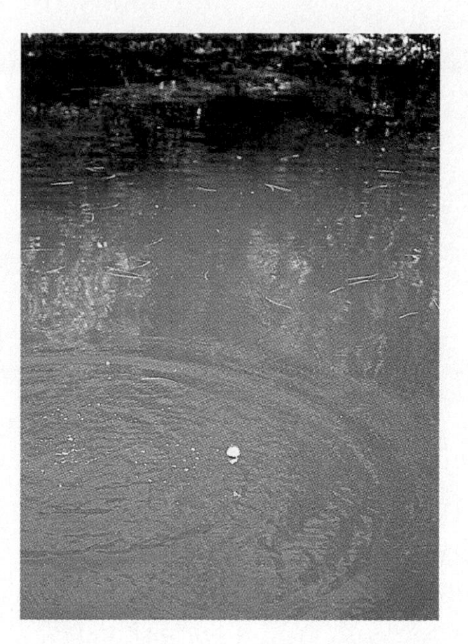

Figure 12
As a wave passes, only energy moves forward. The water particles and the bobber remain in place.

Wave Movement You might have noticed that if you throw a pebble into a pond, a circular wave moves outward from where the pebble entered the water, as shown in **Figure 12.** A bobber on a fishing line floating in the water will bob up and down as the wave passes, but it will not move outward with the wave. Notice that the bobber's position doesn't change.

When you watch an ocean wave, it looks as though the water is moving forward. But unless the wave is breaking onto shore, the water does not move forward. Each molecule of water stays in about the same place as the wave passes. **Figure 13** shows this. Water in a wave moves around in circles. Only the energy moves forward while the water remains in about the same place. Below a depth equal to about half the wavelength, water movement stops. Below that depth, water is not affected by waves. Submarines that travel below this level usually are not affected by surface storms.

Breakers A wave changes shape in the shallow area near shore. Near the shoreline, friction with the ocean bottom slows water at the bottom of the wave. As the wave slows, its crest and trough come closer together. The wave height increases. The top of a wave, not slowed by friction, moves faster than the bottom. Eventually, the top of the wave outruns the bottom and it collapses. The wave crest falls as water tumbles over on itself. The wave breaks onto the shore. **Figure 13** also shows this process. This collapsing wave is a **breaker.** It is the collapse of this wave that propels a surfer and surfboard onto shore. After a wave breaks onto shore, gravity pulls the water back into the sea.

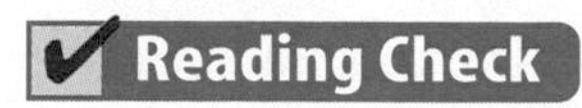

What causes an ocean wave to slow down?

Mini LAB

Modeling Water Particle Movement

Procedure

1. Put a piece of **tape** on the outside bottom of a clear, rectangular **plastic storage box.** Fill the box with **water.**
2. Float a **cork** in the container above the piece of tape.
3. Use a **spoon** to make gentle waves in the container.
4. Observe the movement of the waves and the cork.

Analysis

1. Describe the movement of the waves and the motion of the cork.
2. Compare the movement of the cork in the water with the movement of water particles in a wave.

VISUALIZING WAVE MOVEMENT

Figure 13

As ocean waves move toward the shore, they seem to be traveling in from a great distance, hurrying toward land. Actually, the water in waves moves relatively little, as shown here. It's the energy in the waves that moves across the ocean surface. Eventually that energy is transferred—in a crash of foam and spray—to the land.

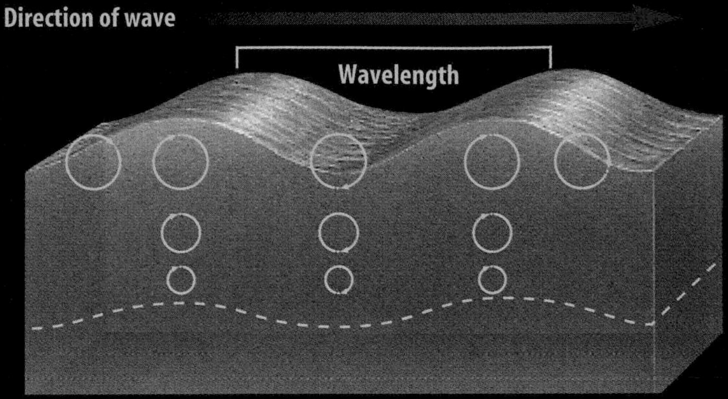

Ⓐ Particles of water move around in circles rather than forward. Near the water's surface, the circles are relatively large. Below the surface, the circles become progressively smaller. Little water movement occurs below a depth equal to about one-half of a wave's length.

Ⓑ The energy in waves, however, does move forward. One way to visualize this energy movement is to imagine a line of dominoes. Knock over the first domino, and the others fall in sequence. As they fall, individual dominoes—like water particles in waves—remain close to where they started. But each transfers its energy to the next one down the line.

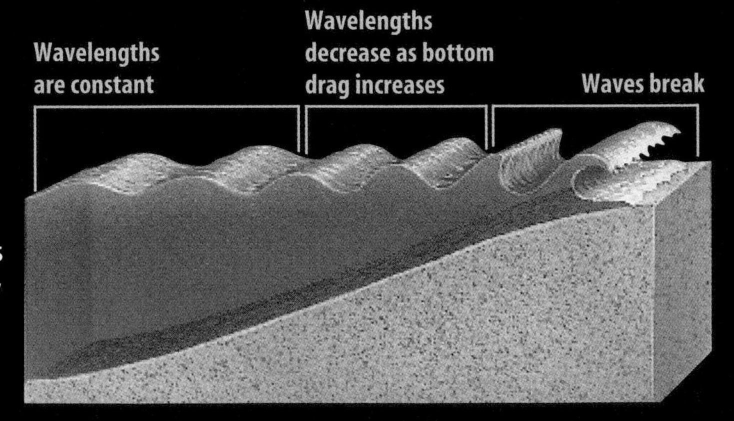

Ⓒ As waves approach shore, wavelength decreases and wave height increases. This causes breakers to form. Where ocean floor rises steeply to beach, incoming waves break quickly at a great height, forming huge arching waves.

Figure 14
Waves formed by storm winds can reach heights of 20 m to 30 m.

How Water Waves Form On a windy day, waves form on a lake or ocean. When wind blows across a body of water, friction causes the water to move along with the wind. If the wind speed is great enough, the water begins to pile up, forming a wave. As the wind continues to blow, the wave increases in height. Some waves reach tremendous heights, as shown in **Figure 14.** Storm winds have been known to produce waves more than 30 m high—taller than a six-story building.

The height of waves depends on the speed of the wind, the distance over which the wind blows, and the length of time the wind blows. When the wind stops blowing, waves stop forming. But once set in motion, waves continue moving for long distances, even if the wind stops. The waves you see lapping at a beach could have formed halfway around the world.

What factors affect the height of waves?

Tides

When you go to a beach, you probably notice the level of the sea rise and fall during the day. This rise and fall in sea level is called a **tide.** A tide is caused by a giant wave produced by the gravitational pull of the Sun and the Moon. This wave has a wave height of only 1 m or 2 m, but it has a wavelength that is thousands of kilometers long. As the crest of this wave approaches the shore, sea level appears to rise. This rise in sea level is called high tide. Later, as the trough of the wave approaches, sea level appears to drop. This drop in sea level is referred to as low tide.

Research Visit the Glencoe Science Web site at **science.glencoe.com** for more information about tides. Communicate to your class what you learn.

Figure 15
A large difference between high tide and low tide can be seen at Mont-Saint-Michel off the northwestern coast of France.

A **Mont-Saint-Michel lies about 1.6 km offshore and is connected to the mainland at low tide.**

B **Incoming tides move very quickly, making Mont-Saint-Michel an island at high tide.**

Tidal Range As Earth rotates, different locations on Earth's surface pass through the high and low positions. Many coastal locations, such as the Atlantic and Pacific coasts of the United States, experience two high tides and two low tides each day. One low-tide/high-tide cycle takes 12 h, 25 min. A daily cycle of two high tides and two low tides takes 24 h, 50 min—slightly more than a day. But because ocean basins vary in size and shape, some coastal locations, such as many along the Gulf of Mexico, have only one high and one low tide each day. The **tidal range** is the difference between the level of the ocean at high tide and low tide. Notice the tidal range in the photos in **Figure 15.**

Extreme Tidal Ranges The shape of the seacoast and the shape of the ocean floor affect the ranges of tides. Along a smooth, wide beach, the incoming water can spread over a large area. There the water level might rise only a few centimeters at high tide. In a narrow gulf or bay, however, the water might rise many meters at high tide.

Most shorelines have tidal ranges between 1 m and 2 m. Some places, such as those on the Mediterranean Sea, have tidal ranges of only about 30 cm. Other places have large tidal ranges. Mont-Saint-Michel, shown in **Figure 15,** lies in the Gulf of Saint-Malo off the northwestern coast of France. There the tidal range reaches about 13.5 m.

The dock shown in **Figure 16** is in Digby, Nova Scotia in the Bay of Fundy. This bay is extremely narrow, which contributes to large tidal ranges. The difference between water levels at high tide and low tide can be as much as 15 m.

Figure 16
The Bay of Fundy has the greatest tidal range in the world. *Was this picture taken at high tide or low tide?*

Tidal Bores In some areas when a rising tide enters a shallow, narrow river from a wide area of the sea, a wave called a tidal bore forms. A tidal bore can have a breaking crest or it can be a smooth wave. Tidal bores tend to be found in places with large tidal ranges. The Amazon River in Brazil, the Tsientang River in China, and rivers that empty into the Bay of Fundy in Nova Scotia have tidal bores.

When a tidal bore enters a river, it causes surface water to reverse its flow. In the Amazon River, the tidal bore rushes 650 km upstream at speeds of 65 km/h, causing a wave more than 5 m in height. Four rivers that empty into the Bay of Fundy have tidal bores. In those rivers, bore rafting is a popular sport.

Limpets are sea snails that live on rocky shores. When the tide comes in, they glide over the rocks to graze on seaweed. When the tide goes out, they use strong muscles to pull their shells tight against the rocks. Find out how other organisms survive in the zone between high and low tides.

The Gravitational Effect of the Moon For the most part, tides are caused by the interaction of gravity in the Earth-Moon system. The Moon's gravity exerts a strong pull on Earth. Earth and the water in Earth's oceans respond to this pull. The water bulges outward as Earth and the Moon revolve around a common center of mass. These events are explained in **Figure 17.**

Two bulges of water form, one on the side of Earth closest to the Moon and one on the opposite side of Earth. The reason two bulges form is because the Moon's gravity pulls harder on parts of Earth closer to the Moon than on parts farther away. You can imagine this if you think about pulling a large ball of dough in the same direction but harder on one side than the other side. The ball of dough will stretch and form two bulges. Earth does the same thing. The ocean bulges are the high tides, and the areas of Earth's oceans that are not toward or away from the Moon are the low tides. As Earth rotates, different locations on its surface pass through high and low tide.

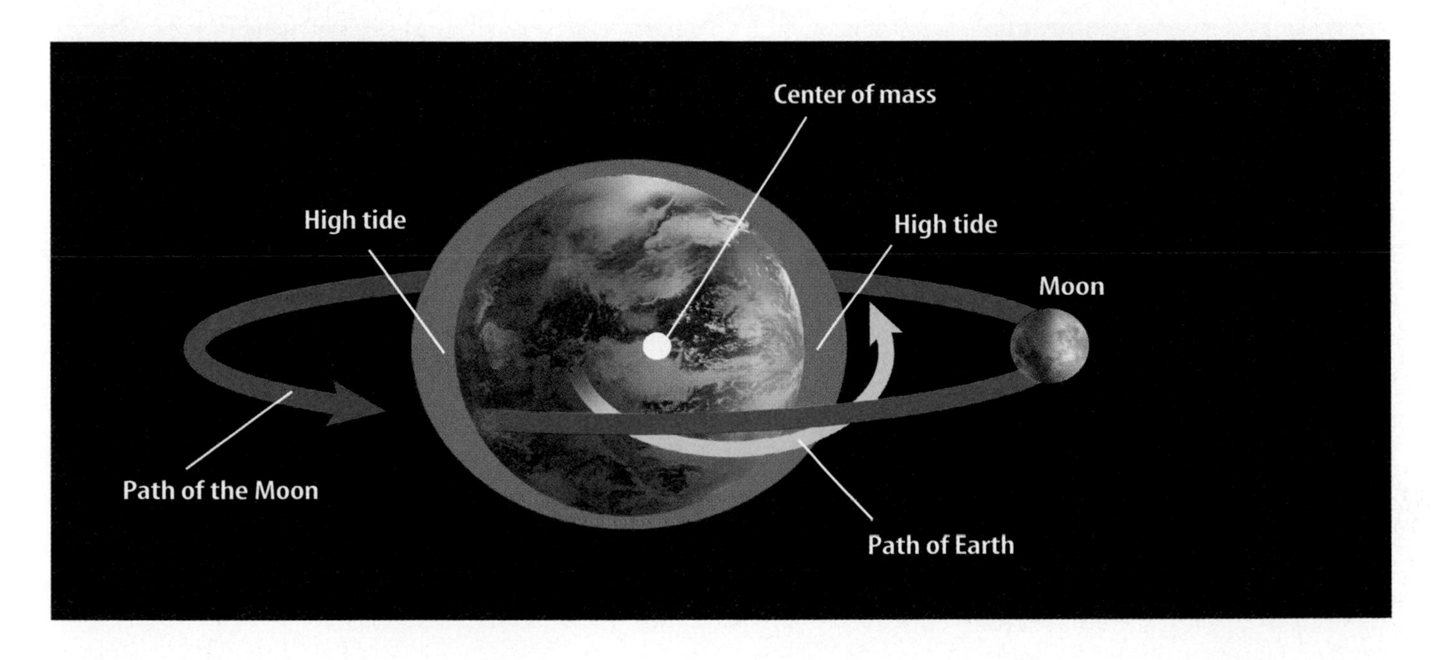

Figure 17
The Moon and Earth revolve around a common center of mass. Because the Moon's gravity pulls harder on parts of Earth closer to the Moon, a bulge of water forms on the side of Earth facing the Moon and the side of Earth opposite the Moon.

Figure 18
The gravitational attraction of the Sun causes spring tides and neap tides.

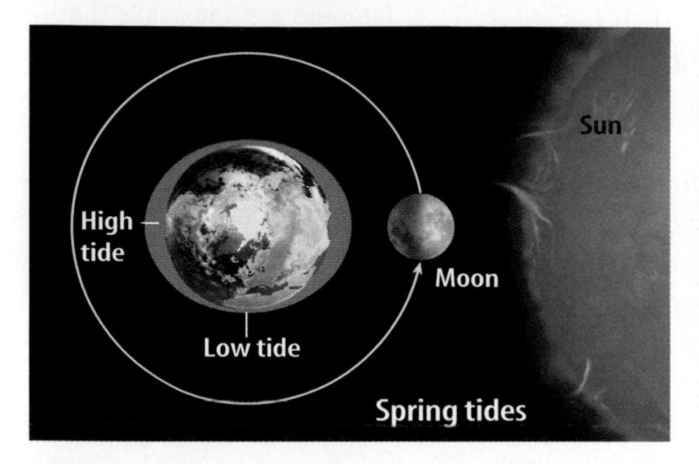

A **When the Sun, the Moon, and Earth are aligned, spring tides occur.**

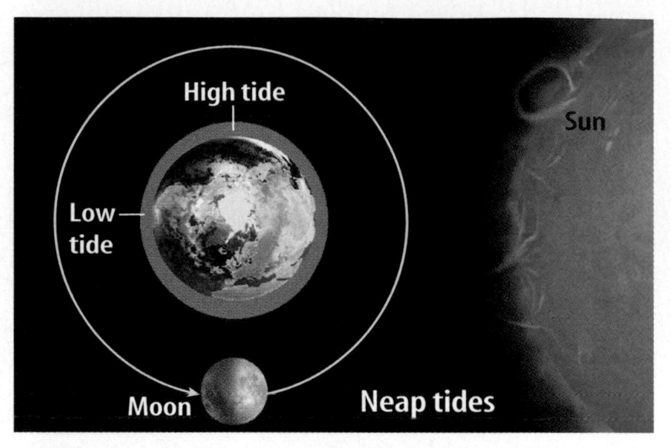

B **When the Sun, Earth, and the Moon form a right angle, neap tides occur.**

The Gravitational Effect of the Sun The Sun also affects tides. The Sun can strengthen or weaken the Moon's effects. When the Moon, Earth, and the Sun are lined up together, the combined pull of the Sun and the Moon causes spring tides, shown in **Figure 18A.** During spring tides, high tides are higher and low tides are lower than normal. The name *spring tide* has nothing to do with the season of spring. It comes from the German word *springen,* which means "to jump." When the Sun, Earth, and the Moon form a right angle, as shown in **Figure 18B,** high tides are lower and low tides are higher than normal. These are called neap tides.

Section  Assessment

1. Describe the parts of an ocean wave.
2. How does wind create water waves?
3. What causes high tides? Spring tides?
4. Compare water and wave movement.
5. **Think Critically** At the ocean, you spot a wave about 200 m from shore. A few seconds later, the wave breaks on the beach. Explain why the water in the breaker is not the same water that was in the wave 200 m away.

Skill Builder Activities

6. **Comparing and Contrasting** Compare and contrast the effects of the Sun and the Moon on Earth's tides. **For more help, refer to the Science Skill Handbook.**
7. **Communicating** Many planets have more than one moon. In your Science Journal, write a description of what tides might be like if Earth had two moons. **For more help, refer to the Science Skill Handbook.**

Activity

Making Waves

Wind generates some waves. The energy of motion is transferred from the wind to the surface water of the ocean. What factors influence the generation of waves?

What You'll Investigate

How do the speed of the wind and the length of time the wind blows affect the height of a wave?

Materials

11" × 14" white paper
3-speed electric fan
gooseneck lamp
clock or watch
rectangular, clear-plastic storage box
water
metric ruler

Goals

- **Observe** how wind speed and duration affect wave height.

Safety Precautions

Do not allow any part of the light or cord to come in contact with the water.

Procedure

1. Position the box on white paper beside the lamp.
2. Fill the plastic box with water to within 3 cm of the top. Direct light from the lamp onto the box.
3. Place the fan at one end of the box to create waves. Start the fan on its slowest speed. Keep the fan on during measuring.
4. After 3 min, measure the height of the waves caused by the fan. Record your observations in a table similar to the one shown. Through the plastic box, observe the shadows of the waves on the white paper.

5. After 5 min, measure the wave height and record your observations.
6. Repeat steps 3 to 5 with the fan on medium, then on high.
7. Turn off the fan. Unplug it. Observe what happens.

Wave Data

Fan Speed	Time (min)	Wave Height (mm)	Observations
Low	3		
Low	5		
Medium	3		
Medium	5		
High	3		
High	5		

Conclude and Apply

1. **Analyze** your data to determine whether the wave height is affected by the length of time that the wind blows. Explain.
2. **Analyze** your data to determine whether the height of the waves is affected by the speed of the wind. Explain.

Activity Design Your Own Experiment

Sink or Float

As you know, ocean water contains many dissolved salts. How does this affect objects within the oceans? Why do certain objects float on top of the ocean's waves, while others sink directly to the bottom? Density is a measurement of mass per volume. You can use density to determine whether an object will float within a certain volume of water of a specific salinity. In this activity you will investigate the effect of salinity on whether an object floats or sinks.

Recognizing the Problem

How does salinity affect whether a potato will float or sink?

Form a Hypothesis

Based on what you know so far about salinity, why things float or sink, and the density of a potato, plus what it looks and feels like, formulate a hypothesis. Do you think the salinity of water has any effect on objects that are floating in water? What kind of effect? Will they float or sink? How would a dense object like a potato be different from a less dense object like a cork?

Goals

- **Design** an experiment to identify how increasing salinity affects the ability of a potato to float in water.

Possible Materials

small, uncooked potato
teaspoon
salt
large glass bowl
water
balance
large graduated cylinder
metric ruler

Safety

Test Your Hypothesis

Plan

1. As a group, agree upon and write your hypothesis statement.
2. Devise a method to test how salinity affects whether a potato floats in water.
3. **List** the steps you need to take to test your hypothesis. Be specific, describing exactly what you will do at each step.
4. Read over your plan for testing your hypothesis.
5. How will you determine the densities of the potato and the different water samples? How you will measure the salinity of the water? How will you change the salinity of the water? Will you add teaspoons of salt one at a time?
6. How you will measure the ability of an object to float? Could you somehow measure the displacement of the water? Perhaps you could draw a line somewhere on your bowl and see how the position of the potato changes.
7. **Design** a data table where you can record your results. Include columns/rows for the salinity and float/sink measurements. What else should you include?

Do

1. Make sure your teacher approves your plan before you start.
2. Carry out the experiment.
3. While conducting the experiment, record your data and any observations that you or other group members make in your Science Journal.

Analyze Your Data

1. **Compare** how the potato floated in water with different salinities.
2. How does the ability of an object to float change with changing salinity?

Draw Conclusions

1. Did your experiment support the hypothesis you made?
2. A heavily loaded ship barely floats in the Gulf of Mexico. Based on what you learned, infer what might happen to the ship if it travels into the freshwater of the Mississippi River.

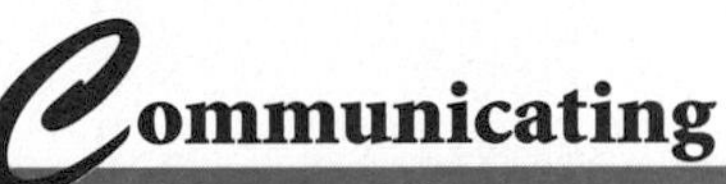

Prepare a chart showing the results of your experiment. Share the chart with members of your class. **For more help, refer to the Science Skill Handbook.**

Science and Language Arts

"The Jungle of Ceylon"

from Passions and Impressions

by Pablo Neruda

Respond to the Reading

1. What were his impressions of the island on arrival?
2. What words does the author choose to describe waves?
3. How would you describe the climate of Ceylon?

The following passage is part of a travel chronicle describing the Chilean poet Pablo Neruda's visit to the island of Ceylon, now called Sri Lanka, which is located southeast of India. The author considered himself so connected to Earth that he wrote in green ink.

Felicitous[1] shore! A coral reef stretches parallel to the beach; there the ocean interposes in its blues the perpetual white of a rippling ruff[2] of feathers and foam; the triangular red sails of sampans[3]; the unmarred line of the coast on which the straight trunks of the coconut palms rise like explosions, their brilliant green Spanish combs nearly touching the sky.

... In the deep jungle, there is a silence like that of libraries: abstract and humid.

1 Happy
2 round collar made of layers of lace
3 East Asian boats

Understanding Literature

Imagery Imagery is a series of words that evoke pictures to the reader. Poets use imagery to connect images to abstract concepts. The poet, here, wants to capture a particular feature of the reef and does so by describing it as a "ruff of feathers and foam," invoking the image of a gentle place, without the author saying so. Imagery also gives the reader more information about the story or chronicle. The poet further describes the shore as "happy", which helps us learn that the poet is arriving on the island on a clear, calm day.

Where else in the poem does the poet use imagery to convey a mood or feeling?

Science Connection In the poem there are several indicators that the wind, which causes waves and currents, is light and the waves are small, using imagery as discussed above.

Sri Lanka, however, often is plagued by monsoons, which affect ocean conditions and local climate. Monsoons are seasonal reversals of the regional winds. During the wet season, moist winds blow in from the sea, causing storms and producing waves. During the dry season, winds blow from the land and sunny days are common.

Linking Science and Writing

Weather Report Write a weather report for fishers and others who work at sea. Pick a geographic location to focus on. If possible, do research on wave conditions in this area. Include the times for low and high tides in your report. Add any additional weather information that you think might be important to people who work at or around the sea.

Career Connection

Oceanographer

Oceanographer Dr. Robert D. Ballard is an American oceanographer who revolutionized deep-sea archaeology. He developed several high-tech vessels that can explore ocean bottoms previously out of reach. Dr. Ballard discovered the location of the wreckage of the *Titanic, Lusitania,* and *Bismark.* He also discovered the wreckage of eight ancient ships in the Mediterranean Sea. Dr. Ballard has degrees in chemistry and geology as well as doctorate degrees in marine geology and geophysics.

SCIENCE*Online* To learn more about careers in oceanography, visit the Glencoe Science Web site at **science.glencoe.com.**

Chapter 16 Study Guide

Reviewing Main Ideas

Section 1 Ocean Water

1. Earth's ocean water might have originated from water vapor released from volcanoes. Over millions of years, the water condensed and rain fell, filling basins.
2. The oceans are a mixture of water, dissolved salts, and dissolved gases that are in constant motion.
3. Groundwater and rivers weather rock and dissolve some minerals to form ions. The ions are carried to the oceans where they give seawater its salty taste. *What kind of salt, shown here, makes up most of the salt left when seawater is evaporated?*

Section 2 Ocean Currents

1. Wind causes surface currents. Surface currents are affected by the Coriolis effect. The Coriolis effect turns currents north of the equator clockwise and turns currents south of the equator counterclockwise.
2. Surface currents can greatly affect climate and economic activity such as fishing. Upwelling brings deep, cold water to the ocean's surface.
3. Cool currents off western coasts originate far from the equator. Warmer currents along eastern coasts begin near the equator.
4. Differences in temperature and salinity between water masses in the oceans set up circulation patterns called density currents. *How do density currents originate in the Mediterranean Sea, shown here?*

Section 3 Ocean Waves and Tides

1. A wave is a rhythmic movement that carries energy. The crest is the highest point of a wave. The trough is the lowest point.
2. In a wave, energy moves forward while water particles move around in small circles.
3. Wind causes water to pile up and form most water waves. Tides are not caused by wind. *What causes the high and low tides shown here?*

After You Read

FOLDABLES Reading & Study Skills

Under the bottom tab of your Foldable, write about the effects of ocean motion on climate, world economics, and ocean water.

Visualizing Main Ideas

Complete the following concept map on ocean motions.

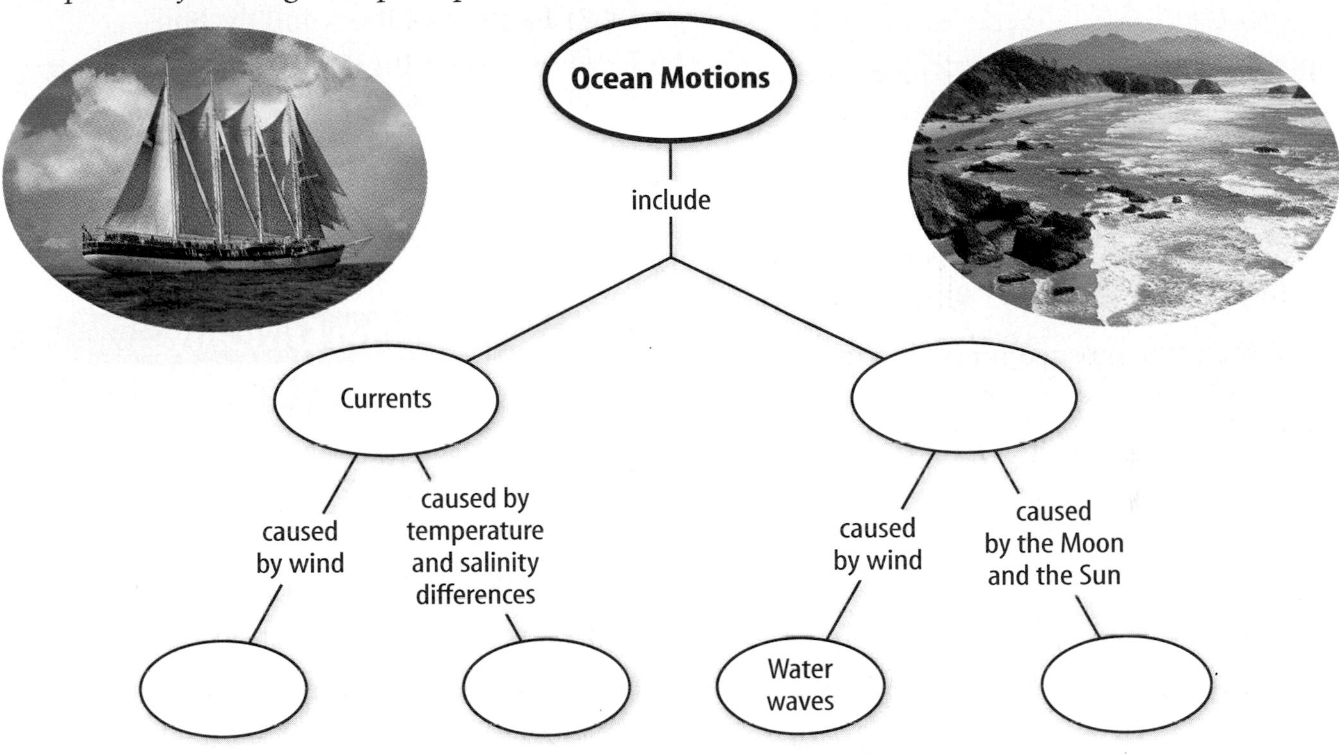

Vocabulary Review

Vocabulary Words

a. basin
b. breaker
c. Coriolis effect
d. crest
e. density current
f. salinity
g. surface current
h. tidal range
i. tide
j. trough
k. upwelling
l. wave

Study Tip

After you've read a chapter, go back to the beginning and speed-read through what you've just read. This will help you better remember what you have read.

Using Vocabulary

Replace the underlined words with the vocabulary words that have the same meaning.

1. The amount of dissolved salts in seawater has stayed about the same for hundreds of millions of years.
2. An area where nutrient-rich water comes to the surface is a good place to catch fish.
3. Wind creates a horizontal current at the top of the ocean.
4. Along most ocean beaches, a rise and fall of the ocean related to gravitational pull is easy to see.
5. Wind pushes on water to make a movement of energy through the water.

Chapter 16 Assessment

Checking Concepts

Choose the word or phrase that best answers the question.

1. Where might ocean water have originated?
   A) salt marshes
   B) volcanoes
   C) basins
   D) surface currents

2. How does chlorine enter the oceans?
   A) volcanoes
   B) rivers
   C) density currents
   D) groundwater

3. What is the most common ion found in ocean water?
   A) chloride
   B) calcium
   C) boron
   D) sulfate

4. What causes most surface currents?
   A) density differences
   B) the Gulf Stream
   C) salinity
   D) wind

5. What is the highest point on a wave called?
   A) wave height
   B) trough
   C) crest
   D) wavelength

6. In the ocean, what is the rhythmic movement that carries energy through seawater?
   A) current
   B) wave
   C) crest
   D) upwelling

7. Which of the following causes the density of seawater to increase?
   A) a decrease in temperature
   B) a decrease in salinity
   C) an increase in temperature
   D) a decrease in pressure

8. In which direction does the Coriolis effect cause currents in the northern hemisphere to turn?
   A) east
   B) south
   C) counterclockwise
   D) clockwise

9. Tides are affected by the positions of which celestial bodies?
   A) Earth and the Moon
   B) Earth, the Moon, and the Sun
   C) Venus, Earth, and Mars
   D) the Sun, Earth, and Mars

10. What affects surface currents?
    A) crests
    B) upwellings
    C) the Coriolis effect
    D) tides

Thinking Critically

11. If a sealed bottle is dropped into the ocean off the coast of Florida, where do you think it might wash up? Explain.

12. Why do silicon and calcium remain in seawater for a shorter time than sodium?

13. Describe the Antarctic density current.

14. How would the density of seawater at the mouth of the Mississippi River and in the Mediterranean Sea compare? Explain.

15. Refer to the graph below. On which day is the high tide highest? Lowest? On which day(s) is the low tide lowest? Highest? On which day(s) would Earth, the Moon, and the Sun be lined up? On which day(s) would the Moon, Earth, and the Sun form a right angle?

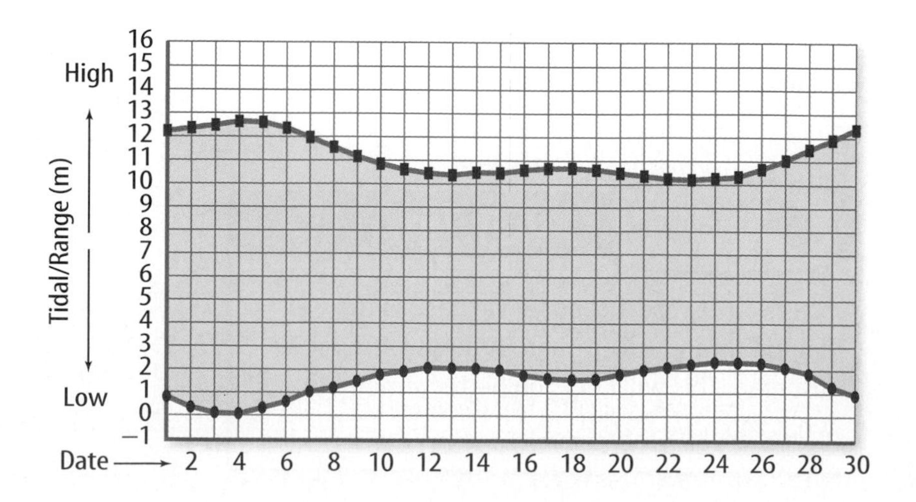

Developing Skills

16. Recognizing Cause and Effect What causes upwelling? What effect does it have? What can happen when upwelling stops?

17. Comparing and Contrasting Compare and contrast ocean waves and ocean currents.

18. Predicting Predict how drift bottles that are dropped into the ocean at points A and B will move. Explain.

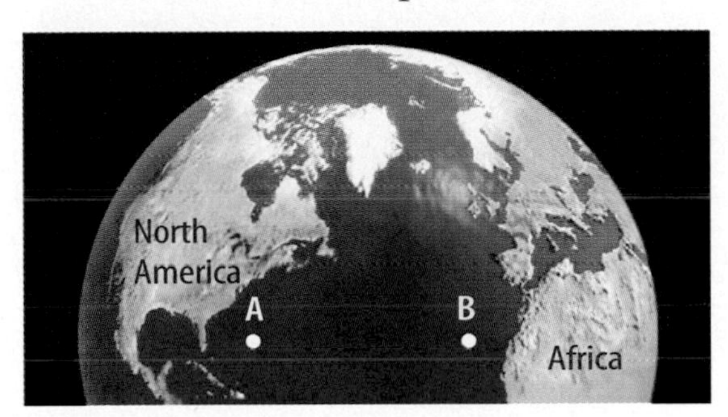

19. Recognizing Cause and Effect In the Mediterranean Sea, a density current forms because of the high rate of evaporation of water from the surface. How can evaporation cause a density current?

Performance Assessment

20. Invention Design a method for desalinating water that does not use solar energy. Draw it, and display it for your class.

21. Design and Perform an Experiment Create an experiment to test the density of water at different temperatures.

Technology

Go to the Glencoe Science Web site at **science.glencoe.com** or use the **Glencoe Science CD-ROM** for additional chapter assessment.

Test Practice

A marine scientist used the following graphic to support a lecture about ocean currents.

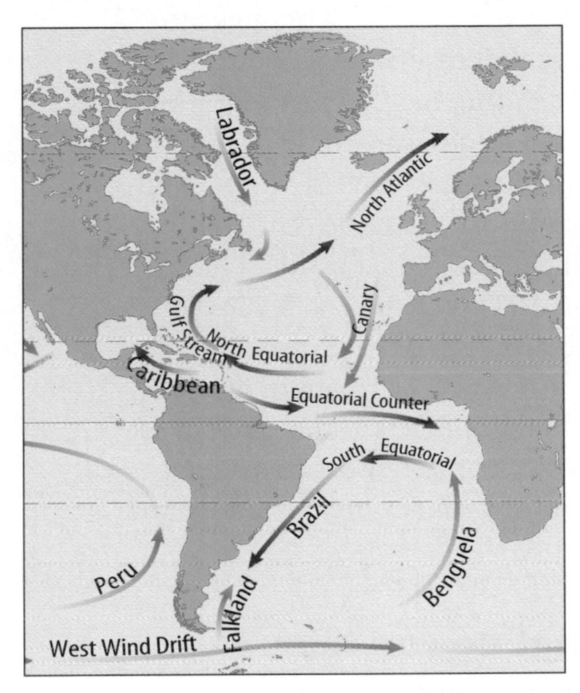

Study the graphic and answer the following questions.

1. The direction of ocean currents in the northern hemisphere is _____.

A) counterclockwise
B) north to south only
C) clockwise
D) east to west only

2. A reasonable conclusion based on the information in the graphic is that _____.

F) the ocean's currents only flow in one direction
G) the ocean's waters are constantly in motion
H) the Gulf Stream flows east to west
J) the Atlantic Ocean is deep

Earth in Space

What do you know about Earth's shape? What do you think happens to cause the seasons? Why do you suppose *maria*, the Latin word for seas, is used to describe some parts of the mostly dry and lifeless Moon? And what do you imagine is on the "other" side—the far side—of the Moon? This chapter will help you find answers to these questions.

You'll also learn about the Sun and other planets and moons in the solar system. They formed, along with the brilliant meteor shown here, billions of years ago.

What do you think?

Science Journal Look at the picture below with a classmate. Discuss what you think this might be or what is happening. Here's a hint: *Have you ever wished upon a falling star?* Write your answer or best guess in your Science Journal.

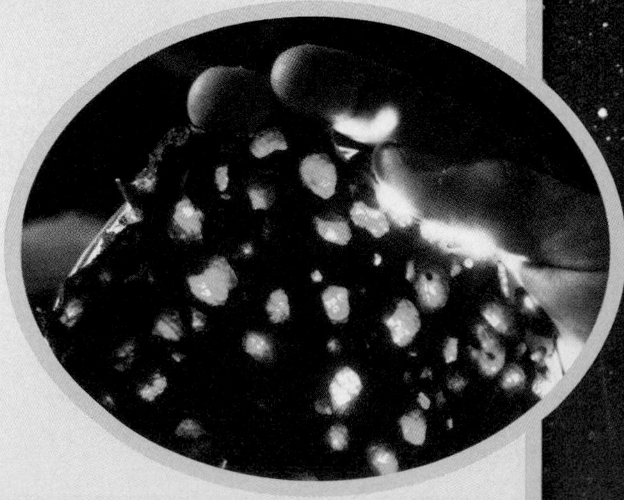

Could you prove that Earth is round? If you look out across a field or a large body of water, Earth seems to be flat, but you know that Earth is shaped like a large ball. What can you see from Earth's surface that proves Earth is round?

Model Earth's shape

1. Cut a strip of cardboard about 8 cm long and 8 cm tall into the shape of a sailboat. Fold up about 2 cm of the cardboard at the boat's base. Tape this 2-cm section to a basketball so that the peak of the sailboat sticks straight up.
2. Roll the basketball on a table at eye level so that the sailboat sticks out horizontally, parallel to the table, and points opposite from your view. Look at the top of the ball. The curving edge of the basketball can be compared to a horizon.
3. Roll the ball toward you slowly so that the sail comes into view over the top of the ball. Stop when you can see the entire paper sailboat.
4. Record everything you observe in your Science Journal.

Observe

In your Science Journal, write a paragraph that explains how the shape of the basketball affected your view of the sailboat. Interpret how this can be considered a model of Earth's shape with you looking out over the sea.

Before You Read

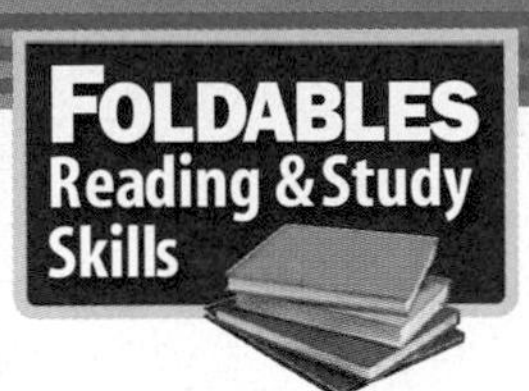

Making a Compare and Contrast Study Fold **Make the following Foldable to help you see how Earth and moon are similar and different.**

1. Place a sheet of paper in front of you so the long side is at the top. Fold the paper in half from the left side to the right side. Fold top to bottom and crease. Then unfold.
2. Through the top thickness of paper, cut along the middle fold line to form two tabs as shown.
3. Label "Alike" and "Different" across the front of the paper as shown.
4. Before you read the chapter, list all the ways you think Earth and the Moon are alike and different under the tabs.
5. As you read the chapter, add to or change the information you wrote under the tabs.

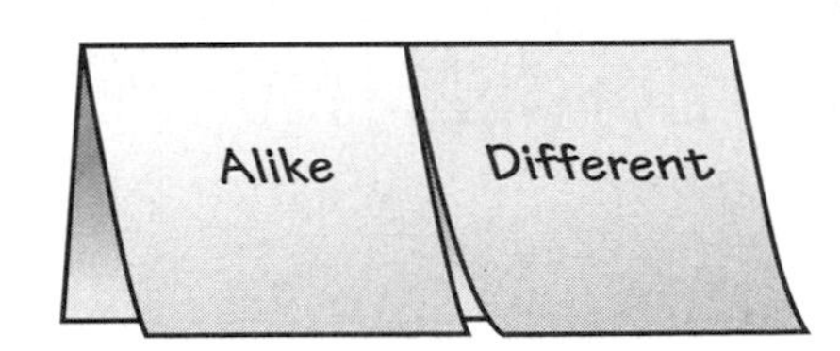

SECTION 1

Earth's Motion and Seasons

As You Read

What You'll Learn

- **Identify** Earth's shape and other physical properties.
- **Compare and contrast** Earth's rotation and revolution.
- **Explain** the causes of Earth's seasons.

Vocabulary

axis
rotation
revolution
orbit
solstice
equinox

Why It's Important

Movements of Earth cause changes from day to night and from one season to another.

Earth's Physical Data

Think about the last time you saw a beautiful sunset. Late in the day, you may have noticed the Sun sinking lower and lower in the western sky. Eventually, as the Sun went below the horizon, the sky became darker. Was the Sun actually traveling out of view, or were you?

In the past, some people thought that the Sun, the Moon, and other objects in space moved around Earth each day. Now it is known that some of the motions of these objects, as observed from Earth, are really caused by Earth's movements.

Also, many people used to think that Earth was flat. They thought that if you sailed far enough out to sea, you eventually would fall off. It is now known that this is not true. What general shape does Earth have?

Spherical Earth As shown in **Figure 1,** pictures from space show that Earth is shaped like a ball, or a sphere. A sphere (SFIHR) is a three-dimensional object whose surface at all points is the same distance from its center. What other evidence can you think of that reveals Earth's shape?

Figure 1
Earth's nearly spherical shape was first observed directly by images taken from spacecraft.
What observations on Earth's surface also suggest that it is spherical?

Evidence for Earth's Shape Have you ever stood on a dock and watched a sailboat come in? If so, you may have noticed that the first thing you see is the top of the boat's sail. This occurs because Earth's curved shape hides the rest of the boat from view until it is closer to you. As the boat slowly comes closer to you, more and more of its sail is visible. Finally, the entire boat is in view.

More proof of Earth's shape is that Earth casts a curved shadow on the Moon during a lunar eclipse, like the one shown in **Figure 2.** Something flat, like a book, casts a straight shadow, whereas objects with curved surfaces cast curved shadows.

Reading Check ***What object casts a shadow on the Moon during a lunar eclipse?***

Influence of Gravity The spherical shape of Earth and other planets is because of gravity. Gravity is a force that attracts all objects toward each other. The farther away the objects are, the weaker the pull of gravity is. Also, the larger an object is, the larger its gravitational pull is. A large object in space is spherical because gravity attracted particles toward its center while it was in a liquid or gaseous state. A spherical shape decreases the distance between particles in the object and its center. In this way, the potential energy due to gravity is less, and a stable shape results.

Even though Earth is round, it may seem flat to you. This is because Earth's surface is so large compared to your size.

Figure 2
Earth's spherical shape also is indicated by the curved shadow it casts on the Moon during a partial lunar eclipse.

Table 1 Physical Properties of Earth

Property	Value
Diameter (pole to pole)	12,714 km
Diameter (equator)	12,756 km
Circumference (poles) *(distance around Earth through N and S poles)*	40,008 km
Circumference (equator) *(distance around Earth through equator)*	40,075 km
Mass	5.98×10^{24} kg
Average Density *(average mass per unit volume)*	5.52 g/cm^3
Average distance from the Sun	149,600,000 km
Period of rotation (1 day) *(spin on axis)*	23 h, 56 min
Period of revolution (1 year) *(path around the Sun)*	365 days, 6 h, 9 min

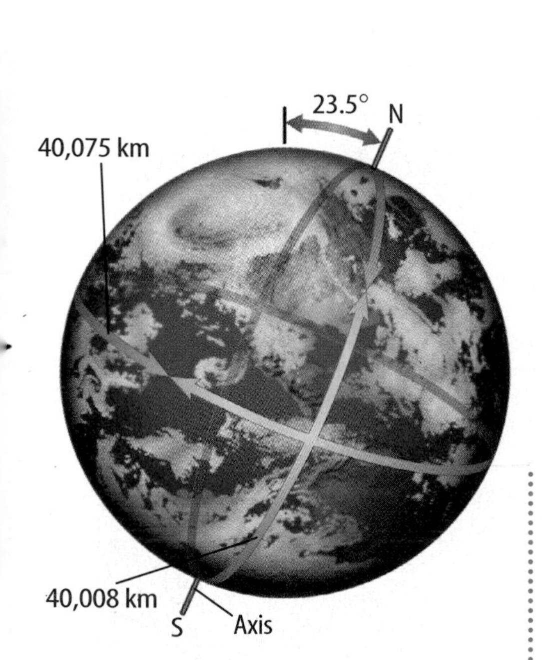

Figure 3
Earth is almost a sphere, but its circumference measurements vary slightly. The north-south circumference of Earth is smaller than the east-west circumference.

Almost a Sphere Earth is shaped like a sphere, but not a perfect one. It bulges slightly at the equator and is somewhat flattened around the poles. As shown in **Figure 3,** this causes Earth's circumference at the equator to be a bit larger than Earth's circumference as measured through the north and south poles. The circumference of Earth and some other physical properties are listed in **Table 1.**

Motions of Earth

Why the Sun appears to set each day and why the Moon and other objects in the sky appear to move from east to west is illustrated in **Figure 4.** Earth's geographic poles are located at the north and south ends of Earth's axis. Earth's **axis** is the imaginary line drawn from the north geographic pole through Earth to the south geographic pole. Earth spins around this imaginary line. The spinning of Earth on its axis, called **rotation,** causes you to experience day and night.

Reading Check ***What imaginary line runs through Earth's north and south geographic poles?***

Earth's Orbit Earth has another type of motion. As it rotates on its axis each day, Earth also moves along a path around the Sun. This motion of Earth around the Sun, shown in **Figure 4,** is called **revolution.** How many times does Earth rotate on its axis during one complete revolution around the Sun? Just as day and night are caused by rotation, what happens on Earth that is caused by its revolution?

Research Visit the Glencoe Science Web site at **science.glencoe.com** for more information about Earth's rotation and revolution.

Seasons A new year has begun. As days and weeks pass, you notice that the Sun remains in the sky later and later each day. You look forward to spring when you will be able to stay outside longer in the evening because the days become longer and longer. What is causing this change?

You learned earlier that Earth's rotation causes day and night. Earth also moves around the Sun, completing one revolution each year. Earth is really a satellite of the Sun, moving around it along a curved path called an **orbit.** The shape of Earth's orbit is an ellipse, which is rounded like a circle but somewhat flattened. As Earth moves along in its orbit, the way in which the Sun's light strikes Earth's surface changes.

Earth's elliptical orbit causes it to be closer to the Sun in January and farther from the Sun in July. But, the total amount of energy Earth receives from the Sun changes little during a year. However, the amount of energy that specific places on Earth receive varies quite a lot.

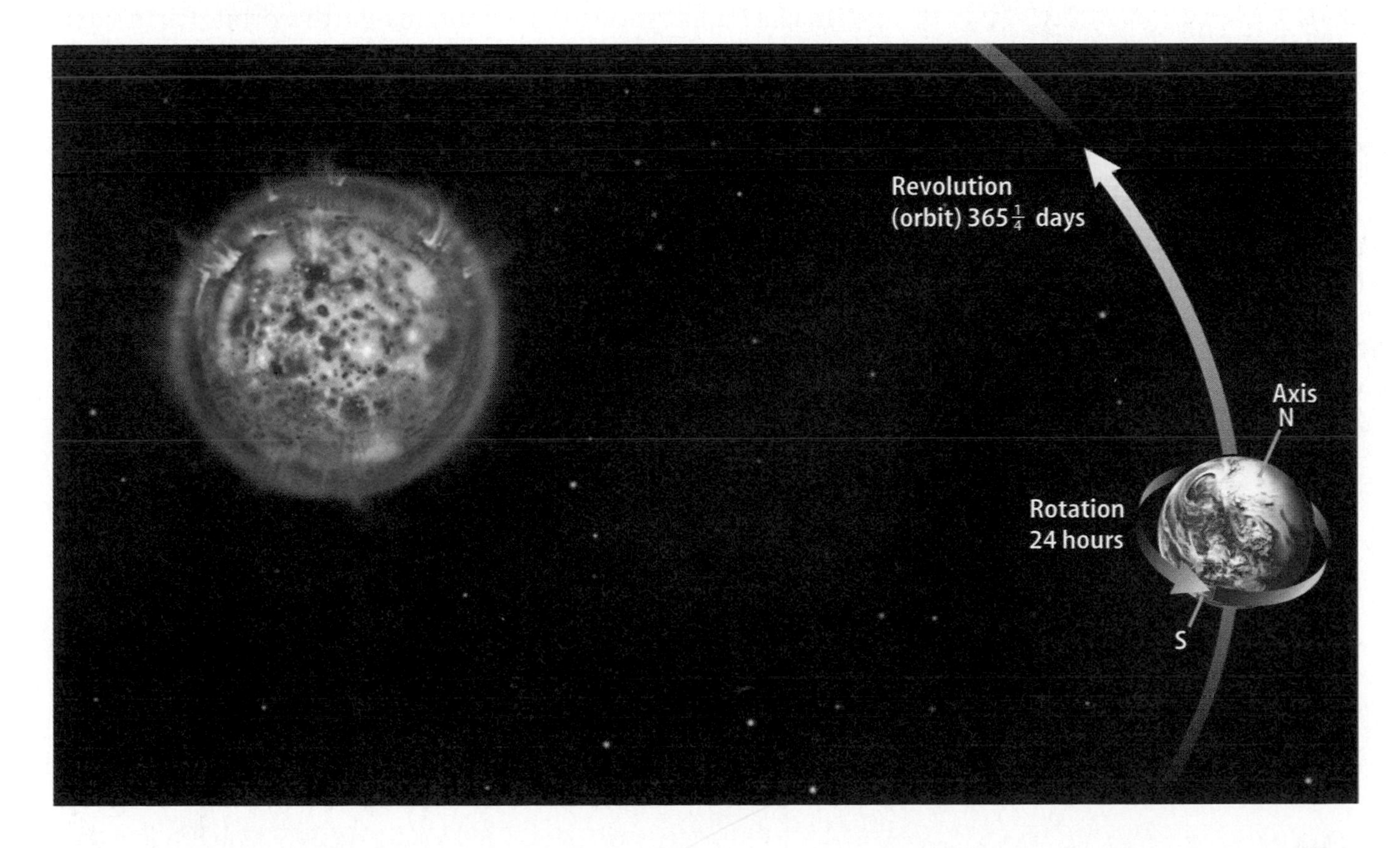

Figure 4
Earth's counterclockwise motions cause day and night and the seasons.

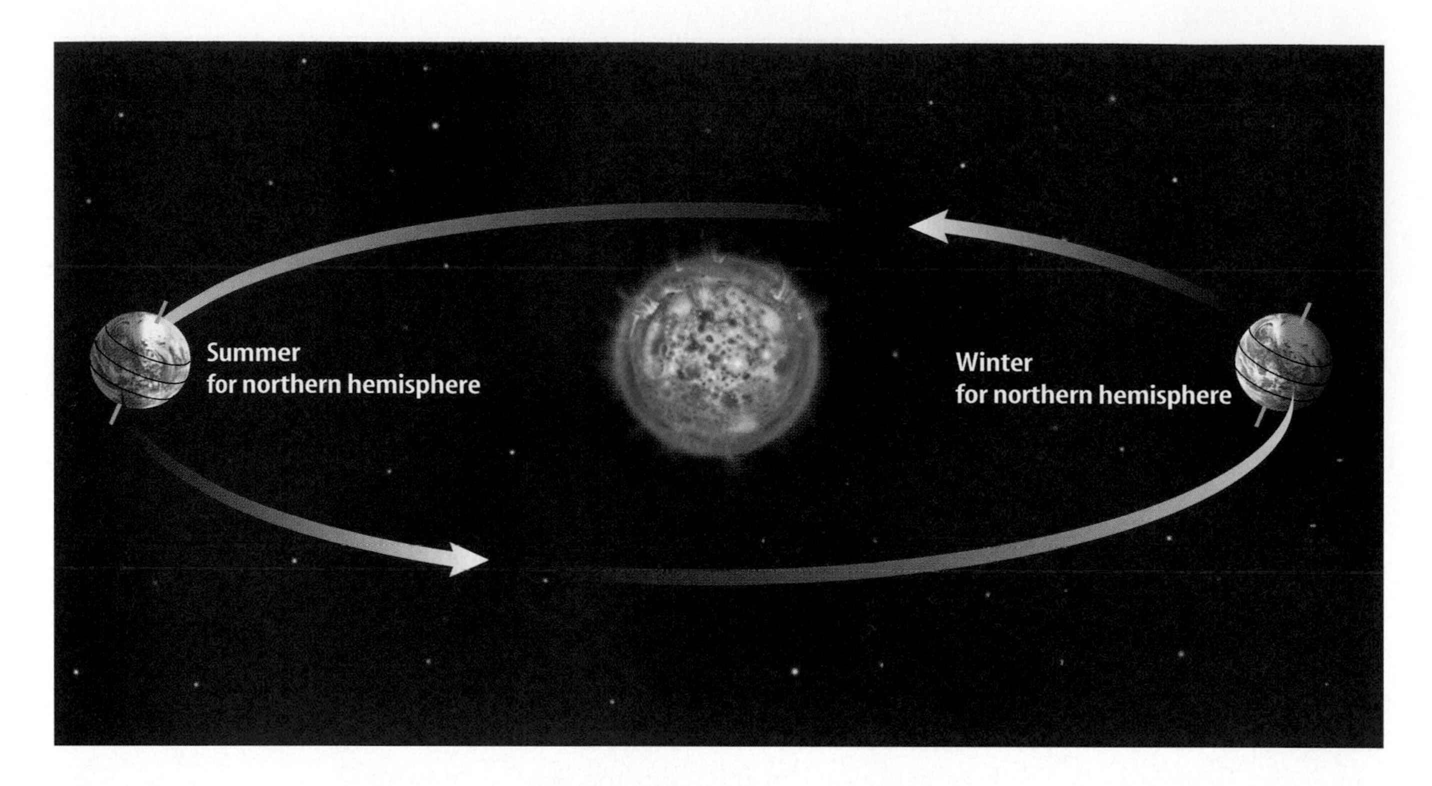

Figure 5
When the northern hemisphere is tilted toward the Sun, it experiences summer. *Why are days longer during the summer?*

Earth's Tilt You can observe one reason why the amount of energy from the Sun varies by moving a globe of Earth slowly around a light source. If you keep the globe tilted in one direction, you will see that the top half of the globe is tilted toward the light during part of its orbit and tilted away from the light during another part of its orbit.

Imagine a flat surface that contains Earth's orbit. Earth's axis forms a 23.5-degree angle with this imaginary surface, and always points to the North Star. Because of this, just as in your model, daylight hours are longer for the half of Earth, or hemisphere, tilted toward the Sun. Also, the Sun's rays hit that hemisphere more directly, at a higher angle. In other words, the Sun is higher in the sky for longer periods of time. Think again about when it gets dark outside at different times of the year. During which season do you notice longer days and shorter nights? As shown in **Figure 5,** this happens during summer.

Solstices Because of the tilt of Earth's axis, the Sun's position relative to Earth's equator changes. At two times during the year, the Sun reaches its greatest distance north or south of the equator and is directly over the Tropic of Cancer or the Tropic of Capricorn, as shown in **Figure 6.** These times are known as the summer and winter **solstices.** Summer solstice, which is about the longest day of the year, happens on June 21 or 22 for the northern hemisphere and on December 21 or 22 for the southern hemisphere. The opposite of this for each hemisphere is winter solstice, which is about the shortest day of the year.

Equinoxes At **equinox,** (EE kwuh nahks) when the Sun is directly above Earth's equator, the lengths of day and night are nearly equal all over the world. During equinox, Earth's tilt is not toward or away from the Sun. In the northern hemisphere, spring equinox is March 21 or 22 and fall equinox is September 21 or 22. As you saw in **Table 1,** the time it takes for Earth to revolve around the Sun is not a whole number of days. Because of this, the dates for solstices and equinoxes change slightly over time.

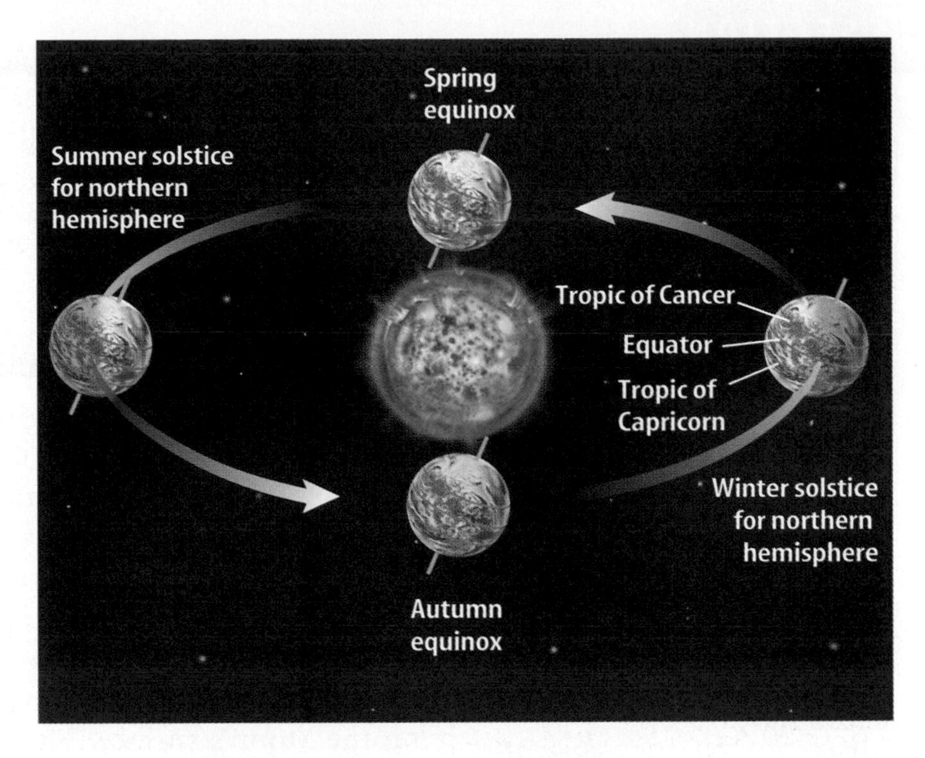

Figure 6
When the Sun is directly above the equator, day and night have nearly equal lengths.

Earth's Place in Space Earth is shaped much like a sphere. As Earth rotates on its axis, the Sun appears to rise and set in the sky. Earth's tilt and revolution around the Sun cause seasons to occur. In the next section, you will learn about Earth's nearest neighbor in space, the Moon. Later, you will learn about other planets in our solar system and how they compare with Earth.

Section  Assessment

1. Which Earth motion causes a point on Earth to experience different amounts of energy from the Sun at different times of the year?
2. How many daylight hours do people in different places on Earth experience during an equinox?
3. Why is Earth's shape considered to be nearly spherical?
4. How do the rotation and revolution of Earth differ?
5. **Think Critically** In **Table 1,** why is Earth's distance from the Sun reported as an average distance?

Skill Builder Activities

6. **Making Models** Use a globe and an unshaded light source to illustrate how the tilt of the Earth on its axis, as it rotates and revolves around the Sun, causes changes in the length of a day. **For more help, refer to the Science Skill Handbook.**
7. **Communicating** Using data in **Table 1,** write a brief paragraph in your Science Journal that describes Earth's shape, size, and motions. Research the same physical properties of Venus. Write another paragraph that compares and contrasts Venus's properties with those of Earth. **For more help, refer to the Science Skill Handbook.**

SECTION

Earth's Moon

As You Read

What You'll Learn

- **Identify** the Moon's surface features and interior.
- **Explain** the Moon's phases.
- **Explain** the causes of solar and lunar eclipses.
- **Compare** possible origins of the Moon.

Vocabulary

crater
moon phase
solar eclipse
lunar eclipse

Why It's Important

The Moon is Earth's closest neighbor in space.

The Moon's Surface and Interior

Take a good look at the surface of the Moon during the next full moon. You can see some of its large surface features, especially if you use binoculars or a small telescope. You will see dark-colored maria (MAR ee uh) and lighter-colored highland areas, as illustrated in **Figure 7.** Galileo first named the dark-colored regions *maria*, the Latin word for seas. They reminded Galileo of the oceans. Maria probably formed when lava flows from the Moon's interior flooded into large, bowl-like regions on the Moon's surface. These depressions may have formed early in the Moon's history. Collected during *Apollo* missions and then analyzed in laboratories on Earth, rocks from the maria are about 3.2 billion to 3.7 billion years old. They are the youngest rocks found on the Moon thus far.

The oldest moon rocks analyzed so far—dating to about 4.4 billion years old—were found in the lunar highlands. The lunar highlands are areas of the lunar surface with an elevation that is several kilometers higher than the maria. Some lunar highlands are located in the south-central region of the Moon.

Figure 7
On a clear night and especially during a full moon, you can observe some of the Moon's surface features. *How can you recognize the maria, lunar highlands, and craters?*

Craters As you look at the Moon's surface features, you will see craters. **Craters,** also shown in **Figure 7,** are depressions formed by large meteorites—space objects that strike the surface. As meteorites struck the Moon, cracks could have formed in the Moon's crust, allowing lava flows to fill in the large depressions. Craters are useful for determining how old parts of a moon's or a planet's surface are compared to other parts. The more abundant the craters are in a region, the older the surface is.

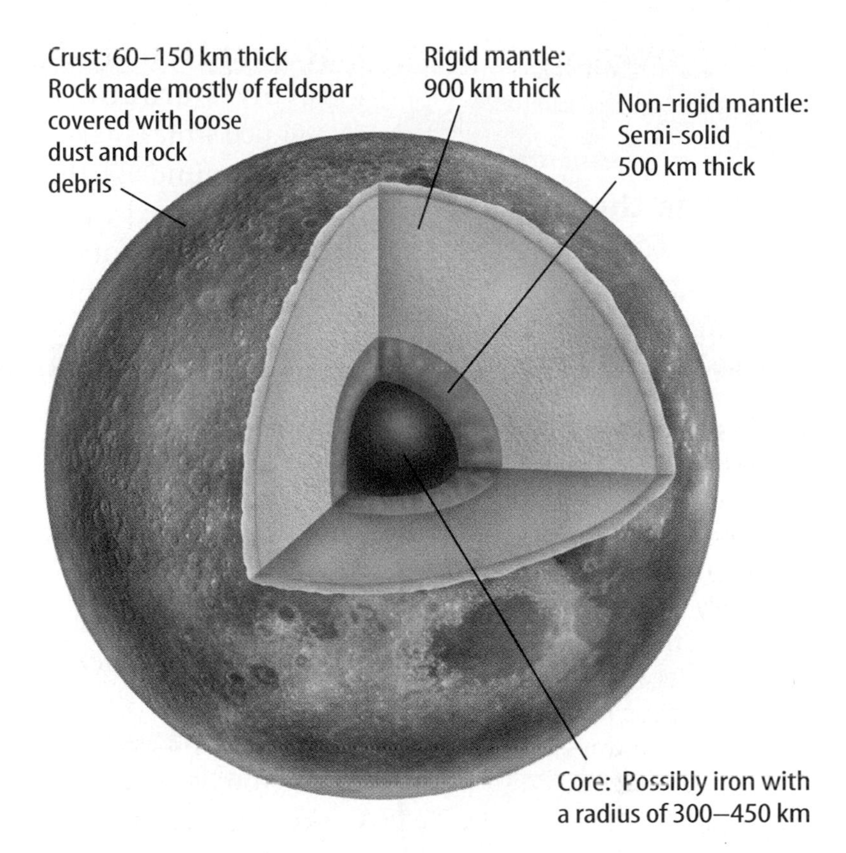

Figure 8
The small size of the Moon's core suggests it formed from a part of the crust and mantle of Earth.

The Moon's Interior During the *Apollo* space program, astronauts left several seismographs (SIZE muh grafs) on the Moon. A seismograph is an instrument that detects tremors, or seismic vibrations. On Earth, seismographs are used to measure earthquake activity. On the Moon, they are used to study moonquakes. Based on the study of moonquakes, a model of the Moon's interior has been proposed, as illustrated in **Figure 8.** The Moon's crust is about 60 km thick on the side facing Earth and about 150 km thick on the far side. The difference in thickness is probably the reason fewer lava flows occurred on the far side of the Moon. Below the crust, a solid layer called the mantle may extend 900 km to 950 km farther down. A soft layer of mantle may continue another 500 km deeper still. Below this may be an iron-rich, solid core with a radius of about 300–450 km.

Like the Moon, Earth also has a dense, iron core. However, the Moon's core is small compared to its total volume. Compared with Earth, the Moon is most like Earth's outer two layers—the mantle and the crust—in density. This supports a hypothesis that the Moon may have formed from material ejected from Earth's mantle and crust.

Research from space probes indicates that conditions on some areas of the Moon might make Moon colonies possible someday. In your Science Journal, write a brief summary of what might have been discovered and how it would be useful to a Moon colony.

Motions of the Moon

The same side of the Moon is always facing Earth. You can verify this by examining the Moon in the sky night after night. You'll see that bright and dark surface features remain in the same positions. Does this mean that the Moon doesn't turn on an axis as it moves around Earth? Next, explore why the same side of the Moon always faces Earth.

Modeling the Moon's Rotation

Procedure

1. Use **masking tape** to place a large X on a **basketball** that will represent the Moon.
2. Ask two students to sit in **chairs** in the center of the room.
3. Place other students around the outer edge of the room.
4. Slowly walk completely around the two students in the center while holding the basketball so that the side with the X always faces the two students.

Analysis

1. Ask the two students in the center whether they think the basketball turned around as you circled them. Then ask several students along the outer edge of the room whether they think the basketball turned around.
2. Based on these observations, infer whether or not the Moon rotates as it moves around Earth. Explain your answer.

Revolution and Rotation of the Moon The Moon revolves around Earth at an average distance of about 384,000 km. It takes 27.3 days for the Moon to complete one orbit around Earth. The Moon also takes 27.3 days to rotate once on its axis. Because these two motions of the Moon take the same amount of time, the same side of the Moon is always facing Earth. Examine **Figure 9** to see how this works.

✔ Reading Check *Why does the same side of the Moon always face Earth?*

However, these two lunar motions aren't exactly the same during the Moon's 27.3-day rotation-and-revolution period. Because the Moon's orbit is an ellipse, it moves faster when it's closer to Earth and slower when it's farther away. During one orbit, observers are able to see a little more of the eastern side of the Moon and then a little more of the western side.

Moon Phases If you ever watched the Moon for several days in a row, you probably noticed how its shape and position in the sky change. You learned that the Moon rotates on its axis and revolves around Earth. Motions of the Moon cause the regular cycle of change in the way the Moon looks to an observer on Earth.

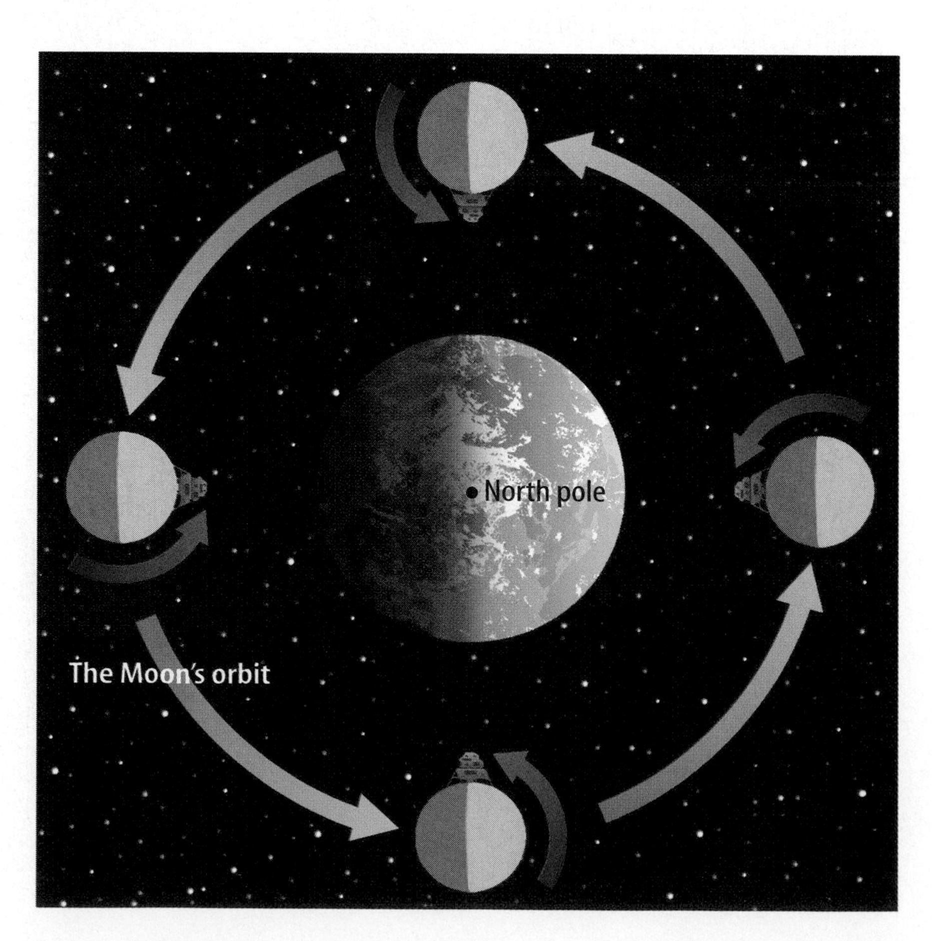

Figure 9
Observers viewing the Moon from Earth always see the same side of the Moon. This is caused by two separate motions of the Moon that take the same amount of time.

The Sun Lights the Moon You see the Moon because it reflects sunlight. As the Moon revolves around Earth, the Sun always lights one half of it. However, you don't always see the entire lighted part of the Moon. What you do see are phases, or different portions of the lighted part. **Moon phases,** illustrated in **Figure 10,** are the changing views of the Moon as seen from Earth.

New Moon and Waxing Phases New moon occurs when the Moon is positioned between Earth and the Sun. You can't see any of a new moon, because the lighted half of the Moon is facing the Sun. The new moon rises and sets with the Sun and never appears in the night sky.

Figure 10
The amount of the Moon's surface that looks bright to observers on Earth changes during a complete cycle of the Moon's phases. *What makes the Moon's surface appear so bright?*

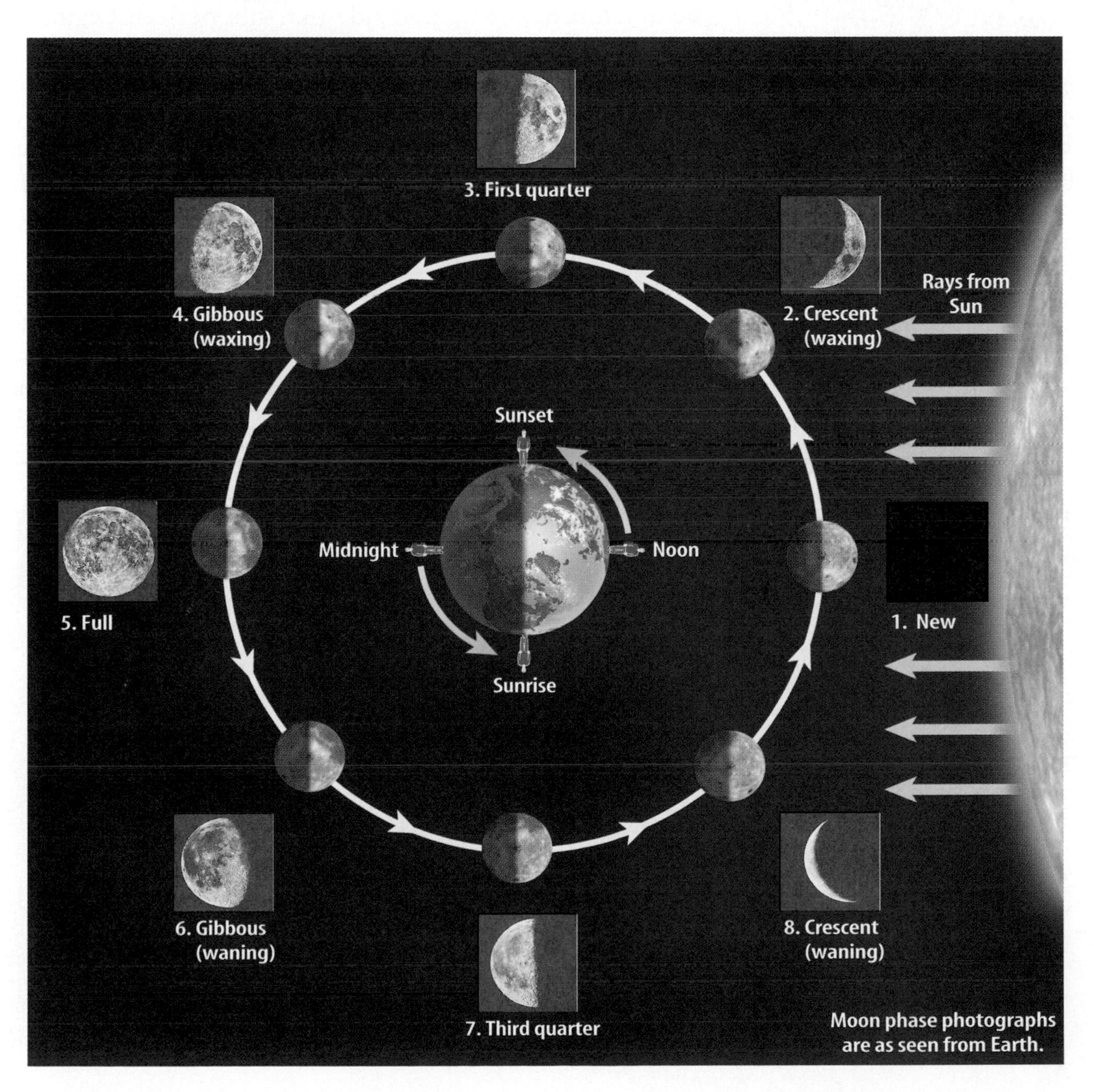

SCIENCE Online

Research Visit the Glencoe Science Web site at **science.glencoe.com** for more information about phases of the Moon.

Waxing Moon

Shortly after new moon, more and more of its lighted side faces Earth and becomes visible. The phases are said to be waxing, or growing in size. About 24 hours after new moon, you can see a thin sliver of the lighted side. This phase is called waxing crescent. As the Moon continues its trip around Earth, you eventually can see half of the lighted side. This phase is first quarter and occurs about a week after new moon.

The Moon's phases continue to wax through waxing gibbous (GIHB us) and then full moon—the phase when you can see all of the lighted side. At full moon, Earth is between the Sun and the Moon.

Full Moon and Waning Phases

After passing full moon, the amount of the lighted side that can be seen begins to decrease. Now the phases are said to be waning. Waning gibbous occurs just after full moon. Next comes third quarter when you can see only half of the lighted side. The phases continue to wane, and waning crescent occurs just before another new moon. Once again you see a small slice of the lighted side.

✔ Reading Check *What are the waning phases of the Moon?*

The complete cycle of the Moon's phases takes about 29.5 days. However, you will recall that the Moon takes only 27.3 days to revolve once around Earth. **Figure 11** explains the time difference between these two lunar cycles. Earth's revolution around the Sun causes the time lag. It takes the Moon about two days longer to align itself again between Earth and the Sun at new moon.

Figure 11
It takes longer for a complete Moon phase cycle than a complete Moon revolution. The revolution of Earth and the Moon around the Sun lengthens the path needed for the Moon, Earth, and the Sun to line up again at new moon. ***Which takes longer, a complete cycle of lunar phases or a complete orbit of the Moon around Earth? How does the revolution of the Earth-Moon system around the Sun cause this difference in time?***

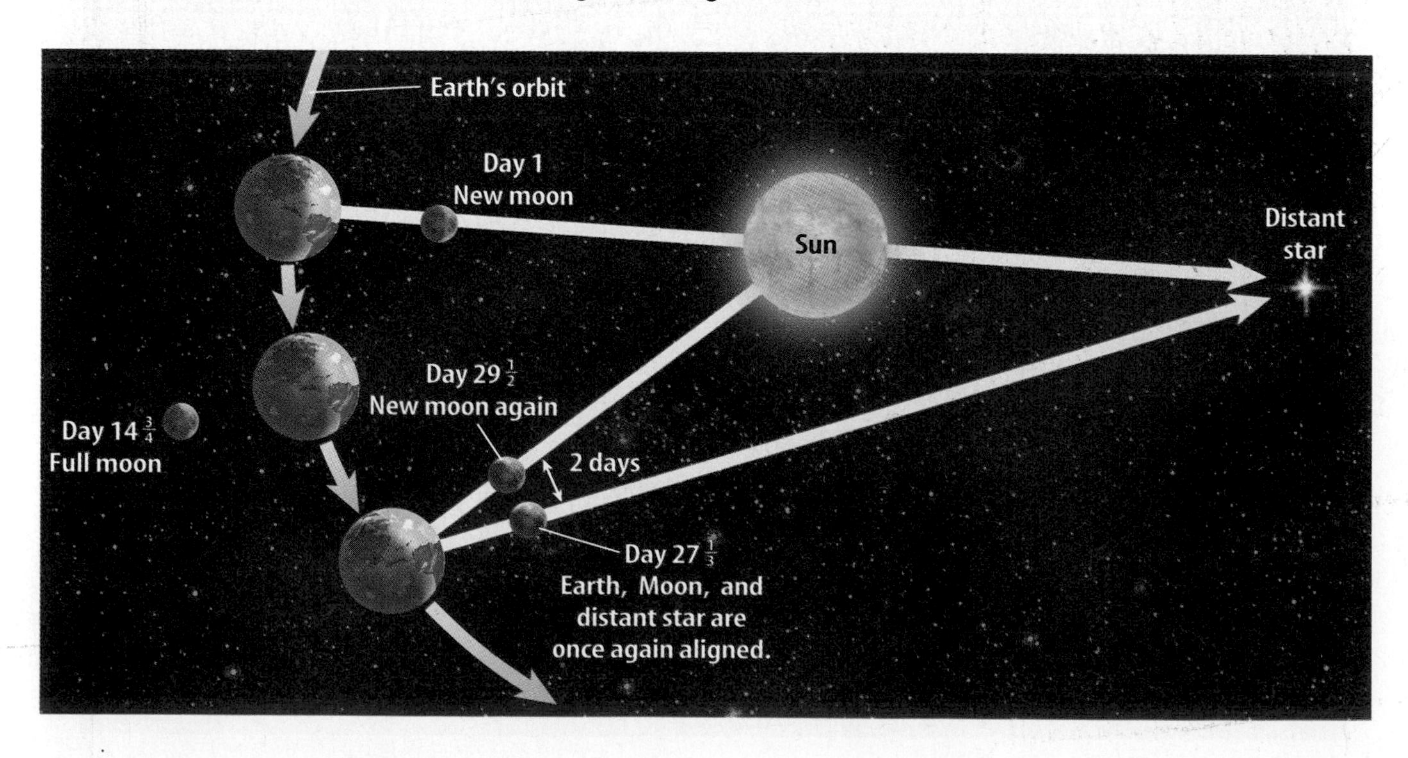

Eclipses

You can see other effects of the Moon's revolution than just the changes in its phases. Sometimes during new and full moon, shadows cast by one object will fall on another. While walking along on a sunny day, have you ever noticed how a passing airplane can cast a shadow on you? On a much larger scale, the Moon can do this too, when it lines up directly with the Sun. When this happens, the Moon can cast its shadow all the way to Earth. Earth also can cast its shadow onto the Moon during a full moon. When shadows are cast in these ways, eclipses occur.

Eclipses occur only when the Sun, the Moon, and Earth are lined up perfectly. Because the Moon's orbit is tilted at a different angle than Earth's orbit, the Moon's shadow most often misses Earth, and eclipses happen only a few times each year.

Solar Eclipses During new moon, if Earth moves into the Moon's shadow, a **solar eclipse** occurs. As shown in **Figure 12,** the Moon blocks sunlight from reaching a portion of Earth's surface. Only areas on Earth in the Moon's umbra, or the darkest part of its shadow, experience a total solar eclipse. Those areas in the penumbra, or lighter part of the shadow, experience a partial solar eclipse. During a total solar eclipse, the sky becomes dark and stars can be seen easily. Because Earth rotates, a solar eclipse lasts only a few minutes in any one location.

A

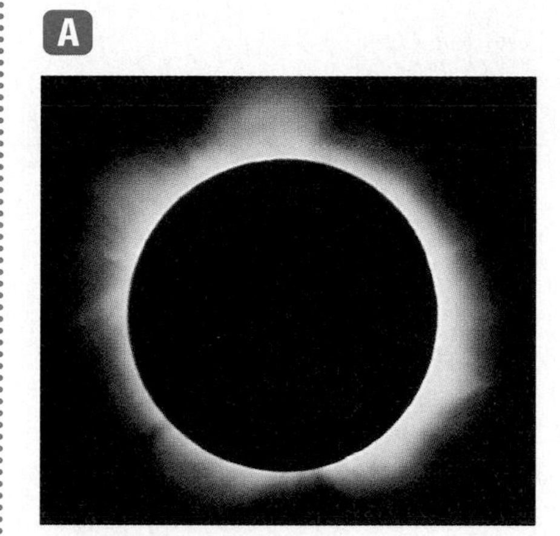

Figure 12
A Solar eclipses occur when the Sun, the Moon, and Earth are lined up in a specific way.
B Eclipses happen because the Moon casts a shadow on Earth.
WARNING: *Never look directly at a solar eclipse. Only observe solar eclipses indirectly.*

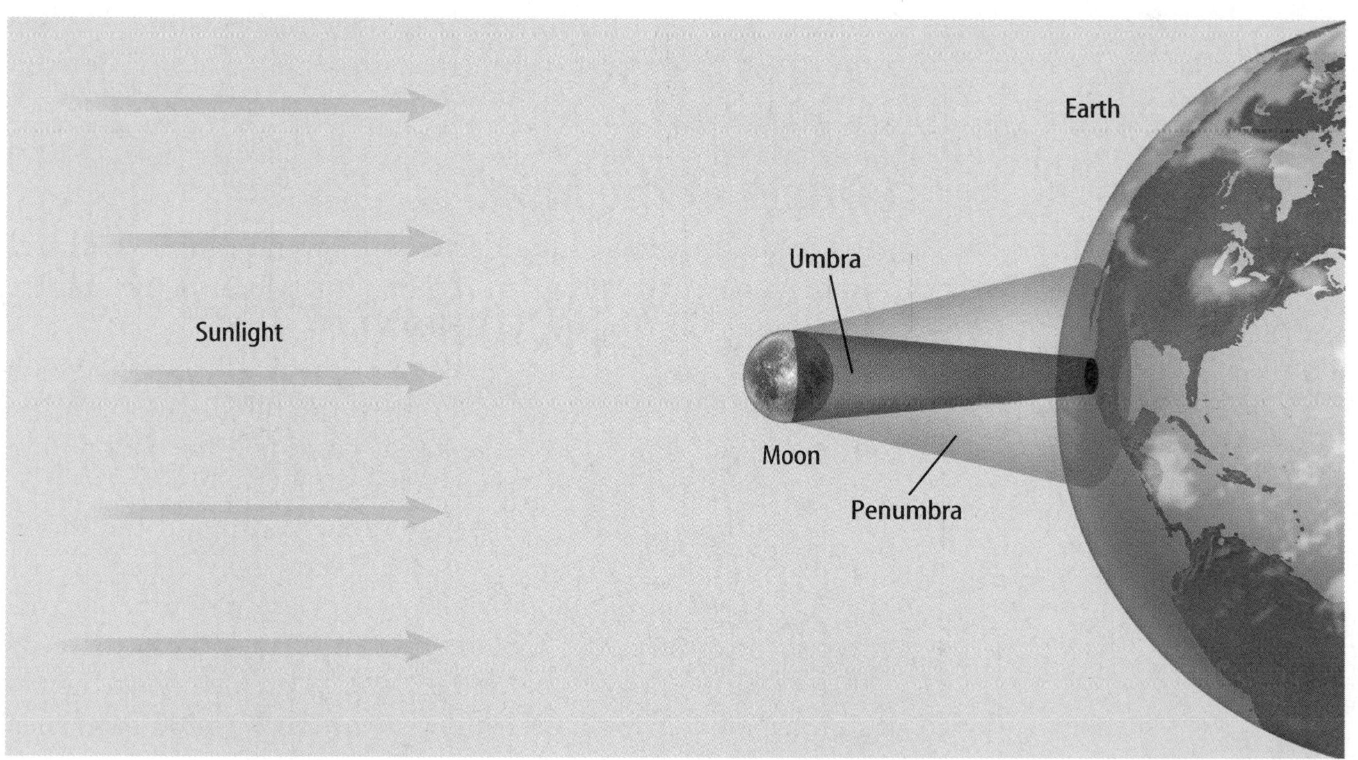

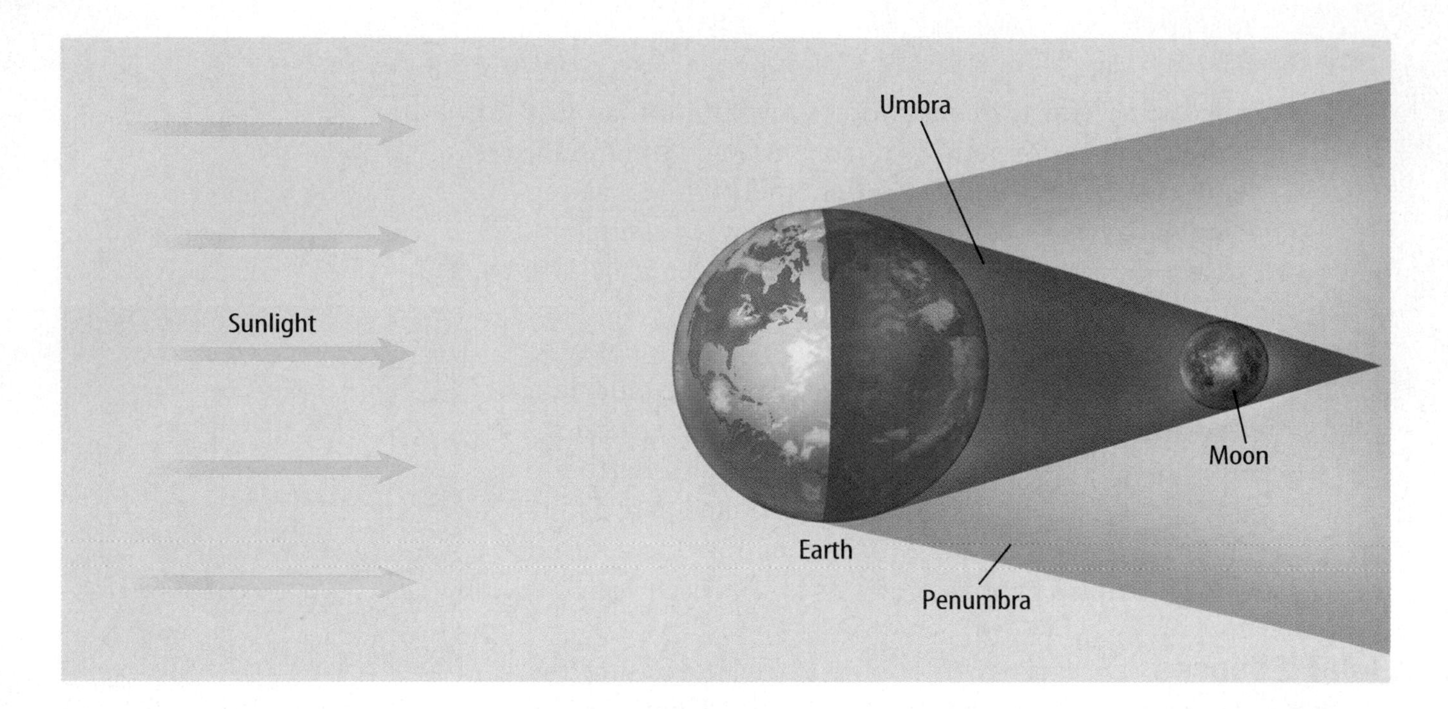

Figure 13
Lunar eclipses occur when the Sun, the Moon, and Earth line up so that Earth's shadow is cast upon a full moon. When the entire Moon is eclipsed, anyone on Earth who can see a full moon can see the lunar eclipse.

Lunar Eclipses A **lunar eclipse,** illustrated in **Figure 13,** occurs when the Sun, Earth, and the Moon are lined up such that the full moon moves into Earth's shadow. Direct sunlight is blocked from reaching the Moon. When the Moon is in the darkest part of Earth's shadow, a total lunar eclipse occurs.

During a total lunar eclipse, the full moon darkens. Because some sunlight refracts through Earth's atmosphere, the Moon appears to be deep red. As the Moon moves out of the umbra and into the penumbra, or lighter shadow, you can see the curved shadow of Earth move across the Moon's surface. When the Moon passes partly through Earth's umbra, a partial lunar eclipse occurs.

Origin of the Moon

Before the *Apollo* space program, several early hypotheses were proposed to explain the origin of the Moon. Some of these hypotheses are illustrated in **Figure 14.**

The co-formation hypothesis states that Earth and the Moon formed at the same time and out of the same material. One problem with this hypothesis is that Earth and the Moon have somewhat different densities and compositions.

According to the capture hypothesis, Earth and the Moon formed at different locations in the solar system. Then Earth's gravity captured the Moon as it passed close to Earth. The fission hypothesis states that the Moon formed from material thrown off of a rapidly spinning Earth. A problem with the fission hypothesis lies in determining why Earth would have been spinning so fast.

NATIONAL GEOGRAPHIC VISUALIZING HOW THE MOON FORMED

Figure 14

Scientists have proposed several possible explanations, or hypotheses, to account for the formation of Earth's Moon. As shown below, these include the co-formation, fission, capture, and collision hypotheses. The latter—sometimes known as the giant impact hypothesis—is the most widely accepted today.

▲ CO-FORMATION Earth and the Moon form at the same time from a vast cloud of cosmic matter that condenses into the bodies of the solar system.

▲ CAPTURE Earth's gravity captures the Moon into Earth orbit as the Moon passes close to Earth.

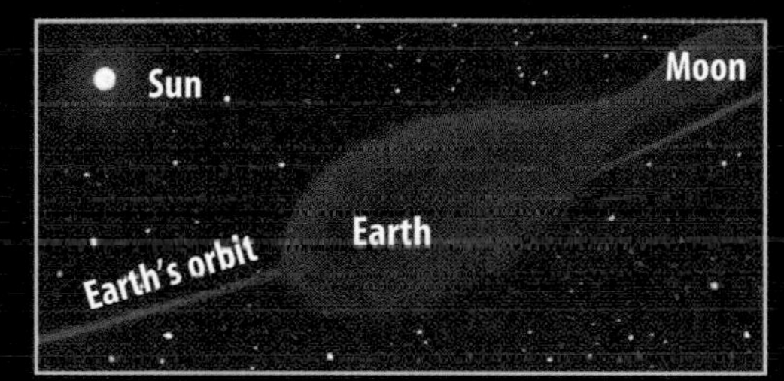

▲ FISSION A rapidly spinning molten Earth tears in two. The smaller blob of matter enters into orbit as the Moon.

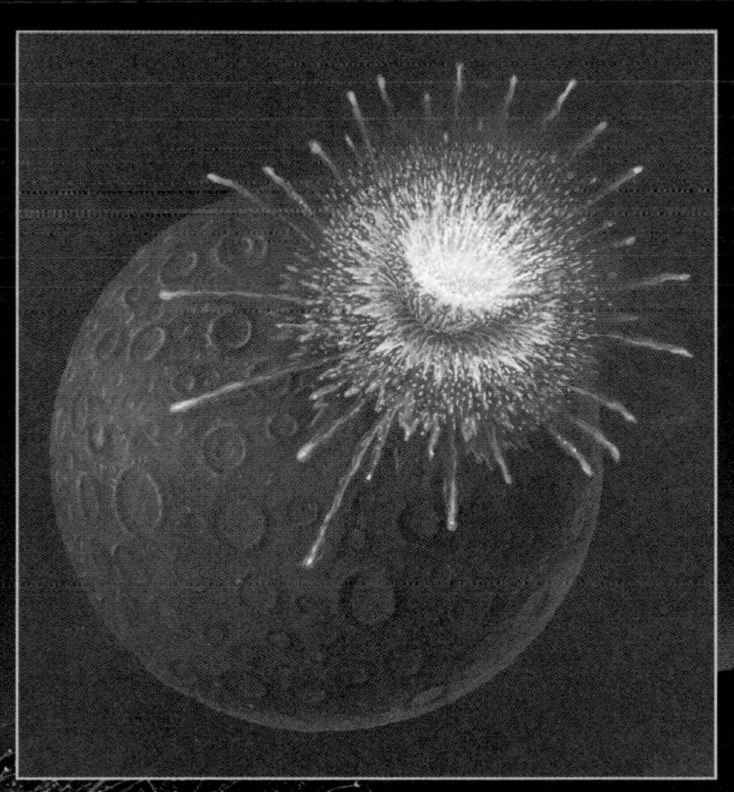

▲ COLLISION A Mars-sized body collides with the primordial Earth. The colossal impact smashes off sufficient debris from Earth to form the Moon.

Figure 15
Moon rocks collected during the *Apollo* space program provide clues about how the Moon formed.

Collision Hypothesis A lot of uncertainty still exists about the origin of the Moon. However, the collection and study of moon rocks, shown in **Figure 15,** brought evidence to support one recent hypothesis. This hypothesis, summarized in **Figure 14,** involves a great collision. When Earth was about 100 million years old, a Mars-sized space object may have collided with Earth. Such an object would have broken through Earth's crust and plunged toward the core. This collision would have thrown large amounts of gas and debris into orbit around Earth. Within about 1,000 years the gas and debris then could have condensed to form the Moon. The collision hypothesis is strengthened by the fact that Earth and the Moon have different densities. The Moon's density is similar to material that would have been thrown off Earth's mantle and crust when the object collided with Earth.

Earth is the third planet from the Sun. Along with the Moon, Earth could be considered a double planet. In the next section you will learn about other planets in the solar system. Some have properties similar to Earth's—others are different from Earth.

Section  Assessment

1. Which phase of the Moon occurs when Earth is located between the Moon and the Sun?
2. Describe the arrangement of the Moon, the Sun, and Earth during a solar eclipse. How is this different from the arrangement during a lunar eclipse?
3. Why are more maria found on the side of the Moon facing Earth than on the opposite side?
4. Describe evidence that supports the collision hypothesis on how the Moon formed.
5. **Think Critically** Explain why more people observe a total lunar eclipse than a total solar eclipse.

Skill Builder Activities

6. **Recognizing Cause and Effect** Answer these questions about the Moon orbiting Earth. **For more help, refer to the Science Skill Handbook.**
   a. Which type of eclipse may occur when the Moon moves between the Sun and Earth?
   b. How does the Moon's orbit around Earth cause the observed cyclical phases of the Moon? What role does the Sun play?
7. **Using an Electronic Spreadsheet** Using a spreadsheet, make a table comparing hypotheses about the origin of the Moon. Include strengths and weaknesses for each hypothesis. **For more help, refer to the Technology Skill Handbook.**

Activity

Viewing the Moon

The position of the Moon in the sky varies as the phases of the Moon change. Do you know when you might be able to see the Moon during daylight hours? How will viewing the Moon through a telescope be different from viewing it with the unaided eye?

What You'll Investigate

What features of the Moon are visible when viewed through a telescope?

Materials

telescope
drawing pencils
drawing paper

Goals

- **Determine** when you may be able to observe the Moon during the day.
- Use a telescope to observe the Moon.
- **Draw** a picture of the Moon's features as seen through the telescope.

Safety Precautions

Never look directly at the Sun. It can damage your eyes.

Procedure

1. Using your own observations, books about astronomy, or other resource materials, determine when the Moon may be visible to you during the day. You will need to find out during which phases the Moon is up during daylight hours, and where in the sky you likely will be able to view it. You will also need to find out when the Moon will be in those phases in the near future.
2. **Observe** the Moon with your unaided eye. Draw the features that you are able to see.
3. Using a telescope, observe the Moon again. Adjust the focus of the telescope so that you can see as many features as possible.
4. **Draw** a new picture of the Moon's features.

Conclude and Apply

1. **Describe** what you learned about when the Moon is visible in the sky. If a friend wanted to know when to try to see the Moon during the day next month, what would you say?
2. **Describe** the differences between how the Moon looked with the naked eye and through the telescope. Did the Moon appear to be the same size when you looked at it both ways?
3. What features were you able to see through the telescope that were not visible with the unaided eye?
4. Was there anything else different about the way the Moon looked through the telescope? Explain your answer.
5. **Identify** some of the types of features that you included in your drawings.

The next time you notice the Moon when you are with your family or friends, talk about when the Moon is visible in the sky and the different features that are visible.

SECTION 3

Our Solar System

As You Read

What You'll Learn

- **List** the important characteristics of inner planets.
- **Identify** how other inner planets compare and contrast with Earth.
- **List** the important characteristics of outer planets.

Vocabulary

solar system
astronomical unit
asteroid
comet
nebula

Why It's Important

Learning about other planets helps you understand Earth and the formation of our solar system.

Size of the Solar System

Measurements in space are difficult to make because space is so vast. Even our own solar system is extremely large. Our **solar system,** illustrated in **Figure 16,** is composed of the Sun, planets, asteroids, comets, and other objects in orbit around the Sun. How would you begin to measure something this large? If you are measuring distance on Earth, kilometers work fine, but not for measuring huge distances in space. Earth, for example, is about 150,000,000 km from the Sun. This distance is referred to as 1 **astronomical unit,** or 1 AU. Jupiter, the largest planet in the solar system, is more than 5 AU from the Sun. Astronomical units can be used to measure distances between objects within the solar system. Even larger units are used to measure distances between stars.

Located at the center of the solar system is a star you know as the Sun. The Sun is an enormous ball of gas that produces energy by fusing hydrogen into helium in its core. More than 99 percent of all matter in the solar system is contained in the Sun.

The asteroid belt is composed of rocky bodies that are smaller than the planets. *About how many AU are between the Sun and the middle of the asteroid belt?*

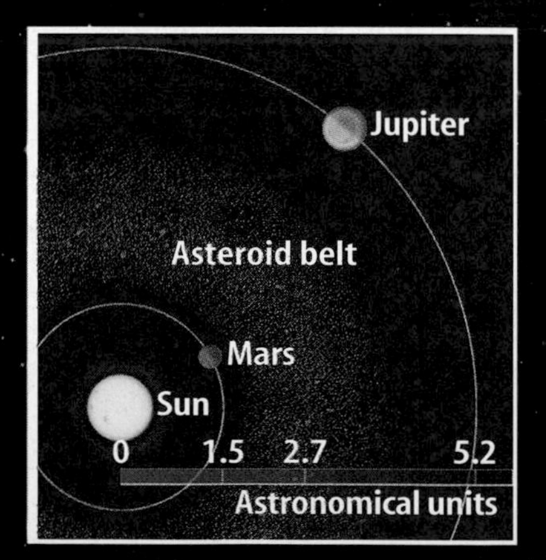

Figure 16
Our solar system is composed of the Sun, planets and their moons, and smaller bodies that revolve around the Sun, such as asteroids and comets.

An Average Star Although the Sun is important to life on Earth, it is much like many other stars in the universe. The Sun is middle-aged and about average in the amount of light it gives off.

The Planets

The planets in our solar system can be classified as inner or outer planets. Inner planets have orbits that lie inside the orbit of the asteroid belt. The inner planets are mostly solid, rocky bodies with thin atmospheres compared with the atmospheres of outer planets. Outer planets have orbits that lie outside the orbit of the asteroid belt. Four of the outer planets are gaseous giants, and one is a small ice/rock planet that seems to be out of place.

Inner Planets

The inner planets are Mercury, Venus, Earth, and Mars. Known as the terrestrial planets, after the Latin word *terra,* they are similar in size to Earth and are made up mainly of rock.

Mercury Mercury is the closest planet to the Sun. It is covered by craters formed when meteorites crashed into its surface. The surface of Mercury also has cliffs, as shown in **Figure 17,** some of which are 3 km high. These cliffs may have formed when Mercury's molten, iron-rich core cooled and contracted, causing the outer solid crust to shrink. The planet seems to have shrunk about 2 km in diameter.

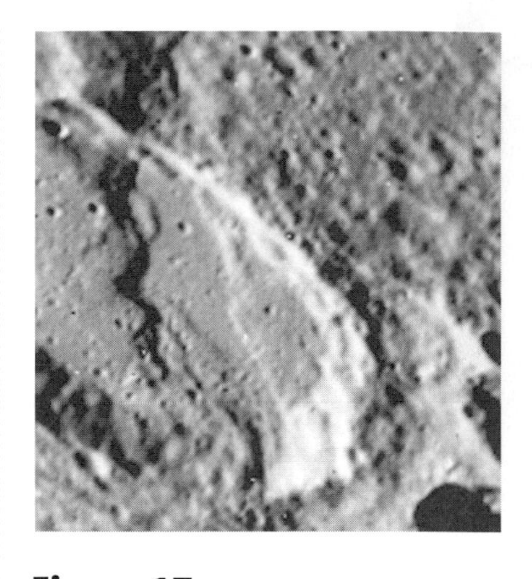

Figure 17
The Discovery Rupes Scarp is a huge cliff that may have formed as Mercury cooled and contracted. *How do craters on Mercury, like craters on the Moon, form?*

Activity: Model and Invent

The Slant of the Sun's Rays

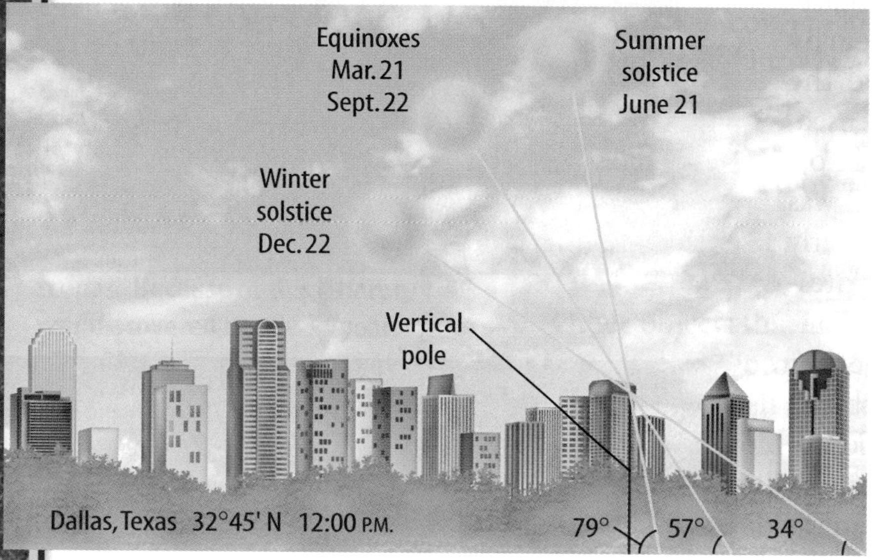

Equinoxes
Mar. 21
Sept. 22
Summer
solstice
June 21
Winter
solstice
Dec. 22
Vertical
pole
Dallas, Texas 32°45' N 12:00 P.M.
79°
57°
34°

During winter in the northern hemisphere, the north pole is positioned away from the Sun. This causes the angle of the Sun's rays striking Earth to be smaller in winter than in summer, and there are fewer hours of sunlight. The reverse is true during the summer months. The Sun's rays strike Earth at higher angles—concentrating more radiation on the surface.

Recognize the Problem

How does the angle of the Sun's rays affect Earth's surface temperature?

Thinking Critically

- How does the angle of the Sun's rays determine seasonal changes?
- How might temperatures on Earth change if Earth were tilted at a different angle?

Angle of the Sun's Rays at Noon at Latitude 32° 45' N (Dallas)

Date	Angle
December 22 (winter solstice)	34°
January 22	37°
February 22	46°
March 21 (vernal equinox)	57°
April 21	69°
May 21	77°
June 21 (summer solstice)	79°

Goals

- **Design** a model for simulating the effect of changing angles of the Sun's rays on Earth's surface temperatures.

Possible Materials

shallow baking pans lined with cardboard
* *paper, boxes, or box lids*
thermometers (non-mercury)
wood blocks,
* *bricks, or textbooks*
protractor
clock
* *stopwatch*

**Alternate materials*

Safety Precautions

Use thermometers as directed by teacher. Do not use "shake down" lab thermometers.

Data Source

Consult the data table providing angles of the Sun's rays for Dallas, Texas, in the northern hemisphere during different months of the year.

Planning the Model

1. **Design** a model that will duplicate the angle of the Sun's rays during different seasons of the year.
2. Choose the materials you will need to construct your model. Be certain to provide identical conditions for each angle of the Sun's rays that you seek to duplicate.

Window

Winter solstice

Sun's rays

34°

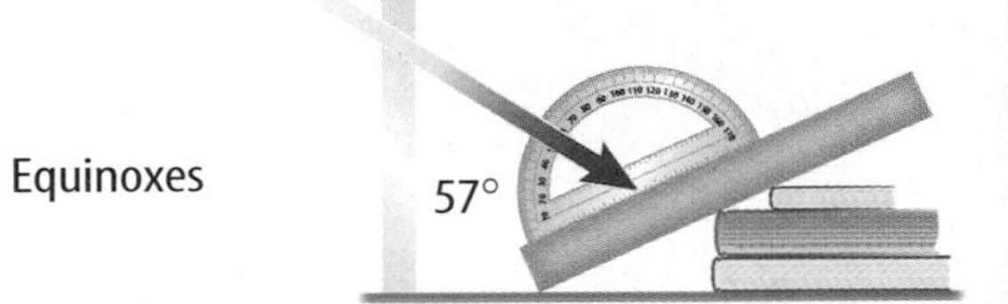

Check Model Plans

1. **Present** your model design to the class in the form of diagrams, poster, slide show, or video. Ask your classmates how your group's model design could be adjusted to make it more accurate.
2. Decide on a location that will provide direct sunlight and will allow your classmates to easily observe your model.

Safety Precautions 

WARNING: *Never look directly at the Sun at any time during your experiment.*

Making the Model

1. Create a model that demonstrates the effects different angles of the Sun's rays have on the temperature of Earth's surface.
2. **Demonstrate** your model during the morning, when the Sun's rays will hit the flat tray at an angle similar to the Sun's rays during winter solstice. Measure the angle of the Sun's rays by laying the protractor flat on the tray. Then sight the angle of the Sun's rays with respect to the tray.
3. Tilt other trays forward to simulate the Sun's rays striking Earth at higher angles during different times of the year.

Analyzing and Applying Results

1. Which angle had the greatest effect on the surface temperature of your trays? Which angle had the least effect?
2. Predict how each of the seasons in your area would change if Earth's axis tilt changed suddenly from 23.5 degrees to 40 degrees.

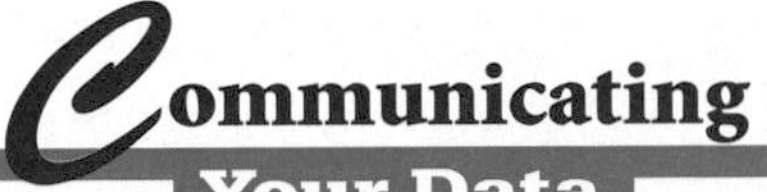

Demonstrate your model for your class. **Explain** how your model replicated the angle of the Sun's rays for each of the four seasons in Dallas, Texas.

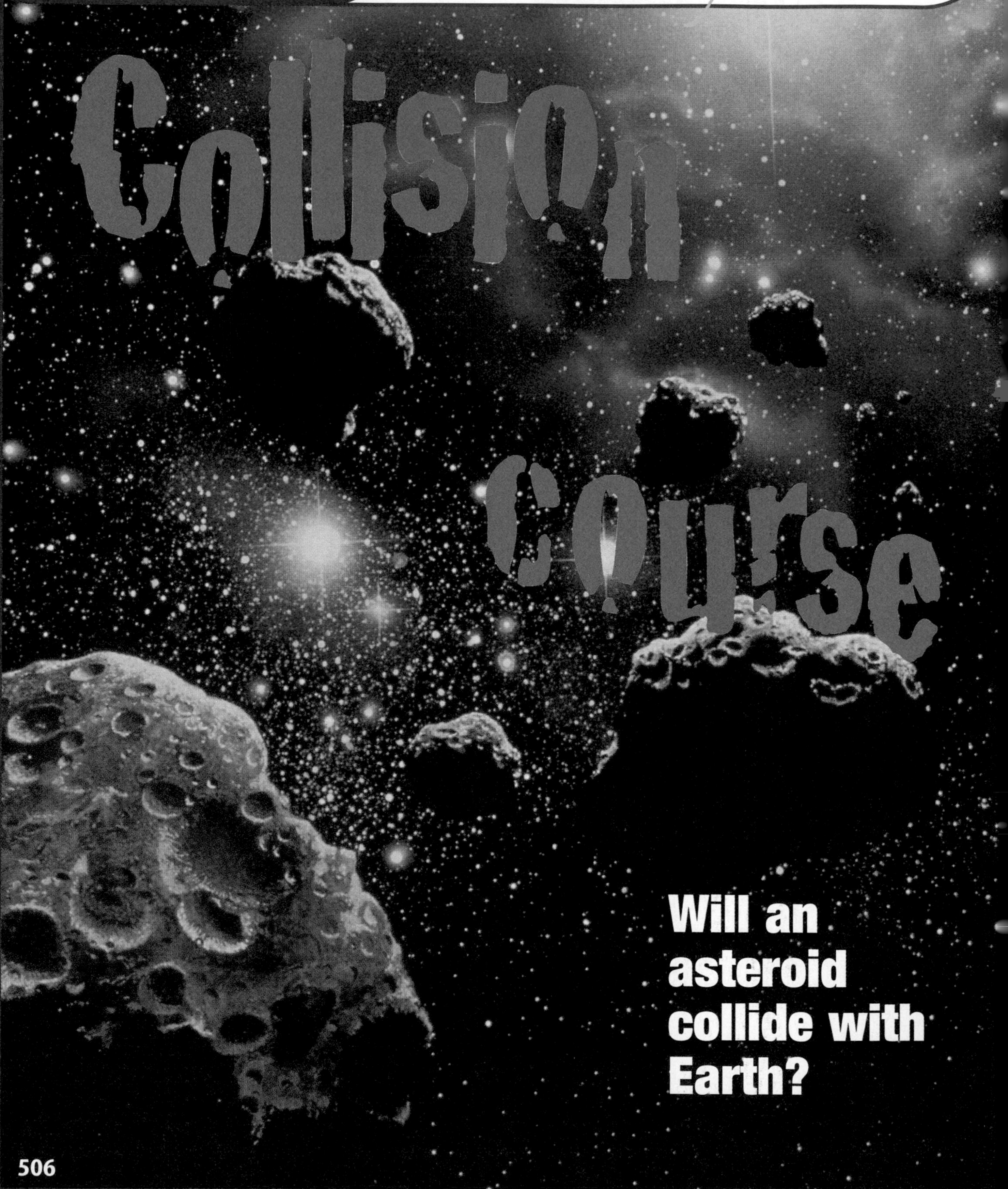

Collision Course

Will an asteroid collide with Earth?

Asteroids—those giant rocks hurtling through space—have been the basis for several disaster movies. But are asteroids really threats? "Absolutely!" say many scientists who study space. In fact, they believe the question is not can—but when—will Earth be hit by an asteroid. Because asteroids already have hit our planet many times since it was formed, it makes sense that an asteroid collision will happen again.

Earth is scarred with about 120 recognizable craters that are visible in many parts of the world. But due to erosion, plant growth, and other processes that cover up Earth's surface features, there are sure to be many craters that have disappeared from sight. However, visitors to Meteor Crater, Arizona, can see a 1.2-km-wide depression caused by an asteroid that impacted Earth about 49,000 years ago. A much older crater lies in Mexico's Yucatan Peninsula. The depression is about 195 km wide and was created about 65 million years ago.

Some scientists believe the Yucatan asteroid created a giant dust cloud that blocked the Sun's rays from reaching Earth. This may have caused the planet to turn dark for about six months and may have led to freezing temperatures. This would have put an end to much of Earth's early plant life. And, scientists hypothesize, this was the asteroid that led to the extinction of the dinosaurs and about half of the other species that once inhabited Earth.

Meteor Crater, near Winslow, Arizona, was formed about 49,000 years ago. It is about 200 m deep.

Rocks in Space

With space crowded with giant chunks of rock, some the size of mountains, is there anything we can do to protect ourselves from such an impact? Astronomer/geologist Eugene Shoemaker thought so and is responsible for alerting the world to the dangers of asteroid impact. In 1973, he and geologist Eleanor Helin began the first Near Earth Objects (NEO) watch at the Mount Palomar Observatory in California. Scanning the sky, they sought out objects that might be on a collision course with Earth. But to the team's disappointment, few people were concerned.

Then, in 1996, all that changed. An asteroid, about 0.5 km wide, came within 450,800 km of Earth. Scientists said this was a close call! Today, groups of scientists are working on creating systems to track NEOs. As of 2000, they recorded 1,082 NEOs. Of those, 407 were about 0.75 km in diameter or more!

Some physicists and astronomers are working on ways to defend our planet from NEOs. One idea is to send a warhead-tipped rocket to try to change a dangerous asteroid's orbit away from Earth.

Don't worry, though: A collision isn't in the foreseeable future. In fact, there is little chance of an asteroid hitting Earth anytime soon. The enormous pressures and temperatures that are generated when an asteroid hits Earth's atmosphere usually vaporize it altogether. So, you can't use the excuse, "An asteroid is coming!" to put off doing your homework or cleaning your room!

CONNECTIONS **Brainstorm** **Working in small groups, come up with as many ways as you can to blast an asteroid to pieces or make it change course before hitting Earth. Present your reports to the rest of the class.**

SCIENCE *Online*

For more information, visit science.glencoe.com

Chapter 17 Study Guide

Reviewing Main Ideas

Section 1 Earth's Motion and Seasons

1. Earth's shape is nearly spherical. *What observations confirm Earth's shape?*

2. Earth's motions include rotation, or spinning on its axis, and revolution, or movement around the Sun in its orbit.
3. Day and night are caused by Earth spinning on its axis. Earth's tilt and revolution cause seasons to occur.

Section 2 Earth's Moon

1. Surface features on the Moon include maria, craters, and lunar highlands. *How can you recognize these surface features from Earth?*

2. The Moon rotates once and revolves around Earth once in 27.3 days. The Moon's orbit is tilted with respect to Earth's orbit.
3. Phases of Earth's Moon, solar eclipses, and lunar eclipses are caused by the Moon's revolution around Earth.
4. One hypothesis concerning the origin of Earth's Moon is that a Mars-sized body collided with Earth, throwing off material that later condensed to form the Moon.

Section 3 Our Solar System

1. The solar system includes the Sun, the planets, moons, asteroids, comets, and meteoroids.
2. Planets in our solar system can be classified as inner or outer planets.
3. Inner planets are small and rocky and have thin atmospheres. Outer planets are generally large and gaseous and have thick, dense atmospheres. *What objects are concentrated in a belt between the inner and outer planets?*

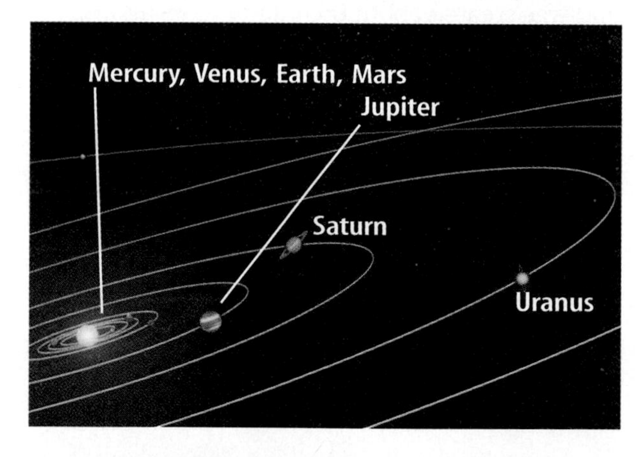

4. One hypothesis on the origin of the solar system states that it formed by the condensation of a large cloud of gas, ice, and dust.

After You Read

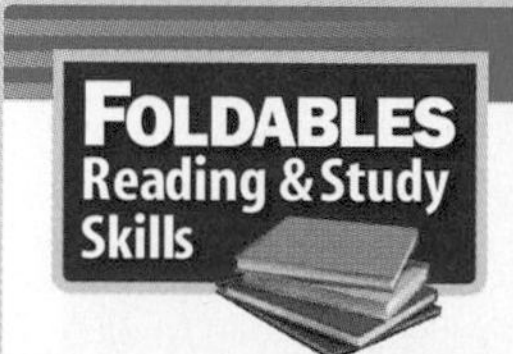

Look at your Foldable and explain what characteristics make Earth a perfect place for life and the Moon an impossible place for life to exist.

Chapter 17 Study Guide

Visualizing Main Ideas

Using the following phrases, fill in the concept map below to complete the moon phase cycle: full moon, waning gibbous, waning crescent, and first quarter.

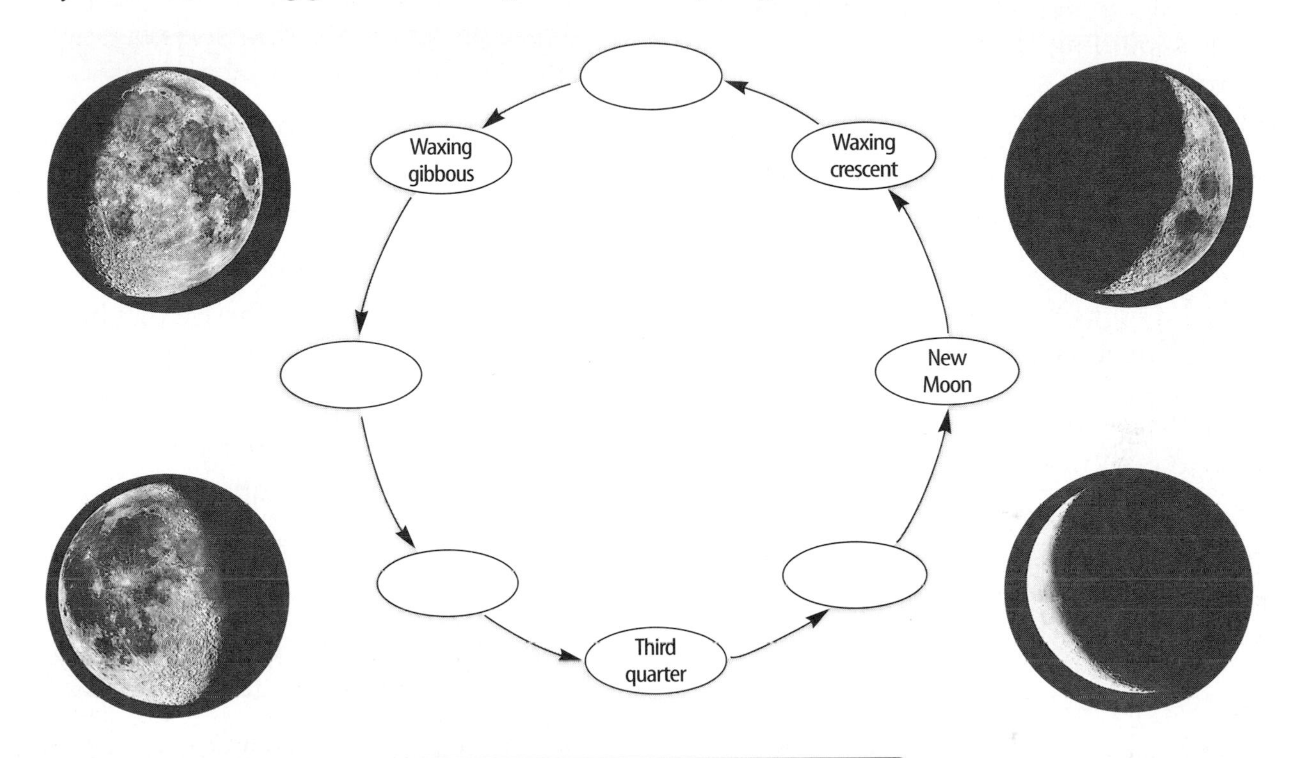

Vocabulary Review

Vocabulary Words

a. asteroid
b. astronomical unit
c. axis
d. comet
e. crater
f. equinox
g. lunar eclipse
h. moon phase
i. nebula
j. orbit
k. revolution
l. rotation
m. solar eclipse
n. solar system
o. solstice

Study Tip

As you read, look up the definition of any prefixes you do not recognize. Once you know the meaning of a prefix, you'll be able to figure out the definition of many new words.

Using Vocabulary

Explain the difference between the vocabulary words in each of the following sets.

1. rotation, revolution
2. orbit, axis
3. solar eclipse, lunar eclipse
4. equinox, solstice
5. comets, craters
6. solar system, solar eclipse
7. Moon phases, lunar eclipse
8. rotation, orbit
9. nebula, solar system
10. asteroid, comet

Chapter 17 Assessment

Checking Concepts

Choose the word or phrase that best answers the question.

1. Earth's spinning on its axis is which motion?
A) rotation
B) waxing
C) revolution
D) waning

2. What occurs when the Sun reaches its greatest distance north or south of the equator?
A) orbit
B) equinox
C) solstice
D) axis

3. What is the imaginary line around which Earth spins called?
A) orbit
B) equinox
C) solstice
D) axis

4. Which moon surface feature probably formed when lava flows filled large basins?
A) maria
B) craters
C) highlands
D) volcanoes

5. Meteorites that strike the Moon's surface cause which surface feature?
A) maria
B) craters
C) highlands
D) volcanoes

6. How long is the Moon's period of revolution?
A) 27.3 hours
B) 29.5 hours
C) 27.3 days
D) 29.5 days

7. How long does it take for the Moon to rotate once on its axis?
A) 27.3 hours
B) 29.5 hours
C) 27.3 days
D) 29.5 days

8. What occurs when the Moon is directly between the Sun and Earth?
A) lunar eclipse
B) solar eclipse
C) full moon
D) waxing crescent

9. Which planet is most like Earth in size and mass?
A) Mercury
B) Mars
C) Saturn
D) Venus

10. Europa is a satellite of which planet?
A) Uranus
B) Saturn
C) Jupiter
D) Mars

Thinking Critically

11. Describe the differences between inner and outer planets.

12. Why are more maria found on the near side of the Moon than on the far side?

13. Why do scientists hypothesize that life might exist on Europa?

14. Explain the Sun's positions relative to the equator during winter and summer solstices and equinox.

15. Describe how Earth's shape is influenced by gravity.

Developing Skills

16. Classifying A new planet is found circling the Sun, and you are given the job of classifying it. The new planet has a thick, dense atmosphere and no apparent solid surface. It lies beyond the orbit of Pluto. How would you classify this newly discovered planet?

17. Making and Using Tables Complete the table of outer planets that shows how many satellites each planet has and what gases are found in each planet's atmosphere.

Planetary Facts

Planet	Number of Satellites	Major Atmospheric Gases
Jupiter		
Saturn		
Uranus		
Neptune		
Pluto		

18. Concept Mapping Complete the following concept maps about Mars and Neptune. Use the following words or phrases: *red, outer planet, rocky, gas giant, blue,* and *inner planet.*

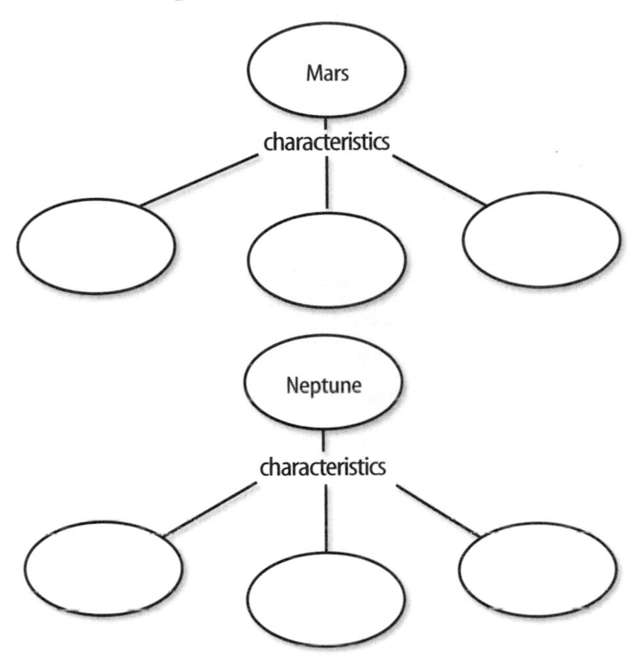

19. Making Models Make a three-dimensional model of a total solar eclipse. Be sure to include all objects involved.

20. Forming Hypotheses Research their characteristics and formulate a hypothesis explaining the origin of Pluto, Charon, and Triton.

21. Poster Use photographs or illustrations of the Sun, the Moon, and Earth. Position each object as it would be located during a lunar and a solar eclipse.

TECHNOLOGY

Go to the Glencoe Science Web site at **science.glencoe.com** or use the **Glencoe Science CD-ROM** for additional chapter assessment.

Test Practice

A student is beginning a study of the inner planets. These planets and some of their characteristics are listed in the table below.

The Inner Planets

Planet	Average Distance From Sun (millions of km)	Diameter (km)	Period of Revolution (Earth days)	Period of Rotation (Earth days-hours)
Mercury	58	4878	88	58-16
Venus	109	12104	225	243-0
Earth	150	12756	365	0-24
Mars	227	6794	686	0-24.5

Study the table and answer the following questions.

1. According to this information, which inner planet takes more than one Earth year to complete its journey around the Sun?

A) Mercury
B) Venus
C) Earth
D) Mars

2. A reasonable hypothesis, based on the data in the table, is that as a planet's distance from the Sun increases ______.

F) the planet's period of revolution increases
G) the planet's period of revolution decreases
H) the planet's period of rotation increases
J) the planet's period of rotation decreases

Standardized Test Practice

Reading Comprehension

Read the passage carefully. Then read the questions that follow the passage. Decide which is the best answer to each question.

Test-Taking Tip Identify whether this passage is fictional or informational.

Earthquakes and Volcanoes

Earthquakes are destructive and potentially fatal natural disasters. Geologists have been working to learn what they can about earthquakes in order to better protect property and save human lives.

Scientists know that many earthquakes occur because tectonic plates interact with one another at plate boundaries. Studies of faults have led to long-term forecasts that earthquakes should occur in certain regions. For example, a major earthquake has been forecast for an area near Parkfield, California, which is along the San Andreas Fault. However, scientists cannot predict exactly when and where an earthquake will occur.

One approach to forecasting earthquakes is known as paleoseismology. Paleoseismology involves the study of past movements of rock and sediment along faults. Motion along a fault results in an earthquake. Therefore, studying this movement is one way to study ancient earthquake occurrences.

This movement can be measured in the field by observing shifted rock and sediment along a fault. If this displaced sediment can be dated, the time at which the earthquake occurred also can be estimated. With information on several past earthquakes, scientists can estimate how long, on average, time intervals are between the earthquakes. This is one way to estimate how many earthquakes might occur along a fault over a period of time.

Although scientists are not yet able to predict earthquakes reliably, the information gained from their research advances the field.

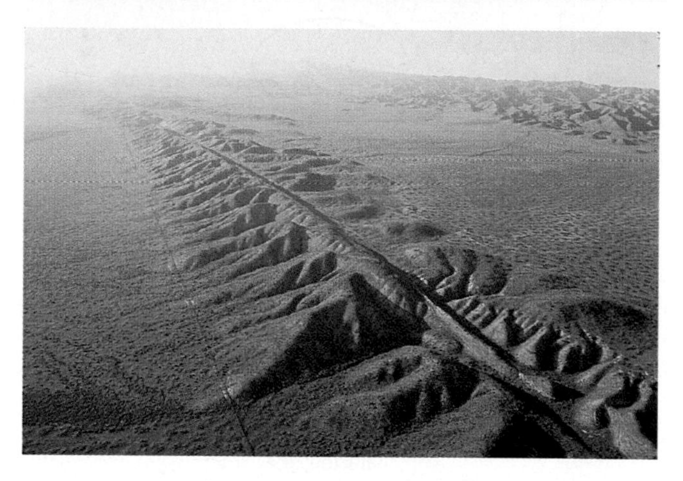

The San Andreas Fault is part of a plate boundary, and many earthquakes occur along it.

1. Based on the information in this passage, what can the reader conclude?

A) Earthquakes always occur during heavy rainstorms.
B) Earthquakes cannot be predicted reliably, but they can be forecast over estimated periods of time.
C) Earthquakes are extremely rare natural occurrences.
D) Earthquakes do not affect property and human lives much.

2. What does the information in this passage suggest?

F) Studies of displacement along faults can provide information for forecasting earthquakes.
G) Studying faults allows scientists to determine the time and date of the next earthquake.
H) Paleoseismology is the study of fossils along the San Andreas Fault.
J) Scientists do not know what causes earthquakes.

Standardized Test Practice

Reasoning and Skills

Read each question and decide which is the best answer.

1. All of the following is evidence used by Alfred Wegener to support the hypothesis of continental drift EXCEPT _____.

A) the puzzlelike fit of the coastlines
B) matching fossils
C) similar rock structures
D) mid-ocean ridges

Test-Taking Tip Think about what evidence Wegener had considered when he presented his hypothesis in 1912.

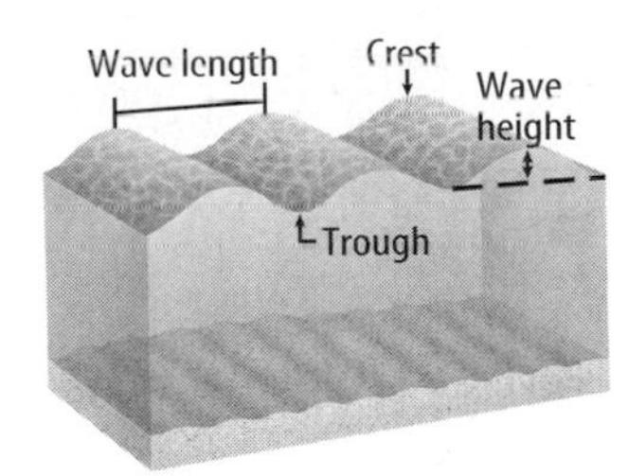

2. Refer to the diagrams above, which show earthquake waves and water waves. The waves are similar in that _____.

F) they travel from the focus in all directions
G) they carry energy through space
H) the material they move through is transported
J) only the energy moves forward; the material does not move forward

Test-Taking Tip Think about and compare the characteristics of seismic and water waves.

3. All of these statements are true about volcanoes EXCEPT _____.

A) three basic types of volcanoes are shield, cinder cone, and composite
B) the amount of water vapor, other gases, and silica in magma influence the kind of eruption that will take place
C) the lower the silica content is in magma, the more explosive an eruption will be
D) a volcano's form depends upon the type of eruption and the composition of the erupted material

Test-Taking Tip Think about the different kinds of volcanoes, how they form, and what factors influence the type of eruption.

Consider this question carefully before writing or sketching your answer on a separate sheet of paper.

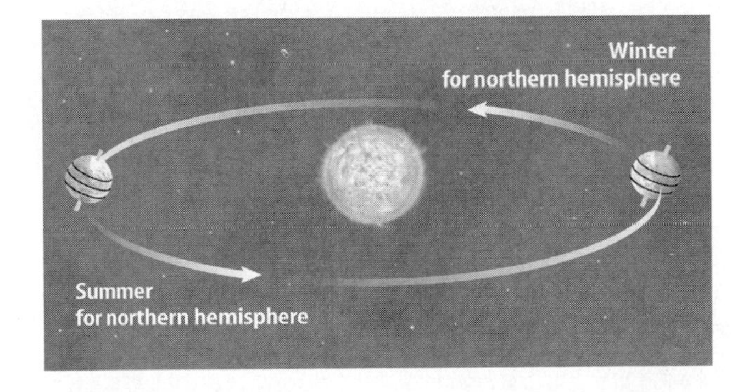

4. The diagram shows the position of Earth relative to the Sun during the northern hemisphere's winter and summer. Describe why the lengths of day and night are different during these two seasons.

Test-Taking Tip Notice that Earth's axis is tilted.

UNIT 6 Building Blocks of Matter

How Are Refrigerators & Frying Pans Connected?

In the late 1930s, scientists were experimenting with a gas that they hoped would work as a new coolant in refrigerators. They filled several metal canisters with the gas and stored the canisters on dry ice. Later, when they opened the canisters, they were surprised to find that the gas had disappeared and that the inside of each canister was coated with a slick, powdery white solid. The gas had undergone a chemical change. That is, the chemical bonds in its molecules had broken and new bonds had formed, turning one kind of matter into a completely different kind of matter. Strangely, the mysterious white powder proved to be just about the slipperiest substance that anyone had ever encountered. Years later, a creative Frenchman obtained some of the slippery stuff and tried applying it to his fishing tackle to keep the lines from tangling. His wife noticed what he was doing and suggested putting the substance on the inside of a frying pan to keep food from sticking. He did, and nonstick cookware was born!

SCIENCE CONNECTION

PHYSICAL AND CHEMICAL CHANGES Working in teams of 3 or 4, look up and write down the definitions of "physical change" and "chemical change." Then brainstorm to compile a list of 10 physical and 10 chemical changes that you might encounter in everyday life. Make flashcards from your list. On each card, write a description of the change on one side and the type of change on the other side. Pair up with another team and use your flashcards to quiz each other.

CHAPTER 18

Matter

What is most striking about this picture—water or ice? What do these things have in common? How are they different? In this chapter, you will find the answers to these questions. You also will learn about the matter that makes up your surroundings and makes up your body.

What do you think?

Science Journal Look at the picture below with a classmate. Discuss what you think these might be. Here's a hint: *They could be part of the landscape of a far-off planet or something in your desk.* Write your answer in your Science Journal.

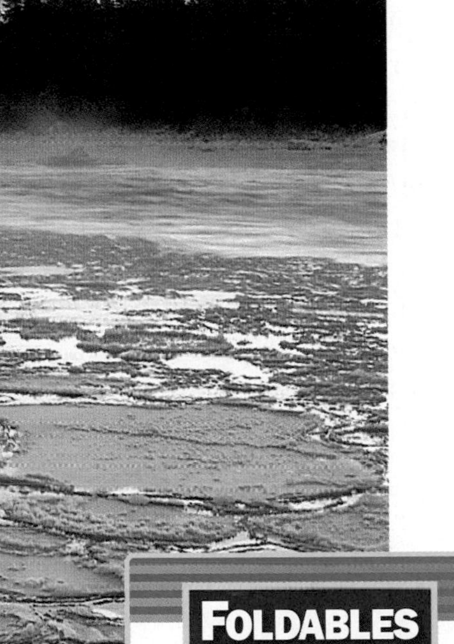

On Earth water is unique because it is found as a solid, liquid, or gas. On a cool autumn morning, you might see gaseous water condensing into fog over a lake or a river whose surface soon will be solid ice. The following activity will help you visualize how matter can change states.

Change the state of water

Safety Precautions

1. Pour 500 mL of water into a 1,000-mL glass beaker.
2. Mark the level of water in the beaker with the bottom edge of a piece of tape.
3. Place the beaker on a hot plate.
4. With the help of an adult, heat the water until it boils for 5 min. Let the water cool.
5. With the help of an adult, compare the level of the water to the bottom edge of the tape.

Observe

What came out of the beaker as the water boiled? Did the amount of water that you started with change? In your Science Journal, explain what happened to the water.

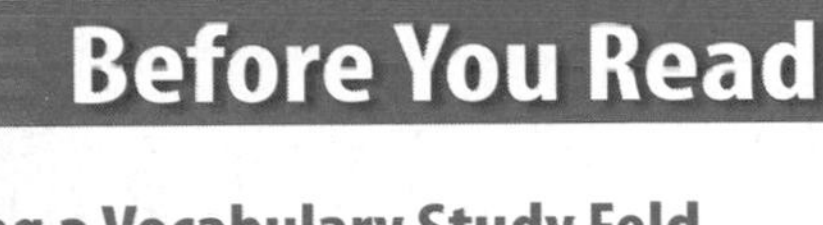

Before You Read

FOLDABLES Reading & Study Skills

Making a Vocabulary Study Fold

Knowing the definition of vocabulary words is a good way to ensure that you understand the content of the chapter.

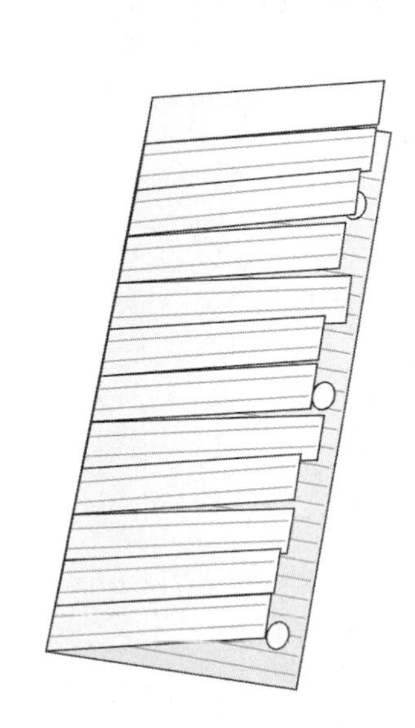

1. Place a sheet of notebook paper in front of you so the short side is at the top and the holes are on the right side. Fold the paper in half from the left side to the right side.
2. Through the top thickness of paper, cut along every third line from the outside edge to the center fold, forming tabs.
3. Before you read, write vocabulary words from each section in this chapter on the front of the tabs. Under each tab, write what you think the word means.
4. As you read the chapter, add to and correct your definitions.

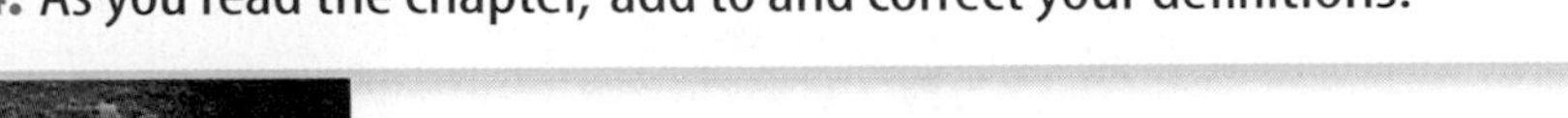

Atoms

As You Read

What You'll Learn

- **Identify** the states of matter.
- **Describe** the internal structure of an atom.
- **Compare** isotopes of an element.

Vocabulary

matter, atom, element, proton, neutron, electron, atomic number, mass number, isotope

Why It's Important

Nearly everything around you—air, water, food, and clothes—is made of atoms.

The Building Blocks of Matter

What do the objects you see, the air you breathe, and the food you eat have in common? They are matter. **Matter** is anything that has mass and takes up space. Heat and light are not matter, because they have no mass and do not take up space. Glance around the room. If all the objects you see are matter, why do they look so different from one another?

Atoms Matter, in its various forms, surrounds you. You can't see all matter as clearly as you see water, which is a transparent liquid, or rocks, which are colorful solids. You can't see air, for example, because air is colorless gas. The forms or properties of one type of matter differ from the properties of another, because matter is made up of tiny particles called **atoms.** The structures of different types of atoms and how they join together determine all the properties of matter that you can observe. **Figure 1** illustrates how small objects, like atoms, can be put together in different ways.

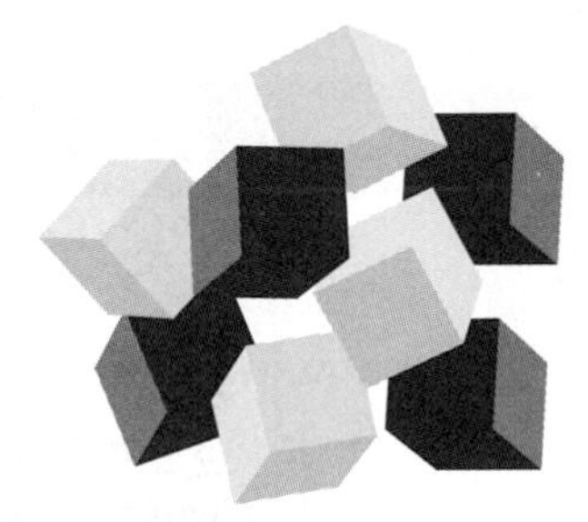

Figure 1
Like atoms, the same few blocks can combine in many ways. *How could this model help explain the variety of matter?*

The Structure of Matter Matter is joined together much like the blocks shown in **Figure 1.** The building blocks of matter are atoms. The types of atoms in matter and how they attach to each other give matter its properties.

Elements When atoms combine, they form many different types of matter. Your body contains several types of atoms combined in different ways. These atoms form the proteins, DNA, tissues, and other matter that make you the person you are. Most other objects that you see also are made of several different types of atoms. However, some substances are made of only one type of atom. **Elements** are substances that are made of only one type of atom and cannot be broken down into simpler substances by normal chemical or physical means.

Elements are useful for making a variety of items you depend on every day. They also combine to make up the minerals that compose Earth's crust. Some minerals, however, are made up of only one element. These minerals, which include copper and silver, are called native elements. **Table 1** shows some common elements and their uses. A table of the elements, called the periodic table of the elements, is included on the inside back cover of this book.

Searching for Elements

Procedure

1. Obtain a copy of the **periodic table of the elements** and familiarize yourself with the elements.
2. Search your house for items made of various elements.
3. Use a **highlighter** to highlight the elements you discovered on your copy of the periodic table.

Analysis

1. Were certain types of elements more common?
2. Infer why you did not find many of the elements.

Table 1 Some Common Uses of Elements

Element	Sulfur	Silver	Copper	Carbon
Native State of the Element	Sulfur	Silver	Copper	Graphite
Uses of the Element	Fertilizer	Tableware	Wire	Ski wax

Figure 2
This model airplane is a small-scale version of a large object.

Modeling the Atom

How can you study things that are too small to be seen with the unaided eye? When something is too large or too small to observe directly, models can be used. The model airplane, shown in **Figure 2,** is a small version of a larger object. A model also can describe tiny objects, such as atoms, that otherwise are difficult or impossible to see.

The History of the Atomic Model The current model of the atom is based on the work of many scientists over hundreds of years. The idea that matter is composed of atoms dates back more than 2,300 years when the Greek philosopher Democritus (dih MAH krih tuhs) proposed that matter is composed of small particles. He called these particles atoms and said that different types of matter were composed of different types of atoms. More than 2,000 years later, John Dalton expanded on these ideas. He theorized that all atoms of an element contain the same type of atom.

Protons and Neutrons In the early 1900s, additional work led to the development of the current model of the atom, shown in **Figure 3.** Three basic particles make up an atom—protons, neutrons (NOO trahnz), and electrons. **Protons** are particles that have a positive electric charge. **Neutrons** have no electric charge. Both particles are located in the nucleus—the center of an atom. With no negative charge to balance the positive charge of the protons, the charge of the nucleus is positive.

Electrons Particles with a negative charge are called **electrons,** and they exist outside of the nucleus. In 1913, Niels Bohr, a Danish scientist, proposed that an atom's electrons travel in orbitlike paths around the nucleus. He also proposed that electrons in an atom have energy that depends on their distance from the nucleus. Electrons in paths that are closer to the nucleus have lower energy, and electrons further from the nucleus have higher energy.

The Current Atomic Model Over the next several decades, research showed that although electrons do have specific amounts of energy, they do not travel in orbitlike paths. Instead, electrons move in an electron cloud surrounding the nucleus. Electrons can be anywhere within the cloud, but evidence suggests that they are located near the nucleus most of the time. To understand how this might work, imagine a beehive. The hive represents the nucleus of an atom. The bees swarming around the hive are like electrons moving around the nucleus. As they swarm, you can't predict their exact location, but they usually stay close to the hive.

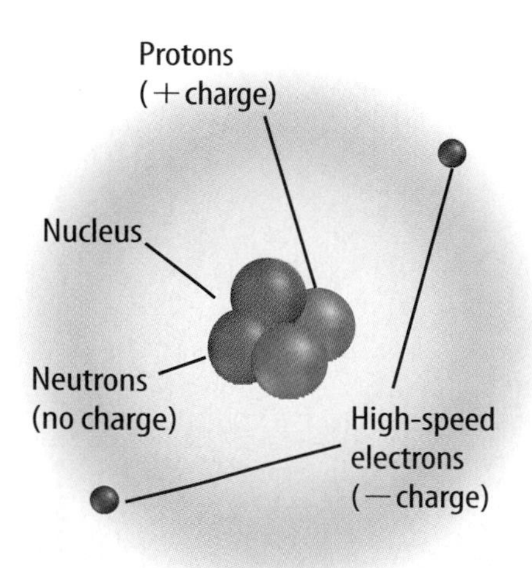

Figure 3
This model of a helium atom shows two protons and two neutrons in the nucleus, and two electrons in the electron cloud.

Counting Atomic Particles

You now know where protons, neutrons, and electrons are located, but how many of each are in an atom? The number of protons in an atom depends on the element. All atoms of the same element have the same number of protons. For example, all iron atoms—whether in train tracks or breakfast cereal—contain 26 protons, and all atoms with 26 protons are iron atoms. The number of protons in an atom is equal to the **atomic number** of the element. This number can be found above the element symbol on the periodic table. Notice that as you go from left to right on the periodic table, the atomic number of the element increases by one.

How many protons are in an atom of gold, which has an atomic number of 79?

How many electrons? In a neutral atom, the number of protons is equal to the number of electrons. This makes the overall charge of the atom zero. Therefore, for a neutral atom:

Atomic number = number of protons = number of electrons

Atoms of an element can lose or gain electrons and still be the same element. When this happens, the atom is no longer neutral. Atoms with fewer electrons than protons have a positive charge, and atoms with more electrons than protons have a negative charge.

How many neutrons? Unlike protons, atoms of the same element can have different numbers of neutrons. The number of neutrons in an atom isn't found on the periodic table. Instead, you need to know the atom's mass number. The **mass number** of an atom is equal to the number of protons plus the number of neutrons. The number of neutrons is determined by subtracting the atomic number from the mass number. In **Figure 4,** the number of neutrons can be determined by counting the blue spheres and the number of protons by counting orange spheres. Atoms of the same element that have different numbers of neutrons are called **isotopes.** **Table 2** lists useful isotopes of some elements.

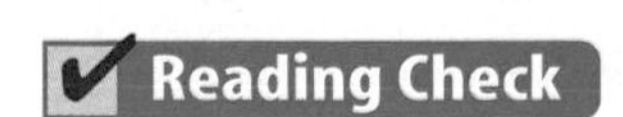

How are isotopes of the same element different?

Health INTEGRATION

Some isotopes of elements are radioactive. Physicians can introduce these isotopes into a patient's circulatory system. The low-level radiation they emit allows the isotopes to be tracked as they move throughout the patient's body. Explain how this would be helpful in diagnosing a disease.

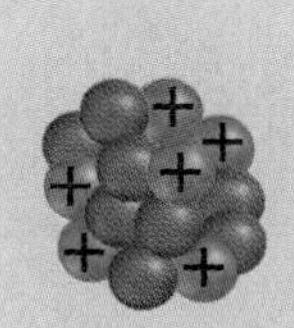

Figure 4
This carbon atom is common in organic material. *What is this atom's mass number?*

Table 2 Some Useful Isotopes

Isotope	Number of Protons	Number of Neutrons	Number of Electrons	Atomic Number	Mass Number
Hydrogen-1	1	0	1	1	1
Hydrogen-2	1	1	1	1	2
Hydrogen-3	1	2	1	1	3
Carbon-12	6	6	6	6	12
Carbon-14	6	8	6	6	14
Uranium-234	92	142	92	92	234
Uranium-235	92	143	92	92	235
Uranium-238	92	146	92	92	238

Uses of Isotopes Scientists have found uses for isotopes that benefit humans. For example, medical doctors use radioactive isotopes to treat certain types of cancer, such as prostate cancer. Geologists use isotopes to determine the ages of some rocks and fossils.

As you continue to investigate matter in this chapter, you will explore how atoms of different elements combine to form the materials around you.

Section  Assessment

1. How does the air you breathe fit the definition of matter?
2. What are the basic particles found in the nucleus of an atom?
3. How do isotopes of an element differ from one another?
4. What is an element?
5. **Think Critically** Oxygen-16 and oxygen-17 are isotopes of oxygen. The numbers 16 and 17 represent their mass numbers, respectively. If the element oxygen has an atomic number of 8, how many protons and neutrons are in these two isotopes?

Skill Builder Activities

6. **Comparing and Contrasting** Review the material in Section 1. How do atoms and elements differ? How are they similar? **For more help, refer to the Science Skill Handbook.**
7. **Solving One-Step Equations** The mass number of a nitrogen atom is 14. Find its atomic number in the periodic table shown on the inside back cover. Then determine the number of neutrons in its nucleus by subtracting the atomic number from the mass number. **For more help, refer to the Math Skill Handbook.**

SECTION

Combinations of Atoms

Interactions of Atoms

When you take a shower, eat your lunch, or do your homework on the computer, you probably don't think about elements. But everything you touch, eat, or use is made from them. Elements are all around you and in you.

There are about 90 naturally occurring elements on Earth. When you think about the variety of matter in the universe, you might find it difficult to believe that most of it consists of combinations of these same elements. How could so few elements produce so many different things? This happens because elements can combine in countless ways. For example, the same oxygen atoms that you breathe also might be found in many other objects, as shown in **Figure 5.** As you can see, each combination of atoms is unique. How do these combinations form and what holds them together?

As You Read

What You'll Learn

- **Describe** ways atoms combine to form compounds.
- **List** differences between compounds and mixtures.

Vocabulary

compound
ion
mixture
heterogeneous mixture
homogeneous mixture
solution

Why It's Important

On Earth, most matter exists as compounds or mixtures.

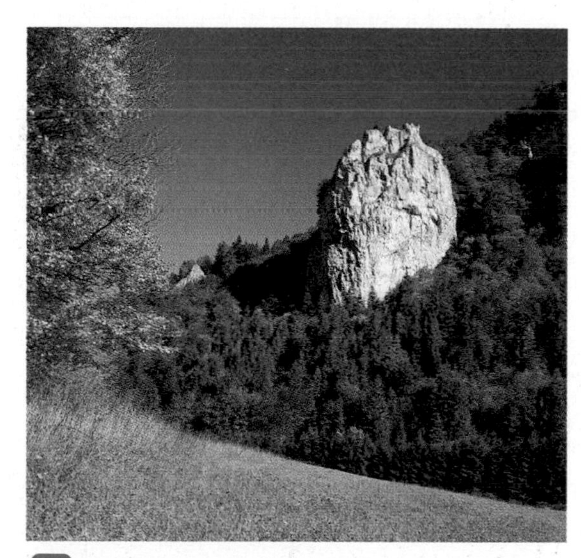

A **Solid limestone has oxygen within its structure.**

B **Oxygen also is present in the juices of these apples.**

C **This canister contains pure oxygen gas.**

Figure 5
Oxygen is a common element found in many different solids, liquids, and gases. *How can the same element, made from the same type of atoms, be found in so many different materials?*

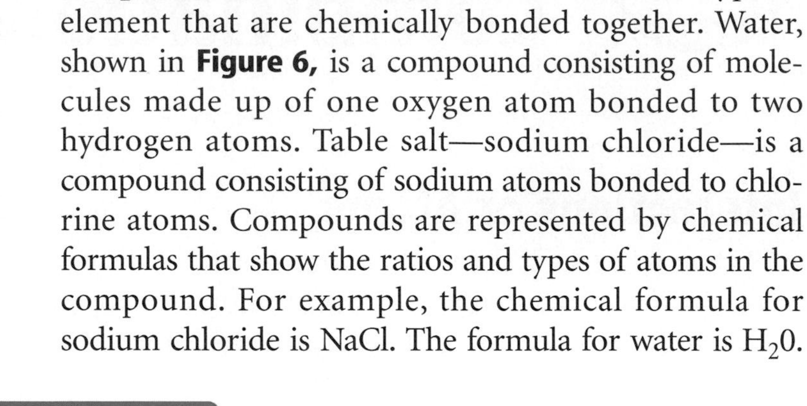

Compounds When atoms of more than one element combine, they form a compound. A **compound** contains atoms of more than one type of element that are chemically bonded together. Water, shown in **Figure 6,** is a compound consisting of molecules made up of one oxygen atom bonded to two hydrogen atoms. Table salt—sodium chloride—is a compound consisting of sodium atoms bonded to chlorine atoms. Compounds are represented by chemical formulas that show the ratios and types of atoms in the compound. For example, the chemical formula for sodium chloride is NaCl. The formula for water is H_20.

Figure 6
The water you drink is a compound consisting of hydrogen and oxygen atoms.

✓ Reading Check *How many atoms does a water molecule have?*

The properties of compounds often are very different from the properties of the elements that combine to form them. Sodium is a soft, silvery metal, and chlorine is a greenish, poisonous gas, but the compound they form is the white, crystalline table salt you use to season food. Under normal conditions on Earth, the hydrogen and oxygen that form water are gases. Water can be solid ice, liquid water, or gaseous vapor. Which state do you think is most common for water at Earth's south pole?

Chemical Properties A property that describes a change that occurs when one substance reacts with another is called a chemical property. For example, one chemical property of water is that it changes to hydrogen and oxygen gas when an electric current passes through it. The chemical properties of a substance depend on what elements are in that substance and how they are arranged. Iron atoms in the mineral biotite will react with water and oxygen to form iron oxide, or rust, but iron mixed with chromium and nickel in stainless steel resists rusting.

Bonding

The forces that hold the atoms in compounds together are called chemical bonds. Some atoms are reactive and form bonds easily. Other atoms are much less reactive. For example, atoms that have eight electrons in the outermost portion of their electron cloud are not likely to combine with other atoms. If an atom has fewer than eight electrons in the outermost portion of its electron cloud, it is unstable and is more likely to combine with other atoms. One exception to this rule is the element helium, with only two electrons in its electron cloud. This atom is stable and does not react easily.

Research Visit the Glencoe Science Web site at **science.glencoe.com** for more information about bonding and the periodic table. Communicate to your class what you learn.

Covalent Bonds Atoms can combine to form compounds in two different ways. One way is by sharing the electrons in the outer portion of their electron clouds. The type of bond that forms by sharing outer electrons is a covalent bond. A group of atoms connected by covalent bonds is called a molecule. For example, two atoms of hydrogen can share outer electrons with one atom of oxygen to form a molecule of water, as shown in **Figure 7.** Each of the hydrogen atoms has one outer electron and the oxygen has six outer electrons. This arrangement causes hydrogen and oxygen atoms to bond together. Each of the hydrogen atoms becomes stable by sharing one electron with the oxygen atom, and the oxygen atom becomes stable by sharing two electrons with the two hydrogen atoms.

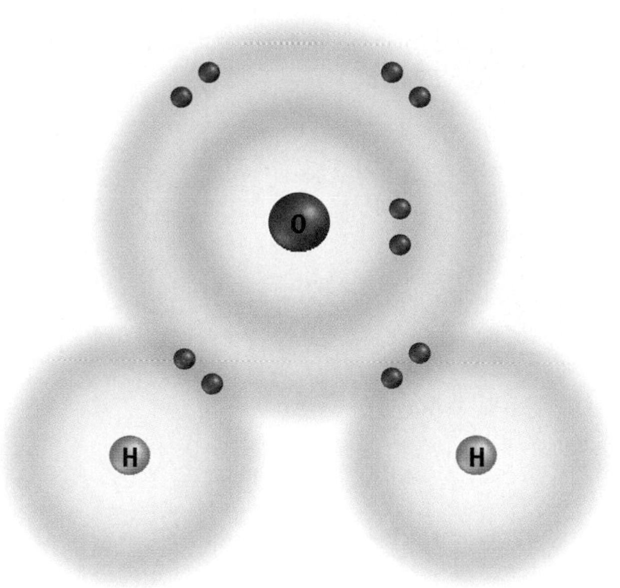

Figure 7
A molecule of water consists of two atoms of hydrogen that share outer electrons with one atom of oxygen.

Ionic Bonds In addition to sharing electrons, atoms also combine if they become positively or negatively charged. This type of bond is called an ionic bond. Atoms can be neutral, or under certain conditions, atoms can lose or gain electrons. When an atom loses electrons, it has more protons than electrons, so the atom is positively charged. When an atom gains electrons, it has more electrons than protons, so the atom is negatively charged. Electrically charged atoms are called **ions.**

Ions are attracted to each other when they have opposite charges. This is similar to the way magnets behave. If the ends of a pair of magnets have the same type of pole, they repel each other. Conversely, if the ends have opposite poles, they attract one another. Ions form electrically neutral compounds when they join. The mineral halite, commonly used as table salt, forms in this way. A sodium (Na) atom loses an outer electron and becomes a positively charged ion. As shown in **Figure 8,** if the sodium ion comes close to a negatively charged chlorine (Cl) ion, they attract each other and form the salt you use on french fries or popcorn.

Figure 8
Table salt forms when a sodium ion and a chlorine ion are attracted to one another. *What kind of bond holds ions together?*

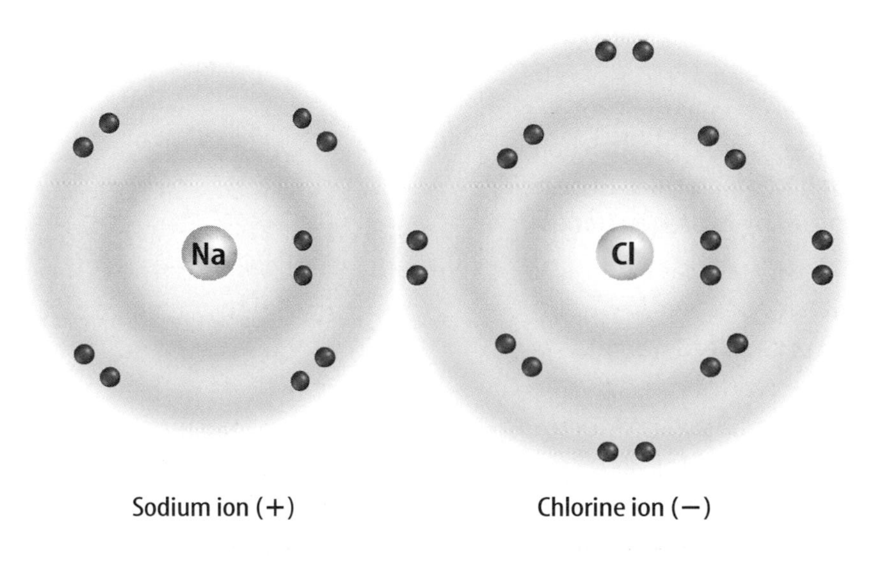

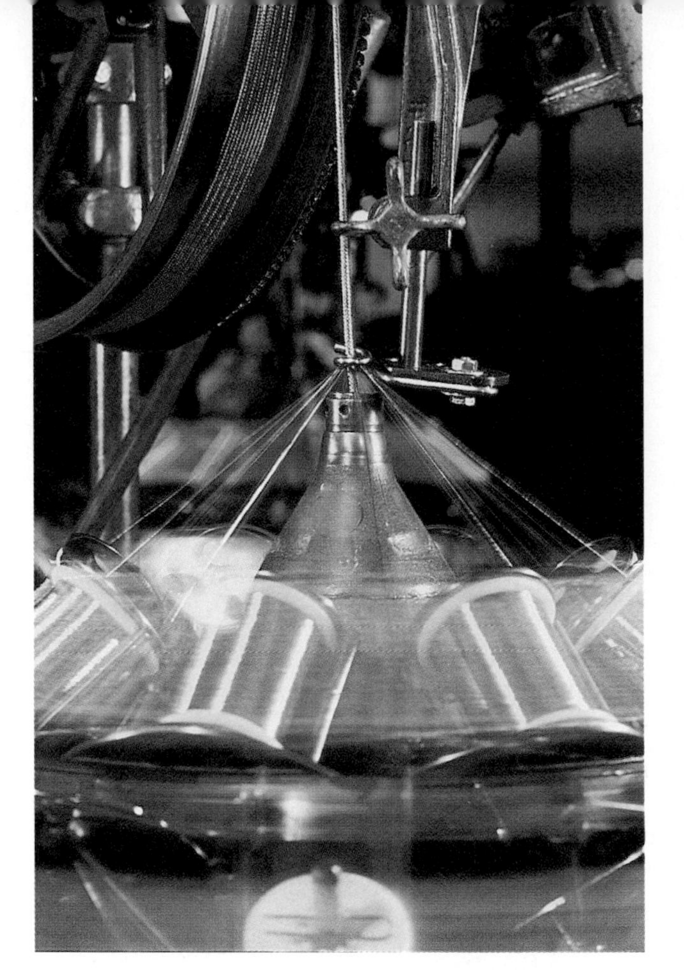

Figure 9
Electrons move freely between the copper atoms in this wire.
What type of bond holds copper atoms together?

Metallic Bonds Metallic bonds are found in metals such as copper, gold, aluminum, and silver. In this type of bond, electrons are free to move from one positively charged ion to another. This free movement of electrons is responsible for key characteristics of metals. The movement of electrons, or conductivity, allows metals like copper, shown in **Figure 9,** to pass an electric current easily.

Hydrogen Bonds Some types of bonds, such as hydrogen bonds, can form without the interactions of electrons. The arrangement of hydrogen and oxygen atoms in water molecules causes them to be polar molecules. A polar molecule has a positive end and a negative end. This happens because the atoms do not share electrons equally. When hydrogen and oxygen atoms form a molecule with covalent bonds, the hydrogen atoms produce an area of partial positive charge and the oxygen atom produces an area of partial negative charge. The positive end of one molecule is attracted to the negative end of another molecule, as shown in **Figure 10,** and a weak hydrogen bond is formed. The different parts of the water molecule are slightly charged, but as a whole, the molecule has no charge. This type of bond is easily broken, indicating that the charges are weak.

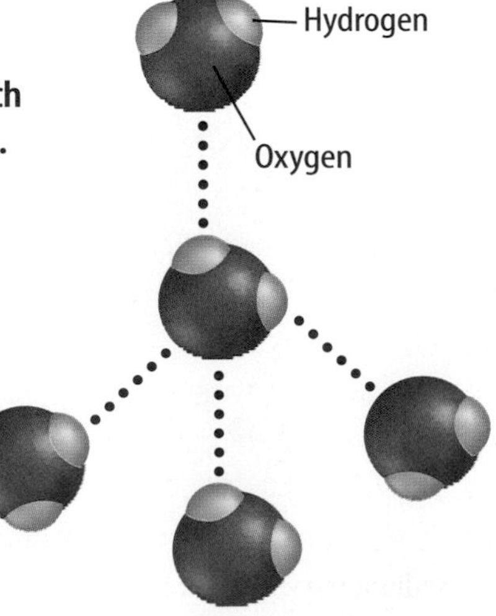

Figure 10
The ends of polar molecules, such as water, have opposite charges. This allows molecules to be held together by hydrogen bonds.

Hydrogen bonds are responsible for several properties of water, some of which are unique. Cohesion is the attraction between water molecules that allows them to form raindrops and to form beads on flat surfaces. Hydrogen bonds cause water to exist as a liquid, rather than a gas, at room temperature. As water freezes, hydrogen bonds force water molecules apart, into a structure that is less dense than liquid water.

Figure 11
This rock contains a variety of minerals that together form a mixture.

Mixtures

Sometimes compounds and elements mix together but do not combine chemically. A **mixture** is composed of two or more substances that are not chemically combined. There are two different types of mixtures—heterogeneous and homogeneous. The components of a **heterogeneous mixture** are not mixed evenly and each component retains its own properties. Maybe you've seen a rock like the one in **Figure 11.** Several different minerals are mixed together, but if you were to examine the minerals separately, you would find that they have the same properties and appearance as they have in the rock.

The components of a **homogeneous mixture** are evenly mixed throughout. You can't see the individual components. Another name for a homogeneous mixture is a **solution**. The properties of the components of this type of mixture often are different from the properties of the mixture. Ocean water is an example of a liquid solution that consists of salts mixed with liquid water.

✔ **Reading Check** *What is a solution?*

Separating Mixtures and Compounds

The components of a mixture can be separated by physical means. For example, you can sit at your desk and pick out the separate items in your backpack, or you can let the water evaporate from a saltwater mixture and the salt will remain.

Separating the components of a mixture is a relatively easy task compared to separating those of a compound. The substances in a compound must be separated by chemical means. This means that an existing compound can be changed to one or more new substances by chemically breaking down the original compound. For example, a drop of dilute hydrochloric acid (HCl) can be placed on calcium carbonate ($CaCO_3$) and carbon dioxide (CO_2) is released. To break down most compounds, several steps usually are required.

Mini LAB

Classifying Forms of Matter

Procedure

1. Make a chart with columns titled Mixtures, Compounds, and Elements.
2. Classify each of these items into the proper column on your chart: **air, sand, hydrogen, muddy water, sugar, ice, sugar water, water, salt, oxygen, copper.**
3. Make a solution using two or more of the items listed above.

Analysis

1. How does a solution differ from other types of mixtures?
2. How does an element differ from a compound?

Figure 12
The ocean is a mixture of many different forms of matter. The ocean water itself is a solution.

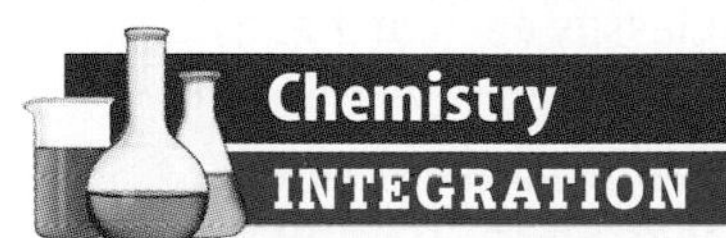

Chemistry INTEGRATION

When one substance dissolves in another, some of the properties of the dissolving substance change. When salt dissolves in water, decide whether its chemical or physical properties change.

Exploring Matter

Air, sweetened tea, salt water, and the contents of your backpack are examples of mixtures. The combination of rocks, fish, and coral shown in **Figure 12** also is a mixture. In each case, the materials within the mixture are not chemically combined. The individual components are made of compounds, or elements. The atoms that make up these compounds lost their individual properties when they combined. Even though atoms are known as the building blocks of matter, they are composed of protons, neutrons, and electrons, which are even smaller. As you continue to explore matter, apply what you've learned about atoms, elements, compounds, mixtures, and solutions to your studies.

Section  2 Assessment

1. How do atoms or ions combine to form compounds?
2. Why is sweetened tea considered to be a solution rather than a compound?
3. Describe the chemical property of iron that results in rust.
4. Explain what makes a solution different from a heterogeneous mixture.
5. **Think Critically** How can you determine whether salt water is a solution or a compound?

Skill Builder Activities

6. **Comparing and Contrasting** How are solutions and compounds similar? How are they different? **For more help, refer to the Science Skill Handbook.**
7. **Communicating** Design an investigation that would show whether sugar water is a mixture or a compound. Discuss your design with your teacher. Perform the investigation and write the results in your Science Journal. **For more help, refer to the Science Skill Handbook.**

Activity

Scales of Measurement

How would you describe some of the objects in your classroom? Perhaps your desktop is about one-half the size of a door. Measuring physical properties in a laboratory experiment will help you make better observations.

What You'll Investigate

How are physical properties of objects measured?

Materials

triple beam balance
100-mL graduated cylinder
metersticks (2)
non-mercury thermometers (3)
stick or dowel
rock sample
string
globe
water

Goals

- **Measure** various physical properties in SI.
- **Determine** sources of error.

Safety Precautions

Never "shake down" lab thermometers.

Procedure

1. Go to every station and determine the measurement requested. Record your observations in a data table and list sources of error.
   - **a.** Use a balance to determine the mass, to the nearest 0.1 g, of the rock sample.
   - **b.** Use a graduated cylinder to measure the water volume to the nearest 0.5 mL.
   - **c.** Use three thermometers to determine the average temperature, to the nearest 0.5°C, at a selected location in the room.
   - **d.** Use a meterstick to measure the length, to the nearest 0.1 cm, of the stick or dowel.
   - **e.** Use a meterstick and string to measure the circumference of the globe. Be accurate to the nearest 0.1 cm.

Measurement and Error

Sample at Station	Value of Measurement	Causes of Error
a.	mass = ____ g	
b.	volume = ____ mL	
c. (location)	average temp. = ____ °C	
d.	length = ____ cm	
e.	circumference = ____ cm	

Conclude and Apply

1. **Compare** your results with those of other students who measured the same objects. Review the values provided by your teacher. How do the values you obtained compare with those provided by your teacher and other students?
2. **Calculate** your percentage of error in each case. Use this formula.

$$\% \text{ error} = \frac{\text{your value} - \text{teacher's value}}{\text{teacher's value}} \times 100$$

3. Decide what percentage of error will be acceptable. Generally, being within five percent to seven percent of the correct value is considered good. If your values exceed ten percent error, what could you do to improve your results and reduce error? What was the most common source of error?

Communicating Your Data

Compare your conclusions with those of other students in your class. **For more help, refer to the Science Skill Handbook.**

Properties of Matter

As You Read

What You'll Learn

- **Distinguish** between chemical and physical properties.
- **List** the four states of matter.

Vocabulary
density

Why It's Important
You can recognize many substances by their physical properties.

Physical Properties of Matter

In addition to the chemical properties of matter that you have already investigated in this chapter, matter also has other properties that can be described. You might describe a pair of blue jeans as soft, blue, and about 80 cm long. A sandwich could have two slices of bread, lettuce, tomato, cheese, and turkey. These descriptions can be made without altering the sandwich or the blue jeans in any way. The properties that you can observe without changing a substance into a new substance are physical properties.

One physical property that you will use to describe matter is density. **Density** is a measure of the mass of an object divided by its volume. Generally, this measurement is given in grams per cubic centimeter (g/cm^3). For example, the average density of liquid water is about 1 g/cm^3. So 1 cm^3 of pure water has a mass of about 1 g.

An object that's more dense than water will sink in water. On the other hand, an object that's not as dense as water will float in water. When oil spills occur on the ocean, as shown in **Figure 13,** the oil floats on the surface of the water and washes up on beaches. Because the oil floats, even a small spill can spread out and cover large areas.

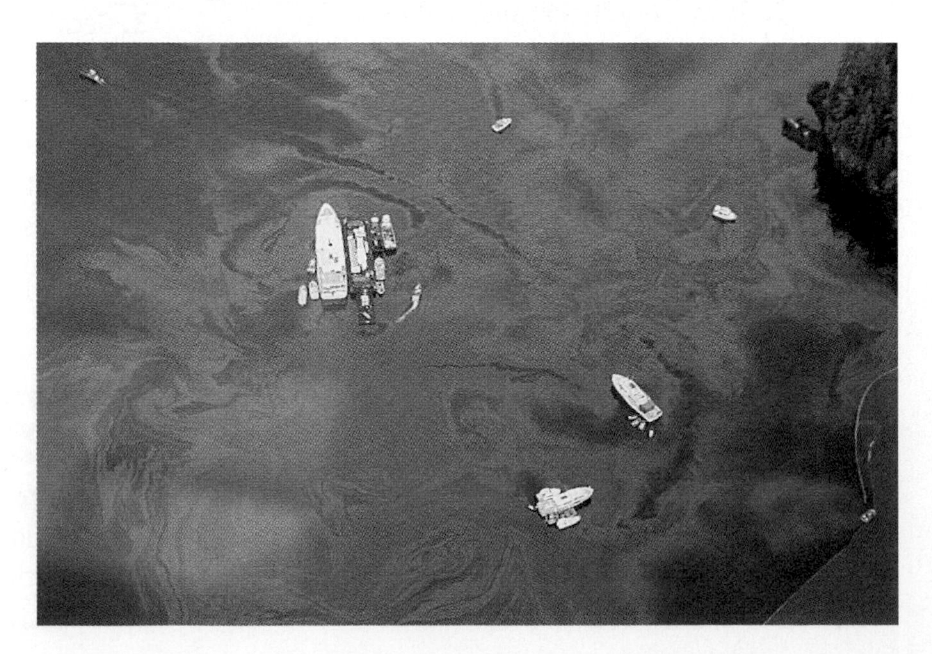

Figure 13
Oil spills on the ocean spread across the surface of the water.
How does the density of oil compare to the density of water?

States of Matter

On Earth, matter occurs in four physical states. These four states are solid, liquid, gas, and plasma. You might have had solid toast and liquid milk or juice for breakfast this morning. You breathe air, which is a gas. A lightning bolt during a storm is an example of matter in its plasma state. What are the differences among these four states of matter?

Solids The reason some matter is solid is that its particles are in fixed positions relative to each other. The individual particles vibrate, but they don't switch positions with each other. Solids have a definite shape and take up a definite volume.

Suppose you have a puzzle that is completely assembled. The pieces are connected so one piece cannot switch positions with another piece. However, the pieces can move a little. For example, you can push on one end of the puzzle and move each individual puzzle piece, but the pieces of the puzzle stay attached to one another. The puzzle pieces in this model represent particles of a substance in a solid state. Such particles are strongly attracted to each other and resist being separated.

Math Skills Activity

Calculating Density

You want to find the density of a small cube of an unknown material. It measures 1 cm × 1 cm × 2 cm. It has a mass of 8 g.

Solution

1. *This is what you know:* mass: $m = 8$ g
   volume: $v = 1 \text{ cm} \times 1 \text{ cm} \times 2 \text{ cm} = 2 \text{ cm}^3$
2. *This is what you need to find:* density: d
3. *This is the equation you need to use:* $d = m/v$
4. *Substitute the known values:* $d = 8 \text{ g}/2 \text{ cm}^3$
   $d = 4 \text{ g}/\text{cm}^3$

Check your answer by multiplying by the volume. Do you calculate the same mass that was given? Explain.

Practice Problem

You discover a gold bar while exploring an old shipwreck. It measures 10 cm × 5 cm × 2 cm. It has a mass of 1,930 g. Find the density of gold.

For more help refer to the Math Skill Handbook.

Research Visit the Glencoe Science Web site at **science.glencoe.com** for information about the four states of matter. Communicate to your class what you learn.

Liquids Particles in a liquid are attracted to each other, but are not in fixed positions as they are in the solid shown in **Figure 15A.** This is because liquid particles have more energy than solid particles. This energy allows them to move around and change positions with each other.

When you eat breakfast, you might have several liquids at the table such as syrup, juice, and milk. These are substances in the liquid state, even though one flows more freely than the others at room temperature. The particles in a liquid can change positions to fit the shape of the container they are held in. You can pour any liquid into any container, and it will flow until it matches the shape of its new container.

Gases The particles that make up gases have enough energy to overcome any attractions between them. This allows them to move freely and independently. Unlike liquids and solids, gases spread out and fill the container in which they are placed. Air fresheners work in a similar way. If an air freshener is placed in a corner, it isn't long before the particles from the air freshener have spread throughout the room. Look at the hot-air balloon shown in **Figure 15C.** The particles in the balloon are evenly spaced throughout the balloon. The balloon floats in the sky, because the hot air inside the balloon is less dense than the colder air around it.

Why do air fresheners work?

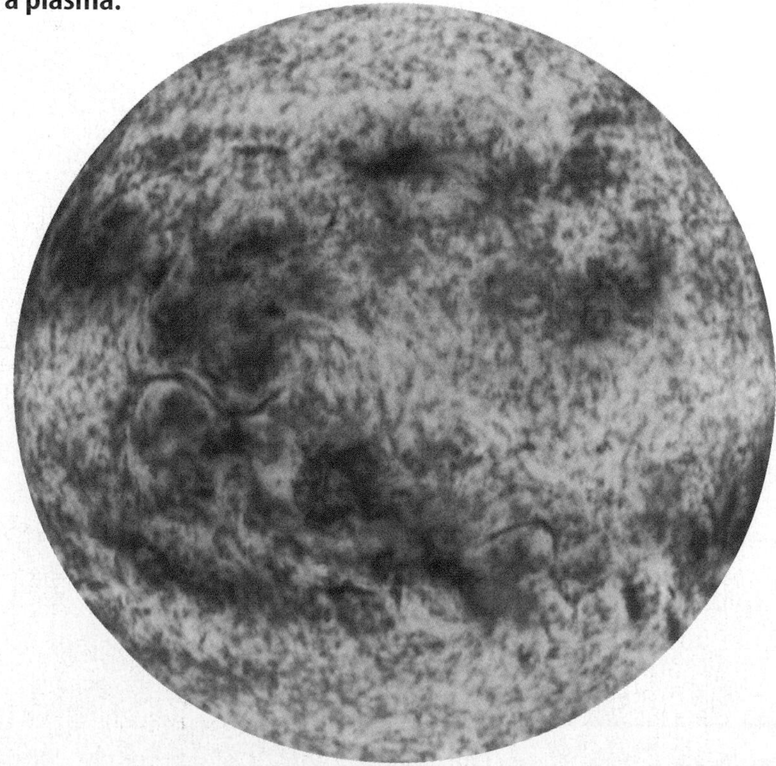

Figure 14
The Sun is an example of a plasma.

Plasma The most common state of matter in the universe is plasma. This state is associated with high temperatures. Can you name something that is in the plasma state? Stars like the Sun, shown in **Figure 14,** are composed of matter in the plasma state. Plasma also exists in Jupiter's magnetic field. On Earth, plasma is found in lightning bolts, as shown in **Figure 15D.** Plasma is composed of ions and electrons. It forms when high temperatures cause some of the electrons normally found in an atom's electron cloud to escape and move outside of the electron cloud.

Figure 15

Matter on Earth exists naturally in four different states—solid, liquid, gas, and plasma—as shown here. The state of a sample of matter depends upon the amount of energy its atoms or molecules possess. The more energy that matter contains, the more freely its atoms or molecules move, because they are able to overcome the attractive forces that tend to hold them together.

D PLASMA Electrically charged particles in lightning are free moving.

A SOLID In a solid such as galena, the tightly packed atoms or molecules lack the energy to move out of position.

B LIQUID The atoms or molecules in a liquid such as water have enough energy to overcome some attractive forces and move over and around one another.

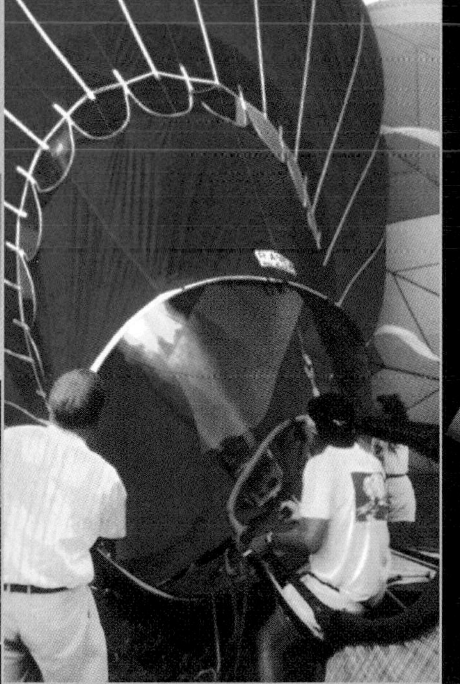

C GAS In air and other gases, atoms or molecules have sufficient energy to separate from each other completely and move in all directions.

Figure 16
A solid metal can be changed to a liquid by adding thermal energy to its molecules. *What is happening to the molecules during this change?*

Changing the State of Matter

Matter is changed from a liquid to a solid at its freezing point and from a liquid to a gas at its boiling point. You may know the freezing and boiling points of water. Water changes from a liquid to a solid at its freezing point of 0°C. It boils at 100°C. Water is the only substance that occurs naturally on Earth as a solid, liquid, and gas. Other substances don't naturally occur in these three states on Earth because of the limited temperature range Earth experiences. For example, temperatures on Earth do not get cold enough for solid carbon dioxide to exist naturally. However, it can be produced by humans.

The attraction between particles of a substance and their rate of movement are factors that determine the state of matter. When thermal energy is added to ice, the rate of movement of its molecules increases. This allows the molecules to move more freely and causes the ice to melt. As **Figure 16** shows, even solid metal can be converted into liquid when enough thermal energy is added.

Changes in state also occur because of increases or decreases in pressure. You can demonstrate this with an ice cube. When subjected to pressure, the ice will change to liquid water when no thermal energy is added. This occurs because the melting temperature of the ice is lowered as more pressure is added. This might explain how the base of a glacier can move around some rock obstacles. It is thought that the pressure of the glacier on the rock melts the ice, creating a thin layer of water. The water then flows around the obstacle and refreezes on the other side.

Figure 17
If ice were more dense than water, lakes would freeze solid from the bottom up. *What effect might this have on the fish?*

Changes in Physical Properties

Chemical properties of matter don't change when the matter changes state, but some of its physical properties change. For example, the density of water changes as water changes state. Ice floats in liquid water, as seen in **Figure 17,** because it is less dense than liquid water. This is unique, because most materials are denser in their solid state than in their liquid state.

✔ Reading Check *Why does ice float in water?*

Some physical properties of substances don't change when they change state. For example, water is colorless and transparent in each of its states.

Changing Mars's Matter

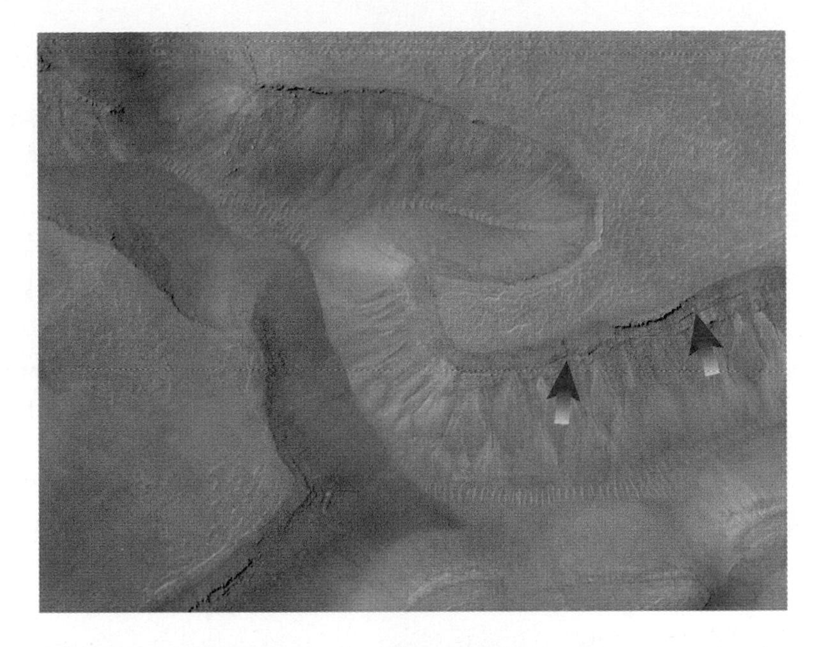

Matter in one state often can be changed to another state by adding or removing thermal energy. Changes in thermal energy might explain why Mars appears to have had considerable water on its surface in the past but now has little or no water on its surface. Recent images of Mars reveal that there might still be some groundwater that occasionally reaches the surface, as shown in **Figure 18.** But what could explain the huge water-carved channels that formed long ago? Much of the liquid water on Mars might have changed state as the planet cooled to its current temperature. Scientists believe that some of Mars's liquid water soaked into the ground and froze, forming permafrost. Some of the water might have frozen to form the polar ice caps. Even more of the water might have evaporated into the atmosphere and escaped to space.

Figure 18
Groundwater might reach the surface of Mars along the edge of this large channel.

Section  Assessment

1. List the four states of matter in order from lowest to highest in terms of amount of particle movement.
2. Compare and contrast the movement of water molecules when water is in a solid, liquid, and gaseous state. How is the movement of water molecules dependent upon temperature?
3. As water freezes, what happens to the water molecules that causes ice to float? Why is this unique?
4. What type of properties of a substance can be observed without changing it into a new substance?
5. **Think Critically** Suppose you blow up a balloon and then place it in a freezer. Later, you find that the balloon has shrunk and has drops of frozen liquid in it. Explain what has happened.

Skill Builder Activities

6. **Classifying** Classify the following items into the four types of matter: *groundwater, lightning, lava, snow, textbook, ice cap, notebook, apple juice, eraser, glass, cotton, helium, iron oxide, lake, limestone,* and *water vapor.* Compare and contrast their characteristics. **For more help, refer to the Science Skill Handbook.**
7. **Using Graphics Software** Research the melting and boiling points in degrees Celsius of several compounds, including water. Use your computer to make a line graph showing the temperatures at which these several compounds change state from solid to liquid to gas. Rearrange your data in order of increasing melting and boiling points. **For more help, refer to the Technology Skill Handbook.**

Activity Design Your Own Experiment

Determining Density

Which has a greater density—a rock or a sponge? Is cork more dense than clay? Density is the ratio of an object's mass to its volume.

Recognize the Problem

How can you determine the densities of several objects in your classroom?

Form a Hypothesis

State a hypothesis about what process you can use to measure and compare the densities of several materials.

Possible Materials

pan
triple-beam balance
100-mL beaker
250-mL graduated cylinder
water
sponge
piece of quartz
piece of clay
small wooden block
small metal block
small cork
rock
ruler

Goals

- **List** some ways that the density of an object can be measured.
- **Design** an experiment that compares the densities of several materials.

Safety Precautions

WARNING: *Be wary of sharp edges on some of the materials and take care not to break the beaker or graduated cylinder. Wash hands thoroughly with soap and water when finished.*

Test Your Hypothesis

Plan

1. As a group, agree upon and write the hypothesis statement.
2. As a group, list the steps that you need to take to test your hypothesis. Be specific, describing exactly what you will do at each step. List your materials.
3. While working as a group, use this equation: density = mass/volume. Devise a method of determining the mass and volume of each material to be tested.
4. **Design** a data table in your Science Journal so that it is ready to use as your group collects data.

Check the Plan

1. Read over your entire experiment to make sure that all steps are in a logical order.
2. Should you run the process more than once for any of the materials?
3. **Identify** any constants, variables, and controls of the experiment.
4. Make sure your teacher approves your plan before you start.

Do

1. Carry out the experiment as planned.
2. While the experiment is going on, write any observations that you make and complete the data table in your Science Journal.

Analyze Your Data

1. Do you observe anything about the way objects with greater density feel compared with objects of lower density?
2. Which of the objects you tested would float in water? Which would sink?

Draw Conclusions

1. Based on your results, would you hypothesize that a cork is more dense, the same density, or less dense than water?
2. Without measuring the density of an object that floats, conclude how you know that it has a density of less than 1.0 g/cm^3.
3. Would the density of the clay be affected if you were to break it into smaller pieces?

Communicating Your Data

Write an informational pamphlet on different methods for determining the density of objects. Include equations and a step-by-step procedure.

Science Stats

Amazing Atoms

Did you know . . .

. . . The diameter of an atom is about 100,000 times as great as the diameter of its nucleus. Suppose that when you sit in a chair, you represent the nucleus of an atom. The nearest electron in your atom would be about 120 km away—nearly half the distance across the Florida peninsula.

. . . Uranium has the greatest mass of the abundant natural elements. One atom of uranium has a mass number that is more than 235 times greater than the mass number of one hydrogen atom, the element with the least mass. However, the diameter of a uranium atom is only about three times the size of a hydrogen atom, similar to the difference between a baseball and a volleyball.

One hundred fourteen Eiffel Towers have a mass of nearly 1 billion kg.

A neutron star speck has about the same mass as 114 Eiffel Towers

. . . The densest material in the universe is found in a neutron star. The core of this type of star is made only of neutrons. Although neutron stars are small, measuring about 10 km to 20 km in diameter, they have a greater mass than the Sun. One tiny pinhead-sized speck of a neutron star would have the same mass as about 114 Eiffel Towers.

. . . The melting point of Cesium is 28.4°C. It would melt in your hand if you held it. You would not want to hold cesium, though, because it would react strongly with your skin. In fact, the metal might even catch fire.

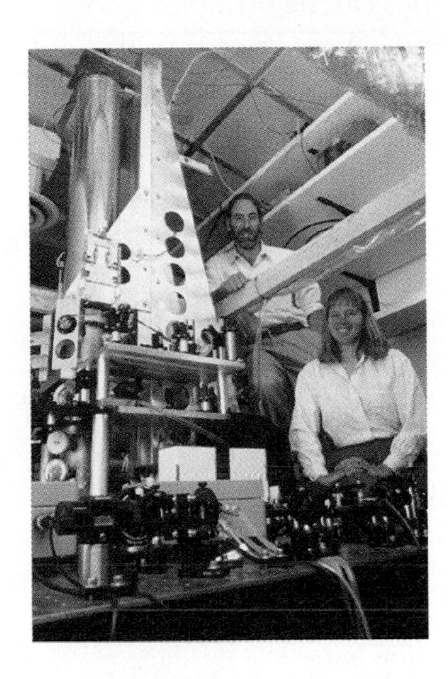

. . . An atomic fountain clock is the world's most accurate timepiece. This timepiece is called an atomic fountain clock because it uses a fountainlike movement of atoms to record time. The atoms are cooled to an extremely low temperature and tossed into a vacuum chamber. The natural vibrations of the atoms are measured. Atomic fountain clocks gain or lose only 1 s in more than 20 million years.

. . . More than ninety elements occur naturally. However, about 98% of Earth's crust consists of only the eight elements shown here.

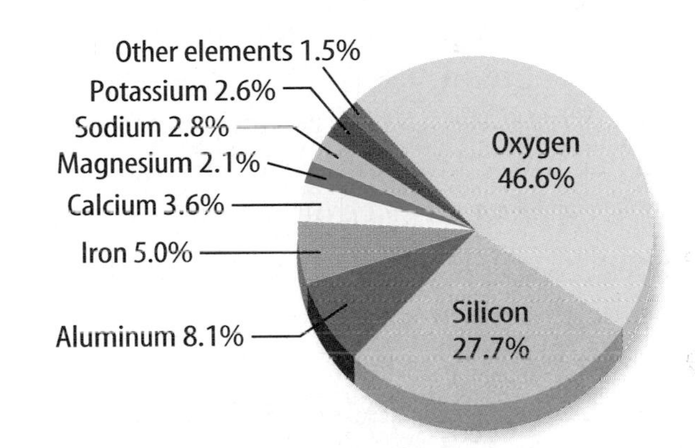

Do the Math

1. Looking at the circle graph, which is the third most abundant element in Earth's crust?
2. How much lower is the melting point of cesium than the average human body temperature of 37°C?
3. The diameter of the Sun is 1,392,000 km. How many neutron stars, each measuring 15 km in diameter, would fit along the Sun's diameter when placed side to side?

Go Further

Do research on the Glencoe Science Web site at **science.glencoe.com** to find out more about atoms and isotopes. What is a radioactive isotope of an element? How are isotopes used in science?

Chapter 18 Study Guide

Reviewing Main Ideas

Section 1 Atoms

1. Matter is anything that has mass and takes up space. *How is matter similar to the snap-together blocks shown below?*

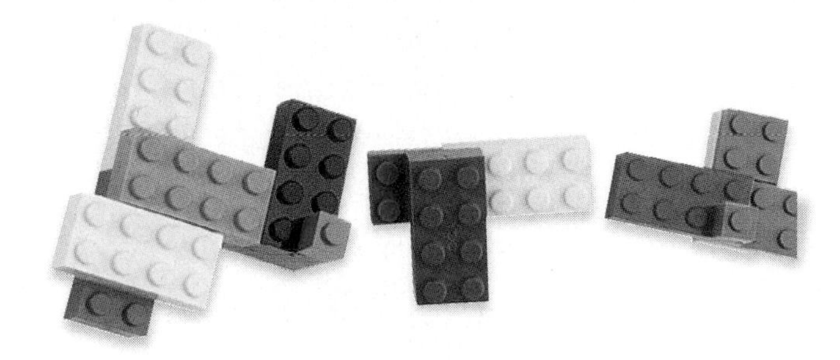

2. Protons and neutrons make up the nucleus of an atom. Protons have a positive charge, and neutrons have no charge. Electrons have a negative charge and surround the nucleus, forming an electron cloud.
3. Isotopes are atoms of the same element that have different numbers of neutrons.

Section 2 Combinations of Atoms

1. Atoms join to form compounds and molecules. A compound is a substance made of two or more elements. The properties of a compound differ from the chemical and physical properties of the elements of which it is composed.
2. A mixture is a substance in which the components are not chemically combined. *Why are the contents of the book bag, shown to the right, considered to be a mixture and not a compound?*

Section 3 Properties of Matter

1. Physical properties can be observed and measured without causing a chemical change in a substance. Chemical properties can be observed only when one substance reacts with another substance.
2. Atoms or molecules in a solid are in fixed positions relative to one another. In a liquid, the atoms or molecules are close together but are freer to change positions. Atoms or molecules in a gas move freely to fill any container. *What makes up plasma, such as the lightning bolts to the right?*

3. Water is the only substance on Earth that occurs naturally as a solid, liquid, and gas. Other substances do not exist in all three states on Earth because of Earth's narrow temperature range.
4. One physical property that is used to describe matter is density. Density is a ratio of the mass of an object to its volume. A material that is less dense will float in a material that is more dense. *Ice floats in liquid water, so which state of water is more dense?*

After You Read

FOLDABLES Reading & Study Skills

Use each vocabulary word on your Vocabulary Study Fold in a sentence about matter and write it next to the definition of the word.

Chapter 18 Study Guide

Visualizing Main Ideas

Complete the following concept map on matter. Use the following terms: liquids, plasma, matter, *and* solids.

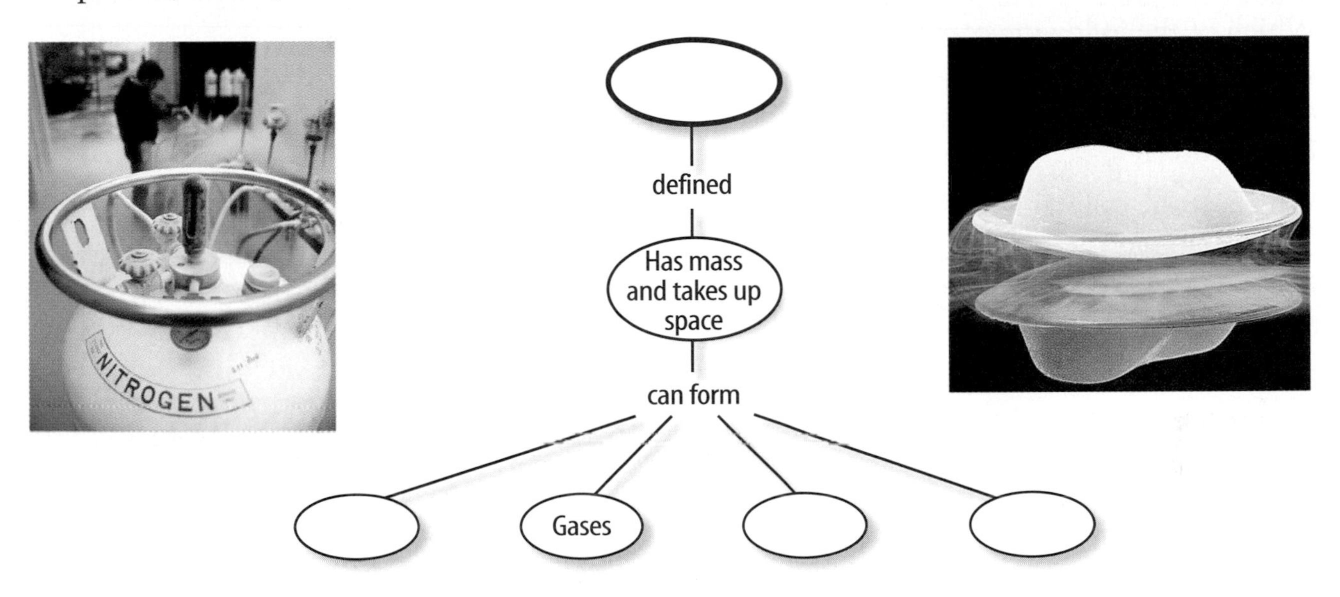

Vocabulary Review

Vocabulary Words

a. atom
b. atomic number
c. compound
d. density
e. electron
f. element
g. ion
h. isotope
i. heterogeneous mixture
j. homogeneous mixture
k. mass number
l. matter
m. mixture
n. neutron
o. proton
p. solution

Study Tip

Take good notes, even during lab. Lab experiments reinforce key concepts, and looking back on these notes can help you better understand what happened and why.

Using Vocabulary

Explain the difference between the vocabulary words in each of the following sets.

1. atom, element
2. mass number, atomic number
3. solution, heterogeneous mixture
4. matter, compound, element
5. heterogeneous mixture, homogeneous mixture
6. proton, neutron, electron
7. isotope, atom
8. atom, ion
9. mixture, compound
10. neutron, mass number

Chapter 18 Assessment

Checking Concepts

Choose the word or phrase that best answers the question.

1. Which of the following contains only one type of atom?
   A) compound **C)** element
   B) mixture **D)** solution

2. Which of the following has a positive electric charge?
   A) electron **C)** neutron
   B) proton **D)** atom

3. In an atom, what forms a cloud around the nucleus?
   A) electrons **C)** neutrons
   B) protons **D)** positively charged particles

4. A carbon atom has a mass number of 12. How many protons and how many neutrons does it have?
   A) 6, 6 **C)** 6, 12
   B) 12, 12 **D)** 12, 6

5. On Earth, oxygen usually exists as which of the following?
   A) solid **C)** liquid
   B) gas **D)** plasma

6. Which of the following isotopes has seven neutrons?
   A) boron-12 **C)** carbon-14
   B) nitrogen-12 **D)** hydrogen-2

7. Which type of bond occurs because of the polar molecules of water?
   A) ionic **C)** metallic
   B) covalent **D)** hydrogen

8. Which of the following are electrically charged?
   A) molecule **C)** isotope
   B) solution **D)** ion

9. What type of property is the color of your clothes?
   A) chemical property
   B) physical property
   C) isotopic property
   D) molecular property

10. Which of the following is not a physical property of water?
    A) transparent
    B) colorless
    C) higher density than ice
    D) changes to hydrogen and oxygen when electricity passes through it

Thinking Critically

11. If an atom has no electric charge, what can be said about the number of protons and electrons it contains?

12. Carbon has six protons and nitrogen has seven protons. Which has the greatest number of neutrons—carbon-13, carbon-14, or nitrogen-14?

13. Would isotopes of the same element have the same number of electrons? Explain.

14. A chlorine atom comes in contact with a lithium atom. Do they combine to form a compound? Why or why not?

15. You pour cooking oil into a glass of water. You briefly stir the materials in the glass. Does the glass contain a mixture? Does it contain a solution? Explain.

Developing Skills

16. Classifying Use the periodic table of the elements, located on the inside back cover, to classify the following substances as elements or compounds: iron, aluminum, carbon dioxide, gold, water, and sugar.

17. Making and Using Graphs Use the following data to make a line graph. For each isotope, plot the mass number along the *y*-axis and the atomic number along the *x*-axis. What is the relationship between mass number and atomic number?

Atomic Number versus Mass Number

Element	Atomic Number	Mass Number
Fluorine	9	19
Lithium	3	7
Carbon	6	12
Nitrogen	7	14
Beryllium	4	9
Boron	5	11
Oxygen	8	16
Neon	10	20

Performance Assessment

18. Song with Lyrics Create a song about how matter changes state by changing the words to a song you know. Include in your song as many states of matter as possible.

TECHNOLOGY

Go to the Glencoe Science Web site at **science.glencoe.com** or use the **Glencoe Science CD-ROM** for additional chapter assessment.

Test Practice

Marlena was instructed by her teacher to find examples of elements and their isotopes. The examples she found are presented in the table below.

Element	Number of Protons	Number of Neutrons
Hydrogen-1	1	0
Hydrogen-2	1	1
Hydrogen-3	1	2
Carbon-12	6	6
Carbon-14	6	8
Oxygen-16	8	8
Oxygen-18	8	10

Study the table and answer the following questions.

1. The hydrogen atom listed first is different from the hydrogen-2 and hydrogen-3 isotopes, because the first hydrogen isotope has _____ .

A) only one neutron
B) fewer neutrons
C) more electrons
D) more neutrons

2. The mass number of an atom is equal to the number of protons and neutrons in its nucleus. According to this definition, which of these has the highest mass number?

F) hydrogen-3
G) carbon-14
H) oxygen-18
J) oxygen-16

CHAPTER 19 Properties and Changes of Matter

At very high temperatures deep within Earth, solid rock melts. When a volcano erupts, the liquid lava cools and turns back to rock. One of the properties of rock is its state—solid, liquid, or gas. Other properties include color, shape, texture, and weight. As lava changes from a liquid to a solid, what happens to its properties? In this chapter, you will learn about physical and chemical properties and changes of matter.

What do you think?

Science Journal Look at the picture below with a classmate. Discuss what you think this might be. Here's a hint: *It can keep your feet smooth.* Write your answer or best guess in your Science Journal.

When a volcano erupts, it spews lava and gases. Lava is hot, melted rock from deep within Earth. After it reaches Earth's surface, the lava cools and hardens into solid rock. The minerals and gases within the lava, as well as the rate at which it cools, determine the characteristics of the resulting rocks. In this activity, you will compare two types of volcanic rock.

Compare properties

1. Obtain samples of the rocks obsidian (uhb SIH dee un) and pumice (PUH mus) of about the same size from your teacher.
2. Compare the colors of the two rocks.
3. Decide which sample is heavier.
4. Look at the surfaces of the two rocks. How are the surfaces different?
5. Place each rock in water and observe.

Observe

What things are different about these rocks? In your Science Journal, make a table that compares the observations you made about the rocks.

Before You Read

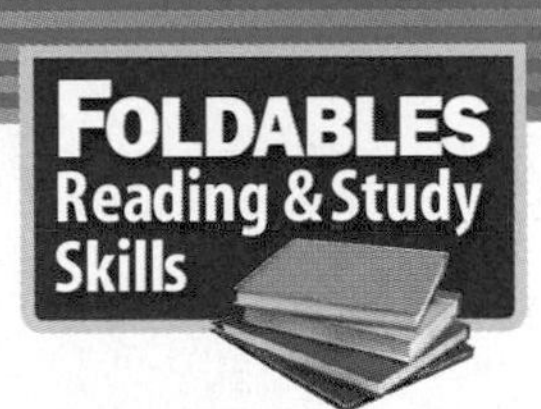

Making an Organizational Study Fold Make the following Foldable to help you organize your thoughts into clear categories about properties and changes.

1. Place a sheet of paper in front of you so the short side is at the top. Fold the paper in half from the top to the bottom two times. Unfold all the folds.
2. Trace over all the fold lines and label the folds *Physical Properties, Physical Changes, Chemical Properties,* and *Chemical Changes* as shown.
3. As you read the chapter, write information about matter's physical and chemical properties and changes on your Foldable.

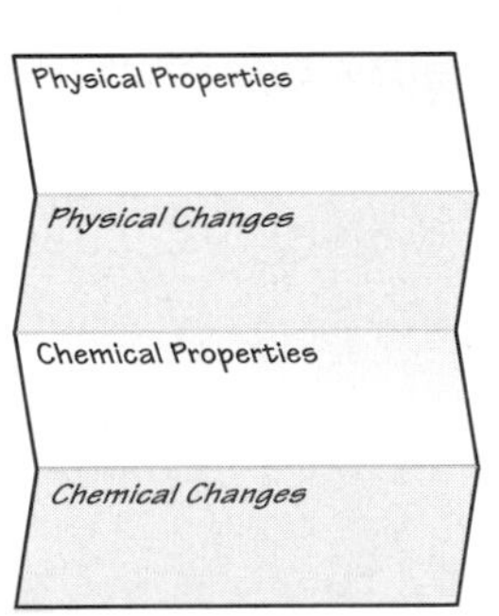

SECTION

Physical and Chemical Properties

As You Read

What You'll Learn

- **Identify** physical and chemical properties of matter.

Vocabulary

physical property
chemical property

Why It's Important

Understanding the different properties of matter will help you to better describe the world around you.

Physical Properties

It's a busy day at the state fair as you and your classmates navigate your way through the crowd. While you follow your teacher, you can't help but notice the many sights and sounds that surround you. Eventually, you fall behind the group as you spot the most amazing ride you have ever seen. You inspect it from one end to the other. How will you describe it to the group when you catch up to them? What features will you use in your description?

Perhaps you will mention that the ride is large, blue, and made of wood. These features are all physical properties, or characteristics, of the ride. A **physical property** is a characteristic you can observe without changing or trying to change the composition of the substance. How something looks, smells, sounds, or tastes are all examples of physical properties. Look at **Figure 1.** You can describe all types of matter and differentiate between them by observing their properties.

Reading Check *What is a physical property of matter?*

Figure 1
All matter can be described by physical properties that can be observed using the five senses.
What types of matter do you think you could see, hear, taste, touch, and smell at the fair?

Using Your Senses Some physical properties describe the appearance of matter. You can detect many of these properties with your senses. For example, you can see the color and shape of the ride at the fair. You can also touch it to feel its texture. You can smell the odor or taste the flavor of some matter. (You should never taste anything in the laboratory.) Consider the physical properties of the items in **Figure 2.**

Figure 2
A Some matter has a characteristic color, such as this sulfur pile. B You can use a characteristic smell or taste to identify these fruits. C Even if you didn't see it, you could probably identify this sponge by feeling its texture.

State To describe a sample of matter, you need to identify its state. Is the ride a solid, a liquid, or a gas? This property, known as the state of matter, is another physical property that you can observe. The ride, your chair, a book, and a pen are examples of matter in the solid state. Milk, gasoline, and vegetable oil are examples of matter in the liquid state. The helium in a balloon, air in a tire, and neon in a sign are examples of matter in the gas state. You can see examples of solids, liquids, and gases in **Figure 3.**

Perhaps you are most familiar with the three states of water. You can drink or swim in liquid water. You use the solid state of water, which is ice, when you put the solid cubes in a drink or skate on a frozen lake. Although you can't see it, water in the gas state is all around you in the air.

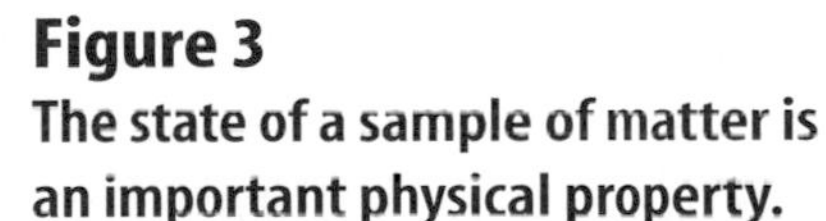

Figure 3
The state of a sample of matter is an important physical property.

A This rock formation is in the solid state.

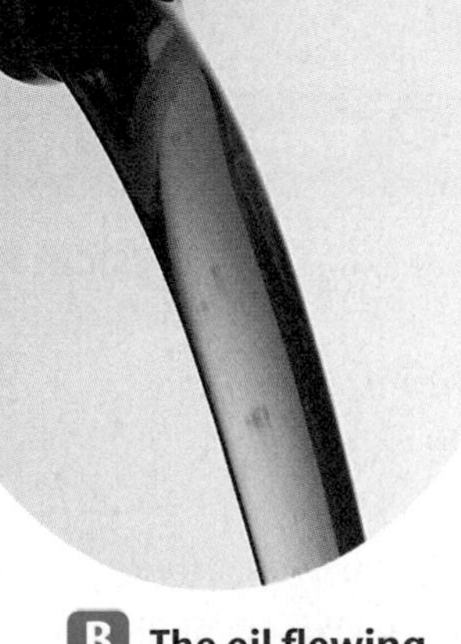

B The oil flowing out of a bottle is in the liquid state.

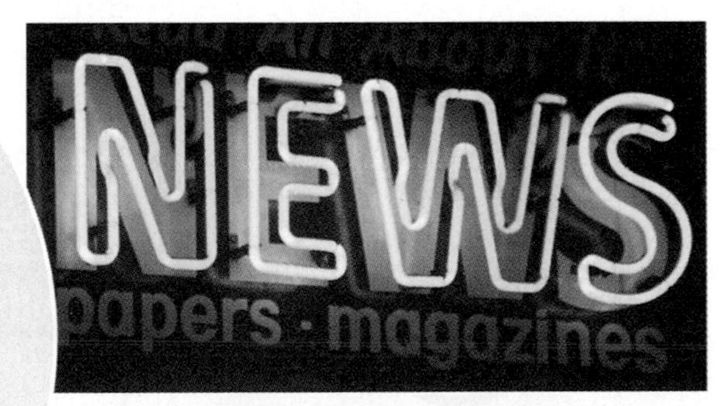

C This colorful sign uses the element neon, which is generally found in the gas state.

Mini LAB

Measuring Properties

Procedure

1. Measure the mass of a **10-mL graduated cylinder.**
2. Fill the graduated cylinder with **water** to the 10-mL mark and measure the mass of the graduated cylinder and water.
3. Determine the mass of the water by subtracting the mass of the graduated cylinder from the mass of the graduated cylinder and water.
4. Determine the density of water by dividing the mass of the water by the volume of the water.

Analysis

1. Why did you need to measure the mass of the empty graduated cylinder?
2. How would your calculated density be affected if you added more than 10-mL of water?

Size Dependent Properties Some physical properties depend on the size of the object. Suppose you need to move a box. The size of the box would be important in deciding if you need to use your backpack or a truck. You can begin by measuring the width, height, and depth of the box. If you multiply them together, you calculate the box's volume. The volume of an object is the amount of space it occupies.

Another physical property that depends on size is mass. Recall that the mass of an object is a measurement of how much matter it contains. A bowling ball has more mass than a basketball. Weight is a measurement of force. Weight depends on the mass of the object and on gravity. If you were to travel to other planets, your weight would change but your size and mass would not. Weight is measured using a spring scale like the one in **Figure 4.**

Size Independent Properties Another physical property, density, does not depend on the size of an object. Density measures the amount of mass in a given volume. To calculate the density of an object, divide its mass by its volume. The density of water is the same in a glass as it is in a tub. Another property, solubility, also does not depend on size. Solubility is the number of grams of one substance that will dissolve in 100 g of another substance at a given temperature. The amount of drink mix that can be dissolved in 100 g of water is the same in a pitcher as it is when it is poured into a glass. Size dependent and independent properties are shown in **Table 1.**

Melting and Boiling Point Melting and boiling point also do not depend upon an object's size. The temperature at which a solid changes into a liquid is called its melting point. The temperature at which a liquid changes into a gas is called its boiling point. The melting and boiling points of several substances, along with some of their other physical properties, are shown in **Table 2.**

Figure 4
A spring scale is used to measure an object's weight.

Table 1 Properties of Matter

	Physical Properties
Dependent on sample size	mass, weight, volume
Independent of sample size	density, melting/boiling point, solubility, ability to attract a magnet, state of matter, color.

Table 2 Physical Properties of Several Substances (at atmospheric temperature and pressure)

Substance	Color	State	Density (g/cm^3)	Melting Point (°C)	Boiling Point (°C)
Bromine	Red-brown	Liquid	3.12	−7	59
Chlorine	Yellowish	Gas	0.0032	−101	−34
Mercury	Silvery-white	Liquid	13.5	−39	357
Neon	Colorless	Gas	0.0009	−249	−246
Oxygen	Colorless	Gas	0.0014	−219	−183
Sodium chloride	White	Solid	2.17	801	1,413
Sulfur	Yellow	Solid	2.07	115	445
Water	Colorless	Liquid	1.00	0	100

Behavior Some matter can be described by the specific way in which it behaves. For example, some materials pull iron toward them. These materials are said to be magnetic. The lodestone in **Figure 5** is a rock that is naturally magnetic.

Other materials can be made into magnets. You might have magnets on your refrigerator or locker at school. The door of your refrigerator also has a magnet within it that holds the door shut tightly.

Reading Check *What are some examples of physical properties of matter?*

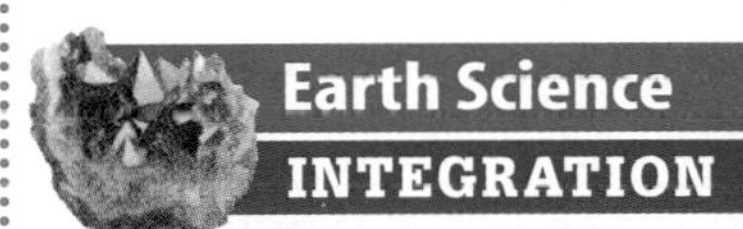

Scientists can learn about the history of the Moon by analyzing the properties of moon rocks. The properties of some moon rocks, for example, are similar to those of rocks produced by volcanoes on Earth. In this way, scientists learned that the Moon once had volcanic activity. Make a list of questions to ask about the properties of a moon rock.

Figure 5
This lodestone attracts certain metals to it. Lodestone is a natural magnet.

Figure 6
Notice the difference between the new matches and the matches that have been burned. The ability to burn is a chemical property of matter.

A

B

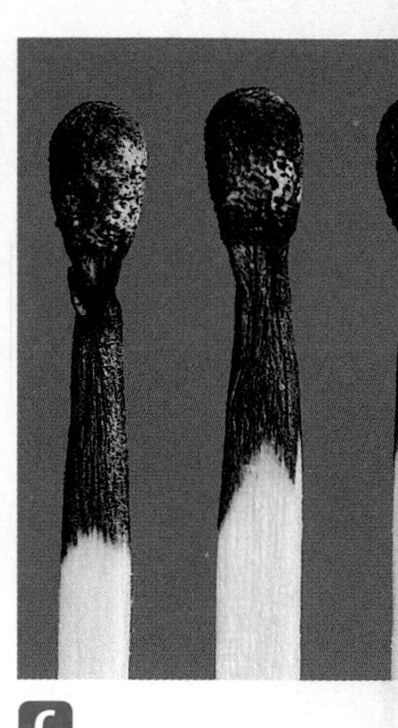
C

SCIENCE Online

Research Visit the Glencoe Science Web site at **science.glencoe.com** for more information about methods of measuring matter. Make a poster showing how to make measurements involving several different samples of matter.

Chemical Properties

Some properties of matter cannot be identified just by looking at a sample. For example, nothing happens if you look at the matches in **Figure 6A.** But if someone strikes the matches on a hard, rough surface they will burn, as shown in **Figure 6B.** The ability to burn is a chemical property. A **chemical property** is a characteristic that cannot be observed without altering the substance. As you can see in **Figure 6C,** the matches are permanently changed after they are burned. Therefore this property can be observed only by changing the composition of the match. Another way to define a chemical property, then, is the ability of a substance to undergo a change that alters its identity. You will learn more about changes in matter in the following section.

Section  1 Assessment

1. What physical properties could you use to describe a baseball?
2. How are your senses important to identifying physical properties of matter?
3. How is density related to mass and volume? Explain and write an equation.
4. Describe a chemical property in your own words. Give an example.
5. **Think Critically** Explain why density and solubility are size-independent physical properties of matter.

Skill Builder Activities

6. **Comparing and Contrasting** How is a chemical property different from a physical property? **For more help, refer to the Science Skill Handbook.**
7. **Solving One-Step Equations** You need to fill a bucket with water. The volume of the bucket is 5 L and you are using a cup with a volume of 50 mL. How many cupfulls will you need? There's a hint: $1\ L = 1000\ mL$. **For more help, refer to the Math Skill Handbook.**

Activity

Finding the Difference

You can identify an unknown object by comparing its physical and chemical properties to the properties of identified objects.

What You'll Investigate

What physical properties can you observe in order to describe a set of objects?

Materials

meterstick
spring scale
block of wood
metal bar or metal ruler
plastic bin
drinking glass
water
rubber ball
paper
carpet
magnet
rock
plant or flower
soil
sand
apple (or other fruit)
vegetable
slice of bread
dry cereal
egg
feather

Goals

- **Identify** the physical properties of objects.
- **Compare and Contrast** the properties.
- **Categorize** the objects based on their properties.

Safety Precautions

Procedure

1. List at least six properties that you will observe, measure, or calculate for each object. Describe how to determine each property.
2. In your Science Journal, create a data table with a column for each property and rows for the objects.
3. Complete your table by determining the properties for each object.

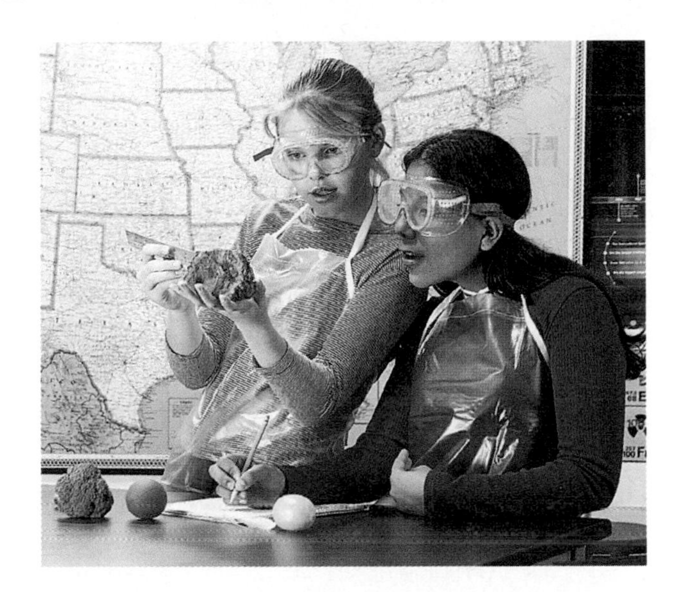

Conclude and Apply

1. Which properties were you able to observe easily? Which required making measurements? Which required calculations?
2. **Compare and contrast** the objects based on the information in your table.
3. Choose a set of categories and group your objects into those categories. Some examples of categories are large/medium/small, heavy/moderate/light, bright/moderate/dull, solid/liquid/gas, etc. Were the categories you chose useful for grouping your objects? Why or why not?

Communicating Your Data

Compare your results with those of other students in your class. **Discuss** the properties of objects that different groups included on their tables. Make a large table including all of the objects that students in the class studied.

SECTION 2

Physical and Chemical Changes

As You Read

What **You'll Learn**

- **Compare** several physical and chemical changes.
- **Identify** examples of physical and chemical changes.

Vocabulary
physical change
chemical change
law of conservation of mass

Why **It's Important**
Physical and chemical changes affect your life every day.

Physical Changes

What happens when the artist turns the lump of clay shown in **Figure 7** into a work of art? The composition of the clay does not change. Its appearance, however, changes dramatically. The change from a lump of clay to a work of art is a physical change. A **physical change** is one in which the form or appearance of matter changes, but not its composition. The lake in **Figure 7** also experiences a physical change. Although the water changes state due to a change in temperature, it is still made of the elements hydrogen and oxygen.

Changing Shape Have you ever crumpled a sheet of paper into a ball? If so, you caused physical change. Whether it exists as one flat sheet or a crumpled ball, the matter is still paper. Similarly, if you cut fruit into pieces to make a fruit salad, you do not change the composition of the fruit. You change only its form. Generally, whenever you cut, tear, grind, or bend matter, you are causing a physical change.

Figure 7
Although each sample looks quite different after it experiences a change, the composition of the matter remains the same. These changes are examples of physical changes.

Dissolving What type of change occurs when you add sugar to iced tea, as shown in **Figure 8?** Although the sugar seems to disappear, it does not. Instead, the sugar dissolves. When this happens, the particles of sugar spread out in the liquid. The composition of the sugar stays the same, which is why the iced tea tastes sweet. Only the form of the sugar has changed.

Figure 8
Physical changes are occurring constantly. The sugar blending into the iced tea is an example of a physical change.

Changing State Another common physical change occurs when matter changes from one state to another. When an ice cube melts, for example, it becomes liquid water. The solid ice and the liquid water have the same composition. The only difference is the form.

Matter can change from any state to another. Freezing is the opposite of melting. During freezing, a liquid changes into a solid. A liquid also can change into a gas. This process is known as vaporization. During the reverse process, called condensation, a gas changes into a liquid. **Figure 9** summarizes these changes.

In some cases, matter changes between the solid and gas states without ever becoming a liquid. The process in which a solid changes directly into a gas is called sublimation. The opposite process, in which a gas changes into a solid, is called deposition.

Figure 9
Look at the photographs below to identify the different physical changes that bromine undergoes as it changes from one state to another.

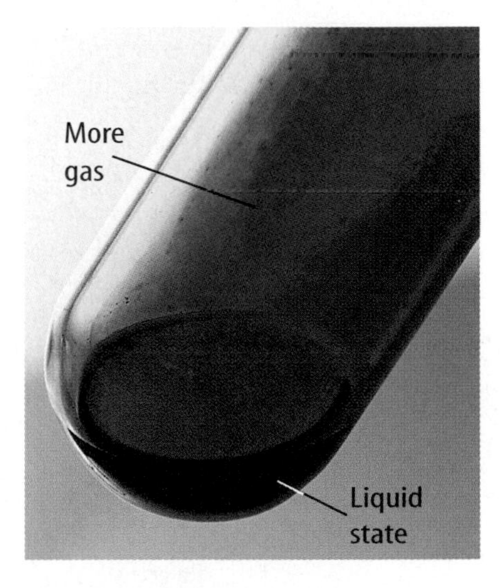

Chemical Changes

It's the Fourth of July in New York City. Brilliant fireworks are exploding in the night sky. When you look at fireworks, such as these in **Figure 10,** you see dazzling sparkles of red and white trickle down in all directions. The explosion of fireworks is an example of a chemical change. During a **chemical change,** substances are changed into different substances. In other words, the composition of the substance changes.

Figure 10
These brilliant fireworks result from chemical changes. *What is a chemical change?*

You are familiar with another chemical change if you have ever left your bicycle out in the rain. After a while, a small chip in the paint leads to an area of a reddish, powdery substance. This substance is rust. When iron in steel is exposed to oxygen and water in air, iron and oxygen atoms combine to form the principle component in rust. In a similar way, coins tarnish when exposed to air. These chemical changes are shown in **Figure 11.**

✔ Reading Check *How is a chemical change different from a physical change?*

Figure 11
Each of these examples shows the results of a chemical change. In each case, the substances that are present after the change are different from those that were present before the change.

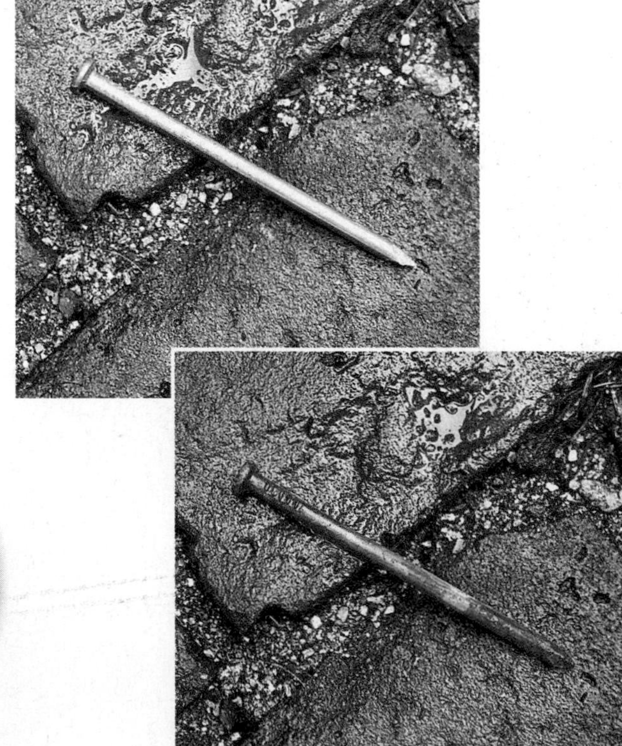

Figure 12
The brilliant colors of autumn result from a chemical change.

Signs of Chemical Changes

Physical changes are relatively easy to identify. If only the form of a substance changes, you have observed a physical change. How can you tell whether a change is a chemical change? If you think you are unfamiliar with chemical changes, think again.

You have witnessed a spectacular change if you have seen the leaves of trees change colors in autumn, but you are not seeing a chemical change. Chemicals called pigments give tree leaves their color. In **Figure 12,** the pigment that is responsible for the green color you see during the summer is chlorophyll (KLOHR uh fihl). Two other pigments result in the colors you see in the red tree. Throughout the spring and summer, chlorophyll is present in much greater amounts than these other pigments, so you see leaves as green. In autumn, however, changes in temperature and rainfall amounts cause trees to stop producing chlorophyll. The chlorophyll that is already present undergoes a chemical change in which it loses its green color. Without chlorophyll, the red and yellow pigments, which are always present, can be seen.

Color Perhaps you have found that a half-eaten apple turns brown. The reason is that a chemical change occurs when food spoils. Maybe you have toasted a marshmallow or a slice of bread and watched them turn black. In each case, the color of the food changes as it is cooked because a chemical change occurs.

SCIENCE *Online*

Research Visit the Glencoe Science Web site at **science.glencoe.com** for more information about how to recognize chemical changes. Choose one example not mentioned in the chapter and present it to the class as a poster or in an oral report.

TRY AT HOME Mini LAB

Comparing Changes

Procedure

1. Separate a piece of **fine steel wool** into two halves.
2. Dip one half in **tap water.**
3. Place each piece of steel wool on a separate **paper plate** and let them sit overnight.
4. Repeat step two with the same half of steel wool for five days.

Analysis

1. Did you observe any changes in the steel wool? If so, describe them.
2. If you observed changes, were they physical or chemical? How do you know?

Figure 13
Cake batter undergoes a chemical change as it absorbs energy during cooking.

Energy Another sign of a chemical change is the release or gain of energy by an object. Many substances must absorb energy in order to undergo a chemical change. For example, energy is absorbed during the chemical changes involved in cooking. When you bake a cake or make pancakes, energy is absorbed by the batter as it changes from a runny mix into what you see in **Figure 13.**

Another chemical change in which a substance absorbs energy occurs during the production of cement. This process begins with the heating of limestone. Ordinarily, limestone will remain unchanged for centuries. But when it absorbs energy during heating, it undergoes a chemical change in which it turns into lime and carbon dioxide.

Energy also can be released during a chemical change. The fireworks you read about earlier released energy in the form of light that you can see. As shown in **Figure 14A,** a chemical change within a firefly releases energy in the form of light. Fuel burned in the camping stove shown in **Figure 14B** releases energy you see as light and feel as heat. You also can see that energy is released when sodium and chlorine are combined and ignited in **Figure 14C.** During this chemical change, the original substances change into sodium chloride, which is ordinary table salt.

Figure 14
A Energy is released when a firefly glows, B when fuel is burned in a camping stove, and C when sodium and chlorine undergo a chemical change to form table salt.

Odor It takes only one experience with a rotten egg to learn that they smell much different than fresh eggs. When eggs and other foods spoil, they undergo chemical change. The change in odor is a clue to the chemical change. This clue can be used to save lives. When you smell an odd odor in foods, such as chicken, pork, or mayonnaise, you know that the food has undergone a chemical change. You can use this clue to avoid eating spoiled food and protect yourself from becoming ill.

Gases or Solids Look at the antacid tablet in **Figure 15A.** You can produce similar bubbles if you pour vinegar on baking soda. The formation of a gas is a clue to a chemical change. What other products undergo chemical changes and produce bubbles?

Figure 15B shows another clue to a chemical change—the formation of a solid. A solid that separates out of a solution during a chemical change is called a precipitate. The precipitate in the photograph forms when a solution containing sodium iodide is mixed with a solution containing lead nitrate.

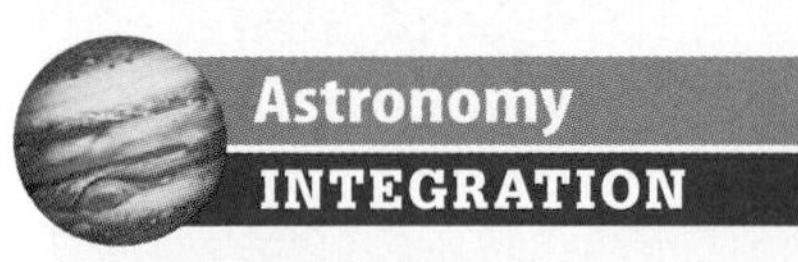

A meteoroid is a chunk of metal or stone in space. Every day, meteoroids enter Earth's atmosphere. When this happens, the meteoroid burns as a result of friction with gases in the atmosphere. A streak of light produced during this chemical change is known as a meteor or shooting star. In your Science Journal, infer why most meteoroids never reach Earth's surface.

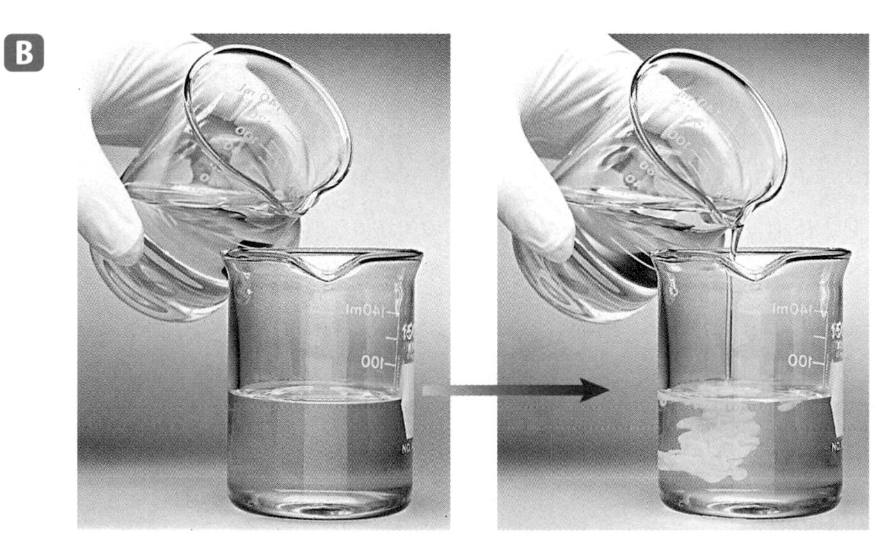

Figure 15
A The bubbles of gas formed when this antacid tablet is dropped into water indicate a chemical change. B The solid forming from two liquids is another sign that a chemical change has taken place.

Figure 16
As wood burns, it turns into a pile of ashes and gases that rise into the air. *Can you turn ashes back into wood?*

Not Easily Reversed How do physical and chemical changes differ from one another? Think about ice for a moment. After solid ice melts into liquid water, it can refreeze into solid ice if the temperature drops enough. Freezing and melting are physical changes. The substances produced during a chemical change cannot be changed back into the original substances by physical means. For example, the wood in **Figure 16** changes into ashes and gases that are released into the air. After wood is burned, it cannot be restored to its original form as a log.

Think about a few of the chemical changes you just read about to see if this holds true. An antacid tablet cannot be restored to its original form after being dropped in water. Rotten eggs cannot be made fresh again, and pancakes cannot be turned back into batter. The substances that existed before the chemical change no longer exist.

Reading Check *What signs indicate a chemical change?*

Math Skills Activity

Converting Temperatures

Fahrenheit is a non-SI temperature scale. Because it is used so often, it is useful to be able to convert from Fahrenheit to Celsius. The equation that relates Celsius degrees to Fahrenheit degrees is:

$(°C \times 1.8) + 32 = °F$

Using this information, what is 15°F on the Celsius scale?

Solution

1. *This is what you know:* temperature = 15°F
2. *This is what you want to find:* temperature in degrees Celsius
3. *This is the equation you need to use:* $(°C \times 1.8) + 32 = °F$
4. *Rearrange the equation to solve for °C.* $(°C \times 1.8) + 32 = °F$
   $°C = (°F - 32)/1.8$
   Then substitute the known value for °F. $°C = (15 - 32)/1.8 = -9.4°\,C$

Check your answer by substituting the Celsius temperature into the original equation. Did you calculate the Fahrenheit temperature that was given in the question?

Practice Problem

Water is being heated on the stove at 156°F. What is this temperature on the Celsius scale?

For help refer to the Math Skill Handbook.

Chemical Versus Physical Change

Now you have learned about many different physical and chemical changes. You have read about several characteristics that you can use to distinguish between physical and chemical changes. The most important point for you to remember is that in a physical change, the composition of a substance does not change and in a chemical change, the composition of a substance does change. When a substance undergoes a physical change, only its form changes. In a chemical change, both form and composition change.

When the wood and copper in **Figure 17** undergo physical changes, the original wood and copper still remain after the change. When a substance undergoes a chemical change, however, the original substance is no longer present after the change. Instead, different substances are produced during the chemical change. When the wood and copper in **Figure 17** undergo chemical changes, wood and copper have changed into new substances with new physical and chemical properties.

Physical and chemical changes are used to recycle or reuse certain materials. **Figure 18** discusses the importance of some of these changes in recycling.

Chemical change

Physical change

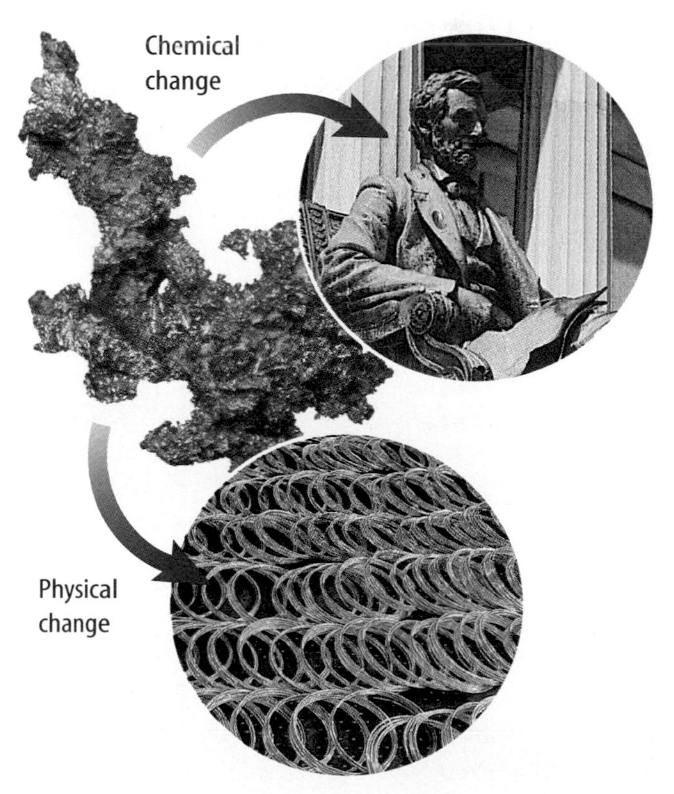

Figure 17
When a substance undergoes a physical change, its composition stays the same. When a substance undergoes a chemical change, it is changed into different substances.

Figure 18

Recycling is a way to separate wastes into their component parts and then reuse those components in new products. In order to be recycled, wastes need to be both physically—and sometimes chemically—changed. The average junked automobile contains about 62 percent iron and steel, 28 percent other materials such as aluminum, copper, and lead, and 10 percent rubber, plastics, and various materials.

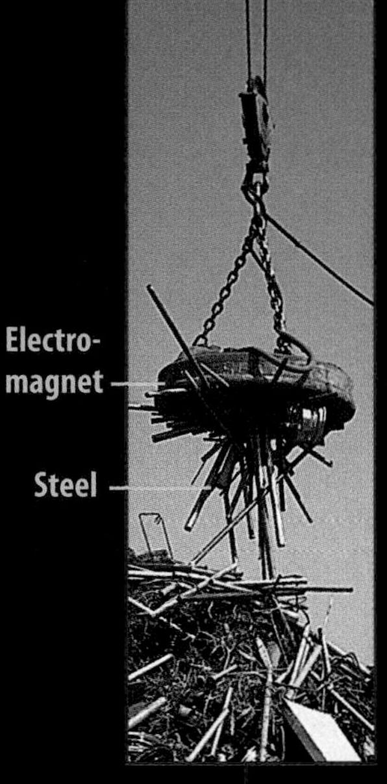

◀ Rubber tires can be shredded and added to asphalt pavement and playground surfaces. New recycling processes make it possible to supercool tires to a temperature at which the rubber is shattered like glass. A magnet can then draw out steel from the tires and parts of the car.

▼ After being crushed and flattened, car bodies are chopped into small pieces. Metals are separated from other materials using physical processes. Some metals are separated using powerful magnets. Others are separated by hand.

◀ Glass can be pulverized and used in asphalt pavement, new glass, and even artwork. This sculpture, named *Groundswell*, was created by artist Maya Lin, using recycled windshield glass.

▲ Some plastics can be melted and formed into new products. Others are ground up or shredded and used as fillers or insulating materials.

Conservation of Mass

During a chemical change, the form or the composition of the matter changes. The particles within the matter rearrange to form new substances, but they are not destroyed and new particles are not created. The number and type of particles remains the same. As a result, the total mass of the matter is the same before and after a physical or chemical change. This is known as the **law of conservation of mass.**

This law can sometimes be difficult to believe, especially when the materials remaining after a chemical change might look different from those before it. In many chemical changes in which mass seems to be gained or lost, the difference is often due to a gas being given off or taken in. The difference, for example, before and after the candle in **Figure 19** is burned is in the gases released into the air. If the gases could be contained in a chamber around the candle, you would see that the mass does not change.

The scientist who first performed the careful experiments necessary to prove that mass is conserved was Antoine Lavoisier (AN twan • luh VWAH see ay) in the eighteenth century. It was Lavoisier who recognized that the mass of gases that are given off or taken from the air during chemical changes account for any differences in mass.

Figure 19
The candle looks as if it lost mass when it was burned. However, if you could trap and measure the gases given up during burning you would find that the mass of the candle and the gases is equal to the mass of the original candle.

Section  Assessment

1. What happens during a physical change?
2. List five physical changes you can observe in your home. Explain how you decided that each change is physical.
3. What kind of change occurs on the surface of bread when it is toasted—physical or chemical? Explain.
4. What does it mean to say that mass is conserved during a chemical change?
5. **Think Critically** A log is reduced to a small pile of ash when it burns. The law of conservation of mass states that the total mass of matter is the same before and after a chemical change. Explain the difference in mass between the log and the ash.

Skill Builder Activities

6. **Classifying** Classify the following changes as physical or chemical: baking a cake, folding towels, burning gasoline, melting snow, grinding beef into a hamburger, pouring milk into a glass, making cookies, and cutting a sheet of paper into paper dolls. **For more help, refer to the Science Skill Handbook.**
7. **Solving One-Step Equations** Magnesium and oxygen undergo a chemical change to form magnesium oxide. How many grams of magnesium oxide will be produced when 0.486 g of oxygen completely react with 0.738 g of magnesium? **For more help, refer to the Math Skill Handbook.**

Activity Design Your Own Experiment

Battle of the Toothpastes

Your teeth are made of a compound called hydroxyapatite (hi DRAHK see A puh tite). The sodium fluoride in toothpaste undergoes a chemical reaction with hydroxyapatite to form a new compound on the surface of your teeth. This compound resists food acids that cause tooth decay, another chemical change. In this activity, you will design an experiment to test the effectiveness of different toothpaste brands. The compound found in your teeth is similar to the mineral compound found in eggshells. Treating hard-boiled eggs with toothpaste is similar to brushing your teeth with toothpaste. Soaking the eggs in food acids such as vinegar for several days will produce similar conditions as eating foods, which contain acids that will produce a chemical change in your teeth, for several months.

Recognize the Problem

Which brands of toothpaste provide the greatest protection against tooth decay?

Form a Hypothesis

Form a hypothesis about the effectiveness of different brands of toothpaste.

Goals

- **Observe** how toothpaste helps prevent tooth decay.
- **Design** an experiment to test the effectiveness of various toothpaste brands.

Safety Precautions

Possible Materials

2 or 3 different brands of toothpaste
drinking glasses or bowls
hard boiled eggs
concentrated lemon juice
apple juice
water
artist's paint brush

Test Your Hypothesis

Plan

1. **Describe** how you will use the materials to test the toothpaste.
2. **List** the steps you will follow to test your hypothesis.
3. **Decide** on the length of time that you will conduct your experiment.
4. **Identify** the control and variables you will use in your experiment.
5. **Create** a data table in your Science Journal to record your observations, measurements, and results.
6. **Describe** how you will measure the amount of protection each toothpaste brand provides.

Do

1. Make sure your teacher approves your plan before you start.
2. **Conduct** your experiment as planned. Be sure to follow all proper safety precautions.
3. **Record** your observations in your data table.

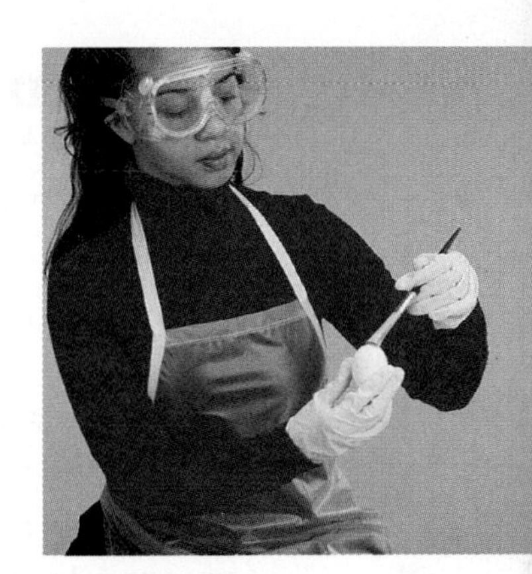

Analyze Your Data

1. **Compare** the untreated eggshells with the shells you treated with toothpaste.
2. **Compare** the condition of the eggshells you treated with different brands of toothpaste.
3. **Compare** the condition of the eggshells soaked in lemon juice and in apple juice.
4. **Identify** unintended variables you discovered in your experiment that might have influenced the results.

Draw Conclusions

1. Did the results support your hypothesis? **Identify** strengths and weaknesses of your hypothesis.
2. **Explain** why the eggs treated with toothpaste were better protected than the untreated eggshells.
3. **Identify** which brands of toothpaste, if any best protected the eggshells from decay.
4. **Evaluate** the scientific explanation for why adding fluoride to toothpaste and drinking water prevents tooth decay.
5. **Predict** what would happen to your protected eggs if you left them in the food acids for several weeks.
6. **Infer** why it is a good idea to brush with fluoride toothpaste.

Compare your results with the results of your classmates. **Create** a poster advertising the benefits of fluoride toothpaste.

Science Stats

Strange Changes

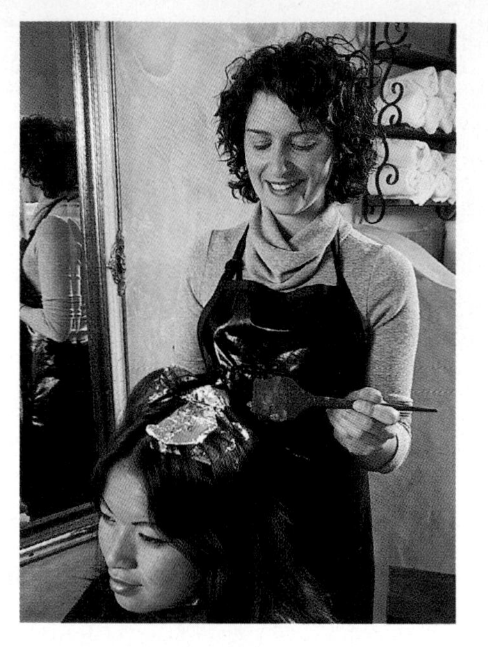

Did you know...

...A hair colorist is also a chemist!

Colorists use hydrogen peroxide and ammonia to swell and open the cuticle-like shafts on your hair. Once these are open, the chemicals in hair dye can get into your natural pigment molecules and chemically change your hair color. The first safe commercial hair color was created in 1909 in France.

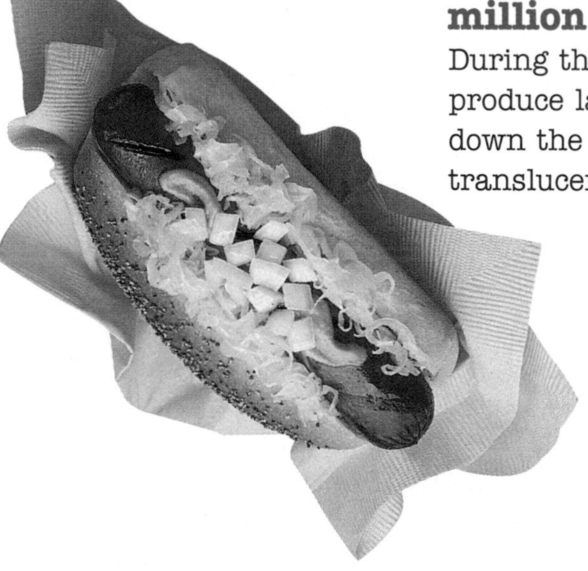

...Americans consume about 175 million kg of sauerkraut each year.

During the production of sauerkraut, bacteria produce lactic acid. The acid chemically breaks down the material in the cabbage, making it translucent and tangy.

...More than 450,000 metric tons of plastic packaging are recycled each year in the U.S.

Discarded plastics undergo physical changes including melting and shredding. They are then converted into flakes or pellets, which are used to make new products. Recycled plastic is used to make clothes, furniture, carpets, and even lumber.

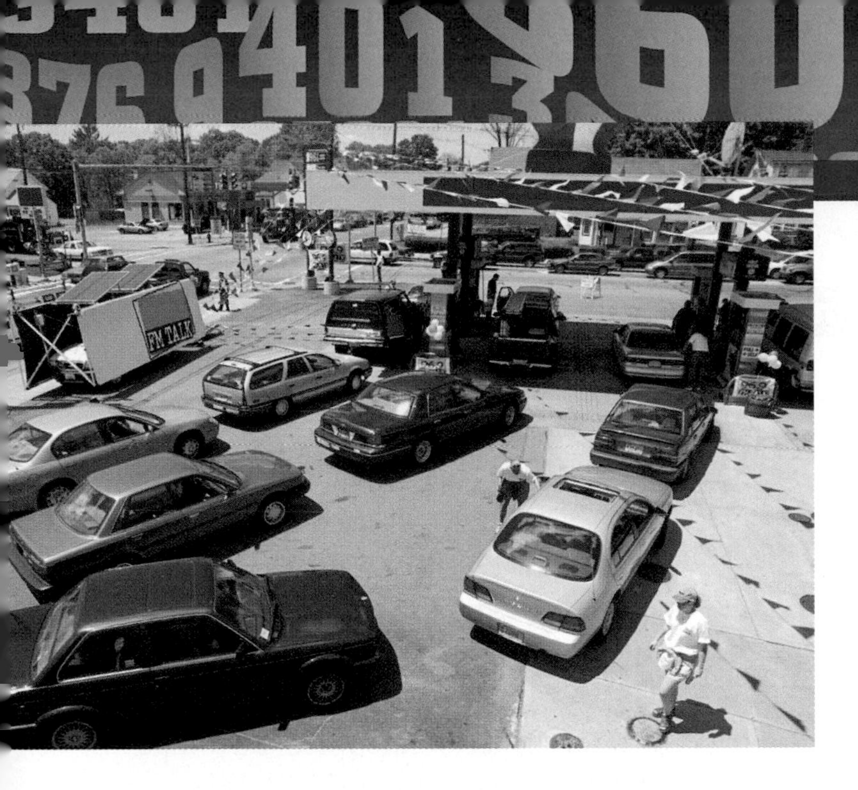

Connecting To Math

...The U.S. population consumes about 1.36 billion L of gasoline each day. When ignited in an enclosed space, such as in an internal combustion engine, gasoline undergoes many chemical changes to deliver the energy needed for an automobile to move.

Projected Recycling Rates by Material, 2000

Material	1995 Recycling	Proj. Recycling
Paper/Paperboard	40.0%	43 to 46%
Glass	24.5%	27 to 36%
Ferrous Metal	36.5%	42 to 55%
Aluminum	34.6%	46 to 48%
Plastics	5.3%	7 to 10%
Yard Waste	30.3%	40 to 50%
Total Materials	**27.0%**	**30 to 35%**

...It is possible to change corn into plastic! Normally plastics are made from petroleum, but chemists have discovered how to chemically convert corn and other plants into plastic. A factory in Nebraska is scheduled to produce about 140,000 metric tons of the corn-based plastic by 2002.

Do the Math

1. There are 275 million people in the United States. Calculate the average amount of sauerkraut consumed by each person in the United States in one year.
2. How many liters of gasoline does the U.S. population consume during a leap year?
3. If chemists can produce 136 kg of plastic from 27 kg of corn, how much corn is needed to make 816 kg of plastic?

Go Further

Every time you cook, you make physical and chemical changes to food. Go to the Glencoe Science Web site at **science.glencoe.com** or to your local or school library to find out what chemical or physical changes take place when cooking ingredients are heated or cooled.

Chapter 19 Study Guide

Reviewing Main Ideas

Section 1 Physical and Chemical Properties

1. Matter can be described by its characteristics, or properties.
2. A physical property is a characteristic that can be observed without altering the composition of the sample.
3. Physical properties include color, shape, smell, taste, and texture, as well as measurable quantities such as mass, volume, density, melting point, and boiling point. *What is the volume of this box?*

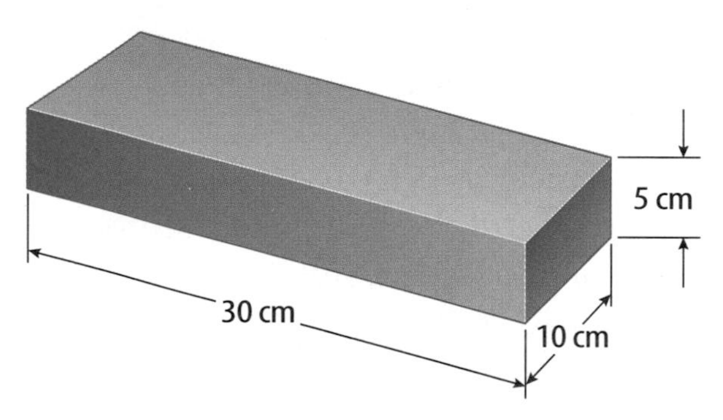

4. Matter can exist in different states—solid, liquid, or gas. The state of matter is a physical property.
5. A chemical property is a characteristic that cannot be observed without changing what the sample is made of. *What chemical property is being shown here?*

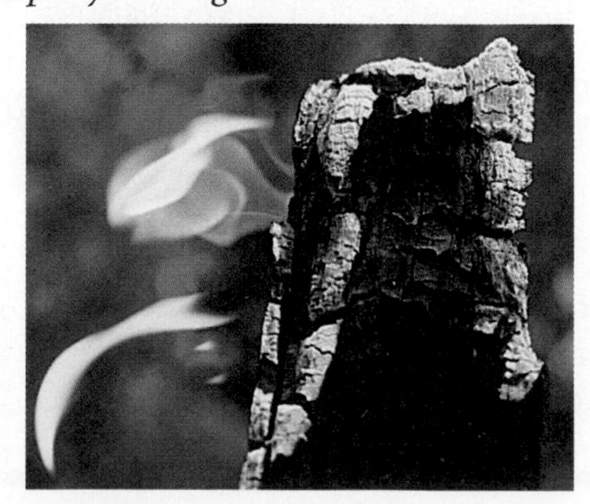

Section 2 Physical and Chemical Changes

1. During a physical change, the composition of matter stays the same but the appearance changes in some way.
2. Physical changes occur when matter is ripped, cut, torn, or bent, or when matter changes from one state to another.
3. A chemical change occurs when the composition of matter changes.
4. Signs of chemical change include changes in energy, color, odor, or the production of gases or solids. *What evidence from this photo indicates a chemical change?*

5. According to the law of conservation of mass, mass cannot be created or destroyed. As a result, the mass of the substances that were present before a physical or chemical change is equal to the mass of the substances that are present after the change.

After You Read

FOLDABLES Reading & Study Skills

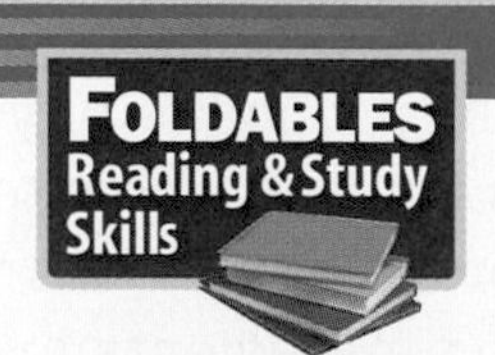

To help you review matter's physical and chemical properties and changes, use the information on your Foldable.

Chapter 19 Study Guide

Visualizing Main Ideas

Complete the following concept map on matter.

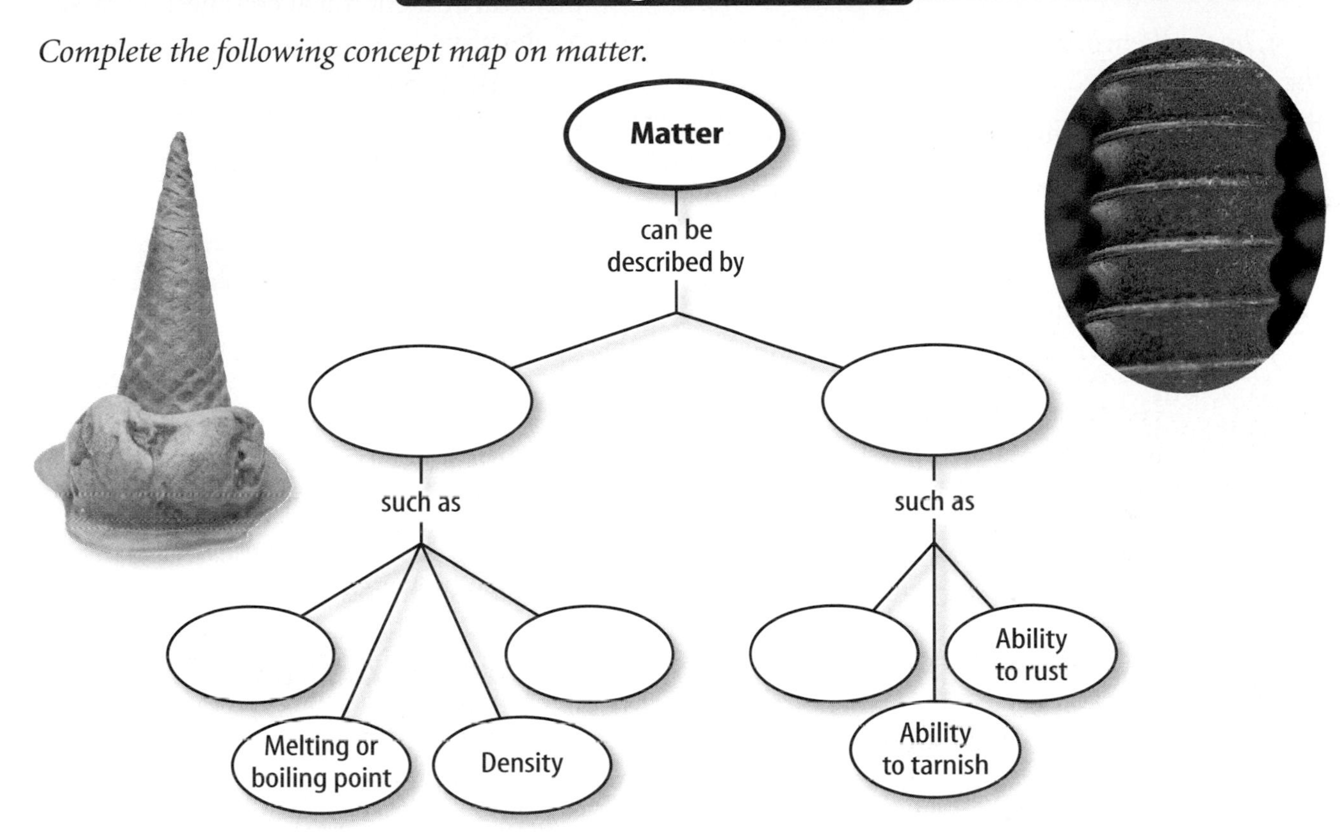

Vocabulary Review

Vocabulary Words

a. chemical change
b. chemical property
c. physical change
d. physical property
e. law of conservation of mass

Study Tip

Study the material you *don't* understand first. It's easy to review the material you know but it's harder to force yourself to learn the difficult concepts. Reread any topics that you find confusing. You should take notes as you read.

Using Vocabulary

Use what you know about the vocabulary words to answer the following questions. Use complete sentences.

1. Why is color a physical property?
2. What is a physical property that does not change with the amount of matter?
3. What happens during a physical change?
4. What type of change is a change of state?
5. What happens during a chemical change?
6. What are three clues that a chemical change has occurred?
7. What is an example of a chemical change?
8. What is the law of conservation of mass?

Chapter 19 Assessment

Checking Concepts

Choose the word or phrase that best answers the question.

1. What changes when the mass of an object increases while volume stays the same?
A) color **C)** density
B) length **D)** height

2. What is the volume of a brick that is 20 cm long, 10 cm wide, and 3 cm high?
A) 33 cm^3 **C)** 600 cm^3
B) 90 cm^3 **D)** 1,200 cm^3

3. What is the density of an object with a mass of 50 g and a volume of 5 cm^3?
A) 45 g/cm^3 **C)** 10 g/cm^3
B) 55 g/cm^3 **D)** 250 g/cm^3

4. What word best describes the type of materials that attract iron?
A) magnetic **C)** mass
B) chemical **D)** physical

5. Which is an example of a chemical property?
A) color **C)** density
B) mass **D)** ability to burn

6. Which is an example of a physical change?
A) metal rusting **C)** water boiling
B) silver tarnishing **D)** paper burning

7. What characteristic best describes what happens during a physical change?
A) composition changes
B) composition stays the same
C) form stays the same
D) mass is lost

8. Which is an example of a chemical change?
A) water freezes **C)** bread is baked
B) wood is carved **D)** wire is bent

9. Which is NOT a clue that could indicate a chemical change?
A) change in color **C)** change in energy
B) change in shape **D)** change in odor

10. What property stays the same during physical and chemical changes?
A) density
B) shape
C) mass
D) arrangement of particles

Thinking Critically

11. When asked to give the physical properties of a painting, your friend says the painting is beautiful. Why isn't this description a true scientific property?

12. The density of gold is 19.3 g/cm^3. How might you use the properties of matter to figure out which of these two samples is gold?

Mineral Samples

Sample	Mass	Volume
A	96.5 g	5 cm^3
B	38.6 g	4 cm^3

13. A jeweler bends gold into a beautiful ring. What type of change is this? Explain.

14. You are told that a sample of matter gives off energy as it changes. Can you conclude which type of change occurred? Why or why not?

15. What happens to mass during chemical and physical changes? Explain.

Developing Skills

16. Classifying Decide whether the following properties are physical or chemical.
a. Sugar can change into alcohol.
b. Iron can rust.
c. Alcohol can vaporize.
d. Paper can burn.
e. Sugar can dissolve.

17. Classifying Decide whether the following changes are physical or chemical.

a. Milk spoils.

b. Dynamite explodes.

c. Eggs and milk are stirred together.

d. Ice cream freezes.

18. Comparing and Contrasting Relate such human characteristics as hair and eye color and height and weight to physical properties of matter. Relate human behavior to chemical properties. Think about how you observe these properties.

19. Interpreting Scientific Illustrations Describe the changes shown below and explain why they are different.

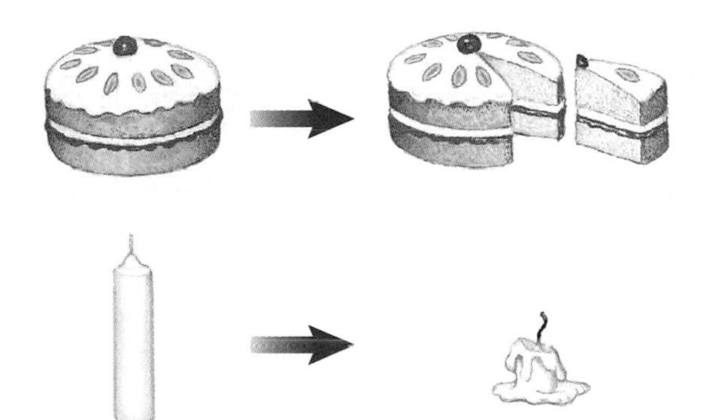

Performance Assessment

20. Write a Story Write a story describing an event that you have experienced. Then go back through the story and circle any physical or chemical properties you mentioned. Underline any physical or chemical changes you included.

TECHNOLOGY

Go to the Glencoe Science Web site at **science.glencoe.com** or use the **Glencoe Science CD-ROM** for additional chapter assessment.

Test Practice

Unknown matter can be identified by taking a sample of it and comparing its physical properties to those of already identified substances.

Physical Properties

Substance	Flash Point (°C)	Density (g/cm³)
Gasoline	– 45.6	0.720
Palmitic Acid	206.0	0.852
Aluminum	645.0	2.702
Methane	– 187.7	0.466

Study the table and answer the following questions.

1. A scientist has a sample of a substance with a density greater than 1 g/cm^3. According to the table, which substance might it be?

A) Gasoline **C)** Aluminum
B) Palmitic acid **D)** Methane

2. Some substances are very combustible and ignite by themselves if they are warm enough. The flashpoint of a substance is the lowest temperature at which it can catch fire spontaneously. If all the substances in the table have to be stored together, at which of the following temperatures should they be stored?

F) 100°C **H)** −100°C
G) 0°C **J)** −200°C

Reading Comprehension

Read the passage. Then read each question that follows the passage. Decide which is the best answer to each question.

Grouping the Elements Using their Properties

By 1860, scientists had discovered a total of 63 chemical elements. Dmitri Mendeleev, a Russian chemist, thought that there had to be some order among the elements.

He made a card for each element. On the card, he listed the physical and chemical properties of the element, such as atomic mass, density, color, and melting point. He also wrote each element's combining power, or its ability to form compounds with other elements.

When he arranged the cards in order of increasing atomic mass, Mendeleev noticed that the elements followed a periodic, or repeating, pattern. After every seven cards, the properties repeated. He placed each group of seven cards in rows, one row under another. He noticed that the columns in his chart formed groups of elements that had similar chemical and physical properties.

In a few places, Mendeleev had to move a card one space to the left or right to maintain the similarities of his groups. This left a few empty spaces. He predicted that they would be filled with elements that were unknown. He even predicted their properties. Fifteen years later, three new elements were discovered and placed in the empty spaces of the periodic table. Their physical and chemical properties agreed with Mendeleev's predictions.

Today there are more than 100 known elements. An extra column has been added for the noble gases, a group of elements that were not yet discovered in Mendeleev's time. Members of this group almost never combine with other elements. As new elements are discovered or are made <u>artificially</u>, scientists can place them in their proper place on the periodic table thanks to Mendeleev.

Test-Taking Tip To answer questions about a sequence of events, make a time line of what happened in each paragraph of the passage.

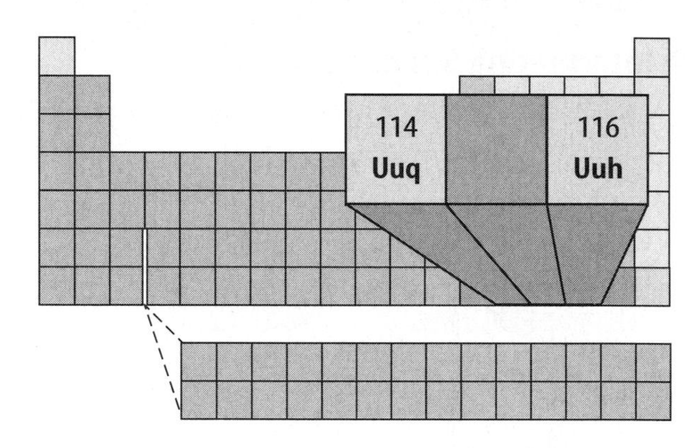

Even the modern periodic table has empty spaces for elements that have not been discovered yet.

1. Which of the following occurred FIRST in the passage?
   - **A)** Three new elements were discovered 15 years after Mendeleev developed the periodic table.
   - **B)** The noble gases were discovered and added to the periodic table.
   - **C)** Mendeleev predicted properties of unknown elements.
   - **D)** New elements were made in the laboratory and added to the periodic table.

2. The word <u>artificially</u> in this passage means _____.
   - **F)** unnaturally
   - **G)** artistically
   - **H)** atomically
   - **J)** radioactively

Standardized Test Practice

Reasoning and Skills

Read each question and choose the best answer.

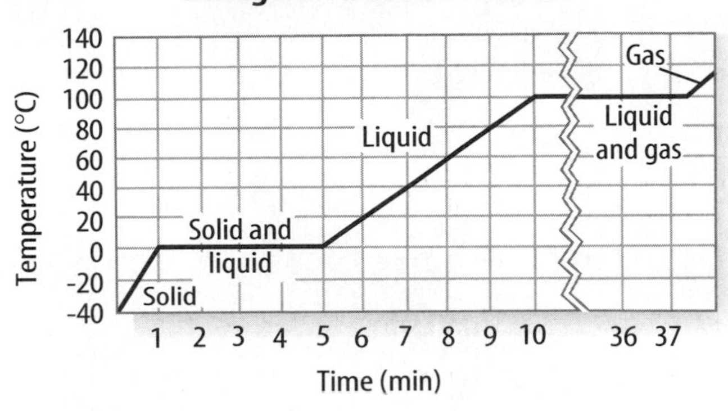

1. The graph shows the change in temperature that occurs as ice changes to water vapor. How much higher than the starting temperature is the boiling point?

A) 40°C
B) 100°C
C) 140°C
D) 180°C

Test-Taking Tip Boiling point is the flat section of the graph where liquid changes to gas.

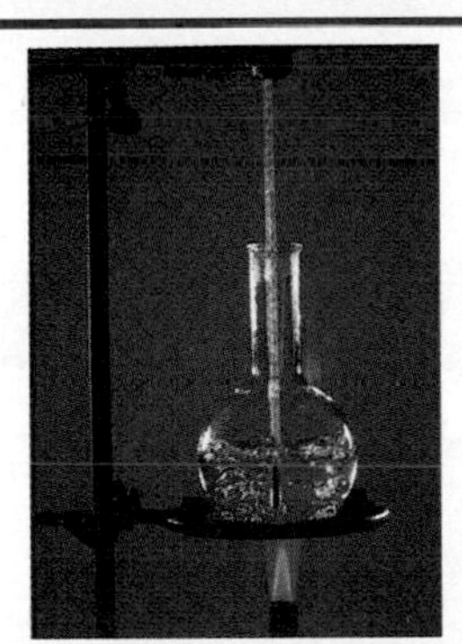

2. What is being measured in the illustration?

F) boiling point
G) melting point
H) density
J) flammability

Test-Taking Tip Think about what you would measure with a thermometer in a liquid that you are heating.

Properties of Selected Pure Substances

Substance	Melting Point (°C)	Boiling Point (°C)	Color
Aluminum	660.4	2,519	silver metallic
Argon	-189.2	-185.7	colorless
Mercury	-38.8	356.6	silver metallic
Water	0	100	colorless

3. Room temperature is about 20°C. In the table, which substance is a solid at room temperature?

A) aluminum
B) argon
C) mercury
D) water

Test-Taking Tip Remember that negative temperatures are below zero.

Read this question carefully before writing your answer on a separate sheet of paper.

4. The density of pure water is 1.00 g/cm^3. Ice floats on water, thus, the density of ice is *less* than that of water. Design an experiment to determine the density of an ice cube. List all the necessary steps.
(Volume = Length × Width × Height;
Density = Mass / Volume)

Test-Taking Tip Consider all the information provided in the question.

UNIT 7

Waves, Sound, and Light

How Are Radar & Popcorn Connected?

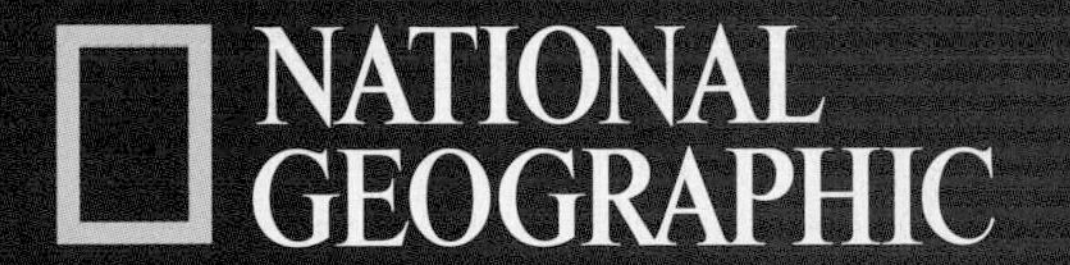

Radar systems—such as the one in this modern air traffic control room—use radio waves to detect objects. In the 1940s, the radio waves used for radar were generated by a device called a magnetron. One day, an engineer working on a radar project was standing near a magnetron when he noticed that the candy bar in his pocket had melted. Intrigued, the engineer got some unpopped popcorn and placed it next to the magnetron. Sure enough, the kernels began to pop. The engineer realized that the magnetron's short radio waves, called microwaves, caused the molecules in the food to move more quickly, increasing the food's temperature. Soon, magnetrons were being used in the first microwave ovens. Today, microwave ovens are used to pop popcorn—and heat many other kinds of food—in kitchens all over the world.

SCIENCE CONNECTION

ELECTROMAGNETIC RADIATION Microwaves and other kinds of radio waves are forms of electromagnetic radiation. So are X rays, gamma rays, infrared radiation, ultraviolet radiation, and visible light. These various forms of electromagnetic radiation differ in the length and frequency of their waves. Find out the wavelengths of the kinds of electromagnetic radiation named above. Then create a diagram that arranges these kinds of radiation in order from longest to shortest.

CHAPTER 20

Waves

On a breezy day in Maui, Hawaii, windsurfers ride the ocean waves. What forces are operating on the windsurfer and his sailboard? The wind catches the sails and helps propel the sailboard, but waves also are at work. Waves carry energy. You can see the ocean waves in this picture, but there are many kinds of waves you cannot see. Microwaves heat your food, radio waves transmit the music you listen to into your home, and sound waves carry that music from the radio to your ears. In this chapter, you will learn about different types of waves and how they behave.

What do you think?

Science Journal Look at the picture below with a classmate. Discuss what you think this might be. Hint: *Some sunglasses have this kind of lens.* Write your answer or best guess in your Science Journal.

It's a beautiful autumn day. You are sitting by a pond in a park. Music blares from a school marching band practicing for a big game. The music is carried by waves. A fish jumps, making a splash. Waves spread past a leaf that fell from a tree, causing the leaf to move. In the following activity, you'll observe how waves carry energy that can cause objects to move.

Observe wave behavior

1. Fill a large, clear plastic plate with 1 cm of water.
2. Use a dropper to release a single drop of water onto the water's surface. Repeat.
3. Float a cork or straw on the water.
4. When the water is still, repeat step 2 from a height of 10 cm, then again from 20 cm.

Observe

In your Science Journal, record your observations. How did the motion of the cork depend on the height of the dropper?

Before You Read

Making a Concept Map Study Fold **Make the following Foldable to organize information by diagramming ideas about waves.**

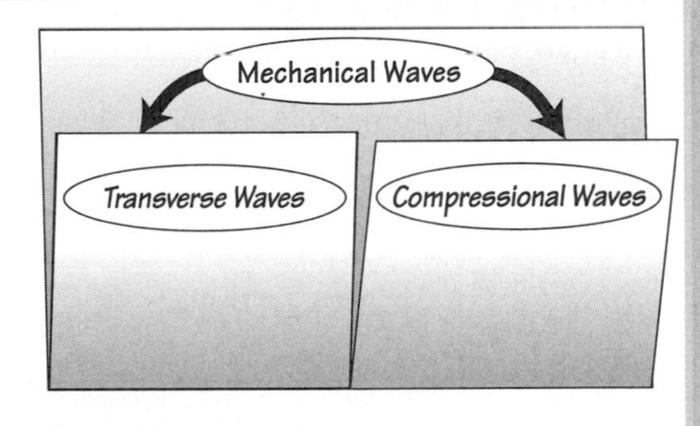

1. Place a sheet of paper in front of you so the long side is at the top. Fold the bottom of the paper to the top, stopping about four centimeters from the top.
2. Draw an oval above the fold. Write *Mechanical Waves* inside the oval.
3. Fold the paper in half from the left side to the right side and then unfold. Through the top thickness of the paper, cut along the fold line to form two tabs.
4. Draw an oval on each tab. Write *Transverse Waves* in one oval and *Compressional Waves* in the other, as shown. Draw arrows from the large oval to the smaller ovals.
5. As you read the chapter, write information about the two types of mechanical waves under the tabs.

SECTION 1

What are waves?

As You Read

***What* You'll Learn**

- **Explain** the relationship among waves, energy, and matter.
- **Describe** the difference between transverse waves and compressional waves.

Vocabulary
wave
mechanical wave
transverse wave
compressional wave
electromagnetic wave

***Why* It's Important**

You can hear music and other sounds because of waves.

What is a wave?

When you are relaxing on an air mattress in a pool and someone does a cannonball dive off the diving board, you suddenly find yourself bobbing up and down. You can make something move by giving it a push or pull, but the person jumping didn't touch your air mattress. How did the energy from the dive travel through the water and move your air mattress? The up-and-down motion was caused by the peaks and valleys of the ripples that moved from where the splash occurred. These peaks and valleys make up water waves.

Waves Carry Energy **Waves** are rhythmic disturbances that carry energy without carrying matter, as shown in **Figure 1A.** You can see the energy of the wave from a speedboat traveling outward, but the water only moves up and down. If you've ever felt a clap of thunder, you know that sound waves can carry large amounts of energy. You also transfer energy when you throw something to a friend, as in **Figure 1B.** However, there is a difference between a moving ball and a wave. A ball is made of matter, and when it is thrown, the matter moves from one place to another. So, unlike the wave, throwing a ball involves the transport of matter as well as energy.

Figure 1
The wave and the thrown ball carry energy in different ways.

A **The waves created by a boat move mostly up and down, but the energy travels outward from the boat.**

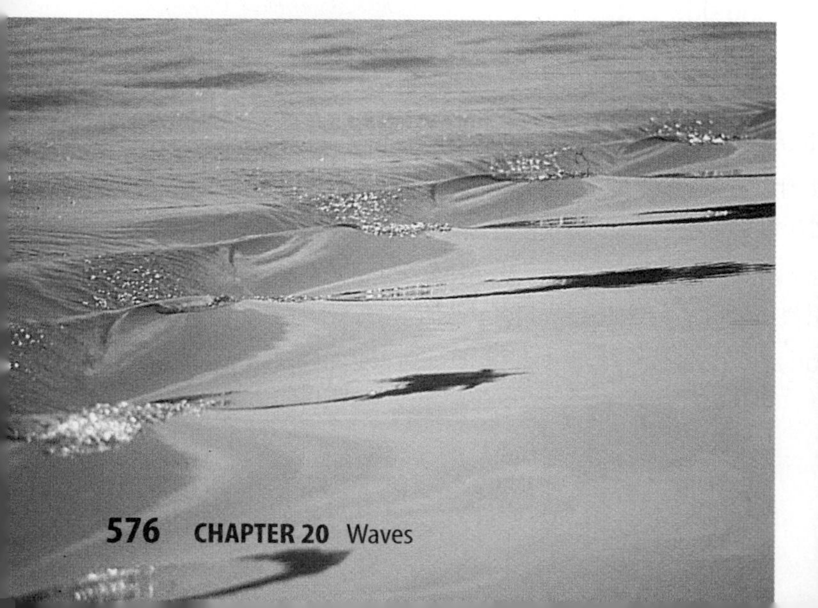

B **When the ball is thrown, the ball carries energy as it moves forward.**

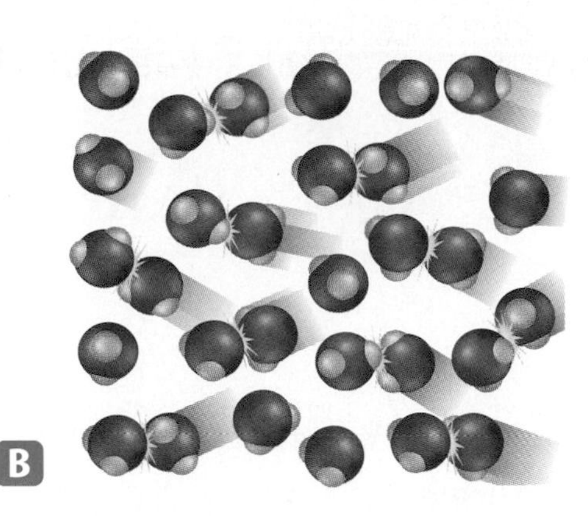

Figure 2
A As the students pass the ball, the students' positions do not change—only the position of the ball changes. B In a water wave, water molecules bump each other and pass energy from molecule to molecule.

A Model for Waves

How does a wave carry energy without transporting matter? Imagine a line of people, as shown in **Figure 2A.** The first person in line passes a ball to the second person, who passes the ball to the next person, and so on. Passing a ball down a line of people is a model for how waves can transport energy without transporting matter. Even though the ball has traveled, the people in line have not moved. In this model, you can think of the ball as representing energy. What do the people in line represent?

Think about the ripples on the surface of a pond. The energy carried by the ripples travels through the water. The water is made up of water molecules. It is the individual molecules of water that pass the wave energy, just as the people in **Figure 2A** pass the ball. The water molecules transport the energy in a water wave by colliding with the molecules around them, as shown in **Figure 2B.**

Reading Check *What is carried by waves?*

Mechanical Waves

In the wave model, the ball could not be transferred if the line of people didn't exist. The energy of a water wave could not be transferred if no water molecules existed. These types of waves, which use matter to transfer energy, are called **mechanical waves.** The matter through which a mechanical wave travels is called a medium. For ripples on a pond, the medium is the water.

A mechanical wave travels as energy is transferred from particle to particle in the medium. For example, a sound wave is a mechanical wave that can travel through air, as well as solids, liquids, and other gases. The sound wave travels through air by transferring energy from gas molecule to gas molecule. Without a medium such as air, you would not hear sounds. In outer space sound waves can't travel because there is no air.

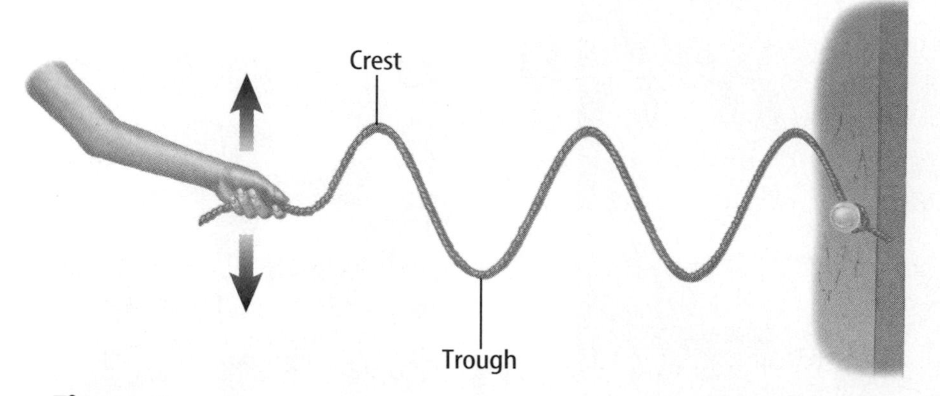

Figure 3
The high points on the wave are called crests and the low points are called troughs.

Transverse Waves In a mechanical **transverse wave,** the wave energy causes the matter in the medium to move up and down or back and forth at right angles to the direction the wave travels. You can make a model of a transverse wave. Stretch a long rope out on the ground. Hold one end in your hand. Now shake the end in your hand back and forth. By adjusting the way you shake the rope, you can create a wave that seems to slide along the rope.

When you first started shaking the rope, it might have appeared that the rope itself was moving away from you. But it was only the wave that was moving away from your hand. The wave energy moves through the rope, but the matter in the rope doesn't travel. You can see that the wave has peaks and valleys at regular intervals. As shown in **Figure 3,** the high points of transverse waves are called crests. The low points are called troughs.

Reading Check ***What are the highest points of transverse waves called?***

Figure 4
A compressional wave can travel through a coiled spring toy.

A **As the wave motion begins, the coils near the string are close together and the other coils are far apart.**

B **The wave, seen in the squeezed and stretched coils, travels along the spring.**

C **The string and coils did not travel with the wave. Each coil moved forward and then back to its original position.**

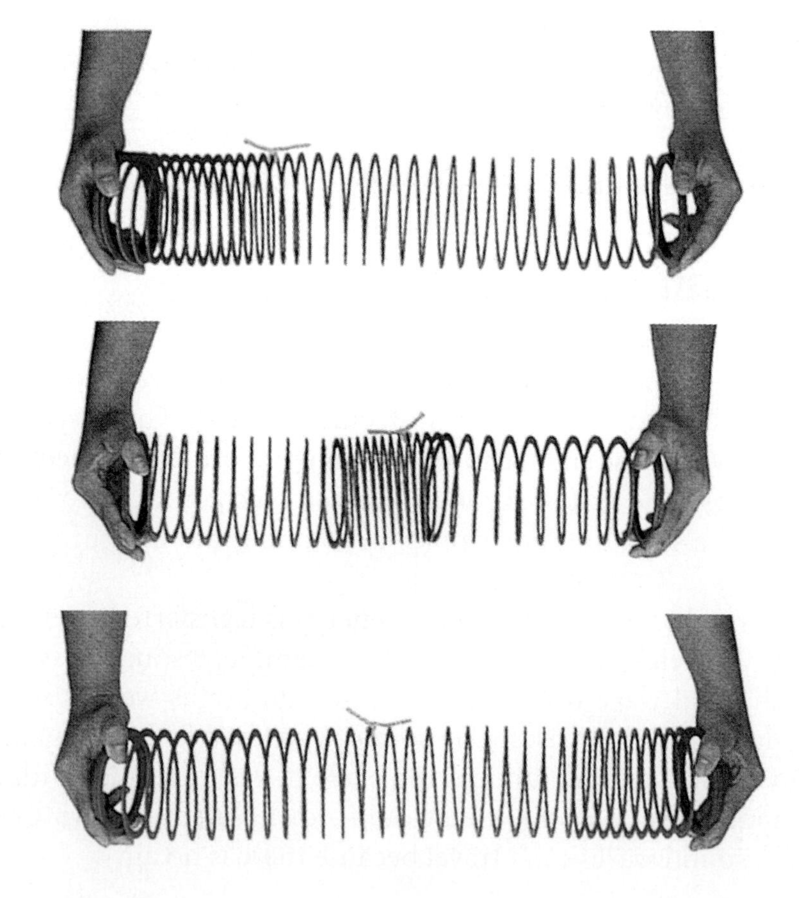

Compressional Waves Mechanical waves can be either transverse or compressional. In a **compressional wave,** matter in the medium moves forward and backward in the same direction that the wave travels. You can make a compressional wave by squeezing together and releasing several coils of a coiled spring toy, as shown in **Figure 4.**

You see that the coils move only as the wave passes. They then return to their original position. So, like transverse waves, compressional waves carry only energy forward along the spring. In this example, the spring is the medium the wave moves through, but the spring does not move along with the wave.

Sound Waves Sound waves are compressional waves. How do you make sound waves when you talk or sing? If you hold your fingers against your throat while you hum, you can feel vibrations. These vibrations are the movements of your vocal cords. If you touch a stereo speaker while it's playing, you can feel it vibrating, too. All waves are produced by something that is vibrating.

Making Sound Waves

How do vibrating objects make sound waves? Look at the drum shown in **Figure 5.** When you hit the drumhead it starts vibrating up and down. As the drumhead moves upward, the molecules next to it are pushed closer together. This group of molecules that are closer together is a compression. As the compression is formed, it moves away from the drumhead, just as the squeezed coils move along the coiled spring toy in **Figure 4.**

When the drumhead moves downward, the molecules near it have more room and can spread farther apart. This group of molecules that are farther apart is a rarefaction (rar uh FAK shun). The rarefaction also moves away from the drumhead. As the drumhead vibrates up and down, it forms a series of compressions and rarefactions that move away and spread out in all directions. This series of compressions and rarefactions is a sound wave.

Comparing Sounds

Procedure

1. Hold a **wooden ruler** firmly on the edge of your **desk** so that most of it extends off the edge of the desk.
2. Pluck the free end of the ruler so that it vibrates up and down. Use gentle motion at first, then pluck with more energy.
3. Repeat step 2, moving the ruler about 1 cm further onto the desk each time until only about 5 cm extend off the edge.

Analysis

1. Compare the loudness of the sounds that are made by plucking the ruler in different ways.
2. Describe the differences in the sound as the end of the ruler extended farther from the desk.

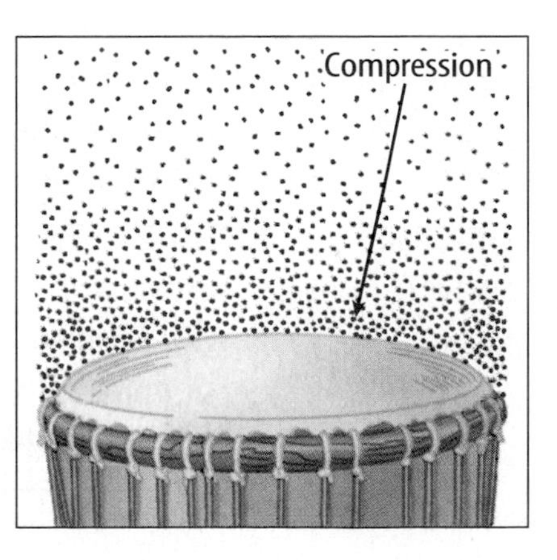

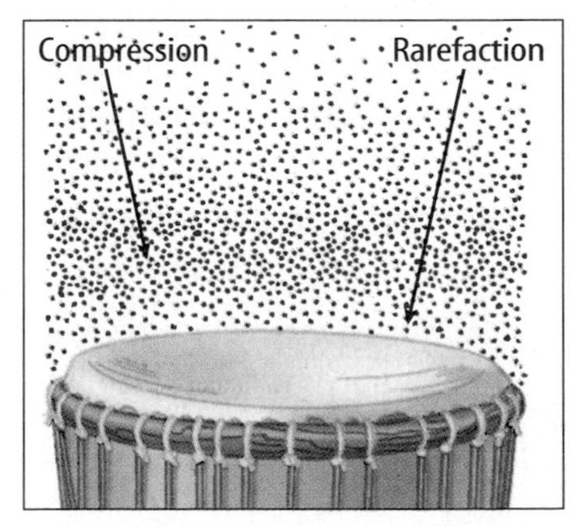

Figure 5
A vibrating drumhead makes compressions and rarefactions in the air. *How are compressions and rarefactions different?*

Electromagnetic Waves

When you listen to the radio, watch TV, or use a microwave oven to cook, you use a different kind of wave—one that doesn't need matter as a medium.

Waves that do not require matter to carry energy are called **electromagnetic waves.** Electromagnetic waves are transverse waves that are produced by the motion of electrically charged particles. Just like mechanical waves, electromagnetic waves also can travel through a medium such as a solid, liquid, or gas. Radio waves are electromagnetic waves that travel through the air from a radio station, and then through the solid walls of your house to reach your radio. However, unlike mechanical waves, electromagnetic waves can travel through outer space or through a vacuum where no matter exists.

Physics INTEGRATION

Maybe you've used a global positioning system (GPS) receiver to determine your location while driving, boating, or hiking. Earth-orbiting satellites send electromagnetic radio waves that transmit their exact locations and times of transmission. The GPS receiver uses information from four of these satellites to determine your location to within about 16 m.

Useful Waves In space, which has no air or any other medium, orbiting satellites beam radio waves to TVs, radios, and cellular phones on Earth's surface. However, radio waves are not the only electromagnetic waves traveling in space. Infrared, visible, and ultraviolet waves travel from the Sun through space before they reach Earth's atmosphere. Infrared waves feel warm when they strike your skin. Without visible light you wouldn't see color or be able to read this page. You use sunscreen to protect yourself from ultraviolet rays. Other useful electromagnetic waves include X rays. X rays are useful not only in medical applications, but also for security checks in airports as luggage is scanned.

Section  Assessment

1. Describe the movement of a floating object on a pond when struck by a wave.
2. Why can't a sound wave travel from a satellite to Earth?
3. Give one example of a transverse wave and one example of a compressional wave. How are they similar and different?
4. What is the difference between a mechanical wave and an electromagnetic wave?
5. **Think Critically** How is it possible for a sound wave to transmit energy but not matter?

Skill Builder Activities

6. **Concept Mapping** Create a concept map that shows the relationships among the following: *waves, mechanical waves, electromagnetic waves, compressional waves,* and *transverse waves.* **For more help, refer to the Science Skill Handbook.**
7. **Using a Word Processor** Use word-processing software to write short descriptions of the waves you encounter during a typical day. **For more help, refer to the Technology Skill Handbook.**

Wave Properties

Amplitude

Can you describe a wave? For a water wave, one way might be to tell how high the wave rises above, or falls below, the normal level. This distance is called the wave's amplitude. The **amplitude** of a transverse wave is one-half the distance between a crest and a trough, as shown in **Figure 6A.** In a compressional wave, the amplitude is greater when the particles of the medium are squeezed closer together in each compression and spread farther apart in each rarefaction.

Amplitude and Energy A wave's amplitude is related to the energy that the wave carries. For example, the electromagnetic waves that make up bright light have greater amplitudes than the waves that make up dim light. Waves of bright light carry more energy than the waves that make up dim light. In a similar way, loud sound waves have greater amplitudes than soft sound waves. Loud sounds carry more energy than soft sounds. If a sound is loud enough, it can carry enough energy to damage your hearing.

As you can see in **Figure 6B,** when a hurricane strikes a coastal area, the resulting water waves can damage almost anything that stands in their path. The large waves caused by a hurricane carry more energy than the small waves or ripples on a pond.

As You Read

What You'll Learn

- **Describe** the relationship between the frequency and wavelength of a wave.
- **Explain** why waves travel at different speeds.

Vocabulary
amplitude
wavelength
frequency

Why It's Important
The energy carried by a wave depends on its amplitude.

Figure 6
A transverse wave has an amplitude.

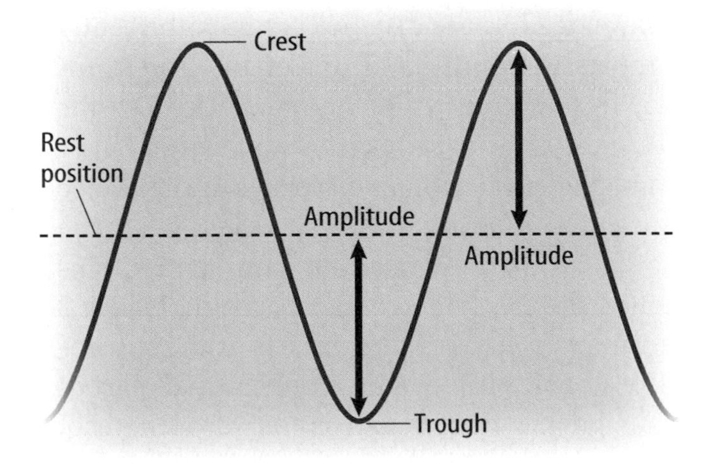

A The amplitude is a measure of how high the crests are or how deep the troughs are.

B A water wave of large amplitude carried the energy that caused this damage.

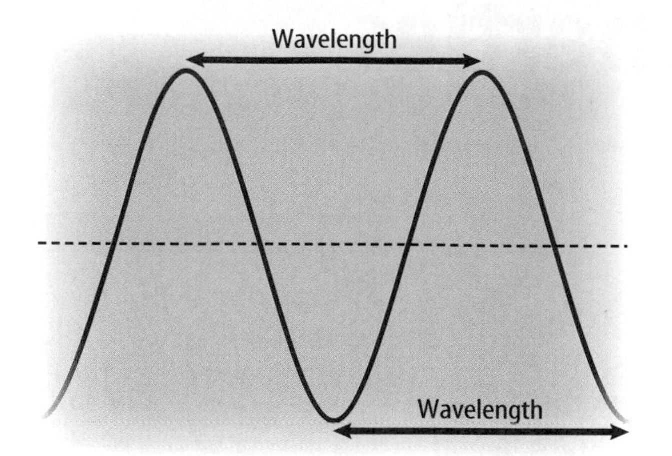

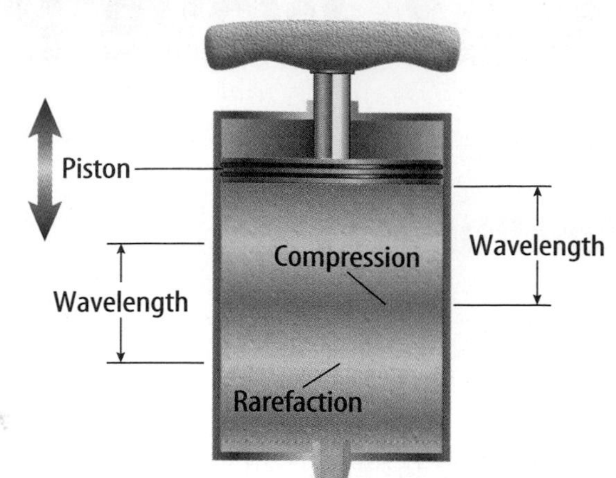

Figure 7
Wavelength is measured differently for transverse and compressional waves.

Figure 8
The wavelengths and frequencies of electromagnetic waves vary.

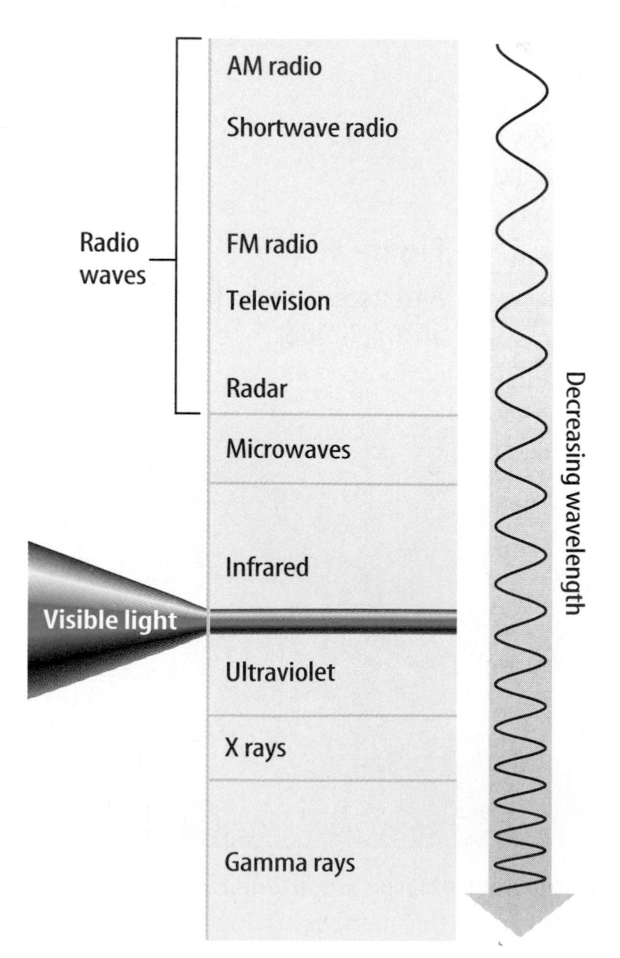

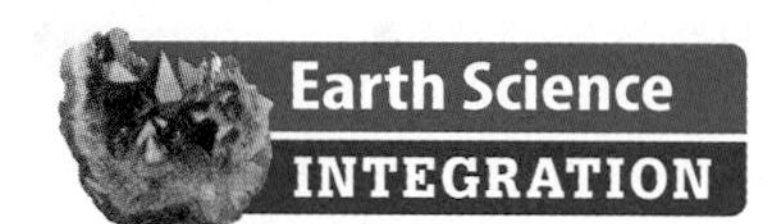

The devastating effect that a wave with large amplitude can have is seen in the aftermath of tsunamis. Tsunamis are huge sea waves that are caused by underwater earthquakes along faults on the seafloor. The movement of the seafloor along the fault produces the wave. As the wave moves toward shallow water and slows down, the amplitude of the wave grows. The tremendous amounts of energy tsunamis carry cause great damage when they move ashore.

Wavelength

Another way to describe a wave is by its wavelength. For a transverse wave, **wavelength** is the distance from the top of one crest to the top of the next crest, or from the bottom of one trough to the bottom of the next trough, as shown in **Figure 7A.** For a compressional wave, the wavelength is the distance between the center of one compression and the center of the next compression, or from the center of one rarefaction to the center of the next rarefaction, as shown in **Figure 7B.**

Electromagnetic waves have wavelengths that range from kilometers, for radio waves, to less than the diameter of an atom, for X rays and gamma rays. This range is called the electromagnetic spectrum. **Figure 8** shows the names given to different parts of the electromagnetic spectrum. Visible light is only a small part of the electromagnetic spectrum. It is the wavelength of visible light waves that determines their color. For example, the wavelength of red light waves is longer than the wavelength of green light waves.

Frequency

The **frequency** of a wave is the number of wavelengths that pass a given point in 1 s. The unit of frequency is the number of wavelengths per second, or hertz (Hz). Recall that waves are produced by something that vibrates. The faster the vibration is, the higher the frequency is of the wave that is produced.

✔ Reading Check *How is the frequency of a wave measured?*

A Sidewalk Model For waves that travel with the same speed, frequency and wavelength are related. To model this relationship, imagine people on two parallel moving sidewalks in an airport, as shown in **Figure 9.** One sidewalk has four travelers spaced 4 m apart. The other sidewalk has 16 travelers spaced 1 m apart.

Now imagine that both sidewalks are moving at the same speed and approaching a pillar between them. On which sidewalk will more people go past the pillar? On the sidewalk with the shorter distance between people, four people will pass the pillar for each one person on the other sidewalk. When four people pass the pillar on the first sidewalk, 16 people pass the pillar on the second sidewalk.

Figure 9
When people are farther apart on a moving sidewalk, fewer people pass the pillar every minute.

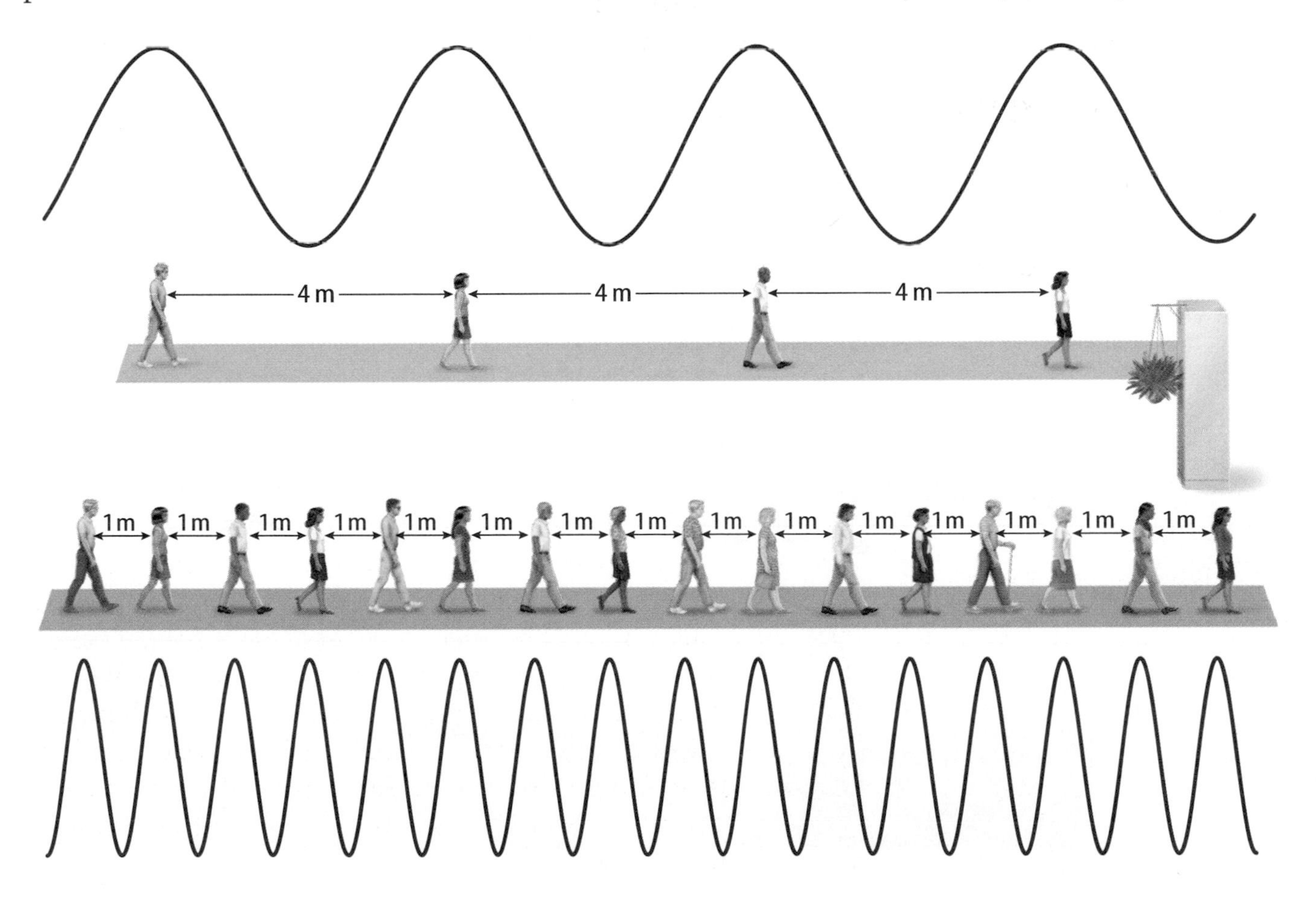

Health INTEGRATION

Sound waves with ultra-high frequencies cannot be heard by the human ear, but are used by medical professionals in several ways. They perform echocardiograms of the heart, produce ultrasound images of internal organs, break up blockages in arteries and kill bacteria and sterilize surgical instruments. *How do the wavelengths of these sound waves compare to sound waves you can hear?*

Frequency and Wavelength Suppose that each person in **Figure 9** represents the crest of a wave. Then the movement of people on the first sidewalk is like a wave with a wavelength of 4 m. For the second sidewalk, the wavelength would be 1 m. On the first sidewalk, where the wavelength is longer, the people pass the pillar *less* frequently. Longer wavelengths result in smaller frequencies. On the second sidewalk, where the wavelength is shorter, the people pass the pillar *more* frequently. Higher frequencies result in shorter wavelengths. This is true for all waves that travel at the same speed. As the frequency of a wave increases, its wavelength decreases.

Color and Pitch Because frequency and wavelength are related, either the wavelength or frequency of a light wave determines the color of the light. For example, blue light has a larger frequency and shorter wavelength than red light.

In a sound wave, either the wavelength or frequency determines the pitch. Pitch is the highness or lowness of a sound. A flute makes musical notes with a high pitch and produces sounds of high frequency. A tuba produces notes with a low pitch and a low frequency. When you sing a musical scale, the pitch and frequency increase from note to note. Wavelength and frequency are also related for sound waves traveling in air. As the frequency of sound waves increases, their wavelength decreases. **Figure 10** shows how the frequency and wavelength change for notes on the musical scale.

Figure 10
The frequency of the notes on a musical scale increases as the notes get higher in pitch, but the wavelength of the notes decreases.

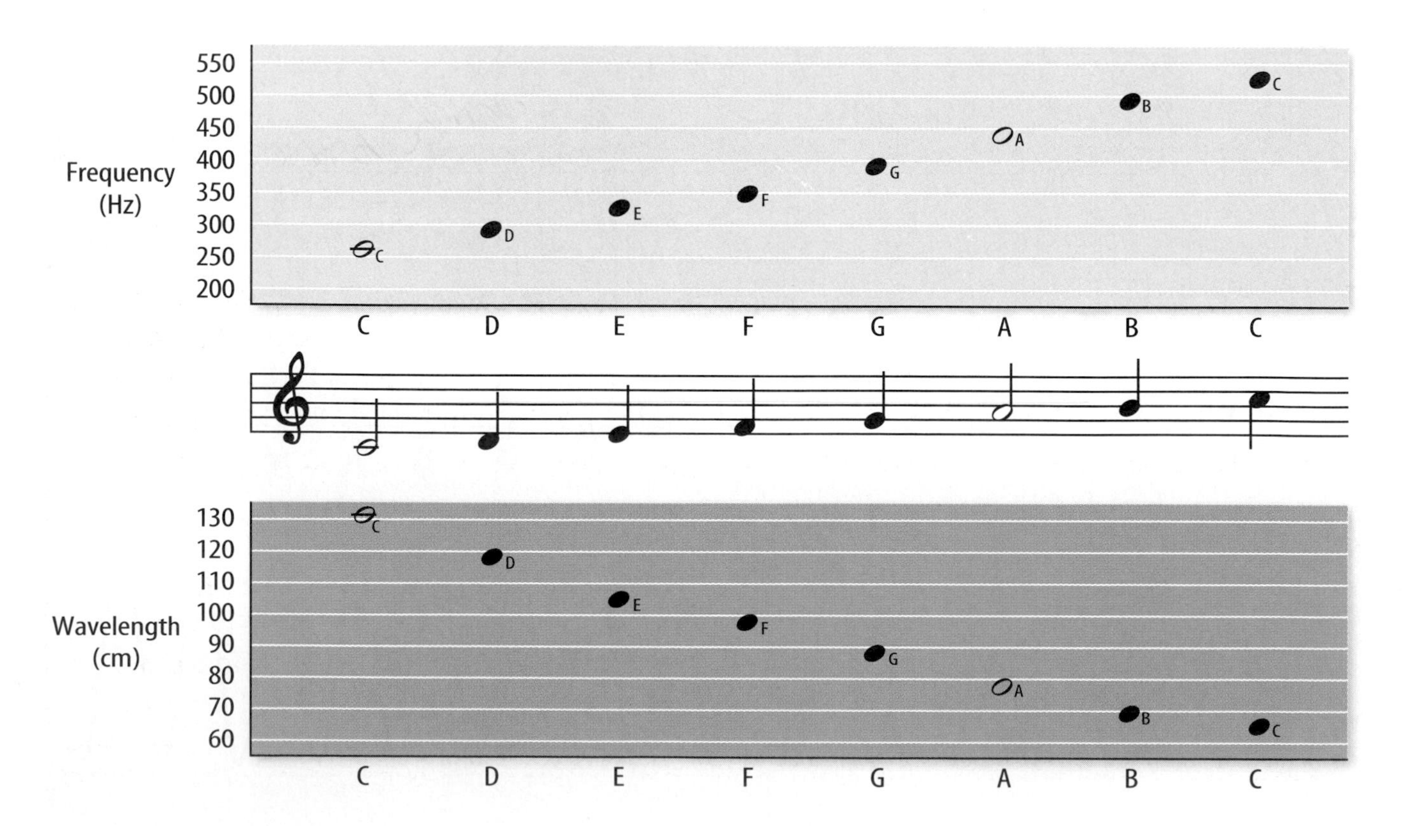

Wave Speed

You've probably watched a distant thunderstorm approach on a hot summer day. You see a bolt of lightning flash between a dark cloud and the ground. Do the sound waves, or thunder, produced by the lightning bolt reach your ears at the same instant you see the lightning? If the thunderstorm is many kilometers away, several seconds will pass between when you see the lightning and when you hear the thunder. This happens because light travels much faster in air than sound does. Light is an electromagnetic wave that travels through air at about 300 million m/s. On the other hand, sound is a mechanical wave that travels through air at about 340 m/s.

Mechanical waves such as sound usually travel faster in a medium in which the atoms that make up the medium are closer together. Sound travels faster in solids than in liquids and faster in liquids than in gases. This is because atoms are closer to each other in a solid than in a liquid, and closer together in a liquid than in a gas.

Electromagnetic waves such as light behave differently than mechanical waves. Unlike mechanical waves, they travel faster in gases than in solids or liquids. You know that you can get to your next class faster if the hallways are nearly empty than if they are filled with other students. Electromagnetic waves behave the same way. If many atoms are in the medium, electromagnetic waves are slowed down. For example, the speed of light is one and a half times faster in air than it is in glass.

SCIENCE *Online*

Research Visit the Glencoe Science Web site at **science.glencoe.com** for information about wave speed in different materials. Make a graph to show the differences.

Section  Assessment

1. How does the frequency of a wave change as its wavelength changes?
2. Why is a sound wave with a large amplitude more likely to damage your hearing than one with a small amplitude?
3. What accounts for the time difference in seeing and hearing a fireworks display?
4. Why is the statement "The speed of light is 300 million m/s" not always correct?
5. **Think Critically** Explain the differences between the waves that make up bright, green light and dim, red light.

Skill Builder Activities

6. **Predicting** A biologist studying bison puts her ear next to the ground. By doing this she knows that the herd is coming toward her. Explain. **For more help, refer to the Science Skill Handbook.**
7. **Solving One-Step Equations** The product of the wavelength and the frequency of a wave is the speed of the wave. If a sound wave traveling through water has a speed of 1,470 m/s and a frequency of 2,340 Hz, what is its wavelength? **For more help, refer to the Math Skill Handbook.**

Activity

Waves on a Spring

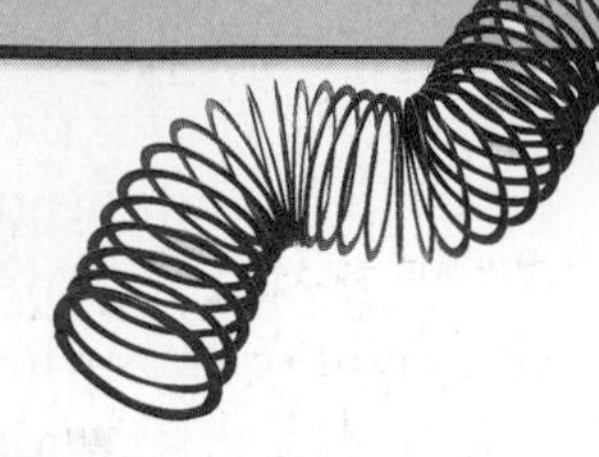

Waves are rhythmic disturbances that carry energy through matter or space. Studying waves can help you understand how the Sun's energy reaches Earth and sounds travel through the air.

What You'll Investigate

What are some of the properties of transverse and compressional waves on a coiled spring?

Materials

long, coiled spring toy
colored yarn (5 cm)
meterstick
stopwatch

Goals

- **Create** transverse and compressional waves on a coiled spring toy.
- **Investigate** wave properties such as speed and amplitude.

Safety Precautions

WARNING: *Avoid overstretching or tangling the spring to prevent injury or damage.*

Procedure

1. **Prepare** a data table such as the one shown.

Wave Data	
Length of stretched spring toy	
Average time for a wave to travel from end to end—step 4	
Average time for a wave to travel from end to end—step 5	

2. Work in pairs or groups and clear a place on an uncarpeted floor about 6 m × 2 m.
3. Stretch the springs between two people to the length suggested by your teacher. Measure the length.
4. Create a wave with a quick, sideways snap of the wrist. Time several waves as they travel the length of the spring. Record the average time in your data table.
5. Repeat step 4 using waves that have slightly larger amplitudes.
6. Squeeze together about 20 of the coils. Observe what happens to the unsqueezed coils. Release the coils and observe.
7. Quickly push the spring toward your partner, then pull it back.
8. Tie the yarn to a coil near the middle of the spring. Repeat step 7, observing the string.

Conclude and Apply

1. **Classify** the wave pulses you created in each step as compressional or transverse.
2. **Calculate** and compare the speeds of the waves in steps 4 and 5.
3. **Classify** the unsqueezed coils in step 6 as a compression or a rarefaction.
4. **Compare and contrast** the motion of the yarn in step 8 with the motion of the wave.

Communicating Your Data

Write a summary paragraph of how this activity demonstrated any of the vocabulary words from the first two sections of the chapter. **For more help, refer to the Science Skill Handbook.**

SECTION

Wave Behavior

Reflection

What causes the echo when you yell across an empty gymnasium or down a long, empty hallway? Why can you see your face when you look in a mirror? The echo of your voice and the face you see in the mirror are caused by wave reflection.

Reflection occurs when a wave strikes an object or surface and bounces off. An echo is reflected sound. Sound reflects from all surfaces. Your echo bounces off the walls, floor, ceiling, furniture, and people. You see your face in a mirror or a still pond, as shown in **Figure 11A,** because of reflection. Light waves produced by a source of light such as the Sun or a lightbulb bounce off your face, strike the mirror, and reflect back to your eyes.

When a surface is smooth and even, the reflected image is clear and sharp. However, when light reflects from an uneven or rough surface, you can't see a sharp image because the reflected light scatters in many different directions, as shown in **Figure 11B.**

Reading Check *What causes reflection?*

As You Read

What You'll Learn

- **Explain** how waves can reflect from some surfaces.
- **Explain** how waves change direction when they move from one material into another.
- **Describe** how waves are able to bend around barriers.

Vocabulary

reflection
refraction
diffraction
interference

Why It's Important

The reflection of waves enables you to see objects around you.

A The smooth surface of a still pond enables you to see a sharp, clear image of yourself.

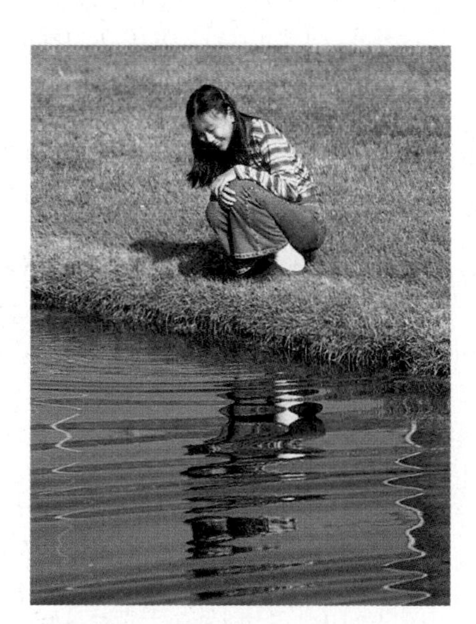

B If the surface of the pond is rough and uneven, your reflected image is no longer clear and sharp.

Figure 11
The image formed by reflection depends on the smoothness of the surface.

TRY AT HOME Mini LAB

Observing How Light Refracts

Procedure

1. Fill a **large, opaque drinking glass or cup** with **water.**
2. Place a **white soda straw** in the water at an angle.
3. Looking directly down into the cup from above, observe the straw where it meets the water.
4. Placing yourself so that the straw angles to your left or right, slowly back away about 1 m. Observe the straw as it appears above, at, and below the surface of the water.

Analysis

1. Describe the straw's appearance from above.
2. Compare the straw's appearance above and below the water's surface in step 4.

Refraction

A wave changes direction when it reflects from a surface. Waves also can change direction in another way. Perhaps you have tried to grab a sinking object when you are in a swimming pool, only to come up empty-handed. Yet you were sure you grabbed right where you saw the object. You missed grabbing the object because the light rays from the object changed direction as they passed from the water into the air. The bending of a wave as it moves from one medium into another is called **refraction.**

Refraction and Wave Speed Remember that the speed of a wave can be different in different materials. For example, light waves travel faster in air than in water. Refraction occurs when the speed of a wave changes as it passes from one substance to another, as shown in **Figure 12.** A line that is perpendicular to the water's surface is called the normal. When a light ray passes from air into water, it slows down and bends toward the normal. When the ray passes from water into air, it speeds up and bends away from the normal. The larger the change in speed of the light wave is, the larger the change in direction is.

You notice refraction when you look down into a fishbowl. Refraction makes the fish appear to be closer to the surface but farther away from you than it is, as shown in **Figure 13.** Light rays reflected from the fish are bent away from the normal as they pass from water to air. Your brain interprets the light that enters your eyes by assuming that light rays always travel in straight lines. As a result, the light rays seem to be coming from a fish that is in a different location.

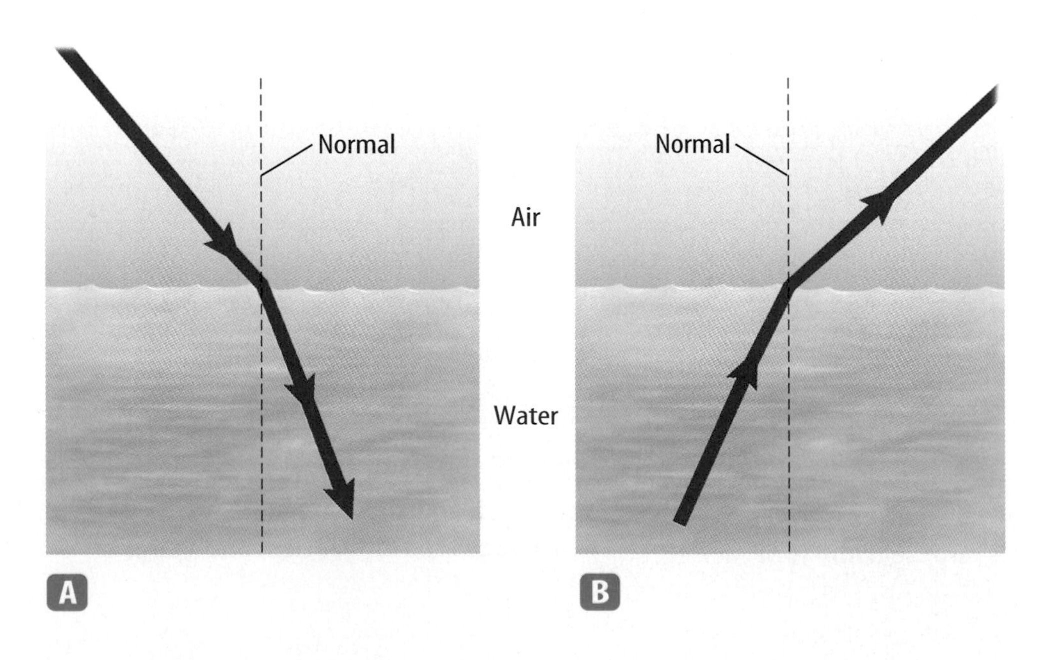

Figure 12
A wave is refracted when it changes speed. **A** As the light ray passes from air to water, it refracts toward the normal. **B** As the light ray passes from water to air, it refracts away from the normal.

Color from Refraction Refraction causes prisms to separate sunlight into many colors and produces rainbows too. **Figure 14** illustrates how refraction and reflection produce a rainbow when light waves from the Sun pass into and out of water droplets in the air.

Reading Check *What produces a rainbow?*

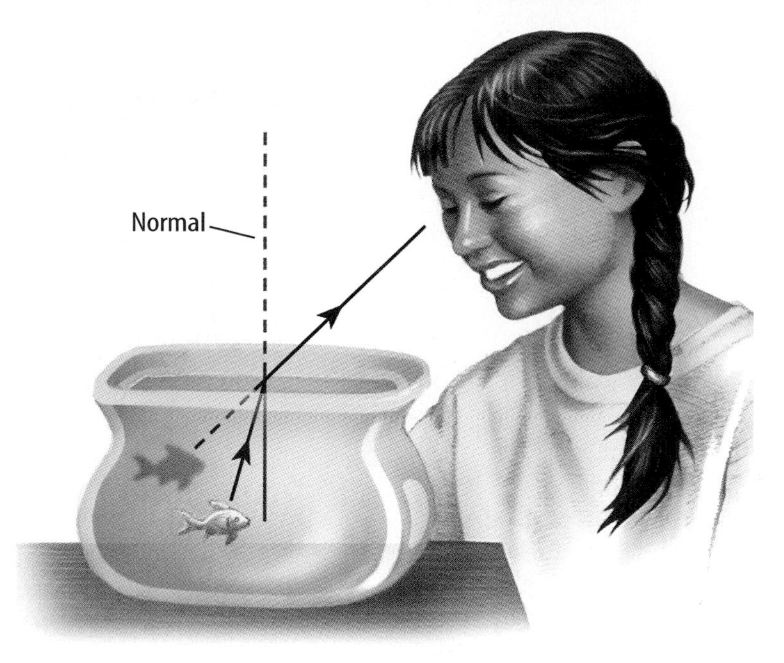

Figure 13
When you look at the goldfish in the water, the fish is in a different position than it appears.

Diffraction

Why can you hear music from the band room when you are down the hall? You can hear the music because the sound waves bend as they pass through an open doorway. This bending isn't caused by refraction. Remember that refraction occurs when waves change speed, but sound waves have the same speed in the band room and in the hallway. Instead, the bending is caused by diffraction. **Diffraction** is the bending of waves around a barrier.

Diffraction of Light Waves Can light waves diffract, too? You can hear your friends in the band room but you can't see them until you reach the open door. Therefore, you know that light waves do not diffract as much as sound waves do.

Are light waves able to diffract at all? Light waves do bend around the edges of an open door. However, for an opening as wide as a door, the amount the light bends is extremely small. As a result, the diffraction of light is far too small to allow you to see around a corner.

Figure 14
Light rays refract as they enter and leave each water drop. Each color refracts at different angles because of their different wavelengths, so they separate into the colors of the visible spectrum.

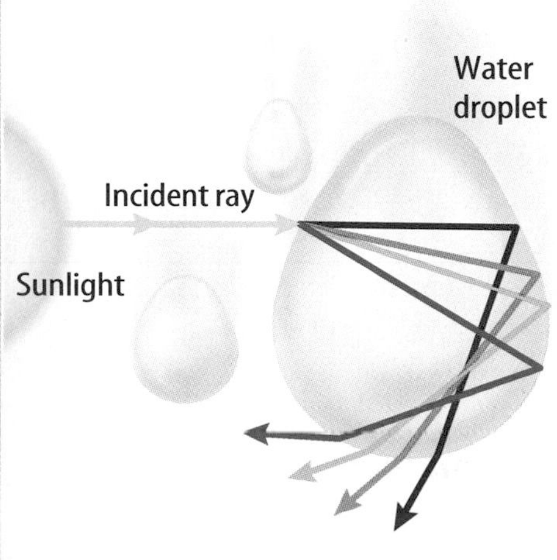

Diffraction and Wavelength The reason that light waves don't diffract much when they pass through an open door is that the wavelengths of visible light are much smaller than the width of the door. Light waves have wavelengths between about 400 and 700 billionths of a meter, while the width of doorway is about one meter. Sound waves that you can hear have wavelengths between a few millimeters and about 10 m. They bend more easily around the corners of an open door. A wave is diffracted more when its wavelength is similar in size to the barrier or opening.

Diffraction of Water Waves Perhaps you have noticed water waves bending around barriers. For example, when water waves strike obstacles such as the islands shown in **Figure 15,** they don't stop moving. Here the size and spacing of the islands is not too different from the wavelength of the water waves. So the water waves bend around the islands, and keep on moving. They also spread out after they pass through openings between the islands. If the islands were much larger than the water wavelength, less diffraction would occur.

What happens when waves meet?

Suppose you throw two pebbles into a still pond. Ripples spread from the impact of each pebble and travel toward each other. What happens when two of these ripples meet? Do they collide like billiard balls and change direction? Waves behave differently from billiard balls when they meet. Waves pass right through each other and continue moving as though the other waves never existed.

Figure 15
Water waves bend or diffract around these islands. More diffraction occurs when the object is closer in size to the wavelength.

Wave Interference While two waves overlap a new wave is formed by adding the two waves together. The ability of two waves to combine and form a new wave when they overlap is called **interference.** After they overlap, the individual waves continue to travel on in their original form.

The different ways waves can interfere are shown in **Figure 16** on the next page. Sometimes when the waves meet, the crest of one wave overlaps the crest of another wave. This is called constructive interference. The amplitudes of these combining waves add together to make a larger wave while they overlap. Destructive interference occurs when the crest of one wave overlaps the trough of another wave. Then the amplitudes of the two waves combine to make a wave with a smaller amplitude. If the two waves have equal amplitudes and meet crest to trough, they cancel each other while the waves overlap.

Waves and Particles Like waves of water, when light travels through a small opening, such as a narrow slit, the light spreads out in all directions on the other side of the slit. If small particles, instead of waves, were sent through the slit, they would continue in a straight line without spreading. The spreading, or diffraction, is only a property of waves. Interference also doesn't occur with particles. If waves meet, they reinforce or cancel each other, then travel on. If particles approach each other, they either collide and scatter or miss each other completely. Interference, like diffraction, is a property of waves, not particles.

SCIENCE Online

Research Visit the Glencoe Science Web site at **science.glencoe.com** for more information about wave interference.

Problem-Solving Activity

Can you create destructive interference?

Your brother is vacuuming and you can't hear the television. Is it possible to diminish the sound of the vacuum so you can hear the TV? Can you eliminate unpleasant sounds and keep the sounds you do want to hear?

Identifying the Problem

It is possible to create a frequency that will destructively interfere with the sound of the vacuum and not the television. The graph shows the waves created by the vacuum and the television.

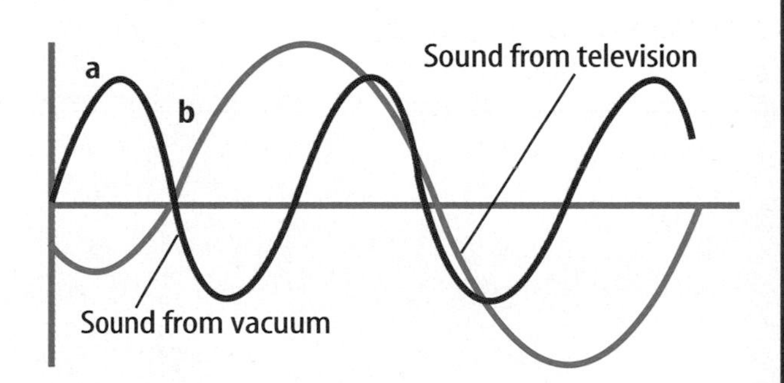

Solving the Problem

1. Can you create the graph of a wave that will eliminate the noise from the vacuum but not the television?
2. Can you create the graph of a wave that would amplify the sound of the television?

NATIONAL GEOGRAPHIC VISUALIZING INTERFERENCE

Figure 16

Whether they are ripples on a pond or huge ocean swells, when water waves meet they can combine to form new waves in a process called interference. As shown below, wave interference can be constructive or destructive.

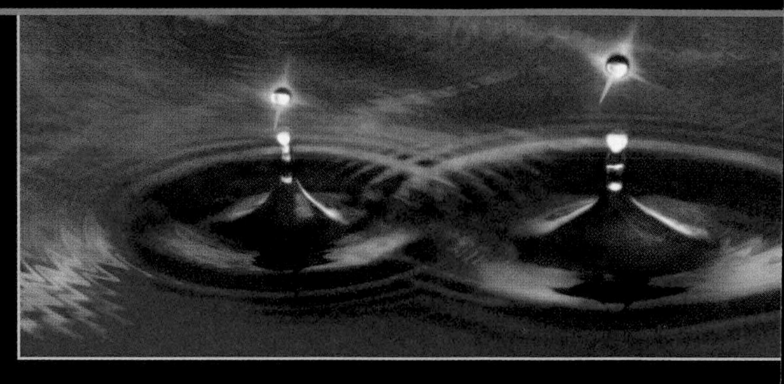

Constructive Interference

In constructive interference, a wave with greater amplitude is formed.

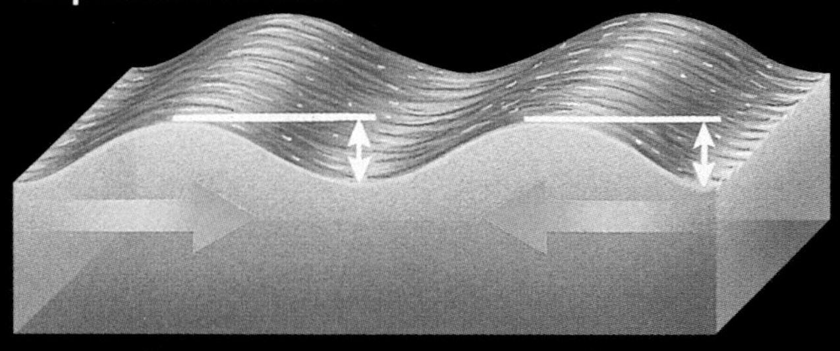

A B

The crests of two waves—A and B—approach each other.

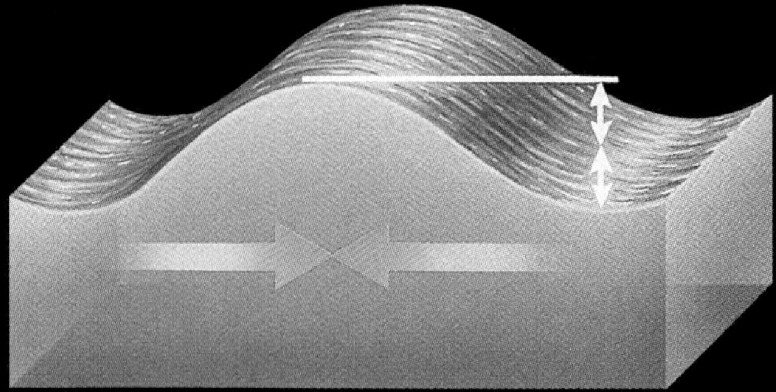

The two waves form a wave with a greater amplitude while the crests of both waves overlap.

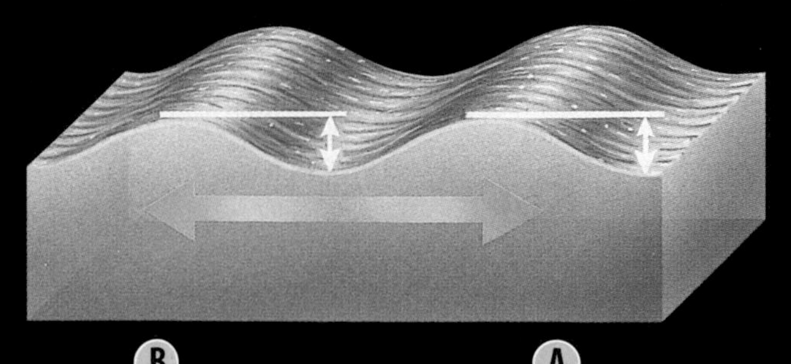

B A

The original waves pass through each other and go on as they started.

Destructive Interference

In destructive interference, a wave with a smaller amplitude is formed.

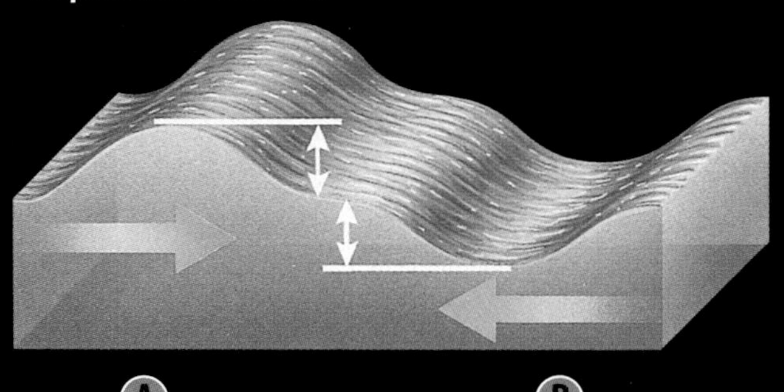

A B

The crest of one wave approaches the trough of another.

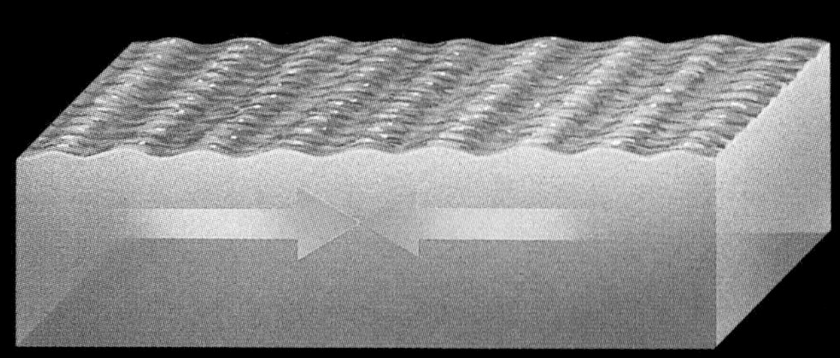

If the two waves have equal amplitude, they momentarily cancel when they meet.

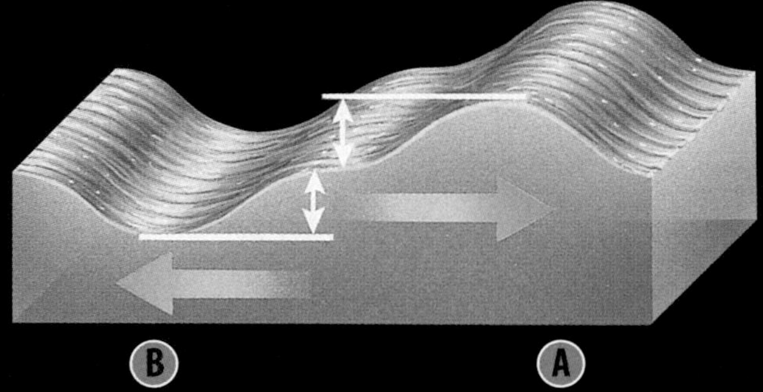

B A

The original waves pass through each other and go on as they started.

Reducing Noise You might have seen someone use a power lawn mower or a chain saw. In the past, many people who performed these tasks damaged their hearing because of the loud noises produced by these machines. Today, specially designed ear protectors absorb the sound from lawn mowers and chain saws. The ear protectors absorb energy and lower the amplitudes of the harmful waves. The waves that reach the ears have smaller amplitudes and won't damage eardrums.

Using Interference Pilots and passengers of small planes have a more complicated problem. They can't use ordinary ear protectors to shut out all the noise of the plane's motor. If they did, the pilots wouldn't be able to hear instructions from air-traffic controllers, and the passengers wouldn't be able to hear each other talk. To solve this problem, engineers invented ear protectors, as shown in **Figure 17,** that have electronic circuits. These circuits detect noise from the aircraft and produce sound frequencies that destructively interfere with the noise. However, the sound frequencies produced do not interfere with human voices, so people can hear and understand normal conversation. In these examples, destructive interference can be a benefit.

Figure 17
Some airplane pilots use special ear protectors that cancel out engine noise but don't block human voices.

Section  Assessment

1. Why don't you see your reflection in a building made of rough, white stone?
2. If you're standing on one side of a building, how are you able to hear the siren of an ambulance on the other side?
3. What behavior of light enables magnifying glasses and contact lenses to bend light rays and help people see more clearly?
4. What is diffraction? How does the amount of diffraction depend on wavelength?
5. **Think Critically** Why don't light rays that stream through an open window into a darkened room spread evenly through the entire room?

Skill Builder Activities

6. **Comparing and Contrasting** When light rays pass from water into a certain type of glass, the rays refract toward the normal. Compare and contrast the speed of light in water and in the glass. **For more help, refer to the Science Skill Handbook.**
7. **Communicating** Watch carefully as you travel home from school or walk down your street. What examples of wave reflection and refraction do you notice? Describe each of these in your Science Journal and explain your reasons. **For more help, refer to the Science Skill Handbook.**

Activity

Design Your Own Experiment

Wave Speed

When an earthquake occurs, it produces waves that are recorded at points all over the world by instruments called seismographs. By comparing the data that they collected from these seismographs, scientists discovered that the interior of Earth must be made of layers of different materials. These data showed that the waves traveled at different speeds as they passed through different parts of Earth's interior.

Recognize the Problem

How can the speed of a wave be measured?

Form a Hypothesis

In some materials, waves travel too fast for their speeds to be measured directly. Think about what you know about the relationship among the frequency, wavelength, and speed of a wave in a medium. Make a hypothesis about how you can use this relationship to measure the speed of a wave within a medium.

Goals

- **Measure** the speed of a wave within a coiled spring toy.
- **Predict** whether the speed you measured will be different in other types of coiled spring toys.

Materials

coiled spring toy
stopwatch
**clock with a second hand*
meterstick
tape

**Alternate materials*

Safety Precautions

Test Your Hypothesis

Plan

1. Make a data table in your Science Journal like the one shown.
2. In your Science Journal, write a detailed description of the coiled spring toy you are going to use. Be sure to include its mass and diameter, the width of a coil, and what it is made of.
3. **Decide** as a group how you will measure the frequency and length of waves in the spring toy. What are your variables? Which variables must be controlled? What variable do you want to measure?
4. Repeat your experiment three times.

Wave Data

	Trial 1	Trial 2	Trial 3
Length spring was stretched (m)			
Number of crests			
Wavelength (m)			
# of vibrations timed			
# of seconds vibrations were timed			
Wave speed (m/s)			

Do

1. Make sure your teacher approves your plan before you start.
2. Carry out the experiment.
3. While you are doing the experiment, record your observations and measurements in your data table.

Analyze Your Data

1. **Calculate** the frequency of the waves by dividing the number of vibrations you timed by the number of seconds you timed them. Record your results in your data table.
2. Use the following formula to calculate the speed of a wave in each trial.

$$\text{wavelength} \times \text{wave frequency} = \text{wave speed}$$

3. Average the wave speeds from your trials to determine the average speed of a wave in your coiled spring toy.

Draw Conclusions

1. Which variables affected the wave speed in spring toys the most? Which variables affected the speed the least? Was your hypothesis supported?
2. What factors caused the wave speed measured in each trial to be different?

Post a description of your coiled spring toy and the results of your experiment on a bulletin board in your classroom. **Compare and contrast** your results with other students in your class.

Science Stats

Waves, Waves, and More Waves

Did you know...

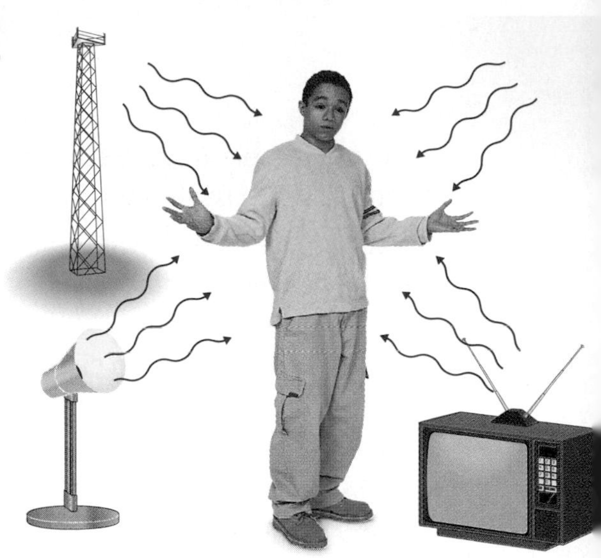

. . . You are constantly surrounded by a sea of waves even when you're on dry land! Electromagnetic waves around us are used to cook our food and transmit signals to our radios and televisions. Light itself is an electromagnetic wave.

. . . The highest recorded ocean wave was 34 meters high, which is comparable to the height of a ten-story building. This super wave was seen in the North Pacific Ocean and recorded by the crew of the naval ship *USS Ramapo* in 1933.

. . . Tsunamis—huge ocean waves— can travel at speeds over 900 km/h.

. . . Waves let dolphins see with their ears! A dolphin sends out ultrasonic pulses, or clicks, at rates of 800 pulses per second. These sound waves are reflected back to the dolphin after they hit another object. This process—echolocation—allows dolphins to recognize obstacles and meals.

Connecting To Math

. . . Earthquakes produce a variety of seismic waves— waves that ripple through Earth after subsurface rock breaks suddenly. The fastest are P and S waves. P waves are compressional waves that travel at about 8 km/s. S waves, which move like ocean waves, travel at about 4.8 km/s.

Electromagnetic Wavelengths

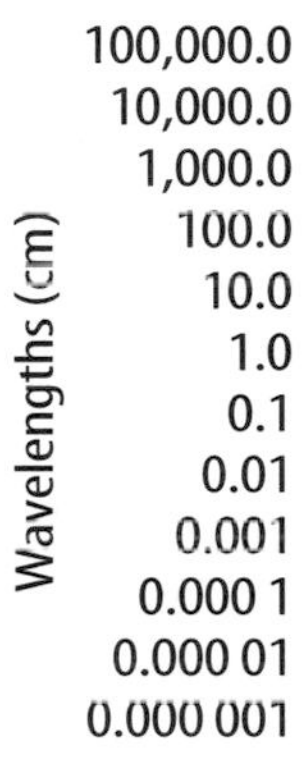
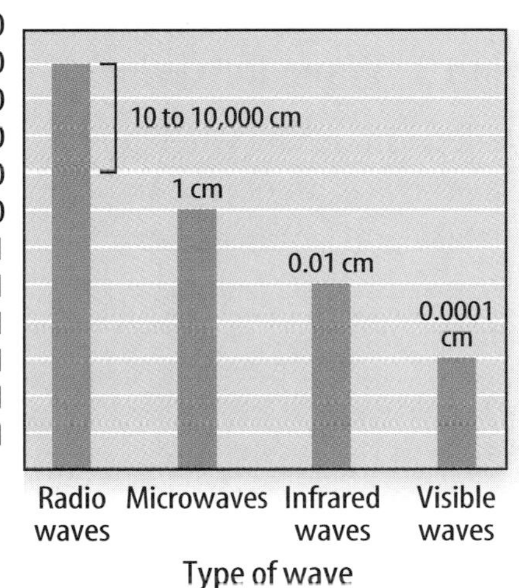

. . . Radio waves from space were discovered in 1932 by Karl G. Jansky, an American engineer. His amazing discovery led to creation of radio astronomy, a field that explores parts of the universe that are hidden by interstellar dust or can't be seen with telescopes.

Do the Math

1. A museum with a dolphin exhibit plays dolphin clicks for its visitors 250 times slower than the rate at which the dolphins emit them. How many clicks do the visitors hear in 10 s?
2. Tsunamis form in the ocean when an earthquake occurs on the ocean floor. How long will it take a tsunami to travel 4,500 km?
3. Make a bar graph to show the speeds of P waves, S waves and tsunamis. Use km/h as your unit of speed.

Go Further

Go to **science.glencoe.com** to learn about discoveries by radio astronomers. Graph the distances of these discoveries from Earth.

Chapter 20 Study Guide

Reviewing Main Ideas

Section 1 What are waves?

1. Waves are rhythmic disturbances that carry energy but not matter.
2. Mechanical waves can travel only through matter. Electromagnetic waves can travel through matter and space.
3. In a mechanical transverse wave, matter in the medium moves back and forth at right angles to the direction the wave travels.
4. In a compressional wave, matter in the medium moves forward and backward in the same direction as the wave. *How does the boat in the picture move as the water wave goes by?*

Section 2 Wave Properties

1. The amplitude of a transverse wave is the distance between the rest position and a crest or a trough.
2. The energy carried by a wave increases as the amplitude increases.
3. Wavelength is the distance between neighboring crests or neighboring troughs.
4. The frequency of a wave is the number of wavelengths that pass a given point in 1 s. *What property of a wave is shown by the figure at the top of the next column?*

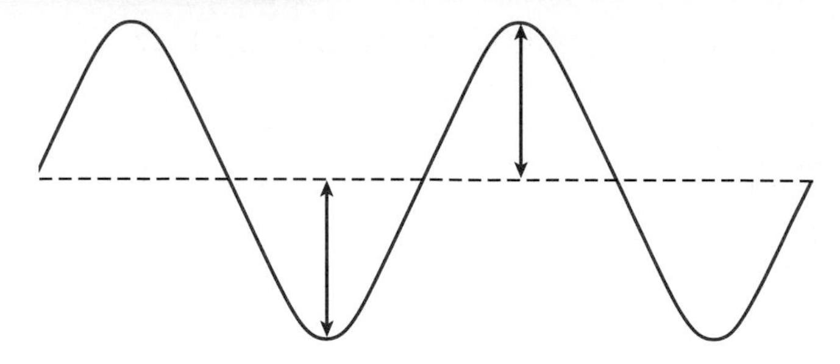

5. Waves travel through different materials at different speeds.

Section 3 Wave Behavior

1. Reflection occurs when a wave strikes an object or surface and bounces off. *Why doesn't the foil show a clear image?*
2. The bending of a wave as it moves from one medium into another is called refraction. A wave changes direction, or refracts, when the speed of the wave changes.
3. The bending of waves around a barrier is called diffraction.
4. Interference occurs when two or more waves combine and form a new wave while they overlap.

FOLDABLES Reading & Study Skills

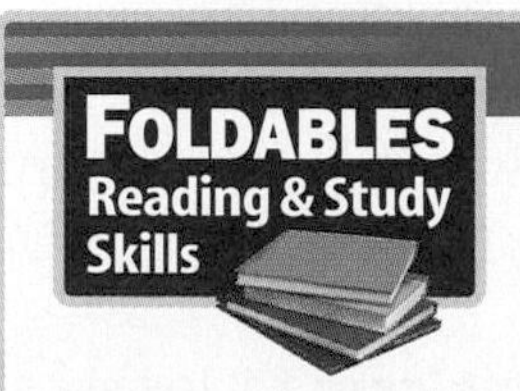

After You Read

Use your Concept Map Study Fold to compare and contrast transverse and compressional mechanical waves.

Visualizing Main Ideas

Complete the following spider map about waves.

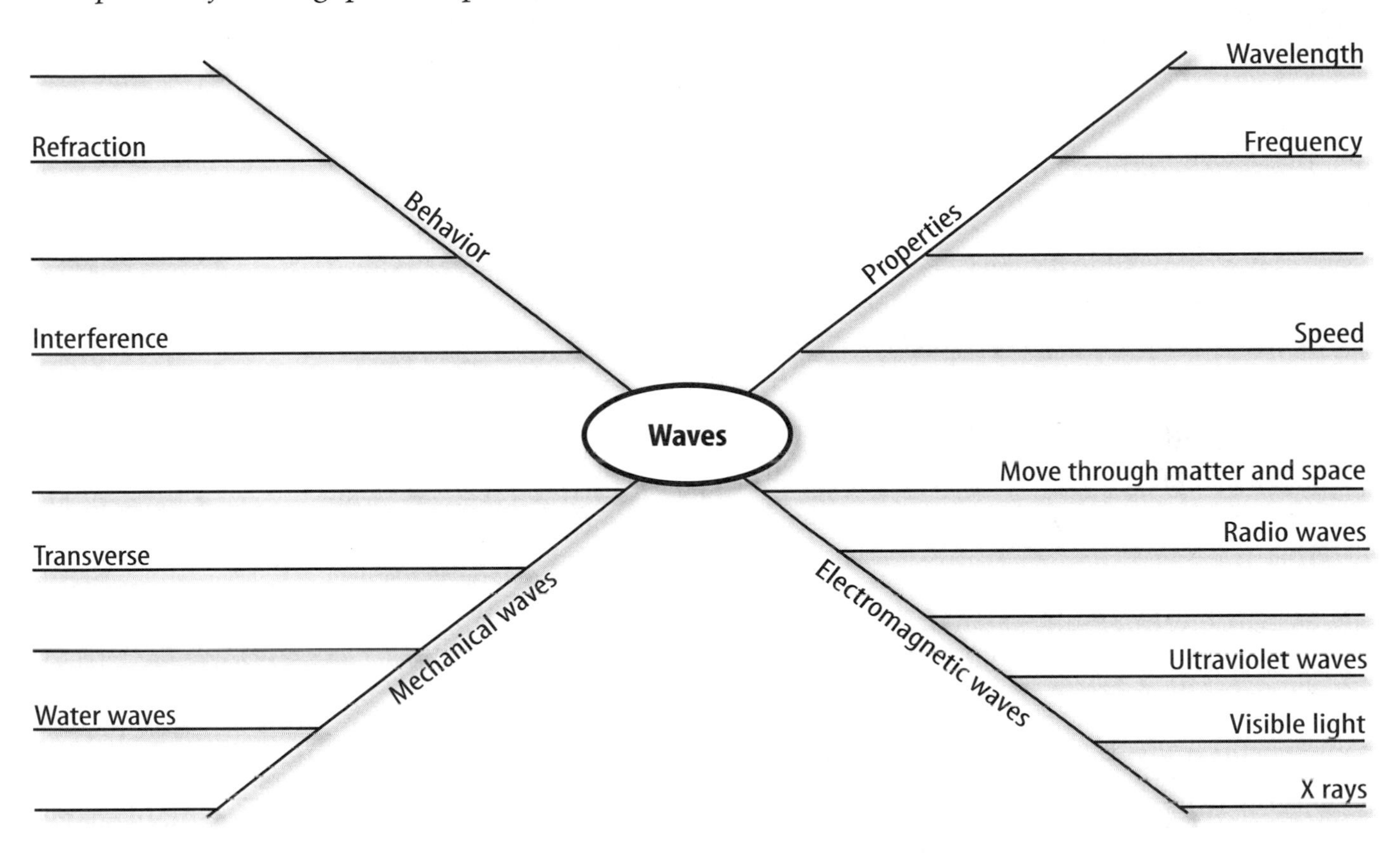

Vocabulary Review

Vocabulary Words

a. amplitude
b. compressional wave
c. diffraction
d. electromagnetic wave
e. frequency
f. interference
g. mechanical wave
h. reflection
i. refraction
j. transverse wave
k. wave
l. wavelength

Study Tip

After you've read a chapter, go back to the beginning and speed-read through what you've just read. This will help your memory.

Using Vocabulary

Using the list, replace the underlined words with the correct vocabulary words.

1. Diffraction is the change in direction of a wave going from one medium to another.
2. The type of wave that has rarefactions is a transverse wave.
3. The distance between two adjacent crests of a transverse wave is the frequency.
4. The more energy a wave carries, the greater its wavelength is.
5. A mechanical wave can travel through space without a medium.

Chapter 20 Assessment

Checking Concepts

Choose the word or phrase that best answers the question.

1. What is the material through which mechanical waves travel?
A) charged particles **C)** a vacuum
B) space **D)** a medium

2. What is carried from particle to particle in a water wave?
A) speed **C)** energy
B) amplitude **D)** matter

3. What are the lowest points on a transverse wave called?
A) crests **C)** compressions
B) troughs **D)** rarefactions

4. What determines the pitch of a sound wave?
A) amplitude **C)** speed
B) frequency **D)** refraction

5. What is the distance between adjacent wave compressions?
A) one wavelength **C)** 1 m/s
B) 1 km **D)** 1 Hz

6. What occurs when a wave strikes an object or surface and bounces off?
A) diffraction
B) refraction
C) a transverse wave
D) reflection

7. What is the name for a change in the direction of a wave when it passes from one medium into another?
A) refraction **C)** reflection
B) interference **D)** diffraction

8. What type of wave is a sound wave?
A) transverse **C)** compressional
B) electromagnetic **D)** refracted

9. When two waves overlap and interfere destructively, what does the resulting wave have?
A) a greater amplitude
B) more energy
C) a change in frequency
D) a lower amplitude

10. What is the difference between blue light and green light?
A) They have different wavelengths.
B) One is a transverse wave and the other is not.
C) They have different pitch.
D) One is mechanical and the other is not.

Thinking Critically

11. Explain what kind of wave—transverse or compressional—is produced when an engine bumps into a string of coupled railroad cars on a track.

12. Is it possible for an electromagnetic wave to travel through a vacuum? Through matter? Explain your answers.

13. Why does the frequency of a wave decrease as the wavelength increases?

14. Why don't you see your reflected image when you look at a white, rough surface?

15. If a cannon fires at a great distance from you, why do you see the flash before you hear the sound?

Developing Skills

16. Solving One-Step Equations An electromagnetic wave travels at the speed of light and has a wavelength of 0.022 m. If the wave speed is equal to the wavelength times the frequency, what is the frequency of the wave?

17. Forming Hypotheses Form a hypothesis that can explain this observation. Waves A and B travel away from Earth through Earth's atmosphere. Wave A continues on into space, but wave B does not.

18. Recognizing Cause and Effect Explain how the object shown below causes compressions and rarefactions as it vibrates in air.

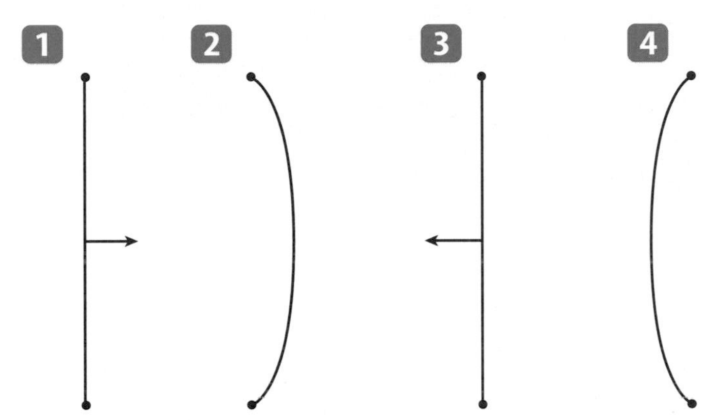

19. Comparing and Contrasting AM radio waves have wavelengths between about 200 m and 600 m, and FM radio waves have wavelengths of about 3 m. Why can AM radio signals often be heard behind buildings and mountains but FM radio signals cannot?

Performance Assessment

20. Making Flashcards Work with a partner to make flashcards for the bold-faced terms in the chapter. Illustrate each term on the front of the cards. Write the term and its definition on the back of the card. Use the cards to review the terms with another team.

Technology

Go to the Glencoe Science Web site at **science.glencoe.com** or use the **Glencoe Science CD-ROM** for additional chapter assessment.

Test Practice

Kamisha's science teacher told her that her remote control sent signals to the TV and VCR by using infrared waves. She decided to do some research about waves. The information she gathered is shown in the diagram below.

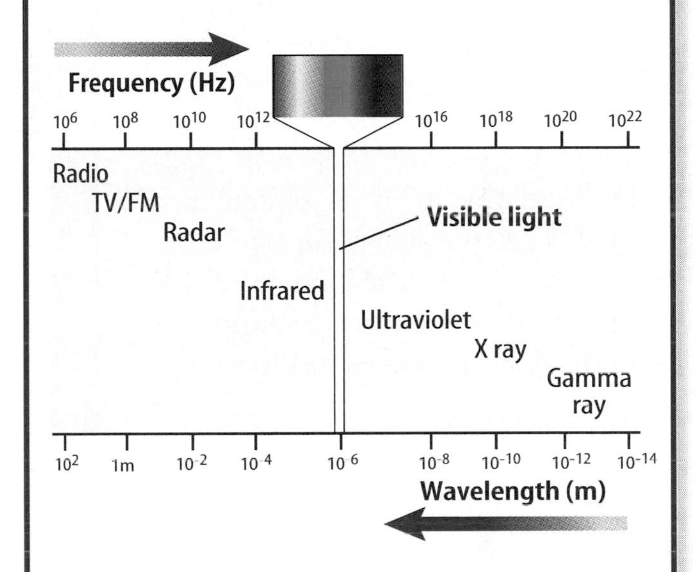

1. According to the diagram, which type of wave has a wavelength greater than 1 m?

A) radio
B) infrared
C) ultraviolet
D) X ray

2. According to the diagram, which type of wave has the HIGHEST frequency?

F) radio
G) ultraviolet
H) X ray
J) gamma ray

CHAPTER

21

Sound

Have you ever experienced complete silence? Unless you have stood in a room like this one, you probably have not. This room is lined with materials that absorb sound waves and eliminate sound reflections. The sounds that you hear are created by vibrations. How do vibrations make sounds with different pitches? What makes a sound loud or soft? In this chapter, you will learn the answers to these questions. You will also learn how musical instruments create sound and how the ear enables you to hear sound.

What do you think?

Science Journal Look at the picture below with a classmate. Discuss what might be happening. Here's a hint: *Sound is caused by vibrations.* Write your answer or best guess in your Science Journal.

When you speak or sing, you push air from your lungs past your vocal cords, which are two flaps of tissue inside your throat. When you tighten your vocal cords, you can make the sound have a higher pitch. Do this activity to explore how you change the shape of your throat to vary the pitch of sound.

Observe throat vibrations

1. Hold your fingers against the front of your throat and say *Aaaah.* Notice the vibration against your fingers.
2. Now vary the pitch of this sound from low to high and back again. How do the vibrations in your throat change? Record your observations.
3. Change the sound to an *Ooooh.* What do you notice as you listen? Record your observations.

Observe

In your Science Journal, describe how the shape of your throat changed the pitch.

Before You Read

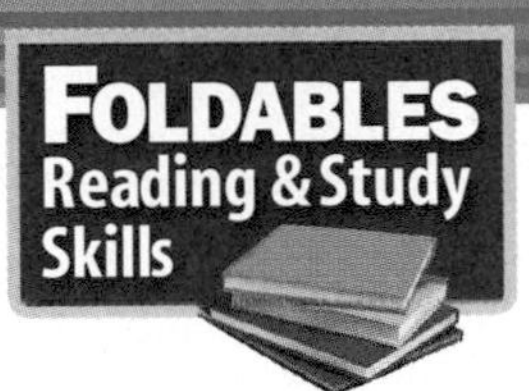

Making a Question Study Fold **Asking yourself questions helps you stay focused so you will better understand sound when you are reading the chapter.**

1. Place a sheet of notebook paper in front of you so the short side is at the top and the holes are on the right side. Fold the paper in half from the left side to the right side.
2. Through the top thickness of paper, cut along every third line from the outside edge to the fold, forming tabs.
3. Before you read the chapter, write a question you have about sound on the front of each tab. As you read the chapter, answer your questions and add more information.

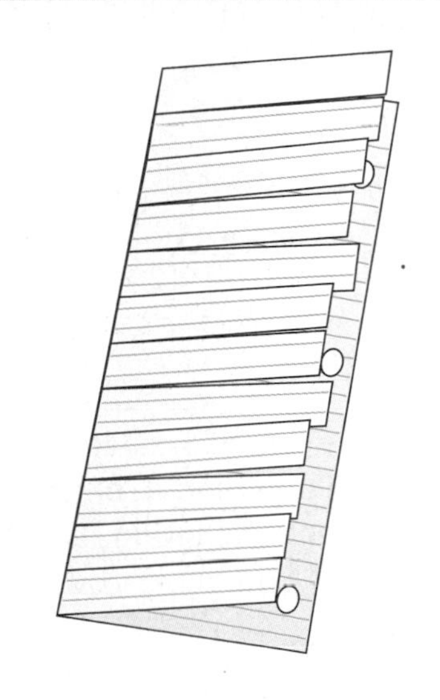

SECTION

What is sound?

As You Read

What You'll Learn
- **Identify** the characteristics of sound waves.
- **Explain** how sound travels.
- **Describe** the Doppler effect.

Vocabulary
loudness
pitch
echo
Doppler effect

Why It's Important
Sound gives important information about the world around you.

Sound and Vibration

Think of all the sounds you've heard since you awoke this morning. Did you hear your alarm clock blaring, car horns honking, or locker doors slamming? Every sound has something in common with every other sound. Each is produced by something that vibrates.

Sound Waves

How does an object that is vibrating produce sound? When you speak, the vocal cords in your throat vibrate. These vibrations cause other people to hear your voice. The vibrations produce sound waves that travel to their ears. The other person's ears interpret these sound waves.

A wave carries energy from one place to another without transferring matter. An object that is vibrating in air, such as your vocal cords, produces a sound wave. The vibrating object causes air molecules to move back and forth. As these air molecules collide with those nearby, they cause other air molecules to move back and forth. In this way, energy is transferred from one place to another. A sound wave is a compressional wave, like the wave moving through the coiled spring toy in **Figure 1.** In a compressional wave, particles in the material move back and forth along the direction the wave is moving. In a sound wave, air molecules move back and forth along the direction the sound wave is moving.

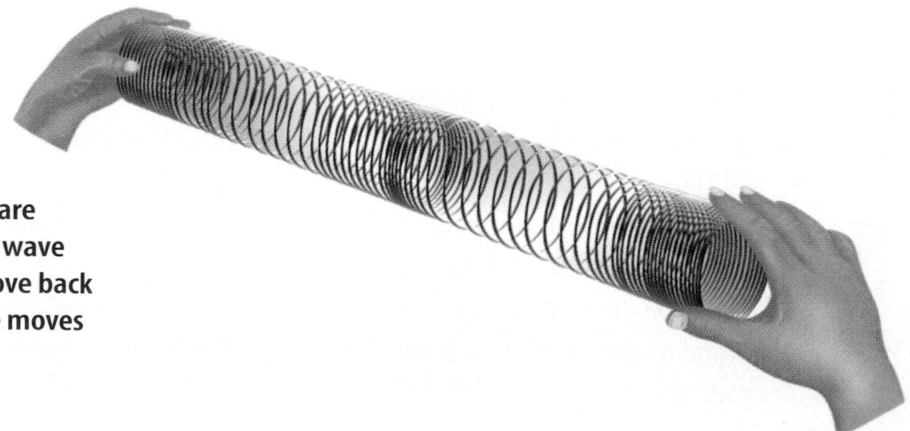

Figure 1
When the coils of a coiled spring toy are squeezed together, a compressional wave moves along the spring. The coils move back and forth as the compressional wave moves past them.

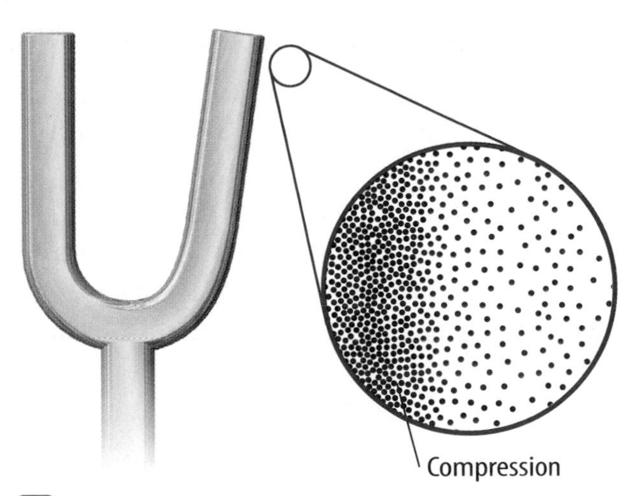

A **When the tuning fork vibrates outward, it forces the air molecules next to it together, creating a region of compression.**

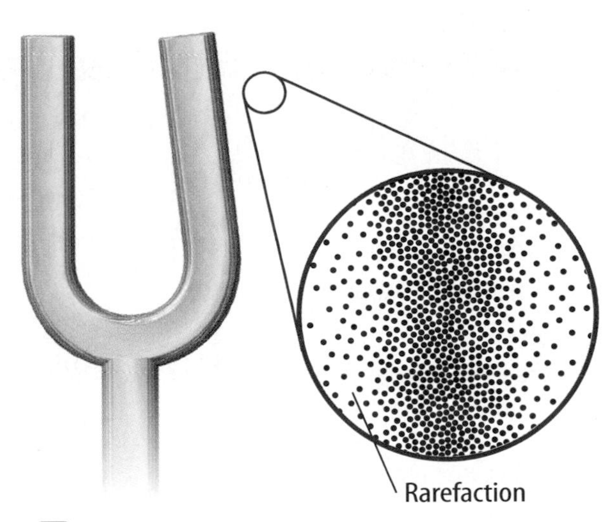

B **When the tuning fork moves back, the air molecules next to it spread apart, creating a region of rarefaction.**

Figure 2
A tuning fork makes a sound wave as the ends of the fork vibrate in the air. *Can a sound wave travel in a vacuum?*

Making Sound Waves When an object vibrates, it exerts a force on the surrounding air. For example, as the end of the tuning fork moves outward into the air, it pushes the air molecules together, as shown in **Figure 2A.** As a result, a region where the air molecules are closer together, or more dense, is created. This region of higher density is called a compression. When the end of the tuning fork moves back, it creates a region of lower density called a rarefaction, as shown in **Figure 2B.** As the tuning fork continues to vibrate, a series of compressions and rarefactions is formed. The compressions and rarefactions move away from the tuning fork as molecules in these regions collide with other nearby molecules.

Like other waves, a sound wave can be described by its wavelength and frequency. The wavelength of a sound wave is shown in **Figure 3.** The frequency of a sound wave is the number of compressions or rarefactions that pass by a given point in one second. An object that vibrates faster forms a sound wave with a higher frequency.

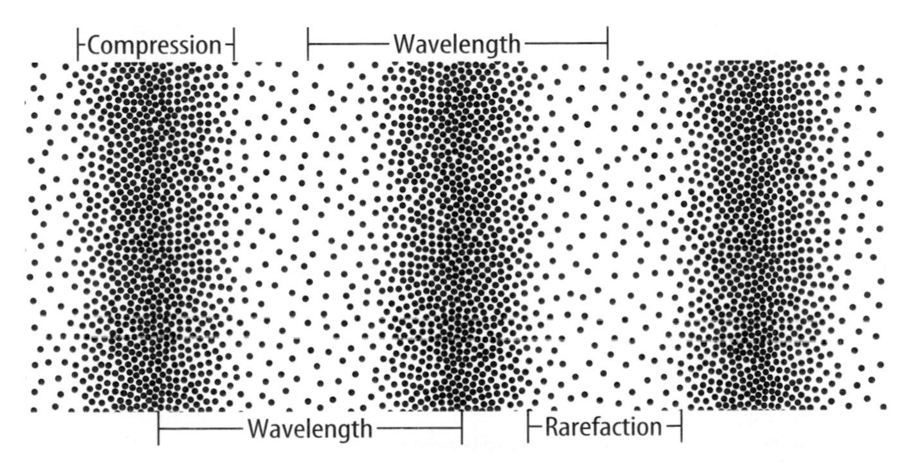

Figure 3
Wavelength is the distance from one compression to another or one rarefaction to another.

TRY AT HOME Mini LAB

Comparing and Contrasting Sounds

Procedure

1. Shake a set of **keys** and listen to the sound they make in air. Then submerge the keys and one ear in **water.** (A **tub** or a wide, deep **bowl** will work.) Again, shake the keys and listen to the sound. Use a **towel** to dry the keys.
2. Tie a **metal spoon** in the middle of a length of **cotton string.** Strike the spoon on something to hear it ring. Now press the ends of the string against your ears and repeat the experiment. What do you hear?

Analysis

1. Did you hear sounds transmitted through water and through string? Describe the sounds.
2. Compare and contrast the sounds in water and in air.

The Speed of Sound

Sound waves can travel through other materials besides air. Even though sound waves travel in the same way through different materials as they do in air, they might travel at different speeds. As a sound wave travels through a material, the particles in the material it is moving through collide with each other. In a solid, molecules are closer together than in liquids or gases, so collisions between molecules occur more rapidly than in liquids or gases. As a result, the speed of sound is usually fastest in solids, where molecules are closest together, and slowest in gases, where molecules are farthest apart. **Table 1** shows the speed of sound through different materials.

The Speed of Sound and Temperature The temperature of the material that sound waves are traveling through also affects the speed of sound. As a substance heats up, its molecules move faster, so they collide more frequently. The more frequent the collisions are, the faster the speed of sound is in the material. For example, the speed of sound in air at 0°C is 331 m/s; at 20°C, it is 343 m/s.

Amplitude and Loudness

What's the difference between loud sounds and quiet sounds? When you play a song at high volume and low volume, you hear the same instruments and voices, but something is different. The difference is that loud sound waves generally carry more energy than soft sound waves do.

Loudness is the human perception of how much energy a sound wave carries. Not all sound waves with the same energy are as loud. Sounds with frequencies between 3,000 Hz and 4,000 Hz sound louder than other sound waves that have the same energy.

Table 1 Speed of Sound Through Different Materials

Material	Speed (m/s)
Air	343
Water	1,483
Steel	5,940
Glass	5,640

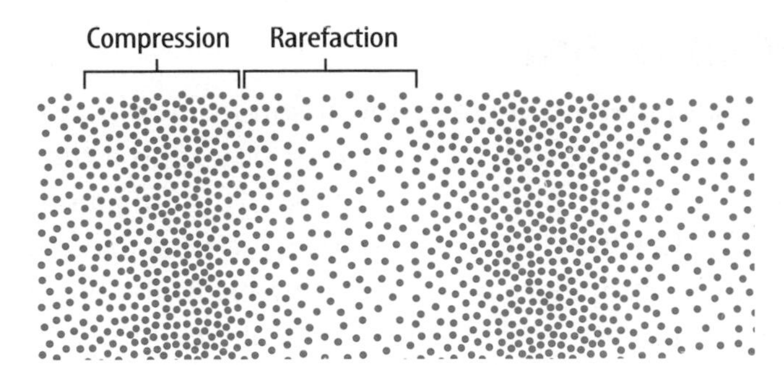

A This sound wave has a lower amplitude.

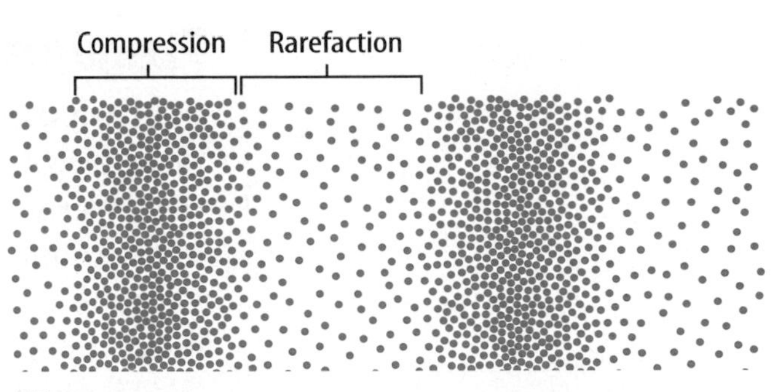

B This sound wave has a higher amplitude. Particles in the material are more compressed in the compressions and more spread out in the rarefactions.

Figure 4
The amplitude of a sound wave depends on how spread out the particles are in the compressions and rarefactions of the wave.

Amplitude and Energy The amount of energy a wave carries depends on its amplitude. For a compressional wave such as a sound wave, the amplitude is related to how spread out the molecules or particles are in the compressions and rarefactions, as **Figure 4** shows. The higher the amplitude of the wave is, the more compressed the particles in the compression are and the more spread out they are in the rarefactions. More energy had to be transferred by the vibrating object that created the wave to force the particles closer together or spread them farther apart. Sound waves with greater amplitude carry more energy and sound louder. Sound waves with smaller amplitude carry less energy and sound quieter.

Reading Check *What determines the loudness of different sounds?*

The Decibel Scale Perhaps an adult has said to you, "Turn down your music, it's too loud! You're going to lose your hearing!" Although the perception of loudness varies from person to person, the energy carried by sound waves can be described by a scale called the decibel (dB) scale. **Figure 5** shows the decibel scale. An increase of 10 dB means that the energy carried by the sound has increased ten times, but an increase of 20 dB means that the sound carries 100 times more energy.

Hearing damage begins to occur at sound levels of about 85 dB. The amount of damage depends on the frequencies of the sound and the length of time a person is exposed to the sound. Some music concerts produce sound levels as high as 120 dB. The energy carried by these sound waves is about 30 billion times greater than the energy carried by sound waves that are made by whispering.

Figure 5
The loudness of sound is measured on the decibel scale.

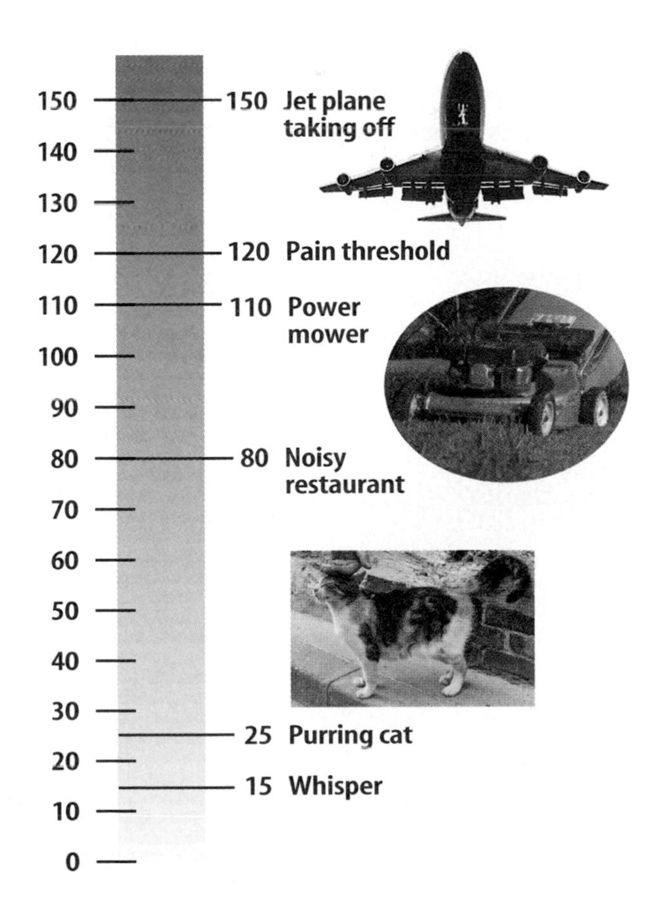

Frequency and Pitch

The **pitch** of a sound is how high or low it sounds. For example, a piccolo produces a high-pitched sound or tone, and a tuba makes a low-pitched sound. Pitch corresponds to the frequency of the sound. The higher the pitch is, the higher the frequency is. A sound wave with a frequency of 440 Hz, for example, has a higher pitch than a sound wave with a frequency of 220 Hz.

The human ear can detect sound waves with frequencies between about 20 Hz and 20,000 Hz. However, some animals can detect even higher and lower frequencies. For example, dogs can hear frequencies up to almost 50,000 Hz. Dolphins and bats can hear frequencies as high as 150,000 Hz, and whales can hear frequencies higher than those heard by humans.

Recall that frequency and wavelength are related. If two sound waves are traveling at the same speed, the wave with the shorter wavelength has a higher frequency. If the wavelength is shorter, then more compressions and rarefactions will go past a given point every second than for a wave with a longer wavelength, as shown in **Figure 6.** Sound waves with a higher pitch have shorter wavelengths than those with a lower pitch.

The Human Voice When you make a sound, you exhale past your vocal cords, causing them to vibrate. The length and thickness of your vocal cords help determine the pitch of your voice. Shorter, thinner vocal cords vibrate at higher frequencies than longer or thicker ones. This explains why children, whose vocal cords are still growing, have higher voices than adults. Muscles in the throat can stretch the vocal cords tighter, letting people vary their pitch within a limited range.

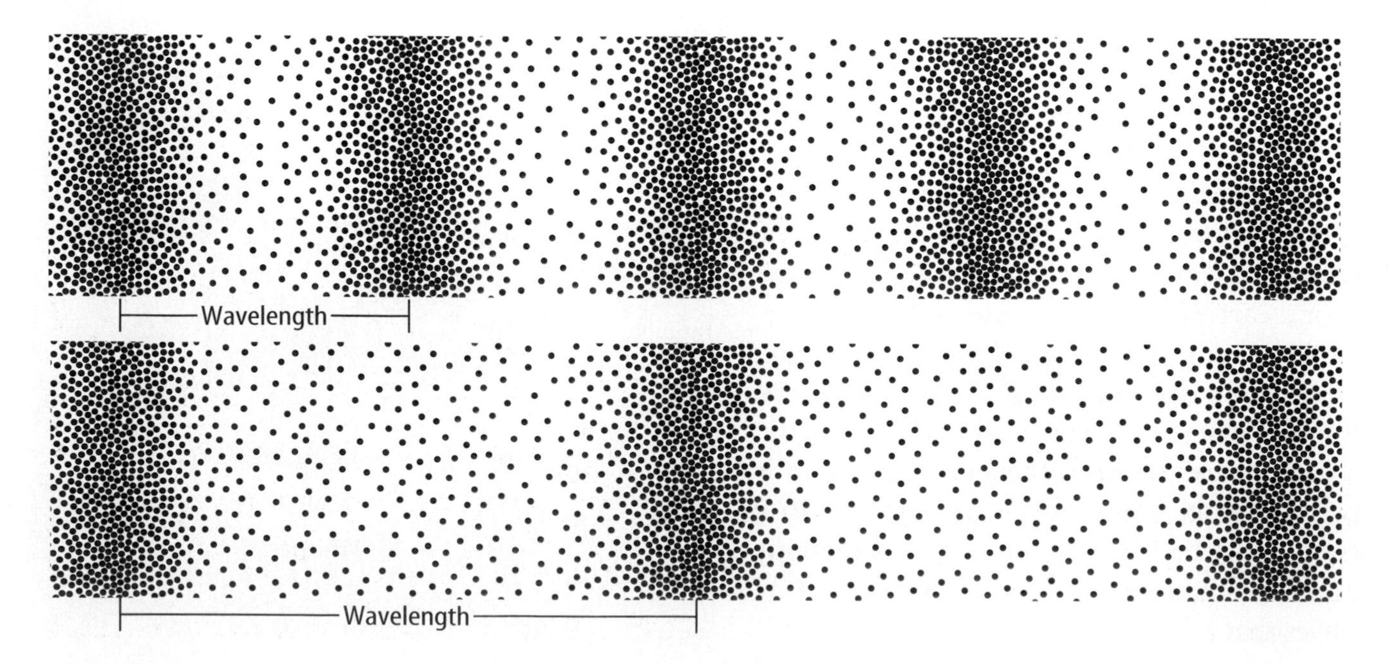

Figure 6
The upper sound wave has a shorter wavelength than the lower wave. If these two sound waves are traveling at the same speed, the upper sound wave has a higher frequency than the lower one. For this wave, more compressions and rarefactions will go past a point every second than for the lower wave. *Which wave has a higher pitch?*

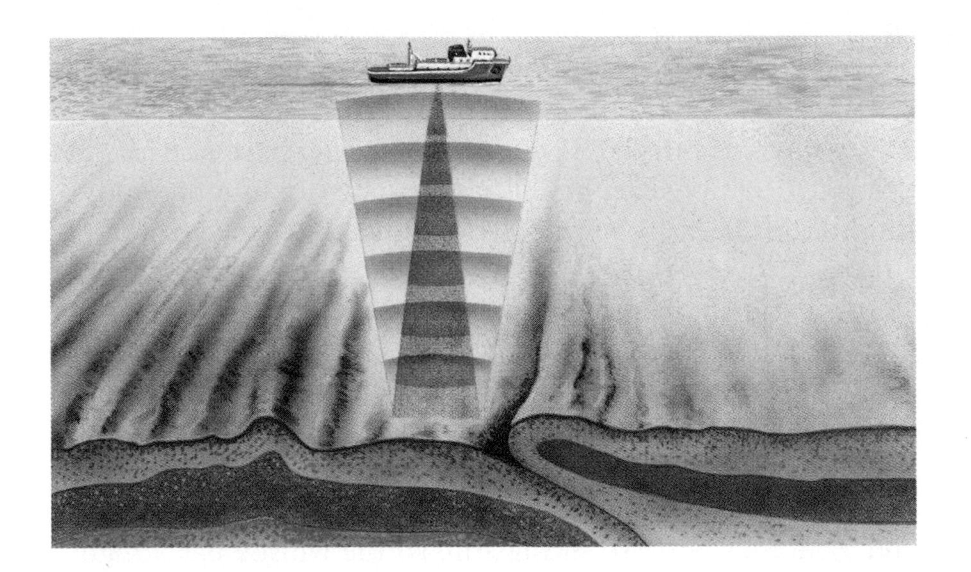

Figure 7
Sonar uses reflected sound waves to determine the location and shape of an object.

Echoes

Sound reflects off of hard surfaces, just like a water wave bounces off the side of a bath tub. A reflected sound wave is called an **echo.** If the distance between you and a reflecting surface is great enough, you might hear the echo of your voice. This is because it might take a few seconds for the sound to travel to the reflecting surface and back to your ears.

Sonar systems use sound waves to map objects underwater, as shown in **Figure 7.** The amount of time it takes an echo to return depends on how far away the reflecting surface is. By measuring the length of time between emitting a pulse of sound and hearing its echo off the ocean floor, the distance to the ocean floor can be measured. Using this method, sonar can map the ocean floor and other undersea features. Sonar also can be used to detect submarines, schools of fish, and other objects.

Research Visit the Glencoe Science Web site at **science.glencoe.com** for more information on how sonar is used to detect objects underwater. Communicate to your class what you learn.

Echolocation Some animals use a method called echolocation to navigate and hunt. Bats, for example, emit high-pitched squeaks and listen for the echoes. The type of echo it hears helps the bat determine exactly where an insect is, as shown in **Figure 8.** Dolphins also use a form of echolocation. Their high-pitched clicks bounce off of objects in the ocean, allowing them to navigate in the same way.

People with visual impairments also have been able to use echolocation. Using their ears, they can interpret echoes to estimate the size and shape of a room, for example.

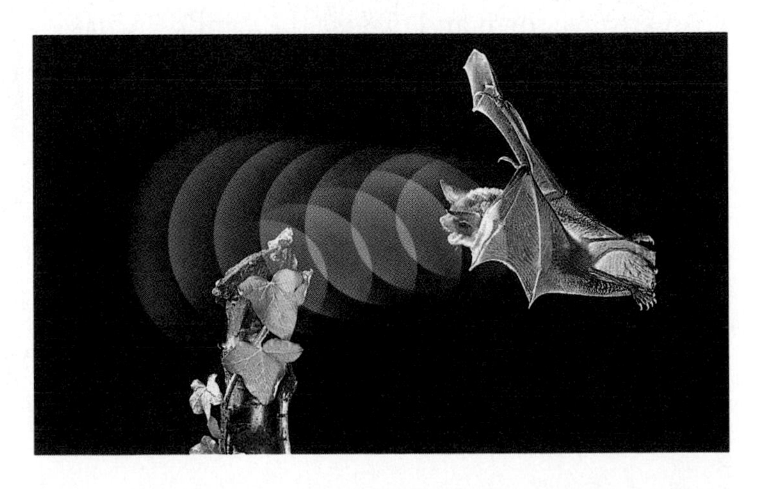

Figure 8
Bats use echolocation to hunt.
Why is this technique good for hunting at night?

The Doppler Effect

The frequency of light waves is also changed by the Doppler shift. If a light source is moving away from an observer, the frequencies of the emitted light waves decrease. Research how the Doppler shift is used by astronomers to determine how other objects in the universe are moving relative to Earth.

Perhaps you've heard an ambulance siren as the ambulance speeds toward you, then goes past. You might have noticed that the pitch of the siren gets higher as the ambulance moves toward you. Then as the ambulance moves away, the pitch of the siren gets lower. The change in frequency that occurs when a source of sound is moving relative to a listener is called the **Doppler effect. Figure 9** shows why the Doppler effect occurs.

The Doppler effect occurs whether the sound source or the listener is moving. If you drive past a factory as its whistle blows, the whistle will sound higher pitched as you approach. As you move closer you encounter each sound wave a little earlier than you would if you were sitting still, so the whistle has a higher pitch. When you move away from the whistle, each sound wave takes a little longer to reach you. You hear fewer wavelengths per second, which makes the sound lower in pitch.

Radar guns that are used to measure the speed of cars and baseball pitches also use the Doppler effect. Instead of a sound wave, the radar gun sends out a radio wave. When the radio wave is reflected, its frequency changes depending on the speed of the object and whether it is moving toward the gun or away from it. The radar gun uses the change in frequency of the reflected wave to determine the object's speed.

Problem-Solving Activity

How does Doppler radar work?

Doppler radar is used by the National Weather Service to detect areas of precipitation and to measure the speed at which a storm moves. Because the wind moves the rain, Doppler radar can "see" into a strong storm and expose the winds. Tornadoes that might be forming in the storm then can be identified.

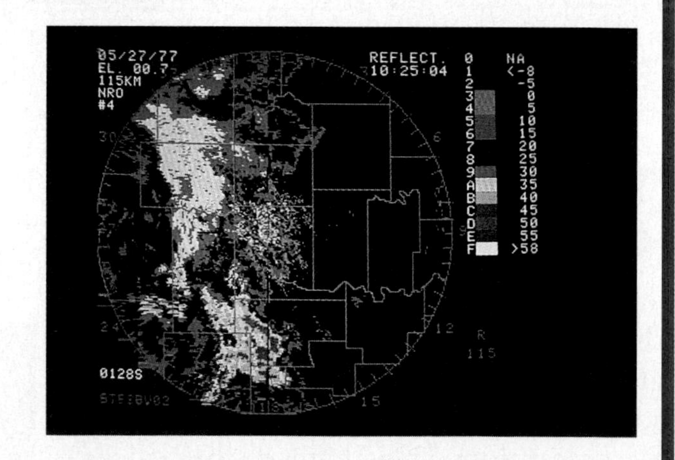

Identify the Problem

An antenna sends out pulses of radio waves as it rotates. The waves bounce off raindrops and return to the antenna at a different frequency, depending on whether the rain is moving toward the antenna or away from it. The change in frequency is due to the Doppler shift.

Solving the Problem

1. If the frequency of the reflected radio waves increases, how is the rain moving relative to the radar station?
2. In a tornado, winds are rotating. How would the radio waves reflected by rotating winds be Doppler-shifted?

NATIONAL GEOGRAPHIC **VISUALIZING THE DOPPLER EFFECT**

Figure 9

You've probably heard the siren of an ambulance as it races through the streets. The sound of the siren seems to be higher in pitch as the ambulance approaches and lower in pitch as it moves away. This is the Doppler effect, which occurs when a listener and a source of sound waves are moving relative to each other.

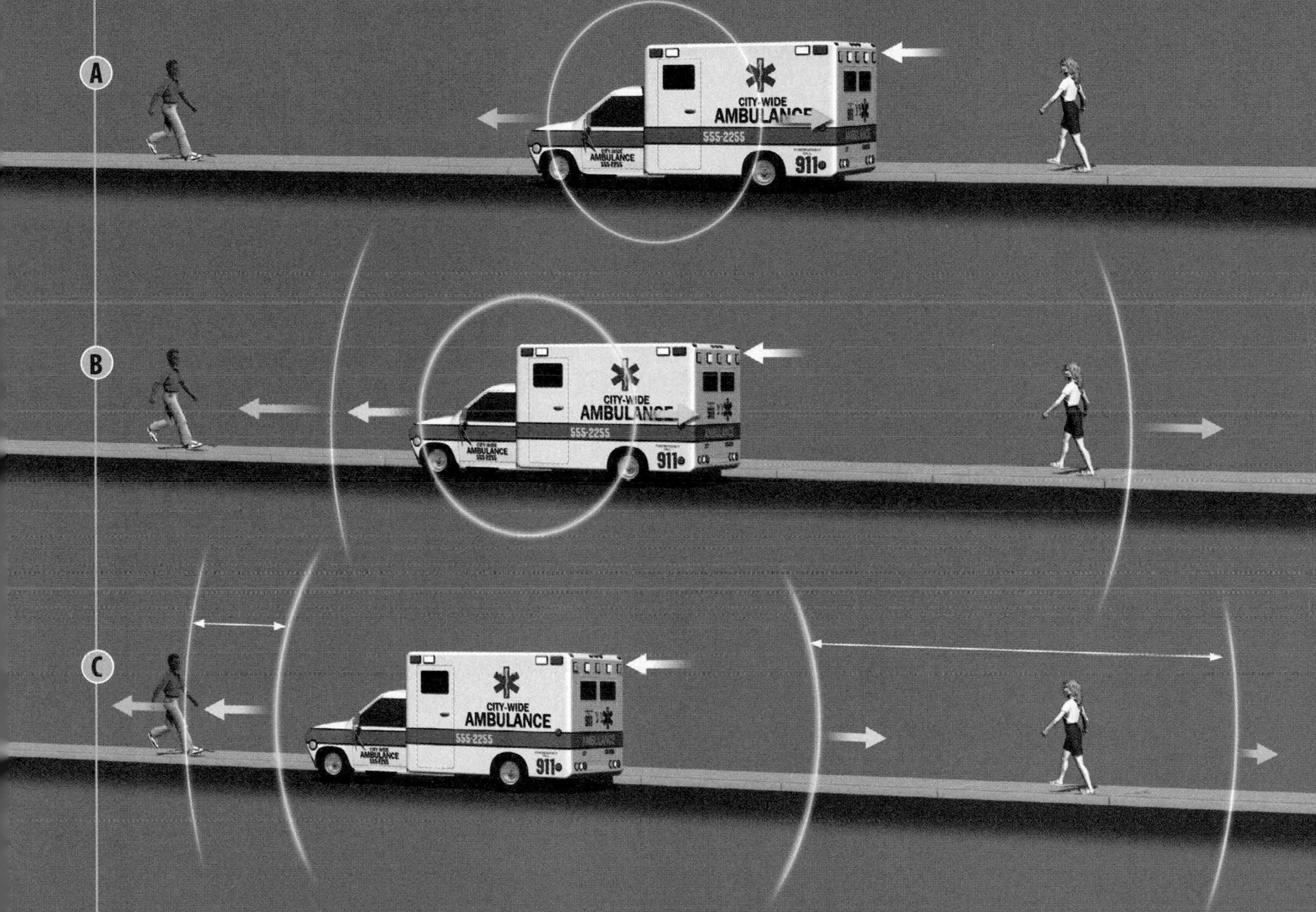

A As the ambulance speeds down the street, its siren emits sound waves. Suppose the siren emits the compression part of a sound wave as it goes past the girl.

B As the ambulance continues moving, it emits another compression. Meanwhile, the first compression spreads out from the point from which it was emitted.

C The waves traveling in the direction that the ambulance is moving have compressions closer together. As a result, the wavelength is shorter and the boy hears a higher frequency sound as the ambulance moves toward him. The waves traveling in the opposite direction have compressions that are farther apart. The wavelength is longer and the girl hears a lower frequency sound as the ambulance moves away from her.

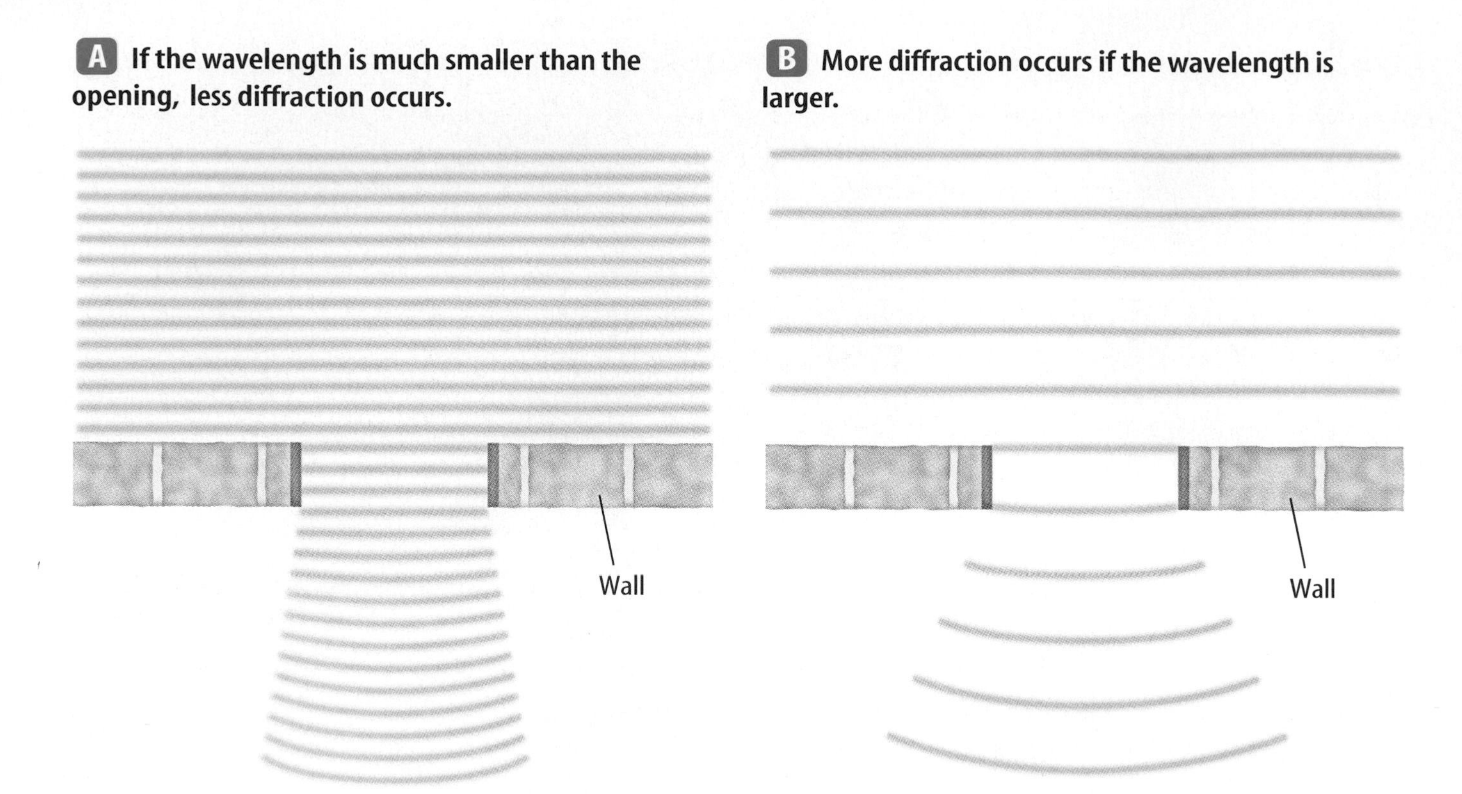

Figure 10
The spreading of a wave by diffraction depends on the wavelength and the size of the opening.

Diffraction of Sound Waves

Like other waves, sound waves diffract. This means they can bend around obstacles or spread out after passing through narrow openings. The amount of diffraction depends on the wavelength of the sound wave compared to the size of the obstacle or opening. If the wavelength is much smaller than the obstacle, almost no diffraction occurs. As the wavelength becomes closer to the size of the obstacle, the amount of diffraction increases.

You can observe diffraction of sound waves by visiting the school band room during practice. If you stand in the doorway, you will hear the band normally. However, if you stand to one side outside the door or around a corner, you will hear the lower-pitched instruments better. **Figure 10** shows why this happens. The sound waves that are produced by the lower-pitched instruments have lower frequencies and longer wavelengths. These wavelengths are closer to the size of the door opening than the higher-pitched sound waves are. As a result, the longer wavelengths diffract more, and you can hear them even when you're not standing in the doorway.

The diffraction of lower frequencies in the human voice allows you to hear someone talking even when the person is around the corner. This is different from an echo. Echoes occur when sound waves bounce off a reflecting surface. Diffraction occurs when a wave spreads out after passing through an opening, or when a wave bends around an obstacle.

Using Sound Waves

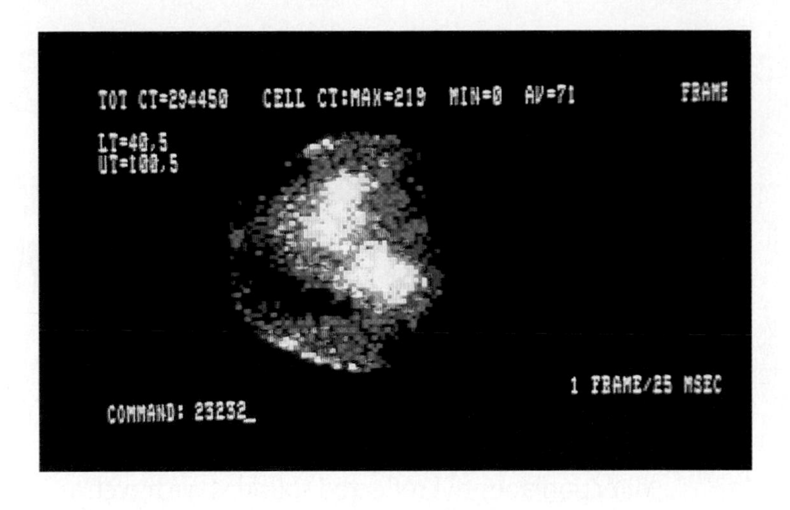

Figure 11
Ultrasound is used to make this image of the heart. *How else is ultrasound used in medicine?*

Sound waves can be used to treat certain medical problems. A process called ultrasound uses high-frequency sound waves as an alternative to some surgeries. For example, some people develop small, hard deposits in their kidneys or gallbladders. A doctor can focus ultrasound waves at the kidney or gallbladder. The ultrasound waves cause the deposits to vibrate rapidly until they break apart into small pieces. Then, the body can get rid of them.

Ultrasound can be used to make images of the inside of the body, just as sonar is used to map the seafloor. One common use of ultrasound is to examine a developing fetus. Also, ultrasound along with the Doppler effect can be used to examine the functioning of the heart. An ultrasound image of the heart is shown in **Figure 11.** This technique can help determine if the heart valves and heart muscle are functioning properly, and how blood is flowing through the heart.

The Doppler effect can be also used with sonar to determine the speed and direction of a detected object, such as a submarine or a school of fish.

Section  Assessment

1. When the amplitude of a sound wave is increased, what happens to the loudness of the sound? The pitch?
2. How does the wavelength of a sound affect the way it moves around corners?
3. How does the temperature of a material affect the speed of sound passing through it? Explain why in terms of the particles within the material.
4. What causes the Doppler effect, and in what ways is it used?
5. **Think Critically** Chemists sometimes use ultrasound machines to clean glassware. How could sound be used to remove particles from glass?

Skill Builder Activities

6. **Using an Electronic Spreadsheet** Think about ten different sounds you've heard today. Make a computer spreadsheet that lists each sound, the vibrating object that made the sound, and how the object was vibrating. **For more help, refer to the Technology Skill Handbook.**
7. **Solving One-Step Equations** If sound travels through water at 1,483 m/s, how far will it travel in 5 s? The speed of sound through air at 20° C is about 343 m/s. How far will sound travel through air in the same amount of time? **For more help, refer to the Math Skill Handbook.**

Activity

Observe and Measure Reflection of Sound

Like all waves, sound waves can be reflected. When sound waves strike a surface, in what direction does the reflected sound wave travel? In this activity, you'll focus sound waves using cardboard tubes to help answer this question.

What You'll Investigate

How do the angles made by incoming and reflected sound waves compare?

Materials

cardboard tubes, 20- to 30-cm-long (2)
watch with a second hand that ticks audibly
protractor

Goals

- **Observe** reflection of sound waves.
- **Measure** the angles incoming and reflected sound waves make with a surface.

Safety Precautions

Procedure

1. Work in groups of three. Each person should listen to the watch—first without a tube and then through a tube. The person who hears the watch most easily is the listener.
2. One person should hold one tube at an angle with one end above a table. Hold the watch at the other end of the tube.
3. The listener should hold the second tube at an angle, with one end near his or her ear and the other end near the end of the first tube that is just above the table. The tubes should be in the same vertical plane.

4. Move the first tube until the watch sounds loudest. The listener might need to cover the other ear to block out background noises.
5. With the tubes held steady, the third person should measure the angle that each tube makes with the table.

Conclude and Apply

1. Are the two angles approximately equal or quite different? How does the angle of reflection compare with the angle made by the incoming wave?
2. Predict how your results would change if the waves reflected from a soft surface instead of a hard surface.

Communicating Your Data

Make a scientific illustration to show how the experiment was done. Describe your results using the illustration. **For more help, refer to the Science Skill Handbook.**

Music

What is music?

What do you like to listen to—rock 'n' roll, country, blues, jazz, rap, or classical? Music and noise are groups of sounds. Why do humans hear some sounds as music and other sounds as noise?

The answer involves sound patterns. Music is a group of sounds that have been deliberately produced to make a regular pattern. Look at **Figure 12.** The sounds that make up music usually have a regular pattern of pitches, or notes. Some natural sounds, such as the patter of rain on a roof, the sound of ocean waves splashing, or the songs of birds can sound musical. On the other hand, noise is usually a group of sounds with no regular pattern. Sounds you hear as noise are irregular and disorganized, such as the sounds of traffic on a city street or the roar of jet aircraft.

However, the difference between music and noise can vary from person to person. What one person considers to be music, another person might consider noise.

Natural Frequencies Music is created by vibrations. When you sing, your vocal cords vibrate. When you beat a drum, the drumhead vibrates. When you play a guitar, the strings vibrate.

If you tap on a bell with a hard object, the bell produces a sound. When you tap on a bell that is larger or smaller or has a different shape you hear a different sound. The bells sound different because each bell vibrates at different frequencies. A bell vibrates at frequencies that depend on its shape and the material it is made from. Every object will vibrate at certain frequencies called its **natural frequencies.**

As You Read

What **You'll Learn**

- **Explain** the difference between music and noise.
- **Describe** how different instruments produce music.
- **Explain** how you hear.

Vocabulary

natural frequency, resonance, fundamental frequency, overtone, reverberation, eardrum

Why **It's Important**

By better understanding how music is produced, you can improve the quality of the sounds you make.

Figure 12
Music and noise have different types of sound patterns.

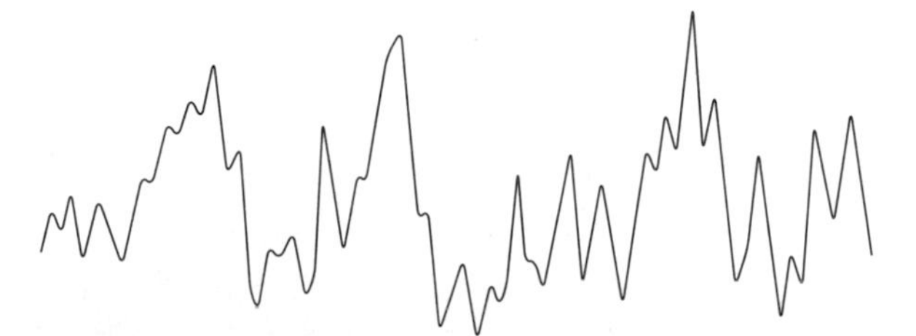

A Noise has no specific or regular sound wave pattern.

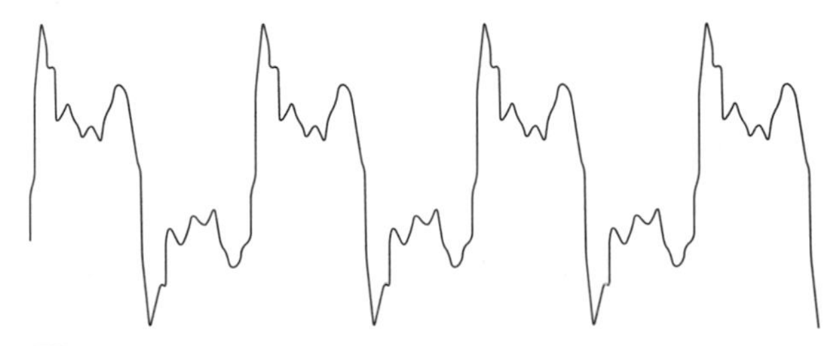

B Music is organized sound. Music has regular sound wave patterns and structures.

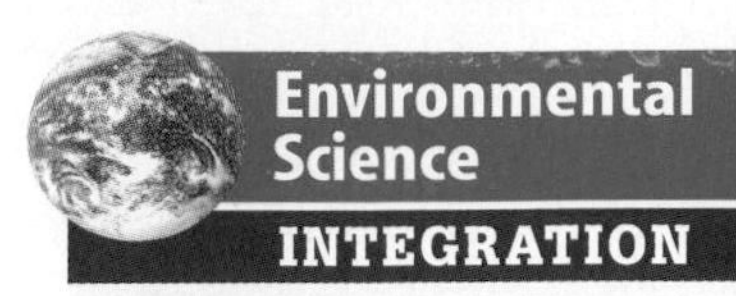

Environmental Science INTEGRATION

Resonance is important in fields outside of music. Earthquake-proof buildings, for example, are designed to resonate at frequencies that are different from those encountered in earthquakes.

Musical Instruments and Natural Frequencies Many objects vibrate at one or more natural frequencies when they are struck or disturbed. Like a bell, the natural frequency of any object depends on the size and shape of the object and the material it is made from. Musical instruments use the natural frequencies of strings, drumheads, or columns of air contained in pipes to produce various musical notes.

✔ Reading Check *What determines the natural frequencies?*

Resonance You may have seen the comedy routine in which a loud soprano sings high enough to shatter glass. Sometimes sound waves cause an object to vibrate. When a tuning fork is struck, it vibrates at its natural frequency and produces a sound wave with the same frequency. Suppose you have two tuning forks with the same natural frequency. You strike one tuning fork, and the sound waves it produces strike the other tuning fork. These sound waves would cause the tuning fork that wasn't struck to absorb energy and vibrate. This is an example of resonance. **Resonance** occurs when an object is made to vibrate at its natural frequencies by absorbing energy from a sound wave or another object vibrating at these frequencies.

Musical instruments use resonance to amplify their sounds. Look at **Figure 13.** The vibrating tuning fork has caused the table to vibrate at the same frequency, or resonate. The combined vibrations of the table and the tuning fork increase the loudness of the sound waves produced.

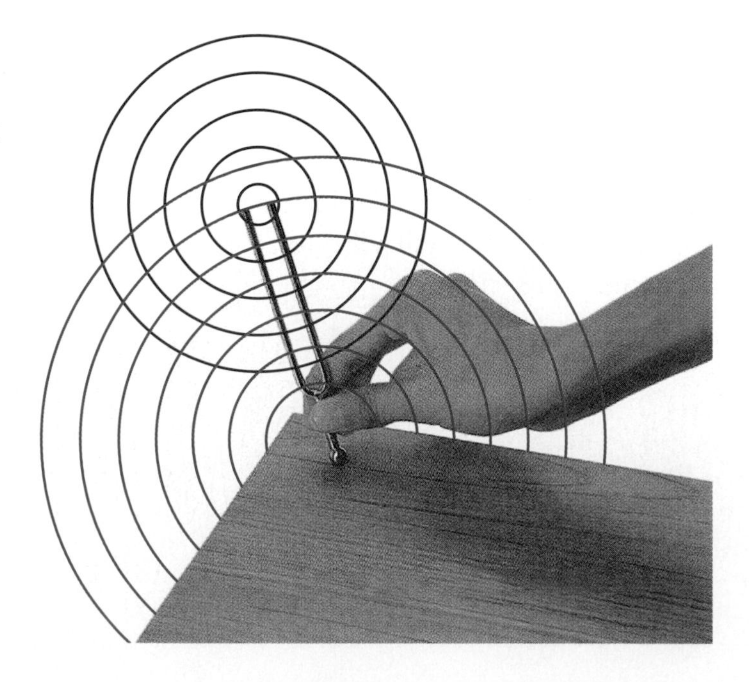

Figure 13
When a vibrating tuning fork is placed against a table, resonance might cause the table to vibrate.

Overtones

Before a concert, all orchestra musicians tune their instruments by playing the same note. Even though the note has the same pitch, it sounds different for each instrument. It also sounds different from a tuning fork that vibrates at the same frequency as the note.

A tuning fork produces a single frequency, called a pure tone. However, the notes produced by musical instruments are not pure tones. Most objects have more than one natural frequency at which they can vibrate. As a result, they produce sound waves of more than one frequency.

If you play a single note on a guitar, the pitch that you hear is the lowest frequency produced by the vibrating string. The lowest frequency produced by a vibrating object is the **fundamental frequency.** The vibrating string also produces higher frequencies. These higher frequencies are **overtones.** Overtones have frequencies that are multiples of the fundamental frequency, as in **Figure 14.** The number and intensity of the overtones produced by each instrument are different and give instruments their distinctive sound quality.

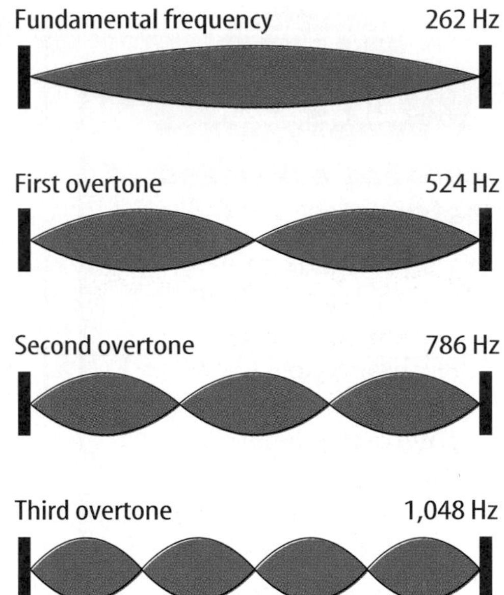

Figure 14
A string vibrates at a fundamental frequency, as well as at overtones. The overtones are multiples of that frequency.

Musical Scales

A musical instrument is a device that produces musical sounds. These sounds are usually part of a musical scale that is a sequence of notes with certain frequencies. For example, **Figure 15** shows the sequence of notes that belong to the musical scale of C. Notice that the frequency produced by the instrument doubles after eight successive notes of the scale are played. Other musical scales consist of a different sequence of frequencies.

How many musical instruments can you name? To find out more about musical instruments, see the **Musical Instruments Field Guide** at the back of the book.

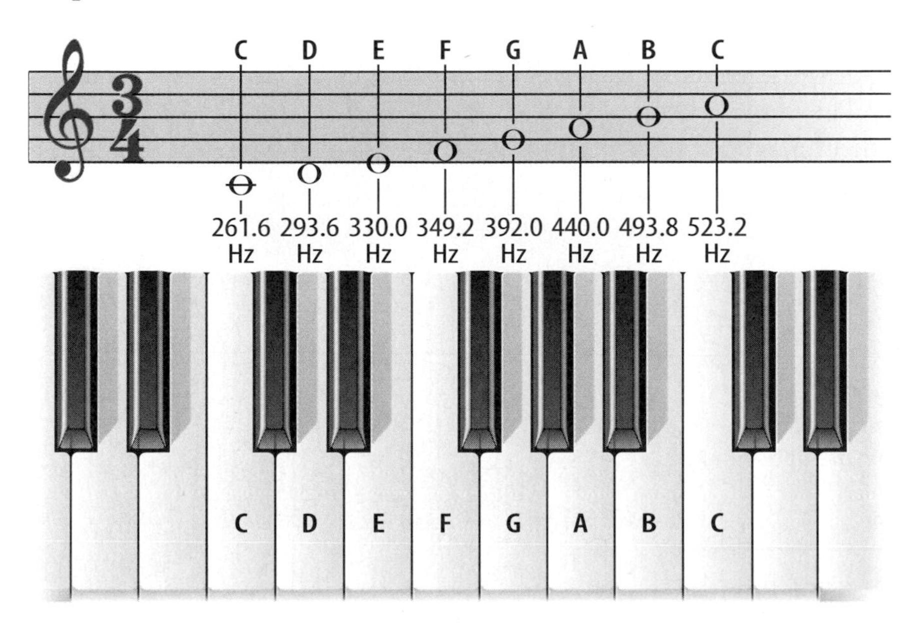

Figure 15
A piano produces a sequence of notes that are a part of a musical scale. *How are the frequencies of the two C notes on this scale related?*

Mini LAB

Modeling a Stringed Instrument

Procedure

1. Stretch a **rubber band** between your fingers.
2. Pluck the rubber band. Listen to the sound and observe the shape of the vibrating band. Record what you hear and see.
3. Stretch the band farther and repeat step 2.
4. Shorten the length of the band that can vibrate by holding your finger on one point. Repeat step 2.
5. Stretch the rubber band over an open box, such as a **shoe box.** Repeat step 2.

Analysis

1. How did the sound change when you stretched the rubber band? Was this what you expected? Explain.
2. How did the sound change when you stretched the band over the box? Did you expect this? Explain.

Stringed Instruments

Stringed instruments, like the cello shown in **Figure 16,** produce music by making strings vibrate. Different methods are used to make the strings vibrate—guitar strings are plucked, piano strings are struck, and a bow is slid across cello strings. The strings often are made of wire. The pitch of the note depends on the length, diameter, and tension of the string—if the string is shorter, narrower, or tighter, the pitch increases. For example, pressing down on a vibrating guitar string shortens its length and produces a note with a higher pitch. Similarly, the thinner guitar strings produce a higher pitch than the thicker strings.

Amplifying Vibrations The sound produced by a vibrating string is soft. To amplify the sound, stringed instruments usually have a hollow chamber, or box, called a resonator, which contains air. The resonator absorbs energy from the vibrating string and vibrates at its natural frequencies. For example, the body of a guitar is a resonator that amplifies the sound that is produced by the vibrating strings. The vibrating strings cause the guitar's body and the air inside it to resonate. As a result, the vibrating guitar strings sound louder, just as the tuning fork that was placed against the table sounded louder.

Figure 16
A cello is a stringed instrument. When strings vibrate, the natural frequencies of the instrument's body amplify the sound.

Percussion

Percussion instruments, such as the drum shown in **Figure 17A,** are struck to make a sound. Striking the top surface of the drum causes it to vibrate. The vibrating drumhead is attached to a chamber that resonates and amplifies the sound.

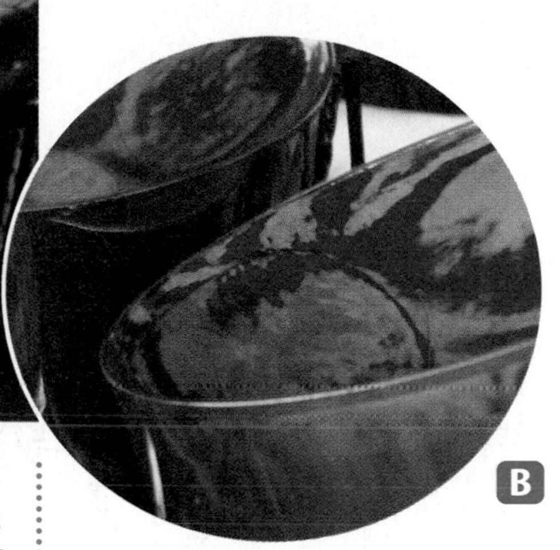

Figure 17
The sounds produced by drums depend on the material that is vibrating. A The vibrating drumhead of this drum is amplified by the resonating air in the body of the drum. B The vibrating steel surface in a steel drum produces loud sounds that don't need to be amplified by an air-filled chamber.

Drums and Pitch Some drums have a fixed pitch, but some can be tuned to play different notes. For example, if the drumhead on a kettledrum is tightened, the natural frequency of the drumhead is increased. As a result, the pitches of the sounds that are produced by the kettledrum get higher. A steel drum, shown in **Figure 17B,** plays different notes in the scale when different areas in the drum are struck. In a xylophone, wood or metal bars of different lengths are struck. The longer the bar is, the lower the note that it produces is.

Brass and Woodwinds

Just as the bars of a xylophone have different natural frequencies, so do the air columns in pipes of different lengths. Brass and woodwind instruments, such as those in **Figure 18,** are essentially pipes or tubes of different lengths that sometimes are twisted around to make them easier to hold and carry. To make music from these instruments, the air in the pipes is made to vibrate at various frequencies.

Different methods are used to make the air column vibrate. A musician playing a brass instrument, such as a trumpet, makes the air column vibrate by vibrating the lips and blowing into the mouthpiece. Woodwinds such as clarinets, saxophones, and oboes contain one or two reeds in the mouthpiece that vibrate the air column when the musician blows into the mouthpiece. Flutes also are woodwinds, but a flute player blows across a narrow opening to make the air column vibrate.

Figure 18
Brass and woodwind instruments produce sounds in a vibrating column of air. *What other instruments make sound this way?*

Figure 19
A flute changes pitch as holes are opened and closed.

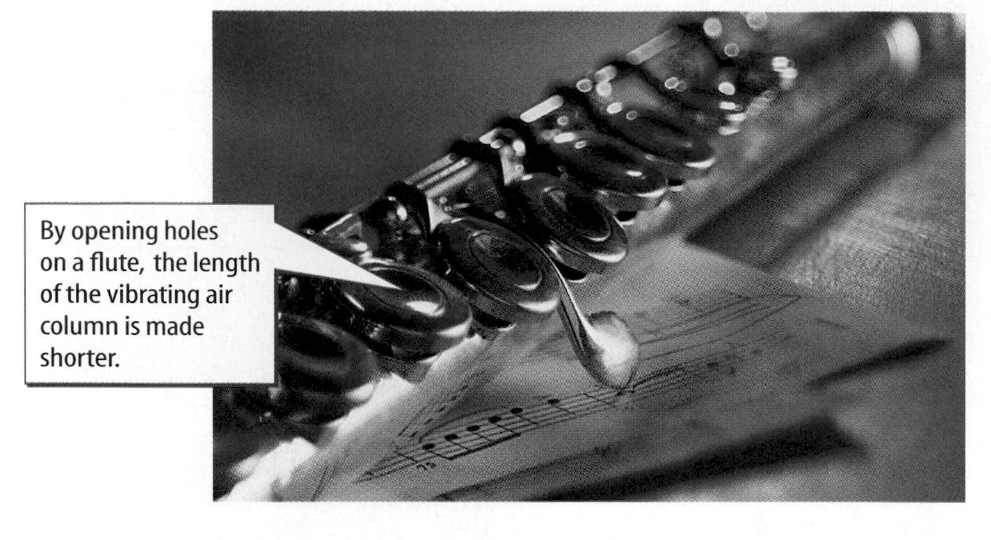

Changing Pitch in Woodwinds To change the note that is being played in a woodwind instrument, a musician changes the length of the resonating column of air. By making the length of the vibrating air column shorter, the pitch of the sound produced is made higher. In a woodwind such as a flute, saxophone, or clarinet, this is done by closing and opening finger holes along the length of the instrument, as shown in **Figure 19.**

Changing Pitch in Brass In brass instruments, musicians vary the pitch in other ways. One is by blowing harder to make the air resonate at a higher natural frequency. Another way is by pressing valves that change the length of the tube.

Beats

When two notes are close in frequency, they interfere in a distinctive way. The two waves combine to form a wave that varies slowly in loudness. This slow variation creates beats. **Figure 20** shows the beats that are produced by the interference of two waves with frequencies of 9 Hz and 12 Hz. The frequency of the beat is the difference in the frequencies —in this case 3 Hz. Listening to two tones at the same time with a frequency difference of 3 Hz, you would hear the sound get louder and softer—a beat—three times each second.

Figure 20
Beats are formed when two frequencies that are nearly the same are played together. The sound wave in A has a frequency of 12 Hz and the sound wave in B has a frequency of 9 Hz. When these two sounds are played together, they interfere and form the wave in C that has a frequency of 3 Hz. You would hear 3 beats each second.

Beats Help Tune Instruments Beats are used to help tune instruments. For example, a piano tuner might hit a tuning fork and then the corresponding key on the piano. Beats are heard when the difference in pitch is small. The piano string is tuned properly when the beats disappear. You might have heard beats while listening to an orchestra tune before a performance. You also can hear beats produced by two engines vibrating at slightly different frequencies.

Reverberation

Sound is reflected by hard surfaces. In an empty gymnasium, the sound of your voice can be reflected back and forth several times by the floor, walls, and ceiling. Repeated echoes of sound are called **reverberation.** In a gym, reverberation makes the sound of your voice linger before it dies out. Some reverberation can make voices or music sound bright and lively. However, reverberation can produce a confusing mess of noise if too many sounds linger for too long. Too little reverberation makes the sound flat and lifeless. Concert halls and theaters, such as the one in **Figure 21,** are designed to produce the appropriate level of reverberation. Acoustical engineers use soft materials to reduce echoes. Special panels that are attached to the walls or suspended from the ceiling are designed to reflect sound toward the audience.

SCIENCE Online

Research Visit the Glencoe Science Web site at **science.glencoe.com** for more information about how concert halls are designed to produce the proper amount of reverberation. Communicate to your class what you learn.

Figure 21
The shape of a concert hall and the materials it contains are designed to control the reflection of sound waves.

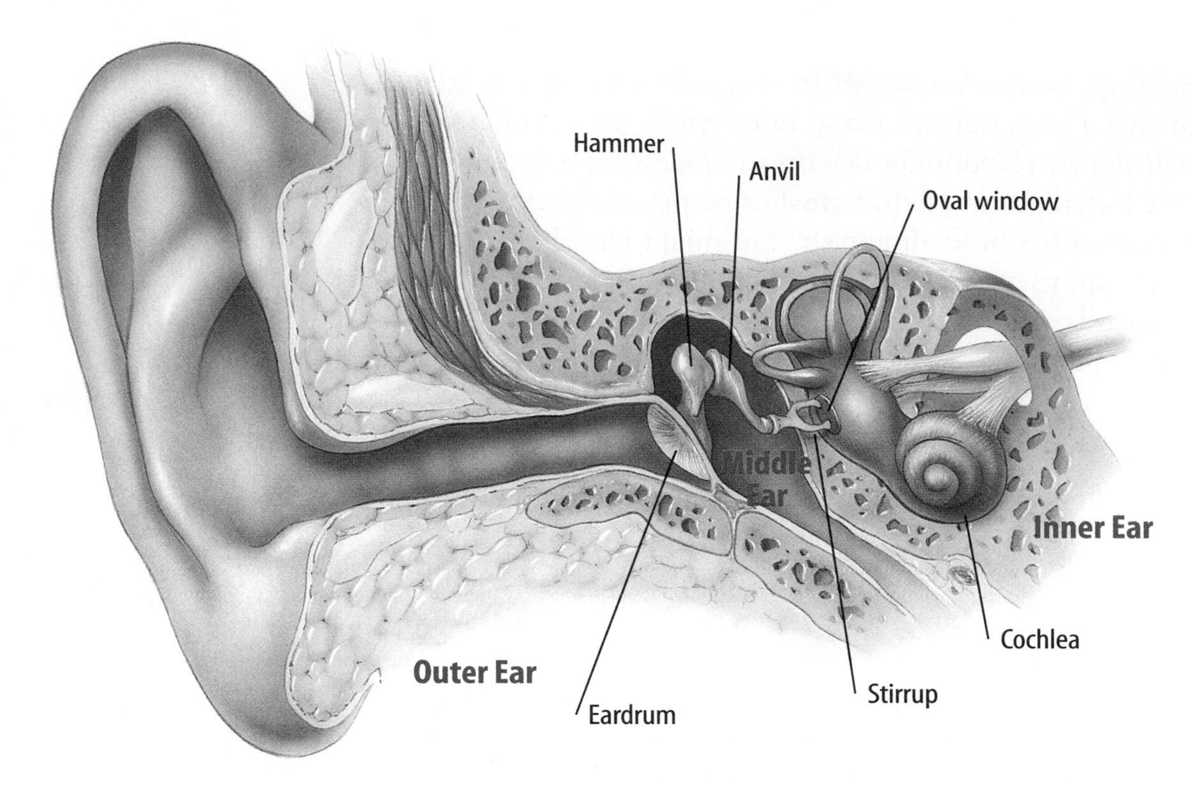

Figure 22
The human ear has three different parts—the outer ear, the middle ear, and the inner ear.

The Ear

Sound is all around you. Sounds are as different as the loud buzz of an alarm clock and the quiet hum of a bee. You hear sounds with your ears. The ear is a complex organ that is able to detect a wide range of sounds. The human ear is illustrated in **Figure 22.** It has three parts—the outer ear, the middle ear, and the inner ear.

Figure 23
Animals, such as rabbits and owls, have ears that are adapted to their different needs.

The Outer Ear—Sound Collector Your outer ear collects sound waves and directs them into the ear canal. Notice that your outer ear is shaped roughly like a funnel. This shape helps collect sound waves.

Animals that rely on hearing to locate predators or prey often have larger, more adjustable ears than humans, as shown in **Figure 23.** A barn owl, which relies on its excellent hearing for hunting at night, does not have outer ears made of flesh. Instead, the arrangement of its facial feathers helps direct sound to its ears. Some sea mammals, on the other hand, have only small holes for outer ears, even though their hearing is good.

The Middle Ear—Sound Amplifier When sound waves reach the middle ear, they vibrate the **eardrum,** which is a membrane that stretches across the ear canal like a drumhead. When the eardrum vibrates, it transmits vibrations to three small connected bones—the hammer, anvil, and stirrup. The bones amplify the vibrations, just as a lever can change a small movement at one end into a larger movement at the other.

The Inner Ear—Sound Interpreter The stirrup vibrates a second membrane called the oval window. This marks the start of the inner ear, which is filled with fluid. Vibrations in the fluid are transmitted to hair-tipped cells lining the cochlea, as shown in **Figure 24.** Different sounds vibrate the cells in different ways. The cells generate signals containing information about the frequency, intensity, and duration of the sound. The nerve impulses travel along the auditory nerve and are transmitted to the part of the brain that is responsible for hearing.

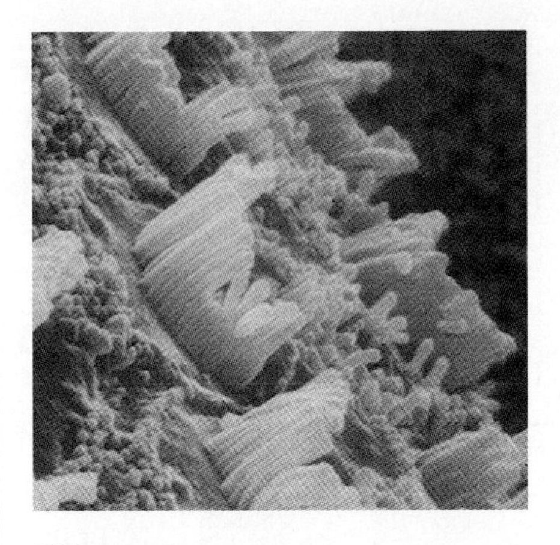

Figure 24
The inner ear contains tiny hair cells that convert vibrations into nerve impulses that travel to the brain.

✔ **Reading Check** ***Where are waves detected and interpreted in the ear?***

Hearing Loss

The ear can be damaged by disease, age, and exposure to loud sounds. For example, constant exposure to loud noise can damage hair cells in the cochlea. If damaged mammalian hair cells die, some loss of hearing results because mammals cannot make new hair cells. Also, some hair cells and nerve fibers in the inner ear degenerate and are lost as people age. It is estimated that about 30 percent of people over 65 have some hearing loss due to aging.

The higher frequencies are usually the first to be lost. The loss of the higher frequencies also distorts sound. The soft consonant sounds, such as those made by the letters *s, f, h, sh,* and *ch,* are hard to hear. People with high-frequency hearing loss have trouble distinguishing these sounds in ordinary conversation.

Section  Assessment

1. How are music and noise different?
2. Two bars on a xylophone are 10 cm and 14 cm long. Which bar will produce a lower pitch when struck? Explain.
3. Why would the sound of a guitar string sound louder when attached to the body of the guitar than when plucked alone?
4. What are the parts of the human ear, and how do they enable you to hear sound?
5. **Think Critically** As the size of stringed instruments increases from violin to viola, cello, and bass, the sound of the instruments becomes lower pitched. Explain.

Skill Builder Activities

6. **Making Models** Illustrate the fundamental and first overtone for a string. **For more help, refer to the Science Skill Handbook.**
7. **Communicating** Imagine that human hearing is much more sensitive than it currently is. Write a story describing a day in the life of your main character. Be sure to describe your setting in detail. For example, does your story take place in a crowded city or a scenic national park? How would life be different? Describe your story in your Science Journal. **For more help, refer to the Science Skill Handbook.**

Activity Design Your Own Experiment

Music

The pitch of a note that is played on an instrument sometimes depends on the length of the string, the air column, or some other vibrating part. Exactly how does sound correspond to the size or length of the vibrating part? Is this true for different instruments?

Recognize the Problem

What causes different instruments to produce different notes?

Form a Hypothesis

Based on your reading and observations, make a hypothesis about what changes in an instrument to produce different notes.

Goals

- **Design** an experiment to compare the changes that are needed in different instruments to produce a variety of different notes.
- **Observe** which changes are made when playing different notes.
- **Measure and record** these changes whenever possible.

Possible Materials

musical instruments
measuring tape
tuning forks

Safety Precautions

Properly clean the mouthpiece of any instrument before it is used by another student.

Test Your Hypothesis

Plan

1. You should do this activity as a class, using as many instruments as possible. You might want to go to the music room or invite friends and relatives who play an instrument to visit the class.
2. As a group, decide how you will measure changes in instruments. For wind instruments, can you measure the length of the vibrating air column? For stringed instruments, can you measure the length and thickness of the vibrating string?
3. Refer to the table of wavelengths and frequencies for notes in the scale. Note that no measurements are given—if you measure C to correspond to a string length of 30 cm, for example, the note G will correspond to two thirds of that length.
4. Decide which musical notes you will compare. Prepare a table to collect your data. List the notes you have selected.

Ratios of Wavelengths and Frequencies of Musical Notes

Note	Wavelength	Frequency
C	1	1
D	8/9	9/8
E	4/5	5/4
F	3/4	4/3
G	2/3	3/2
A	3/5	5/3
B	8/15	15/8
C	1/2	2

Do

1. Make sure your teacher approves your plan before you start.
2. Carry out the experiment as planned.
3. While doing the experiment, record your observations and complete the data table.

Analyze Your Data

1. **Compare** the change in each instrument when the two notes are produced.
2. **Compare and contrast** the changes between instruments.
3. What were the controls in this experiment?
4. What were the variables in this experiment?
5. How did you eliminate bias?

Draw Conclusions

1. How does changing the length of the vibrating column of air in a wind instrument affect the note that is played?
2. **Describe** how you would modify an instrument to increase the pitch of a note that is played.

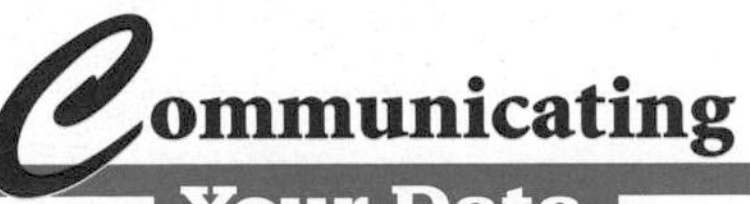

Demonstrate to another teacher or to family members how the change in the instrument produces a change in sound.

It's a Wrap!

No matter how quickly or slowly you open a candy wrapper, it always will make a noise

You're at the movies, and it's the most exciting part of the film. The audience is silent with their eyes riveted to the screen. At that moment, you decide to unwrap the candy you got at the concession stand—"CRACKLE!" The loud noise isn't from the movie. It's from the candy wrapper. Your friends shush you. So you try to open the wrapper more carefully—"POP!" Now you try opening it more slowly—"SNAP!" No matter how you open the candy wrapper—fast or slow—it makes a lot of annoying noise.

Just about everyone has been in that situation at a movie or a concert. And just about everyone has wondered why you can't unwrap candy without making a racket—no matter how hard you try. But now, finally, thanks to the work of a few curious physicists, we know the answer.

To test the plastic problem, researchers took some crinkly wrappers and put them in a silent room. Then the researchers stretched out the wrappers and recorded the sound they made. Next, the crinkling sound was run through a computer. After analyzing the sound, the research team discovered something very interesting—the wrapper didn't make a nonstop, continuous sound. Instead, it made many little separate popping noises. Each of these sound bursts took only a thousandth of a second.

Pop Goes the Wrapper

The researchers found that the loudness of the pops had nothing to do with how fast the plastic was unwrapped. The pops randomly took place. The reason? Little creases in the plastic suddenly snapped into a new position as the wrapper was stretched.

So, if you unwrap candy more slowly, the time between pops will be longer, but the amount of noise made by the pops will be the same. And whether you open the wrapper fast or slow, you'll always hear pops. "And there's nothing you can do about it," said a member of the research team.

Is there another payoff to the candy wrapper research? One scientist said that by understanding what makes a plastic wrapper "snap" when it changes shape, the information can actually help doctors understand molecules in the human body. These molecules, like plastic, can change shape.

But, in the meantime, what are you supposed to do when you absolutely have to open candy in a silent theater? Be considerate of others in the audience. Open the candy as fast as you can, and just get it over with. You can even wait until a noisy part of the movie to hide the crinkle, or open the candy before the film begins.

The pop chart

SOUND LEVEL OVER TIME

The sound that a candy wrapper makes is emitted as a series of pulses or clicks. So, opening a wrapper slowly only increases the length of time in between clicks, but the amount of noise remains the same. (TALLER SPIKES SIGNIFY LOUDER CLICKS)

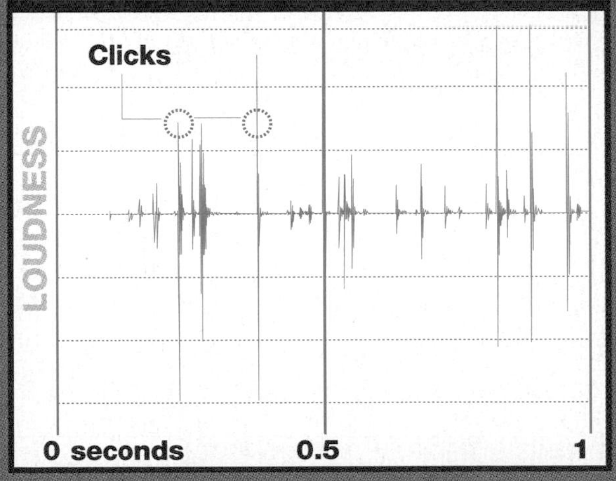

Source: Eric Kramer, Simon's Rock College, 2000

CONNECTIONS Recall and Retell Have you ever opened a candy wrapper in a quiet place? Did it bother other people? If so, did you try to open it more slowly? What happened?

SCIENCE Online
For more information, visit science.glencoe.com

Chapter 21 Study Guide

Reviewing Main Ideas

Section 1 What is sound?

1. Sound is a compressional wave that travels through matter, such as air. Sound is produced by something that vibrates.
2. The speed of sound is different in different materials. In general, sound travels faster in solids than in liquids, and faster in liquids than in gases. *Will the sound of a train travel faster through the air or through these tracks?*

3. The larger the amplitude of a sound wave, the more energy it carries. The loudness of a sound wave increases as its amplitude increases.
4. The pitch of a sound wave corresponds to its frequency. Sound waves can reflect, or bounce, from objects and diffract, or bend around objects.
5. The Doppler effect occurs when the source of sound and the listener are in motion relative to each other. Sound is shifted up or down in pitch. *What happens to the pitch of the train's horn as it approaches the person?*

Section 2 Music

1. Music is made of sounds that are used in a regular pattern. Noise is made of sounds that are irregular and disorganized.
2. Objects vibrate at their natural frequencies. These depend on the shape of the object and the material it's made of.
3. Resonance occurs when an object is made to vibrate by absorbing energy at one of its natural frequencies.
4. Musical instruments produce notes by vibrating at their natural frequencies. Resonance is used to amplify the sound. *How does resonance make this violin sound louder?*

5. Beats occur when two sounds of nearly the same frequency interfere. The beat frequency is the difference in frequency of the sounds.
6. The ear collects sound waves, amplifies the vibrations, and converts the vibrations to nerve impulses.

FOLDABLES Reading & Study Skills

After You Read

Use the library to find answers to any questions remaining on your Question Study Foldable.

Chapter 21 Study Guide

Visualizing Main Ideas

Complete the following concept map on sound.

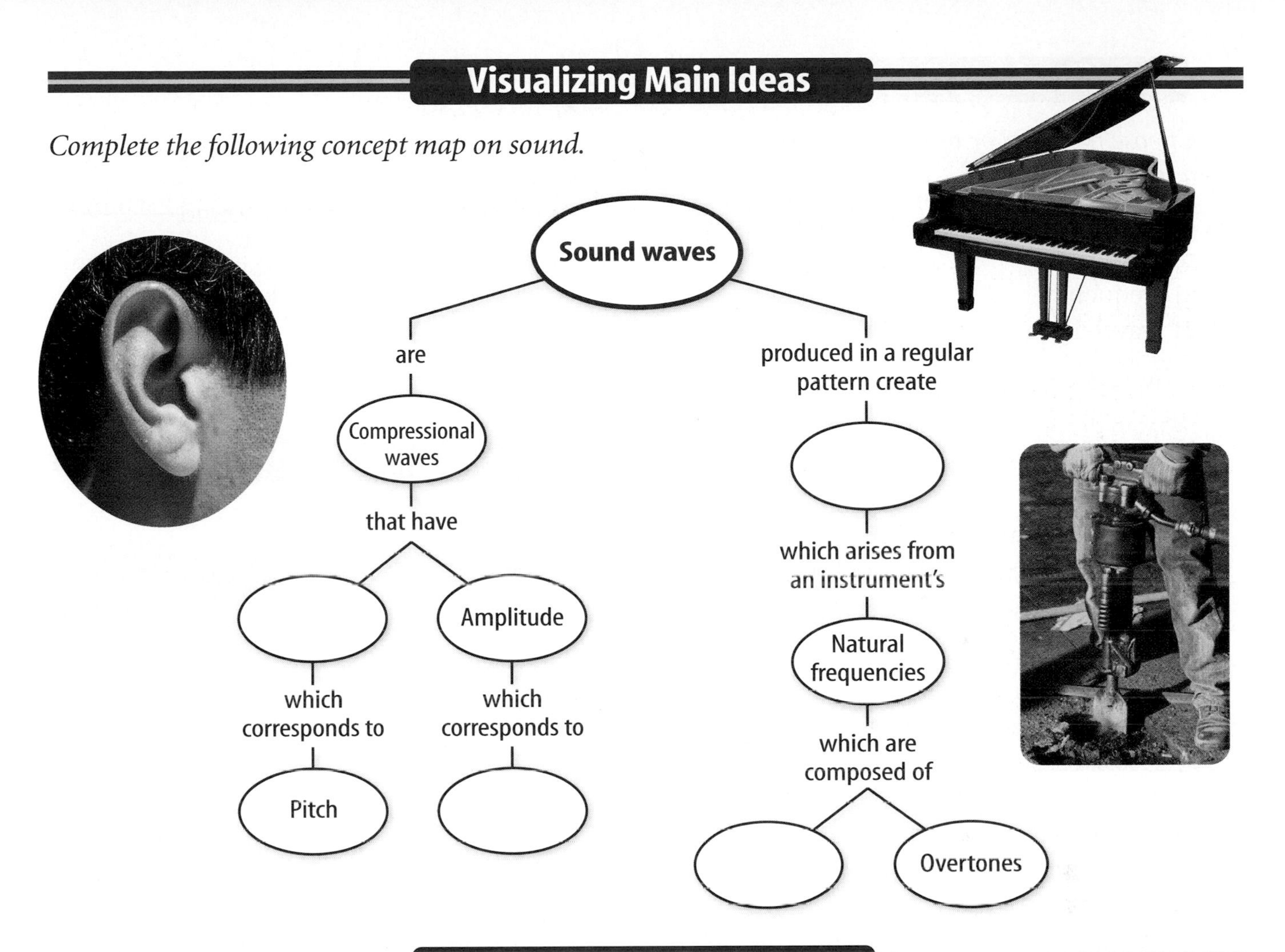

Vocabulary Review

Vocabulary Words

a. Doppler effect
b. eardrum
c. echo
d. fundamental frequency
e. loudness
f. natural frequency
g. overtone
h. pitch
i. resonance
j. reverberation

Study Tip

Recopy your notes from class. As you do, explain each concept in more detail to make sure that you understand it completely.

Using Vocabulary

Distinguish between the terms in each of the following pairs.

1. overtones, fundamental frequency

2. pitch, sound wave

3. pitch, Doppler effect

4. loudness, resonance

5. fundamental, natural frequency

6. loudness, amplitude

7. natural frequency, overtone

8. reverberation, resonance

Chapter 21 Assessment

Checking Concepts

Choose the word or phrase that best answers the question.

1. A tone that is lower in pitch is lower in what characteristic?
   A) frequency
   B) wavelength
   C) loudness
   D) resonance

2. If frequency increases, what decreases if speed stays the same?
   A) pitch
   B) wavelength
   C) loudness
   D) resonance

3. What part of the ear is damaged most easily by continued exposure to loud noise?
   A) eardrum
   B) stirrup
   C) oval window
   D) hair cells

4. What is an echo?
   A) diffracted sound
   B) resonating sound
   C) reflected sound
   D) Doppler-shifted sound

5. A trumpeter depresses keys to make the column of air resonating in the trumpet shorter. What happens to the note that is being played?
   A) The pitch is higher.
   B) The pitch is lower.
   C) It is quieter.
   D) It is louder.

6. When tuning a violin, a string is tightened. What happens to the note that is being played on that string?
   A) The pitch is higher.
   B) The pitch is lower.
   C) It is quieter.
   D) It is louder.

7. If air becomes warmer, what happens to the speed of sound in air?
   A) It increases.
   B) It decreases.
   C) It doesn't change.
   D) It oscillates.

8. Sound is what type of wave?
   A) slow
   B) transverse
   C) compressional
   D) fast

9. What does the middle ear do?
   A) focuses sound
   B) interprets sound
   C) collects sound
   D) transmits and amplifies sound

10. An ambulance siren speeds away from you. What happens to the pitch you hear?
    A) It increases.
    B) It becomes louder.
    C) It decreases.
    D) Nothing happens.

Thinking Critically

11. Some xylophones have open pipes of different lengths hung under each bar. The longer the bar is, the longer the corresponding pipe is. Explain how these pipes amplify the sound of the xylophone.

12. Why don't you notice the Doppler effect for a slow-moving train?

13. Suppose the movement of the bones in the middle ear were reduced. Which would be more affected—the ability to hear quiet sounds or the ability to hear certain frequencies? Explain your answer.

14. Two flutes are playing at the same time. One flute plays a note with frequency 524 Hz. If two beats are heard per second, what are the possible frequencies the other flute is playing?

15. The triangle is a percussion instrument consisting of an open metal triangle hanging from a string. The triangle is struck by a metal rod, and a chiming sound is heard. If the metal triangle is held in the hand rather than by the string, a quiet, dull sound is made when it is struck. Explain why holding the triangle makes it sound quieter.

Developing Skills

16. Predicting If the holes of a flute are all covered while playing, then all uncovered, what happens to the length of the vibrating air column? What happens to the pitch of the note?

17. Identifying and Manipulating Variables and Controls Describe an experiment to demonstrate that sound is diffracted.

18. Making and Using Tables Make a table to show the first three overtones for a note of G, which has a frequency of 392 Hz.

19. Interpreting Scientific Illustrations The picture shows pan pipes. How are different notes produced by blowing on pan pipes?

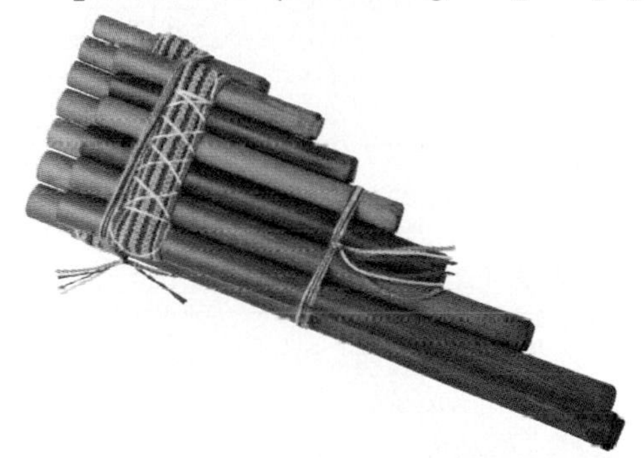

Performance Assessment

20. Recital Perform a short musical piece on an instrument. Explain how your actions changed the notes that were produced.

21. Pamphlet Create a pamphlet describing how a hearing aid works.

Technology

Go to the Glencoe Science Web site at **science.glencoe.com** or use the **Glencoe Science CD-ROM** for additional chapter assessment.

Test Practice

Sound travels in waves that change as the pitch and loudness of the sound vary. These pictures illustrate four recorded sounds.

Q. 

R. 

S. 

T.

Study the pictures and answer the following questions.

1. Which of the four sounds was getting louder while it was recorded?

A) Q
B) R
C) S
D) T

2. Which sound had the highest pitch while it was recorded?

F) Q
G) R
H) S
J) T

Electromagnetic Waves

Wherever you go, you are being bombarded by electromagnetic waves. Some, such as visible light, can be seen. Infrared rays can't be seen but feel warm on your skin. The paint on the tricycle in this picture is being heat cured in an infrared oven. In this chapter, you will learn how electromagnetic waves are formed. You also will learn ways in which electromagnetic waves are used, from cooking to satellite communications.

What do you think?

Science Journal Look at the photograph below with a classmate. Discuss what you think this might be. Here is a hint: *Scientists built this to get a clearer picture.* Write your answer or best guess in your Science Journal.

Light is a type of wave called an electromagnetic wave. You see light every day, but visible light is only one type of electromagnetic wave. Other electromagnetic waves are all around you, but you cannot see them. How can you detect electromagnetic waves that can't be seen with your eyes?

Detecting invisible light

1. Cut a slit 2 cm long and 0.25 cm wide in the center of a sheet of black paper.
2. Cover a window that is in direct sunlight with the paper.
3. Position a glass prism in front of the light coming through the slit so it makes a visible spectrum on the floor or table.
4. Place one thermometer in the spectrum and a second thermometer just beyond the red light.
5. Measure the temperature in each region after 5 min.

Observe

Write a paragraph in your Science Journal comparing the temperatures of the two regions and offer an explanation for the observed temperatures.

Before You Read

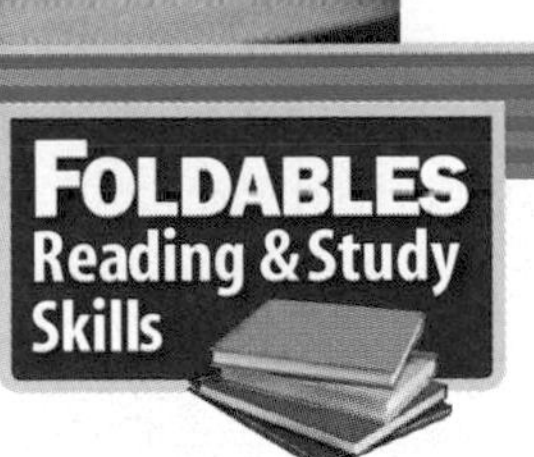

Making a Main Ideas Study Fold **Make the following Foldable to help you identify the major topics about electromagnetic waves.**

1. Stack four sheets of paper in front of you so the short sides are at the top.
2. Slide the top sheet up so about 2 cm of the next sheet shows. Slide each sheet up so about 2 cm of the next sheet shows.
3. Fold the sheets top to bottom to form eight tabs. Staple along the top fold.
4. Label the tabs *Electromagnetic Spectrum, Radio Waves, Microwaves, Infrared Rays, Visible Light, Ultraviolet Light, X Rays,* and *Gamma Rays.*
5. As you read the chapter, list the things you learn about these electromagnetic waves under the tabs.

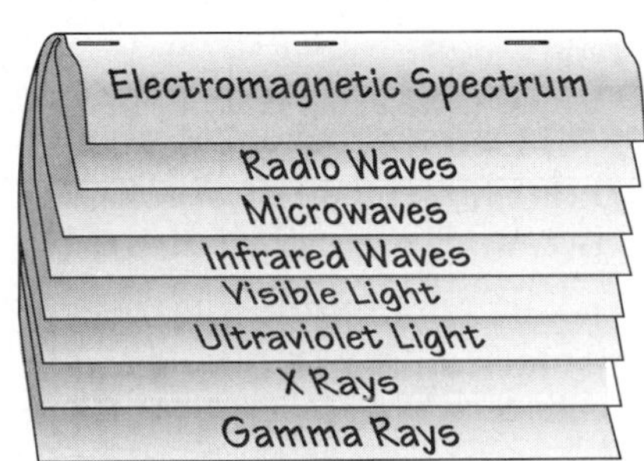

SECTION

The Nature of Electromagnetic Waves

As You Read

What You'll Learn

- **Explain** how electromagnetic waves are produced.
- **Describe** the properties of electromagnetic waves.

Vocabulary

electromagnetic wave
radiant energy

Why It's Important

The energy Earth receives from the Sun is carried by electromagnetic waves.

Waves in Space

On a clear day you feel the warmth in the Sun's rays, and you see the brightness of its light. Energy is being transferred from the Sun to your skin and eyes. Who would guess that the way in which this energy is transferred has anything to do with radios, televisions, microwave ovens, or the X-ray pictures that are taken by a doctor or dentist? Yet the Sun and the objects shown in **Figure 1** use the same type of wave to move energy from place to place.

Transferring Energy A wave transfers energy from one place to another without transferring matter. How do waves transfer energy? Waves, such as water waves and sound waves, transfer energy by making particles of matter move. The energy is passed along from particle to particle as they collide with their neighbors. Mechanical waves are the types of waves that use matter to transfer energy.

How can a wave transfer energy from the Sun to Earth? Mechanical waves, for example, can't travel in the space between Earth and the Sun where no matter exists. Instead, this energy is carried by a different type of wave called an electromagnetic wave. An **electromagnetic wave** is a wave that can travel through empty space and is produced by charged particles that are in motion.

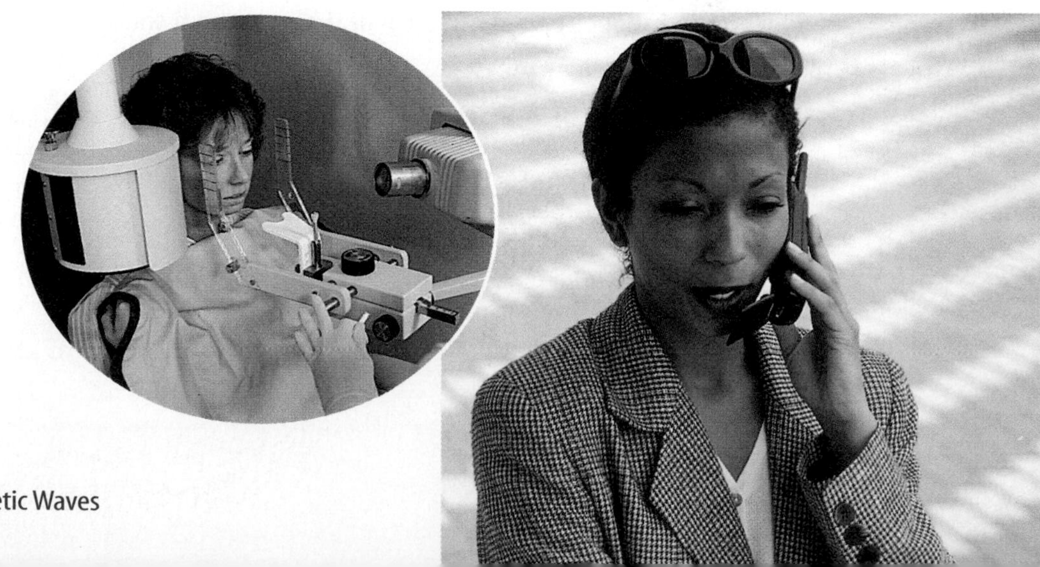

Figure 1
Getting an X ray at the dentist's office and talking on a cell phone are possible because energy is carried through space by electromagnetic waves.

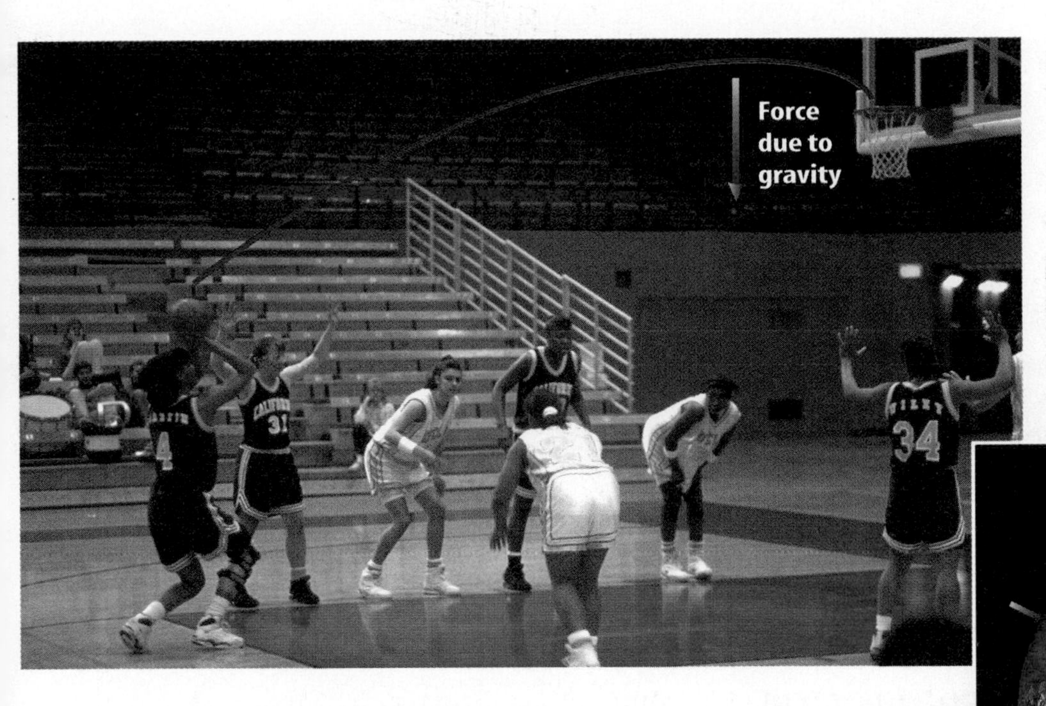

Figure 2
A gravitational field surrounds all objects, such as Earth.

A **When a ball is thrown, Earth's gravitational field exerts a downward force on the ball at every point along the ball's path.**

B **Earth's gravitational field extends out through space, exerting a force on all masses.**

Forces and Fields

An electromagnetic wave is made of two parts—an electric field and a magnetic field. These fields are force fields. A force field enables an object to exert forces on other objects, even though they are not touching. Earth produces a force field called the gravitational field. This field exerts the force of gravity on all objects that have mass.

✔ Reading Check *What force field surrounds Earth?*

How does Earth's force field work? If you throw a ball in the air as high as you can, it always falls back to Earth. At every point along the ball's path, the force of gravity pulls down on the ball, as shown in **Figure 2A.** In fact, at every point in space above or at Earth's surface, a ball is acted on by a downward force exerted by Earth's gravitational field. The force exerted by this field on a ball could be represented by a downward arrow at any point in space. **Figure 2B** shows this force field that surrounds Earth and extends out into space. In fact, it is Earth's gravitational field that causes the Moon to orbit Earth.

Magnetic Fields You know that magnets repel and attract each other even when they aren't touching. Two magnets exert a force on each other when they are some distance apart because each magnet is surrounded by a force field called a magnetic field. Just as a gravitational field exerts a force on a mass, a magnetic field exerts a force on another magnet and on magnetic materials. Magnetic fields cause other magnets to line up along the direction of the magnetic field.

Research In addition to a gravitational field, Earth also is surrounded by a magnetic field. Visit the Glencoe Science Web site at **science.glencoe.com** for more information about Earth's gravitational and magnetic force fields. Place the information you gather on a poster to share with your class.

Figure 3
Force fields surround all magnets and electric charges.

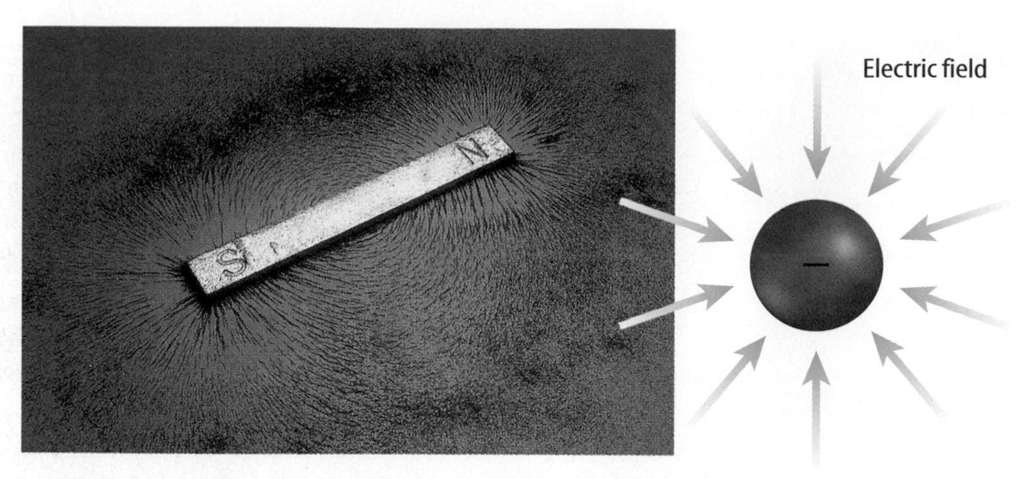

A **A magnetic field surrounds all magnets. The magnetic field exerts a force on iron filings, causing them to line up with the field.**

B **The electric field around an electric charge extends out through space, exerting forces on other charged particles.**

Electric Fields Recall that atoms contain protons, neutrons, and electrons. Protons and electrons have a property called electric charge. The two types of electric charge are positive and negative. Protons have positive charge and electrons have negative charge.

Just as a magnet is surrounded by a magnetic field, a particle that has electric charge, such as a proton or an electron, is surrounded by an electric field, as shown in **Figure 3.** The electric field is a force field that exerts a force on all other charged particles that are in the field.

Making Electromagnetic Waves

An electromagnetic wave is made of electric and magnetic fields. How is such a wave produced? Think about a wave on a rope. You can make a wave on a rope by shaking one end of the rope up and down. Electromagnetic waves are produced by making charged particles, such as electrons, move back and forth, or vibrate.

A charged particle always is surrounded by an electric field. But a charged particle that is moving also is surrounded by a magnetic field. For example, when an electric current flows in a wire, electrons are moving in the wire. As a result, the wire is surrounded by a magnetic field, as shown in **Figure 4.** So a moving charged particle is surrounded by an electric field and a magnetic field.

Figure 4
Electrons moving in a wire produce a magnetic field in the surrounding space.

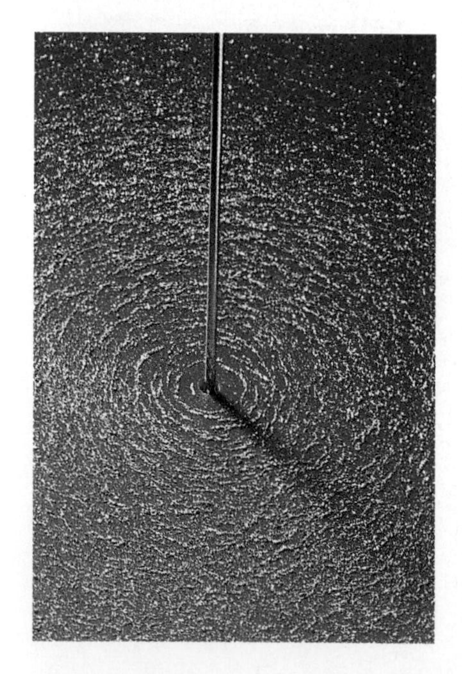

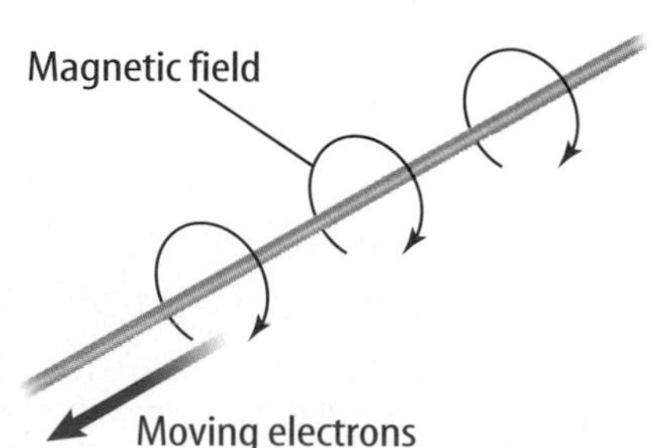

Producing Waves When you shake a rope up and down, you produce a wave that moves away from your hand. As a charged particle vibrates by moving up and down or back and forth, it produces changing electric and magnetic fields that move away from the vibrating charge in all directions. These changing fields traveling in all directions form an electromagnetic wave. **Figure 5A** shows these changing fields along one direction.

Properties of Electromagnetic Waves

Like all waves, an electromagnetic wave has a frequency and a wavelength. When you create a wave on a rope, you move your hand up and down while holding the rope. Look at **Figure 5B.** Frequency is how many times you move the rope through one complete up and down cycle in 1 s. Wavelength is the distance from one crest to the next or from one trough to the next.

Wavelength and Frequency An electromagnetic wave is produced by a vibrating charged particle. When the charge makes one complete vibration, one wavelength is created, as shown in **Figure 5A.** Like a wave on a rope, the frequency of an electromagnetic wave is the number of wavelengths that pass by a point in 1 s. This is the same as the number of times in 1 s that

Observing Electric Fields

Procedure

1. Rub a **hard, plastic comb** vigorously with a **wool sweater or wool flannel shirt.**
2. Turn on a **water faucet** to create the smallest possible continuous stream of water.
3. Hold the comb near the stream of water and observe.

Analysis

1. What happened to the stream of water when you held the comb near it?
2. Explain why the stream of water behaved this way.

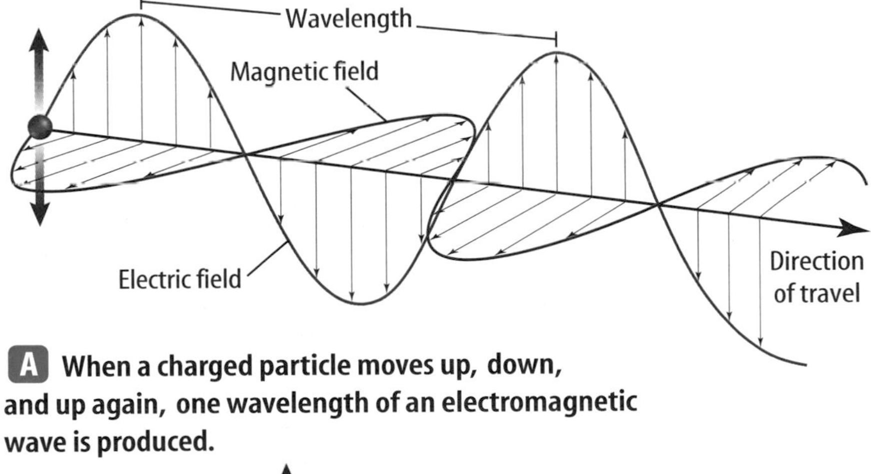

A When a charged particle moves up, down, and up again, one wavelength of an electromagnetic wave is produced.

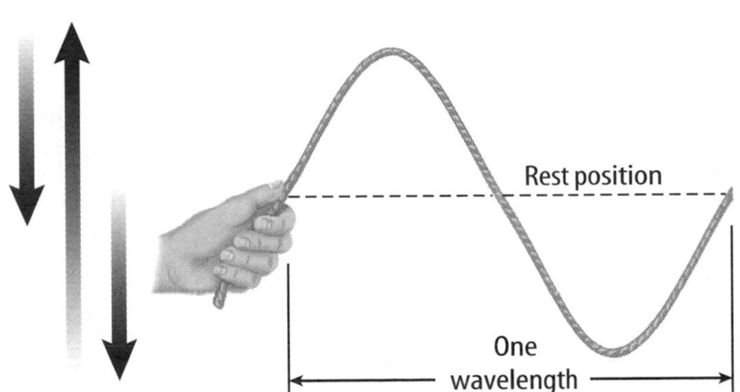

B By shaking the end of a rope down, up, and down again, you make one wavelength.

Figure 5
The vibrating motion of an electric charge produces an electromagnetic wave. One complete cycle of vibration produces one wavelength of a wave.

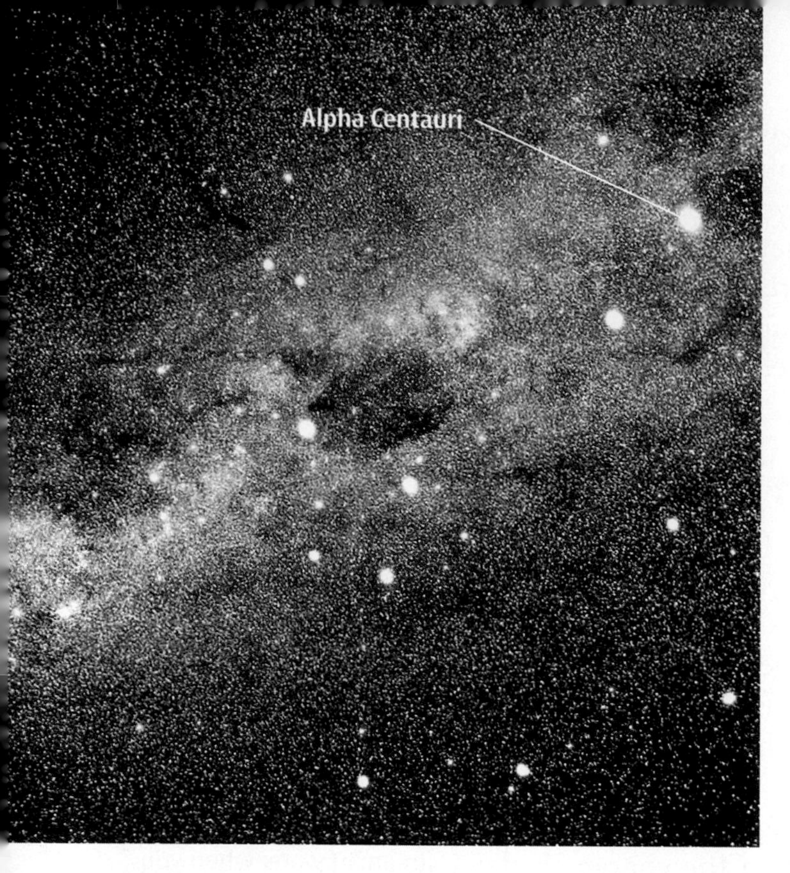

Figure 6
The light that reaches Earth today from Alpha Centauri left the star more than four years ago.

Radiant Energy The energy carried by an electromagnetic wave is called **radiant energy.** What happens if an electromagnetic wave strikes another charged particle? The electric field part of the wave exerts a force on this particle and causes it to move. Some of the radiant energy carried by the wave is transferred into the energy of motion of the particle.

Reading Check *What is radiant energy?*

The amount of energy that an electromagnetic wave carries is determined by the wave's frequency. The higher the frequency of the electromagnetic wave, the more energy it has.

The Speed of Light All electromagnetic waves, such as light, microwaves, and X rays, travel through space at the same speed. This speed has been measured as about 300,000 km/s in space. Because light is an electromagnetic wave, this speed sometimes is called the speed of light. If something could travel at the speed of light, it could travel around the world more than seven times in 1 s. Even though light travels incredibly fast, stars other than the Sun are so far away that it takes years for the light they emit to reach Earth. **Figure 6** shows one of the closest stars to the solar system, Alpha Centauri. This star is more than 40 trillion km from Earth.

Section  Assessment

1. What is an electromagnetic wave?
2. How are electromagnetic waves produced?
3. What two fields surround a moving charged particle?
4. How does the amount of energy carried by a low-frequency wave compare to the amount carried by a high-frequency wave?
5. **Think Critically** Unlike sound waves, electromagnetic waves can travel through a vacuum. What observations can you make to support this statement?

Skill Builder Activities

6. **Comparing and Contrasting** How are electromagnetic waves similar to mechanical waves? How are they different? **For more help, refer to the Science Skill Handbook.**
7. **Calculating Ratios** To go from Earth to Mars, light takes 4 min and a spacecraft takes four months. To go to the nearest star, light takes four years. How long would the same spacecraft take to travel to the nearest star? **For more help, refer to the Math Skill Handbook.**

SECTION 2

The Electromagnetic Spectrum

Electromagnetic Waves

The room you are sitting in is bathed in a sea of electromagnetic waves. These electromagnetic waves have a wide range of wavelengths and frequencies. For example, TV and radio stations broadcast electromagnetic waves that pass through walls and windows. These waves have wavelengths from about 1 m to over 500 m. Light waves that you see are electromagnetic waves that have wavelengths more than a million times shorter than the waves broadcast by radio stations.

Classifying Electromagnetic Waves The wide range of electromagnetic waves with different frequencies and wavelengths is called the **electromagnetic spectrum. Figure 7** shows the electromagnetic spectrum. Though many different types of electromagnetic waves exist, they all are produced by electric charges that are moving or vibrating. The faster the charge moves or vibrates, the higher the energy of the resulting electromagnetic waves is. Electromagnetic waves carry radiant energy that increases as the frequency increases. For waves that travel with the same speed, the wavelength increases as frequency decreases. So the energy carried by an electromagnetic wave decreases as the wavelength increases.

As You Read

***What* You'll Learn**

- **Explain** differences among kinds of electromagnetic waves.
- **Identify** uses for different kinds of electromagnetic waves.

Vocabulary

electromagnetic spectrum
radio wave
infrared wave
visible light
ultraviolet radiation
X ray
gamma ray

***Why* It's Important**

Electromagnetic waves are used to cook food, to send and receive information, and to diagnose medical problems.

Figure 7
Electromagnetic waves have a spectrum of different frequencies and wavelengths.

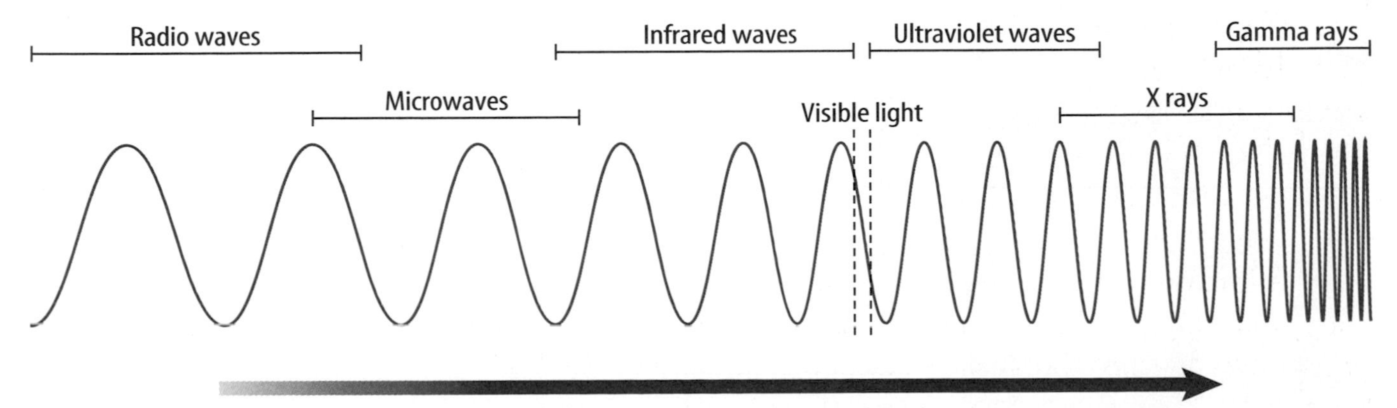

Figure 8
Antennas are useful in generating and detecting radio waves.

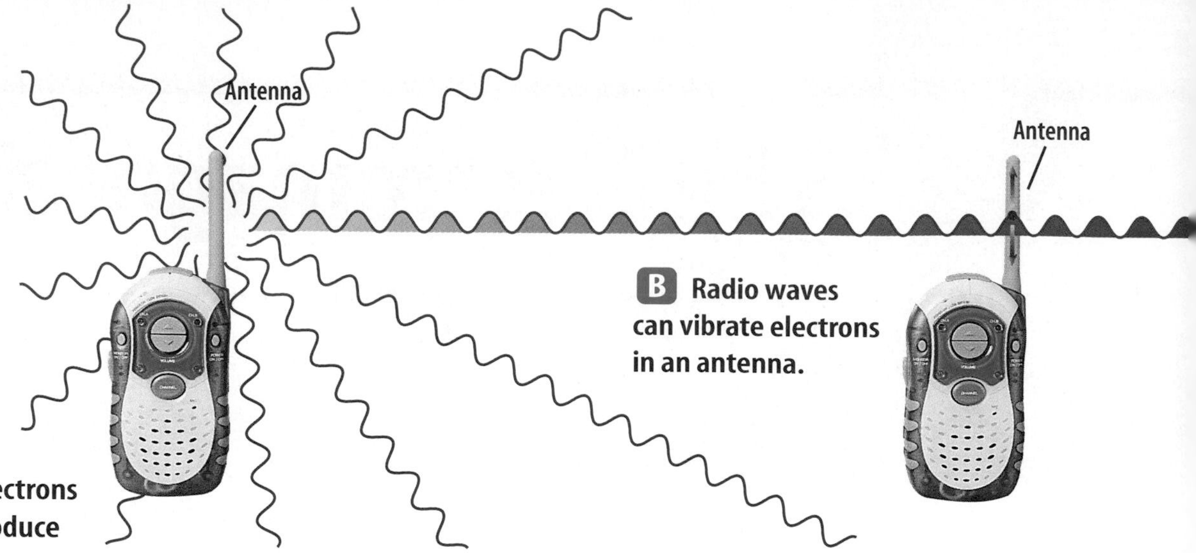

Antenna

Antenna

B Radio waves can vibrate electrons in an antenna.

A Vibrating electrons in an antenna produce radio waves.

Radio Waves

Electromagnetic waves with wavelengths longer than about 0.3 m are called radio waves. **Radio waves** have the lowest frequencies of all the electromagnetic waves and carry the least energy. Television signals, as well as AM and FM radio signals, are types of radio waves. Like all electromagnetic waves, radio waves are produced by moving charged particles. One way to make radio waves is to make electrons vibrate in a piece of metal, as shown in **Figure 8A.** This piece of metal is called an antenna. By changing the rate at which the electrons vibrate, radio waves of different frequencies can be produced that travel outward from the antenna.

Detecting Radio Waves These radio waves can cause electrons in another piece of metal, such as another antenna, to vibrate, as shown in **Figure 8B.** As the electrons in the receiving antenna vibrate, they form an alternating current. This alternating current can be used to produce a picture on a TV screen and sound from a loudspeaker. Varying the frequency of the radio waves broadcast by the transmitting antenna changes the alternating current in the receiving antenna. This produces the different pictures you see and sounds you hear on your TV.

Figure 9
Towers such as the one shown here are used to send and receive microwaves.

Microwaves Radio waves with wavelengths between about 0.3 m and 0.001 m are called microwaves. They have a higher frequency and a shorter wavelength than the waves that are used in your home radio. Microwaves are used to transmit some phone calls, especially from cellular and portable phones. **Figure 9** shows a microwave tower.

Microwave ovens use microwaves to heat food. Microwaves produced inside a microwave oven cause water molecules in your food to vibrate faster, which makes the food warmer.

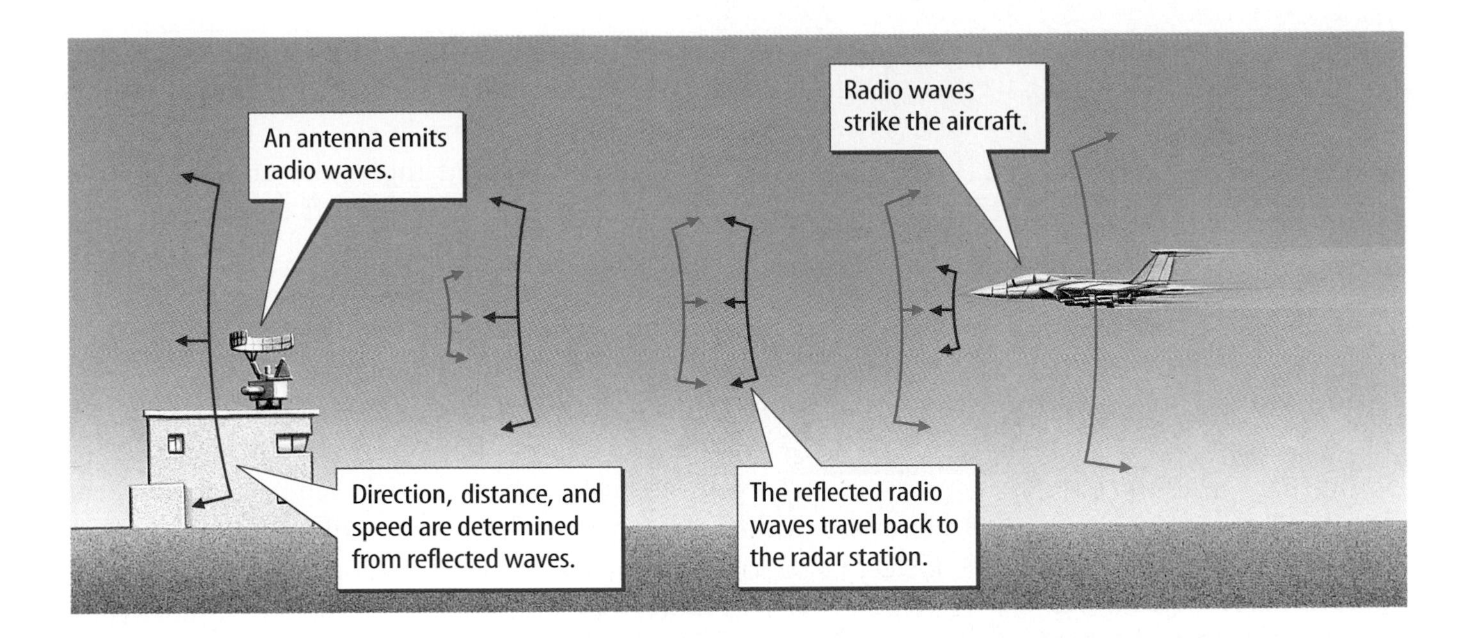

Figure 10
Radar stations determine direction, distance, and speed of aircraft.

Radar You might be familiar with echolocation, in which sound waves are reflected off an object to determine its size and location. Some bats and dolphins use echolocation to navigate and hunt. Radar, an acronym for RAdio Detecting And Ranging, uses electromagnetic waves to detect objects in the same way. Radar was first used during World War II to detect and warn of incoming enemy aircraft.

Reading Check *What does radar do?*

A radar station sends out radio waves that bounce off an object such as an airplane. Electronic equipment measures the time it takes for the radio waves to travel to the plane, be reflected, and return. Because the speed of the radio waves is known, the distance to the airplane can be calculated from the following formula.

$$\text{distance} = \text{speed} \times \text{time}$$

An example of radar being used is shown in **Figure 10.** Because electromagnetic waves travel so quickly, the entire process takes only a fraction of a second.

Infrared Waves

You might know from experience that when you stand near the glowing coals of a barbecue or the red embers of a campfire, your skin senses the heat and becomes warm. Your skin may also feel warm near a hot object that is not glowing. The heat you are sensing with your skin is from electromagnetic waves. These electromagnetic waves are called **infrared waves** and have wavelengths between about one thousandth and 0.7 millionths of a meter.

Observing the Focusing of Infrared Rays

Procedure

1. Place a **concave mirror** 2 m to 3 m away from an **electric heater.** Turn on the heater.
2. Place the palm of your hand in front of the mirror and move it back until you feel heat on your palm. Note the location of the warm area.
3. Move the heater to a new location. How does the warm area move?

Analysis

1. Did you observe the warm area? Where?
2. Compare the location of the warm area to the location of the mirror.

Figure 11
A pit viper hunting in the dark can detect the infrared waves that the warm body of its prey emits.

Detecting Infrared Waves Infrared rays are emitted by almost every object. In any material the atoms and molecules are in constant motion. Electrons in the atoms and molecules also move and vibrate. As a result, they give off electromagnetic waves. Most of the electromagnetic waves given off by an object at room temperature are infrared waves and have a wavelength of about 0.000 01 m, or one hundred thousandth of a meter.

Infrared detectors can detect objects that are warmer or cooler than their surroundings. For example, areas covered with vegetation, such as forests, tend to be cooler than their surroundings. Using infrared detectors on satellites, the areas covered by forests and other vegetation, as well as water, rock, and soil, can be mapped. Some types of night vision devices use infrared detectors that enable objects to be seen in nearly total darkness.

Animals and Infrared Waves Some animals also can detect infrared waves. Snakes called pit vipers, such as the one shown in **Figure 11,** have a pit located between the nostril and the eye that detects infrared waves. Rattlesnakes, copperheads, and water moccasins are pit vipers. These pits help pit vipers hunt at night by detecting the infrared waves their prey emits.

Visible Light

As the temperature of an object increases, the atoms and molecules in the object move faster. The electrons also vibrate faster, and produce electromagnetic waves of higher frequency and shorter wavelength. If the temperature is high enough, the object might glow, as in **Figure 12.** Some of the electromagnetic waves that the hot object is emitting are now detectable with your eyes. Electromagnetic waves you can detect with your eyes are called **visible light.** Visible light has wavelengths between about 0.7 and 0.4 millionths of a meter. What you see as different colors are electromagnetic waves of different wavelengths. Red light has the longest wavelength (lowest frequency), and blue light has the shortest wavelength (highest frequency).

Most objects that you see do not give off visible light. They simply reflect the visible light that is emitted by a source of light, such as the Sun or a lightbulb.

Figure 12
When objects are heated, their electrons vibrate faster. When the temperature is high enough, the vibrating electrons will emit visible light.

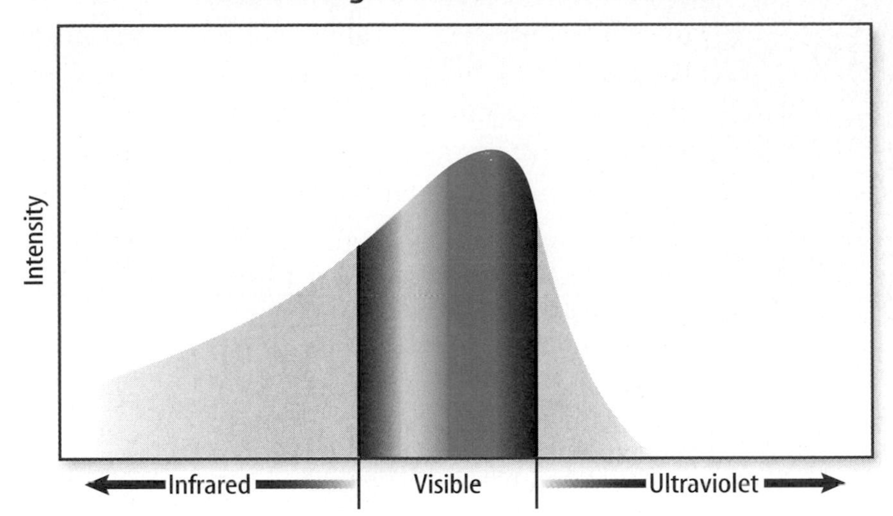

Figure 13
Electromagnetic waves from the Sun have a range of frequencies centered about the visible region. *Which frequencies of light is the Sun brightest in?*

Ultraviolet Radiation

Ultraviolet radiation is higher in frequency than visible light and has even shorter wavelengths—between 0.4 millionths of a meter and about ten billionths of a meter. Ultraviolet radiation has higher frequencies than visible light and carries more energy. The radiant energy carried by an ultraviolet wave can be enough to damage the large, fragile molecules that make up living cells. Too much ultraviolet radiation can damage or kill healthy cells.

Figure 13 shows the electromagnetic waves emitted by the Sun, some of which are in the ultraviolet region. Too much exposure to those ultraviolet waves can cause sunburn. Exposure to these waves over a long period of time can lead to early aging of the skin and possibly skin cancer. You can protect yourself from receiving too much ultraviolet radiation by wearing sunglasses and sunscreen, and staying out of the Sun when it is most intense.

Beneficial Uses of UV Radiation A few minutes of exposure each day to ultraviolet radiation from the Sun enables your body to produce the vitamin D it needs. Most people receive that amount during normal activity. The body's natural defense against too much ultraviolet radiation is to tan. However, a tan can be a sign that overexposure to ultraviolet radiation has occurred.

Ultraviolet radiation's cell-killing effect has led to its use as a disinfectant for surgical equipment in hospitals. In some high school chemistry labs, ultraviolet rays are used to sterilize goggles, as shown in **Figure 14.**

Figure 14
Sterilizing devices, such as this goggle sterilizer, use ultraviolet waves to kill organisms on the equipment.

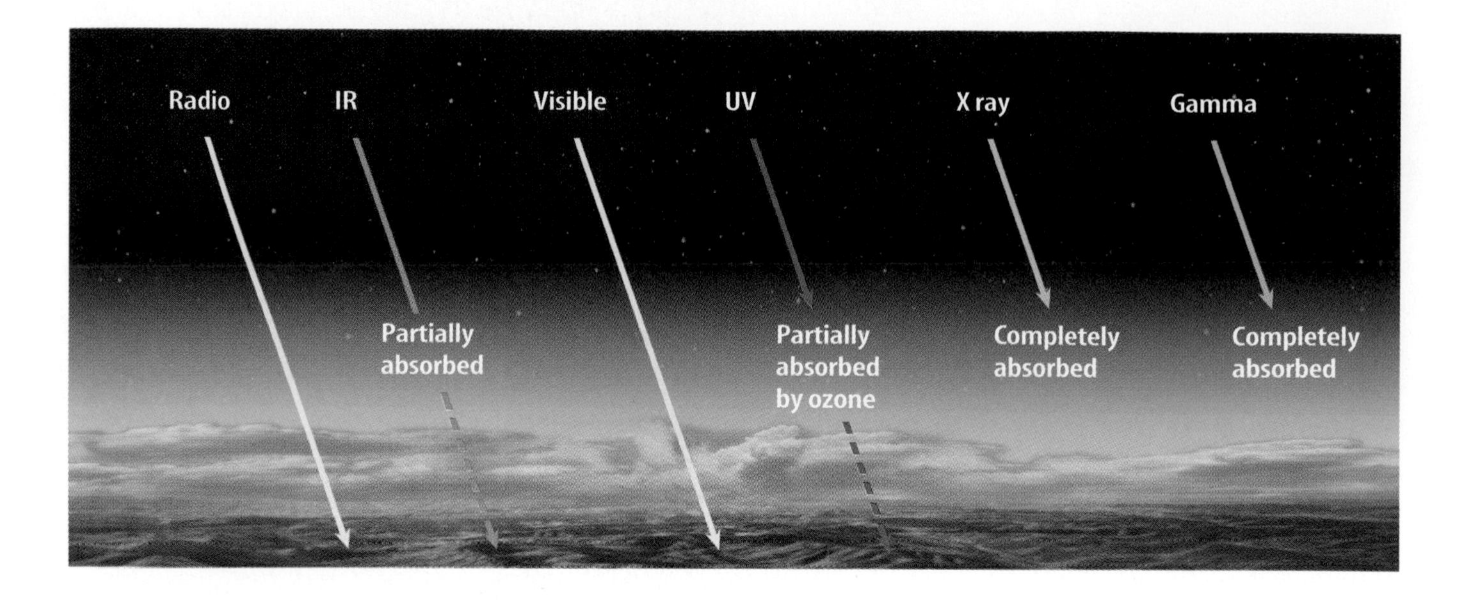

Figure 15
Earth's atmosphere serves as a shield to block certain types of electromagnetic waves from reaching the surface.

Warm-blooded animals, such as mammals, produce their own body heat. Cold-blooded animals, such as reptiles, absorb heat from the environment. Brainstorm the possible advantages of being either warm-blooded or cold-blooded. Which animals would be easier for a pit viper to detect?

The Ozone Layer Much of the ultraviolet radiation arriving at Earth is absorbed in the upper atmosphere by ozone, as shown in **Figure 15.** Ozone is a molecule that has three oxygen atoms and is formed high in Earth's atmosphere.

However, chemical compounds called CFCs, which are used in air conditioners and refrigerators, can react chemically with ozone. This reaction causes ozone to break down and increases the amount of ultraviolet radiation that penetrates the atmosphere. To prevent this, the use of CFCs is being phased out.

Ultraviolet radiation is not the only type of electromagnetic wave absorbed by Earth's atmosphere. Higher energy waves of X rays and gamma rays also are absorbed. The atmosphere is transparent to radio waves and visible light and partially transparent to infrared waves.

X Rays and Gamma Rays

Ultraviolet rays can penetrate the top layer of your skin. **X rays,** with an even higher frequency than ultraviolet rays, have enough energy to go right through skin and muscle. A shield made from a dense metal, such as lead, is required to stop X rays.

Gamma rays have the highest frequency and, therefore, carry the most energy. Gamma rays are the hardest to stop. They are produced by changes in the nuclei of atoms. When protons and neutrons bond together in nuclear fusion or break apart from each other in nuclear fission, enormous quantities of energy are released. Some of this energy is released as gamma rays.

Just as too much ultraviolet radiation can hurt or kill cells, too much X ray or gamma radiation can have the same effect. Because the energy of the waves is so much higher, the exposure that is needed to cause damage is much less.

Using High-Energy Electromagnetic Radiation The fact that X rays can pass through the human body makes them useful for medical diagnosis, as shown in **Figure 16.** X rays pass through the less dense tissues in skin and other organs. These X rays strike a film, creating a shadow image of the denser tissues. X-ray images help doctors detect injuries and diseases, such as broken bones and cancer. A CT scanner uses X rays to produce images of the human body as if it had been sliced like a loaf of bread.

Although the radiation received from getting one medical or dental X ray is not harmful, the cumulative effect of numerous X rays can be dangerous. The operator of the X-ray machine usually stands behind a shield to avoid being exposed to X rays. Lead shields or aprons are used to protect the parts of the patient's body that are not receiving the X rays.

Using Gamma Rays Although gamma rays are dangerous, they also have beneficial uses, just as X rays do. A beam of gamma rays focused on a cancerous tumor can kill the tumor. Gamma radiation also can cleanse food of disease-causing bacteria. More than 1,000 Americans die each year from *Salmonella* bacteria in poultry and *E. coli* bacteria in meat. Although gamma radiation has been used since 1963 to kill bacteria in food, this method is not widely used in the food industry.

Astronomy Across the Spectrum

Some astronomical objects produce no visible light and can be detected only through the infrared and radio waves they emit. Some galaxies emit X rays from regions that do not emit visible light. Studying stars and galaxies like these using only visible light would be like looking at only one color in a picture. **Figure 17** shows how different electromagnetic waves can be used to study the universe.

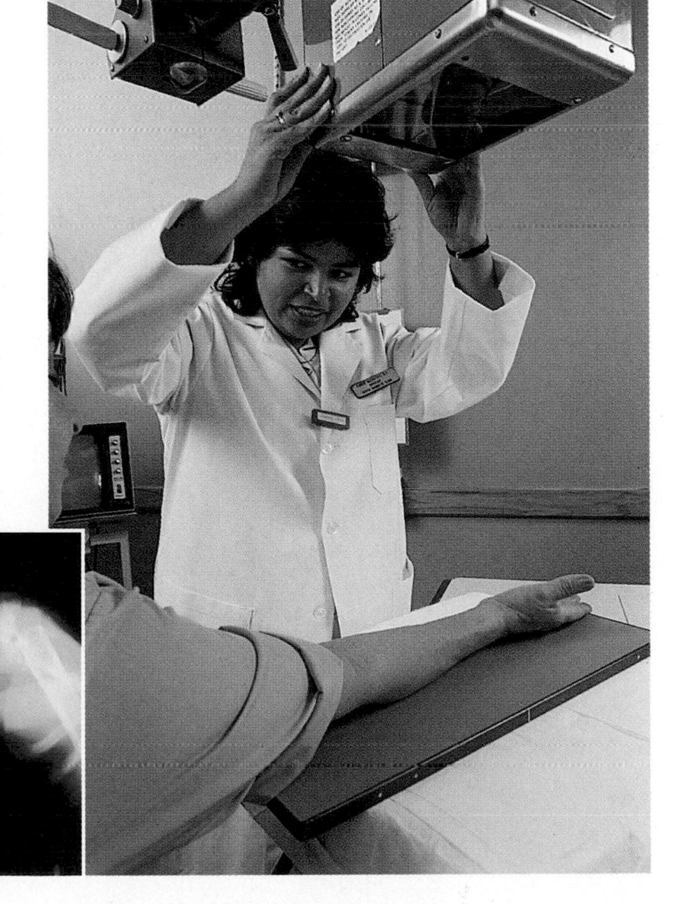

Figure 16
Dense tissues such as bone absorb more X rays than softer tissues do. Consequently, dense tissues leave a shadow on film that can be use to diagnose medical and dental conditions.

NATIONAL GEOGRAPHIC **VISUALIZING THE UNIVERSE**

Figure 17

For centuries, astronomers studied the universe using only the visible light coming from planets, moons, and stars. But many objects in space also emit X rays, ultraviolet and infrared radiation, and radio waves. Scientists now use telescopes that can "see" these different types of electromagnetic waves. As these images of the Sun reveal, the new tools are providing remarkable views of objects in the universe.

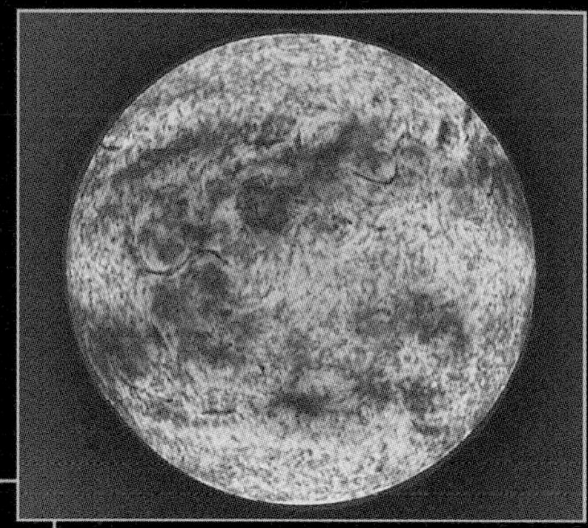

▲ INFRARED RADIATION An infrared telescope reveals that the Sun's surface temperature is not uniform. Some areas are hotter than others.

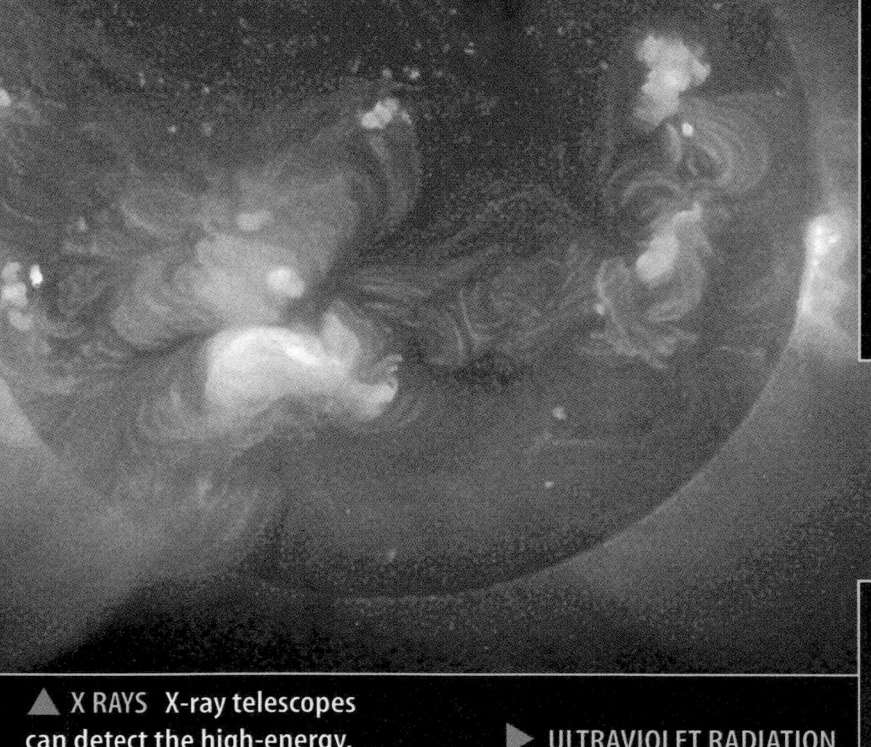

▲ X RAYS X-ray telescopes can detect the high-energy, short-wavelength X rays produced by the extreme temperatures in the Sun's outer atmosphere.

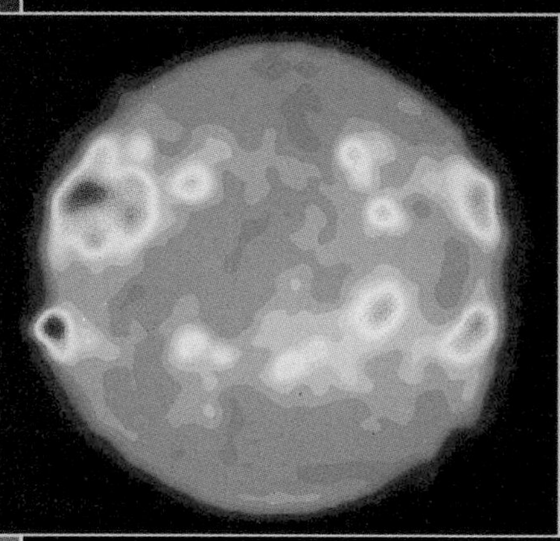

▲ RADIO WAVES Radio telescopes detect radio waves given off by the Sun, which have much longer wavelengths than visible light.

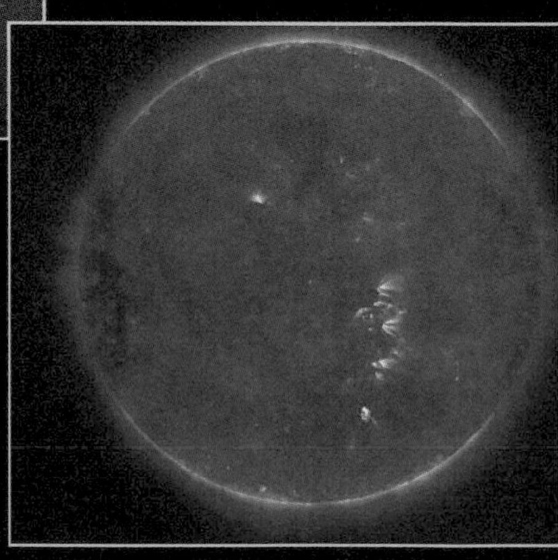

▶ ULTRAVIOLET RADIATION Telescopes sensitive to ultraviolet radiation—electromagnetic waves with shorter wavelengths than visible light—can "see" the Sun's outer atmosphere.

Figure 18
Launching satellite observatories above Earth's atmosphere is the only way to see the universe at electromagnetic wavelengths that are absorbed by Earth's atmosphere.

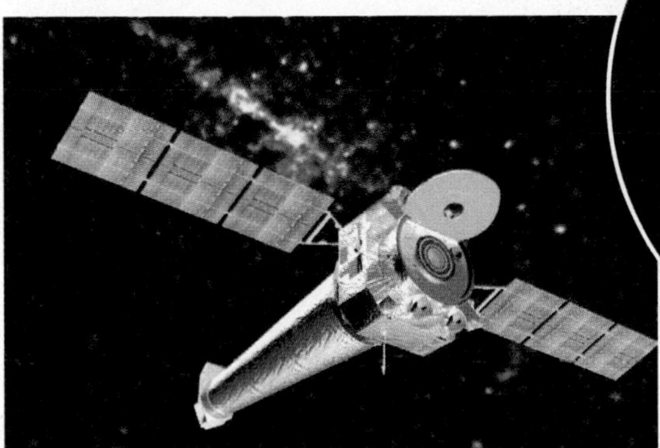

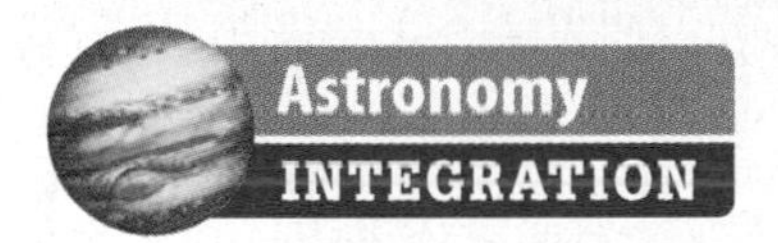

Satellite Observations Recall from **Figure 15** that Earth's atmosphere blocks some parts of the electromagnetic spectrum. For example, X rays, gamma rays, most ultraviolet rays, and some infrared rays cannot pass through. However, telescopes in orbit above Earth's atmosphere can obtain more information than can be obtained at Earth's surface about stars, galaxies, and other objects in the universe. **Figure 18** shows three such satellites—the Extreme Ultraviolet Explorer, the Chandra X-Ray Observatory, and the Infrared Space Observatory.

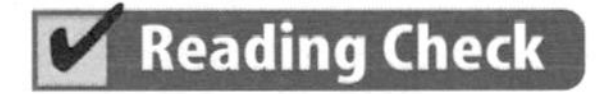

Why are telescopes sent into space on artificial satellites?

Section  Assessment

1. List three types of electromagnetic waves produced by the Sun.
2. Why is ultraviolet light more damaging to cells than infrared light is?
3. Give an application of infrared waves.
4. Describe the difference between X rays and gamma rays.
5. **Think Critically** Why does Earth emit mainly infrared waves and the Sun emit visible light and ultraviolet waves?

Skill Builder Activities

6. **Recognizing Cause and Effect** If visible light is the effect, what is the cause? Do the different colors of light have different causes? **For more help, refer to the Science Skill Handbook.**
7. **Using a Database** What do images of the same object look like if different wavelengths are detected? Use a database to research this topic and present a report to your class. **For more help, refer to the Technology Skill Handbook.**

Activity

Prisms of Light

Do you know what light is? Many would answer that light is what you turn on to see at night. However, white light is made of many different frequencies of the electromagnetic spectrum. A prism can separate white light into its different frequencies. You see different frequencies of light as different colors. What colors do you see when light passes through a prism?

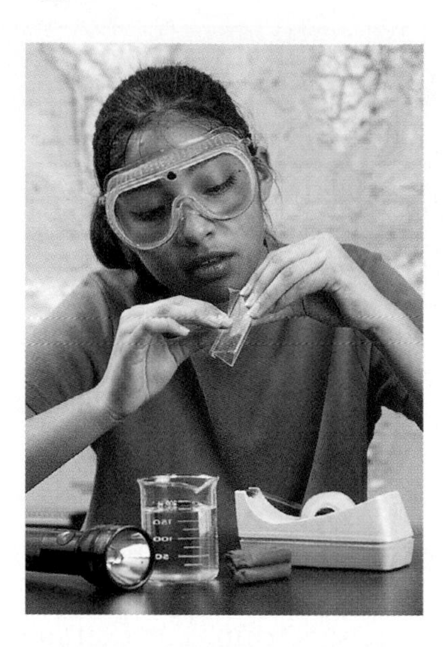

What You'll Investigate

What happens to visible light as it passes through a prism?

Goals

- **Construct** a prism and observe the different colors that are produced.
- **Infer** how the bending of light waves depends on their wavelength.

Materials

microscope slides (3)
transparent tape
clay
flashlight
water

Safety Precautions

Procedure

1. Carefully tape the three slides together on their long sides so they form a long prism.
2. Place one end of the prism into a softened piece of clay so the prism is standing upright.
3. Fill the prism with water and put it on a table that is against a dark wall.
4. Shine a flashlight beam through the prism so the light becomes visible on the wall.

Conclude and Apply

1. What was the order of the colors you saw on the wall?
2. How does the position of the colors on the wall change as you change the direction of the flashlight beam?
3. How does the order of colors on the wall change as you change the direction of the flashlight beam?
4. After passing through the water prism, which color light waves have changed direction, or have been bent, the most? Which color has been bent the least?
5. How does the amount of bending of a light wave depend on its wavelength? How does it depend on the frequency?

Communicating Your Data

Compare your conclusions with those of other students in your class. **For more help, refer to the Science Skill Handbook.**

Using Electromagnetic Waves

Telecommunications

In the past week, have you spoken on the phone, watched television, done research on the Internet, or listened to the radio? Today you can talk to someone far away or transmit and receive information over long distances almost instantly. Thanks to telecommunications, the world is becoming increasingly connected through the use of electromagnetic waves.

Using Radio Waves

Radio waves usually are used to send and receive information over long distances. Using radio waves to communicate has several advantages. For example, radio waves pass through walls and windows easily. Radio waves do not interact with humans, so they are not harmful to people like ultraviolet rays or X rays are. So most telecommunication devices, such as TVs, radios, and telephones, use radio waves to transmit information such as images and sounds. **Figure 19** shows how radio waves can be used to transmit information—in this case transmitting information that enables sounds to be reproduced at a location far away.

As You Read

***What* You'll Learn**

- **Explain** different methods of electronic communication.
- **Compare and contrast** AM and FM signals.

Vocabulary

Carrier wave
Global Positioning System

***Why* It's Important**

Telecommunication enables people to contact others and collect information worldwide.

Figure 19
Transmitting sounds by radio waves uses conversions among sound, electrical, and radiant energies.

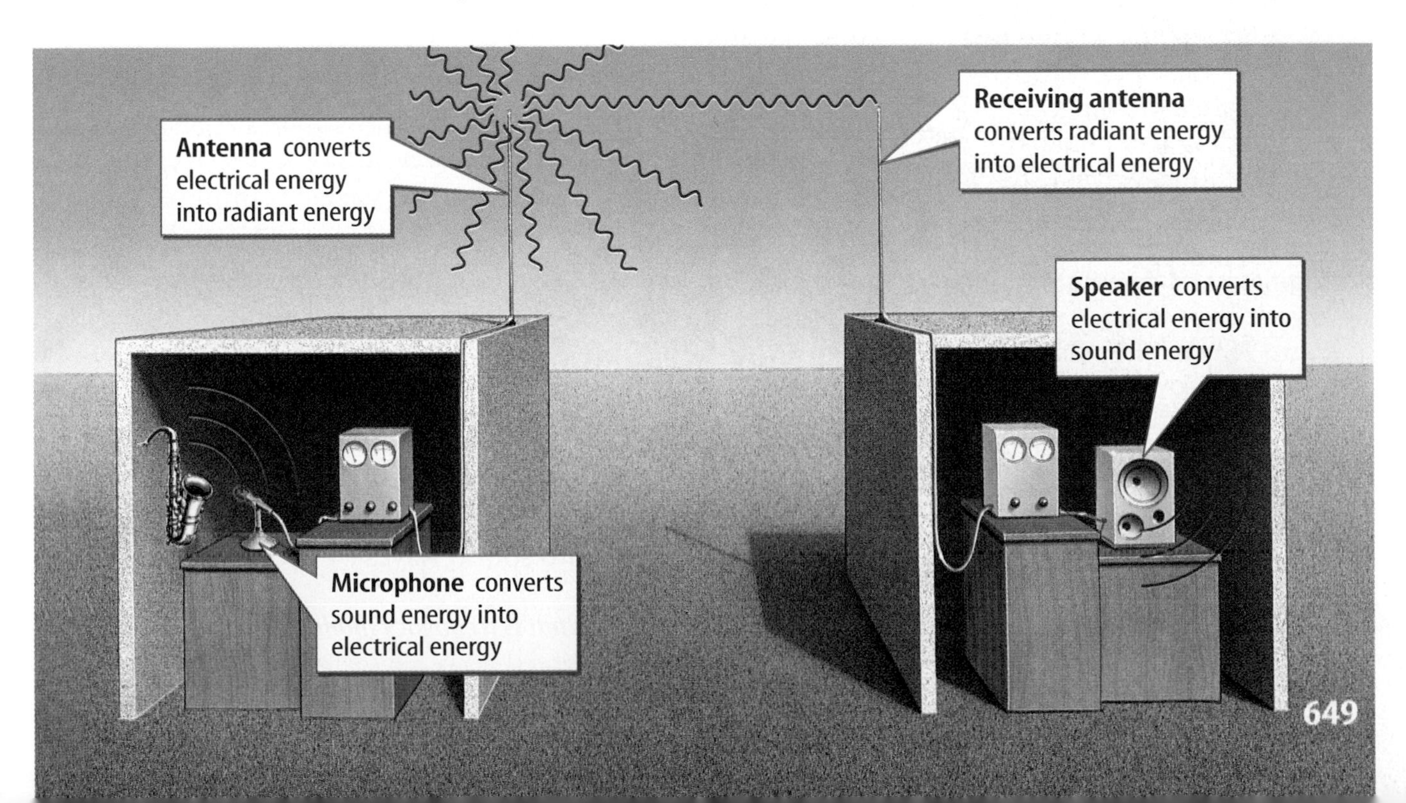

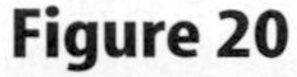

A signal can be carried by a carrier wave in two ways—amplitude modulation or frequency modulation.

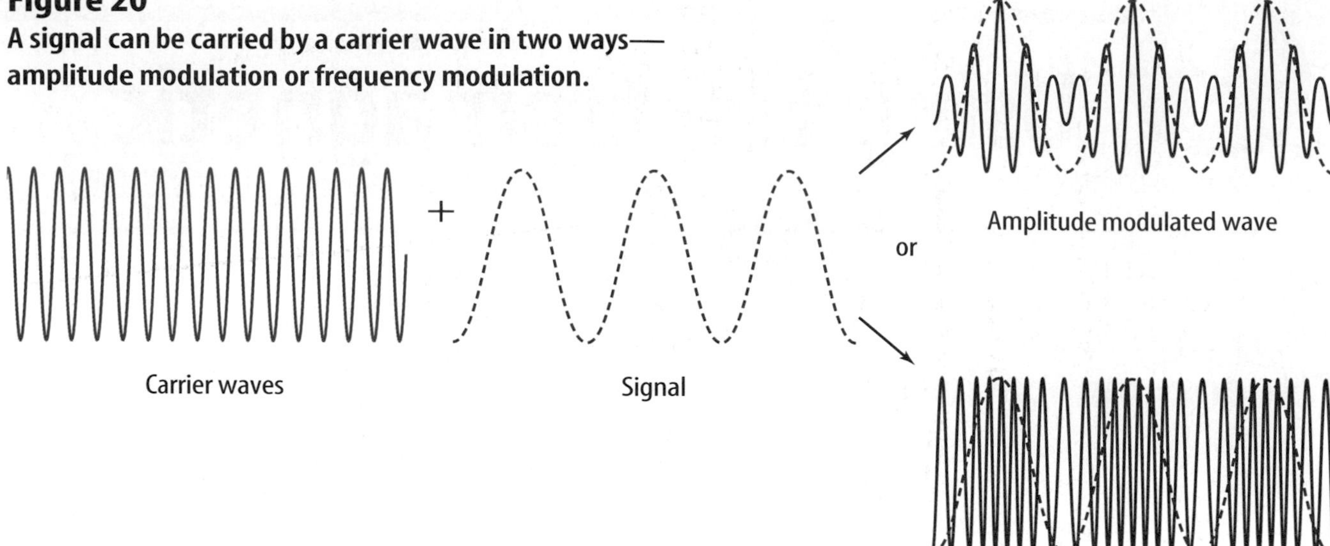

Radio Transmission How is information, such as images or sounds, broadcast by radio waves? Each radio and television station is assigned a particular frequency at which it broadcasts radio waves. The radio waves broadcast by a station at its assigned frequency are the **carrier waves** for that station. To listen to a station you tune your radio or television to the frequency of the station's carrier waves. To carry information on the carrier wave, either the amplitude or the frequency of the carrier wave is changed, or modulated.

Amplitude Modulation The letters *AM* in AM radio stand for amplitude modulation, which means that the amplitude of the carrier wave is changed to transmit information. The original sound is transformed into an electrical signal that is used to vary the amplitude of the carrier wave, as shown in **Figure 20.** Note that the frequency of the carrier wave doesn't change—only the amplitude changes. An AM receiver tunes to the frequency of the carrier wave. In the receiver, the varying amplitude of the carrier waves produces an electric signal. The radio's loudspeaker uses this electric signal to produce the original sound.

Pulsars are astronomical objects that emit periodic bursts of radio waves. The pattern of pulses is regular. Investigate how pulsars originate and communicate to your class what you learn. Why might pulsars have seemed to be signals from intelligent life?

Frequency Modulation FM radio works in much the same way as AM radio, but the frequency instead of the amplitude is modulated, as shown in **Figure 20.** An FM receiver contains electronic components that use the varying frequency of the carrier wave to produce an electric signal. As in an AM radio, this electric signal is converted into sound waves by a loudspeaker.

 ***What is frequency modulation?***

Telephones

A telephone contains a microphone in the mouthpiece that converts a sound wave into an electric signal. The electric signal is carried through a wire to the telephone switching system. There, the signal might be sent through other wires or be converted into a radio or microwave signal for transmission through the air. The electric signal also can be converted into a light wave for transmission through fiber-optic cables.

At the receiving end, the signal is converted back to an electric signal. A speaker in the earpiece of the phone changes the electric signal into a sound wave.

What device converts sound into an electric signal?

Math Skills Activity

Calculating the Wavelength of Radio Frequencies

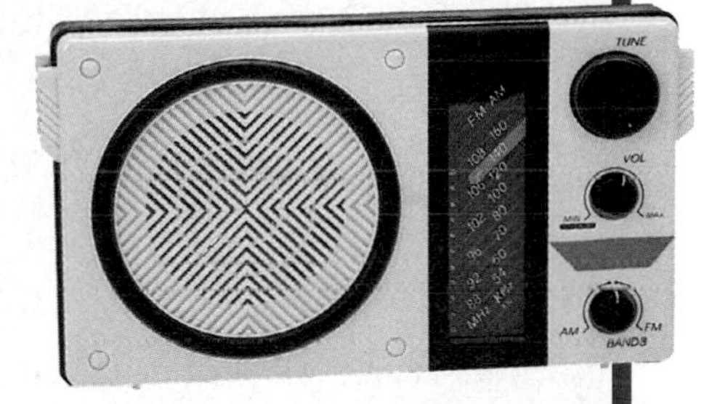

Example Problem

You are listening to an FM station with a frequency of 94.9 MHz or 94,900,000 Hz. How long are the wavelengths that strike the antenna? For any wave, the wavelength equals the wave speed divided by the frequency. The speed of radio waves is 300,000,000 m/s. The SI unit of frequency, Hz, is equal to l/s.

Solution

1. *This is what you know:* frequency = 94,900,000 Hz; wave speed = 300,000,000 m/s
2. *This is what you need to find:* wavelength
3. *This is the equation you need to use:* wavelength = wave speed / frequency
4. *Substitute the known values:* wavelength = (300,000,000 m/s) / (94,900,000 Hz) = 3.16 m

Check your answer by multiplying the units. Do you calculate a unit of distance for your answer?

Practice Problems

1. Your friend prefers an AM radio station at 1,520 kHz (1,520 thousand vibrations each second). What is the wavelength of this frequency? Which has a longer wavelength, AM or FM radio waves?
2. An AM radio station operates at 580 kHz (580 thousand vibrations each second). What is the wavelength of this frequency? What is the relationship between frequency and wavelength?

For more help, refer to the Math Skill Handbook.

Figure 21
Electromagnetic waves make using telephones easier.

A **Cordless phones use radio waves to allow users to talk from anywhere in the house.**

B **Radio waves enable cell phone users to send or receive calls without using wires.**

Remote Phones A telephone does not have to transmit its signal through wires. In a cordless phone, the electrical signal produced by the microphone is transmitted through an antenna in the handset to the base. **Figure 21A** shows how incoming signals are transmitted from the base to the handset. A cellular phone uses an antenna to broadcast and receive information between the phone and a base station, as shown in **Figure 21B.** The base station uses radio waves to communicate with other stations in a network.

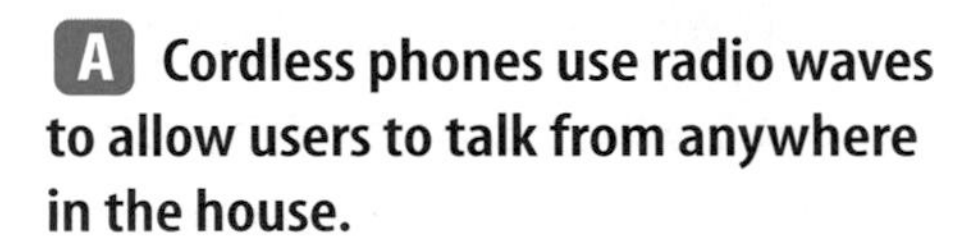

SCIENCE Online

Research Visit the Glencoe Science Web site at **science.glencoe.com** for more information about how satellites are used in around-the-world communications. Summarize what you learn in an informational handout.

Pagers The base station also is used in a pager system. When you dial a pager, the signal is sent to a base station. From there, an electromagnetic signal is sent to the pager. The pager beeps or vibrates to indicate that someone has called. With a touch-tone phone, you can transmit numeric information, such as your phone number, which the pager will receive and display.

Communications Satellites

How do you send information to the other side of the world? Radio waves can't be sent directly through Earth. Instead, radio signals are sent to satellites. The satellites can communicate with other satellites or with ground stations. Some communications satellites are in geosynchronous orbit, meaning each satellite remains above the same point on the ground.

The Global Positioning System

Satellites also are used as part of the **Global Positioning System,** or GPS. GPS is used to locate objects on Earth. The system consists of satellites, ground-based stations, and portable units with receivers, as illustrated in **Figure 22.**

A GPS receiver measures the time it takes for radio waves to travel from several satellites to the receiver. This determines the distance to each satellite. The receiver then uses this information to calculate its latitude, longitude, and elevation. The accuracy of GPS receivers ranges from a few hundred meters for hand-held units, to several centimeters for units that are used to measure the movements of Earth's crust.

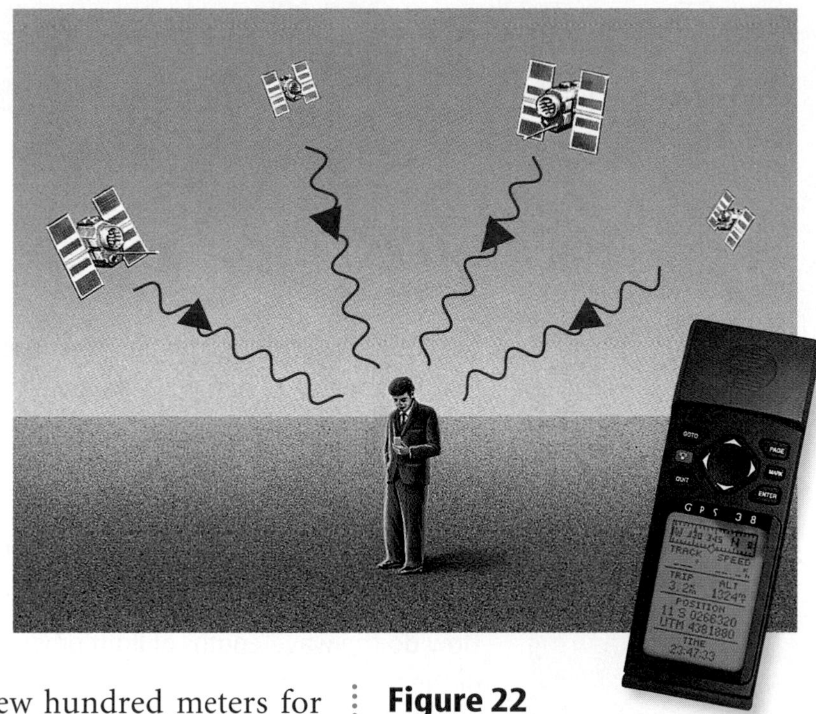

Figure 22
The signals broadcast from GPS satellites enable portable, hand-held receivers to determine the position of an object or person.

✔ Reading Check *What is GPS used for?*

Many of these forms of communication have been developed over the past few decades. For example, an Internet connection transfers images and sound using the telephone network, just as a television signal transfers images and sound using radio waves. What forms of telecommunications do you think you'll be using a few decades from now?

Section  Assessment

1. What is a modulated radio signal?
2. What does a microphone do? What does a speaker do?
3. What types of information does a GPS receiver provide for its user?
4. What is a communications satellite?
5. **Think Critically** Make a diagram showing how a communication satellite could be used to relay information from a broadcasting station in New York to a receiving station in London.

Skill Builder Activities

6. **Researching Information** Find out more about a form of telecommunications, such as email or shortwave radio. **For more help, refer to the Science Skill Handbook.**
7. **Communicating** Think of a story you have enjoyed about a time before telecommunications or one in which telecommunication was not possible. How would telecommunications have changed the story? **For more help, refer to the Science Skill Handbook.**

Activity Design Your Own Experiment

Spectrum Inspection

You've heard the term "red-hot" used to describe something that is unusually hot. When a piece of metal is heated it may give off a red glow or even a yellow glow. All objects emit electromagnetic waves. How do the wavelengths of these waves depend on the temperature of the object?

Recognize the Problem

How do the wavelengths of light produced by a lightbulb depend on the temperature of the lightbulb?

Form a Hypothesis

The brightness of a lightbulb increases as its temperature increases. Form a hypothesis describing how the wavelengths emitted by a lightbulb will change as the brightness of a lightbulb changes.

Goals

- **Design** an experiment that determines the relationship between brightness and the wavelengths emitted by a lightbulb.
- **Observe** the wavelengths of light emitted by a lightbulb as its brightness changes.

Safety Precautions

WARNING: Be sure all electrical cords and connections are intact and that you have a dry working area. Do not touch the bulbs as they may be hot.

Possible Materials

diffraction grating
power supply with variable resistor switch
clear, tubular lightbulb and socket
red, yellow, and blue colored pencils

Test Your Hypothesis

Plan

1. **Decide** how you will determine the effect of lightbulb brightness on the colors of light that are emitted.
2. As shown in the photo at the right, you will look toward the light through the diffraction grating to detect the colors of light emitted by the bulb. The color spectrum will appear to the right and to the left of the bulb.
3. **List** the specific steps you will need to take to test your hypothesis. Describe precisely what you will do in each step. Will you first test the bulb at a bright or dim setting? How many settings will you test? (Try at least three.) How will you record your observations in an organized way?
4. **List** the materials you will need for your experiment. Describe exactly how and in which order you will use these materials.
5. **Identify** any constants and variables in your experiment.

Do

1. Make sure your teacher approves your plan before you start.
2. **Perform** your experiment as planned.
3. While doing your experiment, write down any observations you make in your Science Journal.

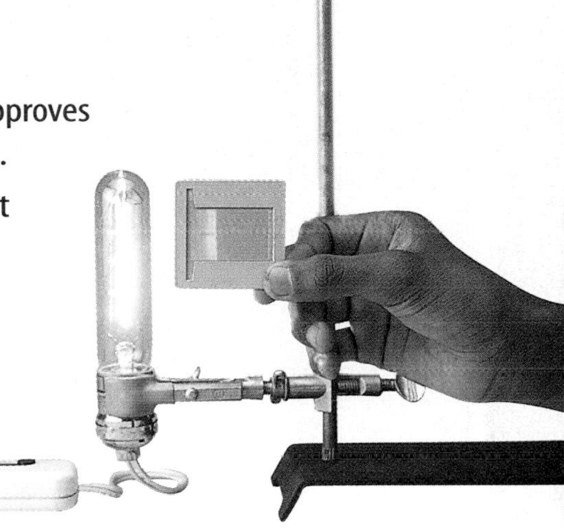

Analyze Your Data

1. Use the colored pencils to draw the color spectrum emitted by the bulb at each brightness.
2. Which colors appeared as the bulb became brighter? Did any colors disappear?
3. How did the wavelengths emitted by the bulb change as the bulb became brighter?
4. Infer how the frequencies emitted by the lightbulb changed as it became hotter.

Draw Conclusions

1. If an object becomes hotter, what happens to the wavelengths it emits?
2. How do the wavelengths that the bulb emits change if it is turned off?
3. From your results, infer whether red stars or yellow stars are hotter.

Compare your results with others in your class. How many different colors were seen?

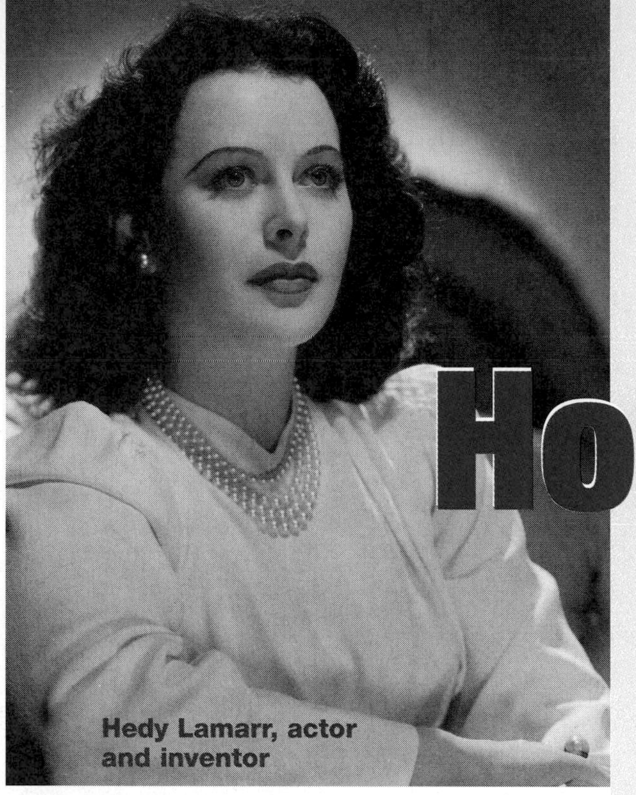
Hedy Lamarr, actor and inventor

Hopping the

Ringgggg. There it is—that familiar beep! Out come the cellular phones—from purses, pockets, book bags, belt clips, and briefcases. At any given moment, a million wireless signals are flying through the air—and not just cell phone signals. With radio and television signals, Internet data, and even Global Positioning System information coming at us, the air seems like a pretty crowded place. How do all of these signals get to the right place? How does a cellular phone pick out its own signal from among the clutter? The answer lies in a concept developed in 1940 by Hedy Lamarr.

Lamarr was born in Vienna, Austria. In 1937, she left Austria to escape Hitler's invading Nazi army. Lamarr left for another reason, as well. She was determined to pursue a career as an actor. And she became a famous movie star.

In 1940, Lamarr came up with an idea to keep radio signals that guided torpedoes from being jammed. Her idea, called frequency hopping, involved breaking the radio signal that was guiding the torpedo into tiny parts and rapidly changing their frequency. The enemy would not be able to keep up with the frequency changes and thus would not be able to divert the torpedo from its target. Lamarr worked with a partner who helped her figure out how to make the idea work. They were awarded a patent for their idea in 1942.

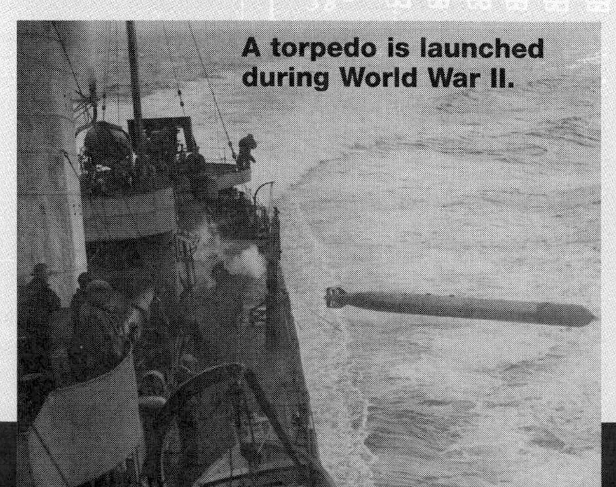
A torpedo is launched during World War II.

Spread Spectrum

Lamarr's idea was ahead of its time. The digital technology that allowed efficient operation of her system wasn't invented until decades later. However, after 1962, frequency hopping was adopted and used in U.S. military communications. It was the development of cellular phones, however, that benefited the most from Lamarr's concept. Cellular phones and other wireless technologies operate by breaking their signals into smaller parts, called packets. The packets are encoded in a certain way for particular receivers and are spread across bands of the electromagnetic spectrum. In this way, millions of users can use the same frequencies at the same time.

Frequencies

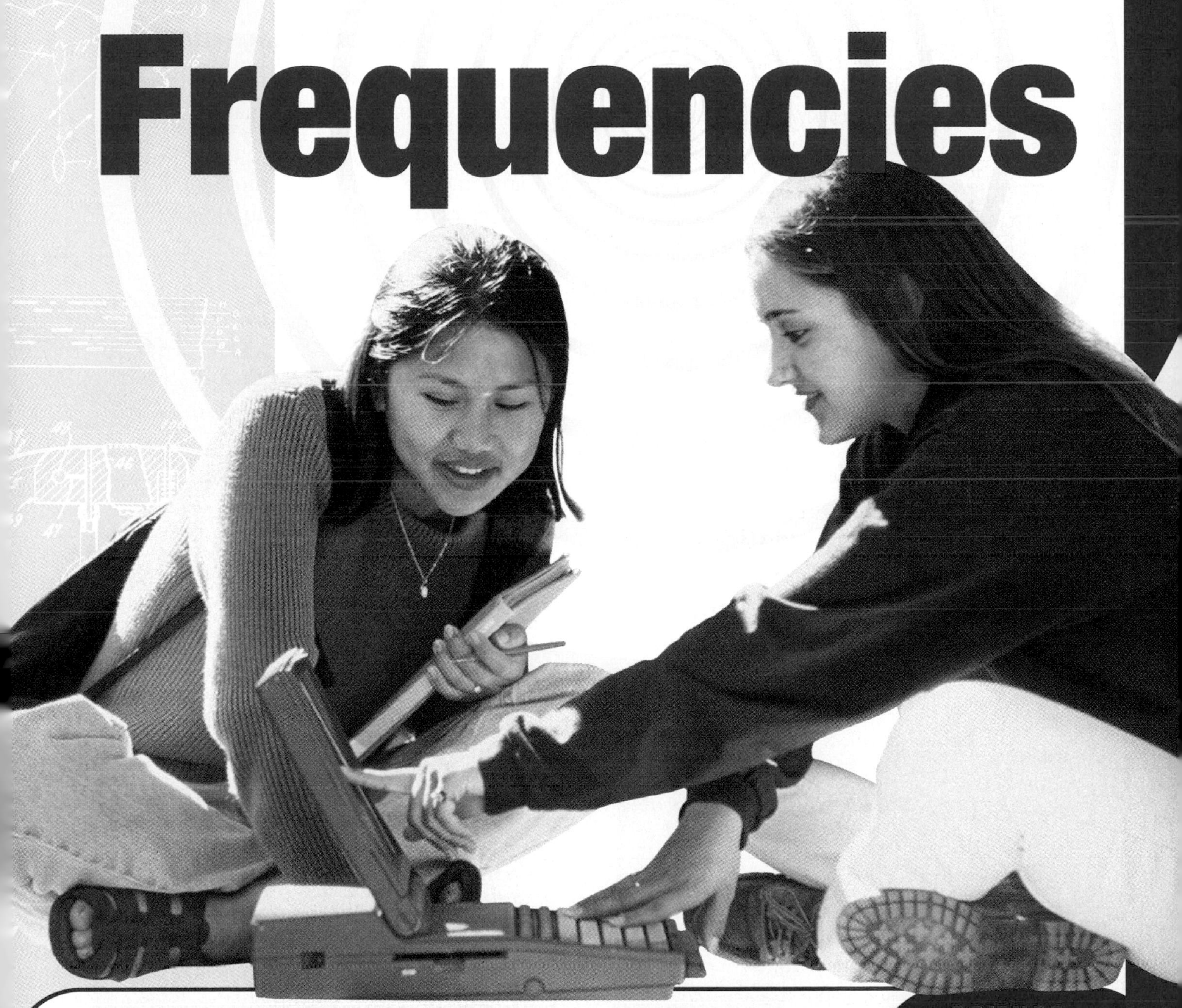

CONNECTIONS Brainstorm How are you using wireless technology in your life right now? List ways it makes your life easier. Are there drawbacks to some of the uses for wireless technology? What are they?

SCIENCE Online
For more information, visit science.glencoe.com

Chapter 22 Study Guide

Reviewing Main Ideas

Section 1 The Nature of Electromagnetic Waves

1. Vibrating charges generate vibrating electric and magnetic fields. These vibrating fields travel through space and are called electromagnetic waves.
2. Electromagnetic waves, like all waves, have wavelength, frequency, amplitude, and carry energy. *How are ocean waves similar to electromagnetic waves?*

Section 2 The Electromagnetic Spectrum

1. Radio waves have the longest wavelength and lowest energy. Radar uses radio waves to locate objects.
2. All objects emit infrared waves. Most objects you see reflect the visible light emitted by a source of light. *How could a person be seen in total darkness?*

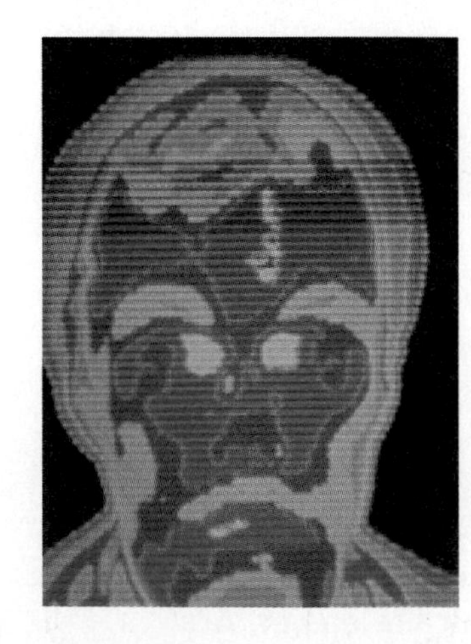

3. Ultraviolet waves have a higher frequency and carry more energy than visible light.
4. X rays and gamma rays are highly penetrating and can be dangerous to living organisms.

Section 3 Using Electromagnetic Waves

1. Communications systems use visible light, radio waves, or electrical signals to transmit information.
2. Radio and TV stations use modulated carrier waves to transmit information.
3. Electromagnetic waves are used in telephone technologies to make communication easier and faster. *What is one way an electromagnetic wave is used in telephone communication?*

4. Communications satellites relay information from different points on Earth so a transmission can go around the globe. The Global Positioning System uses satellites to determine the position of an object on Earth.

After You Read

FOLDABLES Reading & Study Skills

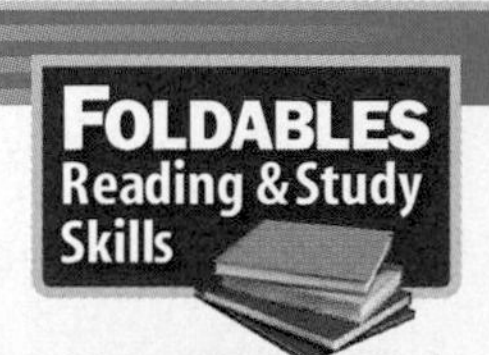

Using the information on your Foldable, compare and contrast visible and invisible waves that form the electromagnetic spectrum.

Visualizing Main Ideas

Complete the following spider map about electromagnetic waves.

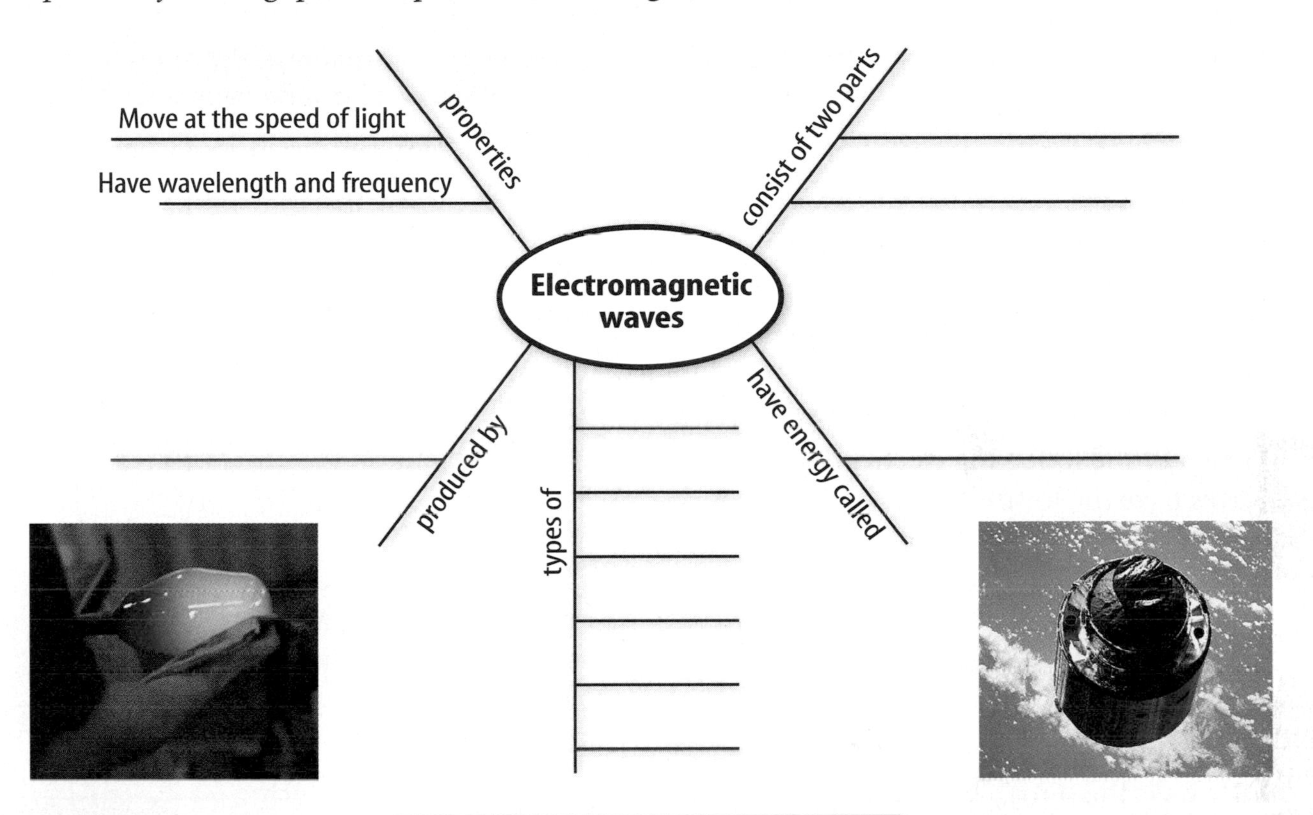

Vocabulary Review

Vocabulary Words

- **a.** carrier wave
- **b.** electromagnetic spectrum
- **c.** electromagnetic wave
- **d.** gamma ray
- **e.** Global Positioning System
- **f.** infrared wave
- **g.** radiant energy
- **h.** radio wave
- **i.** ultraviolet radiation
- **j.** visible light
- **k.** X ray

Study Tip

After you read a chapter, write ten questions that it answers. Wait one day and then try to recall the answers. Look up what you can't remember.

Using Vocabulary

Explain the difference between the terms in each of the following pairs.

1. infrared wave, radio wave
2. radio wave, carrier wave
3. communications satellite, Global Positioning System
4. visible light, ultraviolet radiation
5. X ray, gamma ray
6. electromagnetic wave, radiant energy
7. carrier wave, AM radio signal
8. infrared wave, ultraviolet wave

Chapter 22 Assessment

Checking Concepts

Choose the word or phrase that best answers the question.

1. Which type of force field surrounds a moving electron?
   A) electric and magnetic **C)** magnetic
   B) electric **D)** none of these

2. What does a microphone transform?
   A) light waves to sound waves
   B) radio waves to an electrical signal
   C) sound waves to electromagnetic waves
   D) sound waves to an electrical signal

3. Which of the following electromagnetic waves have the lowest frequency?
   A) visible light **C)** radio waves
   B) infrared waves **D)** X rays

4. What happens to the energy of an electromagnetic wave as its frequency increases?
   A) It increases.
   B) It decreases.
   C) It stays the same.
   D) It oscillates up and down.

5. What type of wave can hot objects emit?
   A) radio **C)** visible
   B) infrared **D)** ultraviolet

6. What can detect radio waves?
   A) film **C)** eyes
   B) antenna **D)** skin

7. Which wave can pass through people?
   A) infrared **C)** ultraviolet
   B) visible **D)** gamma

8. Which color has the lowest frequency?
   A) green **C)** yellow
   B) violet **D)** red

9. What is the key device that allows cordless phones to function?
   A) X ray **C)** GPS
   B) satellite **D)** antenna

10. What does *A* in AM stand for?
    A) amplitude **C)** astronomical
    B) antenna **D)** Alpha centauri

Thinking Critically

11. Infrared light was discovered when a scientist placed a thermometer in each band of the light spectrum produced by a prism. Would the area just beyond red have been warmer or cooler than the room? Explain.

12. Astronomers have built telescopes on Earth that have flexible mirrors that can eliminate the distortions due to the atmosphere. What advantages would a space-based telescope have over these?

13. Heated objects often give off visible light of a particular color. Explain why an object that glows bluish-white is hotter than one that glows red.

14. How can an X ray be used to determine the location of a cancerous tumor?

15. Why are many communications systems based on radio waves?

Developing Skills

16. **Calculating Ratios** How far does light travel in 1 min? How does this compare with the distance to the Moon?

17. **Recognizing Cause and Effect** As you ride in the car, the radio alternates between two different stations. How can the antenna pick up two stations at once?

18. **Classifying** List the colors of the visible spectrum in order of increasing frequency.

19. Comparing and Contrasting Compare and contrast ultraviolet and infrared light.

20. Concept Mapping Electromagnetic waves are grouped according to their frequencies. In the following concept map, write each frequency group and one way humans make use of the electromagnetic waves in that group. For example, in the second set of ovals, you might write "X rays" and "to see inside the body."

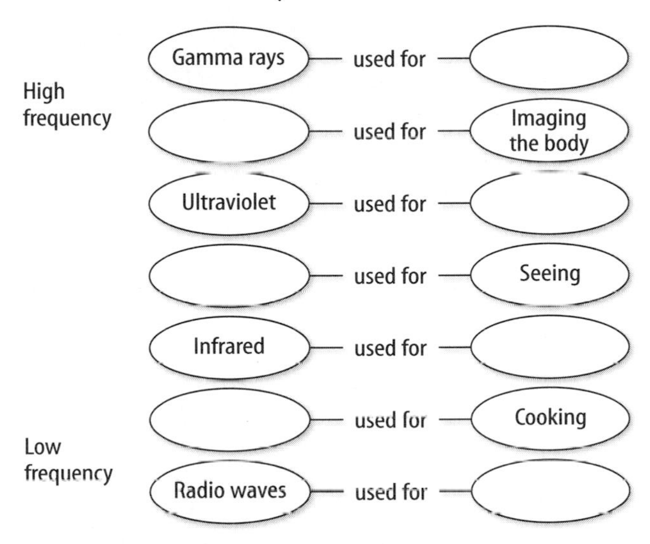

Performance Assessment

21. Oral Presentation Explain to the class how a radio signal is generated, transmitted, and received.

22. Poster Make a poster showing the parts of the electromagnetic spectrum. Show how frequency, wavelength, and energy change throughout the spectrum. How is each wave generated? What are some uses of each?

Technology

Go to the Glencoe Science Web site at **science.glencoe.com** or use the **Glencoe Science CD-ROM** for additional chapter assessment.

Test Practice

Mr. Rubama's class was studying how radio waves are transmitted. An experimental setup involving radio waves and glass is shown below.

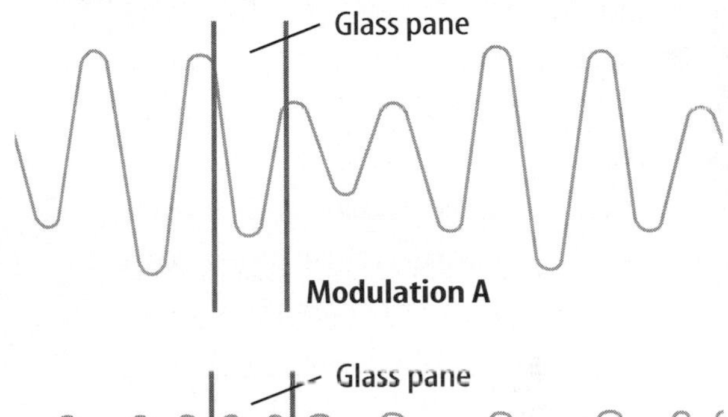

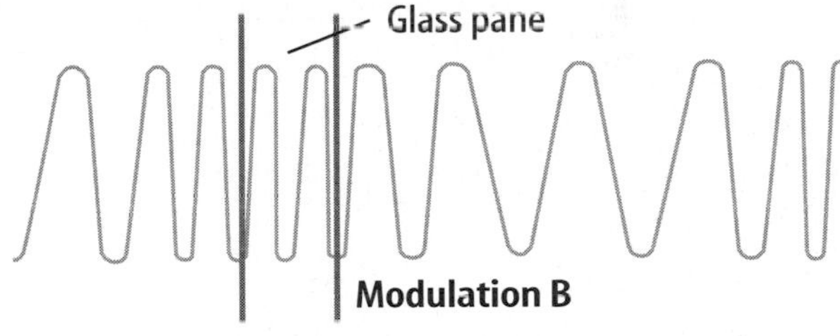

Study the illustrations and answer the following questions.

1. Which of these questions would most likely be answered by this experiment?

A) How fast do radio waves travel through the air?

B) Why do some waves travel more quickly than other waves?

C) Where do radio waves come from?

D) Can radio waves travel through glass?

2. Which of the following describes how the wave in Modulation A is different from the wave in Modulation B?

F) It is a radio wave.

G) It is frequency modulated

H) It is amplitude modulated

J) It is a carrier wave.

CHAPTER

Light, Mirrors, and Lenses

You walk through a door of the fun house and are bombarded by images of yourself. In one mirror, your face seems smashed. You turn around and face another mirror—your chin and neck are gigantic. How do mirrors in a fun house make you look so strange? In this chapter, you'll learn how mirrors and lenses create images. You'll also learn why objects have the colors they have.

What do you think?

Science Journal Look at the picture below with a classmate. Discuss what you think this might be or what is happening. Here's a hint: *It helps you keep in touch.* Write down your answer or your best guess in your Science Journal.

Everything you see results from light waves entering your eyes. These light waves are either given off by objects, such as the Sun and lightbulbs, or reflected by objects, such as trees, books, and people. Lenses and mirrors can cause light to change direction and make objects seem larger or smaller. What happens to light as it passes from one material to another?

Observe the bending of light

1. Place two paper cups next to each other and put a penny in the bottom of each cup.
2. Fill one of the cups with water and observe how the penny looks.
3. Looking straight down at the cups, slide the cup with no water away from you just until you can no longer see the penny.
4. Pour water into this cup and observe what seems to happen to the penny.

Observe

In your Science Journal, record your observations. Did adding water make the cup look deeper or shallower?

Before You Read

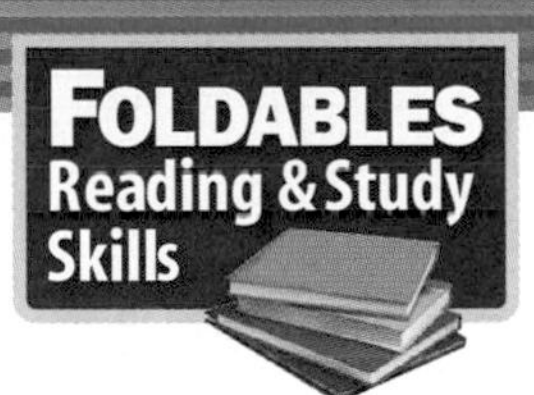

Making a Question Study Fold Asking yourself questions helps you stay focused so you will better understand light, mirrors, and lenses when you are reading the chapter.

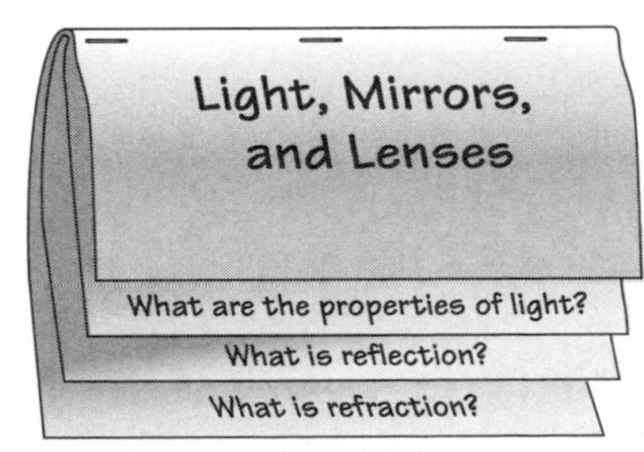

1. Stack two sheets of paper in front of you so the short side of both sheets is at the top.
2. Slide the top sheet up so about 4 cm of the bottom sheet shows.
3. Fold both sheets top to bottom to form four tabs and staple along the fold as shown.
4. Title the Foldable *Light, Mirrors, and Lenses* as shown. Write these questions on the flaps: *What are the properties of light? What is reflection? What is refraction?*
5. Before you read the chapter, try to answer the questions with what you already know. As you read the chapter, add to or correct your answers under the flaps.

SECTION

Properties of Light

As You Read

What You'll Learn

- **Describe** the wave nature of light.
- **Explain** how light interacts with materials.
- **Determine** why objects appear to have color.

Vocabulary

light ray
medium
reflection

Why It's Important

Much of what you know about your surroundings comes from information carried by light waves.

What is light?

Drop a rock on the smooth surface of a pond and you'll see ripples spread outward from the spot where the rock struck. The rock produced a wave much like the one in **Figure 1A.** A wave is a disturbance that carries energy through matter or space. The matter in this case is the water, and the energy originally comes from the impact of the rock. As the ripples spread out, they carry some of that energy.

Light is a type of wave that carries energy. A source of light such as the Sun or a lightbulb gives off light waves into space, just as the rock hitting the pond causes waves to form in the water. But while the water waves spread out only on the surface of the pond, light waves spread out in all directions from the light source. **Figure 1B** shows how light waves travel.

Sometimes, however, it is easier to think of light in a different way. A **light ray** is a narrow beam of light that travels in a straight line. You can think of a source of light as giving off, or emitting, a countless number of light rays that are traveling away from the source in all directions.

Figure 1
Waves carry energy as they travel.

A **Ripples on the surface of a pond are produced by an object hitting the water. As the ripples spread out from the point of impact, they carry energy.**

B **A source of light, such as a lightbulb, gives off light rays that travel away from the light source in all directions.**

Light Travels Through Space There is, however, one important difference between light waves and the water wave ripples on a pond. If the pond dried up and had no water, ripples could not form. Waves on a pond need a material—water—in which to travel. The material through which a wave travels is called a **medium.** Light is an electromagnetic wave and doesn't need a medium in which to travel. Electromagnetic waves can travel in a vacuum, as well as through materials such as air, water, and glass.

Light and Matter

What can you see when you are in a closed room with no windows and the lights are out? You can see nothing until you turn on a light or open a door to let in light from outside the room. Most objects around you do not give off light on their own. They can be seen only if light waves from another source bounce off them and into your eyes, as shown in **Figure 2.** The process of light striking an object and bouncing off is called **reflection.** Right now, you can see these words because light emitted by a source of light is reflecting from the page and into your eyes. Not all the light rays reflected from the page strike your eyes. Light rays striking the page are reflected in many directions, and only some of these rays enter your eyes.

✔ Reading Check *What must happen for you to see most objects?*

TRY AT HOME Mini LAB

Observing Colors in the Dark

Procedure

1. Get six pieces of **paper** that are different colors and about 10 cm × 10 cm.
2. Darken a room and wait 10 min for your eyes to adjust to the darkness.
3. Write on each paper what color you think the paper is.
4. Turn on the lights and see if your night vision detected the colors.

Analysis

1. If the room were perfectly dark, what would you see? Explain.
2. Your eyes contain structures called rods and cones. Rods don't detect color, but need only a little light. Cones detect color, but need more light. Which structure was working in the dark room? Explain.

Figure 2
Light waves are given off by the lightbulb. Some of these light waves hit the page and are reflected. The student sees the page when some of these reflected waves enter the student's eyes.

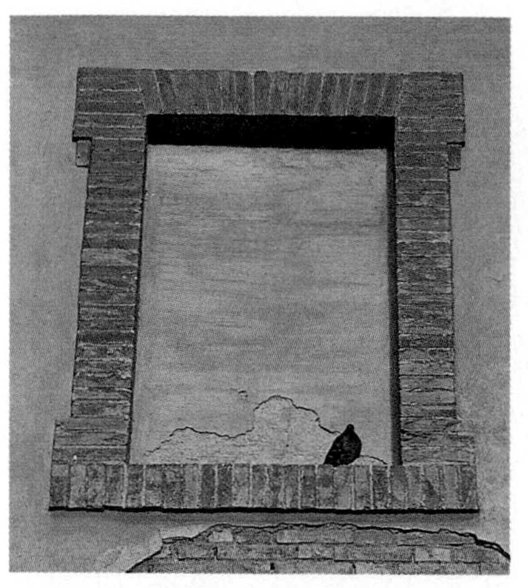

A An opaque object allows no light to pass through it.

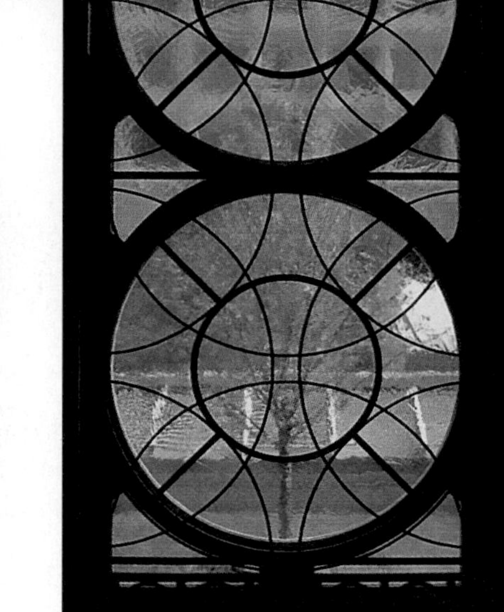

B A translucent object allows some light to pass through it.

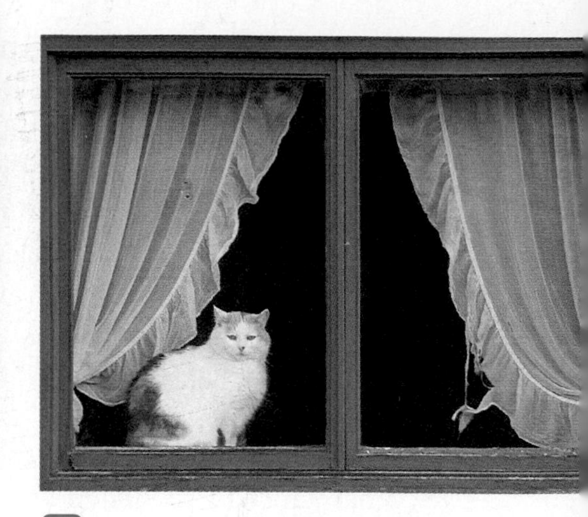

C A transparent object allows almost all light to pass through it.

Figure 3
Materials are opaque, translucent, or transparent depending on how much light passes through them. *Which type of material reflects the least amount of light?*

Opaque, Translucent, and Transparent When light waves strike an object, some of the waves are absorbed by the object, some of the waves are reflected by it, and some of the light waves might pass through it. What happens to light when it strikes the object depends on the material that the object is made of.

All objects reflect and absorb some light waves. Materials that let no light pass through them are opaque (oh PAYK). You cannot see other objects through opaque materials. On the other hand, you clearly can see other objects through materials such as glass and clear plastic that allow nearly all the light that strikes them to pass through. These materials are transparent. A third type of material allows only some light to pass through. Although objects behind these materials are visible, they are not clear. These materials, such as waxed paper and frosted glass, are translucent (trans LEW sent). Examples of opaque, translucent, and transparent objects are shown in **Figure 3.**

Figure 4
A beam of white light passing through a prism is separated into many colors. *What colors can you see emerging from the prism?*

Color

The light from the Sun might look white, but it is a mixture of colors. Each different color of light is a different wavelength. You sometimes can see the different colors of the Sun's light when it passes through raindrops to make a rainbow. As shown in **Figure 4,** white light is separated into different colors when it passes through a prism. The colors in white light range from red to violet. When light waves from all these colors enter the eye at the same time, the brain interprets the mixture as being white.

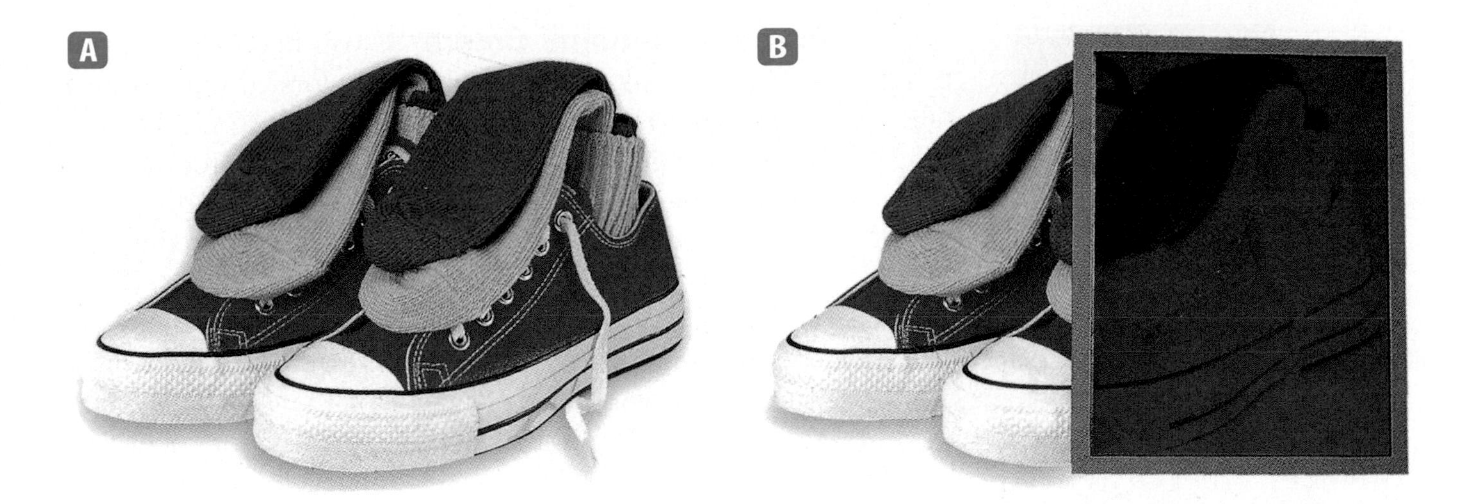

Why do Objects Have Color? Why does grass look green or a rose look red? When a mixture of light waves strikes an object that is not transparent, the object absorbs some of the light waves. Some of the light waves that are not absorbed are reflected. If an object reflects red waves and absorbs all the other waves, it looks red. Similarly, if an object looks blue, it reflects only blue light waves and absorbs all the others. An object that reflects all the light waves that strike it looks white, while one that reflects none of the light waves that strike it looks black. **Figure 5** shows gym shoes and socks as seen under white light and as seen when viewed through a red filter that allows only red light to pass through it.

Figure 5
The color of an object depends on the light waves it reflects.
A Examine the pair of gym shoes and socks as they are seen under white light. *Why do the socks look blue under white light?*
B The same shoes and socks were photographed through a red filter. *Why do the blue socks look black when viewed under red light?*

Primary Light Colors How many colors exist? People often say white light is made up of red, orange, yellow, green, blue, and violet light. This isn't completely true, though. Many more colors than this exist. In reality, most humans can distinguish thousands of colors, including some such as brown, pink, and purple, that are not found among the colors of the rainbow.

Light of almost any color can be made by mixing different amounts of red, green, and blue light. Red, green, and blue are known as the primary colors. Look at **Figure 6.** White light is produced where beams of red, green, and blue light overlap. Yellow light is produced where red and green light overlap. You see the color yellow because of the way your brain interprets the combination of the red and green light striking your eye. This combination of light waves looks the same as yellow light produced by a prism, even though these light waves have only a single wavelength.

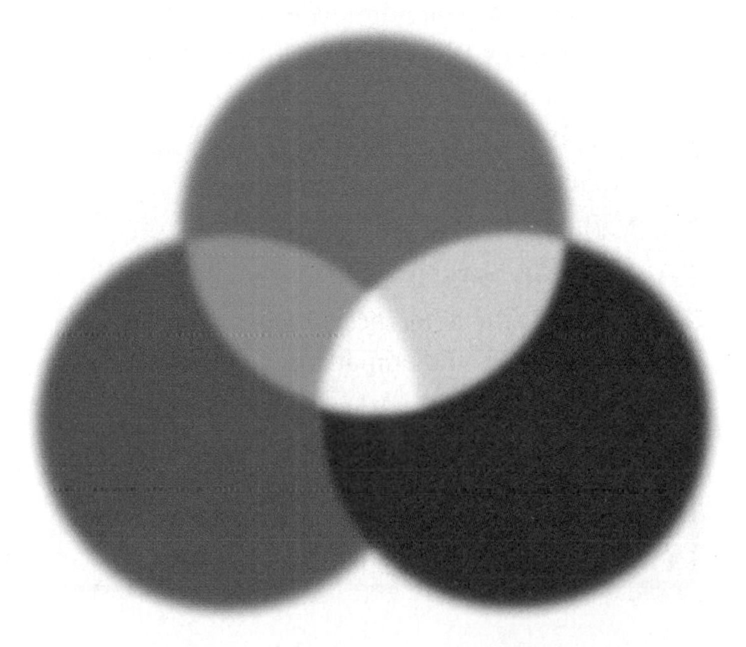

Figure 6
By mixing light from the three primary colors—red, blue, and green—almost all of the visible colors can be made.

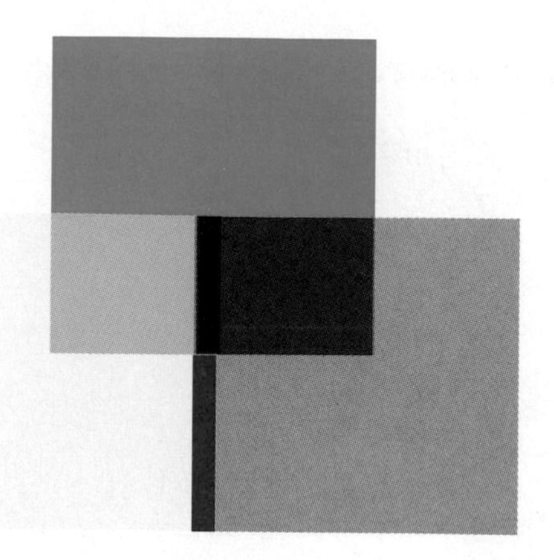

Figure 7
The three primary color pigments—yellow, magenta, and cyan—can form almost all the visible colors when mixed together in various amounts.

Primary Pigment Colors If you like to paint, you might mix two or more different colors to make a new color. Materials like paint that are used to change the color of other objects, such as the walls of a room or an artist's canvas, are called pigments. Mixing pigments together forms colors in a different way than mixing colored lights does.

Like all materials that appear to be colored, pigments absorb some light waves and reflect others. The color of the pigment you see is the color of the light waves that are reflected from it. However, the primary pigment colors are not red, blue, and green—they are yellow, magenta, and cyan. You can make almost any color by mixing different amounts of these primary pigment colors, as shown in **Figure 7.**

✔ Reading Check ***What are the primary pigment colors?***

Although primary pigment colors are not the same as the primary light colors, they are related. Each primary pigment color results when a pigment absorbs a primary light color. For example, a yellow pigment absorbs blue light and it reflects red and green light, which you see as yellow. A magenta pigment, on the other hand, absorbs green light and reflects red and blue light, which you see as magenta. Each of the primary pigment colors is the same color as white light with one primary color removed.

Section  Assessment

1. At night in your room, you are reading a magazine. Describe the path light takes that enables you to see the page.
2. Do the light rays traveling outward from a light source carry energy? Explain.
3. What colors are reflected by an object that appears black? Explain.
4. What is the difference between primary light colors and primary pigment colors?
5. **Think Critically** When you're in direct sunlight, why do you feel cooler if you're wearing light-colored clothes than if you're wearing darker-colored clothes?

Skill Builder Activities

6. **Drawing Conclusions** A white plastic bowl and a black plastic bowl have been sitting in the sunlight. You observe that the black bowl feels warmer than the white bowl. From this information, conclude which of the bowls absorbs and which reflects more sunlight. **For more help, refer to the Science Skill Handbook.**
7. **Communicating** Read an article about the greenhouse effect and draw a diagram in your Science Journal explaining how the greenhouse effect involves absorption. **For more help, refer to the Science Skill Handbook.**

SECTION

Reflection and Mirrors

The Law of Reflection

You've probably noticed your image on the surface of a pool or lake. If the surface of the water was smooth, you could see your face clearly. If the surface of the water was wavy, however, your face might have seemed distorted. The image you saw was the result of light reflecting from the surface and traveling to your eyes. How the light was reflected determined the sharpness of the image you saw.

When a light ray strikes a surface and is reflected as in **Figure 8,** the reflected ray obeys the law of reflection. Imagine a line that is drawn perpendicular to the surface where the light ray strikes. This line is called the normal to the surface. The incoming ray and the normal form an angle called the angle of incidence. The reflected light ray forms an angle with the normal called the angle of reflection. According to the **law of reflection,** the angle of incidence is equal to the angle of reflection. This is true for any surface, no matter what material it is made of.

Reflection from Surfaces

Why can you see your reflection in some surfaces and not others? Why does a piece of shiny metal make a good mirror, but a piece of paper does not? The answers have to do with the smoothness of each surface.

As You Read

What You'll Learn

- **Explain** how light is reflected from rough and smooth surfaces.
- **Determine** how mirrors form an image.
- **Describe** how concave and convex mirrors form an image.

Vocabulary

law of reflection
focal point
focal length

Why It's Important

Mirrors can change the direction of light waves and enable you to see images, such as your own face, that normally would not be in view.

Figure 8
A light ray strikes a surface and is reflected. The angle of incidence is always equal to the angle of reflection. This is the law of reflection.

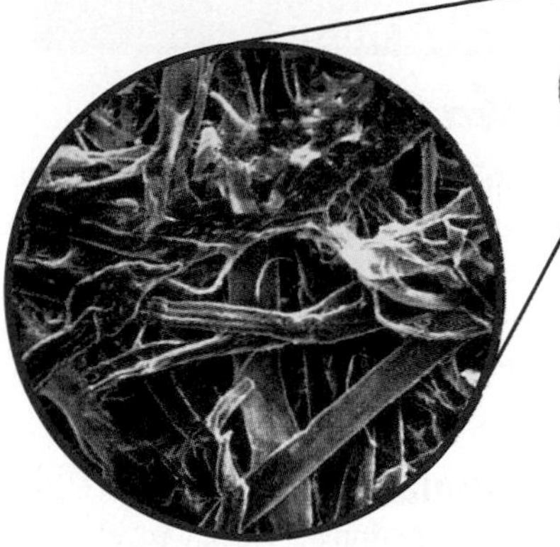
Magnification: 35×

Figure 9
A highly magnified view of the surface of a paper towel shows that the surface is made of many cellulose wood fibers that make it rough and uneven.

Regular and Diffuse Reflection Even though the surface of the paper might seem smooth, it's not as smooth as the surface of a mirror. **Figure 9** shows how rough the surface of a piece of paper looks when it is viewed under a microscope. The rough surface causes light rays to be reflected from it in many directions, as shown in **Figure 10A.** This uneven reflection of light waves from a rough surface is diffuse reflection. The smoother surfaces of mirrors, as shown in **Figure 10B,** reflect light waves in a much more regular way. For example, parallel rays remain parallel after they are reflected from a mirror. Reflection from mirrors is known as regular reflection. Light waves that are regularly reflected from a surface form the image you see in a mirror or any other smooth surface. Whether a surface is smooth or rough, every light ray that strikes it obeys the law of reflection.

✔ Reading Check *Why does a rough surface cause a diffuse reflection?*

Scattering of Light When diffuse reflection occurs, light waves that were traveling in a single direction are reflected, and then travel in many different directions. Scattering occurs when light waves traveling in one direction are made to travel in many different directions. Scattering also can occur when light waves strike small particles, such as dust. You may have seen dust particles floating in a beam of sunlight. When the light waves in the sunbeam strike a dust particle, they are scattered in all directions. You see the dust particle as bright specks of light when some of these scattered light waves enter your eye.

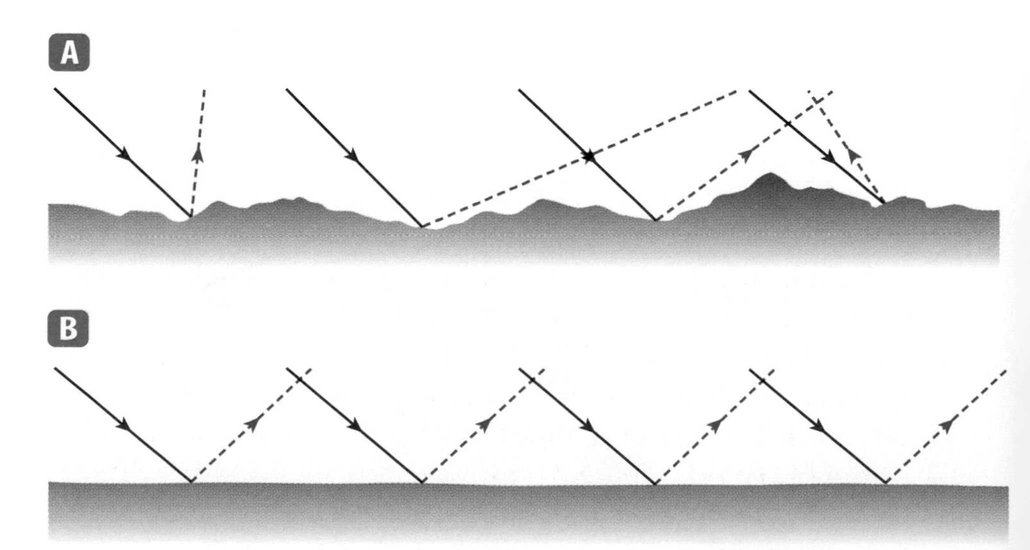

Figure 10
The roughness of a surface determines whether it looks like a mirror. A A rough surface causes parallel light rays to be reflected in many different directions. B A smooth surface causes parallel light rays to be reflected in a single direction. This type of surface looks like a mirror.

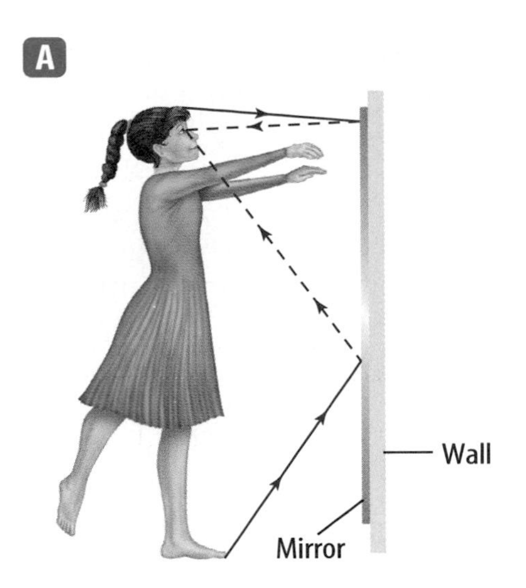

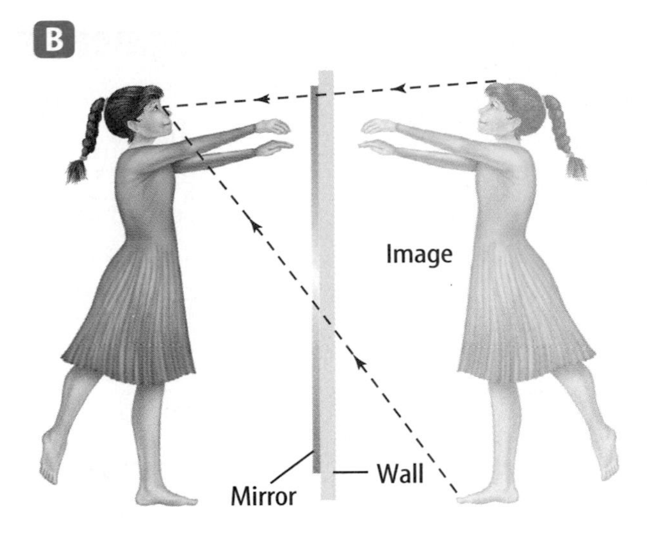

Figure 11
A plane mirror forms an image by changing the direction of light rays. A Light rays that bounce off of a person strike the mirror. Some of these light rays are reflected into the person's eye. B The light rays that are shown entering the person's eye seem to be coming from a person behind the mirror.

Reflection by Plane Mirrors Did you glance in the mirror before leaving for school this morning? If you did, you probably looked at your reflection in a plane mirror. A plane mirror is a mirror with a flat reflecting surface. In a plane mirror, your image looks much the same as it would in a photograph. However, you and your image are facing in opposite directions. This causes your left side and your right side to switch places on your mirror image. Also, your image seems to be coming from behind the mirror. How does a plane mirror form an image?

Reading Check *What is a plane mirror?*

Figure 11 shows a person looking into a plane mirror. Light waves from the Sun or another source of light strike each part of the person. These light rays bounce off of the person according to the law of reflection, and some of them strike the mirror. The rays that strike the mirror also are reflected according to the law of reflection. **Figure 11A** shows the path traveled by a few of the rays that have been reflected off the person and reflected back to the person's eye by the mirror.

The Image in a Plane Mirror Why does the image you see in a plane mirror seem to be behind the mirror? This is a result of how your brain processes the light rays that enter your eyes. Although the light rays bounced off the mirror's surface, your brain interprets them as having followed the path shown by the dashed lines in **Figure 11B.** In other words, your brain always assumes that light rays travel in straight lines without changing direction. This makes the reflected light rays look as if they are coming from behind the mirror, even though no source of light is there. The image also seems to be the same distance behind the mirror as the person is in front of the mirror.

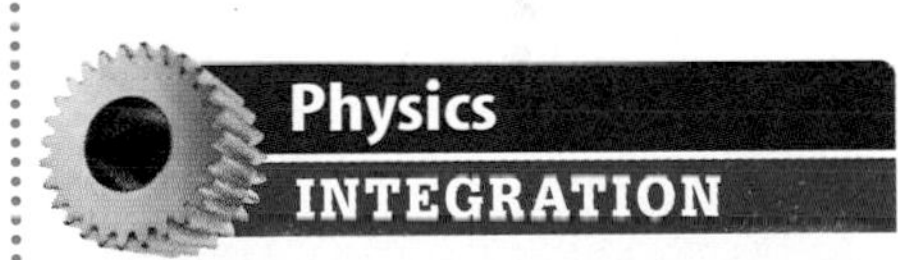

When a particle like a marble or a basketball bounces off a surface, it obeys the law of reflection. Because light also obeys the law of reflection, people once thought that light must be a stream of particles. Today, experiments have shown that light can behave as though it were both a wave and a stream of energy bundles called photons. Read an article about photons and write a description in your Science Journal.

Concave and Convex Mirrors

Some mirrors are not flat. A concave mirror has a surface that is curved inward, like the inside of a spoon. Unlike plane mirrors, concave mirrors cause light rays to come together, or converge. A convex mirror, on the other hand, has a surface that curves outward, like the outside of a spoon. Convex mirrors cause light waves to spread out, or diverge. These two types of mirrors form images that are different from the images that are formed by plane mirrors. Examples of a concave and a convex mirror are shown in **Figure 12.**

✔ **Reading Check** ***What's the difference between a concave and convex mirror?***

Concave Mirrors The way in which a concave mirror forms an image is shown in **Figure 13.** A straight line drawn perpendicular to the center of a concave or convex mirror is called the optical axis. Light rays that travel parallel to the optical axis and strike the mirror are reflected so that they pass through a single point on the optical axis called the **focal point.** The distance along the optical axis from the center of the mirror to the focal point is called the **focal length.**

The image formed by a concave mirror depends on the position of the object relative to its focal point. If the object is farther from the mirror than the focal point, the image appears to be upside down, or inverted. The size of the image decreases as the object is moved farther away from the mirror. If the object is closer to the mirror than one focal length, the image is upright and gets smaller as the object moves closer to the mirror.

A concave mirror can produce a focused beam of light if a source of light is placed at the mirror's focal point, as shown in **Figure 13.** Flashlights and automobile headlights use concave mirrors to produce directed beams of light.

A

B

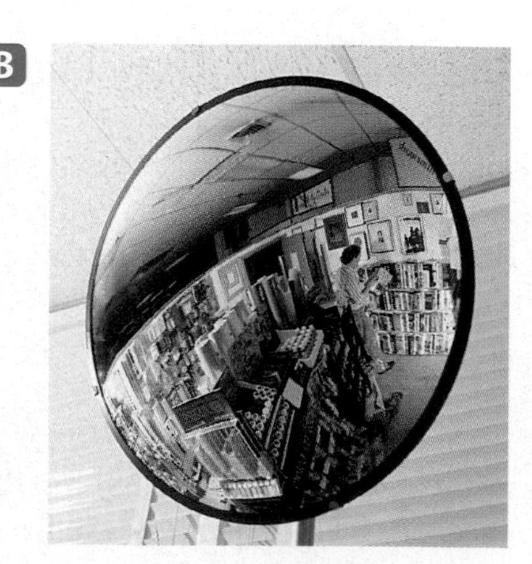

Figure 12
Not all mirrors are flat.
A A concave mirror has a surface that's curved inward. B A convex mirror has a surface that's curved outward.

NATIONAL GEOGRAPHIC VISUALIZING REFLECTIONS IN CONCAVE MIRRORS

Figure 13

Glance into a flat plane mirror and you'll see an upright image of yourself. But look into a concave mirror, and you might see yourself larger than life, right side up, or upside down—or not at all! This is because the way a concave mirror forms an image depends on the position of an object in front of the mirror, as shown here.

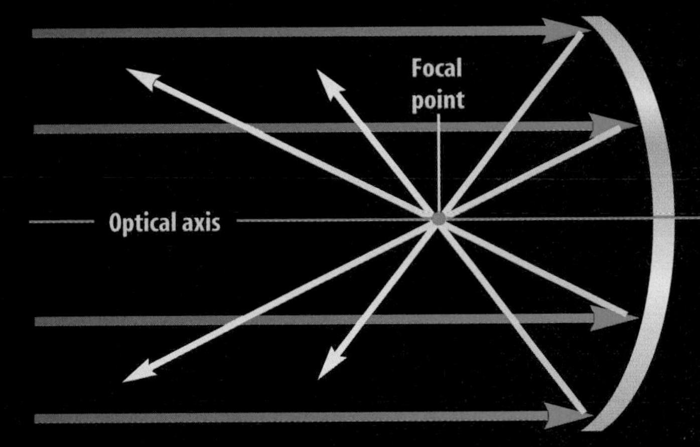

A concave mirror reflects all light rays traveling parallel to the optical axis so that they pass through the focal point.

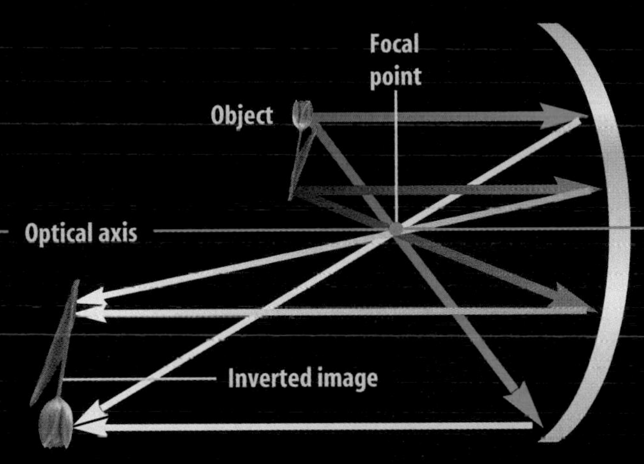

When an object, such as this flower, is placed beyond the focal point, the mirror forms an image that is inverted.

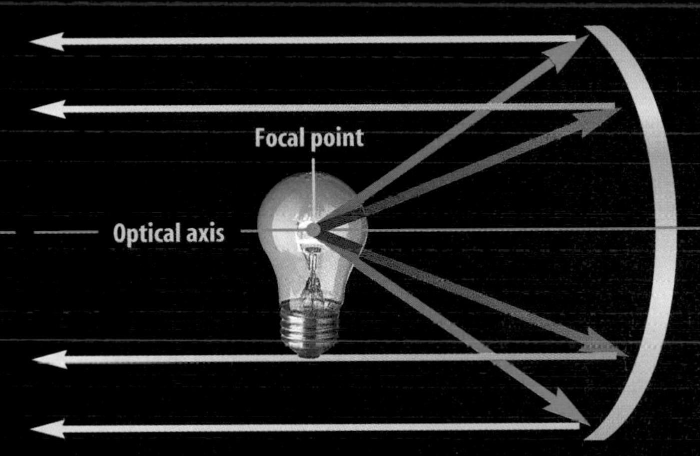

When a source of light is placed at the focal point, a beam of parallel light rays is formed. The concave mirror in a flashlight, for example, creates a beam of parallel light rays.

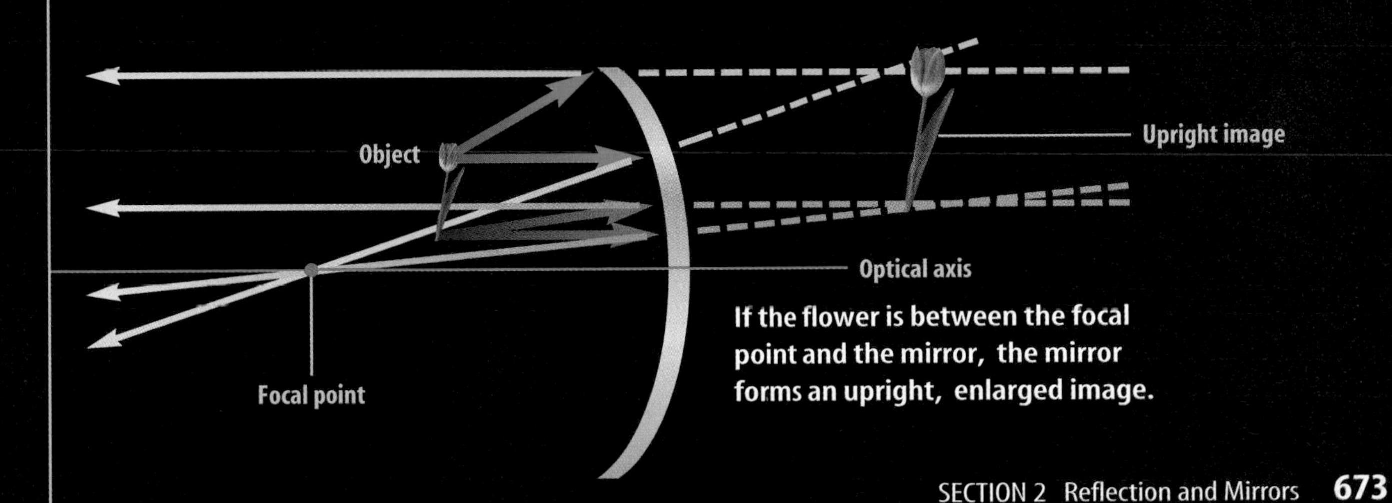

If the flower is between the focal point and the mirror, the mirror forms an upright, enlarged image.

Figure 14
A convex mirror always forms an image that is smaller than the object.

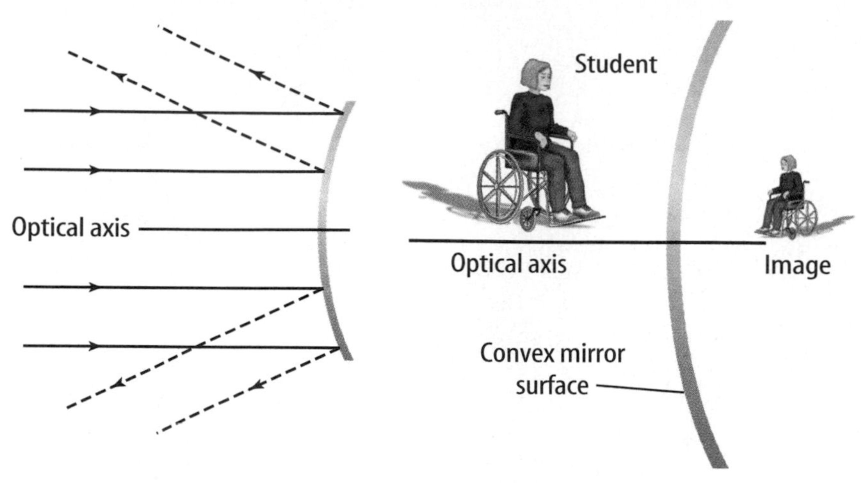

A A convex mirror causes incoming light rays that are traveling parallel to the optical axis to spread apart after they are reflected.

B No matter how far the object is from a convex mirror, the image is always upright and smaller than the object.

Convex Mirrors A convex mirror has a reflecting surface that curves outward. Because the reflecting surface curves outward, a convex mirror causes light rays to spread apart, or diverge, as shown in **Figure 14A.** Like the image formed by a plane mirror, the image formed by a convex mirror seems to be behind the mirror, as shown in **Figure 14B.** Also like a plane mirror, the image formed by a convex mirror is always upright. Unlike a plane mirror or a concave mirror, however, the image formed by a convex mirror is always smaller than the object.

Convex mirrors are used as security mirrors mounted above the aisles in stores and as outside rearview mirrors on cars, trucks and other vehicles. When used in this way, objects in the mirror seem smaller and farther away than they actually are. As a result, you can see a larger area reflected in a convex mirror. A convex mirror is said to have a larger angle of view than a plane mirror.

Section  Assessment

1. Describe how light reflects from rough and smooth surfaces.
2. Why are concave mirrors used in flashlights and automobile headlights?
3. What happens to the image in a concave mirror when an object is closer to the mirror than one focal length?
4. Why do side mirrors on cars carry the warning that objects are closer than they appear to be?
5. **Think Critically** The surface of a car is covered with dust and looks dull. After the car is washed and waxed, you can see your image reflected in the car's surface. Explain.

Skill Builder Activities

6. **Forming Hypotheses** When you look at a window at night, you sometimes can see two images of yourself reflected from the window. Make a hypothesis to explain why two images are seen. **For more help, refer to the Science Skill Handbook.**
7. **Using an Electronic Spreadsheet** Design a table using spreadsheet software to compare the images formed by plane, concave, and convex mirrors. Include in your table how the images depend on the distance of the object from the mirror. **For more help, refer to the Technology Skill Handbook.**

Activity

Reflection from a Plane Mirror

A light ray strikes the surface of a plane mirror and is reflected. Does a relationship exist between the direction of the incoming light ray and the direction of the reflected light ray?

What You'll Investigate

How does the angle of incidence compare with the angle of reflection for a plane mirror?

Materials

flashlight
protractor
metric ruler
scissors
tape
small plane mirror, at least 10 cm on a side
black construction paper
modeling clay
white unlined paper

Goals

- **Measure** the angle of incidence and the angle of reflection for a light ray reflected from a plane mirror.

Safety Precautions

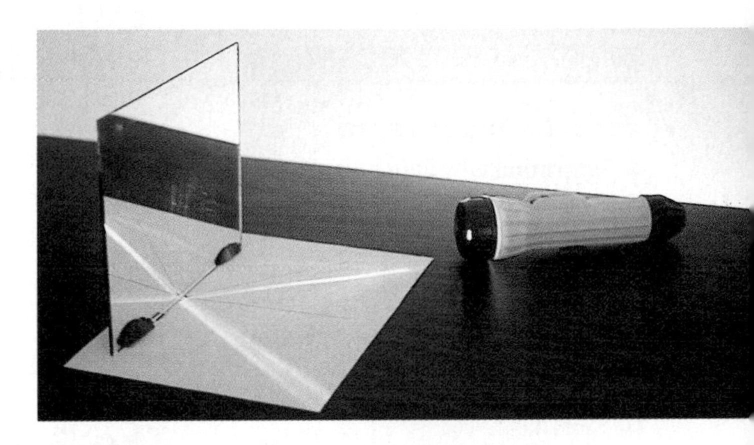

Procedure

1. With the scissors, cut a slit in the construction paper and tape it over the flashlight lens.
2. Place the mirror at one end of the unlined paper. Push the mirror into the lump of clay so it stands vertically, and tilt the mirror so it leans slightly toward the table.
3. **Measure** with the ruler to find the center of the bottom edge of the mirror and mark it. Then use the protractor and the ruler to draw a line on the paper perpendicular to the mirror from the mark. Label this line *P*.
4. Using the protractor and the ruler, draw lines on the paper outward from the mark at the center of the mirror at angles of 30°, 45°, and 60° to line *P*.
5. Turn on the flashlight and place it so the beam is along the 60° line. This is the angle of incidence. Locate the reflected beam on the paper, and measure the angle that the reflected beam makes with line *P*. Record this angle in your data table. This is the angle of reflection. If you cannot see the reflected beam, slightly increase the tilt of the mirror.
6. Repeat step 5 for the 30°, 45°, and *P* lines.

Conclude and Apply

1. What happened to the beam of light when it was shined along line *P*?
2. What can you infer about the relationship between the angle of incidence and the angle of reflection?

Communicating Your Data

Make a poster that shows your measured angles of reflection for angles of incidence of 30°, 45°, and 60°. Write the relationship between the angles of incidence and reflection at the bottom.

Refraction and Lenses

As You Read

What You'll Learn

- **Determine** why light rays refract.
- **Explain** how convex and concave lenses form images.

Vocabulary

refraction
lens
convex lens
concave lens

Why It's Important

Many of the images you see every day in photographs, on TV, and in movies are made using lenses.

Refraction

Objects that are in water can sometimes look strange. A pencil in a glass of water sometimes looks as if it's bent, or as if the part of the pencil in air is shifted compared to the part in water. A penny that can't be seen at the bottom of a cup suddenly appears as you add water to the cup. Illusions such as these are due to the bending of light rays as they pass from one material to another. What causes light rays to change direction?

The Speeds of Light

The speed of light in empty space is about 300 million m/s. Light passing through a material such as air, water, or glass, however, travels more slowly than this. This is because the atoms that make up the material interact with the light waves and slow them down. **Figure 15** compares the speed of light in some different materials.

Figure 15
Light travels at different speeds in different materials.

Air

A **The speed of light through air is about 300 million m/s.**

Water

B **The speed of light through water is about 227 million m/s.**

Glass

C **The speed of light through glass is about 197 million m/s.**

Diamond

D **The speed of light through diamond is about 125 million m/s.**

The Refraction of Light Waves

Light rays from the part of a pencil that is underwater travel through water, glass, and then air before they reach your eye. The speed of light is different in each of these mediums. What happens when a light wave travels from one medium into another in which its speed is different? If the wave is traveling at an angle to the boundary between the two media, it changes direction, or bends. This bending is due to the change in speed the wave undergoes as it moves from one medium into the other. The bending of light waves due to a change in speed is called **refraction.** **Figure 16** shows an example of refraction. The greater the change in speed is, the more the light wave bends, or refracts.

Figure 16
A light ray is bent as it travels from air into water. *In which medium does light travel more slowly?*

Reading Check *What causes light to bend?*

Why does a change in speed cause the light wave to bend? Think about what happens to the wheels of a car as they move from pavement to mud at an angle, as in **Figure 17.** The wheels slip a little in the mud and don't move forward as fast as they do on the pavement. The wheel that enters the mud first gets slowed down a little, but the other wheel on that axle continues at the original speed. The difference in speed between the two wheels then causes the wheel axle to turn, so the car turns a little. Light waves behave in the same way.

Imagine again a light wave traveling at an angle from air into water. The first part of the wave to enter the water is slowed, just as the car wheel that first hit the mud was slowed. The rest of the wave keeps slowing down as it moves from the air into the water. As long as one part of the light wave is moving faster than the rest of the wave, the wave continues to bend.

Convex and Concave Lenses

Do you like photographing your friends and family? Have you ever watched a bird through binoculars or peered at something tiny through a magnifying glass? All of these activities involve the use of lenses. A **lens** is a transparent object with at least one curved side that causes light to bend. The amount of bending can be controlled by making the sides of the lenses more or less curved. The more curved the sides of a lens are, the more light will be bent after it enters the lens.

Figure 17
An axle turns as the wheels cross the boundary between pavement and mud. *How would the axle turn if the wheels were going from mud to pavement?*

Figure 18
A convex lens forms an image that depends on the distance from the object to the lens.

A **Light rays that are parallel to the optical axis are bent so they pass through the focal point.**

B **If the object is more than two focal lengths from the lens, the image formed is smaller than the object and inverted.**

C **If the object is closer to the lens than one focal length, the image formed is enlarged and upright.**

Convex Lenses A lens that is thicker in the center than at the edges is a **convex lens.** In a convex lens, light rays traveling parallel to the optical axis are bent so they meet at the focal point, as shown in **Figure 18A.** The more curved the lens is, the closer the focal point is to the lens, and so the shorter the focal length of the lens is. Because convex lenses cause light waves to meet, they also are called converging lenses.

The image formed by a convex lens is similar to the image formed by a concave mirror. For both, the type of image depends on how far the object is from the mirror or lens. Look at **Figure 18B.** If the object is farther than two focal lengths from the lens, the image seen through the lens is inverted and smaller than the object.

Reading Check ***How can a convex lens be used to make objects appear upside down?***

If the object is closer to the lens than one focal length, then the image formed is right-side up and larger than the object, as shown in **Figure 18C.** A magnifying glass forms an image in this way. As long as the magnifying glass is less than one focal length from the object, you can make the image larger by moving the magnifying glass away from the object.

Research Visit the Glencoe Science web site at **science.glencoe.com** for information about the optical devices that use convex lenses. Prepare a poster or other presentation for your class describing some of these devices.

Concave Lenses A lens that is thicker at the edges than in the middle is a **concave lens.** A concave lens also is called a diverging lens. **Figure 19** shows how light rays traveling parallel to the optical axis are bent after passing through a concave lens.

A concave lens causes light rays to diverge, so light rays are not brought to a focus. The type of image that is formed by a concave lens is similar to one that is formed by a convex mirror. The image is upright and smaller than the object.

Optical axis

Figure 19
A concave lens causes light rays traveling parallel to the optical axis to diverge.

Total Internal Reflection

When you look at a glass window, you sometimes can see your reflection in the window. You see a reflection because some of the light waves reflected from you are reflected back to your eyes when they strike the window. This is an example of a partial reflection—only some of the light waves striking the window are reflected. However, sometimes all the light waves that strike the boundary between two transparent materials can be reflected. This process is called total internal reflection.

The Critical Angle To see how total internal reflection occurs, look at **Figure 20.** Light travels faster in air than in water, and the refracted beam is bent away from the normal. As the angle between the incident beam and the normal increases, the refracted beam bends closer to the air-water boundary. At the same time, more of the light energy striking the boundary is reflected and less light energy passes into the air.

If a light beam in water strikes the boundary so that the angle with the normal is greater than an angle called the critical angle, total internal reflection occurs. Then all the light waves are reflected at the air-water boundary, just as if a mirror were there. The size of the critical angle depends on the two materials involved. For light passing from water to air the critical angle is about 48 degrees.

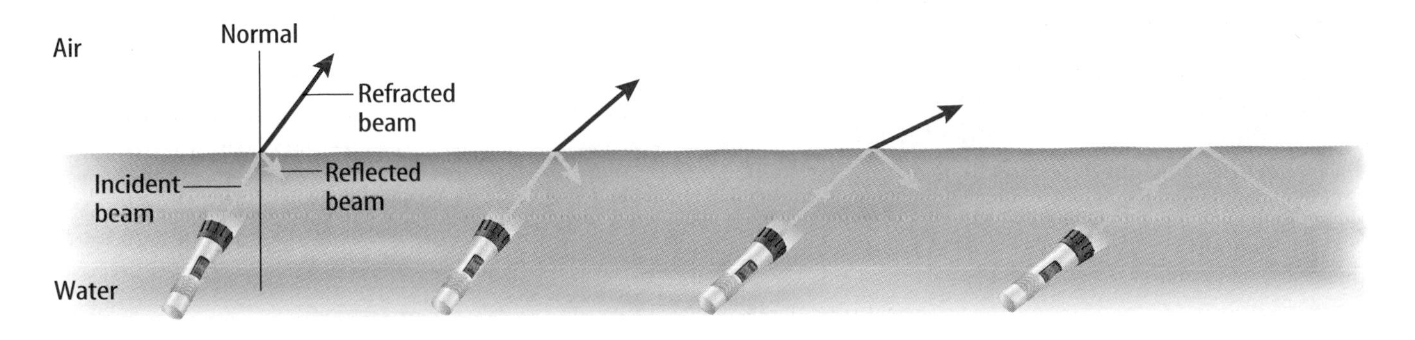

Figure 20
When a light beam passes from one medium to another, some of its energy is reflected (red) and some is refracted (blue). As the incident beam makes a larger angle with the normal, less light energy is refracted, and more is reflected. At the critical angle, all the light is reflected.

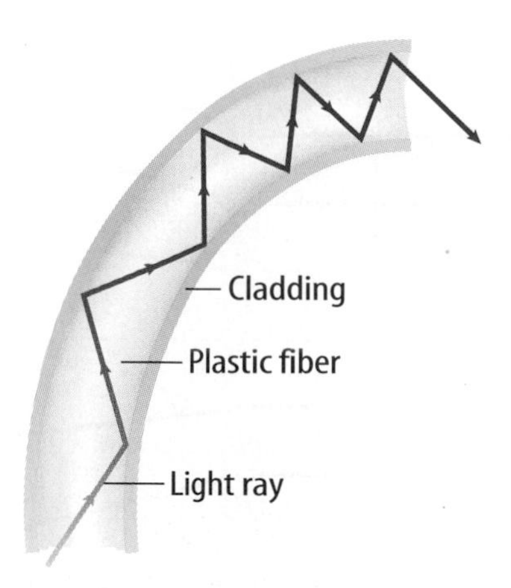

Figure 21
An optical fiber is made of materials that cause total internal reflection to occur. As a result, a light beam can travel for many kilometers through an optical fiber and lose almost no energy.

Optical Fibers

A device called an optical fiber can make a light beam travel in a path that is curved or even twisted. Optical fibers are thin, flexible, transparent fibers. An optical fiber is like a light pipe. Even if the fiber is bent, light that enters one end of the fiber comes out the other end.

Total internal reflection makes light transmission in optical fibers possible. A thin fiber of glass or plastic is covered with another material called cladding in which light travels faster. When light strikes the boundary between the fiber and the cladding, total internal reflection can occur. In this way, the beam bounces along inside the fiber as shown in **Figure 21**.

Using Optical Fibers Optical fibers are used most commonly in communications. For example, television programs and computer information can be coded in light signals. These signals then can be sent from one place to another using optical fibers. Because of total internal reflection, signals can't leak from one fiber to another and interfere with others. As a result, the signal is transmitted clearly. Phone conversations also can be changed into light and sent along optical fibers. One optical fiber the thickness of a human hair can carry thousands of phone conversations.

Section  Assessment

1. How is the image that is formed by a concave lens similar to the image that is formed by a convex mirror?
2. To magnify an object, would you use a convex lens or a concave lens?
3. Describe two ways, using convex and concave lenses, to form an image that is smaller than the object.
4. What are some uses for convex and concave lenses?
5. **Think Critically** A light wave is bent more when it travels from air to glass than when it travels from air to water. Is the speed of light greater in water or glass? Explain.

Skill Builder Activities

6. **Predicting** Air that is cool is more dense than air that is warm. Look at **Figure 15** and predict whether the speed of light is faster in warm air or cool air. **For more help, refer to the Science Skill Handbook.**
7. **Solving One-Step Equations** Earth is about 150 million km from the Sun. Use the formula

$$\text{Distance} = \text{speed} \times \text{time}$$

to calculate how many seconds it takes a beam of light to travel from Earth to the Sun. About how many minutes does it take? About how many hours does it take? **For more help, refer to the Math Skill Handbook.**

SECTION

Using Mirrors and Lenses

Microscopes

For almost 500 years, lenses have been used to observe objects that are too small to be seen with the unaided eye. The first microscopes were simple and magnified less than 100 times. Today a compound microscope like the one in **Figure 22A** uses a combination of lenses to magnify objects by as much as 2,500 times.

Figure 22B shows how a microscope forms an image. An object, such as an insect or a drop of water from a pond, is placed close to a convex lens called the objective lens. This lens produces an enlarged image inside the microscope tube. The light rays from that image then pass through a second convex lens called the eyepiece lens. This lens further magnifies the image formed by the objective lens. By using two lenses, a much larger image is formed than a single lens can produce.

As You Read

What You'll Learn

- **Explain** how microscopes magnify objects.
- **Explain** how telescopes make distant objects visible.
- **Describe** how a camera works.

Vocabulary

refracting telescope
reflecting telescope

Why It's Important

Microscopes and telescopes are used to view parts of the universe that can't be seen with the unaided eye.

Figure 22
A compound microscope uses lenses to magnify objects.

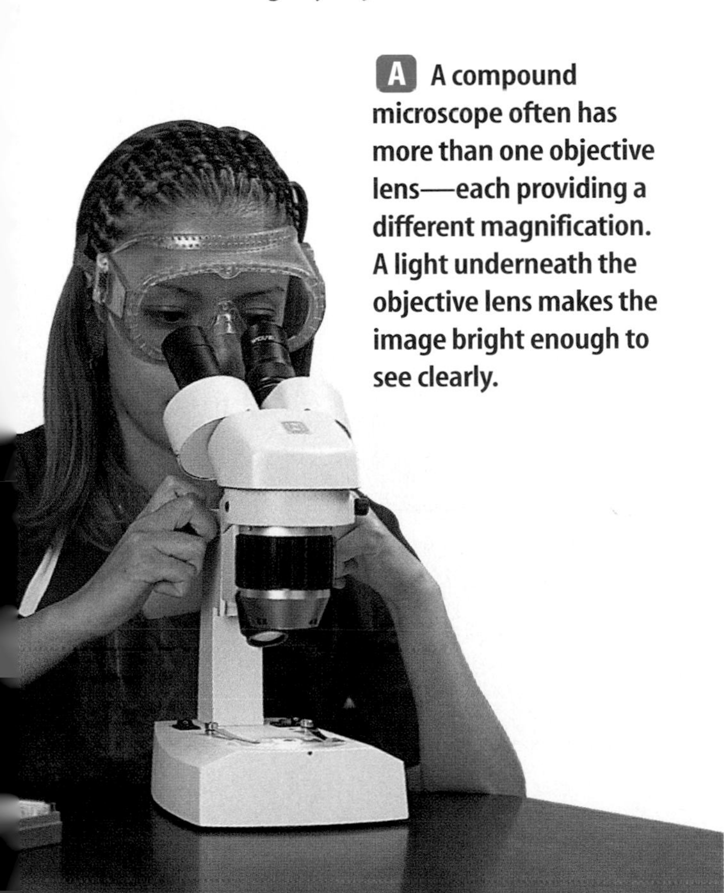

A A compound microscope often has more than one objective lens—each providing a different magnification. A light underneath the objective lens makes the image bright enough to see clearly.

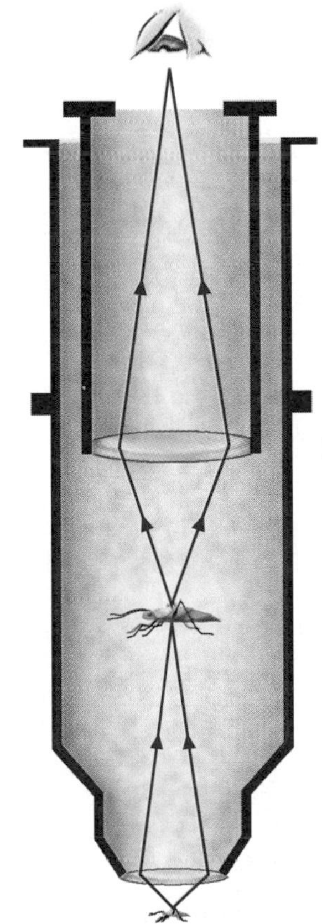

B The objective lens in a compound microscope forms an enlarged image, which is then magnified by the eyepiece lens.

Mini LAB

Forming an Image with a Lens

Procedure

1. Fill a **glass test tube** with **water** and seal it with a **stopper.**
2. Write your name on a **10-cm × 10-cm card.** Lay the test tube on the card and observe the appearance of your name.
3. Hold the test tube about 1 cm above the card and observe the appearance of your name. Record your observations.
4. Observe what happens to your name as you slowly move the test tube away from the card. Record your observations.

Analysis

1. Is the water-filled test tube a concave lens or a convex lens?
2. Compare the image that formed when the test tube was close to the card with the image that formed when the test tube was far from the card.

Telescopes

Just as microscopes are used to magnify very small objects, telescopes are used to examine objects that are very far away. The first telescopes were made at about the same time as the first microscopes. Much of what is known about the Moon, the solar system, and the distant universe has come from images and other information gathered by telescopes.

Refracting Telescopes The simplest **refracting telescopes** use two convex lenses to form an image of a distant object. Just as in a compound microscope, light passes through an objective lens that forms an image. That image is then magnified by an eyepiece, as shown in **Figure 23A.**

An important difference between a telescope and a microscope is the size of the objective lens. The main purpose of a telescope is not to magnify an image. A telescope's main purpose is to gather as much light as possible from distant objects. The larger an objective lens is, the more light that can enter it. This makes images of faraway objects look brighter and more detailed when they are magnified by the eyepiece. With a large enough objective lens, it's possible to see stars and galaxies that are many trillions of kilometers away. **Figure 23B** shows the largest refracting telescope ever made.

Reading Check *How does a telescope's objective lens enable distant objects to be seen?*

Figure 23 Refracting telescopes use a large objective lens to gather light from distant objects.

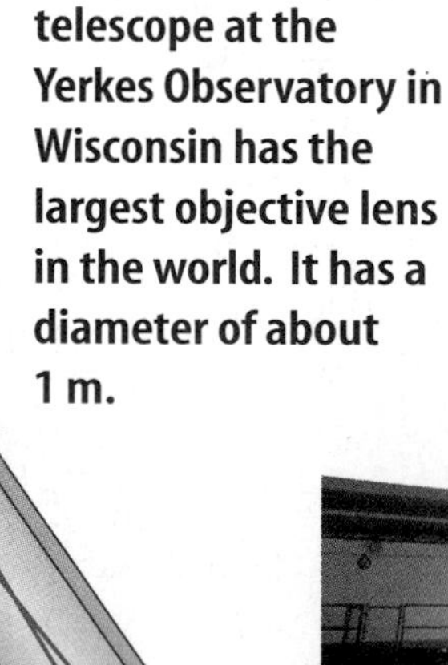

A A refracting telescope is made from an objective lens and an eyepiece. The objective lens forms an image that is magnified by the eyepiece.

B The refracting telescope at the Yerkes Observatory in Wisconsin has the largest objective lens in the world. It has a diameter of about 1 m.

Figure 24
Reflecting telescopes gather light by using a concave mirror.

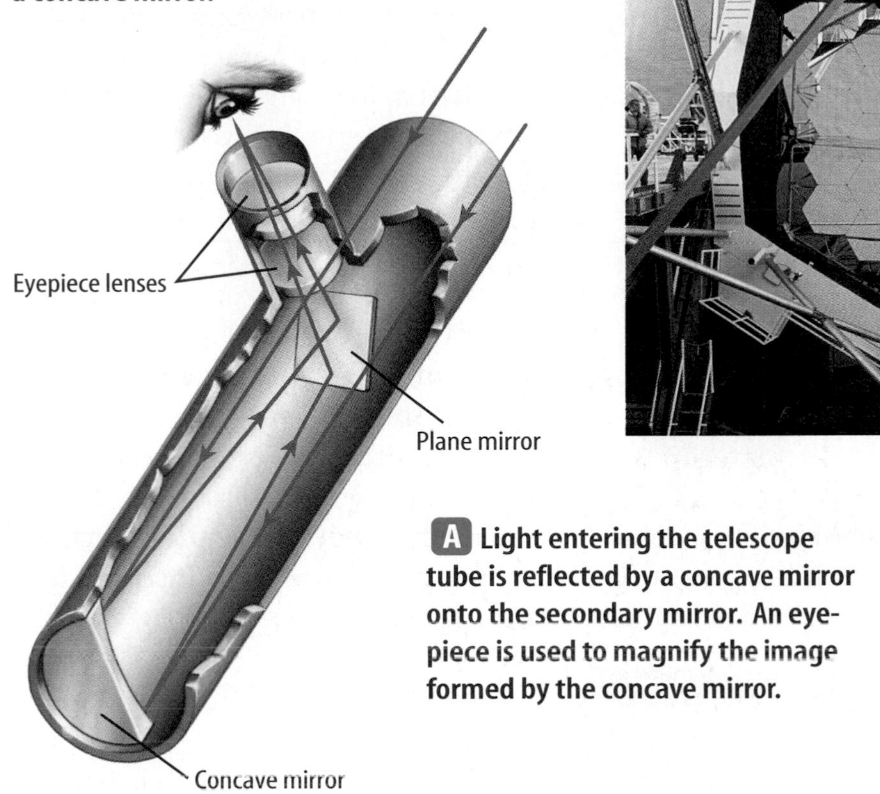

A Light entering the telescope tube is reflected by a concave mirror onto the secondary mirror. An eyepiece is used to magnify the image formed by the concave mirror.

B The Keck telescope in Mauna Kea, Hawaii, is the largest reflecting telescope in the world.

Reflecting Telescopes Refracting telescopes have size limitations. One problem is that the objective lens can be supported only around its edges. If the lens is extremely large, it cannot be supported enough to keep the glass from sagging slightly under its own weight. This causes the image that the lens forms to become distorted.

Reflecting telescopes can be made much larger than refracting telescopes. **Reflecting telescopes** have a concave mirror instead of a concave objective lens to gather the light from distant objects. As shown in **Figure 24A,** the large concave mirror focuses light onto a secondary mirror that directs it to the eyepiece, which magnifies the image.

Because only the one reflecting surface on the mirror needs to be made carefully and kept clean, telescope mirrors are less expensive to make and maintain than lenses of a similar size. Also, mirrors can be supported not only at their edges but also on their back sides. They can be made much larger without sagging under their own weight. The Keck telescope in Hawaii, shown in **Figure 24B,** is the largest reflecting telescope in the world. Its large concave mirror is 10 m in diameter, and is made of 36 six-sided segments. Each segment is 1.8 m in size and the segments are pieced together to form the mirror.

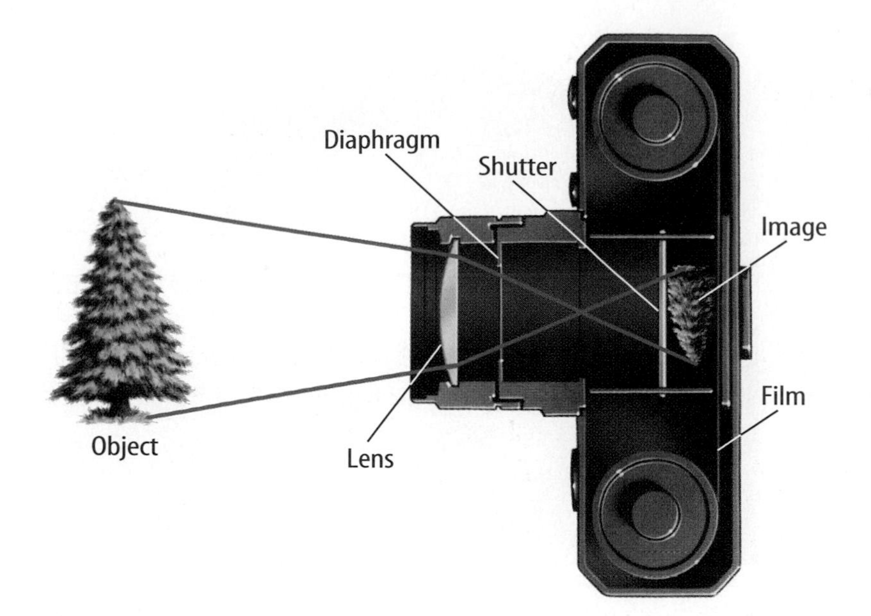

Figure 25
A camera uses a convex lens to form an image on a piece of light-sensitive film. The image formed by a camera lens is smaller than the object and is inverted.

Cameras

You probably see photographs taken by cameras almost every day. A typical camera uses a convex lens to form an image on a section of film, just as your eye's lens focuses an image on your retina. The convex lens has a short focal length so that it forms an image that is smaller than the object and inverted on the film. Look at the camera shown in **Figure 25.** When the shutter is open, the convex lens focuses an image on a piece of film that is sensitive to light. Light-sensitive film contains chemicals that undergo chemical reactions when light hits it. The brighter parts of the image affect the film more than the darker parts do.

Reading Check *What type of lens does a camera use?*

If too much light strikes the film, the image formed on the film is overexposed and looks washed out. On the other hand, if too little light reaches the film, the photograph might be too dark. To control how much light reaches the film, many cameras have a device called a diaphragm. The diaphragm is opened to let more light onto the film and closed to reduce the amount of light that strikes the film.

Lasers

Perhaps you've seen the narrow, intense beams of laser light used in a laser light show. Intense laser beams are also used for different kinds of surgery. Why can laser beams be so intense? One reason is that a laser beam doesn't spread out as much as ordinary light as it travels.

Spreading Light Beams Suppose you shine a flashlight on a wall in a darkened room. The size of the spot of light on the wall depends on the distance between the flashlight and the wall. As the flashlight moves farther from the wall, the spot of light gets larger. This is because the beam of light produced by the flashlight spreads out as it travels. As a result, the energy carried by the light beam is spread over an increasingly larger area as the distance from the flashlight gets larger. As the energy is spread over a larger area, the energy becomes less concentrated and the intensity of the beam decreases.

Figure 26
Laser light is different from the light produced by a lightbulb.

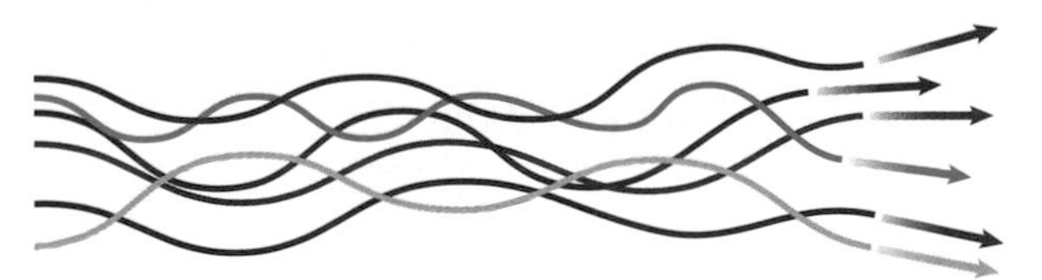

A **The light from a bulb contains waves with many different wavelengths that are out of phase and traveling in different directions.**

B **The light from a laser contains waves with only one wavelength that are in phase and traveling in the same direction.**

Using Laser Light Laser light is different from the light produced by the flashlight in several ways, as shown in **Figure 26.** One difference is that in a beam of laser light, the crests and troughs of the light waves overlap, so the waves are in phase.

Because a laser beam doesn't spread out as much as ordinary light, a large amount of energy can be applied to a very small area. This property enables lasers to be used for cutting and welding materials and as a replacement for scalpels in surgery. Less intense laser light is used for such applications as reading and writing to CDs or in grocery store bar-code readers. Surveyors and builders use lasers to measure distances, angles, and heights. Laser beams also are used to transmit information through space or through optical fibers.

Section  Assessment

1. How is a compound microscope different from a magnifying lens?
2. Compare and contrast reflecting and refracting telescopes. Why aren't refracting telescopes bigger than reflecting telescopes?
3. Why is the objective lens of a refracting telescope bigger than the objective lens of a microscope?
4. Describe how laser light is different from the light produced by a light bulb.
5. **Think Critically** Could a camera with a concave lens instead of a convex lens still take pictures? Explain.

Skill Builder Activities

6. **Communicating** Using words, pictures, or other media, think of a way to explain to a friend how convex and concave lenses work. **For more help, refer to the Science Skill Handbook.**
7. **Solving One-Step Equations** The size of an image is related to the magnification of an optical instrument by the following formula:

$$\text{Image size} = \text{magnification} \times \text{object size}$$

A blood cell has a diameter of about 0.001 cm. How large is the image formed by a microscope with a magnification of 1,000? **For more help, refer to the Math Skill Handbook.**

Activity

Image Formation by a Convex Lens

The type of image formed by a convex lens, also called a converging lens, is related to the distance of the object from the lens. This distance is called the object distance. The location of the image also is related to the distance of the object from the lens. The distance from the lens to the image is called the image distance. What happens to the position of the image as the object gets nearer or farther from the lens?

What You'll Investigate

How are the image distance and object distance related for a convex lens?

Materials

convex lens
modeling clay
meterstick
flashlight
masking tape
20-cm square piece of cardboard with a white surface

Goals

- **Measure** the image distance as the object distance changes.
- **Observe** the type of image formed as the object distance changes.

Safety Precautions

Procedure

1. **Design** a data table to record your data. Make three columns in your table—one column for the object distance, another for the image distance, and the third for the type of image.
2. Use the modeling clay to make the lens stand upright on the lab table.
3. Form the letter *F* on the glass surface of the flashlight with masking tape.
4. Turn on the flashlight and place it 1 m from the lens. Position the flashlight so the flashlight beam is shining through the lens.
5. **Record** the distance from the flashlight to the lens in the object distance column in your data table.
6. Hold the cardboard vertically upright on the other side of the lens, and move it back and forth until a sharp image of the letter *F* is obtained.
7. **Measure** the distance of the card from the lens using the meterstick, and record this distance in the Image Distance column in your data table.
8. **Record** in the third column of your data table whether the image is upright or inverted, and smaller or larger.
9. Repeat steps 4 through 8 for object distances of 0.50 m and 0.25 m and record your data in your data table.

Convex Lens Data

Object Distance (m)	Image Distance (m)	Image Type

Conclude and Apply

1. How did the image distance change as the object distance decreased?
2. How did the image change as the object distance decreased?
3. What would happen to the size of the image if the flashlight were much farther away than 1 m?

Demonstrate this activity to a third-grade class and explain how it works. **For more help, refer to the Science Skill Handbook.**

Oops! Accidents in SCIENCE

SOMETIMES GREAT DISCOVERIES HAPPEN BY ACCIDENT!

Eyeglasses

"It is not yet twenty years since the art of making spectacles, one of the most useful arts on Earth, was discovered. I myself have seen and conversed with the man who made them first."

This quote from an Italian monk dates back to 1306 and is one of the first historical records to refer to eyeglasses. Unfortunately, the monk, Giordano, never actually named the man he met. Thus, the inventor of eyeglasses—one of the most widely used forms of technology today—remains unknown.

The mystery exists, in part, because different cultures in different places used some type of magnifying tool to improve their vision. These tools eventually merged into what today is recognized as a pair of glasses. For example, a rock-crystal lens made by early Assyrians who lived 3,500 years ago in what is now Iraq, may have been used to improve vision. About 2,000 years ago, the Roman writer Seneca looked through a glass globe of water to make the letters appear bigger in the books he read. By the 10th century, glasses were invented in China, but they were used to keep away bad luck, not to improve vision. Trade between China and Europe, however, likely led some unknown inventor to come up with an idea. The inventor fused two metal-ringed magnifying lenses together so they could perch on the nose.

In 1456, the printing press was invented. Suddenly, there was more to read, which, in turn, made the ability to see clearly more important. In Europe, eyeglasses began to appear in paintings of scholars, clergy, and the upper classes—the only people who knew how to read at the time. Although the ability to read spread fairly quickly, eyeglasses were so expensive that only the rich could afford them. In the early 1700s, for example, glasses cost roughly $200, which is comparable to thousands of dollars today. By the mid-1800s, improvements in manufacturing techniques made eyeglasses much less expensive to make, and thus this important invention became widely available to people of all walks of life.

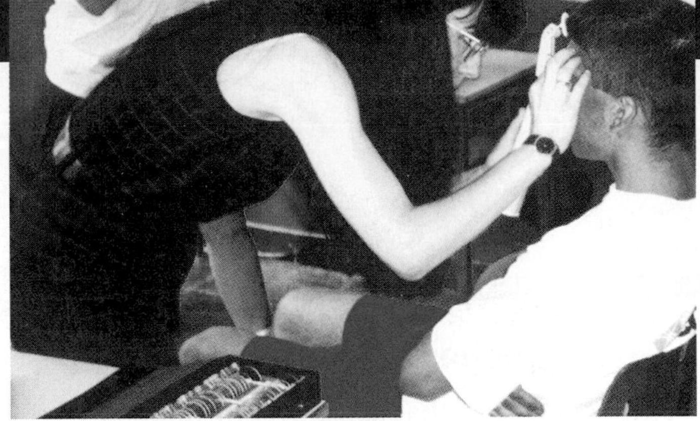

Cheryl Landry at work with a Bosnian teenage soldier

Inventor Unknown

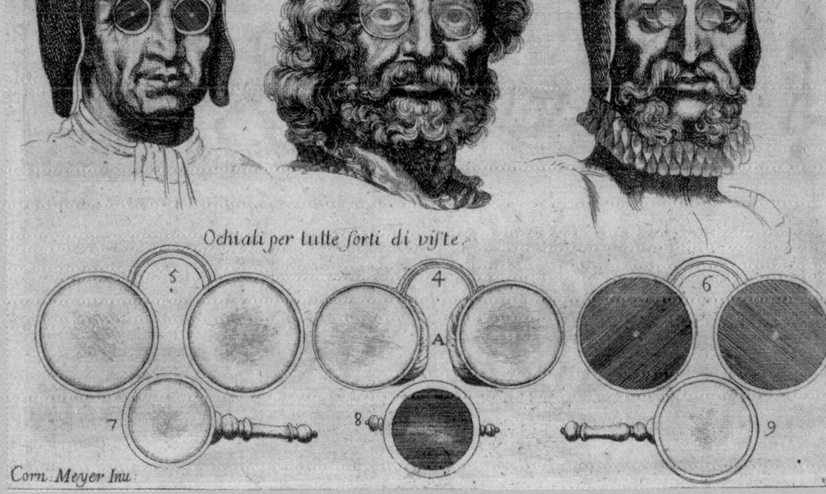

This Italian engraving from the 1600s shows glasses of all strengths.

How Eyeglasses Work

Eyeglasses are used to correct farsightedness and nearsightedness, as well as other vision problems. Farsighted people have difficulty seeing things close up because light rays from nearby objects do not converge enough to form an image on the retina. This problem can be corrected by using convex lenses that cause light rays to converge before they enter the eye. Nearsighted people have problems seeing distant objects because light rays from far-away objects are focused in front of the retina. Concave lenses that cause light rays to diverge are used to correct this vision problem.

CONNECTIONS **Research** In many parts of the world, people have no vision care, and eye diseases and poor vision go untreated. Research the work of groups that bring eye care to people. Start with eye doctor Cheryl Landry, who works with the Bosnian Children's Fund.

SCIENCE Online

For more information, visit science.glencoe.com

Chapter 23 Study Guide

Reviewing Main Ideas

Section 1 Properties of Light

1. Light is a wave that can travel through different materials, including a vacuum.
2. When a light wave strikes an object, some of the light wave's energy is reflected, some is absorbed, and some might be transmitted through the object.
3. The color of a light wave depends on its wavelength. The color of an object depends on which wavelengths of light are reflected by the object. *Why does this flower look red?*

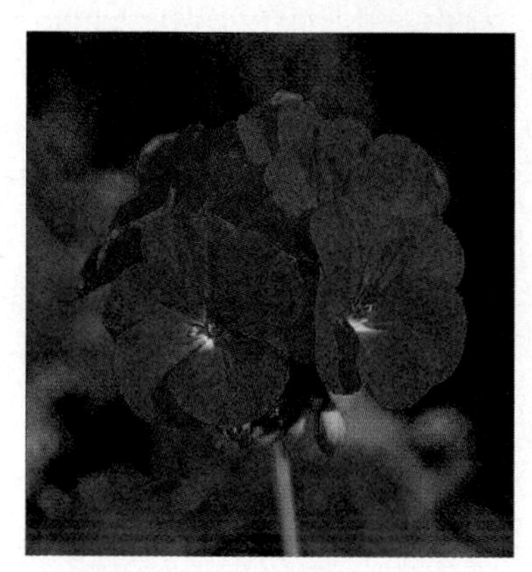

4. Almost any color can be made by mixing the primary light colors or the primary pigment colors.

Section 2 Reflection and Mirrors

1. Light reflected from the surface of an object obeys the law of reflection: the angle of incidence equals the angle of reflection.
2. Diffuse reflection occurs when a surface is rough. Regular reflection occurs from very smooth surfaces and produces a clear, mirrorlike image. *Is the image in the photo a diffuse reflection?*

3. Concave mirrors cause light waves to converge, or meet. Convex mirrors cause light waves to diverge, or spread apart.

Section 3 Refraction and Lenses

1. Light waves can change speed when they travel from one medium to another. The waves bend, or refract, at the boundary between the two media.
2. A convex lens causes light waves to converge, and a concave lens causes light waves to diverge. *What would happen to the image of the insect if the magnifying glass in the photo were moved farther away?*

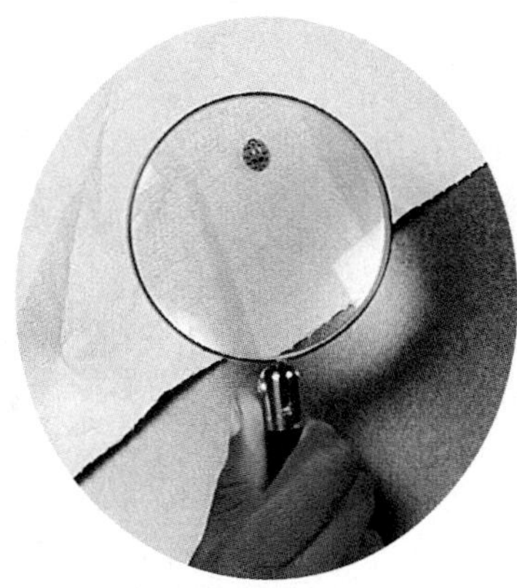

Section 4 Using Mirrors and Lenses

1. A compound microscope is used to enlarge small objects. A convex objective lens forms an enlarged image that is further enlarged by an eyepiece.
2. Most telescopes today are reflecting telescopes, which use a concave mirror to form a real image that is enlarged by an eyepiece.
3. Cameras use a convex lens to form an image on light-sensitive film.

After You Read

FOLDABLES Reading & Study Skills

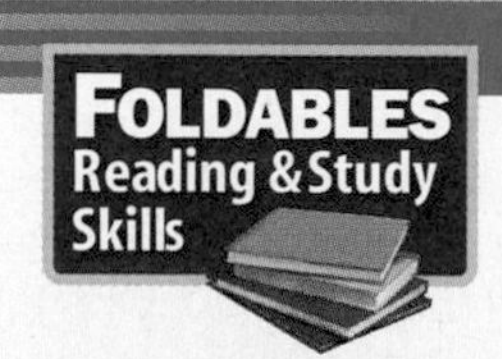

On the back of the top flap of your Making a Question Study Fold, explain why most telescopes are reflecting telescopes.

Visualizing Main Ideas

Complete the following concept map.

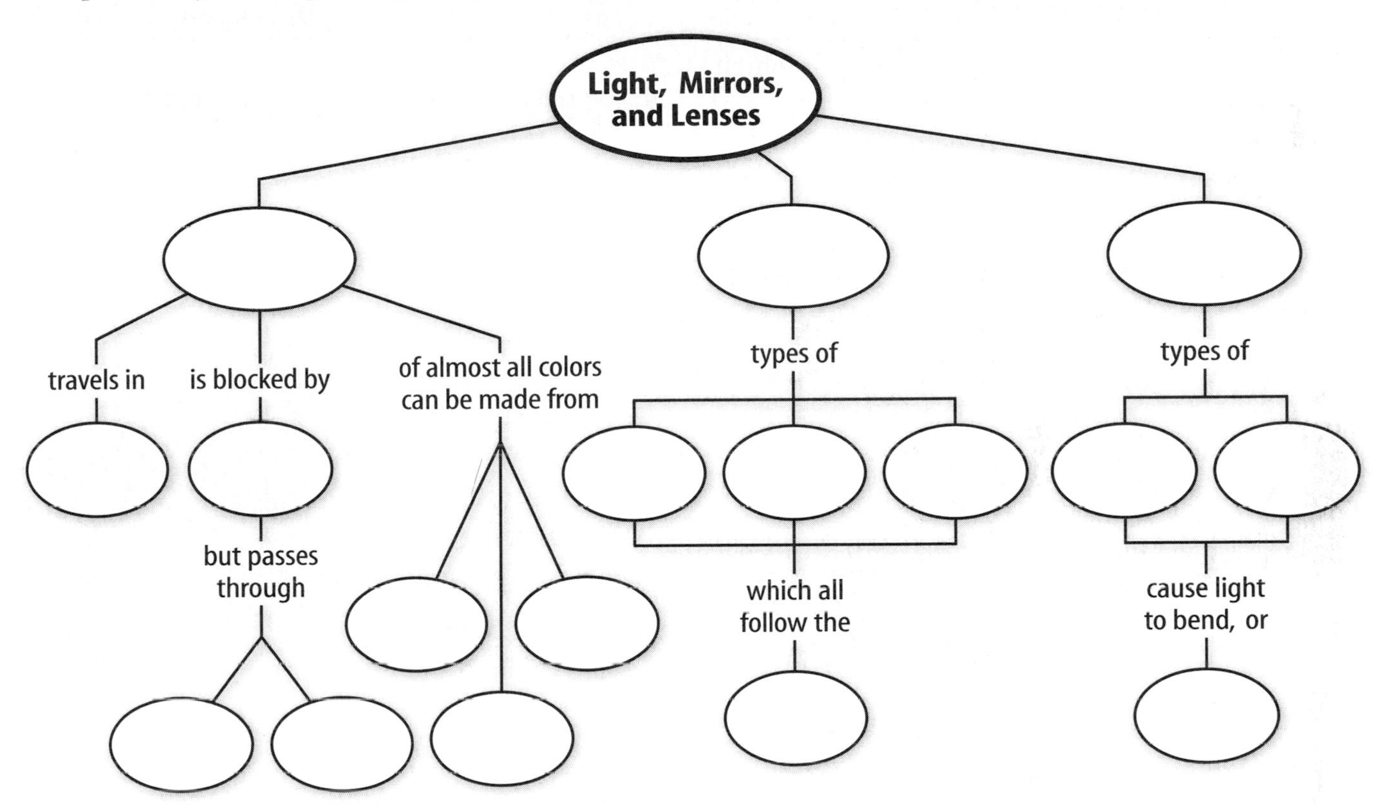

Vocabulary Review

Vocabulary Words

a. concave lens
b. convex lens
c. focal length
d. focal point
e. law of reflection
f. lens
g. light ray
h. medium
i. reflecting telescope
j. reflection
k. refracting telescope
l. refraction

Study Tip

If you're not sure of the relationships between terms in a question, make a concept map of the terms to see how they fit together. Ask your teacher if the relationships you drew are correct.

Using Vocabulary

Explain the differences between the terms in the following sets.

1. reflection, refraction
2. concave lens, convex lens
3. light ray, medium
4. focal length, focal point
5. lens, medium
6. law of reflection, refraction
7. reflecting telescope, refracting telescope
8. focal point, light ray
9. lens, focal length

Chapter 23 Assessment

Checking Concepts

Choose the word or phrase that completes the sentence or answers the question.

1. Light waves travel the fastest through which of the following?
A) air **C)** water
B) diamond **D)** a vacuum

2. What determines the color of light?
A) a prism **C)** its wavelength
B) its refraction **D)** its incidence

3. If an object reflects red and green light, what color does the object appear to be?
A) yellow **C)** green
B) red **D)** purple

4. If an object absorbs all the light that hits it, what color is it?
A) white **C)** black
B) blue **D)** green

5. What type of image is formed by a plane mirror?
A) upright **C)** magnified
B) inverted **D)** all of the above

6. How is the angle of incidence related to the angle of reflection?
A) It's greater. **C)** It's the same.
B) It's smaller. **D)** It's not focused.

7. Which of the following can be used to magnify objects?
A) a concave lens **C)** a convex mirror
B) a convex lens **D)** all of the above

8. Which of the following describes the light waves that make up laser light?
A) same wavelength **C)** in phase
B) same direction **D)** all of the above

9. What is an object that reflects some light and transmits some light called?
A) colored **C)** opaque
B) diffuse **D)** translucent

10. What is the main purpose of the objective lens or concave mirror in a telescope?
A) invert images **C)** gather light
B) reduce images **D)** magnify images

Thinking Critically

11. Do all light rays that strike a convex lens pass through the focal point?

12. Does a plane mirror focus light rays? Why or why not?

13. Explain why a rough surface, such as the road in this photo, is a better reflector when it is wet.

14. If the speed of light were the same in all materials, could lenses be used to magnify objects? Why or why not?

15. A singer is wearing a blue outfit. What color spotlights would make the outfit appear to be black? Explain.

Developing Skills

16. Comparing and Contrasting Compare and contrast plane and convex mirrors.

17. Predicting You see a person's eyes in a mirror. Explain whether he or she can see you.

18. Testing a Hypothesis A convex lens supposedly has a focal length of 0.50 m. Design an experiment that would determine whether the focal length of the lens is 0.50 m.

19. Researching Information Research the ways laser light is used to correct medical and vision problems. Write a page summarizing your results.

20. Making and Using Graphs The graph below shows how the distance of an image from a convex lens is related to the distance of the object from the lens.

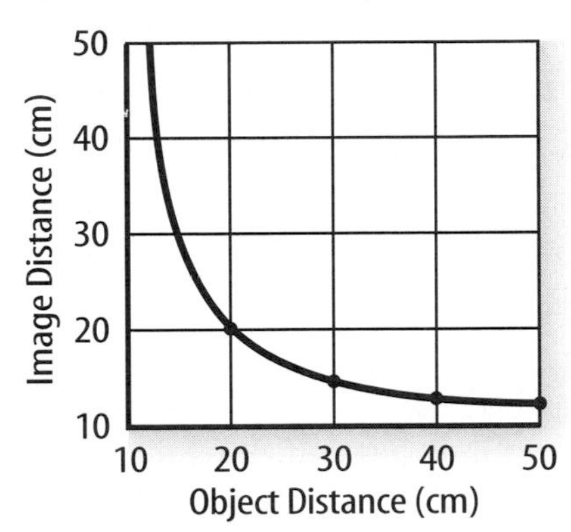

A) How does the image move as the object gets closer to the lens?

B) You can find the magnification of the image with the equation

$$\text{Magnification} = \frac{\text{image distance}}{\text{object distance}}$$

At which object distance is the magnification equal to 2?

Performance Assessment

21. Poster Make a poster describing the difference between the primary colors of light and the primary colors of pigment.

22. Reverse Writing In a plane mirror, images are reversed. With this in mind, write a "backwards" note to a friend and have him or her read it in a mirror.

Technology

Go to the Glencoe Science Web site at **science.glencoe.com** or use the **Glencoe Science CD-ROM** for additional chapter assessment.

Test Practice

Nelson learned that the speed of light is 300,000 km/s. Since he knew that the fastest passenger jet, the SST, flies faster than the speed of sound, he did some research to compare the speeds of sound and light.

The Speed of Sound and Light in Different Media

Medium	Speed of Sound (m/s)	Speed of Light (10^8 m/s)
Glass	5,971	2.0
Water	1,486	2.3
Air	335	3.0
Vacuum	0	3.0

Study the table above and answer the following questions.

1. According to the table, in which medium is the speed of sound the fastest and the speed of light the slowest?

A) glass
B) water
C) air
D) vacuum

2. According to the table, which medium is able to transmit only light?

F) glass
G) water
H) air
J) vacuum

Standardized Test Practice

Reading Comprehension

Read the passage. Then read each question that follows the passage. Decide which is the best answer to each question.

The History of the Telescope: An International Story

Roger Bacon, an English scientist, first wrote about the basic ideas behind the operation of a telescope in the 1200s. It was not until the early 1600s, however, that Hans Lippershey, a Dutchman who made spectacles for people with poor vision, made the first telescope. Lippershey noticed that objects appeared closer if he viewed them through a combination of a concave and a convex lens. He placed the lenses in a tube to hold them more easily. This was the world's first refracting telescope.

A few years later, an Italian scientist, Galileo, was the first to point a telescope toward the stars. Galileo learned of the Dutch invention in 1609. At the time, it was mainly used to see objects on Earth, such as distant ships and enemy armies. This is why the telescope was first called a "spyglass." Galileo made his own telescope and began using it to view the sky. Before this, Galileo had not been particularly interested in astronomy. That quickly changed as he recorded observations of the Moon's surface, spots on the Sun, and four moons circling Jupiter.

Another advance in telescope technology occurred in 1663 when James Gregory, a Scottish scientist, designed the first reflecting telescope. Isaac Newton built the first reflecting telescope 25 years later. The earliest, most valuable contribution to astronomy made by Americans was the construction of the Hooker telescope, a reflecting telescope on Mount Wilson. Completed in 1917, its 254 cm reflecting concave mirror allowed astronomers to see other galaxies clearly for the first time.

Since then, scientists have continued to design and build larger and more powerful telescopes. The development of the modern telescope is the result of many years of work by many scientists across the world.

Test-Taking Tip As you read the passage, make a time line of the history of the telescope.

1. The telescope was first called a "spyglass" because it _____.

A) was helpful in observing the Moon and stars
B) was designed by Roger Bacon
C) could be used to watch other people
D) was first made by a Dutchman

2. According to the passage, scientists often _____.

F) build upon one another's work
G) are slow workers
H) aren't interested in many things
J) never read the work of other scientists

3. The earliest, most valuable contribution to astronomy made by Americans was _____.

A) the first refracting telescope built in the 1600s
B) Roger Bacon's basic ideas about the operation of a telescope in the 1200s
C) the construction of the Hooker telescope on Mount Wilson, which allowed astronomers to see other galaxies clearly for the first time
D) using a telescope to view the Moon's surface, spots on the Sun, and four moons circling Jupiter

Standardized Test Practice

Reasoning and Skills

Power of a Lens		
Lens	**Diopter**	**Focal length (m)**
1	1/4	4
2	1/5	5
3	1/6	6
4	1/7	7
5	1/9	?

1. Diopters are one way to measure the strength of a lens. What is the focal length of lens 5?

A) 5
B) 8
C) 9
D) 10

Test-Taking Tip Study the values for the first four lenses and consider how the diopter value is related to the focal length.

Wavelengths of Electromagnetic Waves	
Type of Wave	**Wavelength**
Radio wave	Greater than 0.3 m
Microwave	0.3 m – 0.001 m
Infrared wave	0.001 m – 0.0000007 m
Visible light	0.0000007 m – 0.0000004 m

2. A wave with a wavelength of 0.03 m would be what type of wave?

F) radio wave
G) microwave
H) infrared wave
J) visible light wave

Test-Taking Tip Read the table's column headings carefully and then reread the question.

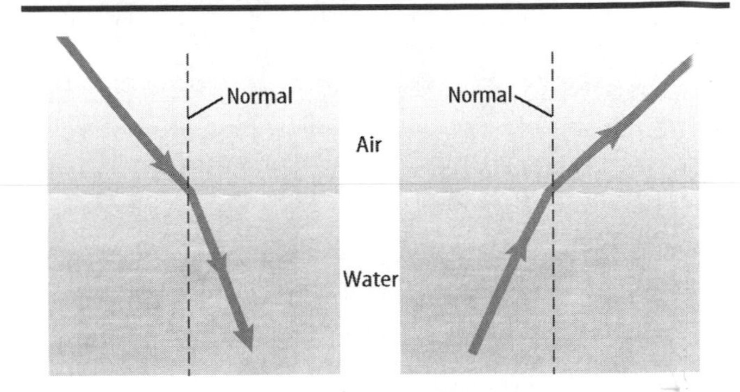

3. What is happening to the wave in this figure?

A) it is being diffracted
B) it is being refracted
C) it is experiencing constructive interference
D) it is experiencing destructive interference

Test-Taking Tip Review the difference between diffraction, refraction, constructive interference, and destructive interference.

4. Explain the relationship between the amplitude of a sound wave and the loudness of a sound. How is the amplitude of the sound wave related to the amount of energy it carries?

Test-Taking Tip Recall the definition of the terms *amplitude* and *loudness*.

Student Resources

CONTENTS

Field Guides 698

Skill Handbooks 710

Reference Handbook 735

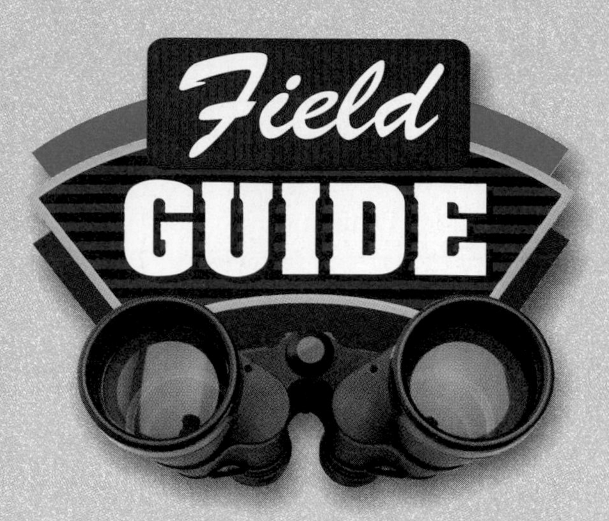

When you hear the word *cone*, you might think of a tasty, edible holder for your favorite ice cream. Maybe you think of the orange cones used on highways and in public places to direct vehicular or pedestrian traffic. However, there's another type of cone in the environment that plays an important role for some plants. These cones are the reproductive organs of a large plant group called the conifers, or cone bearers. The seeds of pines, firs, spruces, redwoods, and other conifers are formed in cones.

Types of Cones

Conifers have two types of cones, male and female. The male cones produce pollen grains and break apart soon after they release pollen. Depending on the species of conifer, the female cones can stay on plants for up to three years. Female cones can be woody or berrylike. Woody cones consist of scales growing from a central stalk and vary in shape and size. Berrylike cones are round and either hard or soft. Each genus of conifers has a different type of female cone. They are so different from one another that you can use them to identify a conifer's genus.

Cones

Cone Characteristics

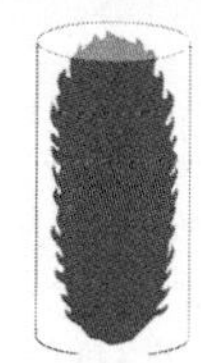

Cylindrical

This cone is shaped like a cylinder and is nearly uniform in size from the base to the tip.

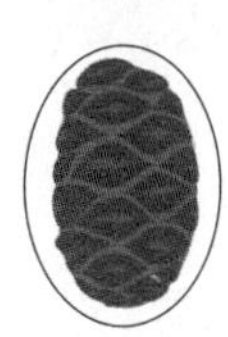

Ovoid

Although this cone is shaped like a cylinder, it is smaller at the ends than in the middle.

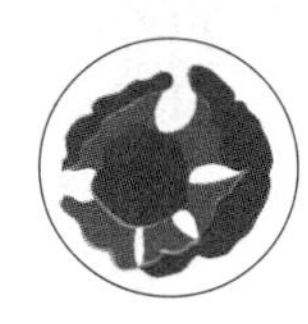

Globose

This cone is rounded like a globe.

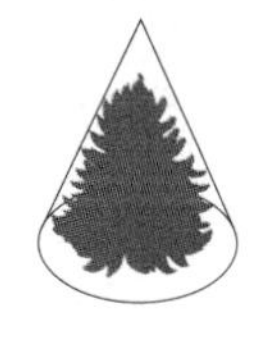

Conic

Shaped like a cone, it decreases in diameter from the base to the tip.

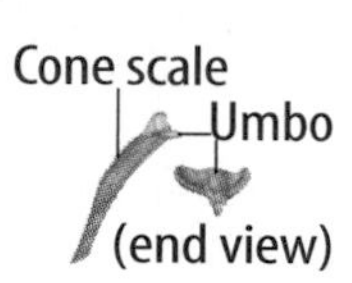

Umbo

A raised, triangular area at the tip of a cone scale varies in size and thickness.

Field Activity

Find three different cones in your neighborhood, a park, around your school, or as part of a craft item. Using this guide, identify the genus of each cone. Go to the Glencoe Science Web site at **science.glencoe.com** if you don't have cones in your neighborhood. Here you can link to different sites about cones. In your Science Journal, sketch each cone and write a description of the plant it came from.

Cone Identification

This field guide contains some of the conifers. Plant features might differ in appearance because of environmental conditions.

Douglas Fir—*Psuedotsuga*

These ovoid cones on short stalks have a three-pointed, papery structure that extends from below each cone scale. The cones range from 5 cm to 10 cm in length.

Douglas fir cone

Juniper berries

Juniper—*Juniperus*

These cones are hard, berrylike structures that stay on the tree or shrub for two to three years. They measure about 1.3 cm in diameter. They are bluish, pale green, reddish, or brown and covered with a white, waxy coating called a bloom.

Spruce—*Picea*

These cones are cylindrical and brown with thin cone scales and tips that usually are pointed. They can be 6 cm to 15 cm long. They stay on the plant for two years and hang from branches on the upper third of the tree. As they mature, they become brittle.

Spruce cones

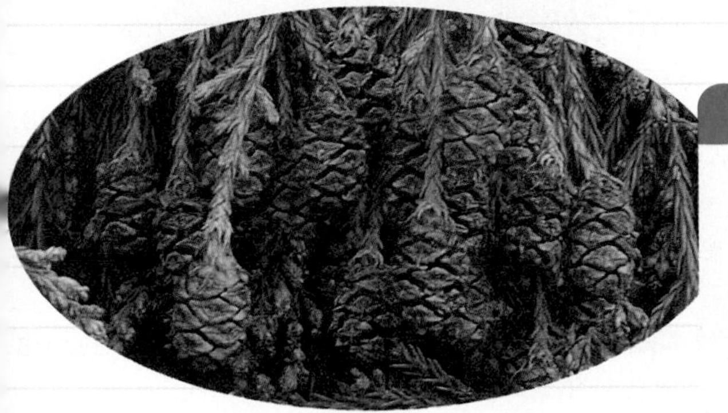

Redwood cones

Redwood—*Sequoia*

Ovoid and reddish brown, these cones hang from the tips of needled twigs. They develop in one year and are small in comparison to the size of the tree—only 1.2 cm to 3 cm. The cone scales are flattened on their ends.

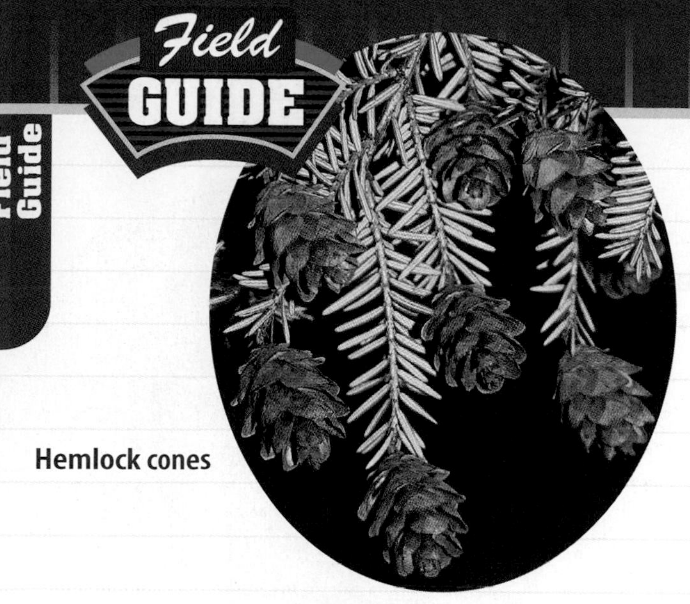

Hemlock cones

Hemlock—*Tsuga*

These cones hang from twigs and are small, ovoid to cylindrical, and 2 cm to 7 cm long. The few cone scales have rounded tips. Although they develop in one year, they usually stay on the tree for more than one year.

Pine—*Pinus*

Each cone has a thick, woody scale tipped with an umbo. The umbo can have a small spine, or prickle. Most pine cones are cylindrical or conic and grow on a small stalk. They vary in length from about 4 cm (scrub pine) to 45 cm (sugar pine) and remain on the tree or shrub for two to three years.

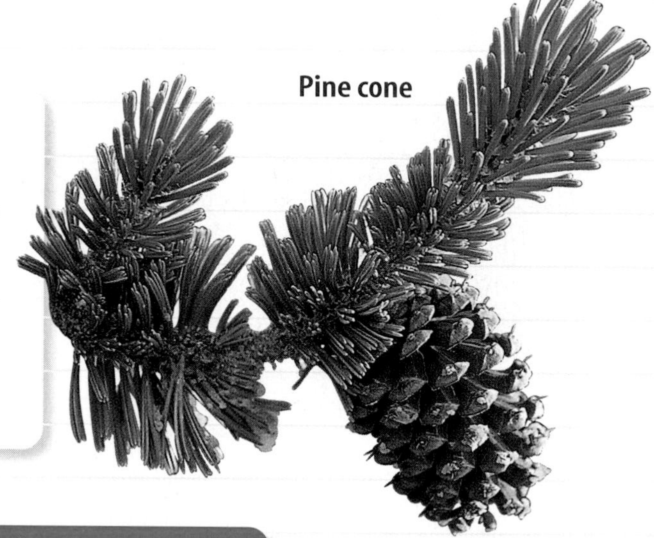

Pine cone

Arborvitae cones

Arborvitae—*Thuja*

These egg-shaped cones are 1.2 cm to 1.5 cm long. They have paired cone scales, usually from six to 12, that are straplike and end in a sharp point. The cones remain attached to the shrub after opening and releasing their seeds.

Cypress—*Cupressus*

These globose cones, which are usually 2 cm to 2.5 cm in diameter, have only six to eight scales. The cone scales have a raised point in the center. They develop in about 18 months and stay closed and attached to the tree.

Cypress cones

False Cypress—*Chamaecyparis*

These small globose cones are only 0.5 cm to 4 cm in diameter with four to ten cone scales. Unlike the cones of the *Cupressus* trees, they open after they are fully developed.

False cypress cones

Swamp or Bald Cypress—*Taxodium*

This globose cone is about 2.5 cm across and develops in one year. The tips of the cone scales are four sided, forming an irregular pattern on the surface of the cone. Trees in this genus are recognized by the projections, called knees, that grow upward from around the base of the tree trunk.

Swamp cypress cones

Fir—*Abies*

Fir cones grow upright on branches and range from 5 cm to 20 cm in length. They are seldom used for identification because the scales drop off when they are developed, leaving only the bare central stalk.

Fir cones

Cedar cones

Cedar—*Cedrus*

These barrel-shaped cones with flattened tips grow upright on branches. They are 5 cm to 10 cm in length and nearly half as wide. After two years, the scales drop off. Cedar trees do not produce these cones until they are 40 to 50 years old.

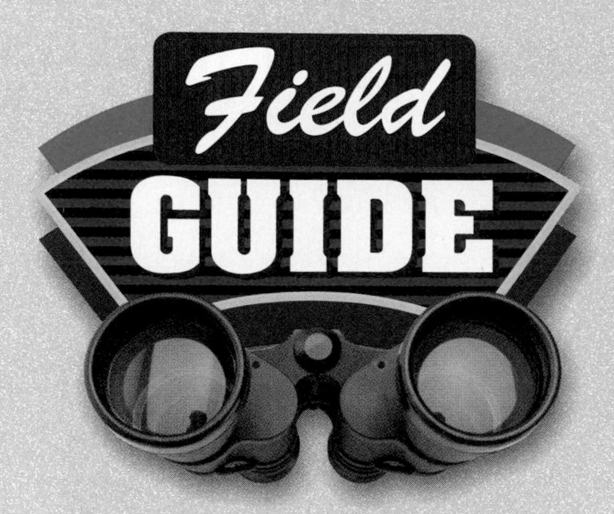

Earth's crust is squeezed and pulled as tectonic plates move. Energy builds up in rocks and when it is released, it can dramatically and visibly change the structure of the land surface. Faults are fractures in Earth produced when sections of Earth's crust move past each other and release energy, sometimes during earthquakes. Small faults can be nearly invisible on the surface. Large faults, caused by intense, long-time crustal movement, can be huge cracks in the ground that extend for hundreds of miles. Types of faults include normal; reverse, or thrust; and strike-slip.

When tectonic plates collide, great mountain chains can be uplifted out of Earth. When rocks are subjected to stress, they do not always break and form faults—under some conditions, rocks will bend and fold. These folded rocks can reach Earth's surface. Upward-arching folds are known as anticlines. Downward-sagging folds are known as synclines. Folds can be large enough to form mountains or small enough to hold in your hand.

Earth's tectonic plates have been moving for hundreds of millions of years. Although they move very slowly—about 10 cm per year—they have moved across great distances over time.

Faults and Folds

Normal Fault: The Rio Grande Rift

Extending across a distance of more than 1,200 km, the Rio Grande Rift stretches from Colorado to Northern Mexico and separates the Colorado Plateau from the Great Plains. The Rio Grande Rift is an example of how normal faults can affect Earth's surface. It is the result of fractures in Earth's crust that are formed as the crust is pulled apart. The Rio Grande Rift is active and the faults in the rift release their built-up energy in the form of frequent, small tremors rather than as large earthquakes. The region also contains many volcanoes.

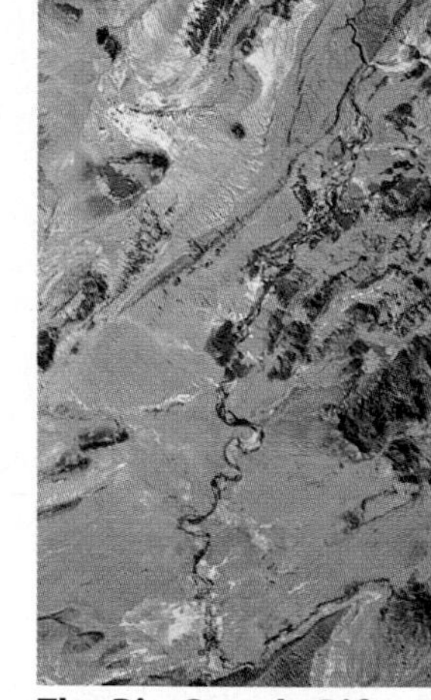

The Rio Grande Rift

Crustal extension

Rift fill

Normal faulting

Upper Mantle

Crustal extension

Formation of rift

Field Activity

Go to **science.glencoe.com** to take an online geology field trip. Click on the links provided to view different rock formations throughout the United States. In your Science Journal, write or draw the overall structure and attitude, or positioning, of the rock formations.

Thrust Fault: The Appalachian Mountains

The Appalachian Mountains were uplifted more than 300 million years ago during a plate collision. As the African and North American Plates collided, huge slices of Earth's crust were moved great distances along thrust faults. These slices of crust make up Earth's surface in some parts of the Appalachian Mountains today.

Thrust fault

Hayward Fault

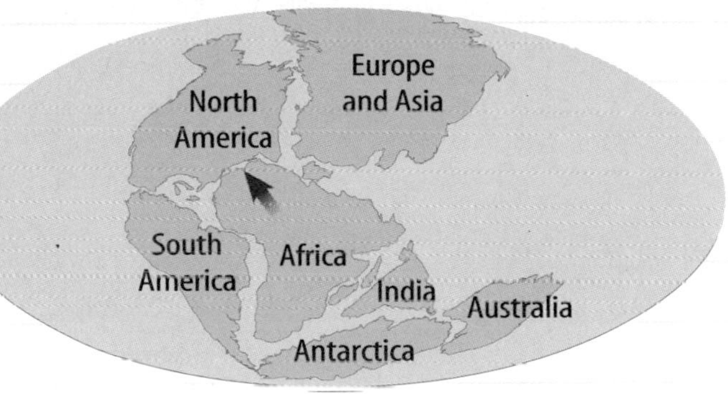

Tectonic Plate collision

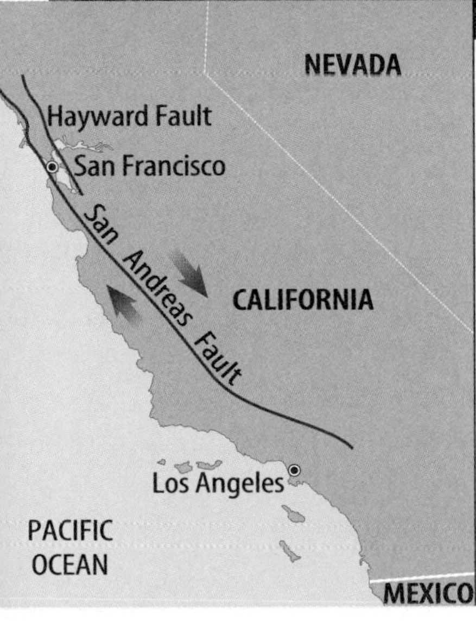

Location of Hayward Fault

Strike-Slip Fault: The Hayward Fault

This strike-slip fault—part of the 900-km San Andreas Fault system—extends about 120 km and is considered to be the state's second-strongest fault. It is a fracture in Earth's crust: the Pacific Plate to the west is moving northwest in relation to the North American Plate. Energy from this movement continues to build until it is released in the form of an earthquake. In 1868, an earthquake occurred along the Hayward Fault, damaging San Francisco.

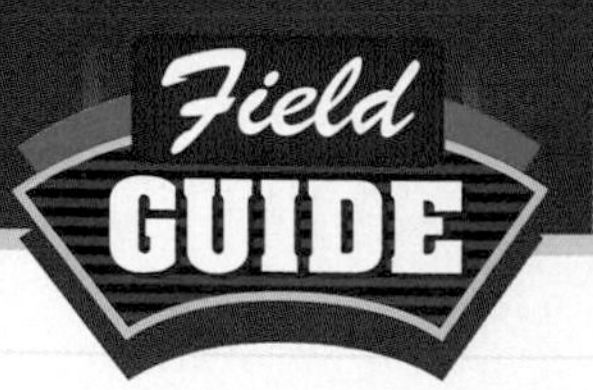

Fold: Anticline at Hartland Quay

Layers of sedimentary rock are folded in this spectacular anticline at Hartland Quay in Devon, England. The anticline has been eroded in the foreground exposing the ends of the tilted rock layers.

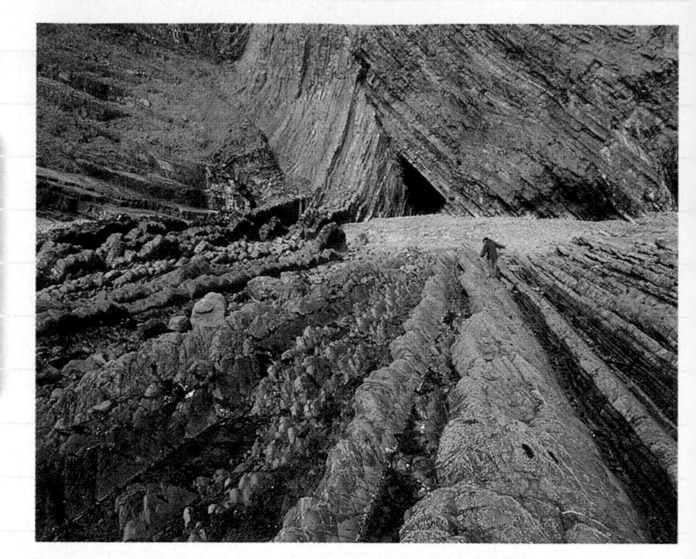

Anticline at Hartland Quay

Anticlines and synclines in sequence

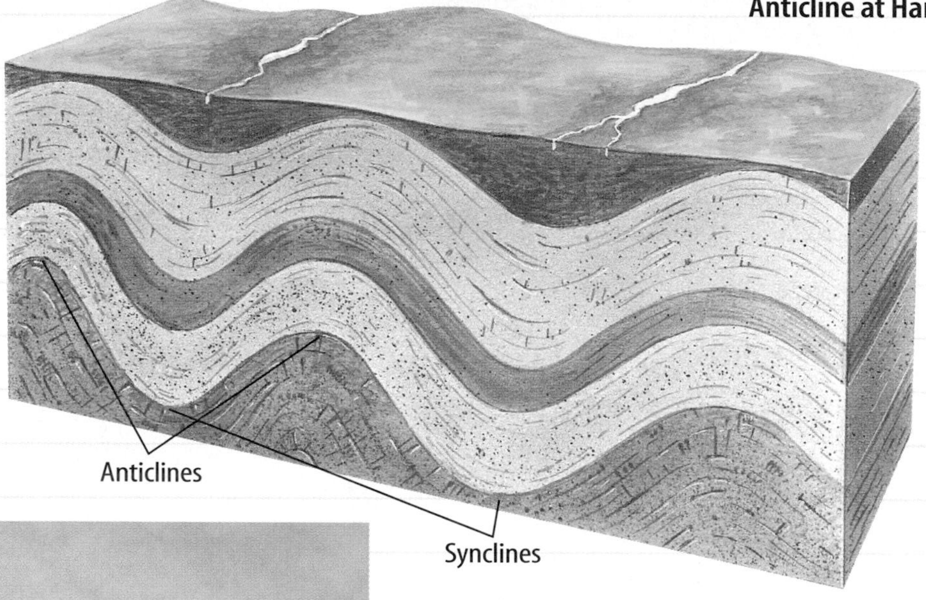

Fold: Syncline on Mount Kidd

This syncline is part of Mount Kidd in Alberta, Canada, in the Kananaskis Valley of the Canadian Rockies. Unlike the American Rockies, which were originally large chunks of granite, the Canadian Rockies began as piles of sediment that were compressed when plates collided.

The syncline on Mount Kidd

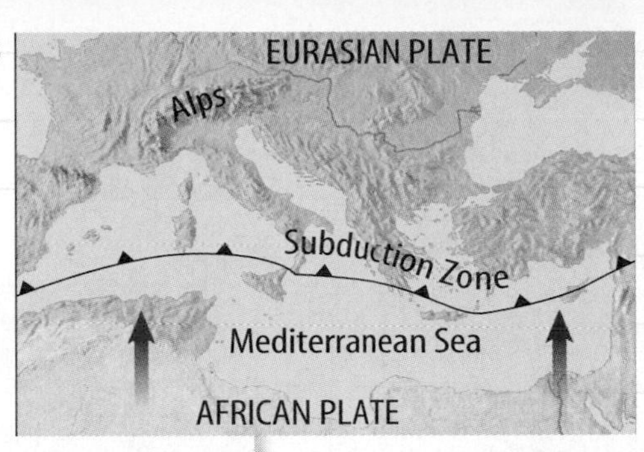

Location of the Alps

Fold: The Alps

The Alps, which are a beautiful example of folded mountains, were formed as the African Plate began to move closer to the Eurasian Plate. The pressure exerted by this movement caused the land to fold, creating these peaks more than 15 million years ago.

The folded Alps

Erosion of the Blue Ridge Mountains

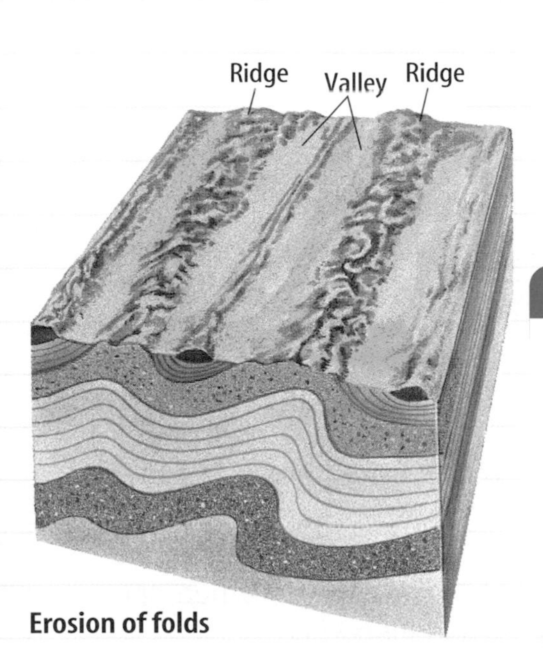

Erosion of folds

Fold: The Blue Ridge Mountains

Folded mountains derive their name from the abundance of folds in the rocks. The Blue Ridge Mountains in the Valley and Ridge region of the Appalachians are an excellent example of mountains formed by rocks that folded, then began to erode. The harder rocks formed ridges, and the softer rocks eroded to form valleys.

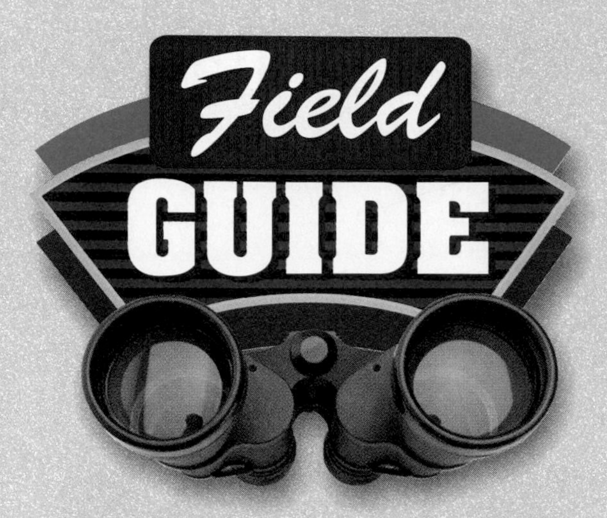

Some people have defined music as "patterns of tones." A tone is a sound with a specific pitch. In music, a tone might also be called a note. Pitch describes how high or low the tone is. Like all sounds, musical tones are produced when an object vibrates. Higher pitches are produced by more vibrations per second, and therefore have a higher frequency.

Most musical instruments use resonance to amplify sounds. To amplify a sound means to increase its volume. Resonance occurs when one object causes another object to vibrate at the same frequency—or pitch.

How Resonance Works

A vibrating object produces sound waves. These waves can affect other objects and cause them to vibrate. As more matter vibrates, a louder sound is produced. For example, resonance is at work in a guitar. When a guitar's strings are plucked or strummed, the strings vibrate. The strings' vibrations make the thin soundboard—in this case, the front of the guitar—vibrate. The soundboard's vibrations make the air inside the guitar's hollow body vibrate. The vibrating air amplifies the sounds that were first produced by the strings.

Musical Instruments

Mandolin

Stringed Instruments

Tones are produced in stringed instruments by making stretched strings vibrate. Each string is tuned to a different pitch. When playing stringed instruments such as the harp, each string produces only one pitch. The player creates different pitches by plucking different strings.

When playing stringed instruments such as the guitar and violin, the player can change the pitch of each string by pressing down on one end and making it shorter. Stringed instruments may be strummed, plucked, or played with a bow.

Harp

Field Activity

Watch an orchestra or band perform in a live concert or on television. In your Science Journal, name all of the different instruments you recognize. Then use this field guide to identify the category in which each instrument is classified.

Wind Instruments—Woodwinds

Woodwind instruments include the clarinet, the saxophone, and the recorder. These instruments are played by blowing into a mouthpiece or across a hole. Some woodwinds, such as clarinets, have a thin flexible reed in the mouthpiece that vibrates. The reed causes the air in the tube to vibrate. As the air vibrates inside the woodwind's hollow tube, tones are produced. Musicians change this instrument's pitch by covering holes with their fingers or by pressing keys that cover holes. Covering a hole changes the length of the column of air inside the tube.

Trombone

Tuba

Wind Instruments—Brass

Brass instruments include the trumpet, the trombone, and the tuba. Their mouthpieces are larger than woodwinds' mouthpieces. Brass instruments are played by pressing the lips against a mouthpiece and blowing so the lips vibrate. Musicians change brass instruments' pitches by tensing or relaxing their lips. With most brass instruments, the pitch also can be changed by pressing valves, which changes the length of the vibrating column of air inside the instrument.

Percussion Instruments—Idiophones

Idiophones vibrate to produce tones. Musicians play them by hitting, shaking, scraping, or plucking them. Idiophones such as cymbals, bells, gongs, music boxes, and xylophone keys play only one pitch. Triangles, clappers, rattles, and cymbals have indefinite pitches—their pitches depend on how they are played and how they are constructed.

Xylophone

Percussion Instruments—Membranophones

Membranophones produce sound when their membranes—the stretched tops of drums or the tiny membranes within kazoos—vibrate. Drums are usually struck with hands, with beaters such as drumsticks, or with knotted cords to produce tones.

Bongo

Electric Instruments

Electric instruments such as the electric guitar and electric violin are played like regular instruments. However, rather than using resonance to amplify their sound, their vibrations are converted to electrical signals that are amplified electronically. The amplified electric signal is then converted to sound by a loudspeaker.

Guitar and Amplifier

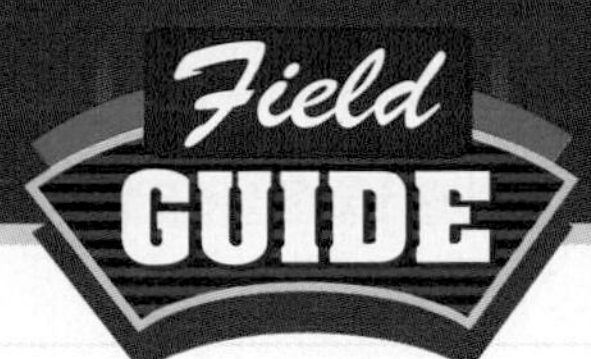

Keyboard Instruments—Piano

Each piano key is attached to a small hammer. When the player presses a key, the hammer hits a string and makes it vibrate. The strings are different lengths and each string produces a different pitch. The piano's body amplifies the tones.

Piano

Keyboards—Pipe Organ

Pressing a pipe organ's key opens a pipe to let air vibrate inside it. The pipes are different lengths, and each produces a different pitch.

Pipe Organ

Electronic Instruments

Unlike all other types of musical instruments, electronic instruments do not rely on vibrations to produce sounds. Instead, these instruments produce electrical signals that a computer then converts to sounds. Even though a synthesizer has a keyboard, it is classified as an electronic instrument because it produces sounds electronically. Today, it is the most widely used electronic instrument.

Synthesizer

Organizing Information

As you study science, you will make many observations and conduct investigations and experiments. You will also research information that is available from many sources. These activities will involve organizing and recording data. The quality of the data you collect and the way you organize it will determine how well others can understand and use it. In **Figure 1,** the student is obtaining and recording information using a microscope.

Putting your observations in writing is an important way of communicating to others the information you have found and the results of your investigations and experiments.

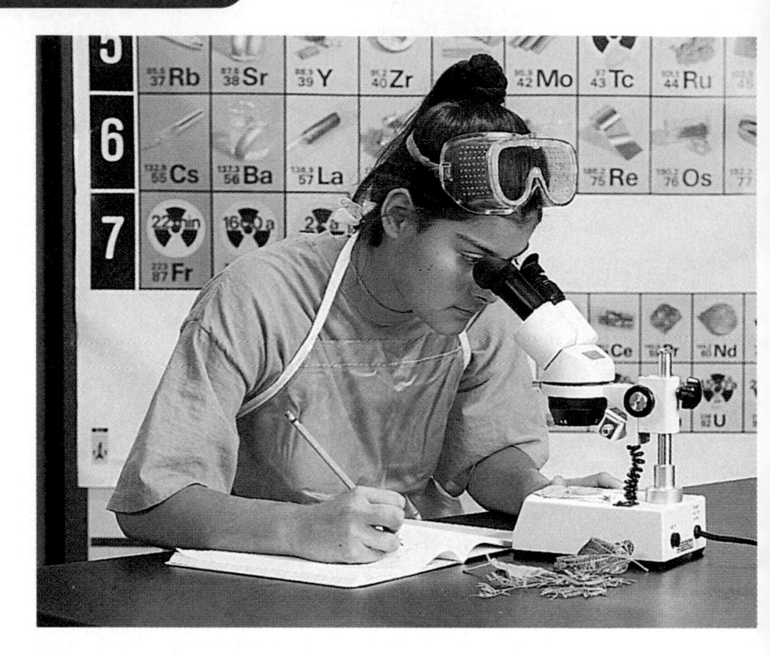

Figure 1
Making an observation is one way to gather information directly.

Researching Information

Scientists work to build on and add to human knowledge of the world. Before moving in a new direction, it is important to gather the information that already is known about a subject. You will look for such information in various reference sources. Follow these steps to research information on a scientific subject:

Step 1 Determine exactly what you need to know about the subject. For instance, you might want to find out what happened to local plant life when Mount St. Helens erupted in 1980.

Step 2 Make a list of questions, such as: When did the eruption begin? How long did it last? How large was the area in which plant life was affected?

Step 3 Use multiple sources such as textbooks, encyclopedias, government documents, professional journals, science magazines, and the Internet.

Step 4 List where you found the sources. Make sure the sources you use are reliable and the most current available.

Evaluating Print and Nonprint Sources

Not all sources of information are reliable. Evaluate the sources you use for information, and use only those you know to be dependable. For example, suppose you want information about the digestion of fats and proteins. You might find two Websites on digestion. One Web site contains "Fat Zapping Tips" written by a company that sells expensive, high-protein supplements to help your body eliminate excess fat. The other is a Web page on "Digestion and Metabolism" written by a well-respected medical school. You would choose the second Web site as the more reliable source of information.

In science, information can change rapidly. Always consult the most current sources. A 1985 source about the human genome would not reflect the most recent research and findings.

Interpreting Scientific Illustrations

As you research a science topic, you will see drawings, diagrams, and photographs. Illustrations help you understand what you read. Some illustrations are included to help you understand an idea that you can't see easily by yourself. For instance, you can't see the bones of a blue whale, but you can look at a diagram of a whale skeleton as labeled in **Figure 2** that helps you understand them. Visualizing a drawing helps many people remember details more easily. Illustrations also provide examples that clarify difficult concepts or give additional information about the topic you are studying.

Most illustrations have a label or a caption. A label or caption identifies the illustration or provides additional information to better explain it. Can you find the caption or labels in **Figure 2?**

Figure 2
A labeled diagram of the skeletal structure of a blue whale.

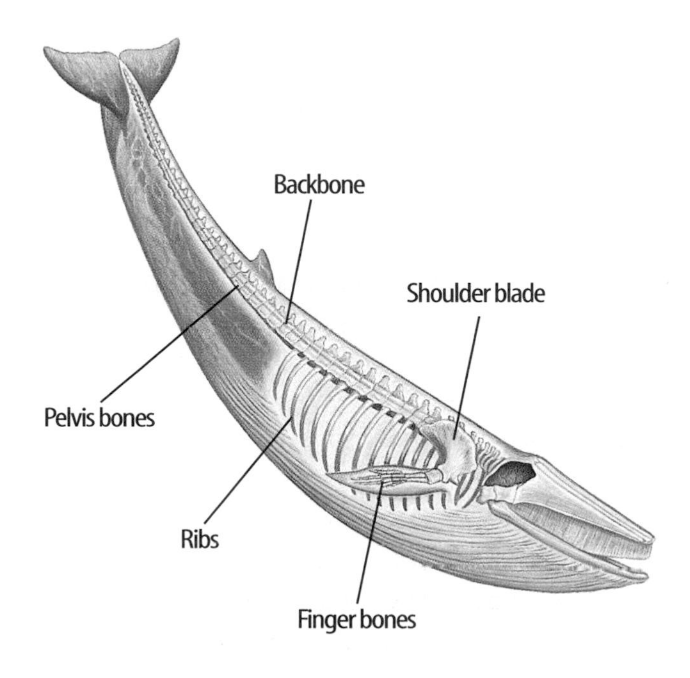

Venn Diagram

Although it is not a concept map, a Venn diagram illustrates how two subjects compare and contrast. In other words, you can see the characteristics that the subjects have in common and those that they do not.

The Venn diagram in **Figure 3** shows the relationship between two categories of organisms, plants and animals. Both share some basic characteristics as living organisms. However, there are differences in the ways they carry out various life processes, such as obtaining nourishment, that distinguish one from the other.

Concept Mapping

If you were taking a car trip, you might take some sort of road map. By using a map, you begin to learn where you are in relation to other places on the map.

A concept map is similar to a road map, but a concept map shows relationships among ideas (or concepts) rather than places. It is a diagram that visually shows how concepts are related. Because a concept map shows relationships among ideas, it can make the meanings of ideas and terms clear and help you understand what you are studying.

Overall, concept maps are useful for breaking large concepts down into smaller parts, making learning easier.

Figure 3
A Venn diagram shows how objects or concepts are alike and how they are different.

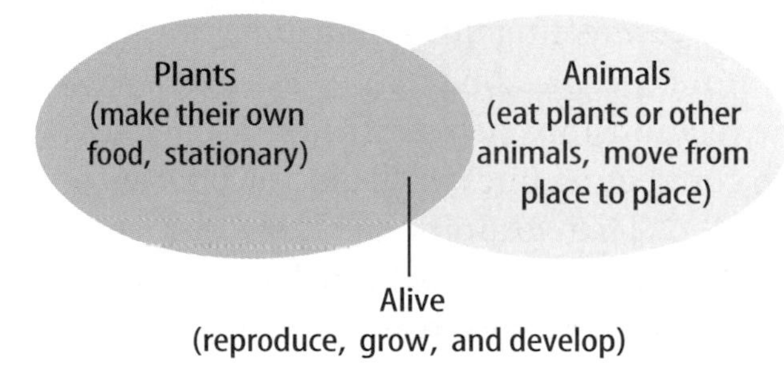

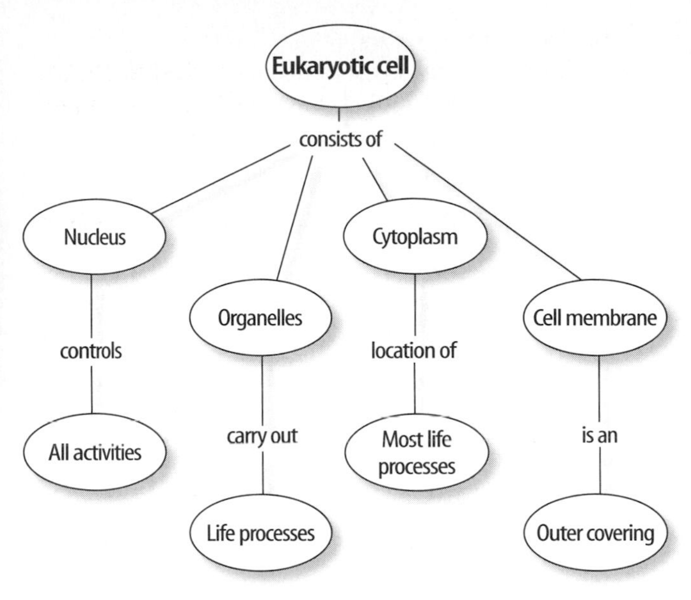

Figure 4
A network tree shows how concepts or objects are related.

Network Tree Look at the network tree in **Figure 4,** that shows details about a eukaryotic cell. A network tree is a type of concept map. Notice how some words are in ovals while others are written across connecting lines. The words inside the ovals are science terms or concepts. The words written on the connecting lines describe the relationships between the concepts.

When constructing a network tree, write the topic on a note card or piece of paper. Write the major concepts related to that topic on separate note cards or pieces of paper. Then arrange them in order from general to specific. Branch the related concepts from the major concept and describe the relationships on the connecting lines. Continue branching to more specific concepts. If necessary, write the relationships between the concepts on the connecting lines until all concepts are mapped. Then examine the network tree for relationships that cross branches, and add them to the network tree.

Events Chain An events chain is another type of concept map. It models the order, or sequence, of items. In science, an events chain can be used to describe a sequence of events, the steps in a procedure, or the stages of a process.

When making an events chain, first find the one event that starts the chain. This event is called the initiating event. Then, find the next event in the chain and continue until you reach an outcome. Suppose you are asked to describe the main stages in the growth of a plant from a seed. You might draw an events chain such as the one in **Figure 5.** Notice that connecting words are not necessary in an events chain.

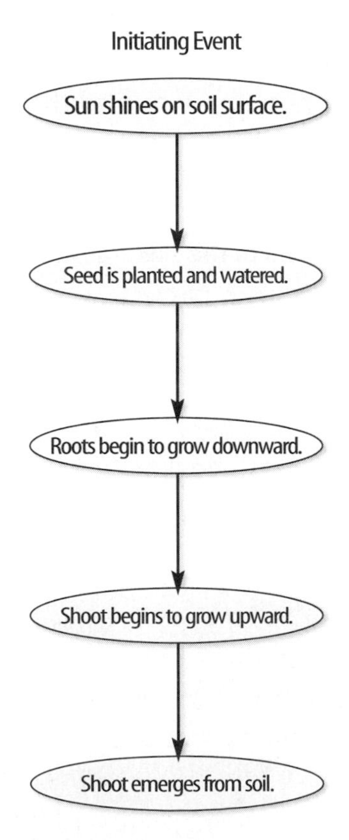

Figure 5
Events chains show the order of steps in a process or event.

Cycle Map A cycle concept map is a specific type of events chain map. In a cycle concept map, the series of events does not produce a final outcome. Instead, the last event in the chain relates back to the beginning event.

You first decide what event will be used as the beginning event. Once that is decided, you list events in order that occur after it. Words are written between events that describe what happens from one event to the next. The last event in a cycle concept map relates back to the beginning event. The number of events in a cycle concept varies but is usually three or more. Look at the cycle map in **Figure 6.**

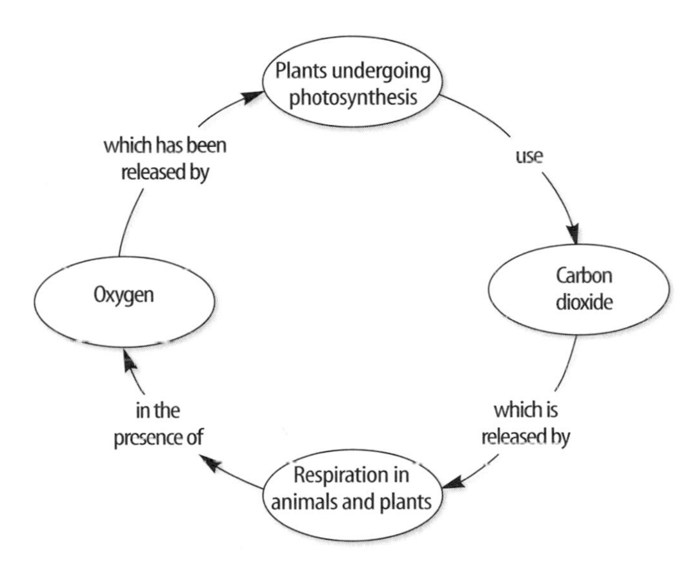

Figure 6
A cycle map shows events that occur in a cycle.

Spider Map A type of concept map that you can use for brainstorming is the spider map. When you have a central idea, you might find you have a jumble of ideas that relate to it but might not clearly relate to each other. The circulatory system spider map in **Figure 7** shows that if you write these ideas outside the main concept, then you can begin to separate and group unrelated terms so they become more useful.

Figure 7
A spider map allows you to list ideas that relate to a central topic but not necessarily to one another.

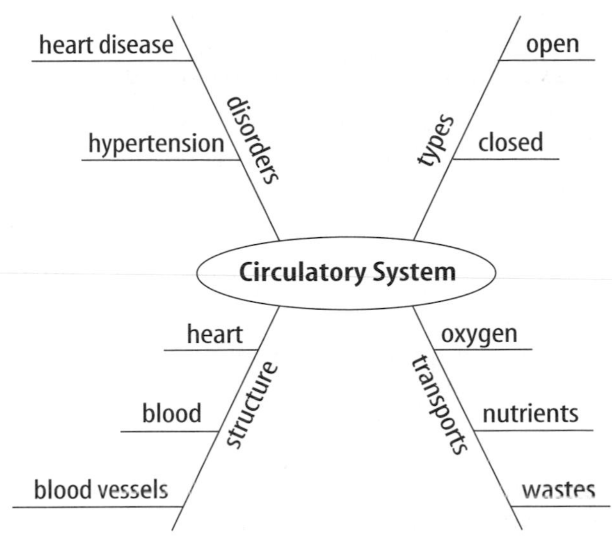

Writing a Paper

You will write papers often when researching science topics or reporting the results of investigations or experiments. Scientists frequently write papers to share their data and conclusions with other scientists and the public. When writing a paper, use these steps.

Step 1 Assemble your data by using graphs, tables, or a concept map. Create an outline.

Step 2 Start with an introduction that contains a clear statement of purpose and what you intend to discuss or prove.

Step 3 Organize the body into paragraphs. Each paragraph should start with a topic sentence, and the remaining sentences in that paragraph should support your point.

Step 4 Position data to help support your points.

Step 5 Summarize the main points and finish with a conclusion statement.

Step 6 Use tables, graphs, charts, and illustrations whenever possible.

Investigating and Experimenting

You might say the work of a scientist is to solve problems. When you decide to find out why one corner of your yard is always soggy, you are problem solving, too. You might observe that the corner is lower than the surrounding area and has less vegetation growing in it. You might decide to see if planting some grass will keep the corner drier.

Scientists use orderly approaches to solve problems. The methods scientists use include identifying a question, making observations, forming a hypothesis, testing a hypothesis, analyzing results, and drawing conclusions.

Scientific investigations involve careful observation under controlled conditions. Such observation of an object or a process can suggest new and interesting questions about it. These questions sometimes lead to the formation of a hypothesis. Scientific investigations are designed to test a hypothesis.

Identifying a Question

The first step in a scientific investigation or experiment is to identify a question to be answered or a problem to be solved. You might be interested in knowing why an animal like the one in **Figure 8** looks the way it does.

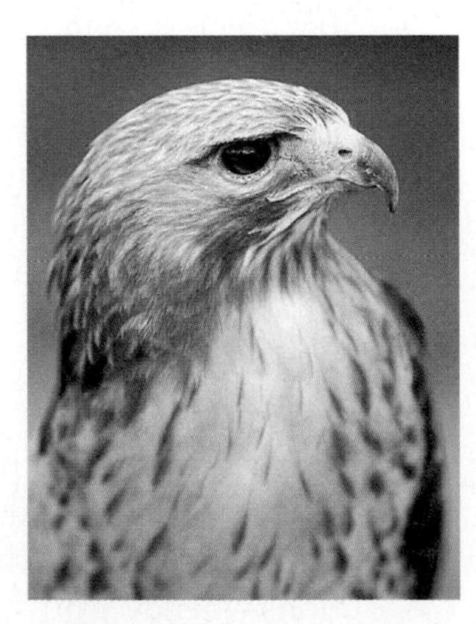

Figure 8
When you see a bird, you might ask yourself, "How does the shape of this bird's beak help it feed?"

Forming Hypotheses

Hypotheses are based on observations that have been made. A hypothesis is a possible explanation based on previous knowledge and observations.

Perhaps a scientist has observed that bean plants grow larger if they are fertilized than if they are not. Based on these observations, the scientist can make a statement that he or she can test. The statement is a hypothesis. The hypothesis could be: *Fertilizer makes bean plants grow larger.* A hypothesis has to be something you can test by using an investigation. A testable hypothesis is a valid hypothesis.

Predicting

When you apply a hypothesis to a specific situation, you predict something about that situation. First, you must identify which hypothesis fits the situation you are considering. People use predictions to make everyday decisions. Based on previous observations and experiences, you might form a prediction that if fertilizer makes bean plants grow larger, then fertilized plants will yield more beans than plants not fertilized. Someone could use this prediction to plan to grow fewer plants.

Testing a Hypothesis

To test a hypothesis, you need a procedure. A procedure is the plan you follow in your experiment. A procedure tells you what materials to use, as well as how and in what order to use them. When you follow a procedure, data are generated that support or do not support the original hypothesis statement.

For example, suppose you notice that your guppies don't seem as active as usual when your aquarium heater is not working. You wonder how water temperature affects guppy activity level. You decide to test the hypothesis, "If water temperature increases, then guppy activity should increase." Then you write the procedure shown in **Figure 9** for your experiment and generate the data presented in the table below.

Procedure

1. Fill five identical glass containers with equal amounts of aquarium water.
2. Measure and record the temperature of the water in the first container.
3. Heat and cool the other containers so that two have higher and two have lower water temperatures.
4. Place a guppy in each container; count and record the number of movements each guppy makes in 5 minutes.

Figure 9
A procedure tells you what to do step by step.

Number of Guppy Movements

Container	Temperature (°C)	Movements
1	38	56
2	40	61
3	42	70
4	36	46
5	34	42

Are all investigations alike? Keep in mind as you perform investigations in science that a hypothesis can be tested in many ways. Not every investigation makes use of all the ways that are described on these pages, and not all hypotheses are tested by investigations. Scientists encounter many variations in the methods that are used when they perform experiments. The skills in this handbook are here for you to use and practice.

Identifying and Manipulating Variables and Controls

In any experiment, it is important to keep everything the same except for the item you are testing. The one factor you change is called the independent variable. The factor that changes as a result of the independent variable is called the dependent variable. Always make sure you have only one independent variable. If you allow more than one, you will not know what causes the changes you observe in the dependent variable. Many experiments also have controls—individual instances or experimental subjects for which the independent variable is not changed. You can then compare the test results to the control results.

For example, in the guppy experiment, you made everything the same except the temperature of the water. The glass containers were identical. The volume of aquarium water in each container and beginning water temperature were the same. Each guppy was like the others, as much as possible. In this way, you could be sure that any difference in the number of guppy movements was caused by the temperature change—the independent variable. The activity level of the guppy was measured as the number of guppy movements—the dependent variable. The guppy in the container in which the water temperature was not changed was the control.

Collecting Data

Whether you are carrying out an investigation or a short observational experiment, you will collect data, or information. Scientists collect data accurately as numbers and descriptions and organize it in specific ways.

Observing Scientists observe items and events, then record what they see. When they use only words to describe an observation, it is called qualitative data. For example, a scientist might describe the color of a bird or the shape of a bird's beak as seen through binoculars. Scientists' observations also can describe how much there is of something. These observations use numbers, as well as words, in the description and are called quantitative data. For example, if a particular dog is described as being "furry, yellow, and short-haired," the data are clearly qualitative. Quantitative data for this dog might include "a mass of 14 kg, a height of 46 cm, and an age of 150 days." Quantitative data often are organized into tables. Then, from information in the table, a graph can be drawn. Graphs can reveal relationships that exist in experimental data.

When you make observations in science, you should examine the entire object or situation first, then look carefully for details. If you're looking at a plant, for instance, check general characteristics such as size and overall structure before using a hand lens to examine the leaves and other smaller structures such as flowers or fruits. Remember to record accurately everything you see.

Scientists try to make careful and accurate observations. When possible, they use instruments such as microscopes, metric rulers, graduated cylinders, thermometers, and balances. Measurements provide numerical data that can be repeated and checked.

Sampling When working with large numbers of objects or a large population, scientists usually cannot observe or study every one of them. Instead, they use a sample or a portion of the total number. To *sample* is to take a small, representative portion of the objects or organisms of a population for research. By making careful observations or manipulating variables within a portion of a group, information is discovered and conclusions are drawn that might apply to the whole population.

Estimating Scientific work also involves estimating. To *estimate* is to make a judgment about the size or the number of something without measuring or counting every object or member of a population. Scientists first count the number of objects in a small sample. Looking through a microscope lens, for example, a scientist can count the number of bacterial colonies in the 1-cm^2 frame shown in **Figure 10.** Then the scientist can multiply that number by the number of cm^2 in the petri dish to get an estimate of the total number of bacterial colonies present.

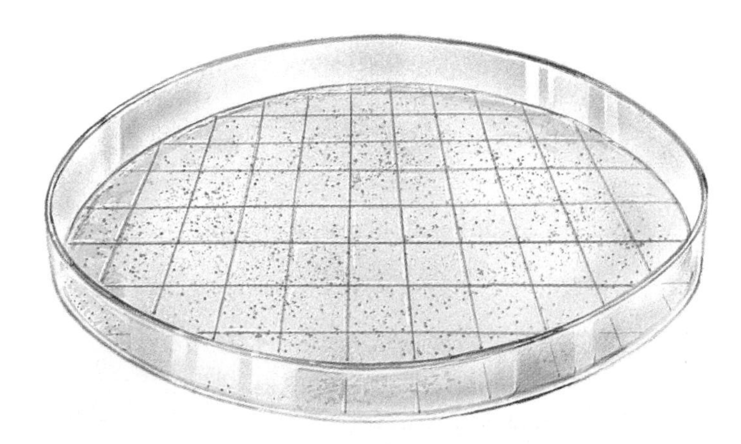

Figure 10
To estimate the total number of bacterial colonies that are present on a petri dish, count the number of bacterial colonies within a 1-cm^2 frame and multiply that number by the number of frames on the dish.

Measuring in SI

The metric system of measurement was developed in 1795. A modern form of the metric system, called the International System, or SI, was adopted in 1960. SI provides standard measurements that all scientists around the world can understand.

The metric system is convenient because unit sizes vary by multiples of 10. When changing from smaller units to larger units, divide by a multiple of 10. When changing from larger units to smaller, multiply by a multiple of 10. To convert millimeters to centimeters, divide the millimeters by 10. To convert 30 mm to centimeters, divide 30 by 10 (30 mm equal 3 cm).

Prefixes are used to name units. Look at the table below for some common metric prefixes and their meanings. Do you see how the prefix *kilo-* attached to the unit *gram* is *kilogram*, or 1,000 g?

Metric Prefixes

Prefix	Symbol	Meaning	
kilo-	k	1,000	thousand
hecto-	h	100	hundred
deka-	da	10	ten
deci-	d	0.1	tenth
centi-	c	0.01	hundredth
milli-	m	0.001	thousandth

Now look at the metric ruler shown in **Figure 11.** The centimeter lines are the long, numbered lines, and the shorter lines are millimeter lines.

When using a metric ruler, line up the 0-cm mark with the end of the object being measured, and read the number of the unit where the object ends. In this instance it would be 4.50 cm.

Figure 11
This metric ruler shows centimeter and millimeter divisions.

Liquid Volume In some science activities, you will measure liquids. The unit that is used to measure liquids is the liter. A liter has the volume of 1,000 cm^3. The prefix *milli-* means "thousandth (0.001)." A milliliter is one thousandth of 1 L and 1 L has the volume of 1,000 mL. One milliliter of liquid completely fills a cube measuring 1 cm on each side. Therefore, 1 mL equals 1 cm^3.

You will use beakers and graduated cylinders to measure liquid volume. A graduated cylinder, as illustrated in **Figure 12,** is marked from bottom to top in milliliters. This graduated cylinder contains 79 mL of a liquid.

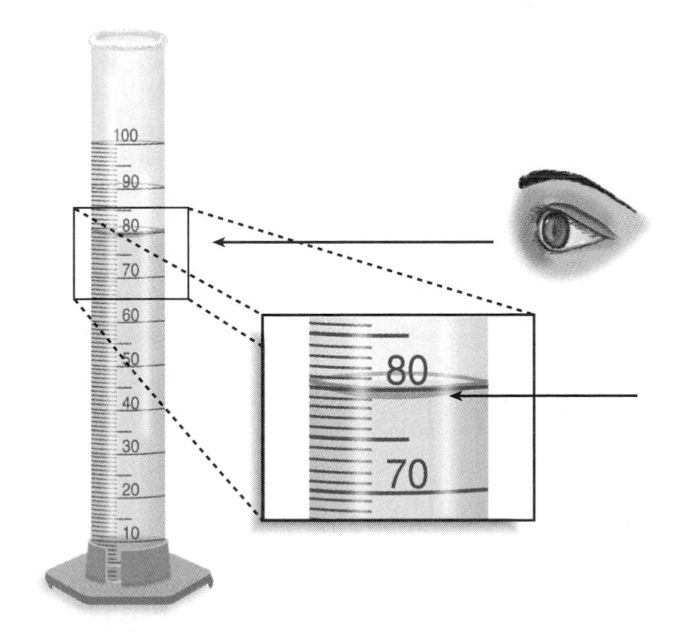

Figure 12
Graduated cylinders measure liquid volume.

Skill Handbooks

Mass Scientists measure mass in grams. You might use a beam balance similar to the one shown in **Figure 13.** The balance has a pan on one side and a set of beams on the other side. Each beam has a rider that slides on the beam.

Before you find the mass of an object, slide all the riders back to the zero point. Check the pointer on the right to make sure it swings an equal distance above and below the zero point. If the swing is unequal, find and turn the adjusting screw until you have an equal swing.

Place an object on the pan. Slide the largest rider along its beam until the pointer drops below zero. Then move it back one notch. Repeat the process on each beam until the pointer swings an equal distance above and below the zero point. Sum the masses on each beam to find the mass of the object. Move all riders back to zero when finished.

Figure 13
A triple beam balance is used to determine the mass of an object.

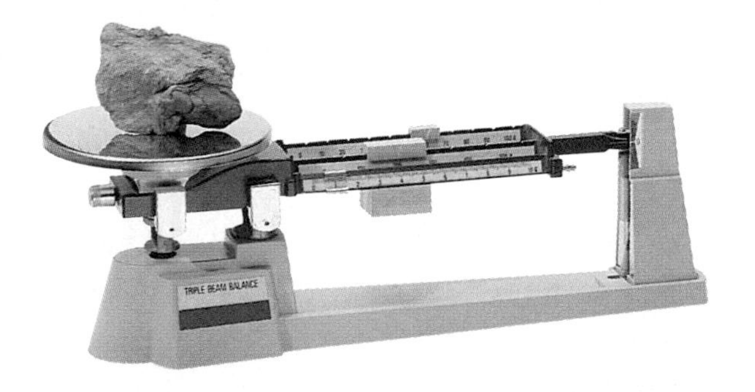

You should never place a hot object on the pan or pour chemicals directly onto the pan. Instead, find the mass of a clean container. Remove the container from the pan, then place the chemicals in the container. Find the mass of the container with the chemicals in it. To find the mass of the chemicals, subtract the mass of the empty container from the mass of the filled container.

Making and Using Tables

Browse through your textbook and you will see tables in the text and in the activities. In a table, data, or information, are arranged so that they are easier to understand. Activity tables help organize the data you collect during an activity so results can be interpreted.

Making Tables To make a table, list the items to be compared in the first column and the characteristics to be compared in the first row. The title should clearly indicate the content of the table, and the column or row heads should tell the reader what information is found in there. The table below lists materials collected for recycling on three weekly pick-up days. The inclusion of kilograms in parentheses also identifies for the reader that the figures are mass units.

Recyclable Materials Collected During Week

Day of Week	Paper (kg)	Aluminum (kg)	Glass (kg)
Monday	5.0	4.0	12.0
Wednesday	4.0	1.0	10.0
Friday	2.5	2.0	10.0

Using Tables How much paper, in kilograms, is being recycled on Wednesday? Locate the column labeled "Paper (kg)" and the row "Wednesday." The information in the box where the column and row intersect is the answer. Did you answer "4.0"? How much aluminum, in kilograms, is being recycled on Friday? If you answered "2.0," you understand how to read the table. How much glass is collected for recycling each week? Locate the column labeled "Glass (kg)" and add the figures for all three rows. If you answered "32.0," then you know how to locate and use the data provided in the table.

Recording Data

To be useful, the data you collect must be recorded carefully. Accuracy is key. A well-thought-out experiment includes a way to record procedures, observations, and results accurately. Data tables are one way to organize and record results. Set up the tables you will need ahead of time so you can record the data right away.

Record information properly and neatly. Never put unidentified data on scraps of paper. Instead, data should be written in a notebook like the one in **Figure 14.** Write in pencil so information isn't lost if your data get wet. At each point in the experiment, record your information and label it. That way, your data will be accurate and you will not have to determine what the figures mean when you look at your notes later.

Figure 14
Record data neatly and clearly so they are easy to understand.

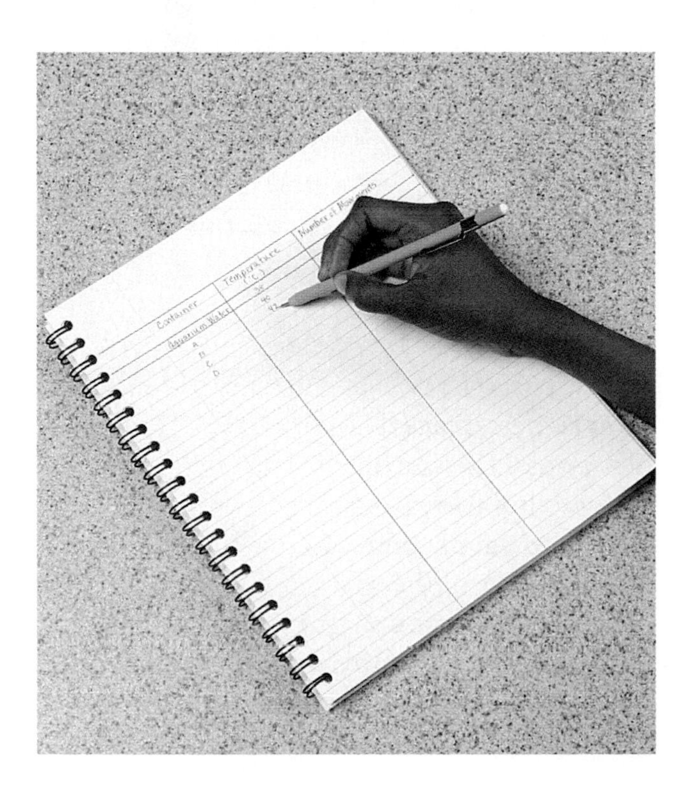

Recording Observations

It is important to record observations accurately and completely. That is why you always should record observations in your notes immediately as you make them. It is easy to miss details or make mistakes when recording results from memory. Do not include your personal thoughts when you record your data. Record only what you observe to eliminate bias. For example, when you record that a plant grew 12 cm in one day, you would note that this was the largest daily growth for the week. However, you would not refer to the data as "the best growth spurt of the week."

Making Models

You can organize the observations and other data you collect and record in many ways. Making models is one way to help you better understand the parts of a structure you have been observing or the way a process for which you have been taking various measurements works.

Models often show things that are very large or small or otherwise would be difficult to see and understand. You can study blood vessels and know that they are hollow tubes. The size and proportional differences among arteries, veins, and capillaries can be explained in words. However, you can better visualize the relative sizes and proportions of blood vessels by making models of them. Gluing different kinds of pasta to thick paper so the openings can be seen can help you see how the differences in size, wall thickness, and shape among types of blood vessels affect their functions.

Other models can be devised on a computer. Some models, such as disease control models used by doctors to predict the spread of the flu, are mathematical and are represented by equations.

Making and Using Graphs

After scientists organize data in tables, they might display the data in a graph that shows the relationship of one variable to another. A graph makes interpretation and analysis of data easier. Three types of graphs are the line graph, the bar graph, and the circle graph.

Line Graphs A line graph like in **Figure 15** is used to show the relationship between two variables. The variables being compared go on two axes of the graph. For data from an experiment, the independent variable always goes on the horizontal axis, called the *x*-axis. The dependent variable always goes on the vertical axis, called the *y*-axis. After drawing your axes, label each with a scale. Next, plot the data points.

A data point is the intersection of the recorded value of the dependent variable for each tested value of the independent variable. After all the points are plotted, connect them.

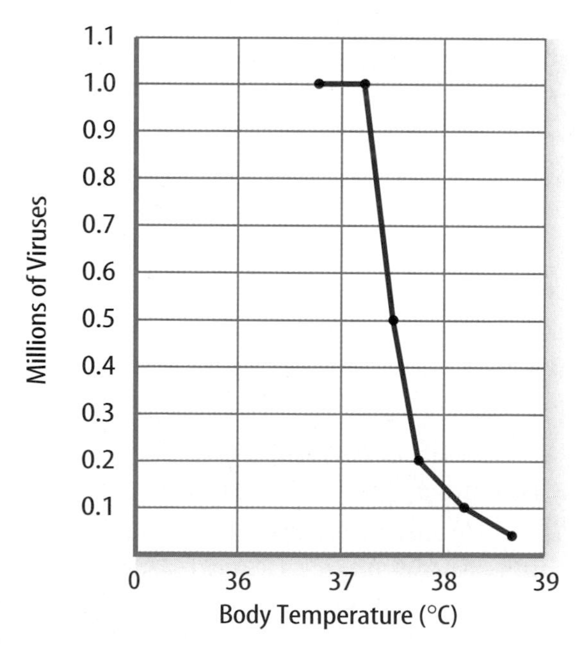

Figure 15
This line graph shows the relationship between body temperature and the millions of infecting viruses present in a human body.

Bar Graphs Bar graphs compare data that do not change continuously. Vertical bars show the relationships among data.

To make a bar graph, set up the *y*-axis as you did for the line graph. Draw vertical bars of equal size from the *x*-axis up to the point on the *y*-axis that represents the value of *x*.

Figure 16
The number of wing vibrations per second for different insects can be shown as a bar graph or circle graph.

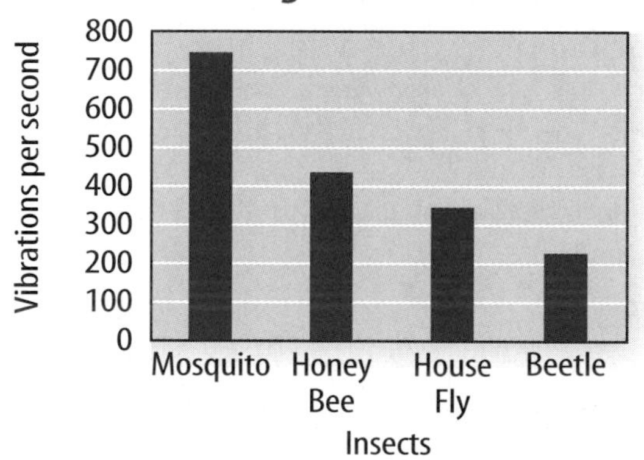

Circle Graphs A circle graph uses a circle divided into sections to display data as parts (fractions or percentages) of a whole. The size of each section corresponds to the fraction or percentage of the data that the section represents. So, the entire circle represents 100 percent, one-half represents 50 percent, one-fifth represents 20 percent, and so on.

Other 1%
Oxygen 21%
Nitrogen 78%

Analyzing and Applying Results

Analyzing Results

To determine the meaning of your observations and investigation results, you will need to look for patterns in the data. You can organize your information in several of the ways that are discussed in this handbook. Then you must think critically to determine what the data mean. Scientists use several approaches when they analyze the data they have collected and recorded. Each approach is useful for identifying specific patterns in the data.

Forming Operational Definitions

An operational definition defines an object by showing how it functions, works, or behaves. Such definitions are written in terms of how an object works or how it can be used; that is, they describe its job or purpose.

For example, a ruler can be defined as a tool that measures the length of an object (how it can be used). A ruler also can be defined as something that contains a series of marks that can be used as a standard when measuring (how it works).

Classifying

Classifying is the process of sorting objects or events into groups based on common features. When classifying, first observe the objects or events to be classified. Then select one feature that is shared by some members in the group but not by all. Place those members that share that feature into a subgroup. You can classify members into smaller and smaller subgroups based on characteristics.

How might you classify a group of animals? You might first classify them by putting all of the dogs, cats, lizards, snakes, and birds into separate groups. Within each group, you could then look for another common feature by which to further classify members of the group, such as size or color.

Remember that when you classify, you are grouping objects or events for a purpose. For example, classifying animals can be the first step in identifying them. You might know that a cardinal is a red bird. To find it in a large group of animals, you might start with the classification scheme mentioned here. You'll locate a cardinal within the red grouping of the birds that you separate from the rest of the animals. A male ruby-throated hummingbird could be located within the birds by its tiny size and the bright red color of its throat. Keep your purpose in mind as you select the features to form groups and subgroups.

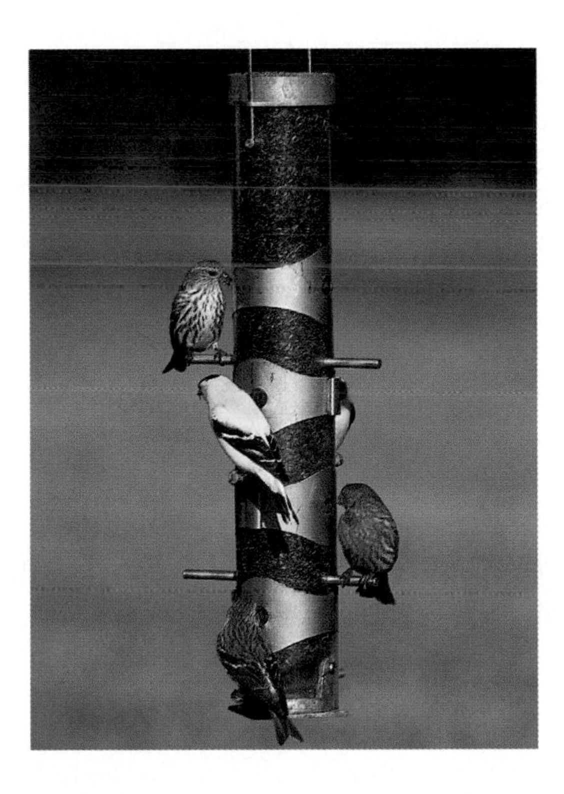

Figure 17
Color is one of many characteristics that are used to classify animals.

Comparing and Contrasting

Observations can be analyzed by noting the similarities and differences between two or more objects or events that you observe. When you look at objects or events to see how they are similar, you are comparing them. Contrasting is looking for differences in objects or events. The table below compares and contrasts the nutritional value of two cereals.

Nutritional Values

	Cereal A	Cereal B
Calories	220	160
Fat	10 g	10 g
Protein	2.5 g	2.6 g
Carbohydrate	30 g	15 g

Recognizing Cause and Effect

Have you ever gotten a cold and then suggested that you probably caught it from a classmate who had one recently? If so, you have observed an effect and inferred a cause. The event is the effect, and the reason for the event is the cause.

When scientists are unsure of the cause of a certain event, they design controlled experiments to determine what caused it.

Interpreting Data

The word *interpret* means "to explain the meaning of something." Look at the problem originally being explored in an experiment and figure out what the data show. Identify the control group and the test group so you can see whether or not changes in the independent variable have had an effect. Look for differences in the dependent variable between the control and test groups.

These differences you observe can be qualitative or quantitative. You would be able to describe a qualitative difference using only words, whereas you would measure a quantitative difference and describe it using numbers. If there are qualitative or quantitative differences, the independent variable that is being tested could have had an effect. If no qualitative or quantitative differences are found between the control and test groups, the variable that is being tested apparently had no effect.

For example, suppose that three pepper plants are placed in a garden and two of the plants are fertilized, but the third is left to grow without fertilizer. Suppose you are then asked to describe any differences in the plants after two weeks. A qualitative difference might be the appearance of brighter green leaves on fertilized plants but not on the unfertilized plant. A quantitative difference might be a difference in the height of the plants or the number of flowers on them.

Inferring Scientists often make inferences based on their observations. An inference is an attempt to explain, or interpret, observations or to indicate what caused what you observed. An inference is a type of conclusion.

When making an inference, be certain to use accurate data and accurately described observations. Analyze all of the data that you've collected. Then, based on everything you know, explain or interpret what you've observed.

Drawing Conclusions

When scientists have analyzed the data they collected, they proceed to draw conclusions about what the data mean. These conclusions are sometimes stated using words similar to those found in the hypothesis formed earlier in the process.

Conclusions To analyze your data, you must review all of the observations and measurements that you made and recorded. Recheck all data for accuracy. After your data are rechecked and organized, you are almost ready to draw a conclusion such as "Plants need sunlight in order to grow."

Before you can draw a conclusion, however, you must determine whether the data allow you to come to a conclusion that supports a hypothesis. Sometimes that will be the case; other times it will not.

If your data do not support a hypothesis, it does not mean that the hypothesis is wrong. It means only that the results of the investigation did not support the hypothesis. Maybe the experiment needs to be redesigned, but very likely, some of the initial observations on which the hypothesis was based were incomplete or biased. Perhaps more observation or research is needed to refine the hypothesis.

Avoiding Bias Sometimes drawing a conclusion involves making judgments. When you make a judgment, you form an opinion about what your data mean. It is important to be honest and to avoid reaching a conclusion if no supporting evidence for it exists or if it was based on a small sample. It also is important not to allow any expectations of results to bias your judgments. If possible, it is a good idea to collect additional data. Scientists do this all the time.

For example, animal behaviorist Katharine Payne made an important observation about elephant communication. While visiting a zoo, Payne felt the air vibrating around her. At the same time, she also noticed that the skin on an elephant's forehead was fluttering. She suspected that the elephants were generating the vibrations and that they might be using the low-frequency sounds to communicate.

Payne conducted an experiment to record these sounds and simultaneously observe the behavior of the elephants in the zoo. She later conducted a similar experiment in Namibia in southwest Africa, where elephant herds roam. The additional data she collected supported the judgment Payne had made, which was that these low-frequency sounds were a form of communication between elephants.

Evaluating Others' Data and Conclusions

Sometimes scientists have to use data that they did not collect themselves, or they have to rely on observations and conclusions drawn by other researchers. In cases such as these, the data must be evaluated carefully.

How were the data obtained? How was the investigation done? Has it been duplicated by other researchers? Did they come up with the same results? Look at the conclusion, as well. Would you reach the same conclusion from these results? Only when you have confidence in the data of others can you believe it is true and feel comfortable using it.

Communicating

The communication of ideas is an important part of the work of scientists. A discovery that is not reported will not advance the scientific community's understanding or knowledge. Communication among scientists also is important as a way of improving their investigations.

Scientists communicate in many ways, from writing articles in journals and magazines that explain their investigations and experiments, to announcing important discoveries on television and radio, to sharing ideas with colleagues on the Internet or presenting them as lectures.

People who study science rely on computers to record and store data and to analyze results from investigations. Whether you work in a laboratory or just need to write a lab report with tables, good computer skills are a necessity.

Using a Word Processor

Suppose your teacher has assigned a written report. After you've completed your research and decided how you want to write the information, you need to put all that information on paper. The easiest way to do this is with a word processing application on a computer.

A computer application that allows you to type your information, change it as many times as you need to, and then print it out so that it looks neat and clean is called a word processing application. You also can use this type of application to create tables and columns, add bullets or cartoon art to your page, include page numbers, and even check your spelling.

Helpful Hints

- If you aren't sure how to do something using your word processing program, look in the help menu. You will find a list of topics there to click on for help. After you locate the help topic you need, just follow the step-by-step instructions you see on your screen.
- Just because you've spell checked your report doesn't mean that the spelling is perfect. The spell check feature can't catch misspelled words that look like other words. If you've accidentally typed *wind* instead of *wing*, the spell checker won't know the difference. Always reread your report to make sure you didn't miss any mistakes.

Figure 18
You can use computer programs to make graphs and tables.

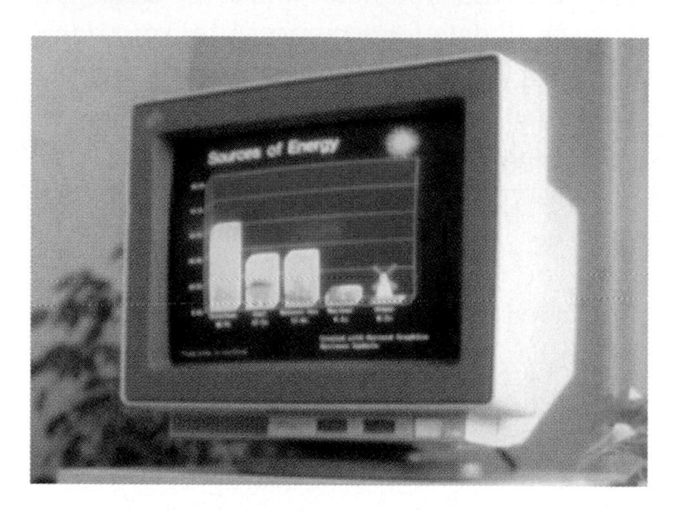

Using a Database

Imagine you're in the middle of a research project busily gathering facts and information. You soon realize that it's becoming more difficult to organize and keep track of all the information. The tool to use to solve information overload is a database. Just as a file cabinet organizes paper records, a database organizes computer records. However, a database is more powerful than a simple file cabinet because at the click of a mouse, the contents can be reshuffled and reorganized. At computer-quick speeds, databases can sort information by any characteristics and filter data into multiple categories.

Helpful Hints

- Before setting up a database, take some time to learn the features of your database software by practicing with established database software.
- Periodically save your database as you enter data. That way, if something happens such as your computer malfunctions or the power goes off, you won't lose all of your work.

Doing a Database Search

When searching for information in a database, use the following search strategies to get the best results. These are the same search methods used for searching Internet databases.

- Place the word *and* between two words in your search if you want the database to look for any entries that have both words. For example, "fox *and* mink" would give you information that mentions both fox and mink.
- Place the word *or* between two words if you want the database to show entries that have at least one of the words. For example "fox *or* mink" would show you information that mentions either fox or mink.
- Place the word *not* between two words if you want the database to look for entries that have the first word but do not have the second word. For example, "canine *not* fox" would show you information that mentions the term *canine* but does not mention the fox.

In summary, databases can be used to store large amounts of information about a particular subject. Databases allow biologists, Earth scientists, and physical scientists to search for information quickly and accurately.

Using an Electronic Spreadsheet

Your science fair experiment has produced lots of numbers. How do you keep track of all the data, and how can you easily work out all the calculations needed? You can use a computer program called a spreadsheet to record data that involve numbers. A spreadsheet is an electronic mathematical worksheet.

Type in your data in rows and columns, just as in a data table on a sheet of paper. A spreadsheet uses simple math to do data calculations. For example, you could add, subtract, divide, or multiply any of the values in the spreadsheet by another number. You also could set up a series of math steps you want to apply to the data. If you want to add 12 to all the numbers and then multiply all the numbers by 10, the computer does all the calculations for you in the spreadsheet. Below is an example of a spreadsheet that records data from an experiment with mice in a maze.

Helpful Hints

- Before you set up the spreadsheet, identify how you want to organize the data. Include any formulas you will need to use.
- Make sure you have entered the correct data into the correct rows and columns.
- You also can display your results in a graph. Pick the style of graph that best represents the data with which you are working.

Figure 19
A spreadsheet allows you to display large amounts of data and do calculations automatically.

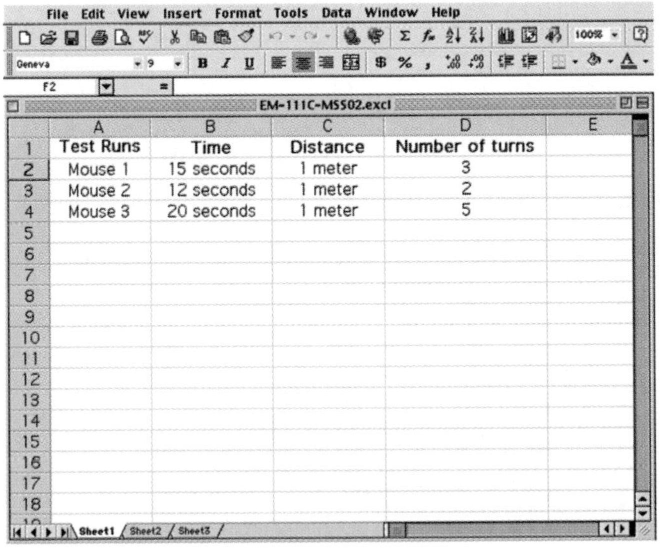

File Edit View Insert Format Tools Data Window Help

EM-111C-MSS02.excl

	A	B	C	D	E
1	Test Runs	Time	Distance	Number of turns	
2	Mouse 1	15 seconds	1 meter	3	
3	Mouse 2	12 seconds	1 meter	2	
4	Mouse 3	20 seconds	1 meter	5	

Sheet1 / Sheet2 / Sheet3

Using a Computerized Card Catalog

When you have a report or paper to research, you probably go to the library. To find the information you need in the library, you might have to use a computerized card catalog. This type of card catalog allows you to search for information by subject, by title, or by author. The computer then will display all the holdings the library has on the subject, title, or author requested.

A library's holdings can include books, magazines, databases, videos, and audio materials. When you have chosen something from this list, the computer will show whether an item is available and where in the library to find it.

Helpful Hints

- Remember that you can use the computer to search by subject, author, or title. If you know a book's author but not the title, you can search for all the books the library has by that author.
- When searching by subject, it's often most helpful to narrow your search by using specific search terms, such as *and*, *or*, and *not*. If you don't find enough sources, you can broaden your search.
- Pay attention to the type of materials found in your search. If you need a book, you can eliminate any videos or other resources that come up in your search.
- Knowing how your library is arranged can save you a lot of time. The librarian will show you where certain types of materials are kept and how to find specific holdings.

Using Graphics Software

Are you having trouble finding that exact piece of art you're looking for? Do you have a picture in your mind of what you want but can't seem to find the right graphic to represent your ideas? To solve these problems, you can use graphics software. Graphics software allows you to create and change images and diagrams in almost unlimited ways. Typical uses for graphics software include arranging clip art, changing scanned images, and constructing pictures from scratch. Most graphics software applications work in similar ways. They use the same basic tools and functions. Once you master one graphics application, you can use any other graphics application relatively easily.

Figure 20
Graphics software can use your data to draw bar graphs.

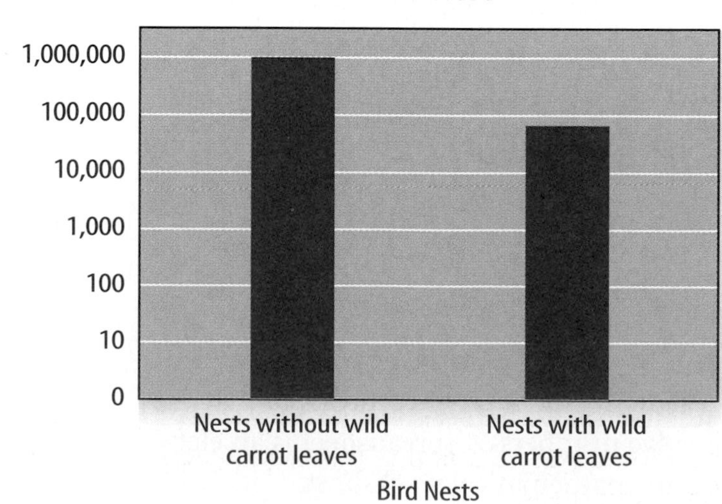

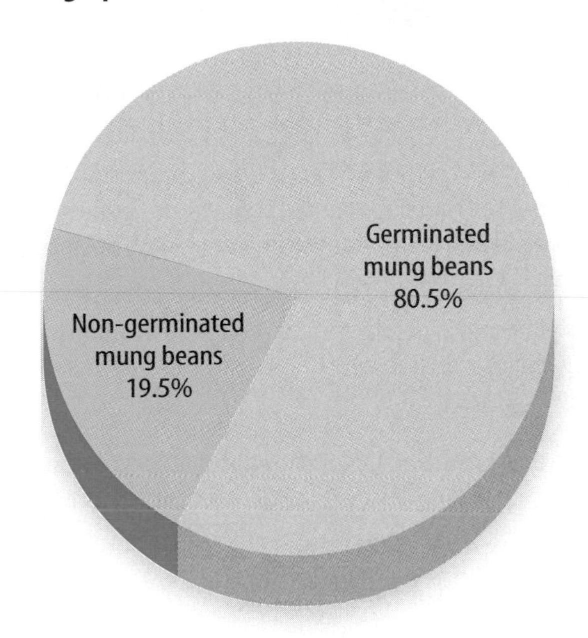

Figure 21
Graphics software can use your data to draw circle graphs.

Helpful Hints

- As with any method of drawing, the more you practice using the graphics software, the better your results will be.
- Start by using the software to manipulate existing drawings. Once you master this, making your own illustrations will be easier.
- Clip art is available on CD-ROMs and the Internet. With these resources, finding a piece of clip art to suit your purposes is simple.
- As you work on a drawing, save it often.

Developing Multimedia Presentations

It's your turn—you have to present your science report to the entire class. How do you do it? You can use many different sources of information to get the class excited about your presentation. Posters, videos, photographs, sound, computers, and the Internet can help show your ideas.

First, determine what important points you want to make in your presentation. Then, write an outline of what materials and types of media would best illustrate those points. Maybe you could start with an outline on an overhead projector, then show a video, followed by something from the Internet or a slide show accompanied by music or recorded voices. You might choose to use a presentation builder computer application that can combine all these elements into one presentation. Make sure the presentation is well constructed to make the most impact on the audience.

Figure 22
Multimedia presentations use many types of print and electronic materials.

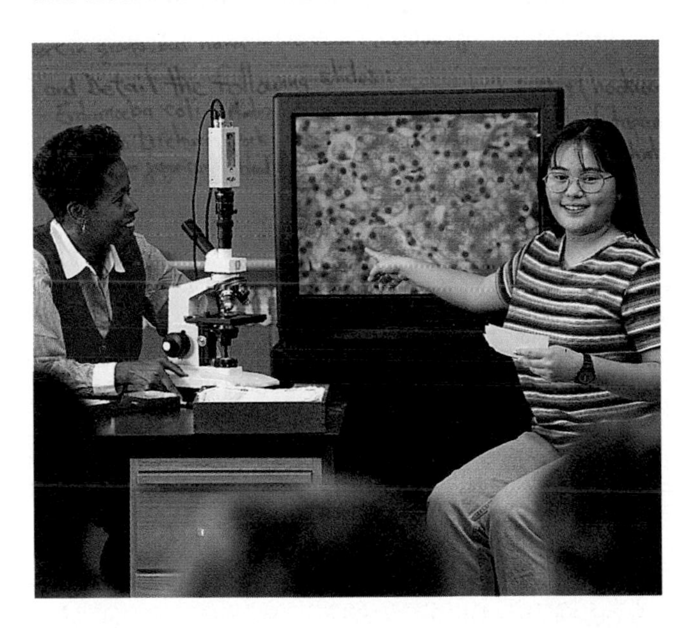

Helpful Hints

- Carefully consider what media will best communicate the point you are trying to make.
- Make sure you know how to use any equipment you will be using in your presentation.
- Practice the presentation several times.
- If possible, set up all of the equipment ahead of time. Make sure everything is working correctly.

Use this Math Skill Handbook to help solve problems you are given in this text. You might find it useful to review topics in this Math Skill Handbook first.

Converting Units

In science, quantities such as length, mass, and time sometimes are measured using different units. Suppose you want to know how many miles are in 12.7 km.

Conversion factors are used to change from one unit of measure to another. A conversion factor is a ratio that is equal to one. For example, there are 1,000 mL in 1 L, so 1,000 mL equals 1 L, or:

$$1{,}000 \text{ mL} = 1 \text{ L}$$

If both sides are divided by 1 L, this equation becomes:

$$\frac{1{,}000 \text{ mL}}{1 \text{ L}} = 1$$

The **ratio** on the left side of this equation is equal to 1 and is a conversion factor. You can make another conversion factor by dividing both sides of the top equation by 1,000 mL:

$$1 = \frac{1 \text{ L}}{1{,}000 \text{ mL}}$$

To **convert units,** you multiply by the appropriate conversion factor. For example, how many milliliters are in 1.255 L? To convert 1.255 L to milliliters, multiply 1.255 L by a conversion factor.

Use the **conversion factor** with new units (mL) in the numerator and the old units (L) in the denominator.

$$1.255 \cancel{\text{L}} \times \frac{1{,}000 \text{ mL}}{1 \cancel{\text{L}}} = 1{,}255 \text{ mL}$$

The unit L divides in this equation, just as if it were a number.

Example 1 There are 2.54 cm in 1 inch. If a meterstick has a length of 100 cm, how long is the meterstick in inches?

Step 1 Decide which conversion factor to use. You know the length of the meterstick in centimeters, so centimeters are the old units. You want to find the length in inches, so inch is the new unit.

Step 2 Form the conversion factor. Start with the relationship between the old and new units.

$$2.54 \text{ cm} = 1 \text{ inch}$$

Step 3 Form the conversion factor with the old unit (centimeter) on the bottom by dividing both sides by 2.54 cm.

$$1 = \frac{2.54 \text{ cm}}{2.54 \text{ cm}} = \frac{1 \text{ inch}}{2.54 \text{ cm}}$$

Step 4 Multiply the old measurement by the conversion factor.

$$100 \cancel{\text{cm}} \times \frac{1 \text{ inch}}{2.54 \cancel{\text{cm}}} = 39.37 \text{ inches}$$

The meterstick is 39.37 inches long.

Example 2 There are 365 days in one year. If a person is 14 years old, what is his or her age in days? (Ignore leap years)

Step 1 Decide which conversion factor to use. You want to convert years to days.

Step 2 Form the conversion factor. Start with the relation between the old and new units.

$$1 \text{ year} = 365 \text{ days}$$

Step 3 Form the conversion factor with the old unit (year) on the bottom by dividing both sides by 1 year.

$$1 = \frac{1 \text{ year}}{1 \text{ year}} = \frac{365 \text{ days}}{1 \text{ year}}$$

Step 4 Multiply the old measurement by the conversion factor:

$$14 \cancel{\text{years}} \times \frac{365 \text{ days}}{1 \cancel{\text{year}}} = 5{,}110 \text{ days}$$

The person's age is 5,110 days.

Practice Problem A cat has a mass of 2.31 kg. If there are 1,000 g in 1 kg, what is the mass of the cat in grams?

Using Fractions

A **fraction** is a number that compares a part to the whole. For example, in the fraction $\frac{2}{3}$, the 2 represents the part and the 3 represents the whole. In the fraction $\frac{2}{3}$, the top number, 2, is called the numerator. The bottom number, 3, is called the denominator.

Sometimes fractions are not written in their simplest form. To determine a fraction's **simplest form,** you must find the greatest common factor (GCF) of the numerator and denominator. The greatest common factor is the largest common factor of all the factors the two numbers have in common.

For example, because the number 3 divides into 12 and 30 evenly, it is a common factor of 12 and 30. However, because the number 6 is the largest number that evenly divides into 12 and 30, it is the **greatest common factor.**

After you find the greatest common factor, you can write a fraction in its simplest form. Divide both the numerator and the denominator by the greatest common factor. The number that results is the fraction in its **simplest form.**

Example Twelve of the 20 corn plants in a field are more than 1.5 m tall. What fraction of the corn plants in the field is 1.5 m tall?

Step 1 Write the fraction.

$$\frac{\text{part}}{\text{whole}} = \frac{12}{20}$$

Step 2 To find the GCF of the numerator and denominator, list all of the factors of each number.

Factors of 12: 1, 2, 3, 4, 6, 12 (the numbers that divide evenly into 12)

Factors of 20: 1, 2, 4, 5, 10, 20 (the numbers that divide evenly into 20)

Step 3 List the common factors.

1, 2, 4.

Step 4 Choose the greatest factor in the list of common factors.

The GCF of 12 and 20 is 4.

Step 5 Divide the numerator and denominator by the GCF.

$$\frac{12 \div 4}{20 \div 4} = \frac{3}{5}$$

In the field, $\frac{3}{5}$ of the corn plants are more than 1.5 m tall.

Practice Problem There are 90 duck eggs in a population. Of those eggs, 66 hatch over a one-week period. What fraction of the eggs hatch over a one-week period? Write the fraction in simplest form.

Skill Handbooks

Calculating Ratios

A **ratio** is a comparison of two numbers by division.

Ratios can be written 3 to 5 or 3:5. Ratios also can be written as fractions, such as $\frac{3}{5}$. Ratios, like fractions, can be written in simplest form. Recall that a fraction is in **simplest form** when the greatest common factor (GCF) of the numerator and denominator is 1.

Example From a package of sunflower seeds, 40 seeds germinated and 64 did not. What is the ratio of germinated to not germinated seeds as a fraction in simplest form?

Step 1 Write the ratio as a fraction.

$$\frac{\text{germinated}}{\text{not germinated}} = \frac{40}{64}$$

Step 2 Express the fraction in simplest form. The GCF of 40 and 64 is 8.

$$\frac{40}{64} = \frac{40 \div 8}{64 \div 8} = \frac{5}{8}$$

The ratio of germinated to not germinated seeds is $\frac{5}{8}$.

Practice Problem Two children measure 100 cm and 144 cm in height. What is the ratio of their heights in simplest fraction form?

Using Decimals

A **decimal** is a fraction with a denominator of 10, 100, 1,000, or another power of 10. For example, 0.854 is the same as the fraction $\frac{854}{1,000}$.

In a decimal, the decimal point separates the ones place and the tenths place. For example, 0.27 means twenty-seven hundredths, or $\frac{27}{100}$, where 27 is the **number of units** out of 100 units. Any fraction can be written as a decimal using division.

Example Write $\frac{5}{8}$ as a decimal.

Step 1 Write a division problem with the numerator, 5, as the dividend and the denominator, 8, as the divisor. Write 5 as 5.000.

Step 2 Solve the problem.

```
   0.625
8)5.000
  48
   20
   16
    40
    40
     0
```

Therefore, $\frac{5}{8} = 0.625$.

Practice Problem Write $\frac{19}{25}$ as a decimal.

Using Percentages

The word *percent* means "out of one hundred." A **percent** is a ratio that compares a number to 100. Suppose you read that 77 percent of all fish on Earth live in the Pacific Ocean. That is the same as reading that the Earth's fish that live in the Pacific Ocean is $\frac{77}{100}$. To express a fraction as a percent, first find an equivalent decimal for the fraction. Then, multiply the decimal by 100 and add the percent symbol. For example, $\frac{1}{2} = 1 \div 2 = 0.5$. Then $0.5 \cdot 100 = 50 = 50\%$.

Example Express $\frac{13}{20}$ as a percent.

Step 1 Find the equivalent decimal for the fraction.

$$\begin{array}{r} 0.65 \\ 20\overline{)13.00} \\ \underline{12\,0} \\ 1\,00 \\ \underline{1\,00} \\ 0 \end{array}$$

Step 2 Rewrite the fraction $\frac{13}{20}$ as 0.65.

Step 3 Multiply 0.65 by 100 and add the % sign.

$0.65 \cdot 100 = 65 = 65\%$

So, $\frac{13}{20} = 65\%$.

Practice Problem In an experimental population of 365 sheep, 73 were brown. What percent of the sheep were brown?

Using Precision and Significant Digits

When you make a **measurement,** the value you record depends on the precision of the measuring instrument. When adding or subtracting numbers with different precision, the answer is rounded to the smallest number of decimal places of any number in the sum or difference. When multiplying or dividing, the answer is rounded to the smallest number of significant figures of any number being multiplied or divided. When counting the number of **significant figures,** all digits are counted except zeros at the end of a number with no decimal such as 2,500, and zeros at the beginning of a decimal such as 0.03020.

Example The lengths 5.28 and 5.2 are measured in meters. Find the sum of these lengths and report the sum using the least precise measurement.

Step 1 Find the sum.

5.28 m	2 digits after the decimal
+ 5.2 m	1 digit after the decimal
10.48 m	

Step 2 Round to one digit after the decimal because the least number of digits after the decimal of the numbers being added is 1.

The sum is 10.5 m.

Practice Problem Multiply the numbers in the example using the rule for multiplying and dividing. Report the answer with the correct number of significant figures.

Solving One-Step Equations

An **equation** is a statement that two things are equal. For example, $A = B$ is an equation that states that A is equal to B.

Sometimes one side of the equation will contain a **variable** whose value is not known. In the equation $3x = 12$, the variable is x.

The equation is solved when the variable is replaced with a value that makes both sides of the equation equal to each other. For example, the solution of the equation $3x = 12$ is $x = 4$. If the x is replaced with 4, then the equation becomes $3 \cdot 4 = 12$, or $12 = 12$.

To solve an equation such as $8x = 40$, divide both sides of the equation by the number that multiplies the variable.

$$8x = 40$$
$$\frac{8x}{8} = \frac{40}{8}$$
$$x = 5$$

You can check your answer by replacing the variable with your solution and seeing if both sides of the equation are the same.

$$8x = 8 \cdot 5 = 40$$

The left and right sides of the equation are the same, so $x = 5$ is the solution.

Sometimes an equation is written in this way: $a = bc$. This also is called a **formula.** The letters can be replaced by numbers, but the numbers must still make both sides of the equation the same.

Example 1 Solve the equation $10x = 35$.

Step 1 Find the solution by dividing each side of the equation by 10.

$$10x = 35 \qquad \frac{10x}{10} = \frac{35}{10} \qquad x = 3.5$$

Step 2 Check the solution.

$$10x = 35 \qquad 10 \times 3.5 = 35 \qquad 35 = 35$$

Both sides of the equation are equal, so $x = 3.5$ is the solution to the equation.

Example 2 In the formula $a = bc$, find the value of c if $a = 20$ and $b = 2$.

Step 1 Rearrange the formula so the unknown value is by itself on one side of the equation by dividing both sides by b.

$$a = bc$$
$$\frac{a}{b} = \frac{bc}{b}$$
$$\frac{a}{b} = c$$

Step 2 Replace the variables a and b with the values that are given.

$$\frac{a}{b} = c$$
$$\frac{20}{2} = c$$
$$10 = c$$

Step 3 Check the solution.

$$a = bc$$
$$20 = 2 \times 10$$
$$20 = 20$$

Both sides of the equation are equal, so $c = 10$ is the solution when $a = 20$ and $b = 2$.

Practice Problem In the formula $h = gd$, find the value of d if $g = 12.3$ and $h = 17.4$.

Using Proportions

A **proportion** is an equation that shows that two ratios are equivalent. The ratios $\frac{2}{4}$ and $\frac{5}{10}$ are equivalent, so they can be written as $\frac{2}{4} = \frac{5}{10}$. This equation is an example of a proportion.

When two ratios form a proportion, the **cross products** are equal. To find the cross products in the proportion $\frac{2}{4} = \frac{5}{10}$, multiply the 2 and the 10, and the 4 and the 5. Therefore **$2 \cdot 10 = 4 \cdot 5$, or $20 = 20$.**

Because you know that both proportions are equal, you can use cross products to find a missing term in a proportion. This is known as **solving the proportion.** Solving a proportion is similar to solving an equation.

Example The heights of a tree and a pole are proportional to the lengths of their shadows. The tree casts a shadow of 24 m at the same time that a 6-m pole casts a shadow of 4 m. What is the height of the tree?

Step 1 Write a proportion.

$$\frac{\text{height of tree}}{\text{height of pole}} = \frac{\text{length of tree's shadow}}{\text{length of pole's shadow}}$$

Step 2 Substitute the known values into the proportion. Let h represent the unknown value, the height of the tree.

$$\frac{h}{6} = \frac{24}{4}$$

Step 3 Find the cross products.

$$h \cdot 4 = 6 \cdot 24$$

Step 4 Simplify the equation.

$$4h = 144$$

Step 5 Divide each side by 4.

$$\frac{4h}{4} = \frac{144}{4}$$

$$h = 36$$

The height of the tree is 36 m.

Practice Problem The proportions of bluefish are stable by the time they reach a length of 30 cm. The distance from the tip of the mouth to the back edge of the gill cover in a 35-cm bluefish is 15 cm. What is the distance from the tip of the mouth to the back edge of the gill cover in a 59-cm bluefish?

Using Statistics

Statistics is the branch of mathematics that deals with collecting, analyzing, and presenting data. In statistics, there are three common ways to summarize the data with a single number—the mean, the median, and the mode.

The **mean** of a set of data is the arithmetic average. It is found by adding the numbers in the data set and dividing by the number of items in the set.

The **median** is the middle number in a set of data when the data are arranged in numerical order. If there were an even number of data points, the median would be the mean of the two middle numbers.

The **mode** of a set of data is the number or item that appears most often.

Another number that often is used to describe a set of data is the range. The **range** is the difference between the largest number and the smallest number in a set of data.

A **frequency table** shows how many times each piece of data occurs, usually in a survey. The frequency table below shows the results of a student survey on favorite color.

Color	Tally	Frequency
red	||||	4
blue	𝍸	5
black	||	2
green	|||	3
purple	𝍸 ||	7
yellow	𝍸 |	6

Based on the frequency table data, which color is the favorite?

Example The high temperatures (in °C) on five consecutive days in a desert habitat under study are 39°, 37°, 44°, 36°, and 44°. Find the mean, median, mode, and range of this set.

To find the mean:

Step 1 Find the sum of the numbers.

$$39 + 37 + 44 + 36 + 44 = 200$$

Step 2 Divide the sum by the number of items, which is 5.

$$200 \div 5 = 40$$

The mean high temperature is 40°C.

To find the median:

Step 1 Arrange the temperatures from least to greatest.

$$36,\ 37,\ \underline{39},\ 44,\ 44$$

Step 2 Determine the middle temperature.

The median high temperature is 39°C.

To find the mode:

Step 1 Group the numbers that are the same together.

44, 44, 36, 37, 39

Step 2 Determine the number that occurs most in the set.

$\underline{44, 44}$, 36, 37, 39

The mode measure is 44°C.

To find the range:

Step 1 Arrange the temperatures from largest to smallest.

44, 44, 39, 37, 36

Step 2 Determine the largest and smallest temperature in the set.

$\underline{44}$, 44, 39, 37, $\underline{36}$

Step 3 Find the difference between the largest and smallest temperatures.

$$44 - 36 = 8$$

The range is 8°C.

Practice Problem Find the mean, median, mode, and range for the data set 8, 4, 12, 8, 11, 14, 16.

Safety in the Science Classroom

1. Always obtain your teacher's permission to begin an investigation.
2. Study the procedure. If you have questions, ask your teacher. Be sure you understand any safety symbols shown on the page.
3. Use the safety equipment provided for you. Goggles and a safety apron should be worn during most investigations.
4. Always slant test tubes away from yourself and others when heating them or adding substances to them.
5. Never eat or drink in the lab, and never use lab glassware as food or drink containers. Never inhale chemicals. Do not taste any substances or draw any material into a tube with your mouth.
6. Report any spill, accident, or injury, no matter how small, immediately to your teacher, then follow his or her instructions.
7. Know the location and proper use of the fire extinguisher, safety shower, fire blanket, first aid kit, and fire alarm.
8. Keep all materials away from open flames. Tie back long hair and tie down loose clothing.
9. If your clothing should catch fire, smother it with the fire blanket, or get under a safety shower. NEVER RUN.
10. If a fire should occur, turn off the gas then leave the room according to established procedures.

Follow these procedures as you clean up your work area

1. Turn off the water and gas. Disconnect electrical devices.
2. Clean all pieces of equipment and return all materials to their proper places.
3. Dispose of chemicals and other materials as directed by your teacher. Place broken glass and solid substances in the proper containers. Make sure never to discard materials in the sink.
4. Clean your work area. Wash your hands thoroughly after working in the laboratory.

First Aid

Injury	Safe Response ALWAYS NOTIFY YOUR TEACHER IMMEDIATELY
Burns	Apply cold water.
Cuts and Bruises	Stop any bleeding by applying direct pressure. Cover cuts with a clean dressing. Apply ice packs or cold compresses to bruises.
Fainting	Leave the person lying down. Loosen any tight clothing and keep crowds away.
Foreign Matter in Eye	Flush with plenty of water. Use eyewash bottle or fountain.
Poisoning	Note the suspected poisoning agent.
Any Spills on Skin	Flush with large amounts of water or use safety shower.

Reference Handbook B

Reference Handbook

SI—Metric/English, English/Metric Conversions			
	When you want to convert:	To:	Multiply by:
Length	inches	centimeters	2.54
	centimeters	inches	0.39
	yards	meters	0.91
	meters	yards	1.09
	miles	kilometers	1.61
	kilometers	miles	0.62
Mass and Weight*	ounces	grams	28.35
	grams	ounces	0.04
	pounds	kilograms	0.45
	kilograms	pounds	2.2
	tons (short)	tonnes (metric tons)	0.91
	tonnes (metric tons)	tons (short)	1.10
	pounds	newtons	4.45
	newtons	pounds	0.22
Volume	cubic inches	cubic centimeters	16.39
	cubic centimeters	cubic inches	0.06
	liters	quarts	1.06
	quarts	liters	0.95
	gallons	liters	3.78
Area	square inches	square centimeters	6.45
	square centimeters	square inches	0.16
	square yards	square meters	0.83
	square meters	square yards	1.19
	square miles	square kilometers	2.59
	square kilometers	square miles	0.39
	hectares	acres	2.47
	acres	hectares	0.40
Temperature	To convert °Celsius to °Fahrenheit		$°C \times 9/5 + 32$
	To convert °Fahrenheit to °Celsius		$5/9\,(°F - 32)$

*Weight is measured in standard Earth gravity.

Care and Use of a Microscope

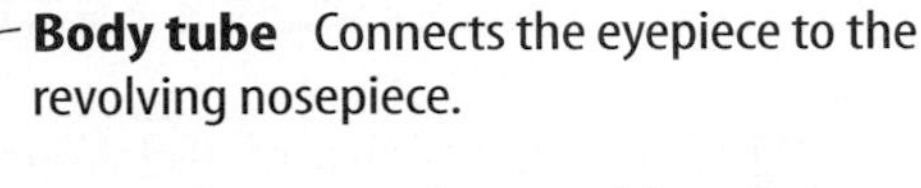

Eyepiece Contains magnifying lenses you look through.

Arm Supports the body tube.

Low-power objective Contains the lens with the lowest power magnification.

Stage clips Hold the microscope slide in place.

Fine adjustment Sharpens the image under high magnification.

Coarse adjustment Focuses the image under low power.

Body tube Connects the eyepiece to the revolving nosepiece.

Revolving nosepiece Holds and turns the objectives into viewing position.

High-power objective Contains the lens with the highest magnification.

Stage Supports the microscope slide.

Light source Provides light that passes upward through the diaphragm, the specimen, and the lenses.

Base Provides support for the microscope.

Caring for a Microscope

1. Always carry the microscope holding the arm with one hand and supporting the base with the other hand.
2. Don't touch the lenses with your fingers.
3. The coarse adjustment knob is used only when looking through the lowest-power objective lens. The fine adjustment knob is used when the high-power objective is in place.
4. Cover the microscope when you store it.

Using a Microscope

1. Place the microscope on a flat surface that is clear of objects. The arm should be toward you.
2. Look through the eyepiece. Adjust the diaphragm so light comes through the opening in the stage.
3. Place a slide on the stage so the specimen is in the field of view. Hold it firmly in place by using the stage clips.
4. Always focus with the coarse adjustment and the low-power objective lens first. After the object is in focus on low power, turn the nosepiece until the high-power objective is in place. Use ONLY the fine adjustment to focus with the high-power objective lens.

Making a Wet-Mount Slide

1. Carefully place the item you want to look at in the center of a clean, glass slide. Make sure the sample is thin enough for light to pass through.
2. Use a dropper to place one or two drops of water on the sample.
3. Hold a clean coverslip by the edges and place it at one edge of the water. Slowly lower the coverslip onto the water until it lies flat.
4. If you have too much water or a lot of air bubbles, touch the edge of a paper towel to the edge of the coverslip to draw off extra water and draw out unwanted air.

Diversity of Life: Classification of Living Organisms

A six-kingdom system of classification of organisms is used today. Two kingdoms—Kingdom Archaebacteria and Kingdom Eubacteria—contain organisms that do not have a nucleus and that lack membrane-bound structures in the cytoplasm of their cells. The members of the other four kingdoms have a cell or cells that contain a nucleus and structures in the cytoplasm, some of which are surrounded by membranes. These kingdoms are Kingdom Protista, Kingdom Fungi, Kingdom Plantae, and Kingdom Animalia.

Kingdom Archaebacteria

one-celled; some absorb food from their surroundings; some are photosynthetic; some are chemosynthetic; many are found in extremely harsh environments including salt ponds, hot springs, swamps, and deep-sea hydrothermal vents

Kingdom Eubacteria

one-celled; most absorb food from their surroundings; some are photosynthetic; some are chemosynthetic; many are parasites; many are round, spiral, or rod-shaped; some form colonies

Kingdom Protista

Phylum Euglenophyta one-celled; photosynthetic or take in food; most have one flagellum; euglenoids

Phylum Bacillariophyta one-celled; photosynthetic; have unique double shells made of silica; diatoms

Phylum Dinoflagellata one-celled; photosynthetic; contain red pigments; have two flagella; dinoflagellates

Phylum Chlorophyta one-celled, many-celled, or colonies; photosynthetic; contain chlorophyll; live on land, in freshwater, or salt water; green algae

Phylum Rhodophyta most are many-celled; photosynthetic; contain red pigments; most live in deep, saltwater environments; red algae

Phylum Phaeophyta most are many-celled; photosynthetic; contain brown pigments; most live in saltwater environments; brown algae

Phylum Rhizopoda one-celled; take in food; are free-living or parasitic; move by means of pseudopods; amoebas

Kingdom Eubacteria
Bacillus anthracis

Phylum Chlorophyta
Desmids

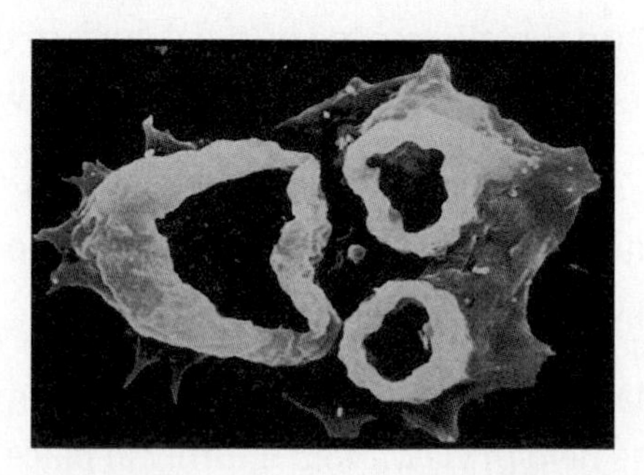

Amoeba

Phylum Zoomastigina one-celled; take in food; free-living or parasitic; have one or more flagella; zoomastigotes

Phylum Ciliophora one-celled; take in food; have large numbers of cilia; ciliates

Phylum Sporozoa one-celled; take in food; have no means of movement; are parasites in animals; sporozoans

Phylum Myxomycota
Slime mold

Phylum Oomycota
Phytophthora infestans

Phyla Myxomycota and Acrasiomycota one- or many-celled; absorb food; change form during life cycle; cellular and plasmodial slime molds

Phylum Oomycota many-celled; are either parasites or decomposers; live in freshwater or salt water; water molds, rusts and downy mildews

Kingdom Fungi

Phylum Zygomycota many-celled; absorb food; spores are produced in sporangia; zygote fungi; bread mold

Phylum Ascomycota one- and many-celled; absorb food; spores produced in asci; sac fungi; yeast

Phylum Basidiomycota many-celled; absorb food; spores produced in basidia; club fungi; mushrooms

Phylum Deuteromycota members with unknown reproductive structures; imperfect fungi; *Penicillium*

Mycophycota organisms formed by symbiotic relationship between an ascomycote or a basidiomycote and green alga or cyanobacterium; lichens

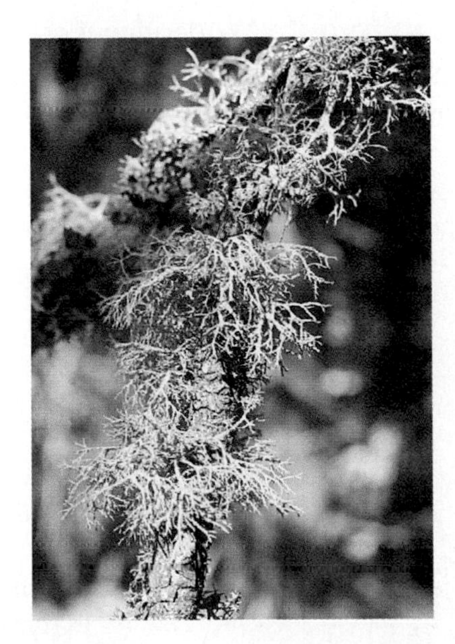

Lichens

Kingdom Plantae

Divisions Bryophyta (mosses), **Anthocerophyta** (hornworts), **Hepatophytal** (liverworts), **Psilophytal** (whisk ferns) many-celled nonvascular plants; reproduce by spores produced in capsules; green; grow in moist, land environments

Division Lycophyta many-celled vascular plants; spores are produced in conelike structures; live on land; are photosynthetic; club mosses

Division Sphenophyta vascular plants; ribbed and jointed stems; scalelike leaves; spores produced in conelike structures; horsetails

Division Pterophyta vascular plants; leaves called fronds; spores produced in clusters of sporangia called sori; live on land or in water; ferns

Division Ginkgophyta deciduous trees; only one living species; have fan-shaped leaves with branching veins and fleshy cones with seeds; ginkgoes

Division Cycadophyta palmlike plants; have large, featherlike leaves; produces seeds in cones; cycads

Division Coniferophyta deciduous or evergreen; trees or shrubs; have needlelike or scalelike leaves; seeds produced in cones; conifers

Division Anthophyta
Tomato plant

Division Gnetophyta shrubs or woody vines; seeds are produced in cones; division contains only three genera; gnetum

Division Anthophyta dominant group of plants; flowering plants; have fruits with seeds

Kingdom Animalia

Phylum Porifera aquatic organisms that lack true tissues and organs; are asymmetrical and sessile; sponges

Phylum Cnidaria radially symmetrical organisms; have a digestive cavity with one opening; most have tentacles armed with stinging cells; live in aquatic environments singly or in colonies; includes jellyfish, corals, hydra, and sea anemones

Phylum Platyhelminthes bilaterally symmetrical worms; have flattened bodies; digestive system has one opening; parasitic and free-living species; flatworms

Division Bryophyta
Liverwort

Phylum Platyhelminthes
Flatworm

Reference Handbook D

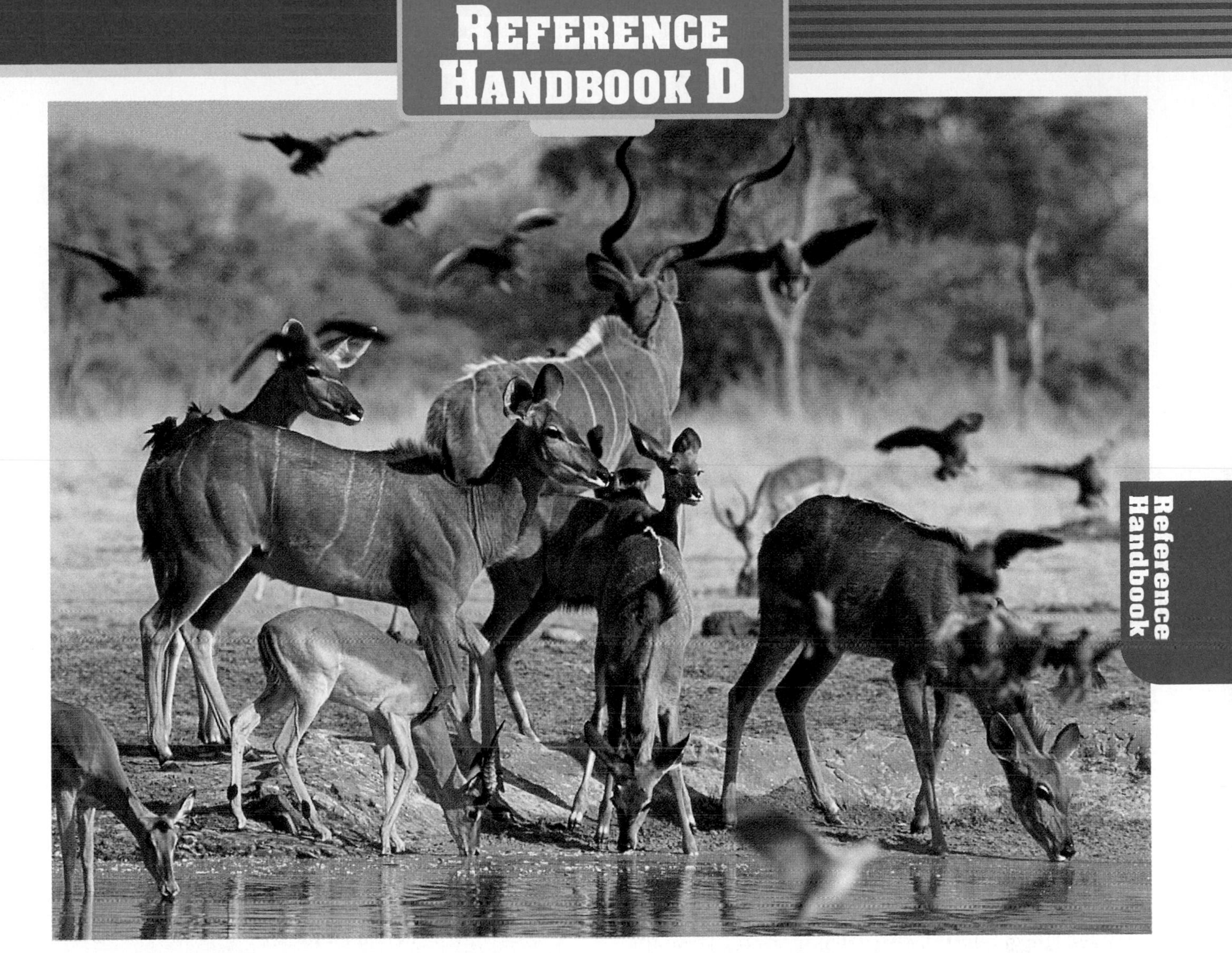

Phylum Chordata

Phylum Nematoda round, bilaterally symmetrical body; have digestive system with two openings; free-living forms and parasitic forms; roundworms

Phylum Mollusca soft-bodied animals, many with a hard shell and soft foot or footlike appendage; a mantle covers the soft body; aquatic and terrestrial species; includes clams, snails, squid, and octopuses

Phylum Annelida bilaterally symmetrical worms; have round, segmented bodies; terrestrial and aquatic species; includes earthworms, leeches, and marine polychaetes

Phylum Arthropoda largest animal group; have hard exoskeletons, segmented bodies, and pairs of jointed appendages; land and aquatic species; includes insects, crustaceans, and spiders

Phylum Echinodermata marine organisms; have spiny or leathery skin and a water-vascular system with tube feet; are radially symmetrical; includes sea stars, sand dollars, and sea urchins

Phylum Chordata organisms with internal skeletons and specialized body systems; most have paired appendages; all at some time have a notochord, nerve cord, gill slits, and a postanal tail; include fish, amphibians, reptiles, birds, and mammals

Topographic Map Symbols

Symbol	Description	Symbol	Description
	Primary highway, hard surface		Index contour
	Secondary highway, hard surface		Supplementary contour
	Light-duty road, hard or improved surface		Intermediate contour
	Unimproved road		Depression contours
	Railroad: single track		
	Railroad: multiple track		Boundaries: national
	Railroads in juxtaposition		State
			County, parish, municipal
	Buildings		Civil township, precinct, town, barrio
cem	Schools, church, and cemetery		Incorporated city, village, town, hamlet
	Buildings (barn, warehouse, etc)		Reservation, national or state
	Wells other than water (labeled as to type)		Small park, cemetery, airport, etc.
	Tanks: oil, water, etc. (labeled only if water)		Land grant
	Located or landmark object; windmill		Township or range line, U.S. land survey
	Open pit, mine, or quarry; prospect		
			Township or range line, approximate location
	Marsh (swamp)		
	Wooded marsh		Perennial streams
	Woods or brushwood		Elevated aqueduct
	Vineyard		Water well and spring
	Land subject to controlled inundation		Small rapids
	Submerged marsh		Large rapids
	Mangrove		Intermittent lake
	Orchard		Intermittent stream
	Scrub		Aqueduct tunnel
	Urban area		Glacier
			Small falls
x7369	Spot elevation		Large falls
670	Water elevation		Dry lake bed

Rocks		
Rock Type	**Rock Name**	**Characteristics**
Igneous (intrusive)	Granite	Large mineral grains of quartz, feldspar, hornblende, and mica. Usually light in color.
	Diorite	Large mineral grains of feldspar, hornblende, and mica. Less quartz than granite. Intermediate in color.
	Gabbro	Large mineral grains of feldspar, augite, and olivine. No quartz. Dark in color.
Igneous (extrusive)	Rhyolite	Small mineral grains of quartz, feldspar, hornblende, and mica, or no visible grains. Light in color.
	Andesite	Small mineral grains of feldspar, hornblende, and mica or no visible grains. Intermediate in color.
	Basalt	Small mineral grains of feldspar, augite, and olivine or no visible grains. No quartz. Dark in color.
	Obsidian	Glassy texture. No visible grains. Volcanic glass. Fracture looks like broken glass.
	Pumice	Frothy texture. Floats in water. Usually light in color.
Sedimentary (detrital)	Conglomerate	Coarse grained. Gravel or pebble size grains.
	Sandstone	Sand-sized grains 1/16 to 2 mm.
	Siltstone	Grains are smaller than sand but larger than clay.
	Shale	Smallest grains. Often dark in color. Usually platy.
Sedimentary (chemical or organic)	Limestone	Major mineral is calcite. Usually forms in oceans, lakes, and caves. Often contains fossils.
	Coal	Occurs in swampy areas. Compacted layers of organic material, mainly plant remains.
Sedimentary (chemical)	Rock Salt	Commonly forms by the evaporation of seawater.
Metamorphic (foliated)	Gneiss	Banding due to alternate layers of different minerals, of different colors. Parent rock often is granite.
	Schist	Parallel arrangement of sheetlike minerals, mainly micas. Forms from different parent rocks.
	Phyllite	Shiny or silky appearance. May look wrinkled. Common parent rocks are shale and slate.
	Slate	Harder, denser, and shinier than shale. Common parent rock is shale.
Metamorphic (non-foliated)	Marble	Calcite or dolomite. Common parent rock is limestone.
	Soapstone	Mainly of talc. Soft with greasy feel.
	Quartzite	Hard with interlocking quartz crystals. Common parent rock is sandstone.

Reference Handbook G

Minerals					
Mineral (formula)	**Color**	**Streak**	**Hardness**	**Breakage Pattern**	**Uses and Other Properties**
Graphite (C)	black to gray	black to gray	1–1.5	basal cleavage (scales)	pencil lead, lubricants for locks, rods to control some small nuclear reactions, battery poles
Galena (PbS)	gray	gray to black	2.5	cubic cleavage perfect	source of lead, used for pipes, shields for X rays, fishing equipment sinkers
Hematite (Fe_2O_3)	black or reddish-brown	reddish-brown	5.5–6.5	irregular fracture	source of iron; converted to pig iron, made into steel
Magnetite (Fe_3O_4)	black	black	6	conchoidal fracture	source of iron, attracts a magnet
Pyrite (FeS_2)	light, brassy, yellow	greenish-black	6–6.5	uneven fracture	fool's gold
Talc ($Mg_3 Si_4O_{10} (OH)_2$)	white, greenish	white	1	cleavage in one direction	used for talcum powder, sculptures, paper, and tabletops
Gypsum ($CaSO_4 \cdot 2H_2O$)	colorless, gray, white, brown	white	2	basal cleavage	used in plaster of paris and dry wall for building construction
Sphalerite (ZnS)	brown, reddish-brown, greenish	light to dark brown	3.5–4	cleavage in six directions	main ore of zinc; used in paints, dyes, and medicine
Muscovite ($KAl_3Si_3 O_{10}(OH)_2$)	white, light gray, yellow, rose, green	colorless	2–2.5	basal cleavage	occurs in large, flexible plates; used as an insulator in electrical equipment, lubricant
Biotite ($K(Mg,Fe)_3 (AlSi_3O_{10}) (OH)_2$)	black to dark brown	colorless	2.5–3	basal cleavage	occurs in large, flexible plates
Halite (NaCl)	colorless, red, white, blue	colorless	2.5	cubic cleavage	salt; soluble in water; a preservative

Reference Handbook G

Minerals

Mineral (formula)	Color	Streak	Hardness	Breakage Pattern	Uses and Other Properties
Calcite ($CaCO_3$)	colorless, white, pale blue	colorless, white	3	cleavage in three directions	fizzes when HCl is added; used in cements and other building materials
Dolomite ($CaMg(CO_3)_2$)	colorless, white, pink, green, gray, black	white	3.5–4	cleavage in three directions	concrete and cement; used as an ornamental building stone
Fluorite (CaF_2)	colorless, white, blue, green, red, yellow, purple	colorless	4	cleavage in four directions	used in the manufacture of optical equipment; glows under ultraviolet light
Hornblende ($(CaNa)_{2-3}(Mg,Al,Fe)_5\text{-}(Al,Si)_2Si_6O_{22}(OH)_2$)	green to black	gray to white	5–6	cleavage in two directions	will transmit light on thin edges; 6-sided cross section
Feldspar ($(KAlSi_3O_8)$ $(NaAlSi_3O_8)$, $(CaAl_2Si_2O_8)$)	colorless, white to gray, green	colorless	6	two cleavage planes meet at 90° angle	used in the manufacture of ceramics
Augite ($((Ca,Na)(Mg,Fe,Al)(Al,Si)_2O_6)$)	black	colorless	6	cleavage in two directions	square or 8-sided cross section
Olivine ($((Mg,Fe)_2SiO_4)$)	olive, green	none	6.5–7	conchoidal fracture	gemstones, refractory sand
Quartz ((SiO_2))	colorless, various colors	none	7	conchoidal fracture	used in glass manufacture, electronic equipment, radios, computers, watches, gemstones

English Glossary

This glossary defines each key term that appears in bold type in the text. It also shows the chapter, section, and page number where you can find the word used.

A

abiotic (ay bi AH tihk): nonliving, physical features of the environment, including air, water, sunlight, soil, temperature, and climate. (Chap. 13, Sec. 1, p. 360)

accuracy: compares a measurement to the true value. (Chap. 2, Sec. 1, p. 41)

active transport: energy-requiring process in which transport proteins bind with particles and move them through a cell membrane. (Chap. 4, Sec. 2, p. 109)

aggression: forceful behavior, such as fighting, used by an animal to control or dominate another animal in order to protect their young, defend territory, or get food. (Chap. 7, Sec. 2, p. 188)

allele (uh LEEL): the different form of a trait that a gene may have. (Chap. 11, Sec. 1, p. 300)

alveoli (al VEE uh li): tiny, thin-walled, grapelike clusters at the end of each bronchiole that are surrounded by capillaries, where carbon dioxide and oxygen exchange takes place. (Chap. 6, Sec. 1, p. 157)

amniotic (am nee AH tihk) **sac:** thin, liquid-filled, protective membrane that forms around the embryo. (Chap. 10, Sec. 3, p. 283)

amplitude: distance a wave rises above or falls below its normal level, which is related to the energy that the wave carries; in a transverse wave, is one-half the distance between a crest and a trough. (Chap. 20, Sec. 2, p. 581)

asexual reproduction: a type of reproduction in which a new organism is produced from one parent and has hereditary material identical to the parent organism. (Chap. 8, Sec. 1, p. 215)

asteroid: small, rocky space object found in the asteroid belt between the orbits of Jupiter and Mars. (Chap. 17, Sec. 3, p. 502)

asthenosphere (as THE nuh sfihr): plasticlike layer of Earth on which the lithospheric plates float and move around. (Chap. 14, Sec. 3, p. 400)

asthma: lung disorder in which the bronchial tubes contract quickly and cause shortness of breath, wheezing, or coughing; may occur as an allergic reaction. (Chap. 6, Sec. 1, p. 162)

astronomical unit: unit of measure used to determine distances between objects in the solar system; 1 AU equals 150,000,000 km. (Chap. 17, Sec. 3, p. 496)

atmosphere: air surrounding Earth; is made up of gases, including 78 percent nitrogen, 21 percent oxygen, and 0.03 percent carbon dioxide. (Chap. 13, Sec. 1, p. 361)

atomic number: the number of protons in an atom. (Chap. 18, Sec. 1, p. 521)

atoms: tiny building blocks of matter, made up of protons, neutrons, and electrons. (Chap. 18, Sec. 1, p. 518)

auxin (AWK sun): plant hormone that causes plant leaves and stems to exhibit positive response to light. (Chap. 5, Sec. 2, p. 140)

axis: imaginary line around which Earth spins, causing day and night; drawn from the north geographic pole through Earth to the south geographic pole. (Chap. 17, Sec. 1, p. 482)

B

bar graph: a type of graph that uses bars of varying sizes to show relationships between variables. (Chap. 2, Sec. 3, p. 54)

basin: low area on Earth in which an ocean formed when the area filled with water from torrential rains. (Chap. 16, Sec. 1, p. 453)

behavior: the way in which an organism interacts with other organisms and its environment; can be innate or learned. (Chap. 7, Sec. 1, p. 180)

biosphere (BI uh sfihr): part of Earth that supports life, including the top portion of Earth's crust, the atmosphere, and all the water on Earth's surface. (Chap. 12, Sec. 1, p. 332)

biotic (bi AHT ik): features of the environment that are alive or were once alive. (Chap. 13, Sec. 1, p. 360)

bladder: elastic, muscular organ that holds urine until it leaves the body. (Chap. 6, Sec. 2, p. 166)

breaker: collapsing ocean wave that forms in shallow water and breaks onto the shore. (Chap. 16, Sec. 3, p. 463)

bronchi (BRAHN ki): two short tubes that branch off the lower end of the trachea and carry air into the lungs. (Chap. 6, Sec. 1, p. 157)

C

carbon cycle: model describing how carbon molecules move between the living and nonliving world. (Chap. 13, Sec. 2, p. 373)

carrier wave: particular transmission frequency assigned to a radio station. (Chap. 22, Sec. 3, p. 650)

carrying capacity: largest number of individuals of a particular species that an ecosystem can support over time. (Chap. 12, Sec. 2, p. 339)

cell membrane: protective outer covering of all cells that is made up of a double layer of fatlike molecules and regulates the interaction between the cell and the environment. (Chap. 3, Sec. 1, p. 70)

cell theory: states that all organisms are made up of one or more cells, the cell is the basic unit of life, and all cells come from other cells. (Chap. 3, Sec. 2, p. 83)

cell wall: rigid structure that encloses, supports, and protects the cells of plants. (Chap. 3, Sec. 1, p. 71)

chemical change: change in which the composition of a substance changes. (Chap. 19, Sec. 2, p. 554)

chemical property: a property of matter that cannot be observed without altering the substance. (Chap. 19, Sec. 1, p. 550)

chemosynthesis (kee moh SIN thuh sus): process in which producers make energy-rich nutrient molecules from chemicals. (Chap. 13, Sec. 3, p. 375)

chlorophyll (KLOR uh fihl): green, light-trapping pigment in plant chloroplasts. (Chap. 5, Sec. 1, p. 130)

chloroplast (KLOR uh plast): green, chlorophyll-containing, plant-cell organelle that converts sunlight, carbon dioxide, and water into sugar. (Chap. 3, Sec. 1, p. 74)

chromosome (KROH muh sohm): structure in a cell's nucleus that contains hereditary material. (Chap. 8, Sec. 1, p. 212)

cinder cone volcano: relatively small volcano formed by moderate to explosive eruptions of tephra. (Chap. 15, Sec. 2, p. 432)

circle graph: a type of graph that shows the parts of a whole; sometimes called a pie graph, each piece of which represents a percentage of the total. (Chap. 2, Sec. 3, p. 54)

climate: average weather conditions of an area over time, including wind, temperature, and rainfall or other types of precipitation. (Chap. 13, Sec. 1, p. 365)

comet: space object made of rocky particles and water ices; forms a tail when orbiting near the Sun and is found mostly in the Kuiper Belt and the Oort Cloud. (Chap. 17, Sec. 3, p. 502)

commensalism (kuh MEN suh lih zum): a type of symbiotic relationship in which one organism benefits and the other organism is not affected. (Chap. 12, Sec. 3, p. 346)

community: all the populations of different species that live in an ecosystem. (Chap. 12, Sec. 1, p. 334)

composite volcano: steep-sided volcano formed from alternating layers of violent eruptions of tephra and quieter eruptions of lava. (Chap. 15, Sec. 2, p. 433)

compound: matter that is made of two or more elements and has physical and chemical properties different from each of the elements that make it up. (Chap. 18, Sec. 2, p. 524)

compressional wave: a type of mechanical wave in which matter in the medium moves forward and backward in the same direction the wave travels. (Chap. 20, Sec. 1, p. 579)

concave lens: lens that is thicker at its edges than in the middle and causes light rays traveling parallel to the optical axis to diverge. (Chap. 23, Sec. 3, p. 679)

condensation: process that takes place when a gas changes to a liquid. (Chap. 13, Sec. 2, p. 369)

conditioning: occurs when the response to a stimulus becomes associated with another stimulus. (Chap. 7, Sec. 1, p. 184)

constant: variable that stays the same during an experiment. (Chap. 1, Sec. 2, p. 21)

consumer: organism that cannot create energy-rich molecules but obtains its food by eating other organisms. (Chap. 12, Sec. 3, p. 345)

continental drift: Wegener's hypothesis that all continents were once connected in a single large landmass that broke apart about 200 million years ago and drifted slowly to their current positions. (Chap. 14, Sec. 1, p. 392)

control: sample that is treated like other experimental groups except that the independent variable is not applied to it. (Chap. 1, Sec. 2, p. 22)

convection current: cycle of heating, rising, cooling, and sinking in Earth's mantle. (Chap. 14, Sec. 3, p. 405)

convex lens: lens that is thicker in the middle than at its edges. (Chap. 23, Sec. 3, p. 678)

Coriolis effect: shifting of winds and surface currents caused by Earth's rotation that turns currents north of the equator clockwise and south of the equator counterclockwise. (Chap. 16, Sec. 2, p. 457)

courtship behavior: behavior that allows males and females of the same species to recognize each other and prepare to mate. (Chap. 7, Sec. 2, p. 189)

crater: depression formed by a large meteorite; the more craters in a region, the older the surface. (Chap. 17, Sec. 2, p. 487)

crest: highest point of a wave. (Chap. 16, Sec. 3, p. 462)

cyclic behavior: innate behavior that occurs in repeated patterns. (Chap. 7, Sec. 2, p. 192)

cytoplasm (SI toh plaz uhm): constantly moving, gel-like mixture inside the cell membrane that contains heredity material. (Chap. 3, Sec. 1, p. 70)

D

day-neutral plant: plant that doesn't require a specific photoperiod and can begin the flowering process over a range of night lengths. (Chap. 5, Sec. 2, p. 142)

density: a physical property of matter that can be determined by dividing the mass of an object by its volume. (Chap. 18, Sec. 3, p. 530)

density current: circulation pattern in the ocean that forms when a mass of more dense seawater sinks beneath less dense seawater. (Chap. 16, Sec. 2, p. 459)

dependent variable: factor that is being measured during an experiment. (Chap. 1, Sec. 2, p. 21)

descriptive research: answers scientific questions through observation. (Chap. 1, Sec. 2, p. 13)

diaphragm (DI uh fram): muscle beneath the lungs that contracts and relaxes to move gases in and out of the body. (Chap. 6, Sec. 1, p. 158)

diffraction: bending of waves around a barrier. (Chap. 20, Sec. 3, p. 589)

diffusion: a type of passive transport in cells in which molecules move from areas where there are more of them to areas where there are fewer of them. (Chap. 4, Sec. 2, p. 107)

diploid (DIH ploid): when a cell has chromosomes in pairs. (Chap. 8, Sec. 2, p. 218)

DNA: deoxyribonucleic acid, which is the genetic material of all organisms, made up of two twisted strands of sugar-phosphate molecules and nitrogen bases. (Chap. 8, Sec. 3, p. 224)

dominant (DAHM uh nunt): describes a trait that covers over, or dominates, another form of that trait. (Chap. 11, Sec. 1, p. 302)

Doppler effect: change in the frequency or pitch of a sound that occurs when the sound source and the listener are in motion relative to each other. (Chap. 21, Sec. 1, p. 610)

E

eardrum: membrane stretching across the ear canal that vibrates when sound waves reach the middle ear. (Chap. 21, Sec. 2, p. 622)

earthquake: movement of the ground that occurs when rocks inside Earth pass their elastic limit, break suddenly, and experience elastic rebound. (Chap. 15, Sec. 1, p. 420)

echo: a reflected sound wave. (Chap. 21, Sec. 1, p. 609)

ecology: study of the interactions that occur among organisms and their environment. (Chap. 12, Sec. 1, p. 333)

ecosystem: all the living organisms that live in an area and the nonliving features of their environment. (Chap. 12, Sec. 1, p. 333)

egg: haploid sex cell formed in the female reproductive organs. (Chap. 8, Sec. 2, p. 218)

electromagnetic spectrum: range of electromagnetic waves with different frequencies and wavelengths. (Chap. 22, Sec. 2, p. 639)

electromagnetic waves: waves that can travel through empty space, have a wide range of wavelengths and frequencies, and are produced by moving charged particles. (Chap. 20, Sec. 1, p. 580) (Chap. 22, Sec. 1, p. 634)

electrons: negatively-charged particles that move around the nucleus of an atom and form an electron cloud. (Chap. 18, Sec. 1, p. 520)

element: substance that contains only one type of atom and cannot be broken down by normal chemical or physical means—for example, oxygen, aluminum, and iron. (Chap. 18, Sec. 1, p. 519)

embryo: zygote that has attached to the wall of the uterus. (Chap. 10, Sec. 3, p. 283)

emphysema (em fuh SEE muh): lung disease in which the alveoli enlarge. (Chap. 6, Sec. 1, p. 161)

endocytosis (en duh si TOH sus): process by which a cell takes in a substance by surrounding it with the cell membrane. (Chap. 4, Sec. 2, p. 110)

endoplasmic reticulum (ER): cytoplasmic organelle that moves materials around in a cell and is made up of a complex series of folded membranes; can be rough (with attached ribosomes) or smooth (without attached ribosomes). (Chap. 3, Sec. 1, p. 75)

energy pyramid: model that shows the amount of energy available at each feeding level in an ecosystem. (Chap. 13, Sec. 3, p. 377)

enzyme: a type of protein that regulates nearly all chemical reactions in cells. (Chap. 4, Sec. 1, p. 103)

epicenter: point on Earth's surface directly above an earthquake's focus. (Chap. 15, Sec. 1, p. 422)

equilibrium: occurs when molecules of one substance are spread evenly throughout another substance. (Chap. 4, Sec. 2, p. 107)

equinox (EE kwuh nahks): twice-yearly time when the Sun is directly above Earth's equator and the length of day equals the length of night worldwide. (Chap. 17, Sec. 1, p. 485)

estimation: method of making an educated guess at a measurement. (Chap. 2, Sec. 1, p. 39)

evaporation: process that takes place when a liquid changes to a gas. (Chap. 13, Sec. 2, p. 368)

exocytosis (ek soh si TOH sus): process by which vesicles release their contents outside the cell. (Chap. 4, Sec. 2, p. 110)

experimental research design: used to answer scientific questions by testing a hypothesis through the use of a series of carefully controlled steps. (Chap. 1, Sec. 2, p. 13)

F

fault: fracture that occurs when rocks change their shape by breaking; can form as a result of compression (reverse fault), being pulled apart (normal fault), or shear (strike-slip fault). (Chap. 15, Sec. 1, p. 421)

fermentation: process by which oxygen-lacking cells and some one-celled organisms release small amounts of energy from glucose molecules and produce wastes such as alcohol, carbon dioxide, and lactic acid. (Chap. 4, Sec. 3, p. 116)

fertilization: in sexual reproduction, the joining of a sperm and egg. (Chap. 8, Sec. 2, p. 218)

fetal stress: can occur during the birth process or after birth as an infant adjusts from a watery, dark, constant-temperature environment to its new environment. (Chap. 10, Sec. 3, p. 286)

fetus: a developing baby after the first two months of pregnancy until birth. (Chap. 10, Sec. 3, p. 284)

focal length: distance along the optical axis from the center of a concave mirror to the focal point. (Chap. 23, Sec. 2, p. 672)

focal point: single point on the optical axis of a concave mirror where reflected light rays pass through. (Chap. 23, Sec. 2, p. 672)

focus: point deep inside Earth where energy is released, causing an earthquake. (Chap. 15, Sec. 1, p. 422)

food web: model that shows the complex feeding relationships among organisms in a community. (Chap. 13, Sec. 3, p. 376)

frequency: number of wavelengths that pass a given point in one second, measured in hertz (Hz). (Chap. 20, Sec. 2, p. 583)

frond: leaf of a fern that grows from the rhizome. (Chap. 9, Sec. 2, p. 246)

fundamental frequency: lowest natural frequency that is produced by a vibrating object. (Chap. 21, Sec. 2, p. 617)

G

gametophyte (guh MEE tuh fite) **stage:** plant life cycle stage that begins when cells in reproductive organs undergo meiosis and produce haploid cells (spores). (Chap. 9, Sec. 1, p. 243)

gamma ray: highest-frequency, most penetrating electromagnetic wave. (Chap. 22, Sec. 2, p. 644)

gene: section of DNA on a chromosome that contains instructions for making specific proteins. (Chap. 8, Sec. 3, p. 226)

genetic engineering: biological and chemical methods to change the arrangement of a gene's DNA to improve crop production, produce large volumes of medicine, and change how cells perform their normal functions. (Chap. 11, Sec. 3, p. 315)

genetics (juh NET ihks): the study of how traits are inherited through the actions of alleles. (Chap. 11, Sec. 1, p. 300)

genotype (JEE nuh tipe): an organism's genetic makeup. (Chap. 11, Sec. 1, p. 304)

germination: series of events that results in the growth of a plant from a seed. (Chap. 9, Sec. 3, p. 258)

Global Positioning System (GPS): uses satellites, ground-based stations, and portable units with receivers to locate objects on Earth. (Chap. 22, Sec. 3, p. 653)

Golgi (GAWL jee) **bodies:** organelles that package cellular materials and transport them within the cell or out of the cell. (Chap. 3, Sec. 1, p. 75)

graph: used to collect, organize, and summarize data in a visual way. (Chap. 2, Sec. 3, p. 53)

H

habitat: place where an organism lives and that provides the types of food, shelter, moisture, and temperature needed for survival. (Chap. 12, Sec. 1, p. 335)

haploid (HA ploid): when a cell has only half the number of chromosomes as body cells. (Chap. 8, Sec. 2, p. 219)

heredity (huh RED ut ee): the passing of traits from parent to offspring. (Chap. 11, Sec. 1, p. 300)

heterogeneous mixture: mixtures which are not mixed evenly and each component retains its own properties. (Chap. 18, Sec. 2, p. 527)

heterozygous (het uh roh ZI gus): describes an organism with two different alleles for a trait. (Chap. 11, Sec. 1, p. 304)

hibernation: cyclic response of inactivity and slowed metabolism that occurs during periods of cold temperatures and limited food supplies. (Chap. 7, Sec. 2, p. 193)

homogeneous mixture: mixtures which are evenly mixed throughout. (Chap. 18, Sec. 2, p. 527)

homozygous (hoh muh ZI gus): describes an organism with two alleles that are the same for a trait. (Chap. 11, Sec. 1, p. 304)

hormone (HOR mohn): chemical produced by the endocrine system and released directly into the bloodstream by ductless glands; affects specific target tissues, and regulate cellular activities. (Chap. 10, Sec. 1, p. 270)

hot cell: living cell in which a virus can actively reproduce or in which a virus can hide until activated by environmental stimuli. (Chap. 3, Sec. 3, p. 84)

hot spot: large, rising body of magma that can force its way through Earth's mantle and crust and may form volcanoes. (Chap. 15, Sec. 3, p. 438)

hybrid (HI brud): an offspring that was given different genetic information for a trait from each parent. (Chap. 11, Sec. 1, p. 302)

hypothesis (hi PAH thuh sus): prediction or statement that can be tested and may be formed by prior knowledge, any previous observations, and new information. (Chap. 1, Sec. 2, p. 21)

I

imprinting: occurs when an animal forms a social attachment to another organism during a specific period following birth or hatching. (Chap. 7, Sec. 1, p. 183)

incomplete dominance: production of a phenotype that is intermediate between the two homozygous parents. (Chap. 11, Sec. 2, p. 308)

independent variable: variable that can be changed during an experiment. (Chap. 1, Sec. 2, p. 21)

infrared wave: electromagnetic wave that is sensed as heat and is emitted by almost every object. (Chap. 22, Sec. 2, p. 641)

innate behavior: behavior that an organism is born with and does not have to be learned, such as a reflex or instinct. (Chap. 7, Sec. 1, p. 181)

inorganic compound: compound, such as H_2O, that is made from elements other than carbon and whose atoms can usually be arranged in only one structure. (Chap. 4, Sec. 1, p. 103)

insight: form of reasoning that allows animals to use past experiences to solve new problems. (Chap. 7, Sec. 1, p. 185)

instinct: complex pattern of innate behavior, such as spinning a web, that can take weeks to complete. (Chap. 7, Sec. 1, p. 182)

interference: ability of two or more waves to combine and form a new wave when they overlap. (Chap. 20, Sec. 3, p. 591)

ion: electrically-charged atom whose charge results from an atom losing or gaining electrons. (Chap. 18, Sec. 2, p. 525)

isotopes: atoms of the same element that have different numbers of neutrons. (Chap. 18, Sec. 1, p. 521)

K

kelvin: SI unit for temperature. (Chap. 2, Sec. 2, p. 50)

kidney: bean-shaped urinary system organ that is made up of about 1 million nephrons and filters blood, producing urine. (Chap. 6, Sec. 2, p. 164)

kilogram (kg): SI unit for mass. (Chap. 2, Sec. 2, p. 49)

L

larynx (LER ingks): airway to which the vocal cords are attached. (Chap. 6, Sec. 1, p. 157)

lava: magma flowing onto Earth's surface. (Chap. 15, Sec. 2, p. 429)

law of conservation of mass: states that the total mass of matter is the same before and after a physical or chemical change. (Chap. 19, Sec. 2, p. 561)

law of reflection: states that the angle of incidence is equal to the angle of reflection. (Chap. 23, Sec. 2, p. 669)

lens: transparent object that has at least one curved surface that causes light to bend. (Chap. 23, Sec. 3, p. 677)

light ray: narrow beam of light traveling in a straight line. (Chap. 23, Sec. 1, p. 664)

limiting factor: anything that can restrict the size of a population, including living and nonliving fea-tures of an ecosystem, such as predators or drought. (Chap. 12, Sec. 2, p. 338)

line graph: a type of graph used to show the relationship between two variables that are numbers on an *x*-axis and a *y*-axis. (Chap. 2, Sec. 3, p. 53)

lithosphere (LIH thuh sfihr): rigid layer of Earth about 100 km thick, made of the crust and a part of the upper mantle. (Chap. 14, Sec. 3, p. 400)

long-day plant: plant that generally requires short nights—less than 10 to 12 hours of darkness—to begin the flowering process. (Chap. 5, Sec. 2, p. 142)

loudness: the human perception of how much energy a sound wave carries. (Chap. 21, Sec. 1, p. 606)

lunar eclipse: occurs during a full moon when the Sun, the Moon, and Earth line up in a specific way and the Moon moves into Earth's shadow. (Chap. 17, Sec. 2, p. 492)

M

magnitude: a measure of the energy released by an earthquake. (Chap. 15, Sec. 1, p. 423)

mass: amount of matter in an object, which is measured in kilograms. (Chap. 2, Sec. 2, p. 49)

mass number: the number of protons plus the number of neutrons in an atom. (Chap. 18, Sec. 1, p. 521)

matter: anything that has mass and takes up space; matter's properties are determined by the structure of its atoms and how they are joined. (Chap. 18, Sec. 1, p. 518)

measurement: way to describe the world with numbers—for example, length, volume, mass, weight, and temperature. (Chap. 2, Sec. 1, p. 38)

mechanical wave: a type of wave that uses matter to transfer energy. (Chap. 20, Sec. 1, p. 577)

medium: material through which a wave can travel. (Chap. 23, Sec. 1, p. 665)

meiosis (mi OH sus): reproductive process that produces four haploid sex cells from one diploid cell and ensures offspring will have the same number of chromosomes as the parent organisms. (Chap. 8, Sec. 2, p. 219)

menstrual cycle: hormone-controlled monthly cycle of changes in the female reproductive system that includes the maturation of an egg and preparation of the uterus to receive a fertilized egg. (Chap. 10, Sec. 2, p. 278)

menstruation (men STRAY shun): monthly flow of blood and tissue cells that occurs when the lining of the uterus breaks down and is shed. (Chap. 10, Sec. 2, p. 278)

metabolism: the total of all chemical reactions in an organism. (Chap. 4, Sec. 3, p. 113)

meter (m): SI unit for length. (Chap. 2, Sec. 2, p. 47)

migration: instinctive seasonal movement of animals to find food or to reproduce in better conditions. (Chap. 7, Sec. 2, p. 194)

mitochondrion: cell organelle that breaks down lipids and carbohydrates and releases energy. (Chap. 3, Sec. 1, p. 74)

mitosis (mi TOH sus): cell process in which the nucleus divides to form two nuclei identical to each other, and identical to the original nucleus, in a series of steps (prophase, metaphase, anaphase, and telophase). (Chap. 8, Sec. 1, p. 212)

mixture: a combination of substances in which the individual substances do not change or combine chemically but instead retain their own individual properties; can be gases, solids, liquids, or any combination of them. (Chap. 4, Sec. 1, p. 101) (Chap. 18, Sec. 2, p. 527)

model: represents something that is too big, too small, too dangerous, too time consuming, too expensive, or happens too quickly or too slowly to observe directly. (Chap. 1, Sec. 2, p. 16)

moon phases: changing views of the Moon as seen from Earth, which are caused by the Moon's revolution around Earth. (Chap. 17, Sec. 2, p. 489)

mutation: any permanent change in a gene or chromosome of a cell. (Chap. 8, Sec. 3, p. 228)

mutualism: a type of symbiotic relationship in which both organisms benefit. (Chap. 12, Sec. 3, p. 346)

N

natural frequency: frequency at which a musical instrument or other object vibrates when it is struck or disturbed; relative to its size, shape, and the material it is made from. (Chap. 21, Sec. 2, p. 615)

nebula (NEB yuh luh): cloud of rotating gases and dust particles from which the solar system may have formed about 5 billion years ago. (Chap. 17, Sec. 3, p. 503)

nephron (NEF rahn): tiny filtering unit of the kidney. (Chap. 6, Sec. 2, p. 165)

neutron: particle without an electrical charge that is located in the nucleus of an atom. (Chap. 18, Sec. 1, p. 520)

niche (NIHCH): the unique ways an organism survives, obtains food and shelter, and avoids danger. (Chap. 12, Sec. 3, p. 347)

nitrogen cycle: model describing how nitrogen moves from the atmosphere to the soil, to living organisms, and then back to the atmosphere. (Chap. 13, Sec. 2, p. 370)

nitrogen fixation: process in which some types of bacteria can form the nitrogen compounds that plants need. (Chap. 13, Sec. 2, p. 370)

nucleus: organelle that controls all the activities of a cell and contains hereditary material made of proteins and DNA. (Chap. 3, Sec. 1, p. 72)

O

orbit: curved path followed by Earth as it moves around the Sun. (Chap. 17, Sec. 1, p. 483)

organ: structure, such as the heart, made up of different types of tissues that all work together. (Chap. 3, Sec. 1, p. 77)

organelle: structure in the cytoplasm of a eukaryotic cell that can act as a storage site, process energy, move materials, or manufacture substances. (Chap. 3, Sec. 1, p. 72)

organic compounds: compounds that always contain hydrogen and carbon; carbohydrates, lipids, proteins, and nucleic acids are organic compounds found in living things. (Chap. 4, Sec. 1, p. 102)

osmosis: a type of passive transport that occurs when water diffuses through a cell membrane. (Chap. 4, Sec. 2, p. 108)

ovary: female reproductive organ that produces eggs and is located in the lower part of the body. (Chap. 10, Sec. 2, p. 277)

ovary: swollen base of an angiosperm's pistil, where egg-producing ovules are found. (Chap. 9, Sec. 3, p. 253)

overtones: multiples of the fundamental frequency. (Chap. 21, Sec. 2, p. 617)

ovulation (ahv yuh LAY shun): monthly process in which an egg is released from an ovary and enters the oviduct, where it can become fertilized by sperm. (Chap. 10, Sec. 2, p. 277)

ovule: in gymnosperms, the female reproductive part that produces eggs. (Chap. 9, Sec. 3, p. 251)

P

Pangaea (pan JEE uh): large, ancient landmass that was composed of all the continents joined together. (Chap. 14, Sec. 1, p. 392)

parasitism: a type of symbiotic relationship in which one organism benefits but the other organism is harmed. (Chap. 12, Sec. 3, p. 346)

passive transport: movement of substances through a cell membrane without the use of cellular energy; includes diffusion, osmosis, and facilitated diffusion. (Chap. 4, Sec. 2, p. 106)

pharynx (FER ingks): tubelike passageway for food, liquid, and air. (Chap. 6, Sec. 1, p. 156)

phenotype (FEE nuh tipe): the way an organism looks and behaves as a result of its genotype. (Chap. 11, Sec. 1, p. 304)

pheromone (FER uh mohn): powerful chemical produced by an animal to influence the behavior of another animal of the same species. (Chap. 7, Sec. 2, p. 189)

photoperiodism (foh toh PIHR ee uh dih zum): a plant's response to the lengths of darkness each day. (Chap. 5, Sec. 2, p. 142)

photosynthesis: process by which plants and many other producers use light energy from the Sun to make sugars, which can be used as food. (Chap. 4, Sec. 3, p. 114) (Chap. 5, Sec. 1, p. 131)

physical change: change in which the form or appearance of matter changes, but not its composition. (Chap. 19, Sec. 2, p. 552)

physical property: characteristic that can be observed, using the five senses, without changing or trying to change the composition of a substance. (Chap. 19, Sec. 1, p. 546)

pistil: female reproductive organ inside the flower of an angiosperm; consists of a sticky stigma, where pollen grains land, and an ovary. (Chap. 9, Sec. 3, p. 253)

pitch: how high or low a sound is. (Chap. 21, Sec. 1, p. 608)

plate: a large section of Earth's oceanic or continental crust and rigid upper mantle that moves around on the asthenosphere. (Chap. 14, Sec. 3, p. 400)

plate tectonics: theory that Earth's crust and upper mantle are broken into plates that float and move around on a plastic-like layer of the mantle. (Chap. 14, Sec. 3, p. 400)

pollen grain: small structure produced by the male reproductive organs of a seed plant; has a water-resistant coat, can develop from a spore, and contains gametophyte parts that will produce sperm. (Chap. 9, Sec. 3, p. 249)

pollination: transfer of pollen grains to the female part of a seed plant by agents such as gravity, water, wind, and animals. (Chap. 9, Sec. 3, p. 249)

polygenic (pahl ih JEHN ihk) **inheritance:** occurs when a group of gene pairs acts together and produces a specific trait, such as human eye color, skin color, or height. (Chap. 11, Sec. 2, p. 310)

population: all the organisms in an ecosystem that belong to the same species. (Chap. 12, Sec. 1, p. 334)

precision: describes how close measurements are to each other. (Chap. 2, Sec. 1, p. 40)

pregnancy: period of development—usually about 38 or 39 weeks in humans—from fertilized egg until birth. (Chap. 10, Sec. 3, p. 282)

producer: organism, such as a green plant or alga, that uses an outside source of energy like the Sun to create energy-rich food molecules. (Chap. 12, Sec. 3, p. 344)

prothallus (proh THA lus): small, green, heart-shaped gametophyte plant form of a fern that can make its own food and absorb water and nutrients from the soil. (Chap. 9, Sec. 2, p. 246)

proton: positively-charged particle that is located in the nucleus of an atom. (Chap. 18, Sec. 1, p. 520)

Punnett (PUN ut) **square:** a tool to predict the probability of certain traits in offspring that shows the different ways alleles can combine. (Chap. 11, Sec. 1, p. 304)

R

radiant energy: energy carried by an electromagnetic wave. (Chap. 22, Sec. 1, p. 638)

radio waves: lowest-frequency electromagnetic waves that carry the least amount of energy and are used in most forms of telecommunications technology—such as TVs, telephones, and radios. (Chap. 22, Sec. 2, p. 640)

rate: amount of change of one measurement in a given amount of time. (Chap. 2, Sec. 2, p. 50)

recessive (rih SES ihv): describes a trait that is covered over, or dominated, by another form of that trait and seems to disappear. (Chap. 11, Sec. 1, p. 302)

reflecting telescope: uses a concave mirror to gather light from distant objects. (Chap. 23, Sec. 4, p. 683)

reflection: occurs when a wave strikes an object or surface and bounces off. (Chap. 20, Sec. 3, p. 587) (Chap. 23, Sec. 1, p. 665)

reflex: simple innate behavior, such as yawning or blinking, that is an automatic response and does not involve a message to the brain. (Chap. 7, Sec. 1, p. 181)

refracting telescope: uses two convex lenses to gather light and form an image of a distant object. (Chap. 23, Sec. 4, p. 682)

refraction: bending of a wave as it moves from one medium into another medium. (Chap. 20, Sec. 3, p. 588) (Chap. 23, Sec. 3, p. 677)

resonance: sound amplification that occurs when an object is vibrated at its natural frequency by absorbing energy from a sound wave or other object vibrating at this frequency. (Chap. 21, Sec. 2, p. 616)

respiration: series of chemical reactions used to release energy stored in food molecules. (Chap. 4, Sec. 3, p. 115) (Chap. 5, Sec. 1, p. 133)

reverberation: repeated echoes of sounds. (Chap. 21, Sec. 2, p. 621)

revolution: the motion of Earth around the Sun, which takes about 365 1/4 days, or one year, to complete. (Chap. 17, Sec. 1, p. 483)

rhizome: underground stem of a fern. (Chap. 9, Sec. 2, p. 246)

ribosome: small structure on which cells make their own proteins. (Chap. 3, Sec. 1, p. 74)

rift: long crack that forms between tectonic plates moving apart at plate boundaries. (Chap. 15, Sec. 3, p. 437)

RNA: ribonucleic acid, which carries codes for making proteins from the nucleus to the ribosomes. (Chap. 8, Sec. 3, p. 226)

rotation: spinning of Earth on its axis, which causes day and night; it takes 24 hours for Earth to complete one rotation. (Chap. 17, Sec. 1, p. 482)

S

salinity (say LIHN ut ee): a measure of the amount of salts dissolved in seawater. (Chap. 16, Sec. 1, p. 454)

science: process used to investigate what is happening around us in order to solve problems or answer questions. (Chap. 1, Sec. 1, p. 6)

scientific methods: ways to solve problems that can include step-by-step plans, making models, and carefully thought-out experiments. (Chap. 1, Sec. 2, p. 13)

seafloor spreading: Hess's theory that new seafloor is formed when magma is forced upward toward the surface at a mid-ocean ridge. (Chap. 14, Sec. 2, p. 397)

seismic safe: describes the ability of structures to stand up against the vibrations caused by an earthquake. (Chap. 15, Sec. 1, p. 427)

seismic waves: earthquake waves, including primary waves, secondary waves, and surface waves. (Chap. 15, Sec. 1, p. 422)

seismograph: instrument used to record seismic waves. (Chap. 15, Sec. 1, p. 423)

semen (SEE mun): mixture of sperm and a fluid that helps sperm move and supplies them with an energy source. (Chap. 10, Sec. 2, p. 276)

sex-linked gene: an allele inherited on a sex chromosome and that can cause human genetic disorders such as color blindness and hemophilia. (Chap. 11, Sec. 2, p. 313)

sexual reproduction: a type of reproduction in which two sex cells, usually an egg and a sperm, join to form a zygote, which will develop into a new organism with a unique identity. (Chap. 8, Sec. 2, p. 218)

shield volcano: large, broad volcano with gently sloping sides that is formed by the buildup of basaltic layers. (Chap. 15, Sec. 2, p. 432)

short-day plant: plant that generally requires long nights—12 or more hours of darkness—to begin the flowering process. (Chap. 5, Sec. 2, p. 142)

SI: International System of Units, related by multiples of ten, that allows quantities to be measured in the exact same way throughout the world. (Chap. 2, Sec. 2, p. 46)

social behavior: interactions among members of the same species, including courtship and mating, getting food, caring for young, and protecting each other. (Chap. 7, Sec. 2, p. 186)

society: a group of animals of the same species that live and work together in an organized way, with each member doing a specific job. (Chap. 7, Sec. 2, p. 187)

soil: mixture of mineral and rock particles, the remains of dead organisms, air, and water that forms the topmost layer of Earth's crust and supports plant growth. (Chap. 13, Sec. 1, p. 362)

solar eclipse: occurs during a new moon, when the Sun, the Moon, and Earth are lined up in a specific way and Earth moves into the Moon's shadow. (Chap. 17, Sec. 2, p. 491)

solar system: extremely large system that includes the Sun, planets, comets, meteoroids and other objects that orbit the Sun. (Chap. 17, Sec. 3, p. 496)

solstice: time when the Sun reaches its greatest distance north or south of the equator, which occurs June 21 or 22 for the northern hemisphere (longest day of the year) and December 21 or 22 for the southern hemisphere (shortest day of the year). (Chap. 17, Sec. 1, p. 484)

solution: a kind of mixture in which one substance is completely and evenly mixed in another substance and is the same throughout. (Chap. 18, Sec. 2, p. 527)

sori: fern structures in which spores are produced. (Chap. 9, Sec. 2, p. 246)

sperm: haploid sex cells formed in the male reproductive organs. (Chap. 8, Sec. 2, p. 218) (Chap. 10, Sec. 2, p. 276)

spores: haploid cells produced in the gametophyte stage that can divide by mitosis and form plant structures or an entire new plant or can develop into sex cells. (Chap. 9, Sec. 1, p. 243)

sporophyte (SPOR uh fite) **stage:** plant life cycle stage that begins when an egg is fertilized by a sperm. (Chap. 9, Sec. 1, p. 243)

stamen: male reproductive organ inside the flower of an angiosperm where pollen grains form. (Chap. 9, Sec. 3, p. 253)

stomata (STOH mut uh): tiny openings in a plant's epidermis through which carbon dioxide and water vapor gases enter and leave a leaf. (Chap. 5, Sec, 1, p. 129)

surface current: wind-powered ocean current that moves water horizontally, parallel to Earth's surface, and moves only the upper few hundred meters of seawater. (Chap. 16, Sec. 2, p. 456)

symbiosis: any close relationship between species, including mutualism, commensalism, and parasitism. (Chap. 12, Sec. 3, p. 346)

T

table: displays information in rows and columns, making it easier to read and understand. (Chap. 2, Sec. 3, p. 53)

technology: application of science to make useful products and tools, such as computers. (Chap. 1, Sec. 1, p. 9)

testes: (TES teez) male organ that produces sperm and testosterone. (Chap. 10, Sec. 2, p. 276)

tidal range: the difference between the level of the ocean at high tide and the level at low tide. (Chap. 16, Sec. 3, p. 466)

tide: daily rise and fall in sea level caused, for the most part, by the interaction of gravity in the Earth-Moon system. (Chap. 16, Sec. 3, p. 465)

tissue: group of similar cells that work together to do one job. (Chap. 3, Sec. 1, p. 77)

trachea (TRAY kee uh): air-conducting tube that connects the larynx with the bronchi, is lined with mucous membranes and cilia, and contains strong cartilage rings. (Chap. 6, Sec. 1, p. 157)

transverse wave: a type of mechanical wave in which the wave energy causes matter in the medium to move up and down or back and forth at right angles to the direction the wave travels. (Chap. 20, Sec. 1, p. 578)

tropism: positive or negative plant response to an external stimulus such as touch, light, or gravity. (Chap. 5, Sec. 2, p. 138)

trough: lowest point of a wave. (Chap. 16, Sec. 3, p. 462)

tsunami: powerful seismic sea wave that begins over an ocean-floor earthquake, can reach 30 m in height when approaching land, and can cause destruction in coastal areas. (Chap. 15, Sec. 1, p. 425)

U

ultraviolet radiation: electromagnetic waves with higher frequencies and shorter wavelengths than visible light. (Chap. 22, Sec. 2, p. 643)

upwelling: circulation in the ocean that brings deep, cold water to the ocean surface. (Chap. 16, Sec. 2, p. 459)

ureter: tube that carries urine from each kidney to the bladder. (Chap. 6, Sec. 2, p. 166)

urethra (yoo REE thruh): tube that carries urine from the bladder to the outside of the body. (Chap. 6, Sec. 2, p. 166)

urinary system: system of excretory organs that rids the blood of wastes, controls blood volume by removing excess water, and balances concentrations of salts and water. (Chap. 6, Sec. 2, p. 163)

urine: wastewater that contains excess water, salts, and other wastes that are not reabsorbed by the body. (Chap. 6, Sec. 2, p. 164)

uterus: hollow, muscular, pear-shaped organ where a fertilized egg develops into a baby. (Chap. 10, Sec. 2, p. 277)

V

vagina (vuh JI nuh): muscular tube that connects the lower end of the uterus to the outside of the body; the birth canal through which a baby travels when being born. (Chap. 10, Sec. 2, p. 277)

virus: a strand of hereditary material surrounded by a protein coating. (Chap. 3, Sec. 3, p. 84)

visible light: electromagnetic waves with wavelengths between 0.7 and 0.4 millionths of a meter that can be seen with your eyes. (Chap. 22, Sec. 2, p. 642)

volcano: cone-shaped hill or mountain formed when hot magma, solids, and gas erupt onto Earth's surface through a vent. (Chap. 15, Sec. 2, p. 429)

W

water cycle: model describing how water moves from Earth's surface to the atmosphere and back to the surface again through evaporation, condensation, and precipitation. (Chap. 13, Sec. 2, p. 369)

wave: rhythmic movement that carries energy through matter or space. (Chap. 16, Sec. 3, p. 462) (Chap. 20, Sec. 1, p. 576)

wavelength: in transverse waves, the distance between the tops of two adjacent crests or the bottoms of two adjacent troughs; in compressional waves, the distance from the centers of adjacent rarefactions. (Chap. 20, Sec. 2, p. 582)

X

X ray: high-energy electromagnetic wave that is highly penetrating. (Chap. 22, Sec. 2, p. 644)

Z

zygote: new diploid cell formed when a sperm fertilizes an egg; will divide by mitosis and develop into a new organism. (Chap. 8, Sec. 2, p. 218)

Spanish Glossary

Este glossario define cada término clave que aparece en negrillas en el texto. También muestra el capítulo, la sección y el número de página en donde se usa dicho término.

A

abiotic / abióticos: factores físicos inanimados del ambiente que incluyen el aire, el agua, la luz solar, el suelo, la temperatura y el clima. (Cap. 13, Sec. 1, pág. 360)

accuracy / exactitud: compara una medida con el verdadero valor. (Cap. 2, Sec. 1, pág. 41)

active transport / transporte activo: proceso que requiere energía en el cual las proteínas de transporte se enlazan con partículas y se mueven a través de la membrana celular. (Cap. 4, Sec. 2, pág. 109)

aggression / agresión: comportamiento enérgico, como las peleas, que usa un animal para controlar o dominar a otro animal con el propósito de proteger sus crías, defender su territorio u obtener alimento. (Cap. 7, Sec. 2, pág. 188)

allele / alelo: formas alternas que un gene puede tener para un sólo rasgo; puede ser dominante o recesivo. (Cap. 11, Sec. 1, pág. 300)

alveoli / alvéolos: racimos minúsculos de paredes finas que se hallan en el extremo de cada bronquiolo y que están rodeados de capilares, donde se lleva a cabo el intercambio de dióxido de carbono y oxígeno. (Cap. 6, Sec. 1, pág. 157)

amniotic sac / saco amniótico: membrana protectora, delgada y llena de líquido, que se forma alrededor del embrión. (Cap. 10, Sec. 3, pág. 283)

amplitude / amplitud: distancia a la cual una onda sube o baja de su nivel normal, la cual se relaciona con la energía que transporta la onda; en una onda transversal, es la mitad de la distancia entre una cresta y un seno. (Cap. 20, Sec. 2, pág. 581)

asexual reproduction / reproducción asexual: tipo de reproducción que comprende la fisión, gemación y regeneración, en el cual un progenitor produce un nuevo organismo que tiene DNA idéntico al organismo progenitor. (Cap. 8, Sec. 1, pág. 215)

asteroid / asteroide: pequeño cuerpo espacial rocoso que se encuentra en el cinturón de asteroides entre las órbitas de Júpiter y Marte. (Cap. 17, Sec. 3, pág. 502)

asthenosphere / astenosfera: capa viscosa de la Tierra en la cual las placas litosféricas flotan y se mueven. (Cap. 14, Sec. 3, pág. 400)

asthma / asma: trastorno pulmonar en el cual los tubos bronquiales se contraen rápidamente y dificultan la respiración y causan estornudo o tos; puede ocurrir como una reacción alérgica. (Cap. 6, Sec. 1, pág. 162)

astronomical unit / unidad astronómica: unidad de medida que se usa para medir distancias entre objetos en el sistema solar; 1 UA equivale a 150 000 000 km. (Cap. 17, Sec. 3, pág. 496)

atmosphere / atmósfera: comprende el aire que rodea la Tierra; la atmósfera está compuesta por gases, entre los cuales se incluye un 78 por ciento de nitrógeno, un 21 por ciento de oxígeno y 0.03 por ciento de dióxido de carbono. (Cap. 13, Sec. 1, pág. 361)

atomic number / número atómico: número de protones en un átomo. (Cap. 18, Sec. 1, pág. 521)

atoms / átomos: partículas diminutas de materia compuestas de protones, neutrones y electrones. (Cap. 18, Sec. 1, pág. 518)

Spanish Glossary

auxin / auxina: hormona vegetal gracias a la cual las hojas y tallos de las plantas exhiben fototropismos positivos. (Cap. 5, Sec. 2, pág. 140)

axis / eje: línea imaginaria alrededor de la cual gira la Tierra, lo cual causa el día y la noche; se traza desde el polo geográfico norte a través de la Tierra hasta el polo geográfico sur. (Cap. 17, Sec. 1, pág. 482)

B

bar graph / gráfica de barras: tipo de gráfica que usa barras de distintos tamaños para mostrar las relaciones entre variables. (Cap. 2, Sec. 3, pág. 54)

basin / cuenca: depresión en la Tierra en donde se formó un océano cuando el área se llenó de agua debido a las lluvias torrenciales. (Cap. 16, Sec. 1, pág. 453)

behavior / comportamiento: la interacción de un organismo con otro organismo y con su ambiente; puede ser innato o adquirido. (Cap. 7, Sec. 1, pág. 180)

biosphere / biosfera: parte de la Tierra que sostiene la vida; incluye la parte superior de la corteza terrestre, la atmósfera y toda el agua sobre la superficie de la Tierra. (Cap. 12, Sec. 1, pág. 332)

biotic / bióticos: los factores del medio ambiente que son seres vivos o seres que una vez estuvieron vivos. (Cap. 13, Sec. 1, pág. 360)

bladder / vejiga: órgano elástico y muscular que retiene la orina hasta que ésta sale del cuerpo por la uretra. (Cap. 6, Sec. 2, pág. 166)

breaker / cachón: una ola oceánica que se forma en aguas poco profundas y la cual rompe en la playa. (Cap. 16, Sec. 3, pág. 463)

bronchi / bronquios: dos conductos cortos que se bifurcan del extremo inferior de la tráquea y por los cuales se introduce el aire en los pulmones. (Cap. 6, Sec. 1, pág. 157)

C

carbon cycle / ciclo del carbono: modelo que describe cómo las células del carbono se movilizan entre el mundo vivo y el inanimado. (Cap. 13, Sec. 2, pág. 373)

carrier wave / onda portadora: frecuencia de transmisión particular asignada a una estación radial. (Cap. 22, Sec. 3, pág. 650)

carrying capacity / capacidad de carga: el número mayor de individuos de una especie en particular que puede mantener un ecosistema de manera prolongada. (Cap. 12, Sec. 2, pág. 339)

cell membrane / membrana celular: cubierta externa protectora de todas las células; formada por una capa doble de moléculas adiposas y controla la interacción entre la célula y el medio ambiente. (Cap. 3, Sec. 1, pág. 70)

cell theory / teoría celular: establece que todos los organismos están formados por una o más células, la célula es la unidad básica de la vida y todas las células provienen de otras células. (Cap. 3, Sec. 2, pág. 83)

cell wall / pared celular: estructura rígida que encierra, sostiene y protege las células vegetales, las células de las algas, de los hongos y de la mayoría de las bacterias. (Cap. 3, Sec. 1, pág. 71)

chemical change / cambio químico: transformación que experimenta la composición de una sustancia. (Cap. 19, Sec. 2, pág. 554)

chemical property / propiedad química: propiedad de la materia que se puede observar cuando dos sustancias reaccionan o cuando una sustancia se convierte en otra sustancia. (Cap. 19, Sec. 1, pág. 550)

chemosynthesis / quimiosíntesis: proceso en el cual los productores elaboran moléculas nutritivas ricas en energía a partir de sustancias químicas. (Cap. 13, Sec. 3, pág. 375)

chlorophyll / clorofila: pigmento verde y absorbente de luz, fijado a los cloroplastos de las plantas y que es importante en el proceso de la fotosíntesis. (Cap. 5, Sec. 1, pág. 130)
chloroplast / cloroplasto: organelo de las células vegetales, de color verde y que contiene clorofila, que convierte la luz solar, el dióxido de carbono y el agua en azúcar. (Cap. 3, Sec. 1, pág. 74)
chromosome / cromosoma: estructura en el núcleo de una célula que contiene el material genético. (Cap. 8, Sec. 1, pág. 212)
cinder cone volcano / volcán de cono de carbonilla: volcán relativamente pequeño que se ha formado debido a erupciones violentas de tefra. (Cap. 15, Sec. 2, pág. 432)
circle graph / gráfica circular: tipo de gráfica que muestra partes de un todo; cada parte de la gráfica es un sector que representa un porcentaje del total. (Cap. 2, Sec. 3, pág. 54)
climate / clima: condiciones meteorológicas promedio de una región durante un período de tiempo, entre las cuales se incluyen el viento, la temperatura y la precipitación pluvial u otro tipo de precipitación como la nieve o la cellisca. (Cap. 13, Sec. 1, pág. 365)
comet / cometa: astro espacial compuesto de partículas rocosas y hielo; forma una cola cuando su órbita lo acerca al Sol y se encuentra principalmente en el cinturón Kuiper y en la Nube de Oort. (Cap. 17, Sec. 3, pág. 502)
commensalism / comensalismo: tipo de relación simbiótica en el cual un organismo se beneficia y el otro organismo no se ve afectado. (Cap. 12, Sec. 3, pág. 346)
community / comunidad: todas las poblaciones de diferentes especies que viven en un ecosistema. (Cap. 12, Sec. 1, pág. 334)
composite volcano / volcán compuesto: volcán de laderas abruptas que se ha formado por la alternación de erupciones violentas de tefra y erupciones más silenciosas de lava. (Cap. 15, Sec. 2, pág. 433)
compound / compuesto: materia que está hecha de dos o más elementos y que tiene propiedades físicas y químicas diferentes a las de los elementos que la formaron. (Cap. 18, Sec. 2, pág. 524)
compressional wave / onda de compresión: tipo de onda mecánica en la cual la materia del medio oscila en la misma dirección en que viaja la onda. (Cap. 20, Sec. 1, pág. 579)
concave lens / lente cóncava: lente que es más gruesa en los bordes que en el medio y que desvía los rayos luminosos que viajan paralelos al eje óptico. (Cap. 23, Sec. 3, pág. 679)
condensation / condensación: proceso que se efectúa cuando un gas se convierte en un líquido. (Cap. 13, Sec. 2, pág. 369)
conditioning / condicionamiento: ocurre cuando la respuesta a un estímulo se asocia con otro estímulo. (Cap. 7, Sec. 1, pág. 184)
constant / constante: variable que permanece igual durante un experimento. (Cap. 1, Sec. 2, pág. 21)
consumer / consumidor: organismo que no puede fabricar moléculas ricas en energía, sino que obtiene su alimento al alimentarse de otros organismos. (Cap. 12, Sec. 3, pág. 345)
continental drift / deriva continental: hipótesis de Wegener que afirmaba que todos los continentes estuvieron unidos en algún momento formando una sola masa continental, la cual se separó hace unos 200 millones de años, haciendo que los continentes derivaran lentamente a sus posiciones actuales. (Cap. 14, Sec. 1, pág. 392)
control / control: muestra que se trata como cualquier otro grupo experimental, excepto que no se le aplica la variable independiente. (Cap. 1, Sec. 2, pág. 22)

convection current / corriente de convección: corriente en el manto terrestre que transfiere energía en el interior de la Tierra y que provee la potencia de la tectónica de placas. (Cap. 14, Sec. 3, pág. 405)

convex lens / lente convexa: lente convergente que es más gruesa en el medio que en los bordes. (Cap. 23, Sec. 3, pág. 678)

Coriolis effect / efecto de Coriolis: cambio de vientos y corrientes superficiales provocados por la rotación de la Tierra; hace que las corrientes al norte del ecuador fluyan en dirección de las manecillas del reloj y las corrientes al sur del ecuador en dirección contraria. (Cap. 16, Sec. 2, pág. 457)

courtship behavior / comportamiento de cortejo: comportamiento que permite que machos y hembras de una especie se reconozcan mutuamente y se preparen para el apareo. (Cap. 7, Sec. 2, pág. 189)

crater / cráter: depresión formada por un meteorito grande; entre más cráteres presente una región, más data en antigüedad. (Cap. 17, Sec. 2, pág. 487)

crest / cresta: punto más alto de una onda. (Cap. 16, Sec. 3, pág. 462)

cyclic behavior / comportamiento cíclico: comportamiento que ocurre en forma de patrones repetidos. (Cap. 7, Sec. 2, pág. 192)

cytoplasm / citoplasma: mezcla gelatinosa en continuo movimiento dentro de la membrana celular que contiene material hereditario y en la cual se lleva a cabo la mayoría de los procesos de una célula. (Cap. 3, Sec. 1, pág. 70)

D

day-neutral plant / planta de día neutro: planta que no necesita un fotoperíodo específico y que puede comenzar el proceso de floración a lo largo de un rango de períodos nocturnos. (Cap. 5, Sec.2, pág. 142)

density / densidad: cambio físico de la materia que puede calcularse dividiendo la masa de un cuerpo entre su volumen. (Cap. 18, Sec. 3, pág. 530)

density current / corriente de densidad: patrón de circulación en el océano que se forma cuando una masa de agua salada más densa se hunde debajo de agua salada menos densa. (Cap. 16, Sec. 2, pág. 459)

dependent variable / variable dependiente: factor que se mide durante un experimento. (Cap. 1, Sec. 2, pág. 21)

descriptive research / investigación descriptiva: responde preguntas científicas a través de la observación. (Cap. 1, Sec. 2, pág. 13)

diaphragm / diafragma: músculo situado debajo de los pulmones que se contrae y se relaja permitiendo así la entrada y salida de gases del cuerpo. (Cap. 6, Sec. 1, pág. 158)

diffraction / difracción: desviación de las ondas alrededor de un obstáculo. (Cap. 20, Sec. 3, pág. 589)

diffusion / difusión: tipo de transporte pasivo celular en el que las moléculas se mueven desde áreas de mayor concentración a áreas de menor concentración. (Cap. 4, Sec. 2, pág. 107)

diploid / diploide: célula cuyos cromosomas se dan en pares. (Cap. 8, Sec. 2, pág. 218)

DNA / DNA: ácido desoxirribonucleico; material genético de todos los organismos y compuesto de dos hebras retorcidas de moléculas de fosfato de azúcar y bases nitrogenadas. (Cap. 8, Sec. 3, pág. 224)

dominant / dominante: describe un rasgo que cubre o domina otra forma de dicho rasgo. (Cap. 11, Sec. 1, pág. 302)

Doppler effect / efecto Doppler: cambio en la frecuencia o el tono de un sonido, el cual ocurre cuando la fuente sonora y el oyente están en movimiento relativo uno del otro. (Cap. 21, Sec. 1, pág. 610)

E

eardrum / tímpano: membrana que se extiende a través del canal auditivo y la cual vibra cuando las ondas sonoras llegan al oído medio. (Cap. 21, Sec. 2, pág. 622)

earthquake / terremoto: movimiento sísmico que ocurre cuando las rocas del interior de la Tierra exceden su límite elástico, se rompen repentinamente y experimentan un rebote elástico. (Cap. 15, Sec. 1, pág. 420)

echo / eco: onda sonora reflejada. (Cap. 21, Sec. 1, pág. 609)

ecology / ecología: estudio de las interacciones que se llevan a cabo entre los organismos y su ambiente. (Cap. 12, Sec. 1, pág. 333)

ecosystem / ecosistema: todos los organismos vivos que habitan en un área y las cosas inanimadas en su ambiente. (Cap. 12, Sec. 1, pág. 333)

egg / óvulo: célula haploide formada en los órganos reproductores femeninos. (Cap. 8, Sec. 2, pág. 218)

electromagnetic spectrum / espectro electromagnético: rango de ondas electromagnéticas, que incluyen las ondas radiales, la luz visible y los rayos X, las cuales poseen distintas frecuencias y longitudes de onda. (Cap. 22, Sec. 2, pág. 639)

electromagnetic waves / ondas electromagnéticas: ondas que pueden viajar a través del espacio vacío, poseen una amplia gama de longitudes de onda y frecuencias y las producen las partículas con carga eléctrica al moverse. (Cap. 20, Sec. 1, pág. 580; Cap. 22, Sec. 1, pág. 634)

electrons / electrones: partículas con carga negativa que se mueven alrededor del núcleo de un átomo y forman la nube de electrones. (Cap. 18, Sec. 1, pág. 520)

element / elemento: sustancia que sólo contiene un tipo de átomo; por ejemplo, el oxígeno, el aluminio y el hierro. (Cap. 18, Sec. 1, pág. 519)

embryo / embrión: óvulo fecundado adherido a la pared uterina. (Cap. 10, Sec. 3, pág. 283)

emphysema / enfisema: enfermedad pulmonar en la cual se produce una dilatación de los alvéolos. (Cap. 6, Sec. 1, pág. 161)

endocytosis / endocitosis: proceso que permite que una célula deje pasar una sustancia al rodearla con la membrana celular. (Cap. 4, Sec. 2, pág. 110)

endoplasmic reticulum (ER) / retículo endoplásmico: organelo citoplásmico que mueve materiales dentro de una célula y que está formado por una serie compleja de membranas plegadas; puede ser áspero (con ribosomas adheridos) o liso (sin ribosomas adheridos). (Cap. 3, Sec. 1, pág. 75)

energy pyramid / pirámide de energía: modelo que muestra la cantidad de energía disponible en cada nivel alimenticio de un ecosistema. (Cap. 13, Sec. 3, pág. 377)

enzyme / enzima: tipo de proteína que regula casi todas las reacciones químicas de las células. (Cap. 4, Sec. 1, pág. 103)

epicenter / epicentro: punto en la superficie terrestre que se halla directamente encima del foco de un terremoto. (Cap. 15, Sec. 1, pág. 422)

equilibrium / equilibrio: ocurre cuando las moléculas de una sustancia se esparcen uniformemente en otra sustancia. (Cap. 4, Sec. 2, pág. 107)

equinox / equinoccio: época que ocurre dos veces el año, cuando el Sol se encuentra directamente encima del ecuador y el número de horas diurnas y nocturnas son iguales en todo el mundo. (Cap. 17, Sec. 1, pág. 485)

estimation / estimación: método de hacer una conjetura razonada de una medida. (Cap. 2, Sec. 1, pág. 39)

evaporation / evaporación: proceso que se lleva a cabo cuando un líquido se convierte en un gas. (Cap. 13, Sec. 2, pág. 368)

exocytosis / exocitosis: proceso a través del cual las vesículas liberan sus contenidos fuera de la célula. (Cap. 4, Sec. 2, pág. 110)

experimental research design / diseño de investigación experimental: se usa para responder preguntas científicas mediante la prueba de una hipótesis usando una serie de pasos cuidadosamente controlados. (Cap. 1, Sec. 2, pág. 13)

F

fault / falla: fractura que ocurre cuando las rocas cambian de forma al fragmentarse; se puede formar debido a una compresión (falla invertida), a la separación (falla normal) o al cizallamiento (falla transformante). (Cap. 15, Sec. 1, pág. 421)

fermentation / fermentación: proceso en que las células carentes de oxígeno y algunos organismos unicelulares liberan pequeñas cantidades de energía de las moléculas de glucosa y producen desechos como el alcohol, el dióxido de carbono y el ácido láctico. (Cap. 4, Sec. 3, pág. 116)

fertilization / fecundación: en la reproducción sexual, la unión del espermatozoide y del óvulo. (Cap. 8, Sec. 2, pág. 218)

fetal stress / estrés fetal: puede presentarse durante el proceso de alumbramiento o después del nacimiento conforme el lactante se ajusta de un entorno acuoso, oscuro y de temperatura constante a su nuevo entorno. (Cap. 10, Sec. 3, pág. 286)

fetus / feto: bebé en desarrollo después de los primeros dos meses de embarazo hasta su nacimiento. (Cap. 10, Sec. 3, pág. 284)

focal length / longitud focal: distancia a lo largo del eje óptico desde el centro de un espejo cóncavo al punto focal. (Cap. 23, Sec. 2, pág. 672)

focal point / punto focal: punto único en el eje óptico de un espejo cóncavo a través del cual pasan los rayos luminosos reflejados. (Cap. 23, Sec. 2, pág. 672)

focus / foco: punto profundo en el interior de la Tierra donde se libera energía, lo cual provoca un terremoto. (Cap. 15, Sec. 1, pág. 422)

food web / red alimenticia: modelo que muestra las complejas relaciones alimenticias entre los organismos de una comunidad. (Cap. 13, Sec. 3, pág. 676)

frequency / frecuencia: número de longitudes de onda que pasan por un punto dado en un segundo; se miden en hertz (Hz). (Cap. 20, Sec. 2, pág. 583)

frond / fronda: hoja de helecho que crece desde el rizoma. (Cap. 9, Sec. 2, pág. 246)

fundamental frequency / frecuencia fundamental: frecuencia natural más baja que produce una cuerda o una columna de aire que vibra. (Cap. 21, Sec. 2, pág. 617)

G

gametophyte stage / etapa gametofita: etapa del ciclo de vida vegetal que comienza cuando las células de los órganos reproductores pasan por la meiosis y producen células haploides. (Cap. 9, Sec. 1, pág. 243)

gamma ray / rayo gama: la onda electromagnética más penetrante y de alta frecuencia. (Cap. 22, Sec. 2, pág. 644)

gene / gene: sección de DNA en un cromosoma que contiene las instrucciones para la elaboración de proteínas específicas. (Cap. 8, Sec. 3, pág. 226)

genetic engineering / ingeniería genética: métodos biológicos y químicos que se usan para cambiar el arreglo del DNA de un gene con el propósito de mejorar la producción de cosechas, producir grandes volúmenes de medicamentos y cambiar el funcionamiento normal de células. (Cap. 11, Sec. 3, pág. 315)

genetics / genética: estudia la manera en que se heredan los rasgos a través de las acciones de los alelos. (Cap. 11, Sec. 1, pág. 300)

genotype / genotipo: la composición genética de un organismo. (Cap. 11, Sec. 1, pág. 304)

germination / germinación: serie de eventos que dan como resultado el crecimiento de una planta a partir de una semilla. (Cap. 9, Sec. 3, pág. 258)

Global Positioning System / Sistema de Posición Global: usa satélites, estaciones terrestres y equipo portátil con receptores para ubicar objetos sobre la Tierra. (Cap. 22, Sec. 3, pág. 653)

Golgi bodies / cuerpos de Golgi: organelos que almacenan materiales celulares y los transportan dentro o fuera de la célula. (Cap. 3, Sec. 1, pág. 75)

graph / gráfica: se utiliza para recopilar, organizar y resumir datos de una manera visual, facilitando de esta manera su uso y comprensión. (Cap. 2, Sec. 3, pág. 53)

H

habitat / hábitat: lugar en donde vive un organismo y que le provee los tipos de alimento, refugio, humedad y temperaturas necesarias para la sobrevivencia. (Cap. 12, Sec. 1, pág. 335)

haploid / haploide: célula que sólo tiene uno de cada tipo de cromosoma. (Cap. 8, Sec. 2, pág. 219)

heredity / herencia: el traspaso de rasgos de los progenitores a la progenie. (Cap. 11, Sec. 1, pág. 300)

heterogeneous / heterogénea: mezclas que se no distribuyen igualmente y en las que cada componente retiene sus propias propiedades. (Cap. 18, Sec. 2, pág. 527)

heterozygous / heterocigoto: describe al organismo que presenta dos alelos distintos para un rasgo. (Cap. 11, Sec. 1, pág. 304)

hibernation / hibernación: respuesta cíclica de inactividad y disminución del metabolismo, la cual ocurre durante períodos de temperaturas frías y abastecimientos limitados de alimentos. (Cap. 7, Sec. 2, pág. 193)

homogeneous / homogénea: mezcla que se distribuyen igualmente en toda la extensión de la mezcla. (Cap. 18, Sec. 2, pág. 527)

homozygous / homocigoto: describe un organismo con dos alelos idénticos para el mismo rasgo. (Cap. 11, Sec. 1, pág. 304)

hormone / hormona: sustancia química que produce el sistema endocrino y la cual se libera directamente en el torrente sanguíneo a través de glándulas sin conductos; actúa en tejidos asignados y puede acelerar o aminorar las actividades celulares. (Cap. 10, Sec. 1, pág. 270)

host cell / célula huésped: célula viva en la cual un virus se puede reproducir activamente o en la cual un virus puede ocultarse hasta que los estímulos ambientales lo activen. (Cap. 3, Sec. 3, pág. 84)

hot spot / foco caliente: masa de magma extensa y ascendente que puede abrirse paso a través del manto y la corteza terrestres y que puede formar volcanes. (Cap. 15, Sec. 3, pág. 438)

hybrid / híbrido: progenie que ha obtenido información genética distinta para un rasgo de cada progenitor. (Cap. 11, Sec. 1, pág. 302)

hypothesis / hipótesis: predicción o enunciado que puede probarse y que se puede deducir del conocimiento previo, cualquier observación previa y de nueva información. (Cap. 1, Sec. 2, pág. 21)

I

imprinting / impronta: ocurre cuando un animal forma un vínculo social con otro organismo durante un período específico después del nacimiento o de salir del cascarón. (Cap. 7, Sec. 1, pág. 183)

incomplete dominance / dominancia incompleta: producción de un fenotipo intermedio al de los dos progenitores homocigotos. (Cap. 11, Sec. 2, pág. 308)

independent variable / variable independiente: variable que puede cambiarse durante un experimento. (Cap. 1, Sec. 2, pág. 21)

infrared wave / onda infrarroja: onda electromagnética que se siente como calor y la cual emiten casi todos los cuerpos. (Cap. 22, Sec. 2, pág. 641)

innate behavior / comportamiento innato: comportamiento con que nace un organismo y el cual no tiene que ser adquirido, como un reflejo o un instinto. (Cap. 7, Sec. 1, pág. 181)

inorganic compound / compuesto orgánico: compuesto cuyos constituyentes son otros elementos, en vez del carbono y cuyos átomos por lo general pueden arreglarse en sólo una estructura, como por ejemplo, el H_2O. (Cap. 4, Sec. 1, pág. 103)

insight / discernimiento: forma de razonamiento que permite a los animales usar las experiencias previas para resolver nuevos problemas. (Cap. 7, Sec. 1, pág. 185)

instinct / instinto: patrón complejo de comportamiento innato, como por ejemplo, tejer una telaraña y el que puede demorar semanas en completarse. (Cap. 7, Sec. 1, pág. 182)

interference / interferencia: capacidad de dos o más ondas de combinarse y formar una nueva onda cuando se traslapan. (Cap. 20, Sec. 3, pág. 591)

ion / ion: átomo con carga eléctrica cuya carga resulta cuando un átomo pierde o gana electrones. (Cap. 18, Sec. 2, pág. 525)

isotopes / isótopos: átomos del mismo elemento con diferentes números de neutrones. (Cap. 18, Sec. 1, pág. 521)

K

kelvin (K) / kelvin (K): unidad de temperatura del SI. (Cap. 2, Sec. 2, pág. 50)

kidney / riñón: órgano del sistema urinario en forma de frijol y que está formado por cerca de 1 millón de nefrones; filtra la sangre produciendo la orina. (Cap. 6, Sec. 2, pág. 164)

kilogram (kg) / kilogramo (kg): unidad de masa del SI. (Cap. 2, Sec. 2, pág. 49)

L

larynx / laringe: vía respiratoria a la cual se encuentran adheridas las cuerdas vocales. (Cap. 6, Sec. 1, pág. 157)

lava / lava: roca derretida que fluye a la superficie terrestre. (Cap. 15, Sec. 2, pág. 429)

law of conservation of mass / conservación de la masa: establece que la masa ni se crea ni se destruye. Como resultado, la masa de las sustancias antes de un cambio físico o químico es igual a la masa de las sustancias después del cambio. (Cap. 19, Sec. 2, pág. 561)

law of reflection / ley de la reflexión: establece que el ángulo de incidencia es igual al ángulo de reflexión. (Cap. 23, Sec. 2, pág. 669)

lens / lente: objeto transparente que tiene por lo menos una superficie que hace que la luz se doble. (Cap. 23, Sec. 3, pág. 677)

light ray / rayo luminoso: rayo angosto de luz que viaja en línea recta. (Cap. 23, Sec. 1, pág. 664)

limiting factor / factor limitativo: cualquier cosa que puede limitar el tamaño de una población, incluye los organismos vivos y los inanimados de un ecosistema, como los depredadores o la sequía. (Cap. 12, Sec. 2, pág. 338)

line graph / gráfica lineal: tipo de gráfica que se utiliza para mostrar la relación entre dos variables, en forma de números, en un eje *x* y un eje *y*. (Cap. 2, Sec. 3, pág. 53)

lithosphere / litosfera: capa rígida de la Tierra de unos 100 km de grosor formada por la corteza y parte del manto superior. (Cap. 14, Sec. 3, pág. 400)

long-day plant / planta de día largo: planta que necesita, por lo general, noches cortas (menos de diez a 12 horas de oscuridad) para comenzar el proceso de floración. (Cap. 5, Sec. 2, pág. 142)

loudness / volumen: de un sonido es el grado de percepción humana de la cantidad de energía que transporta la onda. (Cap. 21, Sec. 1, pág. 606)

lunar eclipse / eclipse lunar: ocurre durante la luna llena cuando el Sol, la Luna y la Tierra se alinean de una manera específica y la Luna se mueve dentro de la sombra de la Tierra. (Cap. 17, Sec. 2, pág. 492)

M

magnitude / magnitud: medida de la energía que libera un terremoto. (Cap. 15, Sec. 1, pág. 423)

mass / masa: cantidad de materia que posee un cuerpo, la cual se mide en kilogramos. (Cap. 2, Sec. 2, pág. 49)

mass number / número de masa: el número de protones más el número de neutrones en un átomo. (Cap. 18, Sec. 1, pág. 521)

matter / materia: cualquier cosa que tiene masa y ocupa espacio; las propiedades de la materia están determinadas según la estructura y enlace de sus átomos. (Cap. 18, Sec. 1, pág. 518)

measurement / medida: manera de describir objetos y eventos con números; por ejemplo: longitud, volumen, masa, peso y temperatura. (Cap. 2, Sec. 1, pág. 38)

mechanical wave / onda mecánica: tipo de onda que sólo puede viajar a través de la materia. (Cap. 20, Sec. 1, pág. 577)

medium / medio: material a través del cual puede viajar una onda. (Cap. 23, Sec. 1, pág. 665)

meiosis / meiosis: proceso reproductor que produce cuatro células sexuales haploides a partir de una célula diploide y asegura que la progenie tenga el mismo número de cromosomas que el organismo progenitor. (Cap. 8, Sec. 2, pág. 219)

menstrual cycle / ciclo menstrual: ciclo de cambios mensual controlado por hormonas del sistema reproductor femenino. Incluye la maduración de un óvulo y la preparación del útero para un posible embarazo. (Cap. 10, Sec. 2, pág. 278)

menstruation / menstruación: descarga mensual de sangre y células tisulares que ocurre cuando el revestimiento uterino se desintegra y se desprende. (Cap. 10, Sec. 2, pág. 278)

metabolism / metabolismo: el total de todas las reacciones químicas en un organismo. (Cap. 4, Sec. 3, pág. 113)

meter (m) / metro (m): unidad de longitud del SI. (Cap. 2, Sec. 2, pág. 47)

migration / migración: movimiento instintivo de ciertos animales de mudarse a lugares nuevos cuando cambian las estaciones, en busca de alimentos o para encontrar condiciones más propicias para el apareo. (Cap. 7, Sec. 2, pág. 194)

mitochondrion / mitocondria: organelo celular que descompone lípidos y carbohidratos y libera energía. (Cap. 3, Sec. 1, pág. 74)

mitosis / mitosis: proceso celular en que el núcleo se divide para formar dos núcleos idénticos uno al otro e idénticos al núcleo original, en una serie de pasos (profase, metafase, anafase y telofase). (Cap. 8, Sec. 1, pág. 212)

mixture / mezcla: combinación de sustancias en que las sustancias individuales no cambian ni se combinan químicamente, sino que retienen sus propiedades individuales; pueden ser gases, sólidos, líquidos o cualquier combinación de estos dos. (Cap. 4, Sec. 1, pág. 101; Cap. 18, Sec. 2, pág. 527)

model / modelo: representa algo que es muy grande, muy pequeño, muy peligroso, que consume mucho tiempo o que es muy costoso para ser observado directamente. (Cap. 1, Sec. 2, pág. 16)

molecules / moléculas: partículas constitutivas de los compuestos que se forman de la unión de átomos. (Cap. 18, Sec. 2, pág. 524)

moon phases / fases lunares: cambio en la apariencia de la Luna, vista desde la Tierra, debido a la revolución de la Luna alrededor de la Tierra. (Cap. 17, Sec. 2, pág. 489)

mutation / mutación: cualquier cambio permanente en un gene o cromosoma de una célula; puede ser beneficioso, perjudicial o puede tener un efecto mínimo en un organismo. (Cap. 8, Sec. 3, pág. 228)

mutualism / mutualismo: tipo de relación simbiótica en que ambos organismos se benefician. (Cap. 12, Sec. 3, pág. 346)

N

natural frequency / frecuencia natural: frecuencia a la cual vibra un instrumento musical u otro objeto cuando se puntea o se perturba, con relación a su tamaño, forma y el material del cual está hecho. (Cap. 21, Sec. 2, pág. 615)

nebula / nebulosa: nube de gases y partículas de polvo que rotan y de la cual puede haberse formado el sistema solar hace unos 5 billones de años. (Cap. 17, Sec. 3, pág. 503)

nephron / nefrón: una diminuta unidad filtradora del riñón. (Cap. 6, Sec. 2, pág. 165)

neutron / neutrón: partícula sin carga eléctrica ubicada en el núcleo del átomo. (Cap. 18, Sec. 1, pág. 520)

niche / nicho: en un ecosistema, se refiere a la manera en particular en que un organismo sobrevive, obtiene alimentos y refugio y evita peligros. (Cap. 12, Sec. 3, pág. 347)

nitrogen cycle / ciclo del nitrógeno: modelo que describe cómo se mueve el nitrógeno de la atmósfera al suelo, pasando luego a los organismos vivos y, finalmente, de regreso a la atmósfera. (Cap. 13, Sec. 2, pág. 370)

nitrogen fixation / fijación del nitrógeno: proceso en el cual algunos tipos de bacterias que se hallan en el suelo convierten el gas de nitrógeno en una forma de nitrógeno que pueden usar las plantas. (Cap. 13, Sec. 2, pág. 370)

nucleus / núcleo: organelo que controla todas las actividades de una célula y que contiene el material hereditario compuesto por proteínas y DNA. (Cap. 3, Sec. 1, pág. 72)

O

orbit / órbita: trayectoria curva que sigue la Tierra en su movimiento alrededor del Sol. (Cap. 17, Sec. 1, pág. 483)

organ / órgano: estructura, como el corazón, compuesta por tipos diferentes de tejidos que funcionan en conjunto. (Cap. 3, Sec. 1, pág. 77)

organelle / organelo: estructura en el citoplasma de una célula eucariota que puede actuar como lugar de almacenamiento, puede procesar energía, mover materiales o elaborar sustancias. (Cap. 3, Sec. 1, pág. 72)

organic compounds / compuestos orgánicos: los compuestos que siempre contienen hidrógeno y carbono; los carbohidratos, los lípidos, las proteínas y los ácidos nucleicos son compuestos orgánicos que se encuentran en los seres vivos. (Cap. 4, Sec. 1, pág. 102)

osmosis / ósmosis: tipo de transporte pasivo que se lleva a cabo cuando el agua se difunde a través de la membrana celular. (Cap. 4, Sec. 2, pág. 108)

ovary / ovario: base hinchada del pistilo de una angiosperma donde se hallan los óvulos productores de huevos. (Cap. 9, Sec. 3, pág. 253)

ovary / ovario: órgano reproductor femenino que produce óvulo; se encuentra ubicado en la parte inferior del cuerpo. (Cap. 10, Sec. 2, pág. 277)

overtones / sobretonos: múltiplos de la frecuencia fundamental. (Cap. 21, Sec. 2, pág. 617)

ovulation / ovulación: proceso mensual en que un ovario libera un óvulo que entra en el oviducto donde un espermatozoide puede fecundarlo. (Cap. 10, Sec. 2, pág. 277)

ovule / óvulo: en las plantas gimnospermas, la parte reproductora femenina, la cual produce huevos y tejidos almacenadores de alimento. (Cap. 9, Sec. 3, pág. 251)

P

Pangaea / Pangaea: masa de tierra extensa y antigua que una vez estuvo formada por el conjunto de todos los continentes. (Cap. 14, Sec. 1, pág. 392)

parasitism / parasitismo: tipo de relación simbiótica en que un organismo se beneficia y el otro organismo es perjudicado. (Cap. 12, Sec. 3, pág. 346)

passive transport / transporte pasivo: movimiento de sustancias a través de la membrana celular que no involucra el uso de energía celular; incluye la difusión, la ósmosis y la difusión facilitada. (Cap. 4, Sec. 2, pág. 106)

pharynx / faringe: región en forma de conducto por donde pasan los alimentos, los líquidos y el aire. (Cap. 6, Sec. 1, pág. 156)

phenotype / fenotipo: apariencia física externa y comportamiento de un organismo. (Cap. 11, Sec. 1, pág. 304)

pheromone / feromona: poderosa sustancia química producida por un animal para influir sobre el comportamiento de otro animal de la misma especie. (Cap. 7, Sec. 2, pág. 189)

photoperiodism / fotoperiodismo: reacción de una planta a la duración de horas de luz y oscuridad cada día. (Cap. 5, Sec. 2, pág. 142)

photosynthesis / fotosíntesis: proceso mediante el cual las plantas y muchos otros productores utilizan la energía luminosa del Sol para elaborar azúcares que pueden usar como alimento. (Cap. 4, Sec. 3, pág. 114; Cap. 5, Sec. 1, pág. 131)

physical change / cambio físico: cambio que experimenta la forma o apariencia de una sustancia pero sin alterar su composición. (Cap. 19, Sec. 2, pág. 552)

physical property / propiedad física: característica que se puede observar usando los cinco sentidos, sin alterar o tratar de alterar la composición de una sustancia. (Cap. 19, Sec. 1, pág. 546)

pistil / pistilo: órgano reproductor femenino que se encuentra dentro de la flor de las angiospermas; consta de un estigma pegajoso (donde aterrizan los granos de polen) y de un ovario. (Cap. 9, Sec. 3, pág. 253)

pitch / tono: el grado de agudeza o gravedad de un sonido. (Cap. 21, Sec. 1, pág. 608)

plate / placa: región extensa del manto superior rígido y de la corteza oceánica o continental de la Tierra que se mueve sobre la astenosfera. (Cap. 14, Sec. 3, pág. 400)

plate tectonics / tectónica de placas: teoría que afirma que la corteza y el manto superior terrestres se separan en placas que flotan y se mueven sobre una capa viscosa del manto. (Cap. 14, Sec. 3, pág. 400)

pollen grain / grano de polen: estructura pequeña producida por los órganos reproductores masculinos de una planta de semilla; posee un revestimiento resistente al agua, se puede desarrollar a partir de una espora y contiene partes gametofitas que producen espermatozoides. (Cap. 9, Sec. 3, pág. 249)

pollination / polinización: traspaso de los granos de polen a la parte femenina de una planta de semilla efectuado por agentes como la gravedad, el agua, el viento y los animales. (Cap. 9, Sec. 3, pág. 249)

polygenic inheritance / herencia poligénica: la que ocurre cuando un grupo de pares de genes actúan en conjunto y producen un rasgo específico; por ejemplo el color de los ojos, el color del cabello, el color de la piel o la estatura de los seres humanos. (Cap. 11, Sec. 2, pág. 310)

population / población: todos los organismos que pertenecen a la misma especie y que viven en una comunidad. (Cap. 12, Sec. 1, pág. 334)

precision / precisión: describe el grado de aproximación de las medidas entre sí y el grado de exactitud con que se tomaron tales medidas. (Cap. 2, Sec. 1, pág. 40)

pregnancy / embarazo: período de desarrollo, generalmente cerca de 38 ó 39 semanas en los seres humanos, a partir de un óvulo fecundado hasta el nacimiento. (Cap. 10, Sec. 3, pág. 282)

producer / productor: organismo que utiliza fuentes externas de energía como el Sol, para fabricar moléculas ricas en energía; por ejemplo, las plantas o algas verdes. (Cap. 12, Sec. 3, pág. 344)

prothallus / protalo: forma vegetal gametofita de un helecho, pequeña, verde y en forma de corazón, capaz de producir su propio alimento y absorber agua y nutrientes del suelo. (Cap. 9, Sec. 2, pág. 246)

proton / protón: partícula con carga positiva ubicada en el núcleo de un átomo. (Cap. 18, Sec. 1, pág. 520)

Punnett square / cuadrado de Punnett: instrumento que se usa para predecir ciertos rasgos en la progenie, que muestra las distintas maneras en que los alelos se pueden combinar. (Cap. 11, Sec. 1, pág. 304)

R

radiant energy / energía radiante: energía que transportan las ondas electromagnéticas. (Cap. 22, Sec. 1, pág. 638)

radio waves / ondas radiales: ondas electromagnéticas de la más baja frecuencia que transportan la menor cantidad de energía y las cuales se utilizan en casi todas las formas de telecomunicaciones, por ejemplo, en los televisores, en los teléfonos y en los radios. (Cap. 22, Sec. 2, pág. 640)

rate / tasa: razón de dos clases distintas de medidas. (Cap. 2, Sec. 2, pág. 50)

recessive / recesivo: describe un rasgo que es cubierto o dominado por otra forma de ese rasgo y, por lo tanto, parece desaparecer. (Cap. 11, Sec. 1, pág. 302)

reflecting telescope / telescopio reflector: usa un espejo cóncavo para recoger la luz de cuerpo distantes. (Cap. 23, Sec. 4, pág. 683)

reflection / reflexión: ocurre cuando una onda choca contra un cuerpo o una superficie y rebota. (Cap. 20, Sec. 3, pág. 587; Cap. 23, Sec. 1, pág. 665)

reflex / reflejo: comportamiento innato simple, como bostezar o parpadear, que es una respuesta automática y que no involucra el envío de un mensaje al encéfalo. (Cap. 7, Sec. 1, pág. 181)

refracting telescope / telescopio refractor: usa dos lentes convexas para recoger la luz y formar una imagen de un cuerpo distante. (Cap. 23, Sec. 4, pág. 682)

refraction / refracción: desviación de una onda a medida que se mueve de un medio a otro. (Cap. 20, Sec. 3, pág. 588; Cap. 23, Sec. 3, pág. 677)

resonance / resonancia: amplificación sonora que ocurre cuando un cuerpo vibra a su frecuencia natural al absorber energía de una onda sonora u otro cuerpo que vibra a esa misma frecuencia. (Cap. 21, Sec. 2, pág. 616)

respiration / respiración: serie de reacciones químicas utilizadas para liberar la energía almacenada en las moléculas de los alimentos. (Cap. 4, Sec. 3, pág. 115; Cap. 5, Sec. 1, pág. 133)

reverberation / reverberación: ecos de sonidos repetidos. (Cap. 21, Sec. 2, pág. 621)

revolution / revolución: movimiento de la Tierra alrededor del Sol, el cual toma unos 365 1/4 días, o sea, un año en completarse. (Cap. 17, Sec. 1, pág. 483)

rhizome / rizoma: tallo subterráneo de un helecho. (Cap. 9, Sec. 2, pág. 246)

ribosome / ribosoma: estructura pequeña en la cual las células producen sus propias proteínas. (Cap. 3, Sec. 1, pág. 74)

rift / dislocación: grieta larga que se forma entre placas tectónicas que se están separando en los límites de placas. (Cap. 15, Sec. 3, pág. 437)

RNA / RNA: ácido ribonucleico que lleva consigo los códigos para la elaboración de proteínas del núcleo a los ribosomas. (Cap. 8, Sec. 3, pág. 226)

rotation / rotación: movimiento giratorio de la Tierra sobre su eje que causa el día y la noche; la Tierra tarda 24 horas en completar una rotación completa. (Cap. 17, Sec. 1, pág. 482)

S

salinity / salinidad: medida de la cantidad de sales disueltas en el agua marina. (Cap. 16, Sec. 1, pág. 454)

science / ciencia: proceso que se usa para investigar lo que ocurre a nuestro alrededor, con el propósito de resolver problemas o responder preguntas; forma parte de la vida cotidiana. (Cap. 1, Sec. 1, pág. 6)

scientific methods / métodos científicos: maneras de resolver problemas que pueden incluir la planificación paso por paso, la confección de modelos y experimentos programados cuidadosamente. (Cap. 1, Sec. 2, pág. 13)

seafloor spreading / expansión del suelo marino: teoría de Hess que afirma que el nuevo suelo marino se forma cuando el magma es forzado a subir a la superficie en una dorsal mediooceánica. (Cap. 14, Sec. 2, pág. 397)

seismic safe / seguridad sísmica: describe la capacidad de las estructuras de soportar las vibraciones que causa un terremoto. (Cap. 15, Sec. 1, pág. 427)

seismic waves / ondas sísmicas: ondas sísmicas, que incluyen las ondas primarias, las secundarias y las ondas de superficie. (Cap. 15, Sec. 1, pág. 422)

seismograph / sismógrafo: instrumento que se utiliza para registrar las ondas sísmicas. (Cap. 15, Sec. 1, pág. 423)

semen / semen: mezcla de espermatozoides y un líquido que ayuda a los espermatozoides a moverse y que les sirve como fuente de energía. (Cap. 10, Sec. 2, pág. 276)

sex-linked gene / gene ligado al sexo: un alelo heredado en un cromosoma del sexo y que puede causar trastornos genéticos, como por ejemplo, el daltonismo o la hemofilia. (Cap. 11, Sec. 2, pág. 313)

sexual reproduction / reproducción sexual: tipo de reproducción en que dos células sexuales, por lo general un óvulo y un espermatozoide, se unen formando un cigoto, el cual se desarrolla en un nuevo organismo con su propia identidad. (Cap. 8, Sec. 2, pág. 218)

shield volcano / volcán de escudo: volcán extenso y ancho y de laderas levemente inclinadas que se forma de la acumulación de capas basálticas. (Cap. 15, Sec. 2, pág. 432)

short-day plant / planta de día corto: planta que necesita, por lo general, noches largas (12 ó más horas de oscuridad) para comenzar el proceso de floración. (Cap. 5, Sec. 2, pág. 142)

SI / SI: Sistema internacional de unidades, relacionado por múltiplos de diez, que permite que las cantidades se midan de la misma manera exacta en todo el mundo. (Cap. 2, Sec. 2, pág. 46)

social behavior / comportamiento social: se dice de las interacciones entre los miembros de la misma especie; incluye el comportamiento de cortejo, el apareo, la obtención de alimentos, el cuidado de las crías y la protección mutua. (Cap. 7, Sec. 2, pág. 186)

society / sociedad: grupo de animales de la misma especie que viven y trabajan juntos de manera organizada en la que cada cual realiza una tarea específica. (Cap. 7, Sec. 2, pág. 187)

soil / suelo: mezcla de partículas minerales y rocosas, restos de organismos muertos, aire y agua que forma la capa superior de la corteza terrestre y que sostiene el crecimiento vegetal. (Cap. 13, Sec. 1, pág. 362)

solar eclipse / eclipse solar: ocurre durante la luna nueva, cuando el Sol, la Luna y la Tierra se encuentran alineados de una forma específica y la Tierra se mueve dentro de la sombra de la Luna. (Cap. 17, Sec. 2, pág. 491)

solar system / sistema solar: sistema extremadamente grande que incluye el Sol, los planetas, los cometas, los meteo roides y otros cuerpos que orbitan el Sol. (Cap. 17, Sec. 3, pág. 496)

solstice / solsticio: época cuando el Sol alcanza su mayor distancia al norte o al sur del ecuador; ocurre el 21 ó el 22 de junio en el hemisferio norte (el día más largo del año) y el 21 ó 22 de diciembre en el hemisferio sur (el día más corto del año en el hemisferio norte). (Cap. 17, Sec. 1, pág. 484)

solution / solución: tipo de mezcla en que una sustancia se mezcla completa y uniformemente con otra y que es idéntica en toda su extensión. (Cap. 18, Sec. 2, pág. 527)

sori / soros: estructuras de los helechos en los cuales se producen las esporas. (Cap. 9, Sec. 2, pág. 246)

sperm / espermatozoide: célula sexual haploide formada en los órganos reproductores masculinos. (Cap. 8, Sec. 2, pág. 218; Cap. 10, Sec. 2, pág. 276)

spores / esporas: células haploides producidas en la etapa gametofita que se puede dividir mediante la mitosis y pueden formar estructuras vegetales o una planta nueva completa o que se puede desarrollar en células sexuales. (Cap. 9, Sec. 1, pág. 243)

sporophyte stage / etapa esporofita: etapa del ciclo de vida vegetal que comienza cuando un espermatozoide fecunda un huevo. (Cap. 9, Sec. 1, pág. 243)

stamen / estambre: órgano reproductor masculino que se encuentra dentro de la flor de las angiospermas; consta de una antera (donde se forman los granos de polen) y de un filamento. (Cap. 9, Sec. 3, pág. 253)

stomata / estomas: aberturas microscópicas de la epidermis de las plantas a través de las cuales los gases de dióxido de carbono y vapor de agua entran y salen de una hoja. (Cap. 5, Sec. 1, pág. 129)

surface current / corriente de superficie: una corriente oceánica accionada por el viento; este tipo de corriente se mueve horizontalmente, paralela a la superficie de la Tierra y solamente mueve unos pocos cientos de los metros superiores del agua marina. (Cap. 16, Sec. 2, pág. 456)

symbiosis / simbiosis: cualquier relación estrecha entre especies, incluye el mutualismo, el comensalismo y el parasitismo. (Cap. 12, Sec. 3, pág. 346)

T

table / tabla: despliega información en hileras y columnas facilitando la lectura y comprensión de los datos. (Cap. 2, Sec. 3, pág. 53)

technology / tecnología: aplicación de la ciencia para fabricar productos y herramientas útiles, como por ejemplo, las computadoras. (Cap. 1, Sec. 1, pág. 9)

testes / testículos: órgano masculino productor de espermatozoides y testosterona. (Cap. 10, Sec. 2, pág. 276)

tidal range / amplitud de la marea: la diferencia entre el nivel del océano en pleamar y su nivel en bajamar. (Cap. 16, Sec. 3, pág. 466)

tide / marea: ascenso y descenso del nivel del mar provocado, en su mayor parte, por la interacción de la gravedad en el sistema Tierra-Luna. (Cap. 16, Sec. 3, pág. 465)

tissue / tejido: grupo de células semejantes que funcionan juntas para efectuar una tarea. (Cap. 3, Sec. 1, pág. 77)

trachea / tráquea: conducto transportador de aire que une la laringe con los bronquios; este conducto está forrado de membranas mucosas y cilios y contiene anillos cartilaginosos resistentes. (Cap. 6, Sec. 1, pág. 157)

transverse wave / onda transversal: tipo de onda mecánica en la cual la energía de la onda hace que la materia del medio suba o baje u oscile formando ángulos rectos con la dirección en que viaja la onda. (Cap. 20, Sec. 1, pág. 578)

tropism / tropismo: reacción positiva o negativa a un estímulo externo como el tacto, la luz o la gravedad. (Cap. 5, Sec. 2, pág. 138)

trough / seno: es el punto más bajo de una onda. (Cap. 16, Sec. 3, pág. 462)

tsunami / tsunami o maremoto: onda marina sísmica de gran intensidad que comienza sobre un terremoto del suelo oceánico; puede alcanzar 30 m de altura cuando se aproxima a tierra y puede causar destrucción en las zonas costeras. (Cap. 15, Sec. 1, pág. 425)

U

ultraviolet radiation / radiación ultravioleta: ondas electromagnéticas con frecuencias más altas y longitudes de onda más cortas que la luz visible. (Cap. 22, Sec. 2, pág. 643)

upwelling / corriente de aguas resurgentes: circulación del océano que lleva agua fría y profunda a la superficie oceánica. (Cap. 16, Sec. 2, pág. 459)

ureter / uréter: conducto que transporta la orina desde los riñones hasta la vejiga. (Cap. 6, Sec. 2, pág. 166)

urethra / uretra: conducto que transporta la orina desde la vejiga y la expulsa del cuerpo. (Cap. 6, Sec. 2, pág. 166)

urinary system / sistema urinario: sistema de órganos excretores que elimina los residuos de la sangre, controla el volumen sanguíneo al eliminar el exceso de agua y mantiene el equilibrio en las concentraciones de sal y agua. (Cap. 6, Sec. 2, pág. 163)

urine / orina: líquido residual que contiene el exceso de agua, sales y otros residuos que el cuerpo no reabsorbe. (Cap. 6, Sec. 2, pág. 164)

uterus / útero: órgano hueco, muscular y con forma de pera donde un óvulo fecundado se desarrolla hasta convertirse en un bebé. (Cap. 10, Sec. 2, pág. 277)

V

vagina / vagina: conducto muscular que conecta el extremo inferior del útero con la parte externa del cuerpo; el canal del nacimiento por el cual pasa el bebé cuando está naciendo. (Cap. 10, Sec. 2, pág. 277)

virus / virus: hebra de material hereditario rodeada por una capa proteíca. (Cap. 3, Sec. 3, pág. 84)

visible light / luz visible: ondas electromagnéticas con longitudes de onda entre 0.4 y 0.7 millonésimas de metro y las cuales se pueden ver a simple vista. (Cap. 22, Sec. 2, pág. 642)

volcano / volcán: montaña o colina cónica que se forma por la emisión de magma caliente, sólidos y gases a la superficie terrestre a través de una chimenea. (Cap. 15, Sec. 2, pág. 429)

W

water cycle / ciclo del agua: modelo que describe cómo se mueve el agua de la superficie de la Tierra a la atmósfera para regresar nuevamente a la superficie a través de la evaporación, condensación y precipitación. (Cap. 13, Sec. 2, pág. 369)

wave / onda: perturbación rítmica que transporta energía pero no materia. (Cap. 16, Sec. 3, pág. 462; Cap. 20, Sec. 1, pág. 576)

wavelength / longitud de onda: en las ondas transversales, es la distancia entre la parte superior de dos crestas adyacentes o la parte inferior de dos senos adyacentes; en las ondas de compresión, es la distancia desde los centros de rarefacciones adyacentes. (Cap. 20, Sec. 2, pág. 582)

X

X ray / rayo X: onda electromagnética de alta frecuencia que es muy penetrante y la cual se usa en el diagnóstico médico. (Cap. 22, Sec. 2, pág. 644)

Z

zygote / cigoto: nueva célula diploide que se forma cuando un espermatozoide fecunda un óvulo; se divide mediante mitosis y se desarrolla en un nuevo organismo. (Cap. 8, Sec. 2, pág. 218)

Index

The index for *Glencoe Science* will help you locate major topics in the book quickly and easily. Each entry in the index is followed by the number of the pages on which the entry is discussed. A page number given in boldfaced type indicates the page on which that entry is defined. A page number given in italic type indicates a page on which the entry is used in an illustration or photograph. The abbreviation *act.* indicates a page on which the entry is used in an activity.

A

B

C

Index

D

E

F

Index

N

O

Index

P

R

Index

U

W

Index

Credits

Art Credits

Glencoe would like to acknowledge the artists and agencies who participated in illustrating this program: Absolute Science Illustration; Andrew Evansen; Argosy; Articulate Graphics; Craig Attebery represented by Frank & Jeff Lavaty; CHK America; Gagliano Graphics; Pedro Julio Gonzalez represented by Melissa Turk & The Artist Network; Robert Hynes represented by Mendola Ltd.; Morgan Cain & Associates; JTH Illustration; Laurie O'Keefe; Matthew Pippin represented by Beranbaum Artist's Representative; Precision Graphics; Publisher's Art; Rolin Graphics, Inc.; Wendy Smith represented by Melissa Turk & The Artist Network; Kevin Torline represented by Berendsen and Associates, Inc.; WILDlife ART; Phil Wilson represented by Cliff Knecht Artist Representative; Zoo Botanica.

Photo Credits

Abbreviation Key: AA=Animals Animals; AH=Aaron Haupt; AMP=Amanita Pictures; BC=Bruce Coleman, Inc.; CB=CORBIS; DM=Doug Martin; DRK=DRK Photo; ES=Earth Scenes; FP=Fundamental Photographs; GH=Grant Heilman Photography; IC=Icon Images; KS=KS Studios; LA=Liaison Agency; MB=Mark Burnett; MM=Matt Meadows; PE=PhotoEdit; PD=PhotoDisc; PQ=PictureQuest; PR=Photo Researchers; SB=Stock Boston; TSA=Tom Stack & Associates; TSM=The Stock Market; VU=Visuals Unlimited.

Cover (l)Corbis, (r)Johnny Johnson/Animal Animals, (bkgd)Chad Ehlers/Stone; **vii** William J. Weber; **ix** Jeremy Woodhouse/DRK; **xi** James H. Robinson; **xiii** Richard T. Nowitz; **xvi** Gary Rosenquist; **xvii** Lappa/Marguart; **xviii** Klaus Guldbrandsen/Science Photo Library/PR; **xix** Georg Custer/PR; **xxi** Timothy Fuller; **xxiii** BMDO/NRL/LLNL/Science Photo Library/PR; **2** PD; **2–3** Wolfgang Kaehler; **3** PD; **4** Moredon Animals Health Ltd./Science Photo Library/PR; **4–5** AP/Wide World Photos; **5** MM; **6** (tr)Stephen Webster, (others)KS; **7 8** AH; **9** (t)Bob Daemmrich, (bl)Paul A. Souders/CB, (br)KS; **10** AH; **11** Geoff Butler; **14** KS; **15** (l)IC, (r)F. Fernandes/Washington Stock Photo; **16** AH; **17** MM; **18** DM; **19** AH; **20** (tr)courtesy IWA Publishing, (others)Patricia Lanza; **21** AMP; **22** Laura Sifferlin; **23** Jeff Greenberg/VU; **24** KS; **25** Jeff Smith/The Image Bank; **26** (tl)Sarita M. James, (tc)AFP/CB, (tr)James D. Wilson/LA, (cl)Bettmann/CB, (c)Bob Rowan/Progressive Image/CB, (cr)Fred Begay, (bl)NASA, (bcl)CB, (bcr)Provident Foundation/CED Photographic Service, (br)Bettmann/CB; **27** The Image Bank; **28** (l)Dominic Oldershaw, (r)Richard Hutchings; **29** Dominic Oldershaw; **30** Gary Retherford/PR; **31** Karl Ammann/CB; **32** (t)MM, (c)Still Pictures/Peter Arnold, Inc., (b)The Image Bank; **33** (l)AH, (r)DM; **34** Geoff Butler; **36** MB; **36–37** Brent Jones/SB; **37** MM; **38** Paul Almasy/CB; **39** AFP/CB; **40** David Young-Wolff/PE; **41** (l)Lowell D. Franga, (tr)The Purcell Team/CB, (br)Len Delessio/Index Stock; **42** Photos by Richard T. Nowitz, imaging by Janet Dell Russell Johnson; **43** MM; **45** MB; **47** Tom Prettyman/PE; **49** (t)Michael Dalton/FP, (tc)David Young-Wolff/PE, (bc)Dennis Potokar/PR, (b)MM; **51** Michael Newman/PE; **53** John Cancalosi/SB; **56** Richard Hutchings; **57** Richard Hutchings; **58** (t)Fletcher & Baylis/PR, (b)Owen Franken/CB; **59** CMCD/PD; **60** Fred Bavendam/Stone; **61** MB; **65** MM; **66–67** Richard Hamilton Smith/CB; **67** (t)courtesy Broan-NuTone, (b)CB; **68** Oliver Meckes/E.O.S./MPI-Tubingen/PR; **68–69** Tim Flach/Stone; **69** MM; **71** Biophoto Associates/PR; **72** (t)Don Fawcett/PR, (b)M. Schliwa/PR; **74** (t)George B. Chapman/VU, (b)P. Motta & T. Naguro/Science Photo Library/PR; **75** (t)Don Fawcett/PR, (b)Biophoto Associates/PR; **79** David M. Phillips/VU; **80** (tl)courtesy Olympus Corporation, (tr)Kathy Talaro/VU, (cl)David M. Phillips/VU, (cr)Mike Abbey/VU, (bl)David M. Phillips/VU, (br)Michael Gabridge/VU; **81** (tl)James W. Evarts, (tr)Bob Krist/CB, (c)Mike Abbey/VU, (bl)Karl Aufderheide/VU, (br)Lawrence Migdale/SB/PQ; **84** (l)Richard J. Green/PR, (c)Dr. J.F.J.M. van der Heuvel, (r)Gelderblom/Eye of Science/PR; **87** Pam Wilson/Texas Dept. of Health; **88 89** MM; **90** (t)Quest/Science Photo Library/PR, (b)courtesy California University; **91** Nancy Kedersha/Immunogen/Science Photo Library/PR; **92** (t)Oliver Meckes/PR, (b)CDC/Science Photo Library/PR; **93** (l)Keith Porter/PR, (r)NIBSC/Science Photo Library/PR; **94** Biophoto Associates/Science Source/PR; **96** Bill Longcore/PR; **96–97** KS; DM; AH; **99** Bob Daemmrich; **101** (t)Runk/Schoenberger from GH, (b)Klaus Guldbrandsen/Science Photo Library/PR; **106** (l)John Fowler, (r)Richard Hamilton Smith/CB; **107** KS; **108** AH; **109** VU; **110** Biophoto Associates/Science Source/PR; **111** Stephen R. Wagner; **112** MM; **113** AH; **114** Craig Lovell/CB; **115** John Fowler; **116** David M. Phillips/VU; **117** (l)GH, (r)Bios (Klein/Hubert)/Peter Arnold, Inc.; **118** (t)Runk/Schoenberger from GH, (b)AH; **119** MM; **120** Lappa/Marguart; 121 The Just/Garcia/Hill Science Web Site; **122** (tl)Michael Pogany/Columbus Zoo, (tr)Bachman/PR, (b)Jim Strawser from GH; **123** CNRI/Science Photo Library/PR; **126** AMP; **126–127** Terry Thompson/Panoramic Images; **127** MM; **129** Dr. Jeremy Burgess/Science Photo Library/PR; **130** (l)John Kieffer/Peter Arnold, Inc., (r)Runk/Schoenberger from GH; **131** M. Eichelberger/VU; **133** (t)Jacques Jangoux/Peter Arnold, Inc., (b)Jeff Lepore/PR; **134** Michael P. Gadomski/PR; **137** Howard Miller/PR; **138** (l)Scott Camazine/PR, (c r)MM; **141** (tl tr)Artville, (cl)Runk/Schoenberger from GH, (c cr)Prof. Malcolm B. Wilkins/University of Glasgow, (bl)Eric Brennan, (br)John Sohlden/VU; **142** Jim Metzger; **144** (t)Ed Reschke/Peter Arnold, Inc., (b)MM; **145** MM; **146** Stone; **147** Kennedy Colombo; **148** (tl)Ed Reschke/Peter Arnold, Inc., (tr)Runk/Schoenberger from GH, (bl)Holt Studios International/Nigel Cattlin/PR, (br)Laura Dwight/Peter Arnold, Inc.; **149** (l)Norm Thomas/PR, (r)S.R. Maglione/PR; **152** Prof. P. Motta, Dept. of Anatomy, University La Sapienza, Rome/Science Photo Library/PR; **152–153** David Madison/Newsport Photography; **153** Geoff Butler; **154** Randy Lincks/CB; **155** Dominic Oldershaw; **156** Bob Daemmrich; **159** Richard T. Nowitz; **161** (l c)SIU/PR, (r)Geoff Butler; **162** Renee Lynn/PR; **164** (l)Science Pictures Ltd./Science Photo Library/PR, (r)SIU/PR; **166** Paul Barton/TSM; **167** (l)Gunther/Explorer/PR, (c r)MB; **168** Richard Hutchings/PR; **169** Cabisco/VU; **170** (t)MM, (b)Larry Mulvehill/PR; **171** MM; **172** Lane Medical Library; **172–173** Science Photo Library/CB; **173** Custom Medical Stock Photo; **174** (t)Biophoto Associates/Science Source/PR, (b)Custom Medical Stock Photo; **175** (l)Ed Beck/TSM, (tr)Gregg Ozzo/VU, (br)Tom & DeeAnn McCarthy; **178** Gary W. Carter/VU; **178–179** Robert Mackinlay/Peter Arnold, Inc.; **179** MB; **180** (l)Michel Denis-Huot/Jacana/PR, (r)Zig Leszczynski/AA; **181** (l)Jack Ballard/VU, (c)Anthony Mercieca/PR, (r)Joe McDonald/VU; **182** (t)Stephen J. Krasemann/Peter Arnold, Inc., (b)Leonard Lee Rue/PR; **183** (t)The Zoological Society of San Diego, (b)Margaret Miller/PR; **186** Michael Fairchild; **187** (t)Bill Bachman/PR, (b)Fateh Singh Rathore/Peter Arnold, Inc.; **188** Jim Brandenburg/Minden Pictures; **189** Michael Dick/AA; **190** (l)Richard Thorn/VU, (c)Arthur Morris/VU, (r)Jacana/PR; **191** (tl)T. Frank/Harbor Branch Oceanographic Institution, (bl bc)Peter J. Herring, (others)Edith Widder/Harbor Branch Oceanographic Institution; **192** Stephen Dalton/AA; **193** Richard Packwood/AA; **194** Ken Lucas/VU; **196** (t)Dave B. Fleetham/TSA, (b)Gary Carter/VU; **197** The Zoological Society of San Diego; **198** Walter Smith/CB; **198–199** Bios (Klein/Hubert)/Peter Arnold, Inc.; **199** courtesy The Seeing Eye; **200** (tl)Norbert Wu/Peter Arnold, Inc., (tr)Fritz Prenzel/AA, (b)Remy Amann-Bios/Peter Arnold, Inc.; **201** (l)Valerie Giles/PR, (r)J & B Photographers/AA; **202** Alan & Sandy Carey/PR; **205** (l)Biophoto Associates/PR, (r)Lee D. Simon/PR; **206** Microworks/PhotoTake NYC; **206–207** Doug Wilson/CB; **208** Biophoto Associates/PR; **208–209** Zig Leszczynski/AA; **209** MM; **210** (l)Dave B. Fleetham/TSA, (r)Cabisco/VU; **212** Cabisco/VU; **213** (t)Michael Abbey/VU, (others)John D. Cunningham/VU; **214** (l)MM, (r)Nigel Cattlin/PR; **215** (l)Barry L. Runk from GH, (r)Runk/Schoenberger from GH; **216** (l)Walker England/PR, (r)TSA; **217** Runk/Schoenberger from GH; **218** David M. Phillips/The Population Council/PR; **219** (tl)Gerald & Buff Corsi/VU, (bl)Susan McCartney/PR, (r)Fred Bruenner/Peter Arnold, Inc.; **221** (l)John D. Cunningham/VU, (c)Jen & Des Bartlett/BC, (r)Breck P. Kent; **222** (tl)Artville, (tr)Tim Fehr, (c)Bob Daemmrich/SB/PQ, (bl)Troy Mary Parlee/Index Stock/PQ, (br)Jeffery Myers/Southern Stock/PQ; **228** Stewart Cohen/Stone; **230** (t)Tom McHugh/PR, (b)file photo; **231** Monica Dalmasso/Stone; **232** Philip Lee Harvey/Stone; **233** Lester V. Bergman/CB; **234** (l)Camille Tokerud/PR, (tr)John Mitchell/PR, (br)David Scharf/Peter Arnold, Inc.; **235** (l)D. Yeske/VU, (r)GH; **238** Noble Proctor/PR; **238–239** Layne Kennedy/CB; **239** DM; **240** (l)Stephen Dalton/PR, (r)MM; **241** (l)Holt Studios/Nigel Cattlin/PR, (r)Inga Spence/VU; **242** (l)H. Reinhard/Okapia/PR, (c)John W. Bova/PR, (r)John D. Cunningham/VU; **244** (l)Biology Media/PR, (c)Andrew Syred/Science Photo Library/PR, (r)Runk/Schoenberger from GH; **246 247** Kathy Merrifield 2000/PR; **248** MM; **249** (l)John Kaprielian/PR, (r)Scott Camazine/Sue Trainor/PR; **250** Dr. Wm. H. Harlow/PR; **251** Christian Grzimek/OKAPIA/PR; **252** (l)M.J. Griffith/PR, (c)Stephen P. Parker/PR, (r)Dan Suzio/PR; **253** (t)Rob Simpson/VU, (c)Gustav Verderber/VU, (b)Alvin E. Staffan/PR; **254** (tl)C. Nuridsany & M. Perennou/Science Photo Library/PR, (tr)Merlin D. Tuttle/PR, (bl)Anthony Mercreca Photo/PR, (bc)Kjell B. Sandved/PR, (br)Holt Studios LTD/PR; **255** William J. Weber/VU; **257** (tl)Kevin Shafer/CB, (c)Darryl Torckler/Stone, (bc)Tom & Pat Leeson, (others)Dwight Kuhn, **258 260** DM; **261** MM; **262** Michael Black/BC; **263** (t)courtesy NIGMS OCPL, (b)Kevin Laubacher/FPG; **264** (tl)Zig Leszczynski/ES, (bl)MM/Peter Arnold, Inc., (r)Tim Davis/PR; **265** (l)Nils Reinhard/OKAPIA/PR, (c)Adrienne T. Gibson/ES, (r)Oliver Meckes/PR; **266** Marcia Griffen/ES; **268** Profs. P.M. Motta & J. Van Blerkom/Science Photo Library/PR; **268–269** Brownie Harris/TSM; **269** John Evans; **270** David Young-Wolff/PE; **272 273** Stephen R. Wagner; **279** Ariel Skelley/TSM; **281** David M. Phillips/PR; **282** (l)Tim Davis/PR, (r)Chris Sorensen/TSM; **283** Science Pictures Ltd./Science Photo Library/PR; **284** Petit Format/Nestle/Science Source/PR; **286** (l)Jeffery W. Myers/SB, (r)Ruth Dixon; **287** (tl b)MB, (tr)AH; **288** KS; **289** (l)NASA/Roger Ressmeyer/CB, (r)AFP/CB; **290** (t)Chris Carroll/CB, (b)Richard Hutchings; **291** MM; **292** (l)John Banagan/The Image Bank, (c)Ron Kimball Photography, (r)SuperStock; **293** (l)Martin B. Withers/Frank Lane Picture Agency/CB, (r)Joe McDonald/CB; **294** (tl)DM, (bl)David M. Phillips/PR, (r)David Woods/TSM; **295** (l)Bob Daemmrich, (r)Maria Taglienti/The Image Bank; **298** David Phillips/Science Source/PR; **298–299** MB; **299** Geoff Butler; **300** Stewart Cohen/Stone; **303** Special Collections, National Agriculture Library, (bkgd)Jane Grushow from GH; **304** Barry L. Runk from GH; **306** Richard Hutchings/PR; **308** (t)Robert Maier/AA, (b)Gemma Giannini from GH; **309** Raymond Gehman/CORBIS; **310** Dan McCoy from Rainbow; **311** (l)Phil Roach/Ipol, Inc., (r)CNRI/Science Photo Library/PR; **312** Gopal Murti/PhotoTake NYC; **313** Tim Davis/PR; **314** (l)Alan & Sandy Carey/PR, (r)Renee Stockdale/AA; **317** Tom Meyers/PR; **318** (t)Runk/Schoenberger from GH, (b)MB; **319** Laura Sefferlin; **321** KS; **322** David R. Frazier Photolibrary; **326** CB; **328–329** Joseph Sohm/ChromoSohm Inc./CB; **329** Andrew A. Wagner; **330** David Cavagnaro/DRK; **330–331** Johnny Johnson/DRK; **331** John D. Cunningham/VU; **332** (tl)Adam Jones/PR, (tc)Tom Van Sant/Geosphere Project, Santa Monica/Science Photo Library/PR, (tr)G. Carleton Ray/PR, (b)Richard Kolar/AA; **333** (t)John W. Bova/PR, (b)David Young/TSA; **335** (t)Zig Leszczynski/AA, (b)Mitsuaki Iwago/Minden Pictures; **339** Joel Sartore from GH; **341** (t)Norm Thomas/PR, (b)Maresa Pryor/ES; **342** (tl)Wyman P. Meinzer, (bl)Wyman P. Meinzer, (r)Bud Neilson/Words & Pictures/PQ; **344** (tl)Michael Abbey/PR, (tr)OSF/AA, (b)Michael P. Gadomski/PR; **345** (tlc)Larry Kimball/VU, (tr)George D. Lepp/PR, (bl)Lynn M. Stone, (blc)Stephen J. Krasemann/Peter Arnold, Inc., (brc)AMP, (others)William J. Weber; **346** (t)Milton Rand/TSA, (c)Marian Bacon/AA, (b)Sinclair Stammers/Science Photo Library/PR; **347** (t)Raymond A. Mendez/AA, (bl)Donald Specker/AA, (br)Joe McDonald/AA; **348** Ted Levin/AA; **349** Richard L. Carlton/PR; **350** (t)Jean Claude Revy/PhotoTake NYC, (b)OSF/AA; **351** Runk/Schoenberger from GH; **353** (l)courtesy US Census, (r)Eric Larravadieu/Stone; **354** (tl)Tui De Roy/Minden Pictures, (bl)Stephen J. Krasemann/DRK, (r)Maslowski/PR; **355** (l)C.K. Lorenz/PR, (r)Hans Pfletschinger/Peter Arnold, Inc.; **356** Scott Camazine/PR; **358** (t)Dr. Jeremy Burgess/Science Photo Library/PR, (b)Jeff Greenberg/VU; **358–359** Steve Bly/International Stock; **360** Kenneth Murray/PR; **361** (t)Jerry L. Ferrara/PR, (b)Art Wolfe/PR; **362** (t)Telegraph Colour Library/FPG, (b)Hal Beral/VU; **363** (l)Fritz Polking/VU, (r)R. Arndt/VU; **364** Tom Uhlman/VU; **368** (l)Jim Grattan, (r)Bruce S. Cushing/VU; **371** (t)Rob & Ann Simpson/VU, (c b)Runk/Schoenberger from GH; **372** Stephen R. Wagner; **374** (l)WHOI/VU, (r)Wolfgang Baumeister/Science Photo Library/PR; **378** (t)MM, (b)Gerald and Buff Corsi/VU; **379** Jeff J. Daly/VU; **380** Gordon Wiltsie/Peter Arnold, Inc.; **382** (tl)Dwight Kuhn, (tr)Stephen J. Krasemann/DRK, (bl)Gregory K. Scott/PR, (br)Simon Battensby/Stone; **383** (l)Soames Summerhay/PR, (r) Tom Uhlman/VU; **388** (t)Ken Lucas/TCL/Masterfile, (b)Patrice Ceisel/SB/PQ; **388–389** Robert Burrington/Index stock; **389** (t)Hal Beral/Photo Network/PQ, (b)Archive Photos/PQ; **390** Francois Gohier/PR; **390–391** Stone; **391** MB; **394** Martin Land/Science Source/PR; **397** Ralph White/CB; **403** Davis Meltzer; **404** Craig Aurness/CB; **406** Craig Brown/Index Stock; **407** Ric Ergenbright/CB; **408** Roger Ressmeyer/CB; **410** Burhan Ozbilici/AP/Wide World Photos; **412** L. Lauber/ES; **413** Courtesy Ed Klimasauskas; **414** (l)courtesy Takeo Suzuki, UCLA, (r)Galen Rowell/CB; **415** (l)Tim Barnwen/SB, (r)Bettmann/CB; **418** JPL/NASA; **418–419** Jim Sugar Photography/CB; **419** AH; **420** KS; **423** (t)Krafft/Explorer/PR, (b)Jean Miele/TSM; **426** (t b)NOAA, (c)Lisa Bigazoli, (bkgd)Galen Rowell/CB; **427** (t)ks, (b) Pacific Seismic Products, Inc.; **428** Roger Ressmeyer/CB; **430** (l)AP/Wide World Photos, (r)Kevin West/AP/Wide World Photos; **432** (t)Breck P. Kent/ES, (b)Dewitt Jones/CB; **433** (t)Lynn Gerig/TSA, (b)Milton Rand/TSA; **435** Otto Hahn/Peter Arnold, Inc.; **436** Spencer Grant/PE; **438** NASA/Peter Arnold, Inc.; **442** AH; **443** AH; **444** (l)Ted Streshinky/CB, (r)Underwood & Underwood/CB; **444–445** Bettmann/CB, (bkgd)Russell D. Curtis/PR; **445** (l)Bettmann/CB, (r)Robert Holmes/CB; **446** (t)Roger Ressmeyer/CB, (b)Tom Walker/Stone; **447** (l)James L. Amos/CB, (c)Michael Collier, (r)Phillip Wallick/TSM; **450** Judy Griesedieck/CB; **450–451** Warren Bolster/Stone; **451** Raven/Explorer/PR; **452** (l)Norbert Wu/Peter Arnold, Inc., (r)Darryl Torckler/Stone; **454** Cathy Church/Picturesque/PQ; **457** Bob Daemmrich; **458** (t)Darryl Torckler/Stone, (b)Raven/Explorer/PR; **462** Jack Fields/PR; **463** Tom & Therisa Stack; **464** (l)Spike Mafford/PD, (r)Douglas Peebles/CB, (bkgd)Stephen R. Wagner; **465** Arnulf Husmo/Stone; **466** (tl)Groenendyk/PR, (tr)Patrick Ingrand/Stone, (b)Kent Knudson/SB; **469** AH; **470** (t)Mark E. Gibson/VU, (b)Timothy Fuller; **471** Timothy Fuller; **473** Seth Resnick/SB/PQ; **474** (t)Worldsat Productions/NRSC/Science Photo Library, (cl)Phillippe Diederich/Contact Press Images/PQ, (cr)S.J. Krasemann/Peter Arnold, Inc., (b)Stephen J. Krasemann/Peter Arnold, Inc.; **475** (l)Carl R. Sams II/Peter Arnold, Inc., (r)Edna Douthat; **478** Johnathan Blair/CB; **478–479** Jerry Schad/PR; **479** MM; **480** NASA/JPL; **481** Jerry Schad/PR; **486 489** Lick Observatory; **491** CB; **493** (bl)Stephen Frisch/SB/PQ, (br)Stephen R. Wagner, (bkgd)NASA, (others)David Meltzer; **494** NASA; **495** Timothy Fuller; **497** NASA/JPL/Northwestern University; **498** (t)NASA/JPL, (b)Dr. Timothy Parker, JPL; **500** (l)CB, (r)Erich Karkoschka, University of Arizona, and NASA; **501** NASA/JPL; **502** (t)Dr. R. Albrecht, ESA/ESO Space Telescope European Coordinating Facility/NASA, (b)Frank Zullo/PR; **506** Kauko Helavuo/The Image Bank; **507** Charles & Josette Lenars/CB; **508** CB; **509** Lick Observatory; **512** Tom Bean/DRK; **514** CB/PQ; **514–515** Stephen Frisch/SB/PQ; **516** IBMRL/VU; **516–517** Roine Magnusson/Stone; **517** MM; **519** (tl)Mark Schneider/Peter Arnold, Inc., (tcl)Dane S. Johnson/VU, (tcr)Ken Lucas/VU, (tr)Mark A. Schneider/PR, (bl)AH, (bcl)AMP, (bcr)Charles D. Winters/PR, (br)AH; **520** John Evans; **523** (l)Herbert Kehrer/OKAPIA/PR, (c)DM, (r)Bruce Hands/Stone; **524** Kenji Kerins; **526** Ken Whitmore/Stone; **527** AMP; **528** Stuart Westmorland; **530** John S. Lough/VU; **532** CB; **533** (t)Storm Pirate Productions/Artville/PQ, (c)Breck P. Kent/ES, (b)CB/PQ; **534** (t)Paul Chesley/Stone, (b)David Muench/CB; **535** NASA/JPL/Malin Space Science Systems; **536** (t)StudiOhio, (b)MM; **537** (t)Tim Courlas, (b)AH; **538** (t)Geoff Butler, (br)StudiOhio, (bl)KS; **539** Geoffrey Wheeler/NIST; **540** (tl)MM, (tr)Keith Kent/Science Photo Library, (b)John Evans; **541** (l)DM, (r)MM; **542** MM; **544** Breck P. Kent/ES; **544–545** Brenda Tharp/TSM; **545** (all)Breck P. Kent/ES; **546** Fred Habegger from GH; **547** (t)David Nunuk/Science Photo Library/PR, (c)AMP, (bl)David Schultz/Stone, (bc)SuperStock, (br)Kent Knudson/PD; **548** KS; **549** Gary Retherford/PR; **550** (l)Peter Steiner/TSM, (c)Tom & DeeAnn McCarthy/TSM, (r)SuperStock; **551** Timothy Fuller; **552** (l lc)A. Goldsmith/TSM, (rc r)Gay Bumgarner/Stone; **553** (t)MM, (others)Richard Megna/FP **554** (t)Ed Pritchard/Stone, (cl, bl)Kip Peticolas/FP, (cr, br) Richard Megna/FP; **555** Rich Iwasaki/Stone; **556** (t)MM, (c)Layne Kennedy/CB, (bl br)Runk/Schoenberger from GH; **557** (tl, tr)AMP, (bl, br)Richard Megna/FP; **558** Anthony Cooper/Ecoscene/CB; **559** (tl)Russell Illig/PD, (tr)SuperStock, (cl)John D. Cunningham/VU, (cr)Coco McCoy/Rainbow/PQ, (bl)Bonnie Kamin/PE, (br)SuperStock; **560** (t)Grantpix/PR, (c)Mark Sherman/Photo Network/PQ, (bl)sculpture by Maya Lin, courtesy Wexner Center for the Arts, The Ohio State University, photo by Darnell Lautt, (br)Rainbow/PQ; **561** MB; **562 563** MM; **564** (t)Michael Newman/PE, (c)Laura Sifferlin, (b)Timothy Fuller; **565** Darren McCollester/Newsmakers; **566** (l)Larry Hamill, (r)AH; **567** (l)C. Squared Studios/PD, (r)Kip Peticolas/FP; **572–573** Matthew Borkoski/SB/PQ; **573** L. Fritz/H. Armstrong Roberts; **574** Jerome Wexler/PR; **574–575** Douglas Peebles/CB; **576** (l)file photo, (r)David Young-Wolff/PE; **577** David Young-Wolff/PE; **578** Mark Thayer; **581** Steven Starr/SB; **586** Ken Frick; **587** MB; **589** Ernst Haas/Stone; **590** Peter Beattie/LA; **592** (t)D. Boone/CB, (b)Stephen R. Wagner; **593** Seth Resnick/SB; **594** (t)Reuters NewMedia, Inc./CB, (b)Timothy Fuller; **596** (t)John Evans, (b)SuperStock; **597** Roger Ressmeyer/CB; **598** Mark Thayer; **602** Paul Silverman/FP; **602–603** Roger Ressmeyer/CB; **603** Timothy Fuller; **607** (t)Joe Towers/TSM, (c)Bob Daemmrich/SB/PQ, (b)Jean-Paul Thomas/Jacana Scientific Control/PR; **609** Stephen Dalton/PR; **610** NOAA; **611** Slim Films; **613** Spencer Grant/PE; **614** Timothy Fuller; **616** Mark Thayer; **618** Dilip Mehta/Contact Press Images/PQ; **619** (tl)CB, (tr)Paul Seheult/Eye Ubiquitous/CB, (b)IC; **620** William Whitehurst/TSM; **621** SuperStock; **622** (t)Geostock/PD, (b)SuperStock; **623** Fred E. Hossler/VU; **624** (t)Ryan McVay/PD, (c)file photo, (b)Oliver Benn/Stone; **626** Douglas Whyte/TSM; **627** (t)Steve Labadessa/Time Inc., (c)courtesy 3M, (b)Bernard Roussel/The Image Bank; **628** (tl)Edmond Van Hoorick/PD, (bl)Kim Steele/PD, (r)Will McIntyre/PR; **629** (tl)The Photo Works/PR, (tr)PD, (b)Gary Braasch; **630** PhotoSpin/Artville/PQ; **631** C. Squared Studios/PD; **632** Stephanie Maze/CB; **632–633** Roger Ressmeyer/CB; **633** IC; **634** (l)Bob Abraham/TSM, (r)Jeff Greenberg/VU; **635** (l)David Young-Wolff/PE, (r)NRSC, Ltd./Science Photo Library/PR; **636** (t)Grantpix/PR, (b)Richard Megna/FP; **638** Luke Dodd/Science Photo Library/PR; **640** (t)MM, (b)Jean Miele/TSM; **642** (t)Gregory G. Dimijian/PR, (b)Charlie Westerman/LA; **643** AH; **645** (l)MM, (r)Bob Daemmrich/The Image Works; **646** (t cl)NASA, (cr)Max Planck Institute for Radio Astronomy/Science Photo Library/PR, (b)ESA/Science Photo Library/PR; **647** (l)NASA/Science Photo Library/PR, (c)Harvard-Smithsonian Center for Astrophysics, (r)ESA; **648** Timothy Fuller; **651** MM; **653** Ken M. Johns/PR; **654** Michael Thomas/Stock South/PQ; **655** Dominic Oldershaw; **656** (t)Culver Pictures, (b)Hulton Getty Library/LA; **657** Aurthur Tilley/FPG; **658** (tl)G. Brad Lewis/LA, (bl)Yoav Levy/Phototake/PQ, (r)George B. Diebold/TSM; **659** (l)Macduff Everton/CB, (r)NASA/Mark Marten/PR; **660** Michael Thomas/Stock South/PQ; **662** Novastock/PE; **662–663** Cary Wolinsky/SB/PQ; **663** MM; **664** Dick Thomas/VU; **665** John Evans; **666** (tl)Bob Woodward/TSM, (tr)SuperStock, (c)Ping Amranand/Pictor, (b)Runk/Schoenberger from GH; **667** Mark Thayer; **670** (l)Dr. Dennis Kunkel/PhotoTake NYC, (r)David Toase/PD; **672** (l)Bill Aron/PE, (r)Paul Silverman/FP; **675** Geoff Butler; **677** Richard Megna/FP; **681** David Young-Wolff/PE; **682 683** Roger Ressmeyer/CB; **685** file photo; **686** (t)MM, (b)Geoff Butler; **687** Geoff Butler; **688–689** Ed Welche's Antiques/Winslow, ME; **689** (t)courtesy Cheryl Landry, (b)The Stapleton Collection/Bridgeman Art Library; **690** (tl)file photo, (bl)Jeremy Horner/Stone, (r)MB**692** Carol Christensen/Stock South/PQ; **696–697** PD; **699** (tl)ES, (tr)Patti Murray/ES, (bl)Larry Ulrich/DRK, (br)Doug Sokell/TSA; **700** (tl)Bill Beatty/ES, (tr)Tom Bean/DRK, (bl)R. Calentine/VU, (br)Gerald and Buff Corsi/VU; **701** (tl)C.C. Lockwood/DRK, (tr)Joseph G. Strauch Jr., (bl)John Frett, (br)Gerald and Buff Corsi/VU; **702** CNES/PR; **703** (l)Lloyd Cluff/CB, (r)Neil E. Johnson; **704** (t)Tom Bean/DRK, (b)Buddy Mays/CB; **705** (t)Joe Cornish/Stone, (b)SuperStock; **706** (t)CB, (b)Artville; **707** (b)Artville, (others)PD; **708** (b)StudiOhio, (others)Artville; **709** (t)PD, (c)Wolfgang Kaehler/CB, (b)CB; **710** Timothy Fuller; **714** MB; **717** Dominic Oldershaw; **718** StudiOhio; **719** MB; **721** Richard Day/AA; **724** Paul Barton/TSM; **727** Charles Gupton/TSM; **737** MM; **738** (l)Dr. Richard Kessel, (c)NIBSC/Science Photo Library/PR, (r)David John/VU; **739** (t)Runk/Schoenberger from GH, (bl)Andrew Syred/Science Photo Library/PR, (br)Rich Brommer; **740** (t)G.R. Roberts, (bl)Ralph Reinhold/ES, (br)Scott Johnson/AA; **741** Martin Harvey/DRK.

Acknowledgments

From the book, *The Everglades: A River of Grass* 50th Anniversary Edition copyright © 1997 by Marjory Stoneham Douglas. Used by permission of Pineapple Press, Inc. The Excerpt from "Sunkissed: An Indian Legend" is reprinted with permission from the publisher of *Tun-ta-ca-tun* (Houston: Arte Publico Press–University of Houston, 1986). "Listening In" by Gordon Judge. Reprinted by permission of the author. "The Jungle of Ceylon" by Pablo Neruda, from PASSIONS AND IMPRESSIONS by Pablo Neruda, translated by Margaret Sayers Peden. Translation copyright © 1983 by Farrar, Straus & Giroux. Reprinted by permission of Farrar, Straus & Giroux. Excerpt from "Tulip" from *Turtle Blessing,* by Penny Harter, published by La Alameda Press, copyright © 1996 by Penny Harter. Reprinted by permission of the author. Excerpt from "The Creatures on my Mind," by Ursula K. Le Guin, from *Harper's* (August 1990). Copyright © 1990 by Ursula K. Le Guin. Reprinted by permisison of the author and the author's agent, Virginia Kidd.